Analog Integrated Circuits

OTHER IEEE PRESS BOOKS

Integrated-Circuit Operational Amplifiers, *Edited by R. G. Meyer*
Modern Spectrum Analysis, *Edited by D. G. Childers*
Digital Image Processing for Remote Sensing, *Edited by R. Bernstein*
Reflector Antennas, *Edited by A. W. Love*
Phase-Locked Loops & Their Application, *Edited by W. C. Lindsey and M. K. Simon*
Digital Signal Computers and Processors, *Edited by A. C. Salazar*
Systems Engineering: Methodology and Applications, *Edited by A. P. Sage*
Modern Crystal and Mechanical Filters, *Edited by D. F. Sheahan and R. A. Johnson*
Electrical Noise: Fundamentals and Sources, *Edited by M. S. Gupta*
Computer Methods in Image Analysis, *Edited by J. K. Aggarwal, R. O. Duda, and A. Rosenfeld*
Microprocessors: Fundamentals and Applications, *Edited by W. C. Lin*
Machine Recognition of Patterns, *Edited by A. K. Agrawala*
Turning Points in American Electrical History, *Edited by J. E. Brittain*
Charge-Coupled Devices: Technology and Applications, *Edited by R. Melen and D. Buss*
Spread Spectrum Techniques, *Edited by R. C. Dixon*
Electronic Switching: Central Office Systems of the World, *Edited by A. E. Joel, Jr.*
Electromagnetic Horn Antennas, *Edited by A. W. Love*
Waveform Quantization and Coding, *Edited by N. S. Jayant*
Communication Satellite Systems: An Overview of the Technology, *Edited by R. G. Gould and Y. F. Lum*
Literature Survey of Communication Satellite Systems and Technology, *Edited by J. H. W. Unger*
Solar Cells, *Edited by C. E. Backus*
Computer Networking, *Edited by R. P. Blanc and I. W. Cotton*
Communications Channels: Characterization and Behavior, *Edited by B. Goldberg*
Large-Scale Networks: Theory and Design, *Edited by F. T. Boesch*
Optical Fiber Technology, *Edited by D. Gloge*
Selected Papers in Digital Signal Processing, II, *Edited by the Digital Signal Processing Committee*
A Guide for Better Technical Presentations, *Edited by R. M. Woelfle*
Career Management: A Guide to Combating Obsolescence, *Edited by H. G. Kaufman*
Energy and Man: Technical and Social Aspects of Energy, *Edited by M. G. Morgan*
Magnetic Bubble Technology: Integrated-Circuit Magnetics for Digital Storage and Processing, *Edited by H. Chang*
Frequency Synthesis: Techniques and Applications, *Edited by J. Gorski-Popiel*
Literature in Digital Processing: Author and Permuted Title Index (Revised and Expanded Edition), *Edited by H. D. Helms, J. F. Kaiser, and L. R. Rabiner*
Data Communications via Fading Channels, *Edited by K. Brayer*
Nonlinear Networks: Theory and Analysis, *Edited by A. N. Willson, Jr.*
Computer Communications, *Edited by P. E. Green, Jr. and R. W. Lucky*
Stability of Large Electric Power Systems, *Edited by R. T. Byerly and E. W. Kimbark*
Automatic Test Equipment: Hardware, Software, and Management, *Edited by F. Liguori*
Key Papers in the Development of Coding Theory, *Edited by E. R. Berkekamp*
Technology and Social Institutions, *Edited by K. Chen*
Key Papers in the Development of Information Theory, *Edited by D. Slepian*
Computer-Aided Filter Design, *Edited by G. Szentirmai*
Laser Devices and Applications, *Edited by I. P. Kaminow and A. E. Siegman*
Integrated Optics, *Edited by D. Marcuse*
Laser Theory, *Edited by F. S. Barnes*
Digital Signal Processing, *Edited by L. R. Rabiner and C. M. Rader*
Minicomputers: Hardware, Software, and Applications, *Edited by J. D. Schoeffler and R. H. Temple*
Semiconductor Memories, *Edited by D. A. Hodges*
Power Semiconductor Applications, Volume II: Equipment and systems, *Edited by J. D. Harnden, Jr. and F. B. Golden*
Power Semiconductor Applications, Volume I: General Considerations, *Edited by J. D. Harnden, Jr and F. B. Golden*
A Practical Guide to Minicomputer Applications, *Edited by F. F. Coury*
Active Inductorless Filters, *Edited by S. K. Mitra*
Clearing the Air: The Impact of the Clean Air Act on Technology, *Edited by J. C. Redmond, J. C. Cook, and A. A. J. Hoffman*

Analog Integrated Circuits

Edited by

Alan B. Grebene
Vice President
Exar Integrated Systems, Inc.

A volume in the IEEE PRESS Selected Reprint Series,
prepared under the sponsorship of the IEEE
Circuits and Systems Society.

The Institute of Electrical and Electronics Engineers, Inc. New York

PRINTED IN THE UNITED STATES OF AMERICA

IEEE International Standard Book Numbers: Clothbound: 0-87942-113-4
Paperbound: 0-87942-114-2

Library of Congress Catalog Card Number 78-059636

Sole Worldwide Distributor (Exclusive of the IEEE):

JOHN WILEY & SONS, INC.
605 Third Ave.
New York, NY 10016

Wiley Order Numbers: Clothbound: 0-471-05211-6
Paperbound: 0-471-05210-8

Contents

Introduction

Since the introduction of the first simple monolithic amplifier in the early 1960's, analog integrated circuits have rapidly evolved into one of the most important fields of electronics. Today, there is hardly a modern electronics system or product that has not been impacted by the low-cost and high-performance advantages of analog IC's.

The evolution of analog IC's from the simple amplifier circuits of the early sixties to the complex signal processing systems of the present day has been a steady one. The first analog IC functional block to gain wide acceptance was the monolithic operational amplifier. Because of its versatility, performance and low cost, the monolithic op amp quickly became a universal building block, and, almost overnight, revolutionized the basic analog system design. The monolithic op amp was then followed by the voltage regulators and broadband amplifiers. In the late sixties, new circuit techniques and functions such as the analog multipliers and phase-locked loops emerged, to extend the use of analog IC's to a wider range of applications, particularly in analog communications. Since the early 1970's one of the most significant areas of growth in analog IC's has been the data acquisition and conversion circuits. This latest growth area is a direct off-shoot of the "microprocessor revolution" which has dramatically changed the digital system design. Over the recent years, the new developments in IC processing technology and precision trimming methods which can be applied to the IC's at the wafer-level have further extended the capabilities of monolithic IC's to precision circuit functions such as 12-bit A/D and D/A conversion which, until recently, could only be achieved in hybrid or discrete (i.e., nonintegrated) form.

The subject of analog integrated circuits is a very broad and general one. A large number of books, technical articles and application notes have been published which cover various aspects of this wide, exciting and rapidly expanding field. This book brings together a selection of important and timely technical articles on the subject of design and applications of analog integrated circuits. All of the articles gathered here have been published, mostly within the recent years, in the recognized technical journals covering the field of solid-state electronics. Each of these articles have been selected from among many others covering their respective areas, because of its technical contribution, timeliness, clarity of expression and the degree of general interest.

The main purpose of this book is to provide an accurate source of design, application and reference information for both the designer and the user of analog integrated circuits by bringing together, under one cover, the key technical articles published on this subject. The papers included in this book are grouped together into eight parts, covering various important aspects of the analog integrated circuit technology. The emphasis, in all cases, is on the circuit design and application rather than on the process or fabrication technology.

A secondary aim of this book is to provide reference for anyone wanting to do extended research or investigation into any one of the specialized product areas in analog IC's. For this purpose, a detailed supplementary bibliography of additional publications is provided at the end of each section.

The first part of this book deals with the fundamental rules, guidelines and building blocks of analog IC design; the remaining parts focus on specific classes or categories of analog circuits. Parts II and III cover the subject of monolithic operational amplifiers, voltage regulators, and voltage references. Part IV presents various techniques for broadband amplifier design; part V covers the subject of analog multipliers and modulators. Parts VI and VII cover the data conversion circuits, i.e., A/D and D/A converters, and a wide range of telecommunication circuits. The final part, Part VIII, discusses some of the advanced trimming techniques for precision monolithic circuits. Each of these parts are preceded by an introduction which gives a perspective and preview of the topics in discussion and briefly summarizes the salient points of each of the papers included. At the end of each part, additional references are included which should prove useful for further reading or investigation in the area. The papers in this book are drawn from several sources; however, the largest number by far were originally published in the IEEE Journal of Solid-State Circuits.

For additional in-depth information on two of the subjects covered in this book, namely operational amplifiers and the data conversion circuits, the reader is referred to two additional companion books, "Integrated Circuit Operational Amplifiers," and "Data Conversion Integrated Circuits," to be published by the IEEE Press, as a part of this series.

Part I
Fundamentals of Analog IC Design

For a linear circuit designer trained in the area of discrete circuits, the basic constraints of integrated circuit technology often pose a difficult challenge. This challenge comes about mainly because of the inherent limitations of monolithic device structures such as poor absolute value tolerances and large temperature coefficients, limited choice of components values and compatible device types, and, last but not the least, lack of monolithic inductors. On the other hand, IC fabrication methods offer a number of unique and powerful advantages to the circuit designer. Some of these are the availability of large numbers of active devices, good matching, and thermal tracking of component values, and the control of the device layout and the geometries during chip design. By making efficient use of these inherent advantages associated with monolithic IC structures, it is often possible to come up with designs which can greatly exceed the performance of similar discrete component circuits.

The purpose of this part of the book is to outline and demonstrate some of the basic design guidelines and fundamentals of analog IC design. Each of the papers included in this part deals with basic design techniques and simple circuit configurations which make use of the advantages of monolithic technology, and manage to circumvent most of its limitations. The basic design approaches and circuit configurations discussed in this part form a set of design guidelines and building blocks and serve as the starting point for the design of more complex functional circuits covered in the later parts of this book.

ADDITIONAL REFERENCES

1. P.R. Gray and R.G. Meyer, *Analysis and Design of Analog Integrated Circuits.* New York: Wiley, 1977.
2. A. B. Grebene, *Analog Integrated Circuit Design.* New York: Van Nostrand Reinhold, 1972.
3. H.R. Camenzind, *Electronic Integrated System Design.* New York: Van Nostrand Reinhold, 1972.
4. D.J. Hamilton and W.G. Howard, *Basic Integrated Circuit Engineering.* New York: McGraw-Hill, 1975.
5. J.A. Connely, *Analog Integrated Circuits.* New York: Wiley, 1975.
6. T.F. Prosser, "An integrated temperature sensor controller," *J. Sol. State Circuits,* vol. SC-1, pp. 13-18, 1966.
7. R.C. Jaeger, "A high output resistance current source," *J. Sol. State Circuits,* vol. SC-9, pp. 192-194, August 1974.
8. A. Bilotti and E. Mariani, "Noise characteristics of current-mirror sinks/sources," *J. Sol. State Circuits,* vol. SC-10, pp. 516-524, December 1975.
9. P.E. Allen, "Graphical analysis of matched transistor current sinks/sources," *J. Sol.State Circuits,* vol. SC-9, pp. 31-35, February 1974.
10. R.A. Blauschild, "High-voltage analog performance from low-voltage digital devices," *ISSCC Digest of Tech. Papers,* pp. 232-233, February 1978.

Some Circuit Design Techniques for Linear Integrated Circuits

R. J. WIDLAR

Abstract—This paper describes some novel circuit design techniques for linear integrated circuits. These techniques make use of elements which can be constructed easily in monolithic form to avoid processing difficulties or substantial yield losses in manufacturing. Methods are shown which eliminate the need for the wide variety of components and tight component tolerances usually required with discrete designs, which substitute parts which can be made simply, and which make use of special characteristics obtainable with monolithic construction.

Introduction

THE COMPONENTS available in integrated circuit processes which are in production today are limited both in type and in range of values. This limitation has been more troublesome with linear circuits than with digital, since many more different kinds of parts are used with conventional designs. Therefore, serious problems have arisen in adapting discrete-component designs to monolithic construction.

In many cases an attempt has been made to develop new technology in order to integrate existing designs. This approach has met with varying degrees of success, but has generally resulted in circuits which are difficult to produce in large volume. To realize a practical microcircuit with any certainty and within a reasonable period of time, it becomes necessary to use existing production technology. Fortunately, with some specialized circuit design techniques, it is possible in certain cases to achieve performance with present technology which is equal to or better than that obtainable with discrete-component circuits. These techniques make it possible to avoid the restrictions imposed by limited types of components, poor tolerances, and restricted range of component values. They make use of the inherent advantages of integrated circuits: close matching of active and passive devices over a wide temperature range, excellent thermal coupling throughout the circuit, the economy of using a large number of active devices, the freedom of selection of active device geometries, and the availability of devices which have no exact discrete-element counterpart.

Some of these techniques will be discussed; an attempt will be made to indicate in a broad sense what can be done rather than going into great detail on particulars. Examples of practical circuits will be given to illustrate important points.

Manuscript received March 8, 1965; revised July 19, 1965.
The author is with Fairchild Semiconductor, Mountain View, Calif.

Biasing Circuits

One of the most basic problems encountered in integrated circuits is bias stabilization of a common emitter amplifier. Conventional methods [1] usually require substantial dc degeneration and a bypass capacitor to reduce the degeneration at the frequencies to be amplified. With integrated circuits, the required bypass capacitors are much too large to be practical. In the past, this problem has been overcome using some sort of differential [2] or emitter-coupled [3] amplifier connection. These solutions have been adequate in many instances, but they suffer from a lack of versatility.

The close matching of components and tight thermal coupling obtained in integrated circuits permit much more radical solutions. An example is given in Fig. 1(a). A current source can be implemented by imposing the emitter-base voltage of a diode-connected transistor operating at one collector current across the emitter-base junction of a second transistor. If the two transistors are identical, their collector currents will be equal; hence, the operating current of the current source can be determined from the resistor (R_1) and the supply voltage (V^+). Experiment has shown that this biasing scheme is stable over a wide temperature range, giving collector current matches between the biasing and operating devices typically better than five percent, even for power dissipations in Q_2 above 100 mW.

An extension of this idea is shown in Fig. 1(b). A transformer with a low resistance secondary can be inserted between the biasing transistor (Q_1) and the second transistor (Q_2). Then Q_2 is stably biased as an amplifier without requiring any bypass elements, and the transformer secondary is coupled to the amplifier without disturbing the bias conditions.

A third and more subtle variation is given in Fig. 1(c). If R_3 and R_4 as well as Q_1 and Q_2 are identical, the collector currents of the two transistors will be equal, since their bases are driven from a common voltage point through equal resistances. The collector current of Q_1 will be given by

$$I_{C1} = \frac{V^+ - V_{BE}}{R_1} - \left(2 + \frac{R_3}{R_1}\right) I_B \tag{1}$$

where a single V_{BE} and I_B term is used since both transistors are identical. For $V_{BE} \ll V^+$ and $I_B \ll I_C$,

$$I_{C1} = I_{C2} \cong \frac{V^+}{R_1}. \tag{2}$$

If $R_2 = \frac{1}{2} R_1$, then

Reprinted from *IEEE Trans. Circuit Theory,* vol. CT-12, pp. 586-590, Dec. 1965.

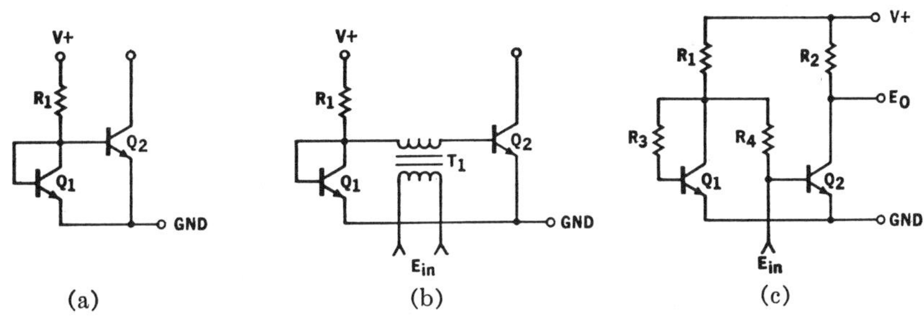

Fig. 1. Biasing techniques applicable to integrated circuitry which take advantage of precise matching and close thermal coupling. (a) Current source. (b) Transformer coupled amplifier. (c) RC-coupled amplifier.

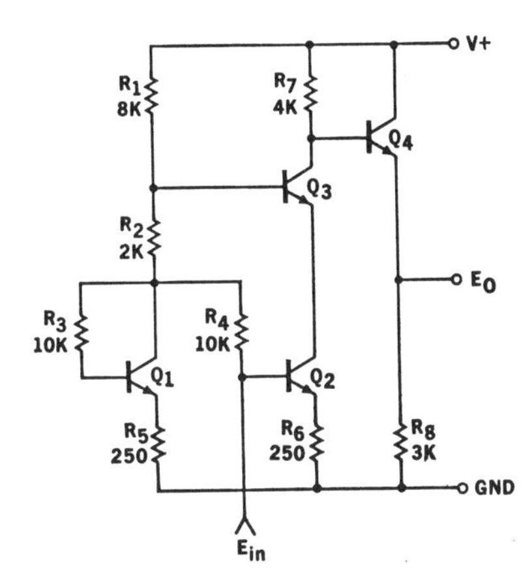

Fig. 2. General purpose wideband amplifier using balanced biasing techniques.

$$E_0 \cong \frac{V^+}{2} \tag{3}$$

which means that the amplifier will be biased at its optimum operating point, at one half the supply voltage, independent of the supply voltage as well as temperature and dependent only on how well the parts within the integrated circuit match.

An evaluation of the effects of mismatch on bias stability is given in the Appendix.

A simple amplifier using the biasing of Fig. 1(c) is illustrated in Fig. 2. An emitter degeneration resistor (R_6) is employed in conjunction with R_5 to control gain and raise input impedance without disturbing the balanced biasing. A cascode connection of Q_2 with Q_3 reduces input capacitance while the emitter follower (Q_4) gives a low output impedance.

It can be seen from the foregoing that the matching characteristics of a monolithic circuit have made possible biasing methods which are superior to those practically attainable with discrete designs. Excellent biasing stability is achieved over a wide temperature range without wasting any of the supply voltage across bias stabilization networks. In addition, no bypass capacitors are required.

The use of a circuit similar to that in Fig. 1(c) as the second stage of differential input, single-ended output amplifiers is described in [4]–[6].

Constant Current Source

The formation of current sources in the microampere current range can be difficult with integrated circuits because of the relatively large resistance values usually required. A circuit is shown in Fig. 3 which makes possible a current source with outputs in the tens of microamperes using resistances of only a few kilohms. It makes use of the predictable difference of the emitter-base voltage of two transistors operating at different collector currents. Its operation can be described as follows: the collector current of a transistor is given as a function of emitter-base voltage by

$$I_C = I_S \exp\cdot\left(\frac{qV_{BE}}{kT}\right), \tag{4}$$

where I_S is the so-called saturation current, q is the charge of an electron, k is Boltzmann's constant and T is the absolute temperature. This expression holds up to high currents where emitter contact and base spreading resistances become important and down to low currents where collector leakage currents cause inaccuracy. It has been shown to be valid, within a percent or so, for operation over at least six decades of collector current with well-made silicon transistors [7]–[9]. This contrasts with similar expressions for diode current and the emitter current of transistors which show substantial error over three decades of current operation.

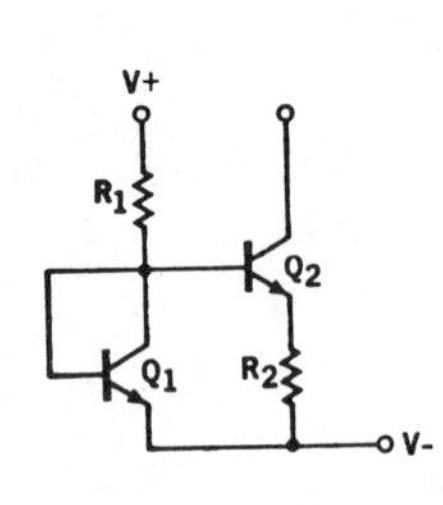

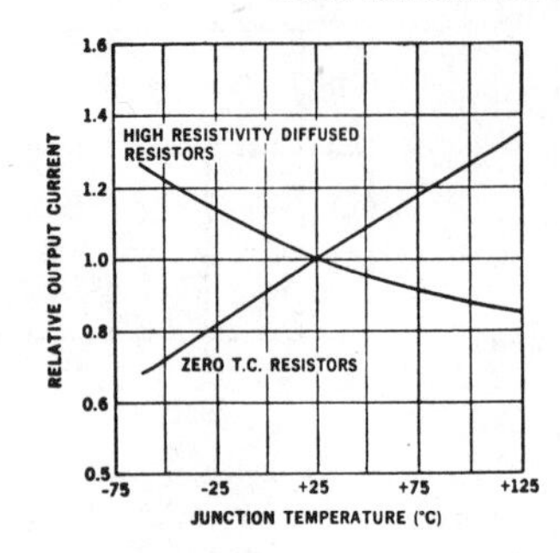

Fig. 3. A current source for generating very small currents using moderate value resistors.

Solving (4) for V_{BE} gives

$$V_{BE} = \frac{kT}{q} \log_e \frac{I_C}{I_S}. \tag{5}$$

This expression can be used to find the emitter-base voltage difference between two transistors:

$$\begin{aligned} \Delta V_{BE} &= V_{BE1} - V_{BE2} \\ &= \frac{kT}{q} \log_e \frac{I_{C1}}{I_{S1}} - \frac{kT}{q} \log_e \frac{I_{C2}}{I_{S2}} \\ &= \frac{kT}{q} \log_e \frac{I_{C1}}{I_{C2}} + \frac{kT}{q} \log_e \frac{I_{S2}}{I_{S1}}. \end{aligned} \tag{6}$$

For equal collector currents, this becomes

$$\Delta V_{BE} = \frac{kT}{q} \log_e \frac{I_{S2}}{I_{S1}}. \tag{7}$$

Considerable testing has shown that for adjacent, identical integrated circuit transistors this term is typically less than 0.5 mV. It is also relatively independent of the current level, as might be expected since I_S should be a constant. Hence, the emitter-base voltage differential between adjacent integrated circuit transistors operating at different collector currents is given by

$$\Delta V_{BE} = \frac{kT}{q} \log_e \frac{I_{C1}}{I_{C2}} \tag{8}$$

within a fraction of a millivolt.

With the circuit in Fig. 3, a collector current which is large by comparison to the desired current-source current is passed through the diode-connected baising transistor Q_1. Its emitter-base voltage is used to bias the current-source transistor Q_2. If the base currents of the transistors are neglected for simplicity, the resistance required to determine the current-source current is given by

$$R_2 = \frac{\Delta V_{BE}}{I_{C2}} = \frac{kT}{qI_{C2}} \log_e \frac{I_{C1}}{I_{C2}}. \tag{9}$$

Or, for the circuit in Fig. 3,

$$R_2 = \frac{kT}{qI_{C2}} \log_e \left[\frac{V^+ - V_{BE}}{R_1 I_{C2}}\right] \tag{10}$$

The effect of nonzero base currents can be easily determined in that they both subtract directly from I_{C1}, and I_{B2} subtracts from I_{C2}.

One interesting feature of this circuit is that for $V^+ \gg V_{BE}$ and $I_{C1} \gg I_{C2}$ the output current will vary roughly as the logarithm of the supply voltage (V^+). Therefore, if the current source is used in such an application as the input stage of an operational amplifier, the operating collector current and voltage gain of the input stage will vary little over an extremely wide range of supply voltages.

From (8) it can be seen that the emitter-base voltage differential is a linear function of absolute temperature. Therefore, it might be expected that the output current of the current source would vary in a similar manner. Such is the case, as illustrated in Fig. 3. The plot is for $I_{C1} \cong 50 I_{C2}$ with both zero temperature coefficient resistors and high-resistivity diffused resistors (bulk impurity concentration less than 10^{17} atoms/cm^3). It is notable that diffused resistors provide overcompensation for this characteristic.

Pinch Resistors

A potentially useful element in integrated circuits which has received much mention but little actual application is the pinch resistor [10]. It is an ordinary diffused (base) resistor, the cross-sectional area of which has been effectively reduced by making an emitter diffusion on top of it [see Fig. 4(a)]. The emitter diffusion raises the sheet resistivity from the usual 100 or 200 Ω/sq to 10 KΩ/sq or higher. This permits rather large resistors to be made in a relatively small area. The pinch resistor, however, has several limiting characteristics. As can be seen from Fig. 4(b), it is linear only for small voltage drops and it has a low breakdown voltage (5 to 10 volts). Neither the linear nor the nonlinear portions of the characteristics can be controlled well and the resistance at the origin can easily vary over a 4-to-1 range in a normal production situation. In addition, the resistor has a very strong positive temperature coefficient, changing by about 3 to 1 over the −55°C to +125°C temperature range.

On the other hand, there is a strong correlation between the pinch resistor values and transistor current gains obtained in manufacture. Within a given process, the sheet resitivity is roughly proportional to the current gains near the current-gain peak (where surface effects have little influence on the current gain). Further, the resistors tend to track with the current gains over temperature. The matching of identical pinch resistors is also nearly as good as base resistors and substantially better than transistor current gains, since the current gains are affected by unpredictable surface phenomena whereas the pinch resistors are not.

Figure 2 provides an example of where pinch resistors can be used effectively. Both R_3 and R_4 have small voltage drops across them, and it would be advantageous to have these resistor values proportional to the transistor current gain to obtain the highest possible input im-

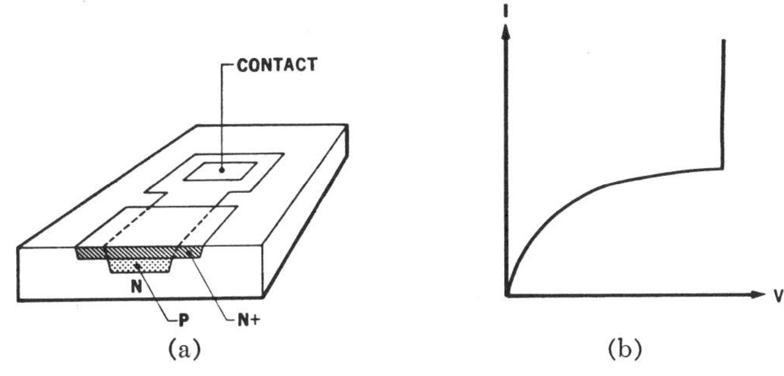

Fig. 4. Pinch resistors. (a) Structure. (b) Characteristics.

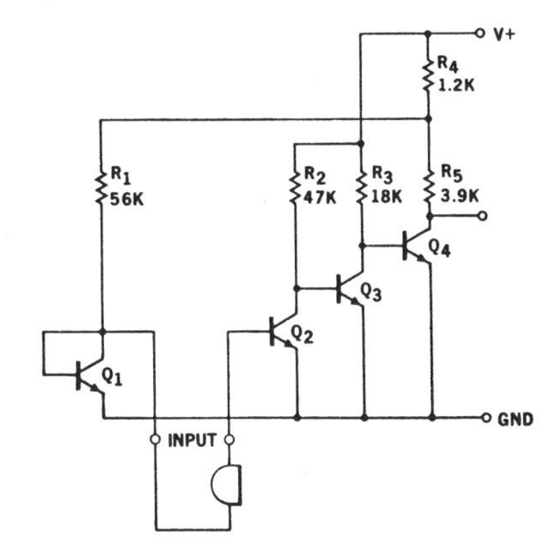

Fig. 5. A low-voltage high-gain microphone preamplifier illustrating the use of pinch resistors.

pedance consistant with satisfactory bias stability.

Another application is the preamplifier shown in Fig. 5 which was designed as part of a hearing aid amplifier. With hearing aids, the maximum supply voltage is 1.55 volts, so the voltage sensitivity and low breakdown of pinch resistors is of little concern. However, power drain is a problem, so large resistances are needed. In this circuit only matching of the pinch resistors (R_1, R_2, and R_3) is required for proper operation. The fact that the pinch resistors correlate with current gain makes the circuit far less sensitive to current gain variations: the pinch resistors and current gains can be varied simultaneously over a range greater than 7 to 1 without any noticeable degradation of performance.

The preceeding examples show that even though pinch resistors have extremely poor characteristics by discrete component standards, certain characteristics, namely good matching, a correlation between resistor values and current gain and high sheet resistivities make them extremely useful elements in circuit design. The pinch resistors can actually function better than precision resistors in certain applications in that they can be used to help compensate for production variations in current gain and the change in gain with temperature.

Conclusions

Biasing techniques have been demonstrated which eliminate the need for the large bypass capacitors required with conventional designs and yield superior performance. A microampere current source using resistance values at least an order of magnitude less than those normally required was also shown. Finally, pinch resistors, devices which can give large resistances in microcircuits, were discussed with respect to both their limitations and their unique properties; examples of how they could be used effectively in circuit design were introduced.

It has been suggested that there are far better approaches available than directly adapting discrete component designs to microcircuits. Many of the restrictions imposed by monolithic construction can be overcome on a circuit design level. This is of particular practical significance because circuit design is a nonrecurring cost in a particular microcircuit while restrictive component tolerances or extra processing steps represent a coutinuing expense in manufacture.

Appendix

The Effect of Mismatches on Balanced Biasing

The assumption that the two transistors in Fig. 1(c) were identical is, of course, not entirely true in a practical microcircuit. This deviation from reality will now be considered.

In (1) of the text, the mismatch in emitter-base voltage and base currents are certainly third-order effects since the absolute values of V_{BE} and I_B are second-order terms. Therefore, (1) will be assumed to hold in this analysis. Equation (2) is, however, strongly affected by the match between Q_1 and Q_2. This influence will be reflected by the equation

$$I_{C2} = I_{C1} + \Delta I_{C2}, \tag{1a}$$

where ΔI_{C2} is the change in collector current of Q_2 due to unequal emitter-base voltages and base currents in Q_1 and Q_2.

Both of the preceding mismatch terms can be combined in

$$\Delta V_{IN} = \Delta V_{BE} + R_4 \Delta I_B \tag{2a}$$

where ΔV_{BE} and ΔI_B are the emitter-base voltage difference and base current difference of the two devices operating at equal collector currents. (These terms are chosen because they are important in the performance of dc amplifiers and because data on the parameters is common, both as transistor pairs and as complete integrated amplifiers.) This ΔV_{IN} will have the same effect on bias stability as a dc input voltage, equal in magnitude, in series with R_4 on an ideally balanced amplifier.

In order to determine the relationship between ΔV_{IN} and ΔI_{C2}, which is the objective of this Appendix, it becomes necessary to introduce the two expressions found in [7],

$$I_C \propto \exp \cdot \left(\frac{q V_{BE}}{kT} \right) \tag{3a}$$

and

$$I_B \propto \exp\cdot\left(\frac{qV_{BE}}{mkT}\right), \tag{4a}$$

which have the same range of validity stated for (4) in the text.

Differentiation of (3a) gives

$$\frac{\Delta I_C}{\Delta V_{BE}} = \frac{qI_C}{kT}. \tag{5a}$$

Similarly, for (4a)

$$\frac{\Delta I_B}{\Delta V_{BE}} = \frac{qI_B}{mkT} \tag{6a}$$

whereas

$$R_{IN} = \frac{mkT}{qI_B}. \tag{7a}$$

From this,

$$\Delta I_{C2} = \left(\frac{R_{IN}\Delta V_{IN}}{R_{IN} + R_4}\right)\left(\frac{qI_{C2}}{kT}\right)$$

$$= \frac{m(\Delta V_{BE} + R_4\Delta I_B)}{\frac{mkT}{q} + R_4 I_{B2}} I_{C2} \tag{8a}$$

where m is electrically significant as the ratio of the ac current gain to the dc current gain at the operating current level (typically 1.6 for low-current operation and 1.2 for operation approaching the current gain peak).

Hence,

$$\Delta I_{C2} \cong \frac{m(\Delta V_{BE} + R_4\Delta I_B)}{\frac{mkT}{q} + R_4 I_B} I_{C1}. \tag{9a}$$

For $R_{IN} \ll R_4$ this becomes

$$\frac{\Delta I_{C2}}{I_{C1}} = \frac{m\Delta I_B}{I_B} = \Delta h_{fe} \tag{10a}$$

where

$$\Delta h_{fe} = \frac{h_{fe2}}{h_{fe1}}.$$

For $R_{IN} \gg R_4$,

$$\frac{\Delta I_{C2}}{\Delta I_{C1}} = \frac{q\Delta V_{BE}}{kT}. \tag{11a}$$

Since R_{IN} varies drastically over the $-55°C$ to $+125°C$ temperature range commonly expected for integrated circuits (10a) can be taken to represent the low-temperature extreme and (11a) the high-temperature extreme. Therefore, a more-than-worst-case solution for full temperature range operation is

$$\frac{\Delta I_{C2}}{I_{C1}} \leq \frac{q\Delta V_{BE}}{kT} + \Delta h_{fe}. \tag{12a}$$

To give some feeling for the results obtainable, the first term in (12a) is typically less than 0.02 while the second term is typically less than 0.07 for full temperature range operation.

References

[1] H. C. Lin and A. A. Barco, "Temperature effects in circuits using junction transistors," *Transistors I*, RCA Laboratories, Princeton, N. J., pp. 369–386, 1956.

[2] W. F. DeBoice and J. F. Bowker, "A general-purpose microelectronic differential amplifier," North American Aviation, Autonetics Div., Downey, Calif., September 1962.

[3] J. A. Narud, N. J. Miller, J. J. Robertson, M. J. Callahan and J. Soloman, "Research on utilization of new techniques and devices in integrated circuits," Motorola Semiconductor Phoenix, Ariz., USAF Contract Rept. AF33(657)-11664.

[4] R. J. Widlar, "A monolithic, high-gain DC amplifier," *1964 Proc. NEC.*, pp. 169–174.

[5] R. J. Widlar, "A fast IC comparator and five ways to use it," *Electrical Design News*, vol. 20, pp. 20–27, May 1965.

[6] R. J. Widlar, "A unique circuit design for a high performance operational amplifier especially suited to monolithic construction," *1965 Proc. NEC*, pp. 169–174.

[7] C. T. Sah, "Effect of surface recombination and channel on *p-n* junction and transistor characteristics," *IRE Trans. on Electron Devices*, vol. ED-9, pp. 94–108, January 1962.

[8] P. J. Coppen and W. T. Matzen, "Distribution of recombination current in emitter-base junctions of silicon transistors," *IRE Trans. on Electron Devices*, vol. ED-9, pp. 75–81, January 1962.

[9] J. E. Iwersen, A. R. Bray and J. J. Kleimack, "Low-current alpha in silicon transistors," *IRE Trans. on Electron Devices*, vol. ED-9, pp. 474–478, November 1962.

[10] G. E. Moore, "Semiconductor integrated circuits," in *Microelectronics*, E. Keonjian, Ed. New York: McGraw-Hill, 1963, ch. V.

An Outline of Design Techniques for Linear Integrated Circuits

HANS R. CAMENZIND, MEMBER, IEEE, AND ALAN B. GREBENE, MEMBER, IEEE

Abstract—**The components available in integrated circuits are well known for their large tolerances and temperature coefficients. This paper is a condensed survey of design principles that make use of the *advantages* offered in monolithic structures: close matching and tracking of parameters, control over component geometries, and the availability of a large number of active devices at little extra cost. With these techniques it is, in many cases, possible to duplicate or even exceed the performance of discrete component linear circuits.**

Manuscript received November 18, 1968; revised February 20, 1969.

The authors are with the Signetics Corp., Sunnyvale, Calif.

INTRODUCTION

THE RAPID advances of the silicon-based integrated-circuit technology during the recent years have posed unique challenges to the linear circuit designer. These challenges arise from the constraints of the monolithic structure, which severely limit the choice of active devices and place large tolerances and temperature coefficients on most parameters. Some of the standard elements of linear circuit design, such as inductors, transformers, and large-value capacitors, are no longer feasible.

Reprinted from *IEEE J. Solid-State Circuits*, vol. SC-4, pp. 110-122, June 1969.

On the other hand, integrated-circuit fabrication methods offer a number of unique advantages to the circuit designer. The most significant of them are the availability of a large number of active devices at little extra cost and the matching and tracking of component parameters due to their close proximity in the silicon chip. This paper presents a condensed description and analysis of some of the major linear-integrated-circuit design ideas that stress the advantages and avoid the shortcomings of the integrated components. These design ideas, though often radically different from discrete component approaches, use the bulk of the established transistor circuit theory as a departing point. With these techniques it is, in many cases, possible to duplicate or even exceed the performance of similar discrete component circuits.

The circuit design techniques covered in this survey are based on the most fully developed integrated-circuit process, the bipolar monolithic approach, which includes both the junction and the dielectric isolation. Structural diagrams of the various components are shown in Fig. 1. Minor modifications in the process permit the inclusion of additional devices, such as junction field effect transistors or substrate p-n-p transistors. In order to make this analysis generally applicable, we have deliberately narrowed the choice of available components.

The performance of these devices depends greatly on the characteristics of the process and how well the process is controlled. Thus the device performance differs somewhat between the various production lines. For our discussion we have compiled a cross section of device characteristics reported in the literature, as shown in Table I.

In this analysis attention was focused on the underlying design principles rather than on a detailed analysis of specific circuits. This was done in the hope that this survey would provide a useful set of tools and guidelines for the circuit designer.

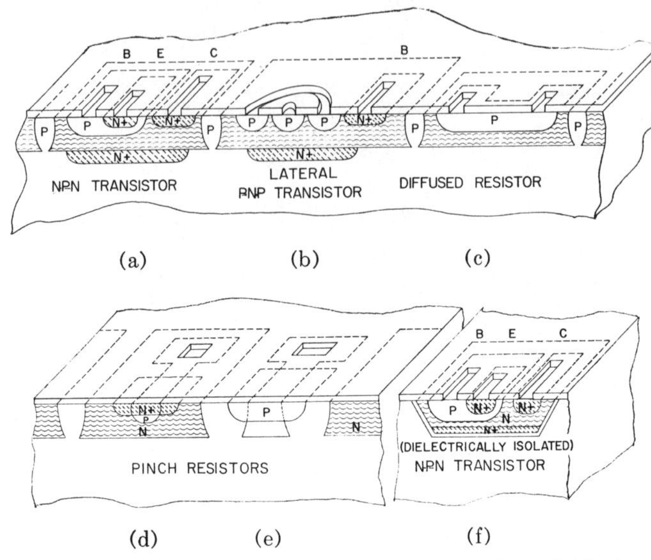

Fig. 1. The integrated components available in monolithic form. Diodes are made from the transistor regions.

TABLE I
TYPICAL PERFORMANCE OF INTEGRATED COMPONENTS

	Symbol	Typical Value	Tolerance	Temperature Coefficient
n-p-n Transistors and Diodes				
Current amplification factor	h_{FE}	30–100	+50, −30 percent	+0.5 percent /°C
Matching of h_{FE} between identical transistors in close proximity	Δh_{FE}	–	±10 percent	+10.5 percent /°C
Base–emitter diode forward voltage drop (low current)	V_{BE}	0.7 volts	±3 percent	−2 mV/°C
Matching of V_{BE} between identical transistors in close proximity	ΔV_{BE}	–	±2 mV	±10 μV/°C
Base–emitter diode reverse breakdown voltage	BV_{EBO}	6–8 volts	±5 percent	+3 mV/°C
Collector–base breakdown voltage	BV_{CBO}	<45 volts	±30 percent	–
Collector–substrate breakdown voltage	BV_{CS}	<60 volts	–	–
Lateral p-n-p Transistors				
Current amplification factor	h_{FE}	0.5–20	+200, −50 percent	+0.5 percent /°C
Base–collector breakdown voltage	BV_{CBO}	<45 volts	–	–
Resistors				
Resistance of diffused resistors (base layer)	R	100 Ω–20 kΩ	±20 percent	±0.15 percent /°C
Resistance of deposited resistors	R	100 Ω–100 kΩ	±18 percent	±0.10 percent /°C
Matching between identical resistors in close proximity	ΔR	–	±3 percent	±0.005 percent /°C
Pinch Resistors				
Sheet resistance (small applied voltage)	R_S	5–10 kΩ/□	+100, −50 percent	+0.3 – 0.5 percent /°C
Matching of identical pinch resistors in close proximity	ΔR	–	±5 percent	–

BIASING

As can be seen from Table I variations and temperature coefficients of most parameters in integrated circuits are very large indeed. The resistance variation of a diffused resistor, for example, is two to four times as large as that of a carbon resistor and its temperature coefficient is larger by an order of magnitude. Neighboring devices, however, have very nearly identical characteristics. This is due to the fact that the properties of diffused and evaporated layers vary only gradually along the surface of the silicon die and there are only minute temperature differences between the devices. It is this property of close matching and tracking that can be utilized for precise biasing of active devices. The first and most simple example [1] using this idea is shown in Fig. 2. Two identical transistors, T_1 and T_2, are employed. T_2 is used as an amplifier in the grounded emitter configuration. T_1, connected as a diode, is placed in parallel with the base–emitter diode of T_2.

If a sufficiently large collector voltage is provided for T_2, both transistors will be in the active region. Consequently, the collector currents of T_1 and T_2 are related to their respective base currents by the common-emitter current gain h_{FE}. Assuming that the close proximity of the two transistors produces identical base–emitter characteristics, then

$$I_{B1} = I_{B2} = I_B. \quad (1)$$

I_{C2}, therefore, is related to I_1 as

$$I_{C2} = \frac{h_{FE2}}{h_{FE1}} (I_1 - 2I_B). \quad (2)$$

Assuming that the current gains are matched and reasonably large (e.g., > 50), we obtain

$$I_{C2} \simeq I_1. \quad (3)$$

If the base–emitter voltage is negligible compared to the supply voltage (or if a diode is connected in series with R_2) we get for the dc output voltage

$$V_{C2} = V_{CC}\left(1 - \frac{R_2}{R_1}\right) \quad (4)$$

This output voltage, therefore, depends on the *ratio* of two resistors, rather than their absolute values, and as can be seen from Table I, this ratio is highly predictable.

In an integrated circuit, matching of parameters is not perfect. Using the worst-case tolerances listed in Table I and assuming that each error increases the deviation of the output dc potential, we get a total deviation of ±18 percent. In practice however, the maximum deviation from the nominal value is somewhat lower due to cancellation of errors.

This approach is not limited to a unity ratio between the two currents of I_1 and I_{C2}. A third transistor of identical size connected in parallel with T_2 would double I_{B2} and therefore I_{C2}. Rather than adding transistors, the size of each transistor (i.e., the effective emitter

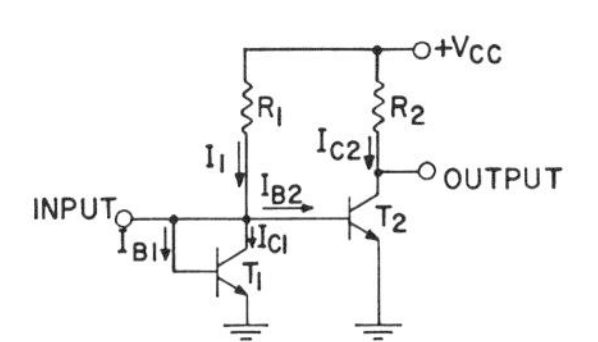

Fig. 2. Diode biasing of transistor.

length) can be chosen to produce the desired current *ratio*. (Thus, by making the emitter of T_2 in the integrated circuit three times the size of T_1, the ratio of the currents I_{C2} to I_1 will be 3:1.)

Transistor Pairs

To provide a large amount of gain two or more transistor stages are generally required. In discrete component circuits the coupling between transistor stages is relatively noncritical, for a variety of active and passive components are available. In integrated circuits the lack of large-value capacitors and high-quality complementary transistors dictates that all stages be direct-coupled. Furthermore, if the gain of this amplifier is to be stable without the use of external components, it must not depend on the absolute values of any of the component parameters. An approach that has been taken from discrete component design and is suited for integrated circuits is the shunt-series feedback pair, Fig. 3(a). Assuming $h_{FE} >> 1$, the current gain of this circuit topology is determined by the ratio of the resistors in the feedback path [2],

$$A_i \simeq \frac{R_2}{R_4}. \quad (5)$$

Thus, since the accurate resistor ratios can be easily realized in integrated circuits, this gain is highly predictable and stable.

In the particular circuit under discussion the base current required to keep the first transistor in its active region is derived from the emitter current of the second transistor. Thus the value of R_4 would be chosen so that the emitter current of T_2 causes a voltage drop of V_{BE} across R_4. Similarly, for a desired dc level at the output, the collector load resistor of T_2, R_3, is determined by the amount of collector current in T_2. Even though R_3 and R_4 might have large variations in their absolute resistances, a predictable and constant *ratio* assures that the amplifier is correctly biased.

The effective dc drift referred to the input, due to V_{BE} variations with temperature, is unaffected by the feedback arrangement, and is of the order of −2 mV/°C. Since extreme resistor ratios are impractical and inaccurate in integrated circuits, the maximum value of A_i obtainable from such a pair is limited to ≤ 1000.

For small-signal operation, the gain of a shunt-series feedback pair can be further increased by utilizing the dynamic impedance, r_d, of a forward-biased diode, D_1, to replace R_4. In integrated form one would use a diode-connected transistor T_3, shown in Fig. 3(b). This configu-

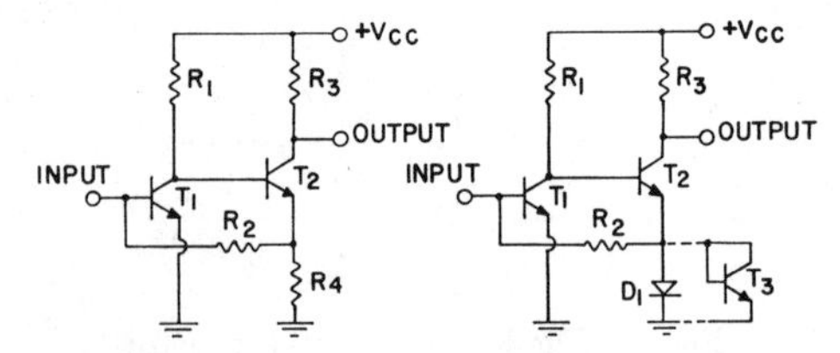

Fig. 3. Shunt-series feedback pairs.

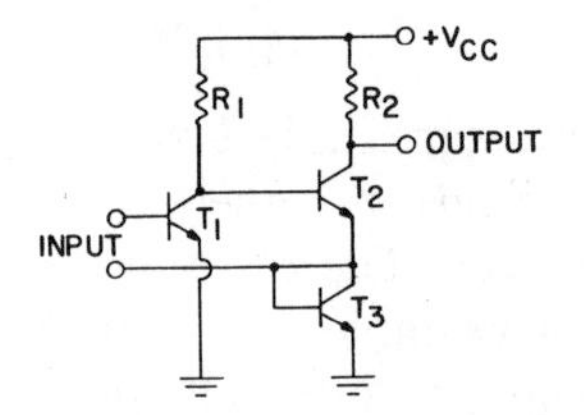

Fig. 4. Series-series feedback pair.

ration closely approximates an ideal diode, since its main ohmic resistance, that of the base region, appears divided by the current gain of the transistor.

An interesting and highly useful property of this scheme is its tolerance tracking ability. If, for example, the resistors are doubled in value, both collector currents would be cut in half. The diode, therefore, runs at half the current and its dynamic resistance is increased by a factor of two. Thus the ratio of R_2 to r_d remains unchanged, providing the same amount of gain for the amplifier.

The shunt-series feedback pairs shown in Fig. 3 exhibit a low input impedance due to the nature of their feedback loop [2]. A much higher input impedance (5 kΩ typical) can be obtained by providing series feedback at the input, as shown in Fig. 4. Here the base current required by the first transistor flows through the source. The source, therefore, must have a reasonably low dc resistance and must be floating above ground. This is usually the case with most transducers and detectors.

This configuration offers the same matching and tracking advantages previously discussed. The voltage gain is again given by a ratio, namely

$$A_V = \frac{R_2}{r_d}. \tag{6}$$

Since the dynamic resistance of D_1 (or T_3) is inversely proportional to the current level through it, the circuits of Figs. 3 (b) and 4 are useful only for small-signal operation. At high signal levels, nonlinear behavior of r_d can result in significant second-harmonic distortion in the output signal.

In a variety of circuit applications, capacitively coupled high input impedance amplifier stages are required. In integrated-circuit design where large bias resistor values are impractical, these requirements can be met by use of a "bootstrap" bias topology as shown in Fig. 5 [3].

The voltage gain of this stage is primarily given by the ratio of collector to emitter resistor,

$$A_V \simeq \frac{R_1}{R_4}. \tag{7}$$

Ignoring the small dynamic emitter resistances in the four transistors, the voltage gain between the input and the collector points of T_3 and T_4 is that of an ideal emitter follower, i.e., exactly unity. Thus R_2 appears to be infinite in value at the input and the input impedance is determined by the parallel combination of R_3 and R_4

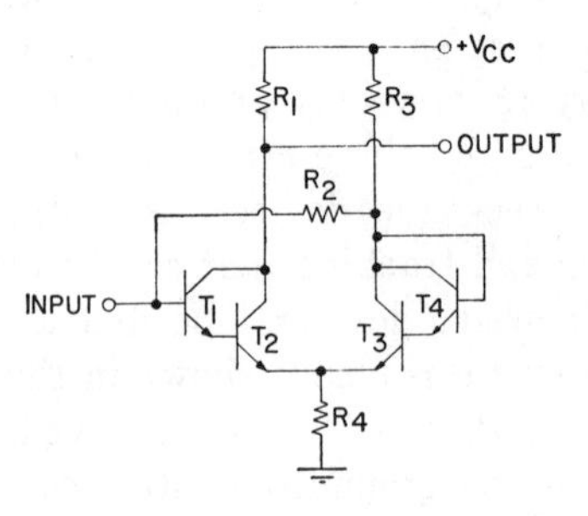

Fig. 5. Self-biasing high-input impedance stage.

multiplied by the current gain of T_1 and T_2. In reality the dynamic emitter resistances of T_1 and T_2 and the dynamic diode resistance of T_3 and T_4 cause this second voltage gain to drop to slightly less than unity. Nevertheless, input impedances up to 1 MΩ can be realized with practical resistor values.

Current Sources and High-Value Resistors

In discrete component circuits current sources are rarely employed. In most cases where a small and relatively constant current is required a high-valued resistor connected to the largest available voltage will suffice. In integrated circuits, however, high-value resistors are costly. Approximate current sources, on the other hand, require only a small amount of surface area, and in most cases provide better performance.

The simplest current source has already been discussed in the previous section (Fig. 2). A current ratio between I_1 and I_{C2} is established by choosing the area ratio of T_1 and T_2. There are, however, two small errors that affect the precision of this current source.

First, I_{C2} is smaller compared to I_1 because it does not include the two base currents, I_{B1} and I_{B2}. Second, base-width modulation causes h_{FE2} to increase slightly with increasing collector voltage. This in turn introduces a voltage dependence [Fig. 7(a)] which is equivalent to a finite shunt conductance.

A significantly improved version [4] of an integrated-current source is shown in Fig. 6. The emitter current of T_2 is fed through a diode-connected transistor, T_3. The voltage drop across T_3 causes a collector current to flow in T_1, which is subtracted from I_1. Thus the error introduced by an increase in h_{FE} in T_2 is reduced through a feedback loop. Referring to Fig. 6, it can be shown by a straightforward node analysis that the controlled current, I_2, is related to the reference current, I_1, as

$$I_2 = I_1 + (2I_B - 2I_{B2}) = \frac{I_1 + 2I_B}{1 + 2/h_{FE2}}, \tag{8}$$

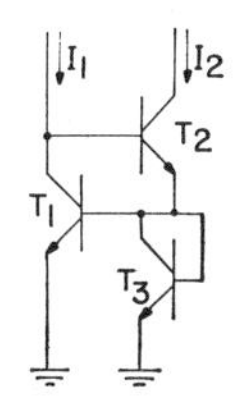

Fig. 6. Improved version of constant current source.

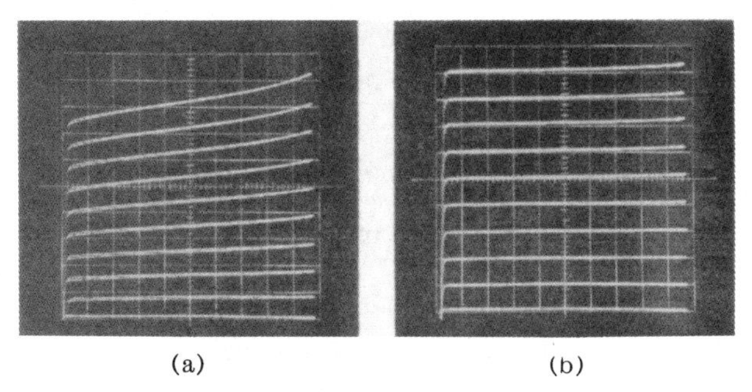

Fig. 7. Current–voltage characteristics of the current sources shown in Fig. 2(a) and Fig. 6(b).

where I_B is the base current drawn by T_1 and T_3. If all three current gains match, then I_2 becomes identically equal to I_1.

Further insight can be gained on the operation of the improved constant current source of Fig. 6 by treating the circuit as a special case of the shunt-series feedback pair topology of Fig. 3 (b), with $R_2 = 0$. This corresponds to unity current feedback condition from the emitter of T_2 to the base of T_1. As a consequence of this feedback arrangement, the output impedance of the resultant current source is greatly increased. This is illustrated in the current–voltage characteristics of Fig. 7. Also, since there is practically no base current error, temperature tracking of I_1 and I_2 is nearly perfect (<0.05 percent/°C).

Although the ratios of currents in both current sources so far discussed can be altered by choosing different sizes for the transistor, an extreme ratio requires a large surace area and reduces the matching advantages. To produce a small constant current from a significantly larger reference current, an emitter resistor can be added as shown in Fig. 8 [1].

Assuming matched device geometries, T_1 operates in this configuration at a higher current density than the base–emitter diode of T_2. The resulting voltage difference (ΔV) appears across R_1.

For an ideal diode under forward bias

$$I = I_s(e^{qV/kT} - 1) \simeq I_s e^{qV/kT} \tag{9}$$

where I_s is the reverse saturation current. Thus

$$\frac{I_1}{I_2} \cdot \frac{I_{s1}}{I_{s2}} = e^{q(\Delta V)/kT} \tag{10}$$

or

$$\frac{\Delta V}{T} = \left(\frac{k}{q}\right) \ln \left[\frac{I_1}{I_2} \cdot \frac{I_{s1}}{I_{s2}}\right] = \frac{I_2 R_1}{T}. \tag{11}$$

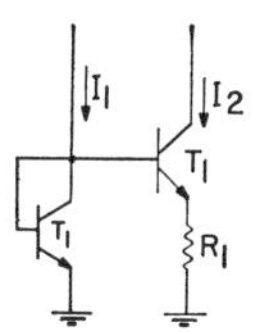

Fig. 8. Current source for small values of current.

Since I_{S1} and I_{S2} have the identical temperature behavior if the transistors are assumed to be matched, $\Delta V/T$ is a constant. To achieve a zero temperature dependence, one must satisfy

$$\frac{dR_1}{R_1} = \frac{dT}{T} \tag{12}$$

This inverse temperature relationship is characteristic of many pure metals. Thus the combination of a thin-film resistor with the proper temperature coefficient with a matched transistor pair produces a current source with a nearly zero temperature coefficient. The temperature behavior in this case is not affected by the absolute magnitudes of resistance and currents or the geometries of the matched transistors.

Pinch Resistors

The combination of the three layers used in integrated circuits can result in devices, which, depending on the operating point, behave as either current sources or high-value resistors.

The pinch resistor structure [5] of Fig. 1 (d) is formed by placing an n⁺-type emitter diffusion over an ordinary p-type diffused resistor. The emitter diffusion greatly reduces the effective cross-sectional area of the p-type resistor, and, consequently, raises the sheet resistivity from approximately 130 Ω/□ to 5 to 10 kΩ/□. Typical current–voltage characteristics of a pinch resistor are shown in Fig. 9 (a). The characteristics are linear only for small voltage drops across the resistor. Application of a higher dc voltage results in an increase of the reverse bias between the p-type resistor body and the surrounding n-type island. This reverse bias in turn causes the junction depletion layer to extend into the resistor structure, and effectively "pinch" the resistor, resulting in a saturation of current. Since the top portion of the pinch resistor is comprised of the heavily doped emitter diffusion, the resistor exhibits a breakdown voltage equal to that of the reverse biased base–emitter diode (6 to 8 volts). Saturation value of the current through the pinch resistor is also temperature-dependent, showing a temperature coefficient that is almost identical to that of the sheet resistance, but opposite in sign [6].

The sheet resistance of a pinched resistor structure exhibits a strong positive temperature coefficient of the order of 3000 ppm/°C to 5000 ppm/°C, as shown in Fig. 9 (b). Absolute values of the pinch resistors are difficult to control in fabrication; however, matching and tracking of

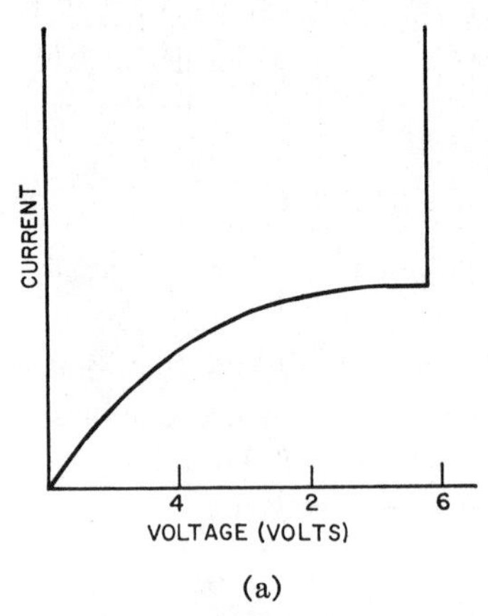

(a)

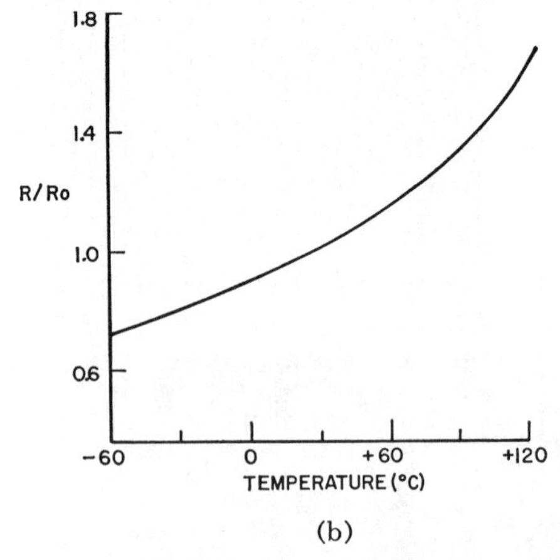

(b)

Fig. 9. Current–voltage characteristics and temperature behavior of the pinch resistor shown in Fig. 1(d).

pinched resistor structures on the same chip can be held to within ±5 percent. Since the effective thickness of the pinched resistor is the same as the base width of the n-p-n transistors on the chip, the variations of the absolute values of the pinch resistors tends to track the variations of the transistor current gain [1].

The buried FET structure [7] of Fig. 1(e) provides an alternate method of obtaining high-value resistors, but without the breakdown voltage limitations of base-diffused pinch resistors. In this case the channel of the buried FET structure, which constitutes the body of the resistor, is formed by the n-type epitaxial region surrounded on all sides by the p-type substrate, isolation, and the base diffusion. Since all p-n junctions forming the device structure are lightly doped, the breakdown voltages associated with the buried FET structure are typically in excess of 40 volts, thus making it well suited for high-voltage applications. It offers a sheet resistance of 3 to 6 kΩ/□, with a temperature coefficient of approximately 4000 ppm/°C. Unlike the pinch resistor, the pinch-off voltage associated with the buried FET structure is a function of the lateral dimensions of the device, as well as the epitaxial layer thickness, and thus can be designed relatively independently of the n-p-n bipolar transistor parameters.

Constant Voltage Sources

In discrete component circuits constant voltage sources are primarily used as voltage references in power supplies. In integrated circuits their importance is greatly increased; not only are voltage sources used as references but also as decoupling elements between stages.

The simplest constant voltage element is the breakdown diode formed by the base–emitter junction. The breakdown voltage of this junction is in the range from 6 to 8 volts, which is not only a convenient potential to work with, but also provides a relatively low temperature coefficient (300 to 800 ppm/°C). However, due to fairly high noise level associated with them, breakdown diodes should not be used at low signal levels.

Another method to obtain a constant reference voltage is the punch-through diode. Its breakdown depends on the depletion of a base region between two reverse-biased junctions [8].

Fig. 10 (a) shows typical current–voltage characteristics of a diode-connected punch-through transistor. The forward current–voltage characteristics of the same transistor are shown in Fig. 10 (b). With the state-of-the-art planar epitaxial technology and existing diffusion schedules, punch-through diodes can be fabricated having reverse breakdown voltages of 2.5 to 5.5 volts, and a typical temperature coefficient of −3 mV/°C. Its noise performance is comparable to that of an alloy-junction Zener diode with the same breakdown voltage.

A similar approach to produce a lower reference voltage with a smaller temperature coefficient [9] is shown in Fig. 11. The base–emitter avalanche diode D_1 is supplied by a current source I_1. A transistor is used as an emitter follower to supply current to a string of two diodes and a tapped resistor. The voltage at the cathode of D_2 has a temperature coefficient of about +7 mV/°C, reflecting the temperature coefficients of D_1, T_1, and D_2. At the other end of this string, D_3 provides a temperature coefficient of about −2 mV/°C. Thus, at a certain point on the tapped resistor the two temperature coefficients cancel exactly and the reference voltage, V_R, is taken from this point. This reference voltage is approximately 1.7 volts.

Stabilization of Substrate Temperature

The close thermal coupling between the components on a monolithic integrated chip allows the circuit designer to control the thermal environment of the circuit by maintaining the substrate at a steady temperature. This is done by incorporating a temperature sensor-controller circuit on the same substrate as the circuit to be stabilized [10], [11]. The function of the temperature regula-

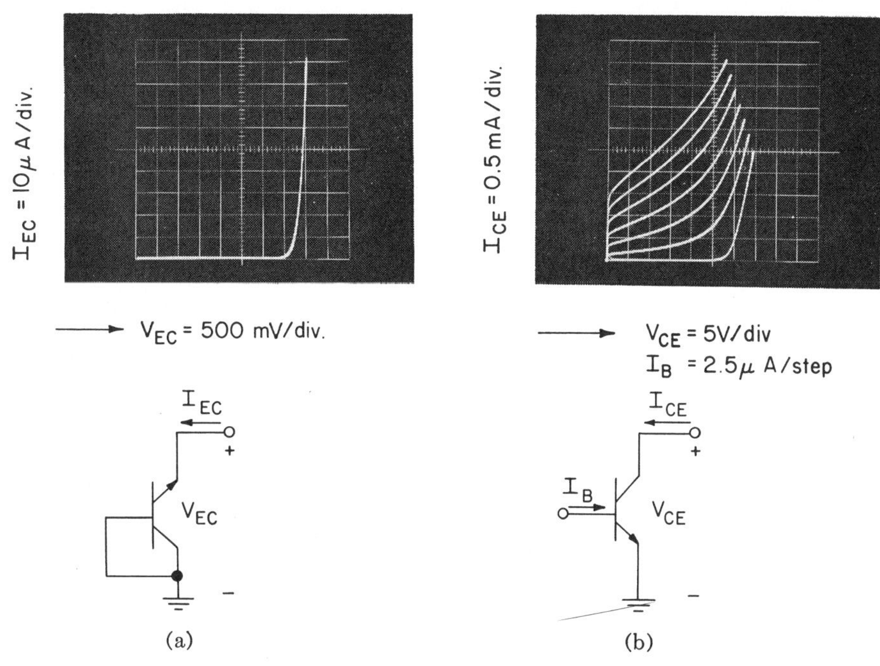

Fig. 10. Current-voltage characteristics of an integrated transistor. (a) Connected as a punch-through diode. (b) In the normal mode.

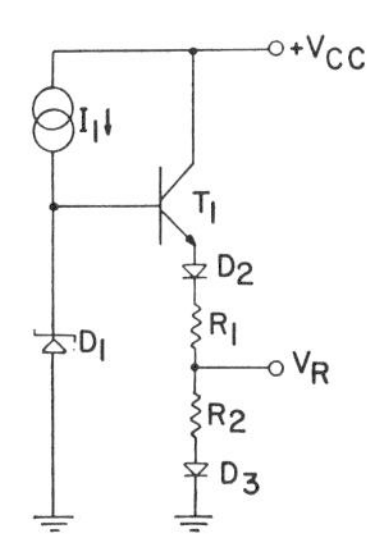

Fig. 11. Temperature-compensated voltage source.

tor circuit is to maintain the substrate at a constant elevated temperature, with minimum dependence on the ambient variations.

Fig. 12 shows a functional block diagram of a temperature sensor-controller circuit for stabilizing the substrate temperature. In practice, predictable temperature dependence of the forward voltage drop or the reverse breakdown voltage of the base–emitter diode is used as the temperature sensing element. A power transistor can be used as a heating element. In order to minimize the thermal gradients through the chip, the circuit to be stabilized is laid out symmetrically about the temperature sensor. To keep the chip at a temperature above the highest ambient temperature with a relatively small amount of additional power dissipation, thermal insulation is provided between the chip and the package.

Using the thermal stabilization technique discussed above, it is possible to control the substrate temperature to better than ±5°C over the entire military temperature range, with an added power dissipation of $\leq$ 500 mW [12]. As a result of regulating the substrate temperature as well as the thermal gradients within the chip, it is possible to design monolithic differential amplifiers with input offset drifts of less than 0.5 μV/°C.

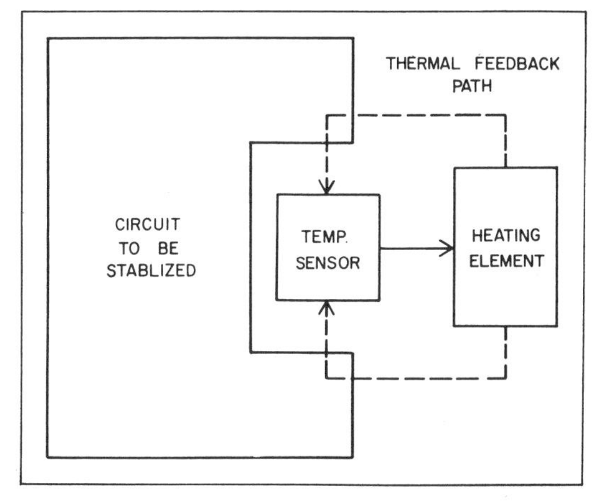

Fig. 12. Block diagram for substrate temperature stabilization.

Amplifier Stage Topologies

DC Level Shifting

In discrete circuit design, direct coupling of successive stages is typically done with complementary transistors. In most monolithic circuit applications, complementary devices (with the exception of lateral p-n-p transistors), are not readily compatible with the conventional integrated-circuit structures. When cascading several gain stages using only n-p-n bipolar transistors as active devices, it is necessary to shift the dc voltage levels within the circuit in order to obtain a desired output voltage swing. In addition, the required shift in dc levels must be obtained with minimum ac signal attenuation.

The use of base–emitter breakdown diode for level shifting may not be suitable for many circuit applications since the breakdown voltage range of these diodes are too restrictive, and they introduce excessive noise into the circuit.

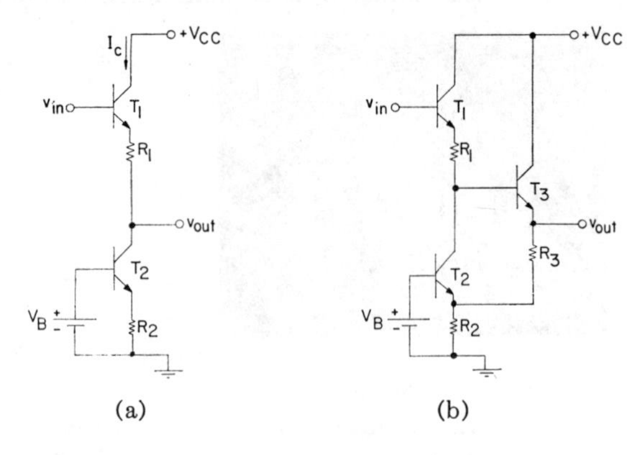

Fig. 13. Level shifting with n-p-n transistors.

A practical level-shifting scheme that meets most integrated-circuit-design requirements is shown in Fig. 13 (a) [13]. In this circuit, T_1 is an emitter follower stage, providing a low impedance source at its emitter, along with a unity voltage transfer ratio between its base and emitter. The transistor T_2 is operated as a constant current source. Assuming that h_{FE} of the transistors is much greater than unity, both transistors carry the same dc current, I_c, as determined by the dc bias on T_2, i.e.,

$$I_c = \frac{V_B - V_{BE}}{R_2} \tag{13}$$

where V_{BE} is the base–emitter voltage drop. Since the collector impedance of T_2 is very large, the ac signal from the emitter of T_1 to the output terminal suffers negligible attenuation. Due to the dc current drawn by T_2, the dc voltage level at the output, however, is more negative with respect to the input, by an amount ΔV, where

$$\Delta V = V_{BE} + (V_B - V_{BE})\left(\frac{R_1}{R_2}\right). \tag{14}$$

The level-shift stage has a relatively high output impedance, as determined by R_1. Therefore, its high-frequency performance is limited by the capacitive loading at the output.

The voltage gain of the simple level-shifting stage of Fig. 13 (a) can be increased above unity by means of a positive feedback scheme [13] shown in Fig. 13 (b). In this configuration, the transistor T_3 provides a low impedance output for the level-shifting stage. Positive feedback is provided to the emitter of T_2 via R_3. The overall voltage gain, A_v, for the circuit of Fig. 13 (b) can be written as

$$A_v = \frac{1}{1 - R_1/R_3}. \tag{15}$$

The frequency response of the circuit is relatively poor ($\simeq$ 200 kHz with $A_v = 4$) due to the narrow-banding effect of the positive feedback.

The positive feedback scheme can result in a negative real part of the input impedance seen at the base of T_1. In order to avoid potential instability, it is necessary that the positive feedback level-shift stage be driven from a low impedance source, such that the effective source impedance, R_s, at the base of T_1 must satisfy the inequality

$$R_s \ll \frac{h_{FE1}R_3}{A_v}. \tag{16}$$

Conventional design practice of cascading complementary gain stages to avoid level shifting can also be applied in integrated circuits, using lateral p-n-p transistors. The circuit performance is hampered, however, by the poor frequency response and high tolerances associated with the lateral p-n-p.

Active Loads

In the design of low-frequency high-gain integrated circuits, it is desirable to obtain the required voltage gain with a minimum number of cascaded gain stages. This is particularly true for the design of operational amplifier-type circuits where excess phase shifts due to added gain stages can result in instability unless sufficient external frequency compensation is provided. To obtain a very high voltage gain from a single transistor stage requires the use of a large collector load resistance. However, the available chip area, power supply values, and dc bias levels make the use of a very high-value load resistance impractical.

A method better suited to obtaining large voltage amplification from a single amplifier stage is the use of active devices as loads. In such an application, the active devices are utilized as current sources to simulate large dynamic resistances.

Active loads can be realized using lateral p-n-p transistors as shown in Fig. 14 (a). In this configuration, T_2 is an ordinary double-diffused n-p-n transistor, and T_1 is a lateral p-n-p, utilizing the device structure shown in Fig. 1. The lateral p-n-p is biased as a constant-current source, and simulates a high dynamic load resistance for the n-p-n transistor.

An alternate approach for obtaining large voltage amplification is shown schematically in Fig. 14 (b) where a pinch resistor (or a p-channel FET) is used as an active load [14]. In this application, the pinch resistor is operated in its saturated region, i.e., beyond pinch-off where the device offers a high dynamic resistance. Since the breakdown voltage of a p-channel pinched resistor is the same as that of a reverse-biased base–emitter diode the use of such a device as an active load is limited to low-voltage low-power circuits.

Fig. 14 (c) shows a high-gain differential amplifier stage which totally eliminates the need for resistive collector loads [7]. In this circuit topology, the p-n-p transistors are again lateral devices. The emitter-coupled transistor pairs T_1/T_3 and T_2/T_4 are equivalent to a common-emitter p-n-p pair with a low-frequency current gain equivalent to that of the n-p-n transistors T_1 and T_2. The p-n-p transistors T_3 and T_4 are biased through a common constant-current source. Assuming that the base–emitter voltage drops of T_5 and T_6 are well-

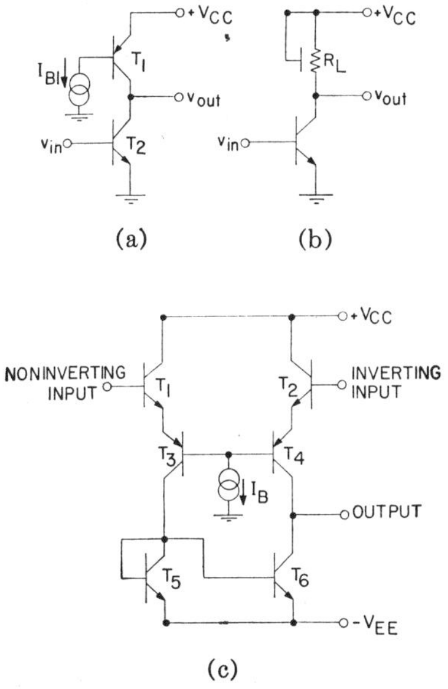

(a) (b)

(c)

Fig. 14. Circuits using active loads.

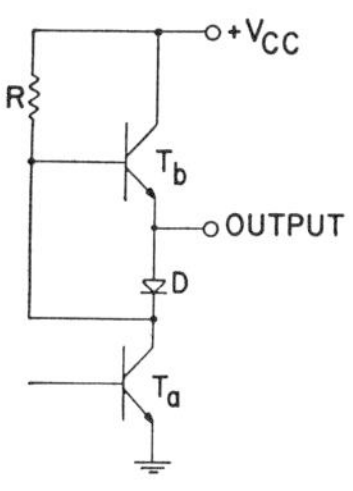

Fig. 15. Class-B output stage.

matched, their collector currents are approximately equal. The differential input signal is applied at the inputs of T_1 and T_2, and causes the currents through T_1/T_3 and T_2/T_4 to vary in opposite directions. However, the diode-connected transistor, T_5, constrains the collector current of T_6 to vary in the same manner as that of T_5. Consequently, T_4 and T_6 function as two opposing ac current sources where each device in turn functions as an active collector load for the other. In this way, the full differential gain of the stage is utilized.

Using active devices as dynamic loads, it is possible to obtain voltage gains in excess of 60 dB from a single amplifier stage. However, since the absolute values of the lateral p-n-p or pinch resistor parameters cannot be closely controlled in the state-of-the-art monolithic IC technology, additional care must be taken in designing the bias networks associated with the active loads. In practical designs, these bias networks rely on some type of feedback arrangement (common-mode or differential) to establish the proper operating points [7].

Class-B Output Stages

One of the few design ideas that have been developed for digital integrated circuits but are also applicable to linear design is shown in Fig. 15 [15]. This class-B output stage splits the signal without requiring transformers and capacitors. For positive swing of the input voltage, T_a is in its active region. The diode drop across D_1 ensures that T_b turns off. Consequently, in this half of the voltage swing the load current is supplied through T_a. The standby current is determined only by the value of R. In this arrangement the two transistors can never conduct simultaneously.

In order to switch conduction from one transistor to the other, the output voltage has to change by two V_{BE} drops. In most linear applications such a large "deadband" results in excessive cross-over distortion. It can be cut in half by connecting another diode between the base of T_b and the collector of T_a, as shown in Fig. 16.

The circuit topology of Fig. 16 also demonstrates a practical method of biasing a class-B stage for maximum output swing. Ignoring the error due to finite base currents, one can show that

$$I_{C3} = nI_{R1} \tag{17}$$

where n is the ratio of effective emitter lengths of T_3 and T_1. If the quiescent output is to be $V_{cc}/2$, then

$$I_{R1} = \left(\frac{V_{cc}}{2} - V_{BE}\right)\frac{1}{R_1}. \tag{18}$$

In the quiescent state, the bias current I_{R1} is supplied by T_2. Similarly the collector current for T_3 is drawn through R_2, or

$$I_{C3} = \left(\frac{V_{cc}}{2} - V_{BE}\right)\frac{1}{R_2}. \tag{19}$$

Thus, choosing the bias resistor values such that

$$\frac{R_2}{R_1} = n \tag{20}$$

the quiescent output voltage can be located at $V_{cc}/2$ over a wide range of supply voltages and resistor values. Within the constraints of integrated component tolerances, the scale factor, n is typically chosen $1 < n < 10$.

In the circuit of Fig. 16, R_1 and T_1 form a shunt-shunt feedback path. As previously discussed, this arrangement can be converted into a series-shunt feedback circuit by feeding the base current for T_3 through the source. Also, a higher input impedance can be achieved when Darlington pairs are substituted for T_1 and T_3.

The p-n-p transistors available in an integrated circuit can occasionally serve to design an effective complementary output stage. Due to their characteristics, however, both frequency performance and output power are impaired.

The p-n-p transistor formed by the substrate, epitaxial layer, and base diffusion can be used in the output stage as an emitter follower with the collector (substrate) connected to the most negative potential. The subtrate p-n-p can be used in combination with an n-p-n emitter follower to form a complementary output stage [16].

Another available approach is the interconnection of a lateral p-n-p with an n-p-n transistor to form a composite p-n-p structure shown in Fig. 17 [17]. This composite p-n-p transistor can be used in connection with

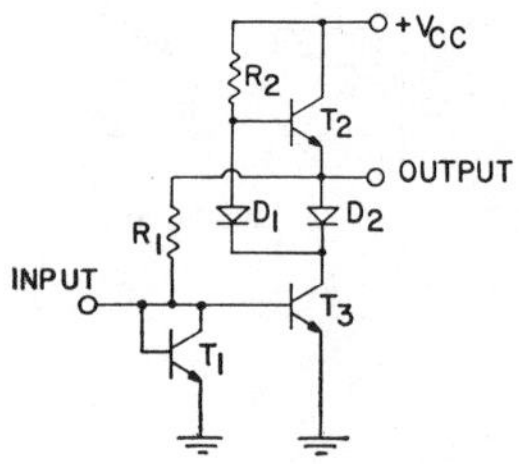

Fig. 16. Class-B output stage with resistance biasing.

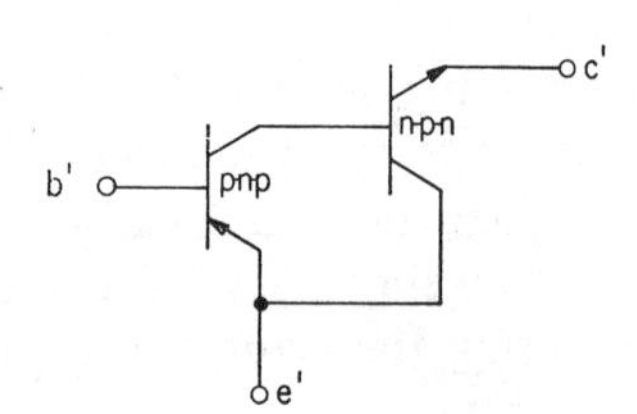

Fig. 17. Composite p-n-p transistor.

additional n-p-n devices to form a class B-output stage. The composite device has the polarity of the p-n-p and the effective current gain of the n-p-n. Its frequency response, however, is severely limited by the lateral p-n-p.

Constant Gain Stages

By making use of the matching and the tracking properties of integrated components, it is possible to design various amplifier stage topologies that offer desensitized gains over a wide range of temperature and parameter variations. The circuit topology of Fig. 18, for example, offers a relatively simple means of obtaining a prescribed voltage gain with 180° phase shift over a broad range of bias conditions, independent of temperature [18]. In this circuit for small signal swings the forward-biased diodes D_1 and D_n serve as a collector load, and provide an effective dynamic resistance, R_L, where

$$R_L = n\frac{kT}{qI_c}, \tag{21}$$

neglecting the effect of the base-spreading resistance. The low-frequency voltage gain of a common-emitter stage can be expressed as

$$\frac{V_{\text{out}}}{V_{\text{in}}} = \frac{-R_L}{kT/qI_c} = \frac{-nI_c}{I_c} \approx -n. \tag{22}$$

Therefore, the circuit topology of Fig. 18 is capable of providing a fixed voltage gain, $-n$, independent of the temperature as long as the transistor and the diodes are kept at the same temperature. This latter condition is very closely approximated in the integrated structure. However, this approach is relatively costly in terms of the chip area since each of the diodes, D_1 through D_n, requires a separate isolation pocket.

Frequency Broad-Banding

High frequency performance of broad-band amplifiers is determined by two dominant factors: gain-bandwidth limitations of active devices, and the parasitic reactances associated with the circuit elements. In the design of in-

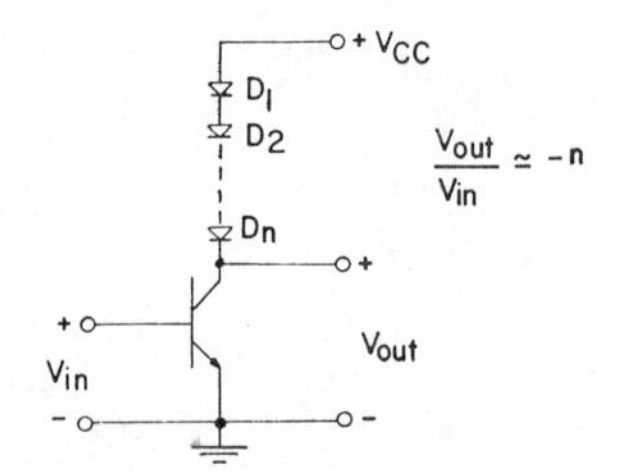

Fig. 18. Constant gain stage.

tegrated circuits, the availability of a large number of closely matched active devices, and the control over the device geometry and structure enables the designer to partially overcome these two limitations. Gain-bandwidth limitations of the active devices can be improved by replacing single devices by compound interconnection of two active devices. Similarly, the effects of the parasitic reactances inherent to the integrated device or circuit structure can be significantly reduced by using various neutralization techniques.

Compound Devices

The gain-bandwidth product of an amplifier stage can be significantly improved by using compound devices in place of single transistors. Fig. 19 shows two such compound device configurations replacing a common-emitter stage. It should be noted that the compound device connections shown in Fig. 19 are not new or novel in transistor circuit design [19]. However, their particular application and usefulness in linear-integrated-circuit design stems from the fact that the added active devices and their associated bias circuitry can be incorporated into the circuit with a minimal increase of circuit cost and complexity.

Fig. 19 (a) shows the ac circuit diagram of a Darlington, or a common-collector–common-emitter stage. In this configuration, T_1 effectively buffers the gain stage T_2 from the input; therefore, the capacitance loading at the input due to the Miller capacitance of T_2 is greatly reduced. The improvement of the frequency performance is particularly significant (several fold) for large values of R_s (i.e., $R_s > 2$ kΩ [20].

The cascode or common-emitter–common-base stage of Fig. 19 (b) also forms a useful compound-device topology, which offers a high-frequency performance superior to that of a common emitter stage. The bandwidth improvement is due to the low input impedance of T_2 as seen by the output (collector) of T_1, which minimizes the narrow-banding influence of the Miller-feedback capacitance associated with T_1. The circuit has a low-frequency current gain equal to that of T_1 since the current gain of transistor T_2 is very nearly unity. Since T_2 buffers the current-gain stage, T_1, from the load, the frequency broad-banding becomes more significant as the value of R_1 is increased. For example, for medium voltage gain amplifier stage design ($R_L > 2$ kΩ), the compound device can offer up to an order of magnitude improvement in the 3-dB bandwidth of the stage.

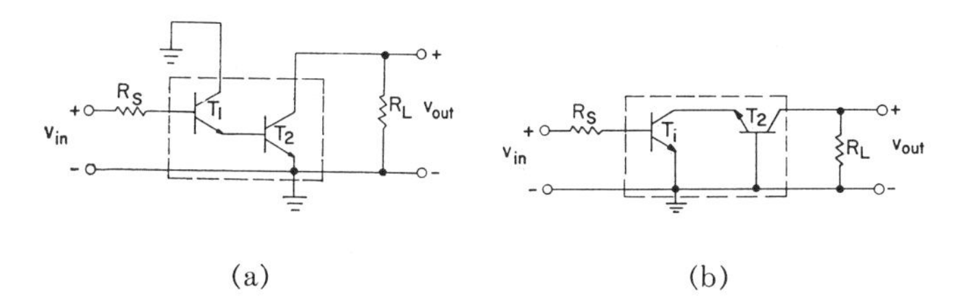

Fig. 19. Ac circuit diagrams for compound device stages. (a) Darlington or common-collector–common-emitter stage. (6) Cascode or common-emitter–common-base stage.

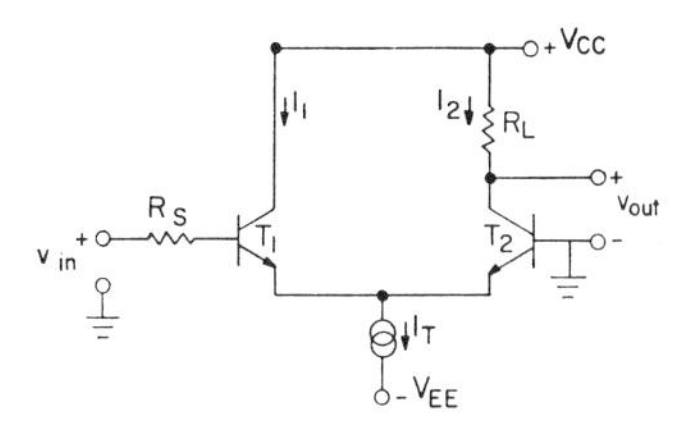

Fig. 20. Emitter-coupled or common-collector–common-base stage.

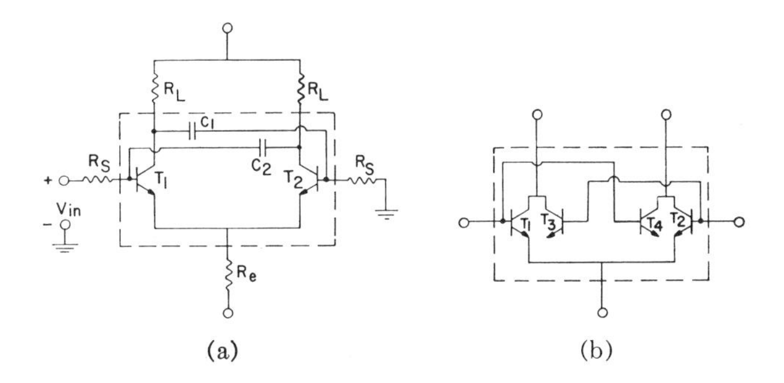

Fig. 21. C_c neutralization.

In both the Darlington and the cascade connections, the reverse transmission through the stage is greatly reduced due to the large impedance mismatches between T_1 and T_2. This effect makes these connections particularly attractive for design of high-frequency tuned amplifier stages where the parasitic cross-coupling between the input and the output tank circuits can make the alignment of the stage difficult, and limits the usable power gains.

In the compound-device connections of Fig. 19, the use of a second transistor adds an additional pole to the overall circuit response. The magnitude of this pole becomes comparable to the dominant pole of the stage as the value of R_s or R_L are raised, respectively. Consequently, the high-frequency response may exhibit a rapid roll-off and excess phase lag. This effect is particularly dominant in the cascade stage, which exhibits a pair of dominant complex conjugate poles for large values of R_L [20]. Therefore, additional care must be taken in applying overall feedback around amplifier stages containing compound devices.

Emitter-Coupled Stages

For high-frequency integrated-circuit applications, the emitter-coupled or the common-collector–common-base amplifier topology, (Fig. 20) also offers considerable advantage over a simple common-emitter stage. In the circuit, the common-collector transistor T_1 provides the necessary current amplification with minimum capacitive loading of the source. T_2, in turn, converts this current gain to a voltage swing across R_L. Similar to the Darlintgon and cascade stages, the emitter-coupled circuit also eliminates the narrow-banding effect of the Miller-feedback capacitance, and greatly reduces the reverse transmission. For low frequencies, the voltage gain of an emitter coupled stage is approximately equal to one half of that available from a similar common-emitter stage, for $I_1 = I_2$, and can be expressed as

$$A_v \simeq \frac{h_{FE}R_L g_m}{g_m R_s + 2h_{FE}}. \qquad (23)$$

The 3-dB bandwidth of the stage, however, is significantly larger than twice the common-emitter bandwidth, especially for large values of R_s and R_L.

Due to common-mode feedback provided by the biasing current source, the emitter-coupled stage offers a very high degree of bias stability. Therefore, it is well suited to direct coupling and can be used as a building block for multistage broad-band amplifier and limiter circuits [21].

Neutralization of the Collector–Base Capacitance

In a differential amplifier stage, the parasitic feedback through the collector–base capacitance, C_c, can be significantly reduced by utilizing a bridge neutralization scheme as shown in Fig. 21 (a) [22]. In a balanced differential stage, the ac voltages at the collectors of T_1 and T_2 have the same amplitude, but differ in phase by 180°. Therefore, the feedback through C_1 and C_2 to the bases of T_2 and T_1 are out of phase with the internal feedback due to C_c of the transistors. By choosing C_1 and C_2 to have the same value as the C_c of the transistors, it is possible to neutralize the parasitic feedback through C_c.

In discrete circuit design applications, this neutralization technique has severe practical limitations because an efficient neutralization of C_c requires that the feedback capacitors C_1 and C_2 be matched by C_c within better than 10 percent in absolute value. In practical devices, C_c is typically less than 1 pF, and its absolute value is predictable to only ±20 percent. Furthermore, discrete capacitors cannot track the variations of C_c as a function of the device operating point, frequency, and temperature.

Since the C_c neutralization scheme requires matching and tracking of the components, rather than accurate absolute values, it is ideally suited for integrated-circuit applications. The feedback capacitances C_1 and C_2 can be realized as the collector–base capacitances of additional transistors T_3 and T_4, as shown in Fig. 21(b). All four transistors have identical geometries, and are fabricated simultaneously. Therefore, the collector–base capacitances of T_3 and T_4 ideally match and track C_c of T_1 and T_2. T_3 and T_4 have their collectors common with T_1 and T_2, respectively; therefore, they do not require

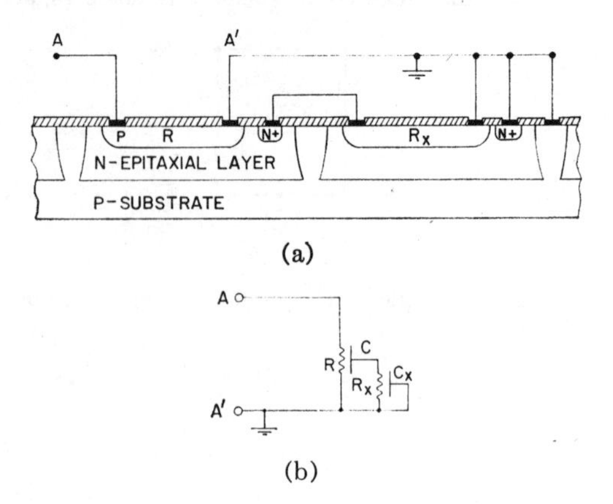

Fig. 22. Diffused resistor with reduced parasitic capacitance.

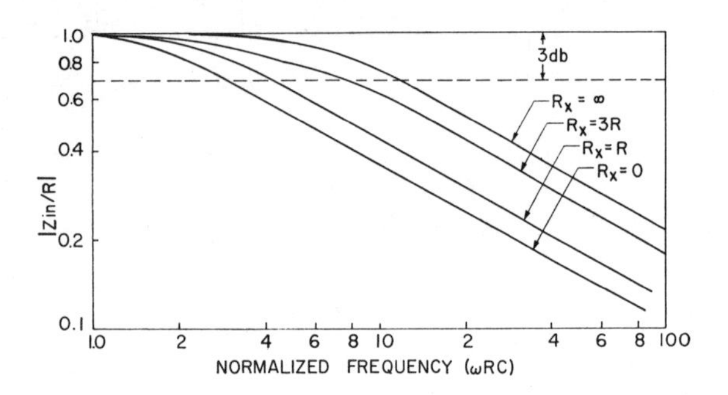

Fig. 23. Frequency response improvement of resistor with reduced parasitic capacitance.

additional isolation. Thus, the increase of circuit chip size due to addition of T_3 and T_4 is negligible.

Reducing the Effects of Parasitic Capacitances

In the design of high-frequency linear-integrated circuits, the distributed parasitic junction capacitance inherent to the diffused resistor structure can be one of the factors that limit speed and performance. In most circuit applications, the n-type pocket, in which the resistor is imbedded, is connected to an ac ground point in the circuit.

The detrimental effects of this parasitic capacitance, as well as the excess phase lag associated with it can be greatly reduced by effectively isolating the n-type pocket from the ac ground point in the circuit as illustrated in Fig. 22 (b) [23]. In the figure, R is the actual circuit resistor, and C is the total distributed capacitance associated with it. An additional resistance, R_x, is used as a means of isolating the resistor pocket from the ac ground point in the circuit. It can be shown that the parasitic capacitances associated with R_x, or with the n-pocket and p-substrate interface have a negligible effect on the frequency response of R. Therefore, these are omitted in the equivalent circuit diagram of Fig. 22 (b).

Fig. 23 gives a calculated plot of normalized driving point impedance for the equivalent circuit of Fig. 22 (b), for various values of R_x. For ease of computation, C is assumed to be uniformly distributed. Examining the family of curves given in the figure, it is seen that R_x does not need to be prohibitively large to provide significant broad-banding of the frequency response of R. For example, for $R_x = 3R$, the 3-dB bandwidth of the diffused resistor is increased by a factor of 2.7.

Since the isolating resistors are at all times in series with reverse biased p-n junctions, they have no effect on the dc performance or the biasing of the circuit, and can be added to an already designed circuit at little additional cost.

Summary

A recurring theme in the above discussion is the exploration of new avenues to solve old problems. The fact that the problems have been previously solved with discrete components is of little value to the integrated-circuit designer and only tends to create confusion. We are apt to believe that the approaches so far developed and widely used are the best ones or even the only ones.

The designer and user of linear-integrated circuits should not forget that during the vacuum tube and transistor eras of electronics the circuits were designed with specific components in mind. Thus during the last fifty years only a few out of many possible approaches were explored and only the device parameters that appeared relevant were examined.

In integrated-circuit design the ground rules have taken on a different form. It comes, therefore, as no surprise that the successful design techniques for linear-integrated circuits are those that are profoundly different from conventional ones.

References

[1] R. J. Widlar, "Some circuit design techniques for linear integrated circuits," *IEEE Trans. Circuit Theory,* vol. CT-12, pp. 586–590, December 1965.
[2] M. S. Ghausi, *Principles* and *Design of Linear Active Circuits.* New York: McGraw-Hill, 1965.
[3] M. J. Hellstrom *et al.,* "An integrated circuit preamplifier with nonlinear bootstrapped input impedance," *Proc. 1967 Natl. Electron. Conf.,* pp. 321–324.
[4] G. R. Wilson, "A monolithic junction FET-NPN operational amplifier," *1968 ISSCC Digest Tech. Papers.*
[5] H. C. Lin, "Nonlinear resistance for microelectronics," *Proc. 1961 Natl. Electron. Conf.,* vol. 17, p. 25.
[6] A. B. Grebene and S. K. Ghandhi, "Behavior of junction-gate field-effect transistors beyond pinch off," *1968 ISSCC Digest Tech. Papers.*
[7] R. J. Widlar, "Monolithic op. amp. with simplified frequency compensation," *EEE,* pp. 58–63, July 1967.
[8] R. H. Rediker, "Narrow base planar punch-through diode," U.S. patent 2975342, March 1961.
[9] R. J. Widlar, "A versatile monolithic voltage regulator," National Semiconductor, Appl. Note AN-1, February 1967.
[10] S. P. Emmons and H. W. Spence, "Very low-drift complementary semiconductor network dc amplifier," *IEEE J. Solid-State Circuits,* vol. SC-1, pp. 13–18, September 1966.
[11] T. F. Prosser, "An integrated temperature sensor-controller," *IEEE J. Solid-State Circuits,* vol. SC-1, pp. 8–13, September 1966.
[12] J. D. Lieux, "A new approach to low drift amplifiers," Fairchild Semiconductor, Appl. Note 20-BR-0021-48, March 1968.
[13] R. J. Widlar, "A monolithic, high gain dc amplifier," *Proc. 1964 Natl. Electron. Conf.,* pp. 169–174.
[14] H. C. Lin, *Integrated Electronics.* San Francisco, Calif.: Holden-Day, 1967, pp. 203–204.
[15] H. R. Camenzind, *Circuit Design for Integrated Electronics.* Reading, Mass.: Addison-Wesley, 1968, p. 155.
[16] R. J. Widlar, "A unique circuit design for a high performance operational amplifier especially suited to monolithic construction," *Proc. 1965 Natl. Electron. Conf.,* vol. 21, pp. 85–89, October 1965.

[17] H. C. Lin, *Integrated Electronics*. San Francisco, Calif.: Holden-Day, 1967, pp. 283–290.
[18] A. J. van Overbeek, "Tunable resonant circuits suitable for integration," *1965 ISSCC Digest Tech. Papers*.
[19] R. D. Thornton, L. L. Searle, D. O. Pederson, R. B. Adler, and E. S. Angelo Jr., *Multistage Transistor Circuits*, SEEC, vol. 5. New York: Wiley, 1965, pp. 172–182.
[20] B. A. Wooley, "Monolithic wideband amplifiers," Electronics Research Lab., University of California, Berkeley, Memo. ERL-M243, March 1968.
[21] "RCA linear integrated circuit fundamentals," Radio Corporation of America, RCA Technical Series IC-40, pp. 231–238, 1966.
[22] G. W. Haines, "C_c compensated transistors," presented at the 1966 IEEE Electron Devices Meeting, Washington, D.C., October 1966.
[23] A B. Grebene, "A practical method for reducing the effects of parasitic capacitances in integrated circuits," *Proc. IEEE* (Letters), vol. 55, pp. 235–236, February 1967; also *ibid.*, pp. 2039–2040, November 1967.

Bipolar Design Considerations for the Automotive Environment

WILLIAM F. DAVIS

Abstract—**In view of the detrimental voltage transients associated with the automotive environment, and the advantages and limitations of the present bipolar integrated circuits (IC) with respect to this environment, new bipolar circuit and device innovations have been specifically developed to economically protect the IC from these destructive transients and improve the IC's functional performance. Many of these techniques prevent the negative voltage transients from injecting electrons into the p-type substrate, insuring that a newly defined parasitic lateral n-p-n transistor cannot degrade the performance of the IC. In addition, the maximum nondestructive supply voltage capability of the IC was extended to be greater than BV_{CEO} of the n-p-n transistors for easier and more economic protection against the automotive's destructive supply voltages. New and effective high-frequency input filtering techniques were also devised on the chip to prevent the high-frequency voltage transients from gaining access to the IC circuitry, especially at the logic inputs.**

I. Introduction

RECENT legislation by the United States government to improve automotive safety and reduce detrimental exhaust emissions has produced a significant demand for massive quantities of the bipolar integrated circuit (IC) to perform the required functions. Present marketing studies indicate that this trend will continue and be even more significant in the future. Several standard bipolar IC's have already been specifically developed for use in recent automotive systems where flexibility and/or a limited development time was of the utmost importance [1]–[4]. Custom bipolar IC's have also been used where system complexity and/or cost considerations prohibits the use of these standard IC's (e.g., seat-belt/starter interlock systems). Many automotive systems require both, and the need for both is apparent. However, in order for either type of bipolar IC to be both technically and economically successful in specific automotive applications, a thorough understanding must be gained not only of the automotive environment, but also of the advantages and disadvantages associated with the IC in this type of environment.

Several potential problems exist for today's automotive bipolar IC. Many positive and negative voltage transients are present in the electrical system and can either have sufficient energy to destroy the IC, or the negative transients, in particular, can cause electrons to be injected into the substrate, activating a parasitic lateral n-p-n transistor (N-epi/P-substrate/N-epi) which can seriously degrade the IC's functional performance. In addition, the system's high-frequency voltages are difficult to filter using present standard components and printed circuit board techniques, and erroneous IC performance can result, especially in view of the excellent high-frequency response offered by today's bipolar technology.

In view of these problems, new bipolar design techniques have been developed which can extend the maximum nondestructive supply voltage capability of present bipolar IC's to values greater than BV_{CE0} of its n-p-n devices, allowing a more economic approach for protecting the IC from destructive automotive supply voltages. Methods were also developed to economically prevent the detrimental substrate injection of electrons, using circuit and device techniques, and to protect the IC's input circuitry from destructive or erroneous high-frequency input voltages, using on-chip high-frequency filtering techniques. In general, these methods can be

Manuscript received May 5, 1973; revised July 18, 1973.
The author is with the Motorola Semiconductor Products Division, Phoenix, Ariz. 85036.

Reprinted from *IEEE J. Solid-State Circuits,* vol. SC-8, pp. 419-426, Dec. 1973.

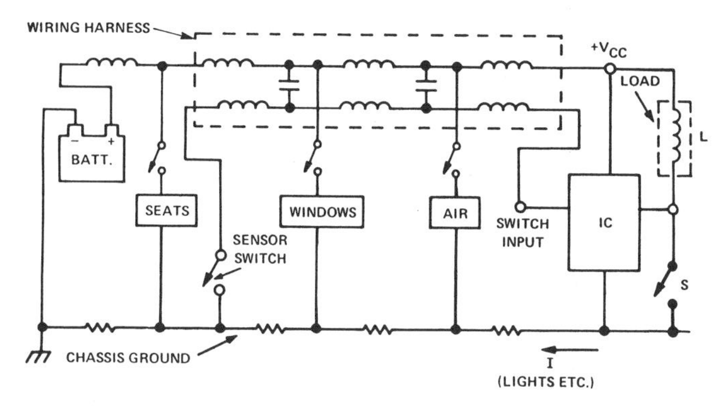

Fig. 1. System model of the automotive electrical environment with respect to an IC.

applied to many automotive systems such as fuel control, ignition, seat-belt/starter interlock, antiskid, and battery charging for optimum protection and functional performance at the lowest possible cost.

II. The Automotive Environment

A. An Automotive System Model

Fig. 1 demonstrates the type of electrical system a bipolar IC may encounter when placed in the automotive environment. The IC's supply voltage $+V_{CC}$ is typically derived from a battery through a long wire that can possess a significant amount of series inductance. Unfortunately, this wire may also distribute power to the air conditioner, power windows, power seats, or any other large current device. Thus, during the high-current switching mode of any of these accessories, large positive inductive voltage transients can develop on the supply wire and be reflected to, and cause possible destruction of, the unprotected IC.

Sometimes it is necessary for the IC's functional inputs to receive information from a remote sensor (e.g., seat-belt switch, a speed, temperature, or pressure sensor) which is referenced to chassis ground but physically located a considerable distance from the IC. Due to the general rust and corrosion of the chassis, and the large circulating currents that can develop within this chassis ground, the electrical ground reference of the sensor can be considerably different than the ground reference of the IC, causing erroneous sensor information to occur at the IC's input. In addition, the sensor wire, which connects the IC to the sensor, may pass through a wiring harness with other wires which possess significant high-voltage transients. As a result of inductive and capacitive coupling between these wires, voltage transients in the harness could be reflected to the IC's sensor input to either produce erroneous sensor information or, if the transient energy is sufficient, destroy the input circuitry.

Sensing a switch closure S to ground for determining the conduction state of an inductive load L referenced to $+V_{CC}$ may also be a requirement of an IC input, as shown. However, immediately after the switch is opened, a large positive inductive voltage transient would be reflected to, and cause possible destruction of, the IC's input. (If this mechanical switch was instead implemented with a bipolar common-emitter IC transistor, the "turn-off" transient voltage could have easily exceeded BV_{CEO}, possibly destroying the device.) Similarly, sensing a switch closure to $+V_{CC}$, to determine the conduction state of an inductive load referenced to ground, can allow large negative inductive voltage transient to be reflected to the IC's input immediately after opening the switch, again causing possible destruction or malfunction of the IC.

Sometimes the automotive IC must withstand other hazards such as a reverse 12-V battery, a cold-temperature "jump start" with 24 V applied, or worse yet, a reverse 24-V battery applied during an attempt at a jump start. Also, the IC must not be destroyed or malfunction during the presence of any high-frequency RF voltages, whether they eminate from sources external to the automobile (radio transmitters, traffic lights, etc.) or from within the automotive environment (two-way communication gear, ignition spark, warning buzzers, etc.). Some electronic systems (e.g., seat-belt/starter interlock) must function continuously without error during the normal cold-temperature cranking mode when the battery voltage may drop to as low as 4 V. Other systems may be located at the rear of the automobile (e.g., antiskid) where the operating voltage, usually derived from lengthy tail or brake light wires, is low due to the resistive voltage drops in the supply line. Although special technical and economic considerations should also be given to the maximum allowable junction temperature and packaging requirements of an IC placed in the hostile under-hood environment, a satisfactory discussion would rival the length of this paper and, for this reason, further discussions will be limited only to the electrical aspects of the environment.

B. The Electrical Charging System

The alternator can be simply represented as a dependent current source I_0, which is proportional to the field current I_F, provided the alternator is driving a relatively low impedance and the shaft is rotating above a given r/min. However, the alternator can also be a source of large positive and negative voltage transients when used in conjunction with the battery, as shown in Fig. 2. Assuming the battery is connected to the system (switch S closed), a large negative voltage transient can develop on the ignition line when the ignition is switched OFF due to current decaying not only in the field coil but also in other inductive loads that may be connected to the ignition line. A large positive voltage transient (commonly referred to as the load dump transient) can occur on the $B+$ line, if the low-impedance battery is disconnected from the system (e.g., switch S opened) while the alternator is providing a given output current. Since the effective instantaneous change in the $B+$ impedance can be rapid

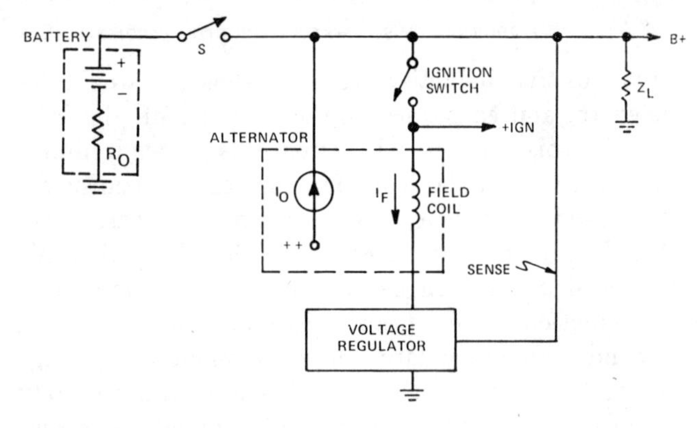

Fig. 2. Basic automotive electrical charging system which can produce large destructive positive and negative voltage transients on the $B+$ line and the ignition line, respectively.

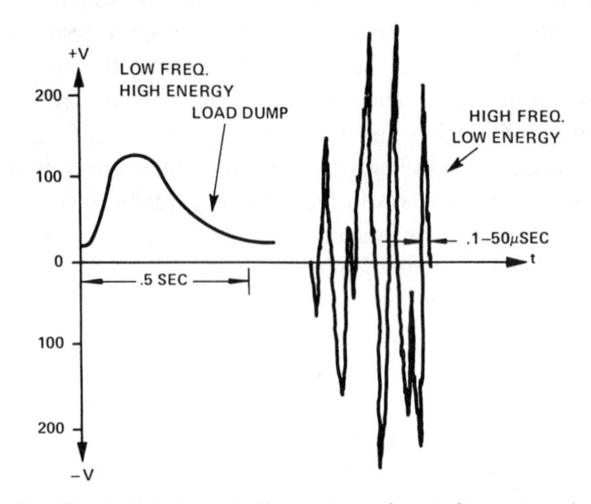

Fig. 3. Characteristics of the automotive voltage transients—high energy, high voltage, and fast rise times.

and significant (from the battery impedance R_0 to the load impedance Z_L), and the alternator output current remains essentially constant during this $B+$ impedance change (the field current cannot change instantaneously), the $B+$ voltage must increase until the open-circuit output voltage of the alternator is reached. The worst case open-circuit output voltage, and thus the maximum load dump voltage, will occur for maximum values of field current, alternator shaft r/min, and load impedance Z_L. This condition corresponds to a full charge on a "dead" battery with all accessories OFF when the battery is disconnected.

C. Automotive Voltage Transient Characteristics

Typical worst case load dump transients range from 80 to 120 V and are the most destructive to IC electronics in view of their associated energy content, as shown in Fig. 3. Although other automotive transients can reach peak voltages in excess of ±200 V, their energy content is considerably less, presenting less of a threat to the IC electronics. However, each of these general automotive transient characteristics—high energy, fast rise times, and large peak negative voltages—presents three particular problems to the bipolar IC and each requires a different method of solution.

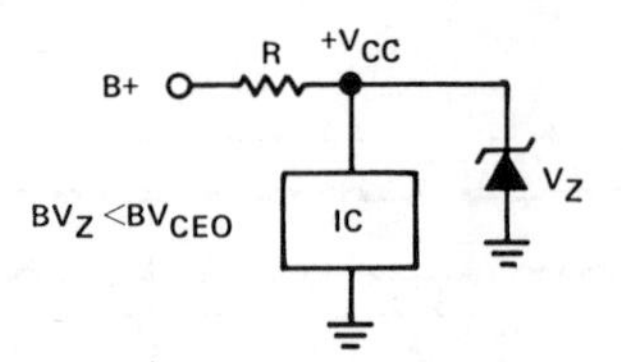

Fig. 4. Conventional protection technique which employs a Zener diode clamp at the $+V_{CC}$ terminal to protect the bipolar IC from destructive automotive $B+$ transients.

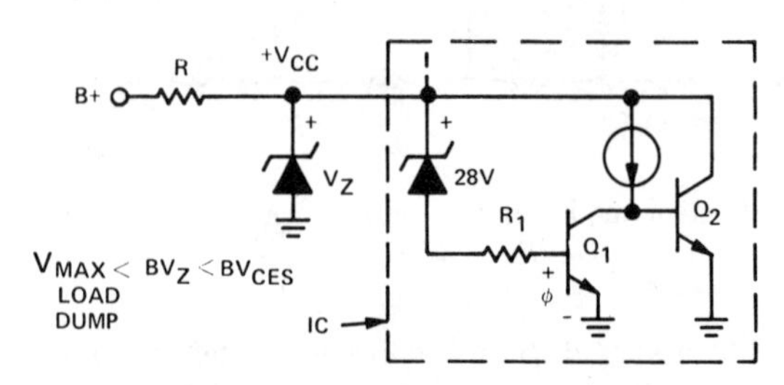

Fig. 5. New method which protects the IC devices from destructive $B+$ voltage transients by extending the nondestructive supply voltage capability of the IC to BV_{CES}, eliminating the need for the external Zener diode to dissipate any part of the load dump energy.

III. Destructive Transient Voltage Protection at $+V_{CC}$

In the past, a bipolar IC was protected from large positive and negative supply voltages by connecting a Zener diode across the IC, as shown in Fig. 4, to limit the suply voltage to less than BV_{CEO} of the IC devices and protect the N-epi/P-substrate junctions during the negative voltage transients (including reverse battery). In many cases, the resistance R must be sufficiently small to insure proper low-voltage cranking mode operation at the IC's required quiescent bias current. However, during the load dump transient, the voltage differential across this resistor can become sufficiently large to cause substantial current flow and thus significant power dissipation in the Zener diode.

The technique of Fig. 5, however, can extend the applied input voltage capability of the IC above the worst case load dump voltage, enabling the use of a less expensive protective Zener diode, since it would only be required to clamp the lower energy higher voltage transients and never dissipate any part of the load dump energy. Basically, all potential BV_{CEO} breakdown voltages are converted into substantially larger BV_{CES} voltages for the duration of the transient. When the supply voltage exceeds approximately $(28 + \phi)$ V, transistor Q_1 conducts and provides a BV_{CES} condition for transistor Q_2 before its collector–emitter voltage can exceed BV_{CEO}. Thus, the maximum nondestructive supply voltage capability of this IC exceeds BV_{CEO} and is defined as BV_{CES}. By choosing the proper epi resistivity and epi thickness, BV_{CES} can be designed to be greater than the specified maximum load dump voltage. In general, this technique can also be economically applied to external high-current power transistors, since the simultaneous requirement for large collector–emitter

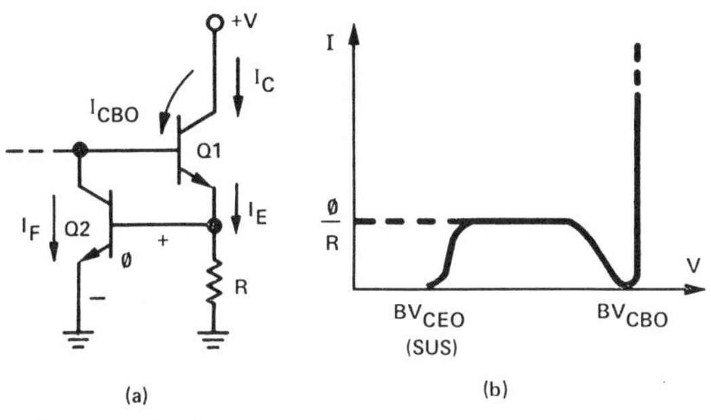

Fig. 6. One method which extends the nondestructive voltage capability of an IC device by limiting the I_{CEO} breakdown current to a nondestructive value.

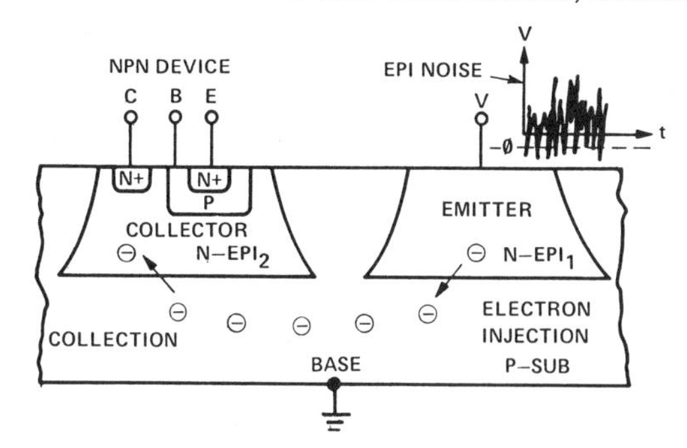

Fig. 7. Parasitic lateral n-p-n transistor. Negative voltages on the N-epi$_1$ region cause injection of electrons into the substrate which are collected by the other reverse biased N-epi$_2$ region, causing possible circuit malfunction.

voltage and collector–emitter current is eliminated during the load dump transient, and only BV_{CES} is required to be greater than the maximum load dump voltage.

A more general technique can be employed, which also extends the nondestructive supply voltage capability, by simultaneously providing the stressed device with a low base impedance and a high emitter impedance to safely limit the I_{CEO} breakdown current when the BV_{CEO} breakdown voltage is exceeded. Usually, this is acomplished by terminating the base of the device to be protected on a voltage source and connecting its emitter to a large valued resistor or controlled current source. Another unique way of implementing this function is shown in Fig. 6. During normal operation [Fig. 6(a)], the dynamic emitter impedance of transistor Q_1 is dominant with respect to the resistance R, and the resistive voltage drop is insignificant with respect to the threshold voltage ϕ of transistor Q_2. However, as the collector voltage of transistor Q_1 exceeds BV_{CEO}, the I_{CEO} breakdown current is limited to ϕ/R [Fig. 6(b)] due to the localized negative feedback through transistor Q_2 which regulates the total base current of transistor Q_1 such that β $(I_{CBO} - I_F) = \phi R$, where I_{CBO} is the collector–base leakage of transistor Q_1, β is the current gain of transistor Q_1, and I_F is the collector–emitter current of transistor Q_2. The maximum allowable value of the I_{CEO} breakdown current is dependent upon the device's ability to withstand both the defined emitter current density and the worst case power dissipation in the collector–base junction. A small-signal minimum-geometry n-p-n IC transistor, for example, with a BV_{CEO} of 35 V, can easily withstand a collector–emitter voltage of 80 V without destruction, provided the collector–emitter breakdown current is limited to 5 mA.

In general, these basic techniques also permit bipolar IC's to be ideally compatible with new automotive surge suppressors (which limit the maximum load dump voltage to approximately 60 V), since the need for any other protection device is eliminated.

IV. The Parasitic Lateral n-p-n Transistor: Substrate Injection of Electrons

Negative voltage transients, applied to any N-epi region with sufficient magnitude to forward bias the N-epi/P-substrate junction, can seriously degrade functional IC performance due to the injection of electrons into the substrate (substrate injection). These minority charge carrier electrons will diffuse throughout the P-substrate region until they recombine or are collected by other reversed biased N-epi regions. If collected, a parasitic lateral n-p-n transistor can be defined, as shown in Fig. 7, where the forward biased N-epi$_1$ region is the emitter, the P-substrate is the base, and the reversed bias N-epi$_2$ region is the collector. In practice, for adjacent N-epi islands with an N-epi thickness and resistivity of 13 μ and 2 $\Omega \cdot$ cm respectively, separated by a 10-$\Omega \cdot$ cm P-substrate base width of 1.4 mil (including out diffusion of the P-type isolation diffusion), the parasitic transistor exhibits a typical emitter–collector current gain α of about 0.3. For an increased N-epi island separation of 30 mil, α becomes approximately 0.001. Unfortunately, N-epi$_2$ may also be the collector of another transistor (or the base of a lateral p-n-p transistor), incorporated, for example, as part of a flip-flop memory or timing circuit or any other circuit where this parasitic current would be intolerable. Thus circuit malfunction could result due to the parasitic collection of substrate injected electrons and illustrates the need to prevent the negative voltages associated with the automotive environment from forward biasing any N-epi region with respect to the P-substrate during normal IC operation.

A. Substrate Injection at the $+V_{CC}$ Supply Terminal

It is common practice in the IC industry to connect N-epi islands to the $+V_{CC}$ supply terminal for biasing purposes. For example, the collector of an n-p-n transistor emitter follower, the N-epi island associated with the P-type resistor, and the N-epi FET resistor are usually all connected to the positive supply $+V_{CC}$. Although a Zener diode is normally connected across the IC to prevent destruction of the N-epi/P-substrate junction as previously mentioned, it would not prevent the N-epi islands from injecting electrons into the substrate, since the diode's high-current forward voltage drop is more than sufficient to forward bias these junctions. To prevent substrate injection, a capacitor could be placed from the supply terminal to ground in order to integrate the negative voltage transients, but the

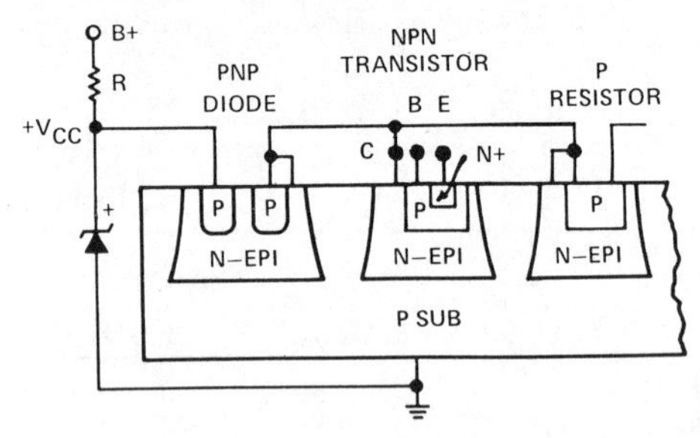

Fig. 8. Prevention of substrate injection at the V_{CC} terminal with a high-voltage IC protection diode. The diode's P-type region is reversed biased during negative B+ voltage transients, preventing the N-epi islands from injecting electrons into the substrate.

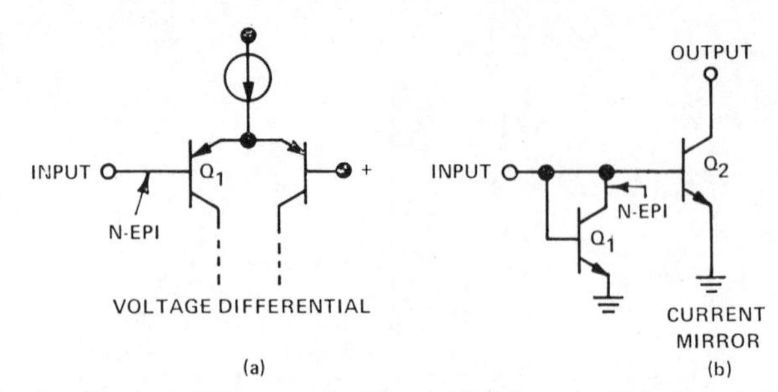

Fig. 9. Typical IC input circuitry which is suseptible to substrate injection. (a) A p-n-p transistor voltage differential stage. (b) An n-p-n transistor current mirror stage.

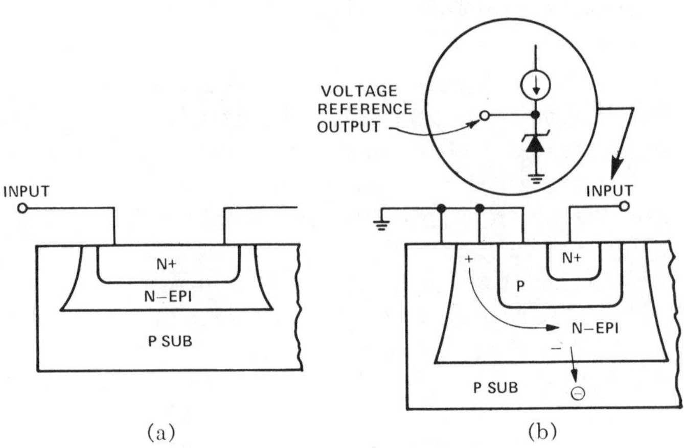

Fig. 10. Other IC devices capable of producing substrate injection of electrons. (a) N+ crossunder connected to an input. (b) Zener reference diode connected to an output.

required capacitance value may be large for the energy involved and prove rather costly. Alternatively, a high-voltage diode could be placed between the IC and the supply terminal, through which the IC's quiescent bias current would normally flow, except during the negative voltage transients when the diode would reverse bias and prevent substrate injection. Aside from the additional cost of the diode, this method is also undesirable, since it directly subtracts one forward biased diode voltage from the operating supply voltage which, for some systems, cannot be tolerated during the low-voltage cranking mode.

A similar technique, shown in Fig. 8, places a high-voltage IC diode (p-n-p) between the appropriate N-epi regions and $+V_{CC}$ terminal, again for the prevention of substrate injection. However, the low-voltage IC performance is not degraded by this technique, since the P-type regions of the remaining functional IC circuitry (such as a P-type resistor in a floating N-island or the P-type emitter of an active p-n-p transistor current source) can also be connected directly to the $+V_{CC}$ terminal, essentially bypassing the p-n-p diode and eliminating the detrimental diode voltage drop without fear of substrate injection.

B. Substrate Injection at the IC's Input–Ouput Terminals

The substrate injection of electrons can also be induced due to negative voltage transients at any of the IC's input–output terminals. Fig. 9, for example, demonstrates two common types of IC input stages which are capable of injecting electrons into the substrate [2], [4]. The input base of the voltage differential stage [Fig. 9(a)] and the collector region associated with the current mirror's input transistor, Q_1 [Fig. 9(b)], are both N-epi regions and thus capable of producing substrate injection if the input voltages are pulled sufficiently negative. Negative voltage transients at the current mirror's output terminal could also produce substrate injection, since the N-epi collector of transistor Q_2 is connected to this terminal.

More subtle examples are shown in Fig. 10 to demonstrate the hazardous nature of the problem. The N+ crossunder, commonly employed for cross metalization [Fig. 10(a)], may be placed in series with an input–output terminal and other IC circuitry, allowing the negative voltages at the terminal to cause substrate injection at the N-epi region. Sometimes an external reference voltage is provided for the system, using an IC Zener diode which is commonly built as an n-p-n transistor with a collector–base–substrate short, as shown in Fig. 10(b). However, negative voltage transients at the reference terminal could cause large currents to flow in the N-epi region, producing a corresponding N-epi voltage drop sufficient to forward bias the N-epi/P-substrate junction and inject electrons into the substrate. In all of these examples, substrate injection could have been eliminated by preventing the input–output terminal voltages from decreasing below ground with the use of some type of voltage clamp. Although IC voltage clamps, such as n-p-n emitter followers, are capable of sufficiently clamping the higher energy negative terminal voltages, they cannot be used in all cases, since the relatively low reverse breakdown voltage of the emitter–base junction ($\approx$7 V) may seriously limit the positive terminal voltage while, at the same time, seriously degrade beta of the device and thus its capability as an emitter follower.

Fortunately, substrate injection can also be prevented by connecting the input–output terminals entirely to P-type regions wherever practical. Input switching information, for example, containing negative voltage transients sufficient to cause substrate injection, can be processed through the IC without activating a parasitic lateral n-p-n transistor if a p-n-p transistor emitter

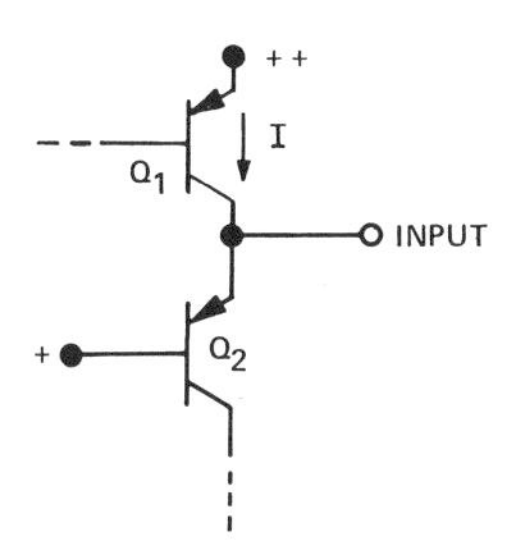

Fig. 11. Ideal input switching stage. Negative input voltages from the switching input cannot inject electrons into the substrate, since the input terminates entirely on P-type regions.

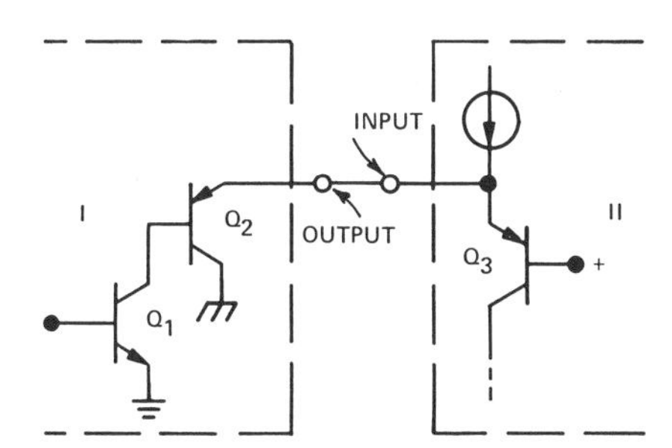

Fig. 12. Ideal input-output interface circuitry for processing switching information between two IC's, since the interconnect terminates entirely on P-type regions.

follower is utilized for the input stage (Fig. 11). Since the input terminates entirely on P-type regions (the P-type emitter of transistor Q_2 and the P-type collector of current source transistor Q_1), the negative voltage transients will only cause the P-type regions to reverse bias in their respective N-epi regions without causing substrate injection. The magnitude of the negative voltage can be allowed to reach the collector-emitter breakdown voltage (e.g., 80 V) of transistor Q_1 before input circuitry protection would be required. This technique can also be incorporated at IC output terminals in order to optimize chip-to-chip input-output interfacing. In Fig. 12, for example, a signal at the base of transistor Q_1 is transferred to the collector of transistor Q_3 through the input-output interface, which terminates entirely on P-type regions for the prevention of substrate injection. Also, since the input-output interconnect impedance is defined by the relatively low emitter impedance of either p-n-p transistor emitter follower, the interconnect is less susceptible to extraneous noise voltages. If the low input impedance of the stage shown in Fig. 11 cannot be tolerated due to the use of a relatively high input sensor impedance, the P-type base of an n-p-n transistor can also be used for an input connection, also without fear of substrate injection, but the nondestructive input voltage capability may not be as good. Analog input information can also be processed through the IC without producing substrate injection provided the same basic concepts of input termination are employed, but the exact methods used tend to be more complex.

Although the techniques thus far described will prevent electrons from being injected into the substrate, complete protection against this detriment cannot be achieved unless special consideration is also given to

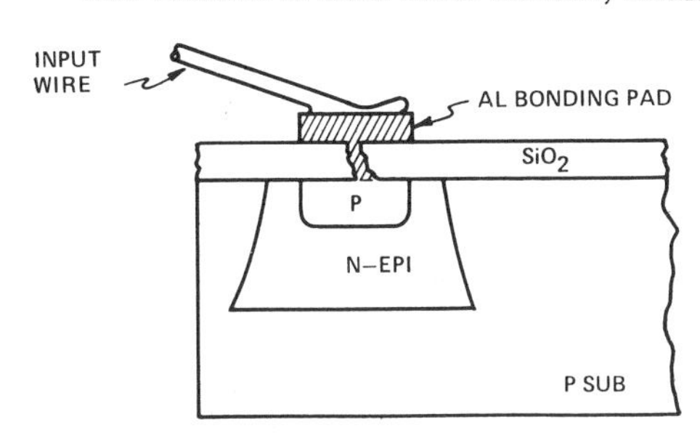

Fig. 13. Substrate injection of electrons prevented by a P-type isolation diffusion directly below the damaged bonding pad.

the bonding pad. In the past, the IC industry has typically placed N-epi regions beneath the aluminum bonding pads in order to prevent the bonding wires from shorting to the substrate due to accidental fracture in the SiO_2 during the wire bond operation. Unfortunately, if an N-epi region actually prevents the bonding wire from shorting to the substrate, a negative voltage transient on the input wire could forward bias the N-epi region and again cause substrate injection. However, by placing a P-type region directly beneath the bonding pad (in the N-epi island), as shown in Fig. 13, the damaged pad would short only to the P-type region, preventing not only a bonding pad short to the substrate, but also the possibility of substrate injection.

The detrimental effects of substrate injection could also be eliminated by employing techniques which would sufficiently reduce the lifetime of the minority carrier electrons in the P-type substrate and in the P-type isolation diffusion, such that α would approach 0—even between adjacent N-epi islands. Further investigation of this technique is required, however, before it can be used in mass production, since it must not degrade other IC parameters during the normal IC fabrication process.

V. Destructive Transient Voltage Protection at the IC's Inputs

IC input information can contain sufficient transient energy, from either positive or negative voltage sources, to destroy the IC's input circuitry unless adequate preventative techniques are employed. Assuming false input information is not degrading to IC performance for the duration of the voltage transient, a first-order method of protection employs an external resistor to limit the input breakdown current to a nondestructive value. Conventional IC Zener diodes can also be used in conjunction with this resistor to clamp all transient input voltages to satisfactory values, but substrate injection could occur during the negative transients as discussed in Section IV. Other methods can incorporate n-p-n and p-n-p transistor emitter followers, which clamp the respective negative and positive input voltages to satisfactory values, again using the external resistors to limit the follower currents. Since the external current limiting resistor must also be dc compatible with the input circuitry during normal operation, the maximum allowable limiting resistance usually determines which type of input protection is employed.

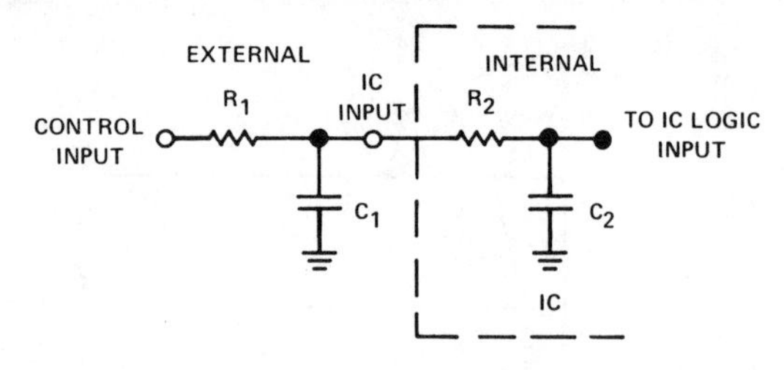

Fig. 14. Basic high-frequency filtering technique for IC logic inputs. The high-frequency on-die IC filtering is required due to the the high-frequency inductive reactance of capacitor C_1 and the actual physical separation between capacitor C_1 and the IC's input pin.

VI. IC Frequency Filtering Techniques

Frequency filtering techniques can be generally applied at the inputs of an IC to minimize high-frequency voltage components, especially at the IC's logic input circuitry, where false input information cannot be tolerated for durations greater than the response time of bipolar IC devices. These techniques can also be generally applied to limit the input terminal voltages below a destructive value for the duration of a destructive input transient. Usually, before external input information is applied to the logic input terminals, extraneous input voltages are filtered with an external single-pole low-pass resistor-capacitor network (R_1C_1), as shown in Fig. 14. Although this filter can be very effective at the lower frequencies, the erroneous high-frequency input voltages are more difficult to filter due to the series inductance typically associated with the capacitor. (An inexpensive 0.1-μF disk ceramic capacitor, for example, will become inductive at frequencies greater than 6 MHz.) In addition, since the R_1C_1 components and the IC would both be placed on a printed circuit board for high volume production, the physical distance between the capacitor C_1 and the IC's input terminal may not be sufficient to allow proper decoupling of all high-frequency voltages at the IC's input.

It is thus desirable to provide high-frequency input filtering techniques on the die, since effective input filtering could then be achieved, not only above the inductive frequency of capacitor C_1, but also at frequencies where the filtering of terminal voltages is unsatisfactory due to the capacitor/input-pin separation. Generally this can be accomplished by establishing a second input pole location (R_2C_2), which is considerably lower in frequency than the upper limit associated with the external filter's effectiveness. Ideally, the R_2C_2 product should be as large as possible in order to reduce or even possibly eliminate the requirements of the external filter. Although a high-valued resistance can be easily achieved using standard "pinched resistor" techniques, the maximum value of resistance, $R_1 + R_2$, is usually limited by the input impedance (input current requirements) of the input stage. Large resistive values can be used, for example, with remote analog low-impedance sensors (e.g., wheel velocity indicators) which typically drive high-impedance operational amplifiers or comparators. Lower values must be employed, however, for systems utilizing the input

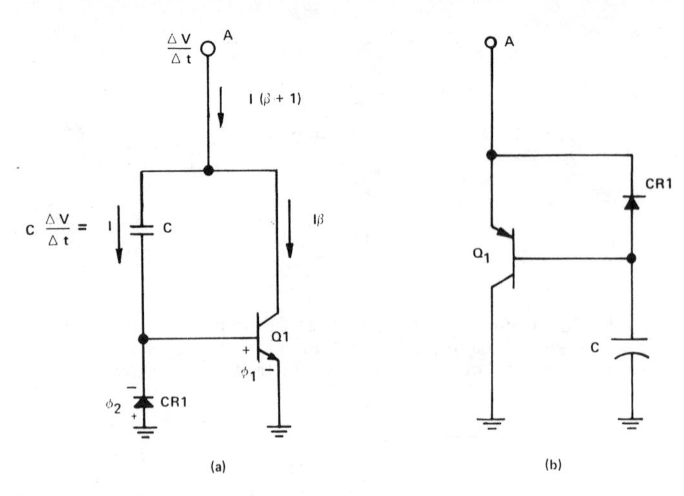

Fig. 15. Capacitance multiplication circuits for positive increasing voltages. (a) Multiplication of capacitance C by the n-p-n transistor's current gain β effective at terminal A. (b) Multiplication of capacitance C by the p-n-p transistor's current gain β effective at terminal A.

switching stage of Fig. 11, or where input sensor switches are used (e.g., seat-belt/starter interlock). The relatively low open-circuit switch impedance to ground (1 MΩ typical due to environmental contamination of switching contacts) requires a minimum input pull-up current, to insure that the open-circuit switched condition will produce an input voltage in excess of the input threshold voltage. The maximum value of the MOS or junction capacitor C_2 is primarily limited by die area constraints.

As a filtering example, it may be required to attenuate IC logic input voltages in the frequency range from 5 kHz to at least 100 MHz in order to prevent erroneous high-frequency triggering of the IC logic stages. Utilizing inexpensive components, an R_1C_1 product of (3 kΩ) (0.01 μF) would yield the 5-kHz pole. An R_2C_2 product of (80 kΩ) (10 pF) would produce a second pole on the die at 200 kHz, providing 55 dB of attenuation at 100 MHz and, more important, a 40-dB attenuation at 20 MHz where the reactance of capacitor C_1 begins to appear inductive.

If, however, filtering is only required for positive increasing voltages, the circuit shown in Fig. 15(a) could be substituted for the capacitor C_2, significantly increasing the effective value of this capacitance by the common-emitter current gain of a transistor to improve the filter's effectiveness, especially at the lower frequencies. Basically, the increasing rates of voltage change (dV/dt) at terminal A will also occur across capacitance C and thus defines the base current I of transistor Q_1. Since the collector current is greater than the base current by the grounded emitter current gain β of transistor Q_1, $I(\beta + 1)$ current will flow from terminal A due to the original dV/dt, thus defining an effective capacitance at terminal A of approximately βC. Diode CR_1 is employed to rapidly discharge the capacitance C to voltage ϕ_2 when the terminal A voltage returns to zero. Since the effective terminal capacitance can only be defined for increasing voltages greater than ($\phi_1 + \phi_2$) above the minimum discharge voltage, the input logic threshold voltage should be significantly above voltage $(\phi_1 + \phi_2)$. Some type of voltage clamp must also be provided at terminal A, since the N-epi collector region of transistor Q_1 may inject electrons into the substrate.

Returning now to the previous filtering example, an equivalent capacitance of 1000 pF can be directly assumed for capacitance C_2 of Fig. 14, by using the original 10-pF capacitor and taking advantage of the circuit of Fig. 15(a) with β equal to 100. If resistor R_2 is 30 kΩ, the required 5-kHz input pole is actually produced on the die for positive increasing voltages. The need for any external capacitor is eliminated. By also allowing resistor R_1 to equal 30 kΩ, the input circuitry can be completely protected from high input voltage transients, as discussed in Section V, in addition to creating a more dominant pole at the input.

To investigate, first order, the high-frequency effectiveness of an IC filter using the circuit of Fig. 15(a), the transfer function of the on-die filter of Fig. 14 can be written as

$$\frac{V_{\text{out}}}{V_{\text{in}}} = \frac{1}{1 + j\omega R_2 C_2} \tag{1}$$

where ω is the radian input frequency. From Fig. 15(a), capacitor C_2 would equal βC, and the frequency dependence of β can be written as

$$\beta = \frac{\beta_0}{1 + j\omega/\omega_\alpha(1 - \alpha_0)} \tag{2}$$

where β_0 is the dc grounded emitter current gain, ω_α is the alpha cutoff frequency, and α_0 is the dc grounded base current gain [5]. Since $1/(1 - \alpha_0)$ is approximately β_0 for $\alpha_0 \approx 1$, (1) can be rewritten as

$$\frac{V_{\text{out}}}{V_{\text{in}}} = \frac{1 + j\omega\beta_0/\omega_\alpha}{1 + j\omega\beta_0(1 + \omega_\alpha RC)/\omega_\alpha} \tag{3}$$

and for values of $\omega \gg \omega_\alpha/\beta_0$

$$\frac{V_{\text{out}}}{V_{\text{in}}} = \frac{1}{1 + \omega_\alpha RC}. \tag{4}$$

Equation (4) demonstrates that the active IC filter attenuation will become a constant for frequencies greater than the beta cutoff frequency, $\omega_\beta = \omega_\alpha/\beta_0$. Since 30×10^8 rad/s is the existing bipolar ω_α, the $\omega_\alpha RC$ product is about 900, and thus a constant attenuation of about 60 dB would be expected for input voltages with rise times corresponding to frequencies greater than ω_β.

The circuitry shown in Fig. 15(b) is conceptually identical with that of Fig. 15(a), and since it exhibits essentially the same basic functional characteristic at terminal A, it can again be directly substituted for capacitor C_2, providing all the basic advantages that were offered by utilizing the circuit of Fig. 15(a). In addition, both circuits can be directly connected to an input terminal and used, in conjunction with an external resistor, to limit the terminal voltages to nondestructive values during positive destructive voltage transients.

Although the frequency filtering techniques thus far described apply only to positive voltage transients with respect to ground, the basic concepts are directly applicable for negative transients with respect to some positive voltage reference and are often implemented, since the need to filter both positive and negative input voltages is often required.

VII. Summary

New bipolar IC techniques have been developed that can significantly extend the nondestructive supply voltage capability of an IC in order to achieve complete protection from large automotive voltage transients with a minimum expense in external protective components. In addition, a parasitic lateral n-p-n transistor has been defined that can be activated with large negative voltage transients by forward biasing any N-epi/P-substrate junction and inject electrons into the substrate. Since this parasitic device can be very detrimental to IC performance, new circuit and device innovations have been developed to prevent the substrate injection of electrons. Finally, methods have been employed which effectively filter, on the die, the destructive and erroneous high-frequency voltage transients at the input terminals of the IC. All of these techniques resulted from a necessary understanding not only of the automotive environment, but also of the advantages and limitations of the present bipolar IC technology with respect to this environment, and allows the bipolar IC to be highly competitive in specific automotive applications, from both an economic and a technical standpoint.

Acknowledgment

The author would like to thank P. Coolman for his collaboration during the investigation of substrate injection.

References

[1] W. F. Davis and T. M. Frederiksen, "A precision monolithic time-delay generator for use in automotive electronic fuel injection system," *IEEE J. Solid-State Circuits,* vol. SC-7, pp. 462–469, Dec. 1972.

[2] T. M. Frederiksen, W. F. Davis, and D. W. Zobel, "A new current-differencing single-supply operational amplifier," *IEEE J. Solid-State Circuits,* vol. SC-6, pp. 340–347, Dec. 1971.

[3] T. M. Frederiksen, W. F. Davis, and R. W. Russel, "Building blocks for an onboard computer," presented at the Society of Automotive Engineers Automotive Engineering Congress, Detroit, Mich., Jan. 10–14, 1972, Paper 720281.

[4] R. W. Russel and T. M. Frederiksen, "Automotive and industrial electronic building blocks," *IEEE J. Solid-State Circuits,* vol. SC-7, pp. 446–454, Dec. 1972.

[5] A. B. Phillips, *Transistor Engineering.* New York: McGraw-Hill, 1962, p. 302.

Analysis and Design of Temperature Stabilized Substrate Integrated Circuits

PAUL R. GRAY, MEMBER, IEEE, DOUGLAS J. HAMILTON, MEMBER, IEEE, AND J. DARRYL LIEUX, MEMBER, IEEE

***Abstract*—A generalized temperature-stabilized substrate integrated circuit system containing heat sources and temperature sensors which are distributed in an arbitrary way in two-dimensions over the surface of the chip is analyzed. A computer-aided technique for optimum placement of these components at the layout stage so as to achieve zero nominal temperature coefficient in the performance parameter of interest is described, and other practical design considerations, such as optimum choice of stabilized chip temperature, are considered. The application of this technique is illustrated with the design of a precision temperature stabilized voltage reference supply. Experimental results from this circuit are presented.**

INTRODUCTION

A MAJOR limitation of linear integrated circuit technology is the difficulty of fabricating diffused circuit components whose parameters have low temperature coefficients. Nondiffused elements, such as metal film resistors, can alleviate the problem in some cases, but require extra processing steps. In many integrated circuits, clever design can cause the critical circuit parameters to depend on ratios of component values rather than on absolute values. Diffused components on the same chip can be made to track within several percent over the military temperature range (−55°C to +125°C). However, many applications require temperature sensitivities orders of magnitude or more smaller than this.

Another approach to the problem involves mounting the chip on a thermal insulator and stabilizing the chip temperature at a fixed value independent of the ambient temperature by means of a heater, sensor, and feedback circuit on the chip itself (temperature stabilized substrate, or TSS). Such an approach always requires an increase in power dissipation in the chip over that required for the original circuit alone, but the effective temperature sensitivity of the critical circuit elements can be reduced by two to three orders of magnitude. Performance superior to discrete component circuits is readily obtainable. Applications include low-drift amplifiers, stable reference voltage sources, oscillators, active filters, etc.

The performance of a TSS system with simple one-dimensional geometry was first analyzed by Matzen *et al.* [1] in 1964. Several practical applications of this technique have been described in the literature [2], [3]. In this paper, a generalized TSS system containing heat sources and temperature sensitive elements of arbitrary geometric shape and location is analyzed, and design techniques are given for optimizing heater, sensor, and circuit geometry. The effect of the TSS system on power dissipation is considered, and results are presented from an experimental integrated circuit die which agree well with theoretical predictions. Finally, the results are applied to the design of a precision TSS voltage reference source. Experimental results from this circuit are also presented.

A SIMPLE EXAMPLE OF A TSS SYSTEM

The analysis of a very simple, nonoptimum TSS configuration is used to clarify the effect of ambient temperature variations on the chip temperature distribution. Consider a chip with a heater at one end, a feedback sensor adjacent to it, and group of temperature sensitive elements which we wish to stabilize at the other end. The configuration is illustrated in Fig. 1 and is similar to that of the Fairchild μA726 temperature stabilized transistor pair which is commercially available.

In Fig. 2(a), the form of the open loop temperature distribution over the chip is shown for various steady-state heater power inputs at a constant ambient temperature. Note that the temperature change along the length of the chip increases with increasing power input. With the insertion of a large amount of gain in the heater-sensor feedback loop, the power dissipation will be regulated so as to maintain the temperature at the feedback sensor essentially constant. In Fig. 2(b), the resultant temperature profiles across the chip are shown for various values of ambient temperature T_a. Note that the temperature of the temperature sensitive elements does not remain constant; for typical chip sizes and insulator thicknesses, the variation would be 1–10°C over the military temperature range. It is obvious that it would be more desirable to locate the temperature sensitive element closer to the feedback sensor where the temperature variation is smaller. A general condition for zero temperature variation in the primary sensor is derived in a later section.

Manuscript received October 3, 1973; revised December 14, 1973. This research was sponsored by the Joint Services Electronics Program, Contract F44620-71-C-0087 to the University of California at Berkeley, and National Science Foundation Grant GK-31906 to the University of Arizona.

P. R. Gray is with the Department of Electrical Engineering and Computer Science, University of California, Berkeley, Calif. 94720.

D. J. Hamilton is with the Department of Electrical Engineering, University of Arizona, Tucson, Ariz. 85720.

J. D. Lieux is with the Fairchild Camera and Instrument Corp., Palo Alto, Calif.

Reprinted from *IEEE J. Solid-State Circuits*, vol. SC-9, pp. 61-69, April 1974.

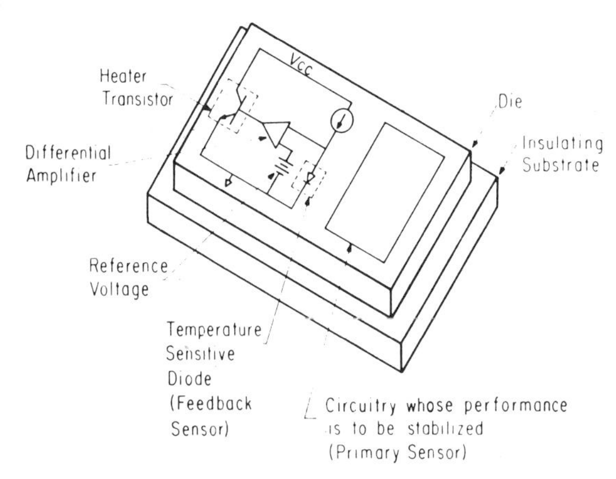

Fig. 1. Simple, nonoptimum TSS cofiguration.

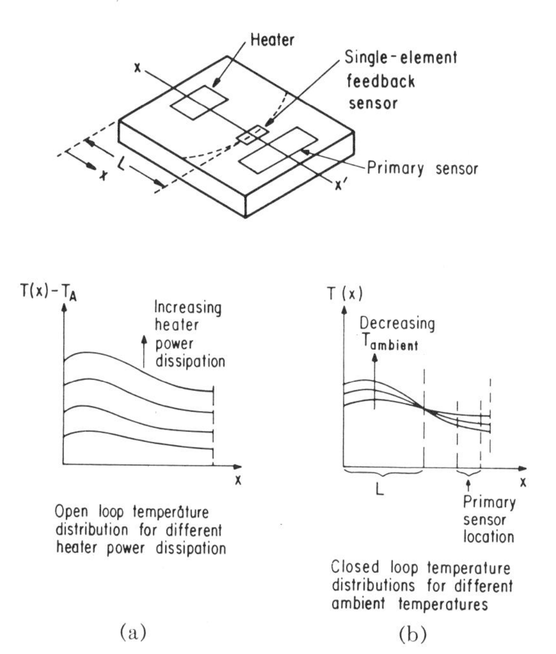

Fig. 2. Open and closed loop temperature distributions for the example chip. For the closed loop case, the loop gain has been assumed to be very large.

Characterization of TSS Systems

A block diagram of a generalized TSS system is shown in Fig. 3. The electrical output of the feedback temperature sensor, ζ_s, is subtracted from an electrical reference ζ_{ref} and the error signal difference is amplified to drive the controlled heat source. The function $F_1(s)$ represents the thermal transfer function between the controlled heat source and the group of interconnected circuit elements, called the primary circuit, whose performance is to be made insensitive to temperature. The parameter ζ_{out} is the parameter of the primary circuit whose temperature coefficient is to be minimized. This could be, for example, the output voltage of a stable reference supply. The function $F_2(s)$ is the thermal transfer function between the heat source and the circuitry making up the feedback sensor. $F_1(s)$ and $F_2(s)$ are dependent upon both the layout of the components on the die, and upon the way they are electrically interconnected. Calculation of $F_1(s)$ and $F_2(s)$ will be discussed later. It should be pointed out that the assumption has been made that the system is linear; this will be discussed in a later section.

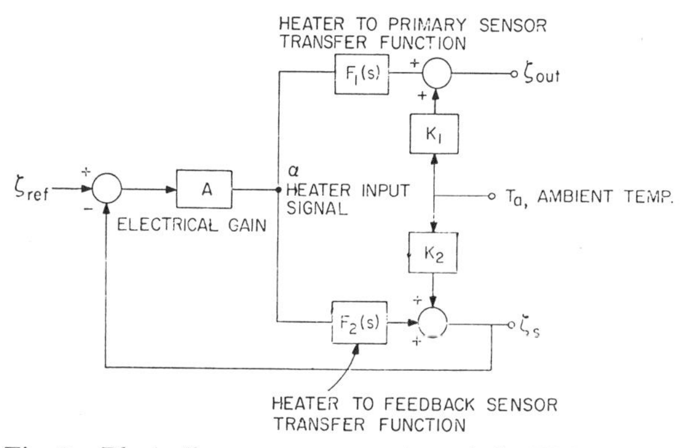

Fig. 3. Block diagram representation of the TSS system.

The constants K_1 and K_2 are the open-loop temperature coefficients of ζ_{out} and ζ_s, or their sensitivity to ambient variations which would be observed if the loop were opened and α was set equal to zero so that the heat source dissipation was zero. The assumption has been made that the variations in ambient temperature are much slower than the longest thermal time constant of the chip structure.

In the next section, expressions for $F_1(s)$, $F_2(s)$, K_1, and K_2 are derived using a simplified physical model for the chip. In a later section, the block diagram of Fig. 1 is used to analyze the performance of the system.

Evaluation of $F_1(s)$, $F_2(s)$, K_1, K_2

The physical model assumed for the TSS is shown in Fig. 4. The rectangular silicon chip is mounted on a thermal insulator, which is in turn attached to the mechanical package. A set of simplifying assumptions is made which allow a straightforward analysis while providing good accuracy in the cases of interest. The assumptions are:

1) In their thermal properties the silicon and insulator are linear, isotropic, homogeneous media.

2) The insulator thermal conductance is sufficiently low that in the silicon

$$\frac{\partial T}{\partial z} \approx 0$$

and

$$T(x, y, z, t) \approx T(x, y, t).$$

3) The insulator thermal conductance is small compared to that of the silicon, and lateral heat flow in the insulator is small compared to that in the silicon.

4) The mounting surface of the mechanical package is isothermal.

5) Heat losses from the chip due to convection and radiation are negligible.

6) Heat losses through bonding wires is negligible.

It has been shown in a previous paper [4] that under the listed assumptions the interactions between the

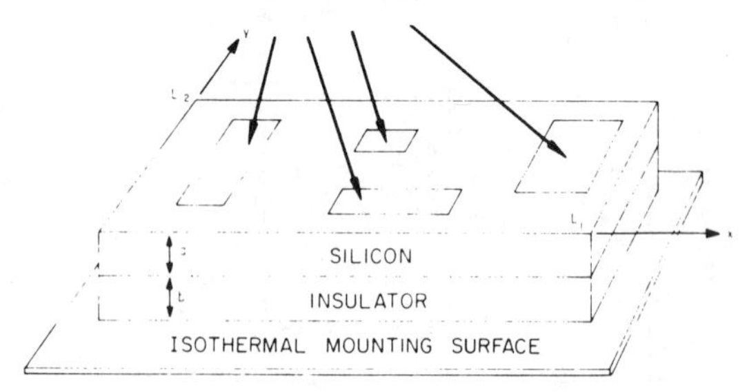

Fig. 4. Physical model for the TSS system.

electrical and thermal properties of integrated circuit components can conveniently be expressed in terms of weighting functions. For a temperature sensitive circuit element or interconnected array of elements with an output signal ζ_0, one can write

$$\zeta_0(t) = \int_0^{L_1}\int_0^{L_2} W_s(x, y)T(x, y, t)\,dx\,dy \tag{1}$$

where the function $W_s(x, y)$ is dependent upon the geometrical layout of the temperature sensor, and if it is made up of more than one circuit element upon the way in which the elements are interconnected. Shown in Fig. 5 is a simple, rectangular diode temperature sensor, and its corresponding weighting function.

Similarly, for a heat dissipating element or group of elements which are controlled by a single input signal $\alpha(t)$, one can write

$$p(x, y, t) = W_p(x, y)\alpha(t) \tag{2}$$

where $p(x, y, t)$ is the power dissipated per unit area at (x, y) and $W_p(x, y)$ is a function describing the geometrical arrangement and electrical interconnection of the heater elements.

In closed loop operation the temperature variations on the chip are small and the assumption is made that the system is linear. Convenient expressions for $F_1(s)|_{s=0}$, $F_2(s)|_{s=0}$, K_1, and K_2 can be written by defining the thermal Green's function for the chip. $G(x', y', x, y)$ is the steady-state temperature distribution $T(x, y)$ which results from a unit input of power at the point (x', y'). The temperature distribution is given by

$$T(x, y) - T_a = \alpha\int_0^{L_1}\int_0^{L_2} U_p(x', y')G(x', y', x, y)\,dx'\,dy' \tag{3}$$

where $U_p(x', y')$ is the heater weighting function and α is the heater input signal. The steady-state feedback sensor output is given by

$$\zeta_s = \int_0^{L_1}\int_0^{L_2} W_{s2}(x, y)T(x, y)\,dx\,dy \tag{4}$$

where $W_{s2}(x, y)$ is the feedback sensor weighting function. Thus, we find that ξ_s is given by

$$\zeta_s = \alpha\int_0^{L_1}\int_0^{L_2} W_{s2}(x, y)$$

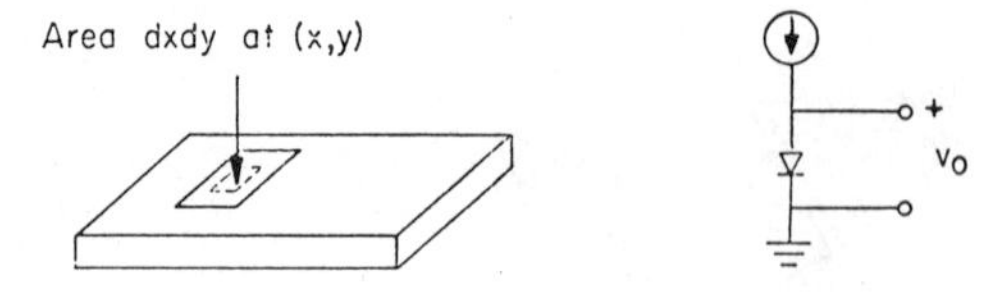

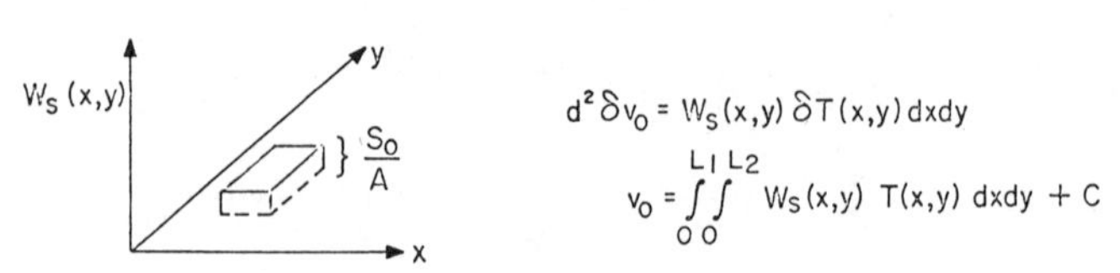

Fig. 5. Simple diode sensor example with its weighting function representation for small temperature variations.

$$\cdot\int_0^{L_1}\int_0^{L_2} U_p(x', y')G(x', y', x, y)\,dx'\,dy'\,dx\,dy$$
$$+\,T_a\int_0^{L_1}\int_0^{L_2} W_{s2}(x, y)\,dx\,dy \tag{5}$$

so that

$$F_2(s)\,|_{s=0} = \int_0^{L_1}\int_0^{L_2} W_{s2}(x, y)$$
$$\cdot\int_0^{L_1}\int_0^{L_2} U_p(x', y')G(x', y', x, y)\,dx'\,dy'\,dx\,dy. \tag{6}$$

$$K_2 = \int_0^{L_1}\int_0^{L_2} W_{s2}(x, y)\,dx\,dy. \tag{7}$$

Similarly, for the primary sensor

$$\zeta_{\text{out}} = \alpha\int_0^{L_1}\int_0^{L_2} W_{s1}(x, y)$$
$$\cdot\int_0^{L_1}\int_0^{L_2} U_p(x', y')G(x', y', x, y)\,dx'\,dy'\,dx\,dy$$
$$+\,T_a\int_0^{L_1}\int_0^{L_2} W_{s1}(x, y)\,dx\,dy. \tag{8}$$

Equation (8) implies that

$$F_1(s)\,|_{s=0} = \int_0^{L_1}\int_0^{L_2} W_{s1}(x, y)$$
$$\cdot\int_0^{L_1}\int_0^{L_2} U_p(x', y')G(x', y', x, y)\,dx'\,dy'\,dx\,dy \tag{9}$$

and

$$K_1 = \int_0^{L_1}\int_0^{L_2} W_{s1}(x, y)\,dx\,dy. \tag{10}$$

The above equations apply only in the steady-state. Thus they give the correct results only for changes in α and T_a which are much slower than the longest time constant of the chip. This is the case of interest, however, so (6)–(10) will be of use in analyzing the effect of the normally slow variations of ambient temperature.

The variations of ζ_{out} result from temperature-induced variations in all the circuit elements that effect it. For small temperature variations, all these elements can be regarded together as a sensor, to be designated the primary sensor. In general, ζ_{out} will be some function of the element values λ_n of the circuit.

$$\zeta_{\text{out}} = f(\lambda_1, \lambda_2, \lambda_3, \cdots \lambda_n). \tag{11}$$

For small variation in temperature, the variation of ζ_{out} can be written as

$$\delta\zeta_{\text{out}} = \frac{\partial f}{\partial \lambda_1}\,\delta\lambda_1 + \frac{\partial f}{\partial \lambda_2}\,\delta\lambda_2 \cdots + \frac{\partial f}{\partial \lambda_n}\,\delta\lambda_n. \tag{12}$$

Each of the elements will be distributed over the chip, and the temperature sensitivity of each element can be described in terms of a weighting function $Z_j(x, y)$

$$\delta\lambda_j = \int_0^{L_1}\int_0^{L_2} Z_j(x, y)\,\delta T(x, y)\,dx\,dy \tag{13}$$

and (12) becomes

$$\delta\zeta_{\text{out}} = \int_0^{L_1}\int_0^{L_2} W_{s1}(x, y)\,\delta T(x, y)\,dx\,dy \tag{14}$$

where

$$W_{s1}(x, y) = \sum_{j=1}^{n} \frac{\partial f}{\partial \lambda_j} Z_j(x, y). \tag{15}$$

Thus the composite weighting function for the complete, multielement sensor is a weighted sum of the individual element weighting functions.

Computer Evaluation of the Thermal Green's Function for the Chip Structure

For the physical model assumed, the thermal Green's function for the chip can be evaluated using finite transform techniques.

$$G(s) = \frac{4\omega_x\omega_y}{ak_s} \cdot \sum_{n=0}^{\infty}\sum_{m=0}^{\infty} \frac{\cos n\pi x \cos m\pi y \cos n\pi x' \cos m\pi y'}{(\delta_n + 1)(\delta_m + 1)[s + (n\pi)^2\omega_x + (m\pi)^2\omega_y + \omega_i]} \tag{16}$$

where

$$\omega_x = \frac{k_s}{\rho_s c_s L_1^2}$$

$$\omega_y = \frac{k_s}{\rho_s c_s L_2^2}$$

$$\omega_i = \frac{k_i}{\rho_s c_s ab}.$$

The expressions for F_1, F_2, K_1, and K_2 then become simply doubly infinite series for the simple model assumed.

$$F_{1,2}(s) = \frac{\sqrt{\omega_x\omega_y}}{ak_s} \cdot \sum_{n=0}^{\infty}\sum_{m=0}^{\infty} \frac{U_{pnm}W_{s1,2nm}}{[\delta_n+1][\delta_n+1][s+(n\pi)^2\omega_x+(m\pi)^2\omega_y+\omega_i]} \tag{17}$$

$$K_{1,2} = \frac{\sqrt{\omega_x\omega_y}}{ak_s}[U_{poo}W_{s1,2oo}] \tag{18}$$

where

$$\omega_x = \frac{k_s}{\rho_s c_s L_1^2} \tag{19}$$

$$\omega_y = \frac{k_s}{\rho_s c_s L_2^2} \tag{20}$$

$$\omega_i = \frac{k_i}{\rho_s c_s ab} \tag{21}$$

$$U_{pnm} = \int_0^{L_1}\int_0^{L_2} U_p(x, y) \cos\frac{n\pi x}{L_1} \cos\frac{m\pi y}{L_2}\,dx\,dy \tag{22}$$

$$W_{snm} = \int_0^{L_1}\int_0^{L_2} W_s(x, y) \cos\frac{n\pi x}{L_1} \cos\frac{m\pi y}{L_2}\,dx\,dy. \tag{23}$$

Equation (17) is easily summed by computer; the numerical evaluation of thermal transfer functions using (17) has been implemented in a program described in an earlier paper [4].

Analysis of TSS System Performance

In this Section the effect of geometrical layout on system performance is analyzed. Referring to Fig. 3, straightforward algebraic manipulation yields the variation of ζ_{out} as a function of ambient temperature.

$$\frac{d\zeta_{\text{out}}}{dT_A} = K_1\left[\frac{1 + AF_1(0)\left[\frac{F_2(0)}{F_1(0)} - \frac{K_2}{K_1}\right]}{1 + AF_2(0)}\right]. \tag{24}$$

This expression together with the results of the last section can be used to analyze the performance of a TSS system of arbitrary geometrical layout.

An objective in the design of TSS systems is the realization of identically zero effective temperature coefficient for the parameter of interest. The possibility of doing this was first pointed out by Matzen [1]. Setting (24) equal to zero, we obtain

$$\left[\frac{K_2}{K_1} - \frac{F_2(0)}{F_1(0)}\right] = \frac{1}{AF_1(0)}. \tag{25}$$

Configurations which satisfy this condition are those for which the gradient-induced "effective" temperature difference between the primary sensor and the feedback sensor just cancels the temperature variations at the feedback sensor. The satisfaction of (25) is thus an objective in TSS design.

The use of (25) as a design criteria will be discussed in two parts. First, the unrealistic case of very large gain ($A \to \infty$) will be considered. The results are then modified to account for the effects of finite gain.

For the special case of very large gain, (25) becomes

$$\frac{F_2(0)}{K_2} = \frac{F_1(0)}{K_1}. \tag{26}$$

The quantities $F_2(0)/K_1$, $F_1(0)/K_1$ are independent of the *magnitude* of the open loop temperature sensitivity of the primary and feedback sensors, and dependent only on the way in which this temperature sensitivity is dis-

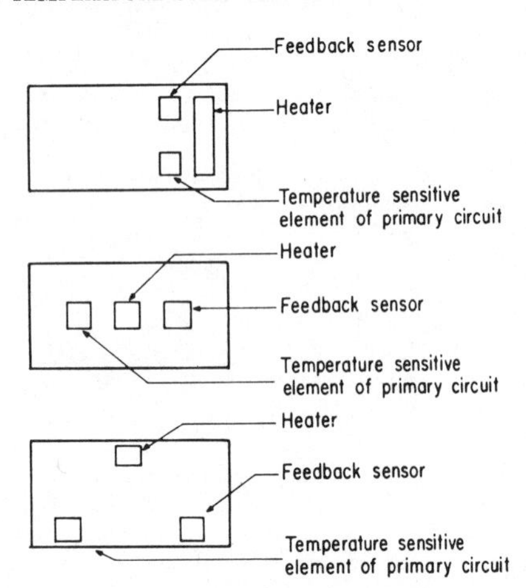

Fig. 6. Example optimum layout configurations for the infinite gain case.

tributed over the surface of the chip. Examination of (17), (18) and (26) shows that a sufficient condition for the satisfaction of (26) is that

$$F_1(s) = CF_2(s) \text{ for all } s, \tag{27}$$

where C is an arbitrary constant. In other words, one way to insure zero effective temperature coefficient for the case of infinite loop gain is to insure by symmetry that $F_1(s)$ and $F_2(s)$ are identical except for a multiplying constant.

Several example chip geometries which accomplish this are shown in Fig. 6. While these configurations are optimum only for the case of large loop gain, they provide a starting point for design purposes. For any particular situation, the validity of the infinite-gain assumption can be checked since in the closed loop case the variation in feedback sensor temperature is equal to the variation in ambient temperature divided by the loop gain. Unfortunately, the characteristics of thermal transfer functions place an upper limit on the dc loop gain that can be used because of the stability problem. For typical insulators, chip sizes, and minimum heater-sensor spacing, this limit is around 60–70 dB. At this value of loop gain, the total variation of temperature in the feedback sensor will be less than 0.1°C over the full military range of ambient temperatures. Stability considerations will usually require that part of the sensor be as close as possible to the heater.

We now turn our attention to the effect of variations in power dissipation of elements other than the heater. Equation (26) implies that any change of power dissipation in the heater must result in equal changes of temperature at the feedback sensor and primary sensor. Thus, by making the variation of feedback sensor temperature zero we also make the variations in primary sensor temperature zero. Exactly the same argument applies to other heat sources on the chip. One must ensure by symmetry that any change of power dissipation in the elements on the chip which occurs during normal operation will result in identical temperature changes at the feedback sensor and primary sensor in the open loop case. Since there may be many elements on the chip whose power dissipation variations are significant, practical constraints may preclude the satisfaction of this condition exactly. For this case, the criteria should be regarded as an ideal to strive for in circuit layout.

Temperature Stabilized Substrates with Finite Loop Gain

In the last section, a criterion was established which ensures that the weighted primary sensor temperature variations will be zero if the loop gain is infinite. However, since in actual TSS systems the loop gain is finite, and since by (25), zero variation of ζ_{out} can be realized with finite loop gain, the results of the last section do not necessarily yield an optimum TSS design. To illustrate this, consider an example TSS system with a line heater at one end ($x = 0$) and a point feedback sensor at the other end ($x = L$). Since the system is linear, the temperature at the sensor is given by

$$T_s = KP_{\text{in}} + T_a, \qquad K = \text{constant}. \tag{28}$$

If a feedback system is constructed such that $P_{\text{in}} = G(T_s - T_0)$, we have

$$T_s = T_0 + \frac{(T_a - T_0)}{1 + KG}. \tag{29}$$

If the loop gain KG is very large, $T_s \cong T_0$ and we have the case discussed in the previous section. However, if it is not, T_s will vary appreciably over the range of ambient temperature. For this simple one-dimensional case, $T(x)$ can be found in closed form as a function of P and T_a.

$$T(x) = AP \cosh\left(\sqrt{\frac{\omega_i}{\omega_x}}\frac{x}{L}\right) + T_a \qquad A = \text{constant}. \tag{30}$$

Combining (30) and (29) one obtains

$$T(x) = T_a\left[1 - \frac{KG \cosh\sqrt{\dfrac{\omega_i}{\omega_x}}\dfrac{x}{L}}{1 + KG}\right] + D \cosh\sqrt{\frac{\omega_i}{\omega_x}}\frac{x}{L}, \qquad D = \text{constant}. \tag{31}$$

We now take the derivative with respect to T_a.

$$\frac{\partial T(x)}{\partial T_a} = \left[1 - \frac{KG \cosh\sqrt{\dfrac{\omega_i}{\omega_x}}\dfrac{x}{L}}{1 + KG}\right]. \tag{32}$$

Note that this derivative can be zero for a single value of x. Since $\omega_i/\omega_x \ll 1$, and $x/L \leq 1$, the hyperbolic cosine can be approximated by a truncated Taylor series. The value of x for which $\partial T(x)/\partial T_a = 0$ is given by

$$\frac{x}{L} \cong \frac{2\omega_x}{KG\omega_i}. \tag{33}$$

Thus, if the temperature sensitive elements of the primary circuit were located at this point, zero variation of ζ_{out} with T_a would be obtained. Note that in order for such a point to exist on the chip

$$KG \geq \frac{2\omega_x}{\omega_i}. \tag{34}$$

Returning to the general case with finite loop gain, (25) is rewritten in terms of the series expansion for $F_1(s)$ and $F_2(s)$. The result is

$$\frac{1}{AW_{s2_{00}}} + \sum_{n=0}^{\infty}\sum_{m=0}^{\infty} \frac{W_{s2_{nm}} U_{p_{nm}}}{W_{s2_{00}}[\delta(n)+1][\delta(m)+1][(n\pi)^2\omega_x+(m\pi)^2\omega_y+\omega_i]}$$
$$= \sum_{n=0}^{\infty}\sum_{m=0}^{\infty} \frac{W_{s2_{nm}} U_{p_{nm}}}{W_{s1_{00}}[\delta(n)+1][\delta(m)+1][(n\pi)^2\omega_x+(m\pi)^2\omega_y+\omega_i]}. \tag{35}$$

This criterion is unfortunately not as easy to interpret as (26). It is a condition on the dependence of W_{s1}, W_{s2}, and U_p on x and y. However, this equation is in a very convenient form for computer evaluation. Also, it is intuitively clear from the simple example given earlier in this section that as one decreases the loop gain, the optimum single-element primary sensor location always moves closer to the heater. This gives rise to a practical design approach as follows:

1) By use of symmetry, design a configuration which satisfies (26) for the infinite loop gain case.

2) Evaluate (35) for that configuration, and then move the primary sensor or some portion thereof closer to the heater in small increments, until (35) is satisfied.

For loop gains on the order of 70 dB, the movement required is only a small fraction of the length of the chip.

This procedure amounts to designing the system for infinite loop gain, and then perturbing it slightly to account for the finite gain. This process has been used in the design of a voltage reference to be described later. Of course, there will be no solution if the loop gain is below a certain minimum value.

The problem of elimination of the effects of varying power dissipation in elements on the chip other than the heater, is more troublesome in the finite gain case. It is important to note that we are concerned with variations in power dissipation in these elements, not steady-state dissipation. In many applications such as an integrated small signal amplifier, these variations may be small compared to the steady-state heater power output, and their effects can be neglected. It can be demonstrated that a criterion for elimination of the effects of other significant power-dissipating elements is to insure by symmetry that the transfer functions from each of those elements to the primary and feedback sensor are identical to those from the heater to the primary and feedback sensor, respectively. In practical circuits, this would be very costly in die area.

TSS performance generally improves with increasing loop gain. The higher the gain, the closer one can approach the simultaneous realization of zero sensitivity to ambient variations and on-chip power variations. Also, for low loop gain, the optimum configuration depends on the insulator characteristics, chip dimensions, and the actual value of the loop gain. All of these are subject to variations from unit to unit. For infinite loop gain, the optimum configurations are independent of these parameters.

Power Dissipation Considerations

As mentioned earlier, stabilization of the chip temperature over a finite range of ambient temperatures always implies a net power dissipation greater than that of the original circuit alone. To investigate this fact more thoroughly, the following quantities are defined:

- P_o Net power dissipated by the circuit to be stabilized.
- P_h Power dissipated by the TSS circuit in controlling the chip temperature.
- T_c Stabilized chip temperature.
- T_a Ambient temperature.
- $T_{a\max}$ Maximum ambient temperature.
- $T_{a\min}$ Minimum ambient temperature.
- θ Net thermal resistance from the chip to the ambient.

These quantities are related by the following equation in the steady-state

$$T_c - T_a = (P_o + P_n)\theta. \tag{36}$$

If the TSS system is properly designed for minimum control power dissipation, the control power dissipation will be zero when the ambient temperature is at its maximum value.

$$T_c - T_{a\max} = P_o\theta \tag{37}$$

where

$$T_c \geq T_{a\max}. \tag{38}$$

Solving this question for θ and inserting the result in (38) one obtains

$$P_h = \left[\frac{T_{a\max} - T_a}{T_c - T_{a\max}}\right] P_0 \qquad T_c \geq T_{a\max} \tag{39}$$

and

$$P_{h\max} = \left[\frac{T_{a\max} - T_{a\min}}{T_c - T_{a\max}}\right] P_0 \qquad T_c \geq T_{a\max}. \tag{40}$$

Equation (39) is a monotonically decreasing function of T_c for all permissible values of $T_{a\max}$ and T_a except $T_a = T_{a\max}$. Thus, in order to minimize the power dissipation penalty imposed by the TSS one should always choose the stabilized chip temperature to be as high as possible, and design the insulator so that

$$\theta = \frac{T_c - T_{a\max}}{P_0}. \tag{41}$$

Unfortunately, the maximum operating temperature for most integrated silicon circuitry is around 125°C. Were it not for this fact, temperature stabilized substrates could be used in many applications with little

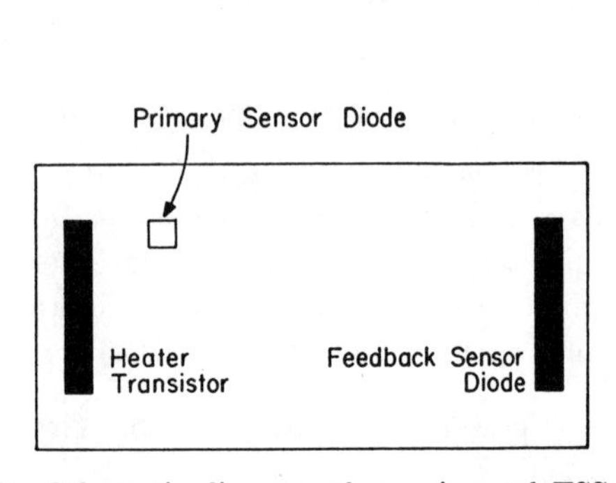

Fig. 7. Schematic diagram of experimental TSS chip.

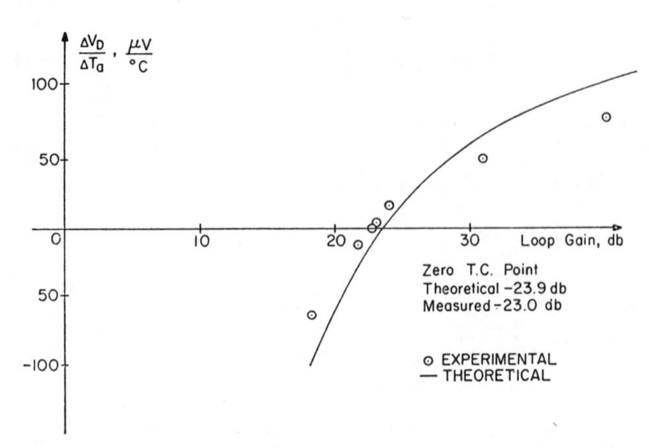

Fig. 8. Theoretically predicted and experimentally observed effective temperature coefficient of primary sensor diode as a function of loop gain, experimental chip.

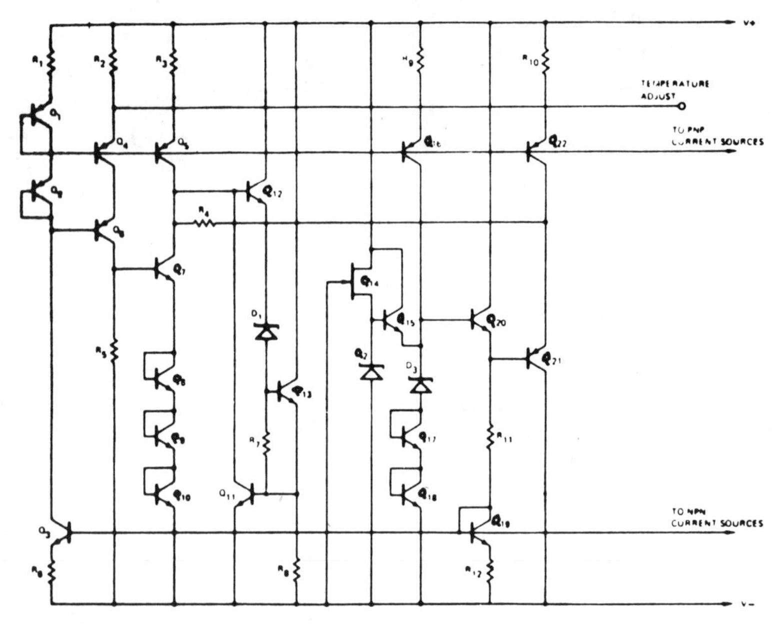

Fig. 9. Temperature regulator circuitry, TSS voltage reference.

penalty in terms of power dissipation. It is apparent from (40) that the maximum excess power dissipation $P_{h\,\max}$ will be small compared to P_o if

$$(T_{a\,\max} - T_{a\,\min}) \ll (T_c - T_{a\,\max}). \tag{42}$$

Thus, if the chip temperature T_c is chosen to be the maximum possible value, the power dissipation penalty involved in stabilizing the substrate will be small if the range of variation of T_a is small compared to the difference :between T_c and $T_{a\,\max}$.

Experimental Results

The comparison of predicted TSS performance with experiment was done in two parts. First, an experimental integrated circuit die was constructed and used together with discrete components to investigate the effective temperature coefficient as a function of loop gain. Second, the TSS design technique described earlier was used in the layout of a TSS voltage reference circuit, with the objective of obtaining a temperature coefficient of less than 5 ppm/°C. Results from both of these circuits are given in this section.

The first experimental die, shown in Fig. 7, contained a number of independent transistors, each of which was contacted separately. A number of different TSS systems of the type shown in Fig. 1 were constructed using an external reference voltage and amplification. In each case the effective temperature coefficient of the primary sensor diode was compared with that predicted by a computer evaluation of (24). A typical result is shown in Fig. 8. The agreement is quite good.

The results regarding TSS design were then applied to the optimum layout of a precision monolithic voltage reference. This circuit was intended to provide a stable reference voltage for use in A/D and D/A conversion, and contains a compensated Zener diode reference, operational amplifier, and temperature regulator circuit. The circuit design aspects of the reference have been described in an earlier paper [3]. The temperature regulator circuit is shown in Fig. 9. Transistors $Q7$–$Q10$ are

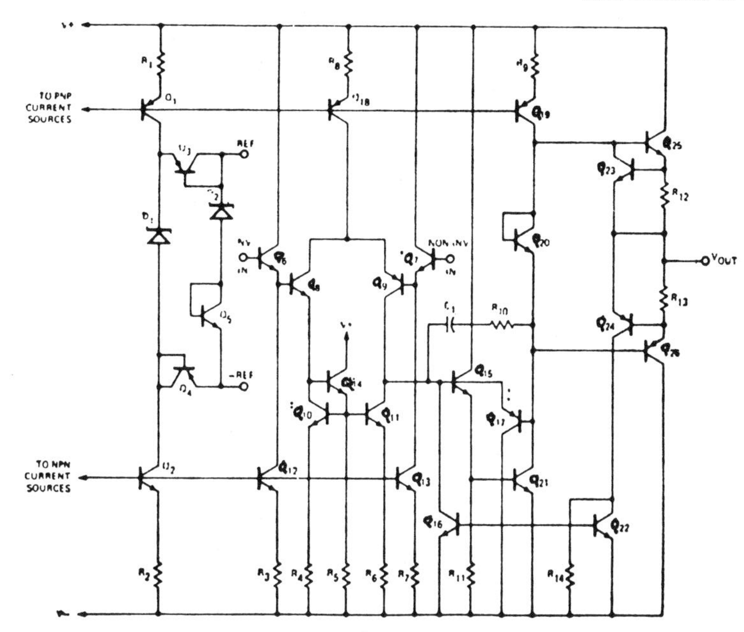

Fig. 10. Voltage reference and buffer amplifier circuit, TSS voltage reference.

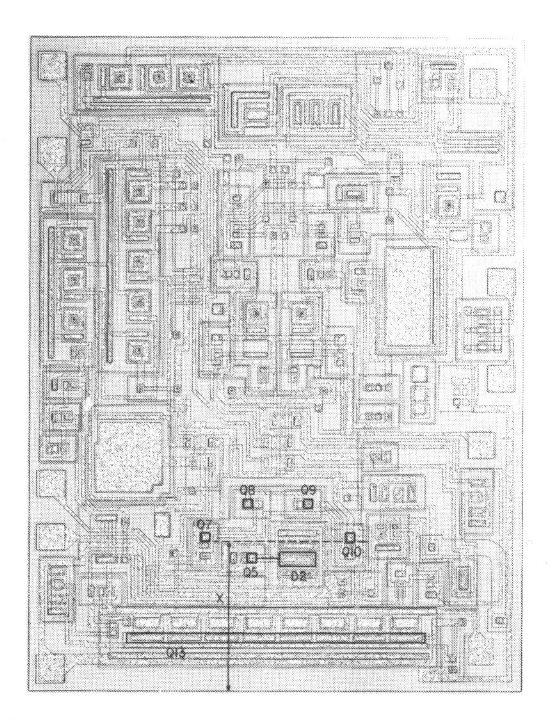

Fig. 11. Die microphotograph showing heater, feedback sensor, and primary sensor, TSS voltage reference.

the feedback sensors, while $Q7$, $Q12$, and $Q13$ provide gain. The resistor $R4$ allows reasonably precise control of the loop gain. Transistor $Q13$ is the heater, while the $Q8$–$Q11$ combination is a startup surge limiter. The remainder of the circuitry is used for biasing. The reference diode and amplifier are shown in Fig. 10. Zener diode $D2$ and compensating diode $Q5$ make up the primary sensor.

The physical location of the various elements on the die is shown in Fig. 11. During the layout process, the locations of $Q7$ and $Q10$ were treated as variables, and (24) was evaluated numerically for a sequence of five values of the distance $X1$. These theoretical results are shown in Fig. 12. The final location was chosen from this theoretical curve. Also shown is the distribution of experimentally observed temperature coefficients from a typical lot of devices. The nominal TC is very near zero, at -0.3 ppm/°C, although random variations in regu-

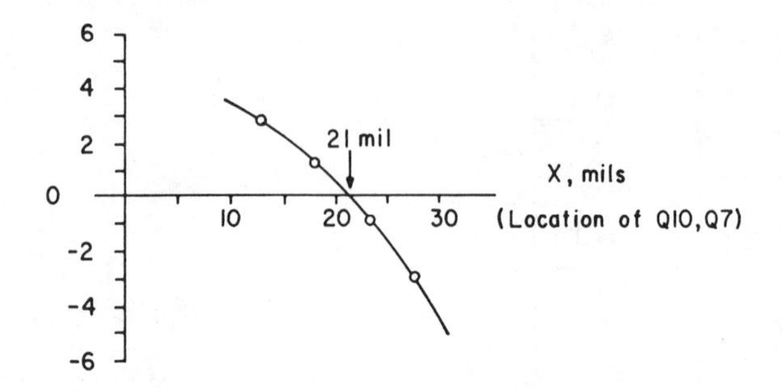

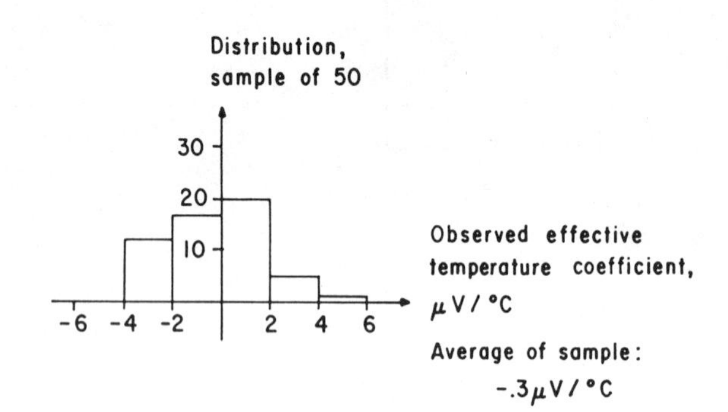

Fig. 12. Theoretically predicted effective reference temperature coefficient as a function of distance x in Fig. 11. Shown on the same axis is a bar graph illustrating the distribution of TC's actually observed.

lator loop gain, die attach uniformity, etc., result in a variation about this nominal value. A typical set of observed TC's from a sample of 50 is shown in Fig. 12.

Summary

A generalized temperature stabilized substrate integrated circuit containing distributed heat sources and temperature sensors of arbitrary shape has been analyzed, and the results used to develop a computer-aided technique for the optimum layout of such circuits. The results have been used to design a precision voltage reference with better than 1 ppm/°C nominal TC. Performance agreed closely with theoretical predictions.

References

[1] W. T. Matzen, R. A. Meadows, J. D. Merryman, and S. P. Emmons, "Thermal techniques as applied to functional electronic blocks," *Proc. IEEE,* vol. 52, pp. 1496–1501, Dec. 1964.
[2] G. Bosch and J. Koster, "Integrated thermostat voltage stabilizer," in *ISSCC Dig. Tech. Papers, 1971,* pp. 144–145.
[3] J. D. Lance, and K. A. Lance, "An ultra-stable voltage referference using a new type of bulk zener diode," in *ISSCC Dig. Tech. Papers, 1972,* pp. 148–149.
[4] P. R. Gray and D. J. Hamilton, "Electrothermal integrated circuits," *IEEE J. Solid-State Circuits,* vol. SC-6, pp. 8–14, Feb. 1971.

Computer Simulation of Integrated Circuits in the Presence of Electrothermal Interaction

KIYOSHI FUKAHORI, STUDENT MEMBER, IEEE, AND PAUL R. GRAY, SENIOR MEMBER, IEEE

Abstract—**A computer program which predicts the dc and transient performance of integrated circuits in the presence of electrothermal interactions on the integrated circuit die is described. The thermal modeling of the die/package structure and the numerical analysis procedure is discussed. Experimental and simulation results are compared for monolithic operational amplifiers, voltage regulators, and a temperature-stabilized voltage reference.**

I. INTRODUCTION

EXISTING electronic circuit simulation programs generally model the integrated circuit as if all the devices within the circuit are at a constant temperature. Actually, localized power dissipation within the elements of the integrated circuit can cause chip temperature gradients and variations which strongly affect the performance of the circuit, particularly in cases where very high accuracy is required or where large power dissipation occurs on the chip. This paper will describe a new electronic circuit simulator which includes the effect of chip temperature gradients and variation on the performance of integrated circuits. In Section II of the paper the significance of electrothermal integrated circuit design is illustrated by means of two examples. In Section III the problem of lumped thermal modeling of the die package structure is discussed. In Section IV the actual operation of the program is described. In Section V experimental results are presented from the simulations of several different integrated circuits.

II. ELECTROTHERMAL INTERACTIONS IN INTEGRATED CIRCUIT DESIGN

Electrothermal interactions are significant in integrated circuit design in two different ways. As a design problem, undesirable thermal feedback can severely degrade the performance of certain types of analog integrated circuits, particularly those in which large power dissipation is experienced on a chip. Secondly, the thermal interactions can be utilized as a design degree of freedom to enhance the performance of certain types of integrated circuits. The significance of electrothermal effects as a design problem is illustrated by the dc transfer characteristic of a commercially available operational amplifier 741 shown in Fig.1 [1]. A heavily distorted dc transfer curve is observed for the amplifier when a load resistor of 1 kΩ is attached from the output to ground, resulting in power dissipation in the output transistors. Fig. 1 also shows this observed dc transfer characteristic for the case of R_L open. Notice that the gain has about the right magnitude but the wrong polarity. This results from undesirable thermal feedback from elements in the circuit other than the output transistors. The computer-predicted dc transfer characteristic with no thermal interaction present is also shown in Fig. 1.

These feedback effects are more clearly illustrated in Fig. 2, in which a simplified version of the 741 circuit is shown, along with a simplified drawing of the die layout of this particular 741 operational amplifier. The output transistors $Q14$ and $Q20$ are located in such a way that power dissipation in those elements results in temperature gradients which cause temperature differences between the critical input pairs $Q1$-$Q2$, $Q3$-$Q4$, and $Q5$-$Q6$. Thus under loaded conditions this thermal coupling mechanism causes the large distortion of the dc transfer characteristic. The p-n-p current source transistor $Q13$, which biases the output stage and acts as the load for the common emitter amplifier $Q17$, has a power dissipation which varies linearly with output voltage. Thus when the amplifier does not have a load attached to the output, this results in the thermal feedback effect which varies linearly with output voltage. Because of the location of $Q13$ the thermal feedback is positive and causes the dc transfer to take on the wrong gain polarity.

The practical effect of the thermal feedback in operational amplifiers is that the effective value of dc open loop gain must be specified at a lower value than that which would be realizable considering electrical effects only. Other types of circuits which are subject to these effects are those circuits which require high precision such as D/A converters, instrumentation amplifers, analog multipliers, and precision voltage references, and those types of circuits which experience large amount of power dissipation such as voltage regulators, audio and servo amplifiers, etc.

Electrothermal interactions are important as a design problem but can also be utilized as a design degree of freedom in certain types of circuits. An example of this is a temperature-stabilized substrate integrated circuit shown in Fig. 3 [2]. The effect of the temperature stabilization system is to hold the chip temperature constant at an elevated value independent of the variations in ambient temperature, so that the effective temperature sensitivity of the circuitry on the chip is reduced below that of an unstabilized chip. The stabilization is accomplished by including temperature sensor, reference source, error amplifier, and controlled heat source on the chip itself and by isolating the chip from its surrounding environment as illustrated in Fig. 3. An important consideration in the layout of such circuits is that when the ambient temperature varies, the power dissipation in the controlled

Manuscript received May 28, 1976; revised July 28, 1976. This work was supported by NSF Grant ENG75-04986.

The authors are with the Department of Electrical Engineering and Computer Sciences and the Electronics Research Laboratory, University of California, Berkeley, CA 94720.

Reprinted from *IEEE J. Solid-State Circuits*. vol. SC-11, pp. 834-846, Dec. 1976.

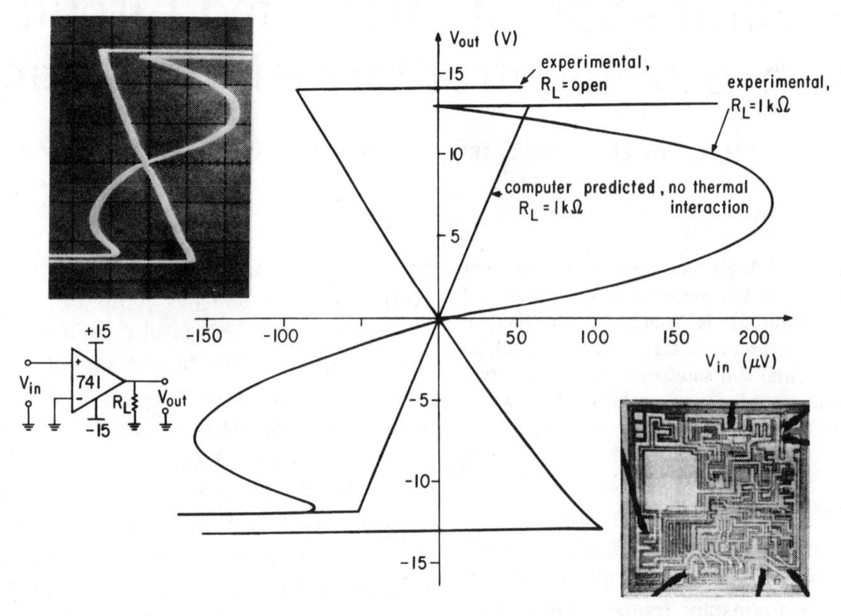

Fig. 1. Experimentally observed dc transfer characteristic, commercially available 741 operational amplifier.

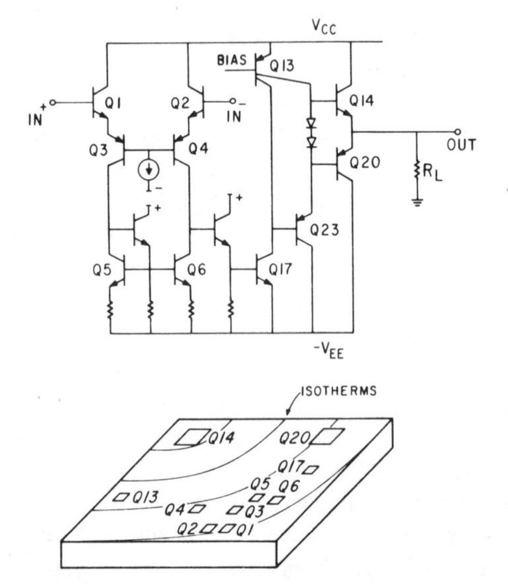

Fig. 2. Simplified schematic and layout of the 741 example of Fig. 1.

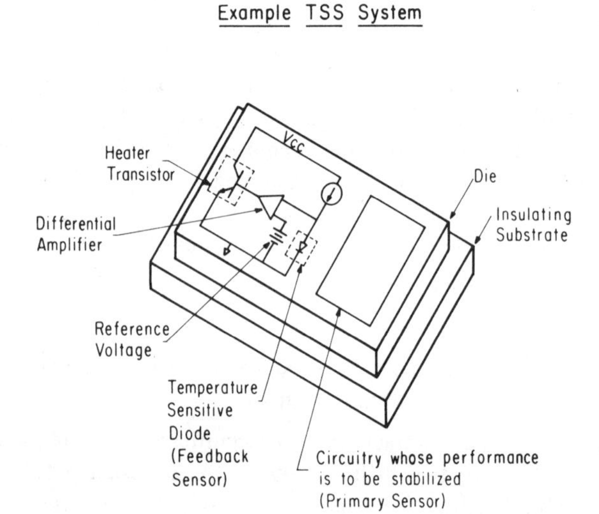

Fig. 3. Temperature-stabilized substrate physical model.

heat source must vary. As a result, temperature gradients across the surface of the chip vary as the ambient temperature changes. These varying gradients can strongly affect the performance of the circuitry whose performance is to be stabilized and can cause the sensitivity of that circuitry to the ambient temperature variation to be much larger than otherwise expected. Other applications of electrothermal interactions to improve circuit performance include thermal shutdown circuits for power voltage regulators and power amplifier, electrothermal multivibrators [3], and electrothermal low-frequency filters [4].

In order to be widely useful as a design tool in such applications, an electrothermal simulator must have certain capabilities as summarized below.

1) It must be capable of accurately modeling the thermal behavior of chip-package structure under dc, transient, and ac conditions.

2) The thermal parameters which must be specified by the user for use in the program must be either easily measured or calculated.

3) The temperature sensitivity of electrical circuit elements within the integrated circuit must be accurately modeled in the program.

4) The capability for simulating the different aspect of electronic circuit behavior must be similar to that of the existing programs.

5) The program should be capable of simulating such anamolies as die-attach voids, flip chip attach bonding, etc.

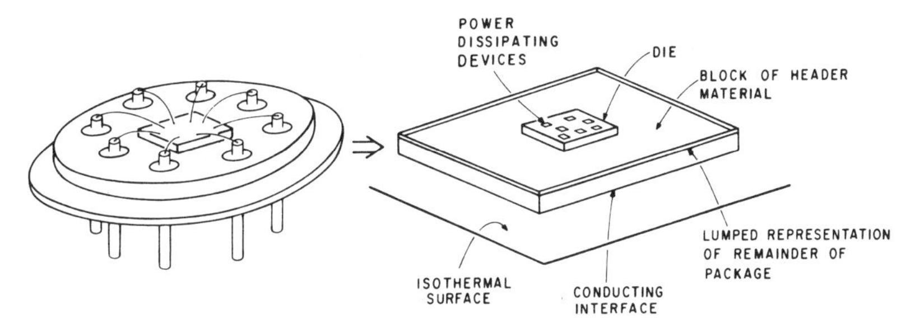

Fig. 4. Die-header structure and simplified physical model.

III. Lumped Thermal Model of the Die-Header Structure

One of the most important problems in the development of a generally useful electrothermal simulation program is that of adequately modeling the thermal behavior of the die-package structure with an effective compromise between accuracy in simulation on one hand and economy in computer simulation time on the other. Shown in Fig. 4(a) is the physical structure whose thermal behavior must be modeled. It consists of the die and the underlying package material, with which it is in intimate thermal contact. The package in turn is subject to a complex set of thermal boundary conditions. The aspects of the thermal behavior of the package which are important to the electrical behavior of the circuit are twofold. First, the package dictates the total thermal resistance between the top surface of the chip and the ambient atomosphere. Second, the presence of the header material underneath the die and in intimate contact with it affects the temperature distribution on the top surface of the die because of a lateral thermal conductance in the header material. In order to properly model those two effects without introducing unnecessary complexity in the modeling of the remote regions of the die-package structure, a simplified model was adopted as shown in Fig. 4(b). This simplified physical model consists of the die which is mounted on the header material of twice the length and width of the die. The rectangular piece of header material is in turn mounted on an isothermal substrate. In order to adjust the net junction to ambient thermal resistance to the proper value, a thermally conducting interface is included between the header material and the conducting substrate. In addition the thermal capacitance of the remainder of the package structure is included in lumped form in a strip around the outside edge or the rectangular block. This simplified physical model provides an adequate compromise between accuracy in simulation and efficient use of computer time. Having selected the simplified physical model, the next problem is the generation of lumped representation of this physical structure.

The usual approach to the lumped modeling of such a rectangular structure is the symmetrical finite difference approach. With this technique the solid which is to be modeled is subdivided into subregions which have a rectangular plan view as shown in Fig. 5. Proper matching of the boundary conditions on the top surface of the chip requires that each active electrical circuit element on the die have one subregion associated with it. This requirement results in the formation of a number of nodes which is greatly in excess of the number of electrical elements on the surface of the die as illustrated in the example of Fig. 5. This excessive number of thermal nodes results in high cost in the computer simulation.

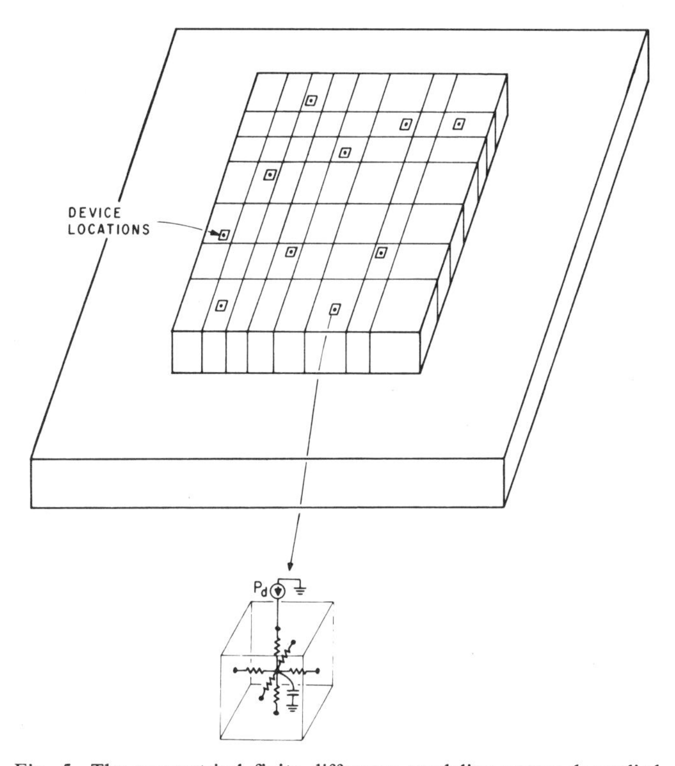

Fig. 5. The symmetrical finite difference modeling approach applied to a typical IC layout.

A better approach is a generalization of the symmetrical finite difference approach [5]. In this method a network of triangles is formed by connecting the thermal nodes corresponding to device locations and some internally created nodes necessary to insure that none of the interior angles of triangles are obtuse. By employing the finite difference approximation to the network thus formed, the values of the thermal resistance and capacitances are easily evaluated. The application of this technique for the example of Fig. 5 is shown in Fig. 6. The proper matching of boundary conditions at the edges of the die require the formation of several additional thermal subregions. We have found that for typical integrated circuit layouts this technique results in the number of thermal nodes which is roughly two to three times the

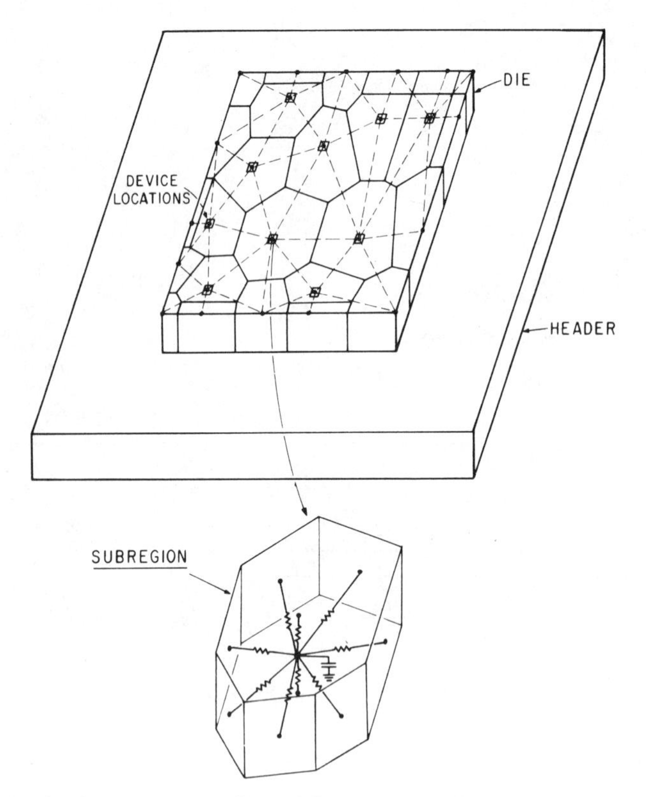

Fig. 6. The asymmetrical finite difference approach applied to a typical IC layout.

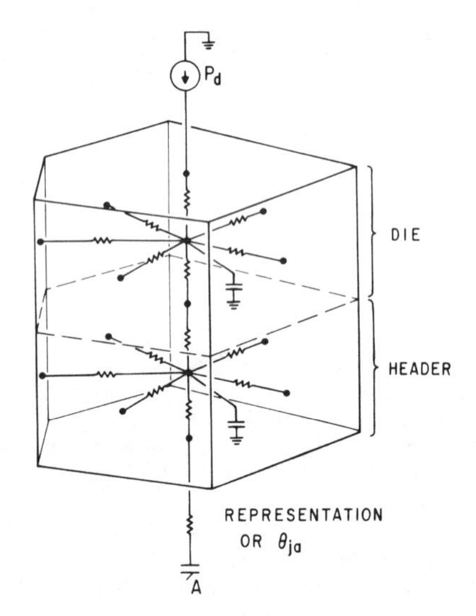

Fig. 7. Subregion of the asymmetrical lumped model.

number of electrical elements on the chip. The rectangular header block is modeled in the same manner. A typical resulting subregion of the die-header sandwich is shown in Fig. 7. In this model the current is analogous to power flow, and voltage to temperature. This current source represents power dissipation within the associated electrical circuit elements. The resistance leading to the ambient represents a portion of the junction to ambient thermal resistance.

Three important limitations arise from this modeling approach. The first is that in circuits which contain only a few circuit elements which are widely spaced physically, the thermal model will not contain enough thermal nodes to accurately predict the chip temperature distribution. To avoid this problem, the model formation algorithm introduces additional nodes to insure that no point on the surface of the chip is more than 10 mils from a thermal node.

A second important limitation is that the heat sources are represented as point sources, and the temperature sensitivity of any element is localized at one point. Device structures are often encountered which are distributed over a large area, large resistors and power transistors being the best examples. These structures can be accurately modeled by the user of the program by representing them as being composed of a number of individual devices connected in parallel or series as appropriate. Each of the individual devices has a different physical location. Such an approach is particularly important in the modeling of power transistors where nonuniform current distribution in the device is an important thermal effect.

A third important limitation is that since the thermal nodes are separated by a distance on the order of the electrical device separation, thermal phenomenon which take place over distances shorter than this are not well modeled. Thus for example, certain self-heating effects which occur in very small-geometry devices operated at high power densities with short pulsewidths will not be accurately predicted. Such phenomenon depend heavily on the temperature distribution with a few tens of microns of the device in the vertical and horizontal directions.

IV. Program Description

The program first reads the data regarding the topology and interconnections of electrical elements as well as the data regarding the location of electrical elements on the die and the thermal data. The lumped thermal model for the die-package structure is then generated as described above. The matrices describing the resulting coupled electrothermal network are then set up and solved by using a standard Newton-Raphson iteration technique. The results are then printed for dc transient and ac conditions.

The technique used for the solution of electrothermal coupled circuit is illustrated in more detail in the block diagram of Fig. 8. The circuit is described by a coupled set of electrothermal equations. In the branch linearization step inherent in the Newton-Raphson method, the branch linearized circuit matrix contains elements corresponding to the electrical circuits, designated Ye, elements corresponding to the thermal circuit, designated Y_{th}, and additional elements corresponding to the coupling between two systems. The latter are temperature-controlled current sources corresponding to the effects of temperature on the electrical performance of the circuit and electrically controlled power sources which indicate the dependence of power dissipation in the circuit on the electrical node voltages. In addition to the usual update logic in an iteration process, a scheme is used in which from one iteration to another the change in temperature and/or power dissipation is limited. Once these elements in the matrix have been included and the limiting process employed, the solution of the circuit proceeds in a

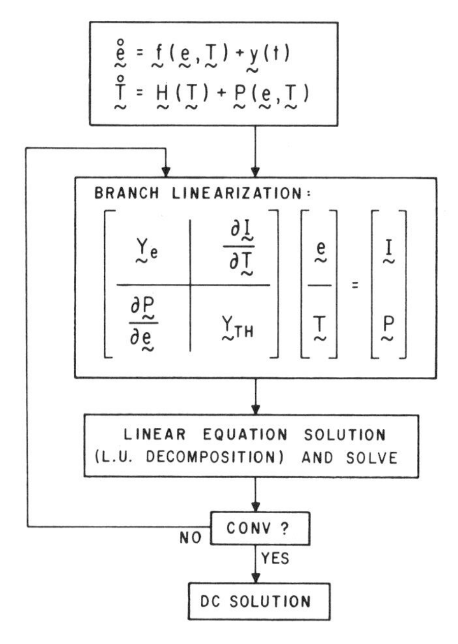

Fig. 8. Numerical analysis approach.

TABLE I

User-specified thermal parameters	
1) Device locations	(x_i, y_i) (mils)
2) Die dimensions and thickness	(L_x, L_y, a_0) (mils)
3) Die thermal parameters	k_s (mW/°C-mil); $\rho_s c_s \left(\frac{\text{mW/s}}{°\text{C-mil}^3}\right)$
4) Thickness of header material	b_0 (mils)
5) Header thermal parameters	k_h (mW/°C-mil); $\rho_h c_h \left(\frac{\text{mW/s}}{°\text{C-mil}^3}\right)$
6) Total thermal capacitance of package	$C_p \left(\frac{\text{mW/s}}{°\text{C}}\right)$
7) Thermal resistance, junction-to-ambient	R_{ja} (°C/W)
Data output options	
1) Standard output options data	
2) Print, plot temp (n) versus time	
3) Print, plot temp (n)-temp (m) versus time	
4) Print, plot temp (x_0) versus y or temp (y_0) versus x	
5) Print, plot temp (n) versus v_{in}	
6) Print, plot temp (n)-temp (m) versus v_{in}	
7) Print, plot temp (n) versus $T_{ambient}$	
8) Print, plot temp (n)-temp (m) versus $T_{ambient}$	
9) Print, plot v_n, i_n versus $T_{ambient}$	

manner similar to existing programs such as SPICE 2 [6]. The transient analysis portion of the program utilizes the trapezoidal integration technique with automatic time step control.

In order to utilize the program the user must specify certain thermal parameters as summarized in Table I. The physical location of the electrical elements are obtained directly from mask drawing or circuit itself. The thermal conductivity and specific heat of the die are assumed to be those of silicon unless otherwise specified. The thermal conductivity, density, and specific heat of the header material can be obtained from the dimensions and material properties of the header. The same is true of the total thermal capacitance of the package structure and the thermal resistance of the package. In addition, the user can insert an element card which specifies the location of a bonding pad and the thermal resistance of the bonding wire. The thermal conduction through bonding wires is not significant for integrated circuits packaged with the usual techniques; conduction through the substrate is vastly larger. However, it is significant in temperature-stabilized substrates where a thermal insulator greatly reduces heat flow through the substrate. The bonding wire is assumed to behave as a thermal resistance terminating at the ambient temperature. This option also permits simulation of flip-chip bonded circuits.

The user can elect all of the output data options which currently exist in most electronic simulators. In addition the user can print or plot the temperature of any device or the temperature difference between any two devices as a function or input voltage for the case of dc transfer curve and time for the case of transient analysis. He can also ask for a plot of temperature variation across the die as a function of the distance for any value of x or y. A typical deck is illustrated in Fig. 9. The addition over and above the standard SPICE deck include x and y locations of each device written in the element card itself and a single thermal card which specifies thermal data required.

V. Experimental Results

The program described above has been utilized to simulate the performance of several different integrated circuits. The program was first used to predict the dc transfer characteristics of the same 741 circuit whose dc transfer characteristic was illustrated in Fig. 1. This integrated circuit die is shown in more detail in Fig. 10. The experimentally observed and

```
***** 9 OCT 75 *************** SPICE 2A ************** 01:00:22 *****

TRANSIENT ANALYSIS OF VOLTAGE REGULATOR ---NATIONAL

***************************************************************************

    VBE(INITIAL)=    4.00000E-01  VBC(INITIAL)=     0.         VTLIM=     2.64500E-02  VRLIM=      1.00000E+00
    TDEL1=       1.00000E+00  TDEL2=       1.00000E+00  FTR=       1.00000E+00  ITLIMIT=    60
    TC1= 1.800E-03      TC2= 0.

Q1 3 2 4 MODN 2.25               X=5.00      Y=16.8
Q2 4 5 6 MODN 2.0                X=12.2      Y=14.6
Q3 7 6 0 MODN 6.0                X=20.0      Y=11.3
Q4 8 7 10 MODN 6.0               X=26.0      Y=11.30
Q5 9 10 11 MODN 7.0              X=35.6      Y=11.3
Q6 17 2 8 MODN 2.25              X=51.0   Y=10.0
Q7 12 9 13 MODN 2.25             X=40.8      Y=11.3
Q8 12 13 0 MODN 2.25             X=43.3      Y=11.3
Q9 14 14 0 MODN 2.25             X=48.0      Y=6
Q10 0 12 15 MODVP 6.15           X=62.0      Y=27.3
Q11 16 3 1 MODLP 3.0             X=50.7      Y=21.0
Q12 3 3 1 MODLP 3.0              X=46.4      Y=21.0
Q15 1 16 17 MODN 17.5            X=77.4      Y=42.0
Q1601 1 17 18 MODN 18.0          X=9.3       Y=42.0
Q1603 1 17 20 MODN 18.0          X=15.8      Y=42.0
Q1605 1 17 22 MODN 18.0          X=22.5      Y=42.0
Q1607 1 17 24 MODN 18.0          X=29.0      Y=42.0
Q1609 1 17 26 MODN 18.0          X=35.6      Y=42.0
Q1611 1 17 28 MODN 18.0          X=51.0      Y=42.0
Q1613 1 17 30 MODN 18.0          X=57.5      Y=42.0
Q1615 1 17 32 MODN 18.0          X=64.1      Y=42.0
Q1617 1 17 34 MODN 18.0          X=70.8      Y=42.0
R1 3 4 30K                       X=10.0      Y=21.0
R2 4 5 1.9K                      X=7.5       Y=11.0
R3 5 7 26.0                      X=16.4      Y=11.0
R4 6 0 1.2K                      X=20.5      Y=8.5
R5 10 0 12.1K                    X=39.0      Y=7.0
R6 11 0 1K                       X=39.0      Y=4.7
R7 8 9 17.7K                     X=25.0      Y=7.0
R8 13 14 4K                      X=48.0      Y=6.0
R9 15 12 4K                      X=53.5      Y=16.0
R10 16 15 850                    X=55.0      Y=20.0
R15 17 2 2K                      X=25.0      Y=26.5
R1601 18 2 3.5                   X=9.3       Y=42.0
R1603 20 2 3.5                   X=15.8      Y=42.0
R1605 22 2 3.5                   X=22.5      Y=42.0
R1607 24 2 3.5                   X=29.0      Y=42.0
R1609 26 2 3.5                   X=35.6      Y=42.0
R1611 28 2 3.5                   X=51.0      Y=42.0
R1613 30 2 3.5                   X=57.5      Y=42.0
R1615 32 2 3.5                   X=64.1      Y=42.0
R1617 34 2 3.5                   X=70.8      Y=42.0

R18 2 0 2.67K                    X=69.0      Y=6.7
VCC 1 0 10
.MODEL MODN NPN BF=100 IS=1.E-15 RB=300 RC=260 VA=200 PE=.65 ME=.5 PC=.5 MC=.3
.MODEL MODLP PNP BF=15 IS=1.5E-15 RB=200 RC=25 RE=270 VA=50 PE=.65 ME=.3 PC=.5
+ MC=.3
.MODEL MODVP PNP BF=50 IS=1.5E-15 RB=100 RC=100 RE=270 VA=90 PE=.65 ME=.3 PC=.5
+ MC=.3
IOUT 2 0 PULSE(1M 1.5 0US 10US 10US 10MS 20MS)
.PRINT TRAN V(2) V(9) V(7) V(7,10) V(10,11) V(11)
.PRINT TRAN T(Q3) T(Q2,Q3) T(Q4,Q3) T(Q5,Q3) T(Q7,Q3) T(Q8,Q3) T(Q6,Q3)
.PLOT TRAN V(2) V(9) V(7) V(11)
.PLOT TRAN T(Q3) T(Q2) T(Q2,Q3)
.PLOT TRAN T(Q2,Q3) T(Q4,Q3) T(Q5,Q3) T(Q7,Q3) T(Q8,Q3) T(Q6,Q3)
.TRAN 100US 10MS
.THRML LX=86 LY=59 A0=7 R0=100 KS=.00223 KH=0.01011 GH=.02857
+TCS=27.7E-9 TCH=56.33E-9 LXHDR=500.0 LYHDR=500.0
.OPTION     ACCT
.END
```

Fig. 9. Typical input deck.

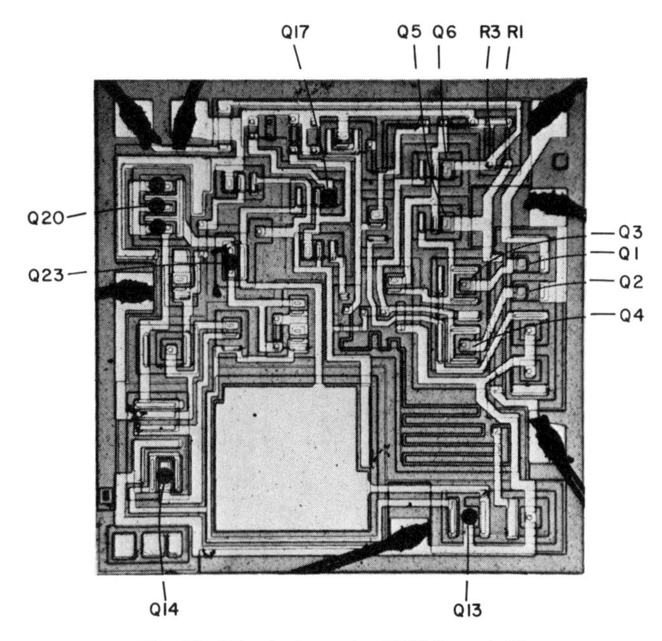

Fig. 10. Die photograph of 741 layout #1.

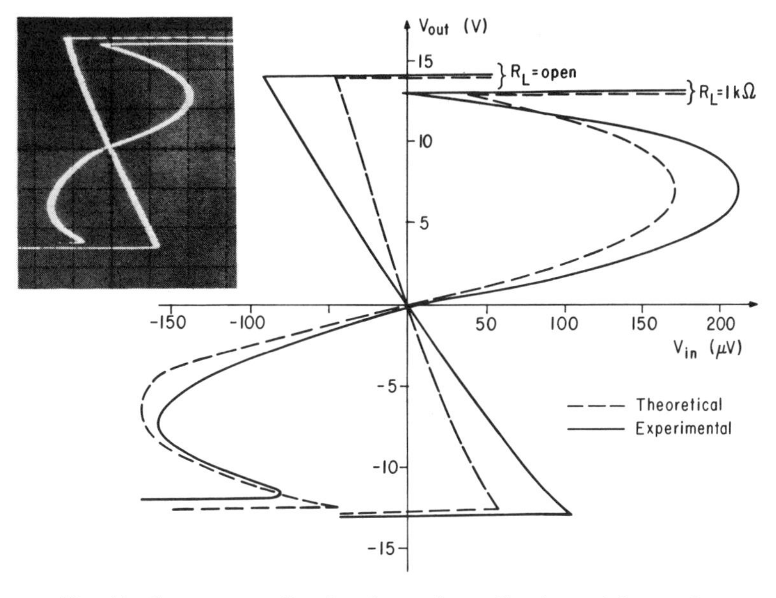

Fig. 11. Computer predicted and experimentally observed dc transfer characteristic of 741 #1.

computer predicted transfer characteristics are shown in Fig. 11. Notice that agreement between experimentally observed and computer predicted performance is good. This device was in dual in line plastic package. The time of simulation was 300 s on a CDC 6400; the same simulation considering electrical effects only requires 25 s.

The program was next applied to a second 741 amplifier which was made by a different manufacturer and had the same electrical circuit but a different layout. The die shown in Fig. 12 and computer predicted and experimentally observed characteristics are shown in Fig. 13 Again the agreement is good. This device was contained in a T0-5 metal package.

The program was used on a third layout of the same 741 circuit fabricated by third manufacturer. A photograph of this die is shown in Fig. 14. A first glance this circuit appears to be one which is much more optimal than the first two in terms of susceptibility to thermal feedback effects. The output transistors $Q14$ and $Q20$ are located symmetrically along the center line of the die as are the critical input pairs $Q1$-$Q2$, $Q3$-$Q4$, and $Q5$-$Q6$, and indeed there is very little thermal interaction between the output transistors and input pairs. However, this circuit is a good example of a case in which a simple symmetrical layout of the obvious large power dissipating elements does not yield ideal thermal performance. Notice that current source $Q13$ is placed so that power dissipa-

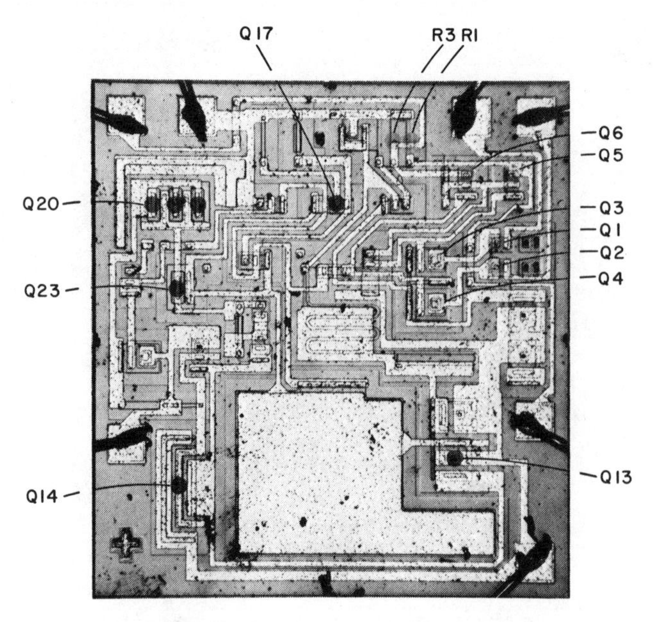

Fig. 12. Die photograph of 741 #2.

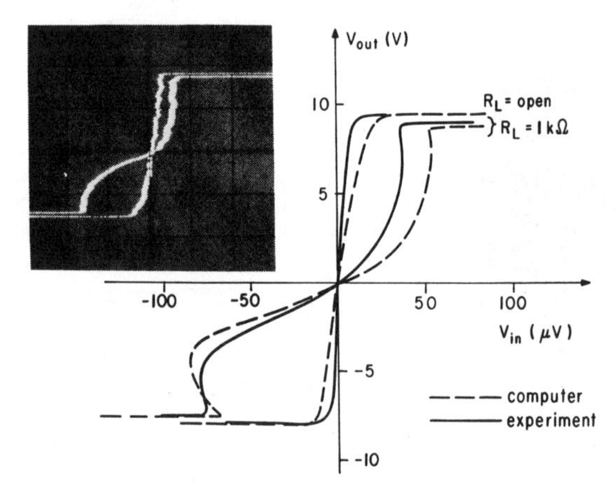

Fig. 13. Computer predicted and experimentally observed dc transfer characteristic of 741 #2.

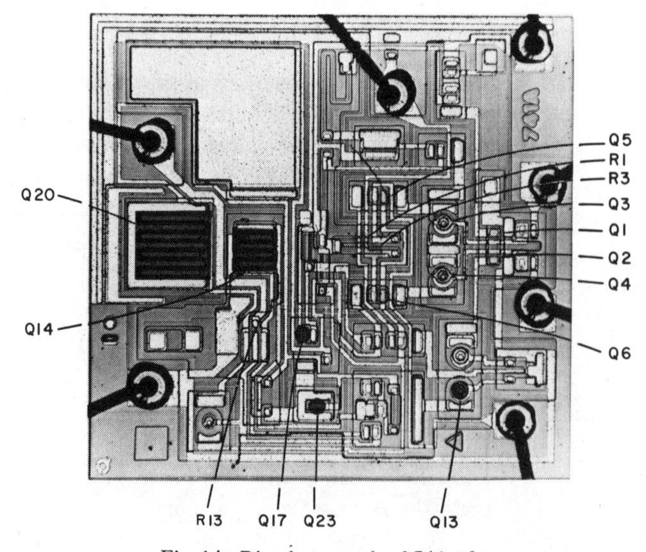

Fig. 14. Die photograph of 741 #3.

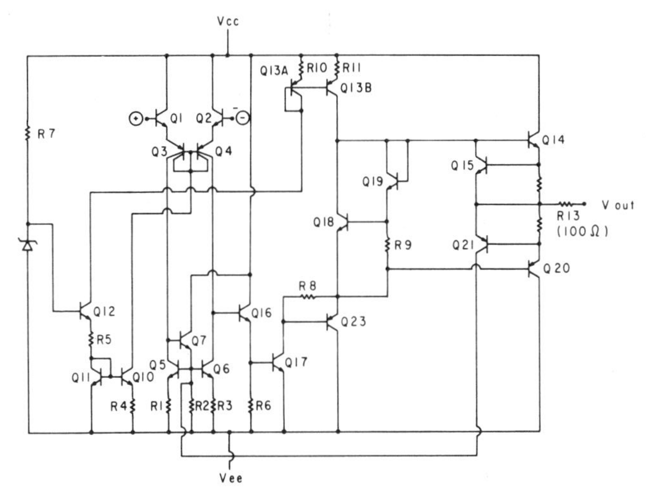

Fig. 15. Schematic of 741 #3 showing 100-Ω buffering resistor in the output lead.

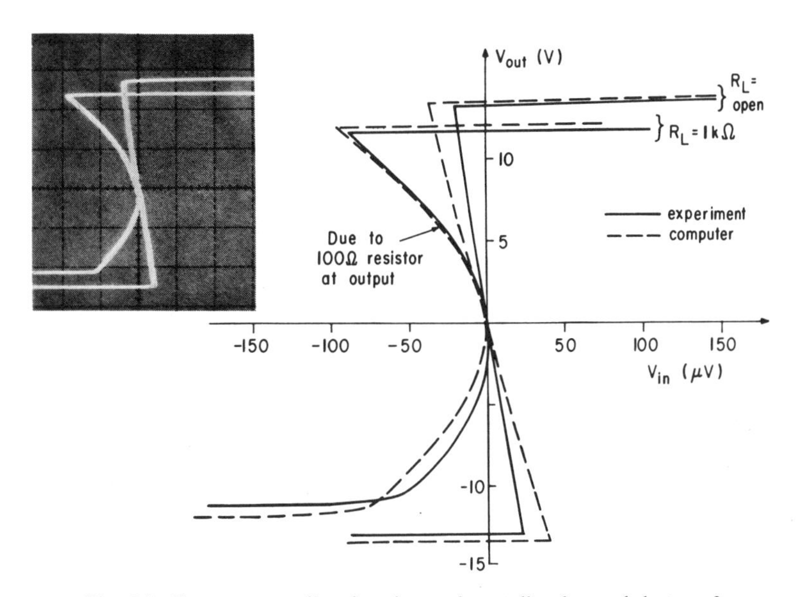

Fig. 16. Computer predicted and experimentally observed dc transfer characteristic of 741 #3.

tion within it can cause temperature differences between the input transistor. As a result, dc transfer curve still shows a gain reversal in the no load condition. In addition, the 100-Ω resistor $R13$ shown in Fig. 15 which appears in series with output lead for buffering purposes is not located along the die centerline, and its dissipation can cause temperature differences between the input devices. The power dissipation in this resistor can be very significant under heavy load conditions. The observed and computer predicted dc transfer characteristics are shown in Fig. 16. Notice that indeed under no load condition the gain reversal still occurs. Notice also under loaded condition the dc transfer characteristic is distorted with a shape which is quite different from the earlier two circuits. The distortion has a shape which indicates that the thermally induced offset voltage is proportional to output current squared. Further computer simulation shows this distortion indeed results from power dissipation in the series buffering resistor $R13$.

As mentioned earlier, the undesired effects of thermal feedback can be most significant in integrated circuits which experience large power dissipation. A good example of such a circuit is a three-terminal voltage regulator illustrated schematically in Fig. 17(a). The circuit usually consists of a reference voltage source, error amplifier, and series pass transistor. When a load resistor is suddenly attached to the output of the regulator the load current increases to a large value and the power dissipation in the pass transistor increases suddenly. As a result, die temperature increases and thermal gradients are generated which propagate from the power transistor across the die. These thermal gradients affect the circuit performance in two important ways. An output waveform which might be observed as a function of time

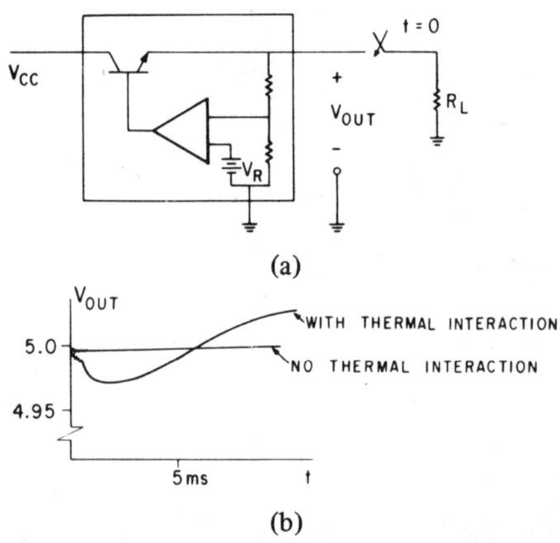

Fig. 17. (a) Block diagram of a typical three-terminal voltage regulator. (b) Typical output voltage waveform in response to a step increase in input current.

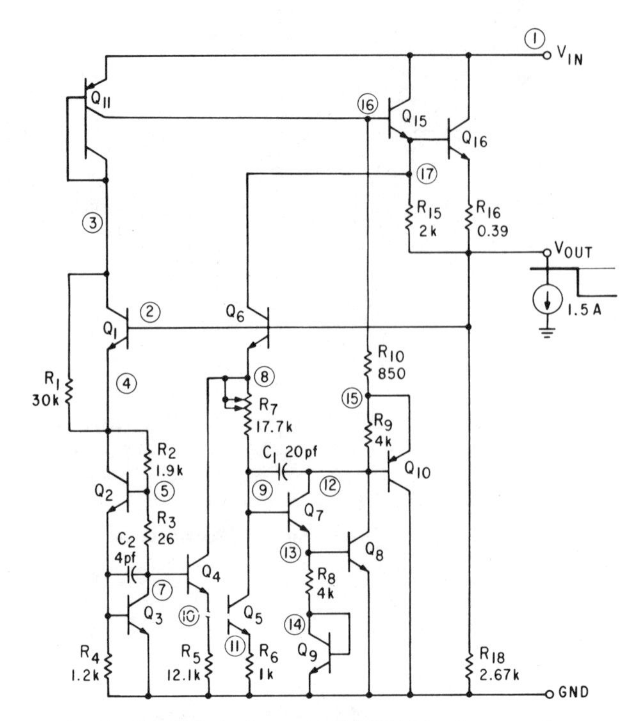

Fig. 18. Simplified schematic of LM140d voltage regulator.

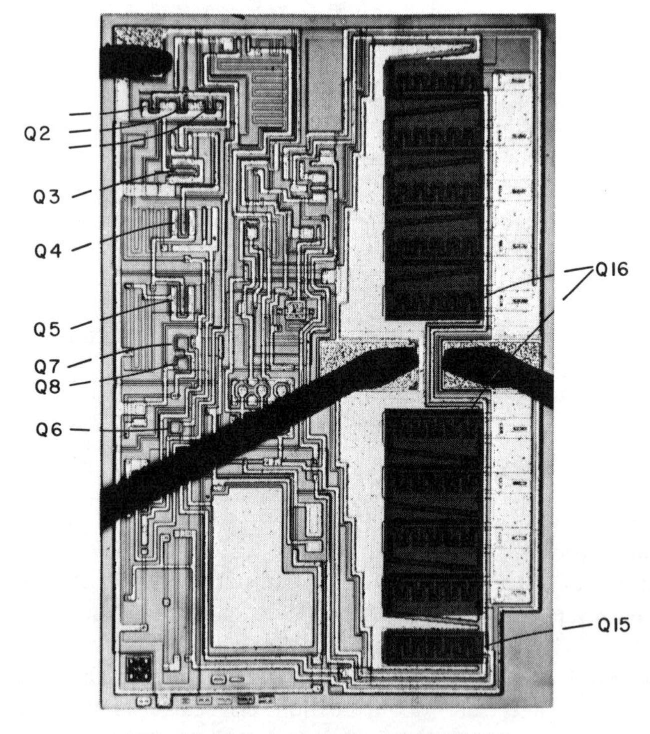

Fig. 19. Die photograph of the LM140d.

after the application of the load at $t = 0$ is illustrated in Fig. 17(b). First, after the thermal gradients have settled to their steady-state value the resulting temperature differences between critical circuit elements can cause the steady-state output voltage to be different from that which existed under no load condition. Secondly, in addition to this phenomena, when the load is applied at $t = 0$, the thermal gradients which are generated do not necessarily reach all the devices within the circuit at the same time. As a result a nonmonotonic transient output waveform can occur due to the resulting transient thermal unbalance. In the frequency domain these time varying thermal gradients have the effect of producing humps and dips in the output impedance and line regulation. Because of the thermal natural frequencies of the die-package structure, they tend to occur at the undesirable values near 60 Hz, 120 Hz, and 180 Hz.

A typical monolithic voltage regulator circuit is shown in simplified form in Fig. 18. The circuit consists of series pass transistor $Q16$ and the band gap circuitry composed of transistor $Q1$ through $Q8$. The output voltage can be seen to be the sum of the base-emitter diodes drops in $Q6$, $Q7$, and $Q8$ plus the voltage drop across R_7. The voltage drop in R_7 is proportional to the voltage $(V_{BE3} + V_{BE2}) - (V_{BE4} + V_{BE5})$ multiplied by a constant which is approximately equal to 17. Thus if any temperature differences exists between transistors $Q2$-$Q3$ and transistors in $Q4$-$Q5$ a large change in output voltage will occur. In addition any difference in temperature between the group of transistors $Q6$, $Q7$, $Q8$, and that of the second group $Q2$, $Q3$, $Q4$, $Q5$ will result in a significant but smaller change in the output voltage.

The die photo of one layout of this particular regulator circuit is shown in Fig. 19. The elements associated with the band gap reference are located down the left side of the die while the large structure on the right side is the output power transistor. The behavior of two different versions of this circuit was simulated using the program described above. In the first version, only the center emitter of the three-emitter transistor $Q2$ was electrically connected so that all of the devices associated with the band gap reference were distributed in a straight line along the left edge of the die. Because the isothermal lines which result from dissipation in the output transistor are curved, it is clear that transistors $Q6$, $Q5$, $Q4$ get hotter than $Q2$ and $Q3$. As a result this circuit displays a large change in the steady state output voltage when power is dissipated in the output transistor. Fig. 20 shows the observed and predicted output voltage waveform, labeled version 1, when a 1.5 A current pulse is applied to the output of the regulator. The solid line is the experimentally observed

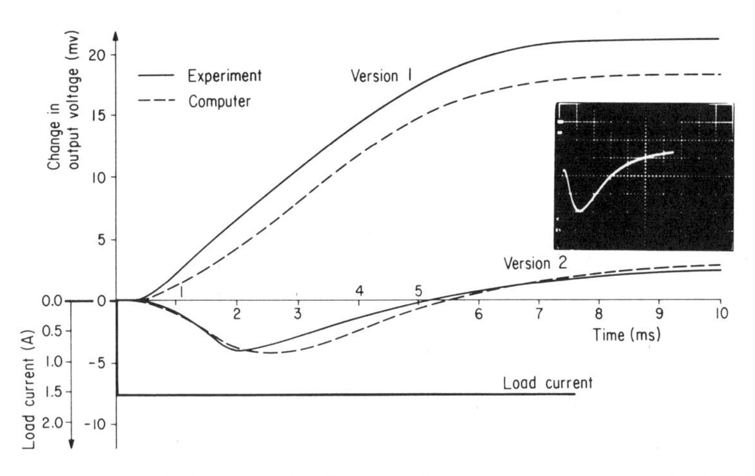

Fig. 20. Observed and computer predicted output voltage waveforms for two versions of the LM140d in response to a 1.5-A output current step.

response and the dotted line is the computer predicted response. This simulation required 800 s of CPU time on the CDC 6400.

A second, modified version of this chip was fabricated in order to attempt to correct this problem in which the emitter of $Q2$ which is closer to the power transistor than in the first version was electrically connected. This was done in an effort to counterbalance the thermal unbalance between the group $Q2$, $Q3$, $Q4$, and $Q5$ with reference to the group $Q6$, $Q7$, and $Q8$ by introducing a compensating thermal unbalance between $Q2$ and $Q3$. By doing this the net steady-state thermal feedback could hopefully be reduced. The resulting transient waveform is also shown in Fig. 20, labeled version 2. Notice that indeed the steady-state output voltage change is smaller than in version 1. However, the change introduced a negative going transient to the output voltage which did not exist before. This results because, although thermal coupling mechanism has been cancelled under steady-state conditions, the thermal imbalance has not been cancelled under transient conditions. The agreement between experimentally observed and computer predicted response is good.

A complete relayout of the circuit was next carried out using the simulator as a tool to check each layout change [7]. The entire set of devices $Q2$, $Q3$, $Q4$, $Q5$, $Q6$, $Q7$, and $Q8$ was placed along an isothermal line as shown in Fig. 21. The resulting simulated and experimentally observed output waveforms are shown in Fig. 22. The transient and steady-state output voltage variations are greatly reduced, and the agreement between predicted and observed response is good.

The program was next applied to a temperature-stabilized substrate integrated circuit system. The objective of this system, illustrated in Fig. 3, is the reduction of effective temperature sensitivity of this integrated circuit by stabilizing the die temperature at constant value independent of the ambient temperature variations. The problem which occurs in the actual behavior of the circuit is that as the ambient temperature varies the heater transistor power must vary, which gives rise to the chip temperature variation. In Fig. 23 the chip temperature distribution which results when the ambient temperature is kept constant and the heater power is varied is illustrated. Notice that the temperature differences between one end of the chip and the other increases as the heater power is increased. If the feedback loop is then closed with a large amount of loop gain, the effect will be to keep the temperature at the feedback sensor constant. However, the temperature variation from one end of the chip to the other as a function of heater power results in a family of temperature curves across the chip as illustrated in Fig. 23(b). All of the

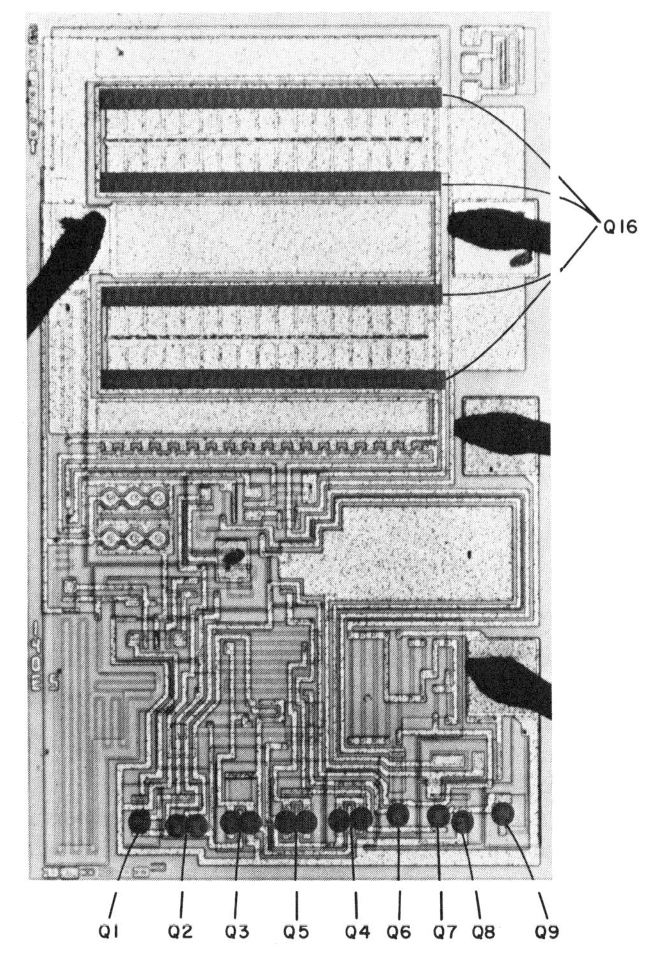

Fig. 21. Photograph of revised LM140d chip.

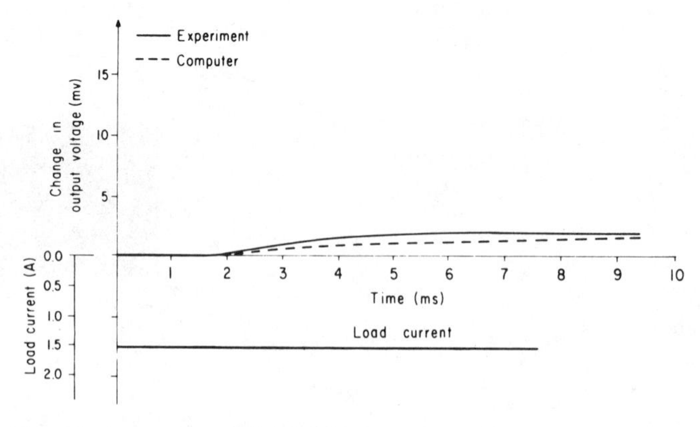

Fig. 22. Computer predicted and experimentally observed transient response of revised chip.

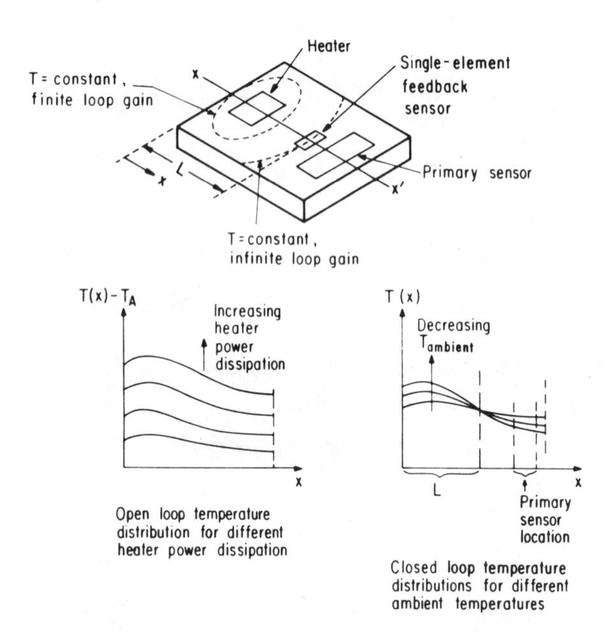

Fig. 23. Temperature distributions within a TSS system.

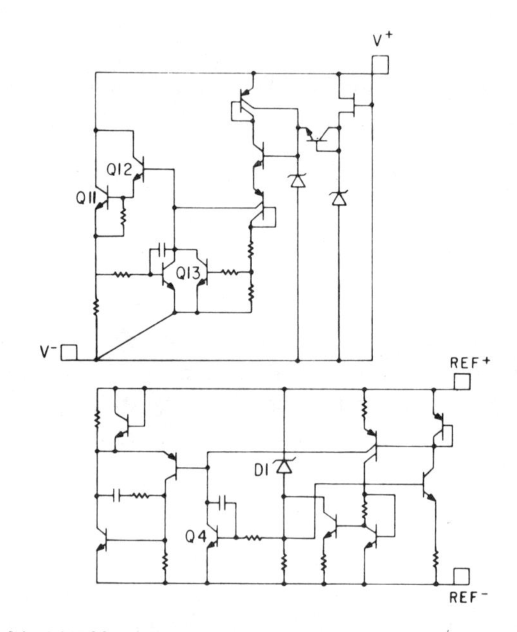

Fig. 24. LM199 temperature-stabilized substrate voltage reference.

devices located along the dotted line passing through the feedback sensor will not experience any temperature variation as the ambient temperature varies. The other locations of the chip will experience either positive or negative temperature variations. In practice a rather moderate amount of loop gain must be used for stability reasons. Because of this the point of zero temperature variation moves closer to the heat source. For optimum performance, the critical elements of the stabilized circuit should be located on this locus or zero temperature variation.

The program was used to simulate the performance of the LM199 temperature-stabilized voltage reference source [8]. The schematic diagram of this circuit is shown in Fig. 24. The upper portion of the circuit is the temperature regulator; $Q11$ is the controlled heat source and $Q13$ is the temperature sensing device. The lower portion of the circuit is the circuit to be stabilized. Zener diode $D1$ and diode $Q4$ together form a temperature-compensated reference diode. The rest of the circuitry is a shunt regulator for the Zener to reduce its incremental impedance. The regulator circuit is set up to stabilize the chip at approximately 90°C.

The LM199 die is shown in Fig. 25. The heater, sensor, and reference diode are labeled. The dark lines on the photograph are computer predicted lines of constant temperature for an ambient temperature variation of -40°C to 90°C. As expected, a line of zero temperature variation does exist, and passes within about 2 mils of the feedback sensor. The regions close to the heater experience a negative shift in temperature as the ambient temperature increases, while the regions far from the heater experience a positive shift in temperature. According to the simulation, the feedback sensor, the Zener and its compensating diode all experience a positive shift of about 1°C. Direct measurement of the Zener diode temperature by measurement of its forward voltage shows that the actual temperature variation is very close to this value. Movement of the Zener diode and the compensating diode closer to the heater would result in a smaller variation in the temperature in these elements. The compensating diode

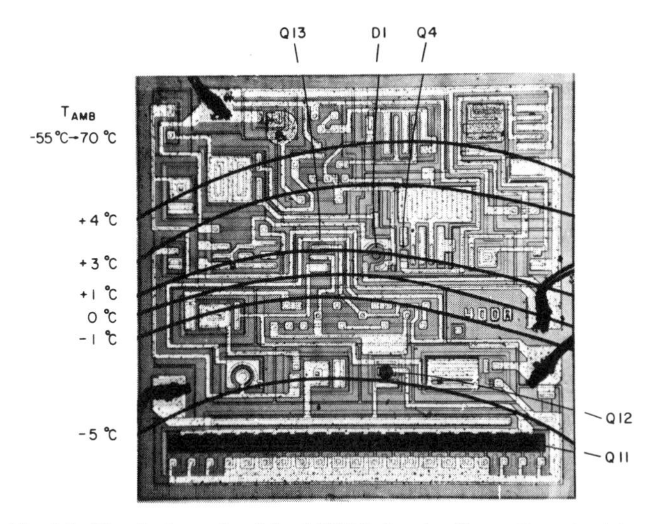

Fig. 25. Die photograph of the LM199 showing lines of constant temperature. The indicated temperatures on the isotherms are the variation in temperature experienced by points on the line for an ambient change from -40°C to 70°C.

temperature compensates the Zener voltage drop so that the unstabilized temperature coefficient of forward drop in the combination is nominally zero with some statistical deviation about this nominal value from unit to unit. The amount of temperature variation at the Zener and compensating diode will determine the amount of deviation from zero temperature coefficient in a sample of stabilized units. Thus the effect of moving the Zener and the compensating diodes closer to the heater would be to tighten the distribution of observed TC's in stabilized units, without affecting the nominal value of zero temperature coefficient. The simulation of this circuit required 330 s of CPU time on the CDC 6400 for 20 ambient temperature points.

VI. Summary

A new circuit simulator has been described which should be useful as a design tool during the layout of a wide variety of analog integrated circuits. Experimental results were presented which indicate that the simulator is capable of predicting the important aspects of electrothermal interactions in several types of analog integrated circuits. Computer simulation run times are typically a factor times ten greater than when electrical effects only are considered.

References

[1] J. E. Solomon, "The monolithic op amp, a tutorial study," *IEEE J. Solid-State Circuits*, vol. SC-9, Dec. 1974.

[2] P. R. Gray, D. J. Hamilton, and D. J. Lieux, "Analysis and design of temperature stabilized integrated circuits," *IEEE J. Solid-State Circuits*, vol. SC-9, Apr. 1974.

[3] P. R. Gray, "Electro-thermal interactions in integrated circuit design," in *Digest Tech. Papers, 1973 IEEE Symp. Circuit and System Theory*, Toronto, Ont., Canada, Apr. 1973.

[4] P. R. Gray and D. J. Hamilton, "Electro-thermal integrated circuits," *IEEE J. Solid-State Circuits*, vol. SC-6, Feb. 1971.

[5] R. H. MacNeal, "Asymmetrical finite difference networks," *Quart. Appl. Math.*, Oct. 1953.

[6] L. W. Nagel, "A computer program to simulate semiconductor circuits," Memorandum ERL-M520, Electronics Res. Lab., Univ. of Calif., Berkeley, CA.

[7] G. Cleveland, National Semiconductor Corp., private communication.

[8] R. C. Dobkin, "Monolithic temperature stabilized voltage reference with .5ppm°C drift," in *Digest Tech. Papers, 1976 Int. Solid-State Circuits Conf.*, Philadelphia, PA, Feb. 1976.

Technology for the Design of Low-Power Circuits

CHARLES A. BITTMANN, GARTH H. WILSON, MEMBER, IEEE, RONALD J. WHITTIER, MEMBER, IEEE, AND ROBERT K. WAITS

***Abstract*—Operation of integrated circuits at micropower levels requires transistors with adequate current gain at collector currents of 1 μA and less and resistors of the order of 1 MΩ within reasonable areas. In this paper the factors affecting current gain at low currents are discussed and design criteria presented that optimize gain at low collector current. A benefit of micropower operation is low-current noise. Factors tending to optimize noise performance are discussed. In order to obtain voltage gain at low collector current, high values of load resistance are required. Both passive and active loads suitable for incorporation in micropower integrated circuits are discussed. As an example of these design principles, a micropower operational amplifier is discussed.**

I. Introduction

THIS PAPER will discuss the design of low-power bipolar transistor circuits [1], [2]. It will review the design of transistors for operation at low collector current and the necessary resistor technology to make use of transistor gain at low collector current, and will give an example of this type of design—a micropower operational amplifier. An example of digital design is given in a companion paper by Foglesong [3].

The principal advantages of micropower operation of operational amplifiers are

low power consumption,
low heat generation,
low input current,
low input offset current,
low input current noise,
low input offset voltage,
high input impedance.

The advantage of low power consumption is obvious. The advantage of low heat generation becomes increasingly important as more and more complex functions are fabricated. It is also important in linear circuits where heat generated in output stages affects the balance and offsets of input stages. Again, in linear integrated circuits, micropower operation is important to give low input current, low input offset current, and low input current noise. In fact, most modern designs of linear integrated circuits are, at least in their input stages, micropower designs. Micropower design also gives low input offset voltage and provides high input impedances.

Manuscript received June 17, 1969; revised August 29, 1969. This work was supported in part by the Air Force Avionics Laboratory, Air Force Systems Command, Wright-Patterson Air Force Base, Ohio, under Contracts AF33(615)-[illegible]010 and F33615-67-C-1484.

C. A. Bittman, R. J. Whittier, and R. K. Waits are with the Fairchild Semiconductor Research and Development Laboratory, Palo Alto, Calif. 94304.

G. H. Wilson was with the Fairchild Semiconductor Research and Development Laboratory, Palo Alto, Calif. He is now with Precision Monolithics, Inc., Santa Clara, Calif. 95050.

Reprinted from *IEEE J. Solid-State Circuits*, vol. SC-5, pp. 29-37, Feb. 1970.

II. Design of Transistors for Operation at Low Collector Current

A. Basic Small-Signal Equivalent Circuit

Fig. 1(a) shows the π equivalent circuit for a junction transistor. At low collector current,

$$r_\pi \doteq r_e h_{fe_0} \gg r_b'$$

$$C_\pi \doteq C_{TE}$$

and the equivalent circuit may be simplified to that shown in Fig. 1(b), where the input capacity $C_i = C_{TE} + C_c(1 + g_m R_L)$. If $R_L \ll r_{22}$,

$$\text{voltage gain} = g_m R_L = \frac{I_c R_L}{0.026}$$

$$\text{current gain} = h_{fe}(I_c)$$

$$\text{unity gain cutoff frequency} = \frac{1}{2\pi}\frac{I_c}{(0.026)C_{TE}}.$$

Thus, the voltage gain is directly proportional to collector current; the current gain is a complex function of collector current (which will be discussed in detail in this paper); and the cutoff frequency is proportional to the collector current and inversely proportional to the emitter transition capacity.

B. Transistor Large-Signal Current Gain

There are two ways of looking at the linearity of the large-signal current gain of transistors as a function of their operating point. One is to plot the current gain as a function of collector current. Ideally, current gain would be independent of collector current. In general, however, current gain will decrease with decreasing collector current. Iwersen *et al.* [4] were among the first to present these data in a more enlightening way, plotting base and collector currents as a function of the base-to-emitter voltage. A plot of this type is shown in Fig. 2. The collector current depends only on the impurity profile in the base and the emitter forward bias [5], increasing exponeitially with qV_{BE}/kT. In order to have perfect beta linearity, the base current must be composed entirely of neutral-region recombination (diffusion) terms, which increase exponentially with qV_{BE}/kT. In general, two types of effects tend to cause deviation from ideal beta linearity. One is recombination in the emitter–base and surface space-charge regions. Such recombination leads to an additional component of base current that has a slope $qV_{BE}/2kT$ (see I_{B_2} in Fig. 2). All means possible must be taken to suppress this additional component of base current. The second deviation from linear beta is even more disastrous. An emitter–base channel can add an additional component of current that will have a slope of approximately $qV_{BE}/4kT$, resulting in very poor linearity (see I_{B_3} in Fig. 2).

Let us elaborate on the factors that have a disastrous effect upon h_{FE} linearity. As shown in Fig. 3, these are

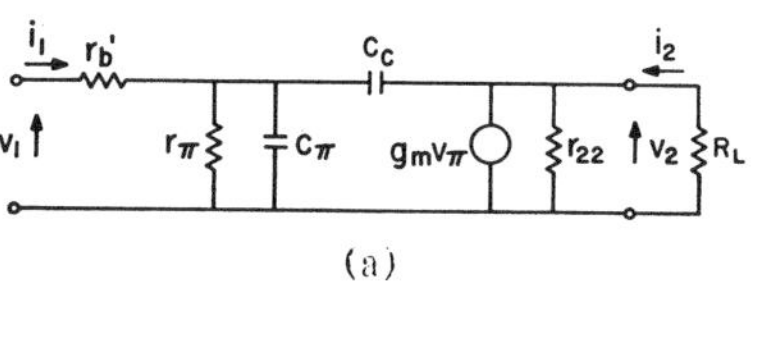

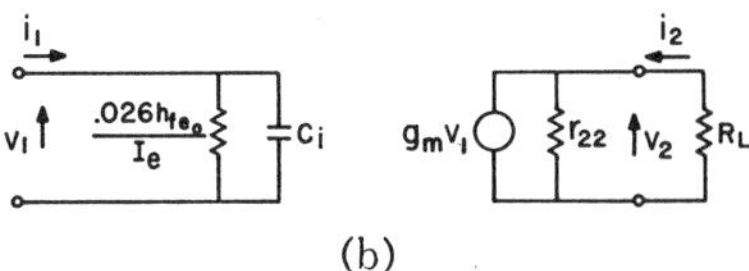

Fig. 1. (a) Small-signal π equivalent circuit of a junction transistor. (b) Simplified small-signal equivalent circuit applicable at low values of collector current.

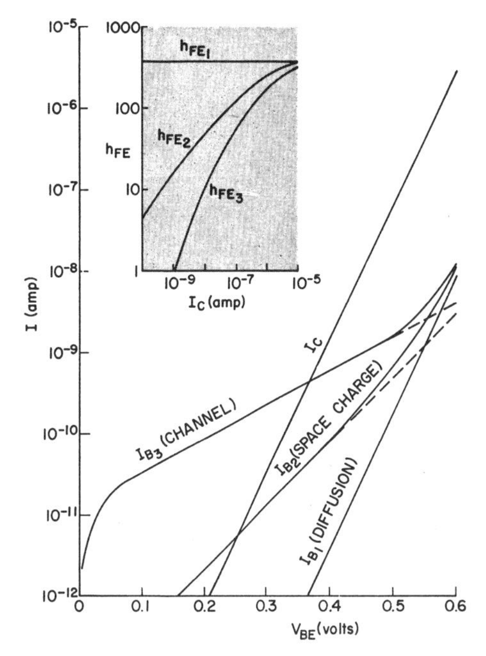

Fig. 2. Collector and base current of a junction transistor as a function of base-to-emitter voltage for three types of transistors. The first type is dominated by neutral-region recombination (I_{B_1}); the second has an additional component of base current due to space-charge region recombination (I_{B_2}); the third is dominated by a channel current (I_{B_3}). Inset shows corresponding plots of h_{FE} as a function of collector current.

related to channels and similar phenomena that produce base-current components that have slopes of the order of exp $(qV_{BE}/4kT)$ [6]–[8]. The most common problem is a base channel that allows a direct path of current flow from the base to the emitter. The amount of this current is limited by pinching-off of the channel. The inversion layer is formed under the base oxide by a fixed charge either on the oxide or within the oxide. Proper design and processing can prevent inversion of the base. The second similar effect is caused by a channel upon the base region due to charge in or on the oxide. The inversion layer does not reach the base contact, but does reach a defect that provides a source of current; again, it limits itself by the channel characteristics. There are two other possibilities involving inversion layers. These both involve inversion of heavily doped layers, leading to depletion regions so thin that significant tunneling can occur through the region. This can be

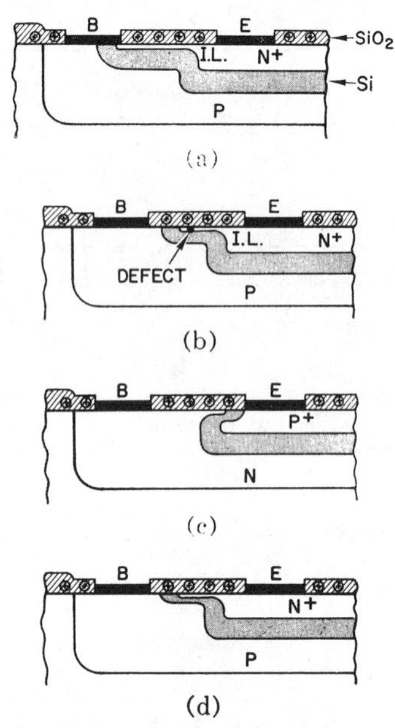

Fig. 3. Factors resulting in extreme degradation of low-current h_{FE}. (a) Emitter–base channel. (b) Base channel with defect. (c) Tunneling in inverted emitter with high C_s. (d) Tunneling in inverted base with high C_s.

due to either an inversion of an emitter layer or inversion of a very heavily doped base layer. In general, the more probable occurrence is the inversion of a p-type base. This additional component can be prevented by reducing the base-surface concentration [7], or by controlling the amount of charge in or on the oxide.

The second class of base-current components that can detract from the low-current gain of transistors is that which affects the linearity of h_{FE} and leads to an I_B component proportional to exp $(qV_{BE}/2kT)$. The physical locations of such recombination are described pictorially in Fig. 4. This component of base current is due to recombination in the emitter–base and surface space-charge regions [9] and is given by

$$I_B = \frac{qn_iA_{MJ}}{2}\left(\frac{W_{EB}}{\tau_0} + s_0\frac{A_s}{A_{MJ}}\right)e^{qV_{BE}/2kT}.$$

In Fig. 4[1] the magnitude of $(1 - \alpha)$, due to space-charge recombination, is plotted for a transistor operated at 1 μA as a function of the lifetime and the surface-recombination velocity. This is a transistor with a 1 × 1 mil emitter. (This transistor structure will be used in examples to follow.) If the surface recombination is negligible, $(1 - \alpha)_{SC}$ will be proportional to 1/lifetime (slope of −1). As the surface-recombination velocity is increased, there will be deviations from lifetime-controlled behavior at very high lifetimes giving a higher space-charge recombination component. This deviation can come about in two ways: either through an increase in s_0 or by a partial depletion of the surface that increases the area at which space-charge recombination can occur.

[1] Alpha defect is defined as $1 - \alpha$, where α is the common-base current gain. Alpha defect terms due to different causes add as they each contribute independent base current.

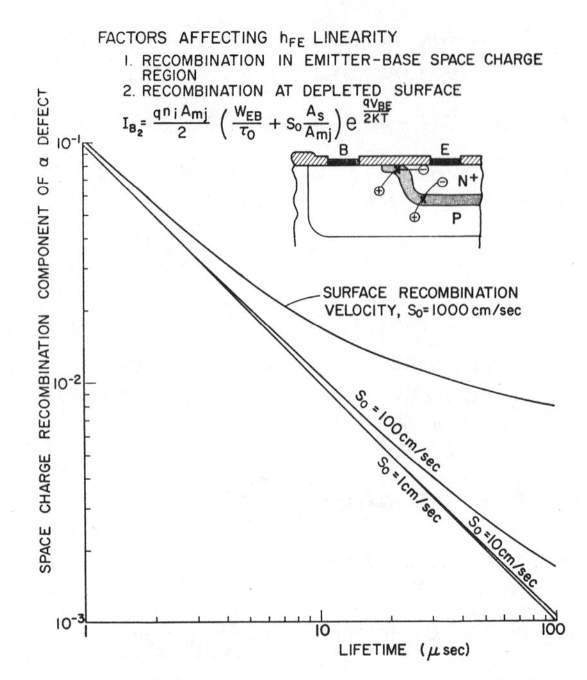

Fig. 4. Space-charge recombination component of alpha defect as a function of lifetime with surface-recombination velocity as a parameter. Inset shows the locations of the recombination process leading to this base-current component.

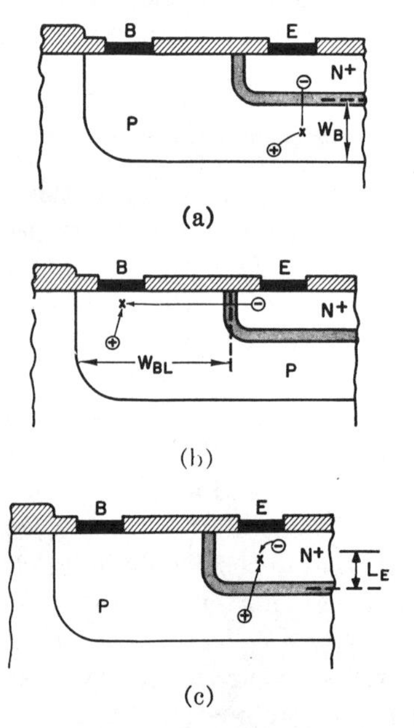

Fig. 5. Factors affecting the magnitude of h_{FE} but not its linearity. (a) Recombination in neutral base, vertical injection. (b) Recombination in neutral base, lateral injection. (c) Recombination in neutral emitter.

Fig. 5 shows the factors that affect the magnitude of h_{FE} but not its linearity. (In other words, one can have ideal beta linearity but just plain low beta.) There are three neutral regions where recombination occurs. There is recombination of carriers traversing the base that are injected vertically into the base. Recombination of this type is negligible in almost all modern double-diffused transistors. There is recombination of carriers that are injected laterally. It is obvious that the distance the carrier would have to travel if injected laterally is

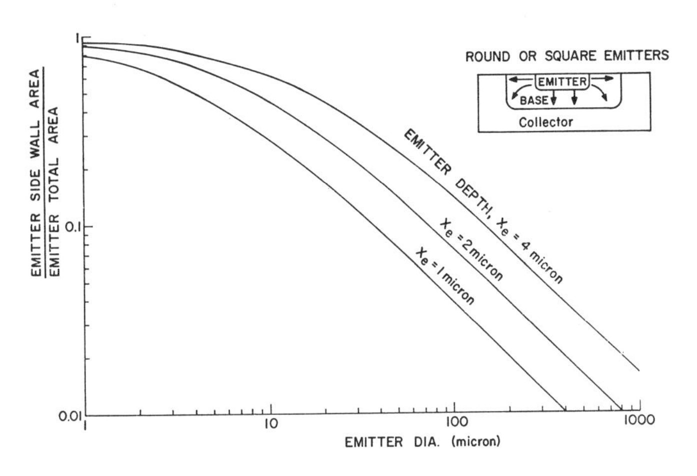

Fig. 6. Fraction of emitter area, which is sidewall (resulting in lateral injection) as a function of the emitter diameter.

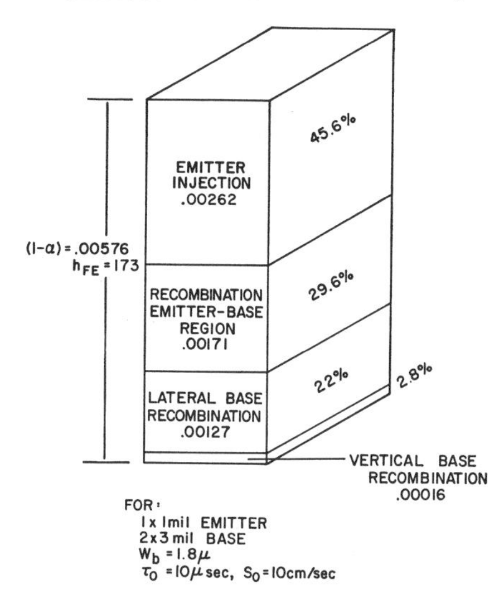

Fig. 7. The relative importance of the various components of alpha defect for a typical transistor.

considerably farther, thus the fraction of the carriers that get across this lateral base will be much less than the fraction that get across the normal (vertical) base. The third location for neutral-region recombination is in the emitter. This term is in the worst shape to handle analytically. Whittier and Downing [10] have postulated that the life path in the emitter is limited by the concentration at which emitter dopants also serve as recombination centers. Thus, the life path will effectively end when the doping reaches the order of 10^{19} cm^{-3} in the emitter. They find the emitter factor Q_E/D_D saturates at a value of $\sim 2 \times 10^{14}$ $cm^{-4} \cdot s$.

Plotted in Fig. 6 is the fraction of the area for emission that is sidewall area that will result in lateral injection as a function of the entire emitter area. For typical diffusion depths of 2 to 4 microns and a typical emitter diameter of 1 mil, 30–40 percent of the area will be sidewall area. For smaller emitters, a greater portion becomes sidewall area so that more and more carriers are lost by recombination of laterally injected carriers. Once an emitter size is chosen, there is nothing that can be done to lower this component of base current except to raise the lifetime high enough so that the carriers will get across the base before recombining.

Fig. 7 is a summary of the effects of various components of base current on alpha defect for a fairly typical transistor: a 1×1 mil emitter, a 2×3 mil base, a 1.8-micron basewidth, a lifetime of 10 μs, and a surface-recombination velocity of 10 cm/s. The surface potential is assumed to be at the flatband condition so that there is no widening or narrowing of the space charge at the surface. In this particular case, 3 percent of the alpha defect—the current that does not get to the collector and that results in base current—is due to recombination of the carriers going across the vertical base; 22 percent is due to carriers laterally injected and dying on the way across; 30 percent is recombination in the emitter–base space-charge region; and 46 percent is due to recombination of carriers that are injected into the emitter. If the lifetime were increased, it would lower all of these components, but it would have a particularly large effect upon the emitter–base space-charge recombination and the lateral injection recombination. An increase in the surface-recombination velocity would cause a fairly small decrease in $(1 - \alpha)$, as the space-charge recombination term is only about 20 percent surface recombination and the rest is recombination in the junction-depletion region.

In summary, to obtain high h_{FE} at low current:

use high-bulk-lifetime material;
have a low-surface-recombination velocity;
use the largest emitter diameter possible.

Thus a tradeoff is required; increasing the emitter diameter increases the emitter transition capacity and lowers the already low cutoff frequency of the device. Surface inversion or depletion must be prevented as either will cause a disastrous loss in low-current h_{FE}. An example of the degree of h_{FE} linearity that can be achieved following these design requirements is shown in Fig. 8, a kit part pair from a linear integrated circuit.

Once good low-current h_{FE} has been achieved, the next problem is to stabilize it so that it will not drift during life. The approaches that traditionally have been taken are to provide field plates, which lock the surface potential. In general, in integrated circuits, there is not enough room to allow for field plates. Phosphorous glass coatings over silicon dioxide are an effective barrier to sodium, the principal ionic contaminant. They also, to some extent, remove ionic contamination already there. The third possibility is silicon nitride coatings over silicon dioxide. These are a completely impervious barrier to sodium and are stable. They are the best overall approach in stabilizing low-current h_{FE}.

We will just mention the special problems of p-n-p's in linear integrated circuits. Fig. 9 shows the current gain as a function of collector current for a typical double-diffused n-p-n transistor, a lateral p-n-p transistor, and a substrate p-n-p. It may be seen that although the h_{FE} of both types of p-n-p is low, the linearity of h_{FE} is not particularly bad. Also, the high-frequency cutoff at low

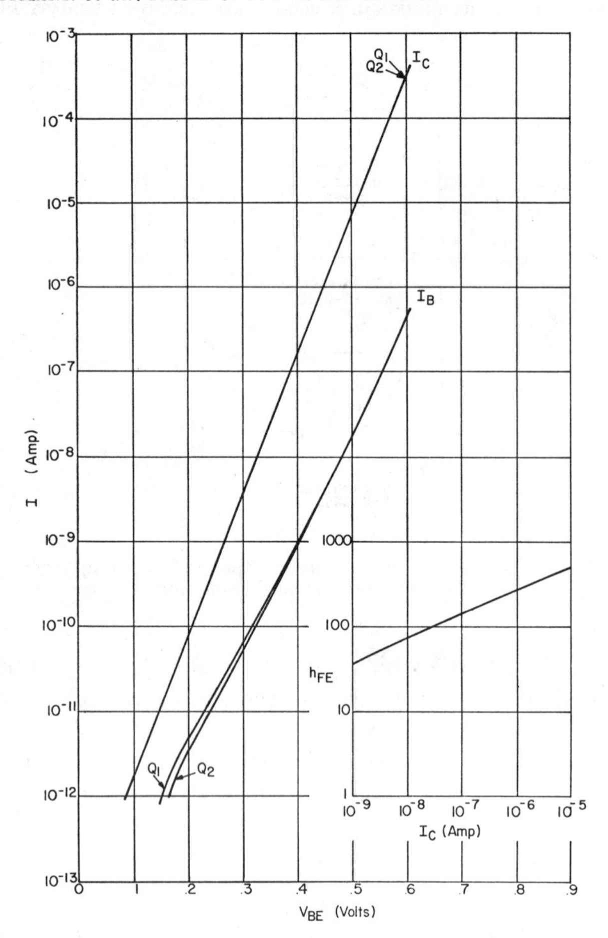

Fig. 8. Collector current and base current for an input pair of transistors of a linear integrated circuit as a function of base-to-emitter voltage. Inset shows corresponding plot of h_{FE} as a function of collector current.

values of collector current is not much worse for p-n-p's than for n-p-n's due to the fact that this cutoff frequency is dominated by the emitter time constant. Therefore, in the design of integrated circuits at micropower, p-n-p's can be used more freely than at conventional power levels. In a sense, the problem of optimizing a p-n-p is opposite that of a double-diffused n-p-n. First, recombination in the base is the dominant rather than the least important term. Second, the design should be made so as to maximize emission from the sidewall and minimize the emission from the bottom of the emitter.

III. Noise Considerations

The spot noise figure of the micropower amplifier will, in general, be required to meet stringent specifications. For example, the spot noise figure may be required to be less than 3 dB for all frequencies in the range of 10 Hz to 20 kHz. To meet such specifications, a considerable knowledge is required in the areas of noise properties of transistors, noise mechanisms, and processing effects on noise.

A standard method of representing the noise behavior of bipolar transistors is to refer all noise to the input of the device [11]. When this is done, it results in the equivalent circuit shown in Fig. 10.

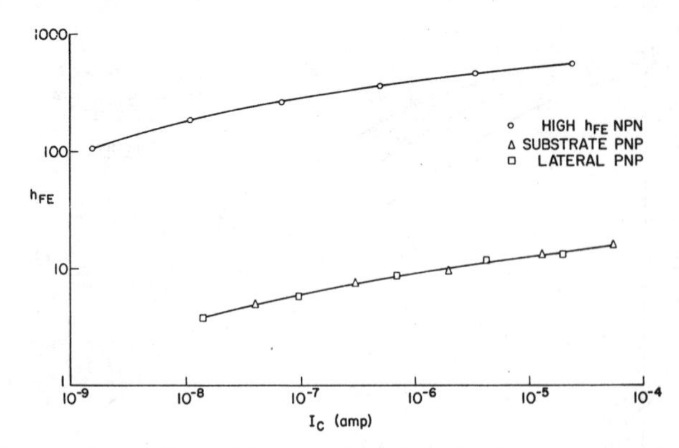

Fig. 9. Comparison of h_{FE} as a function of collector current for a double-diffused vertical n-p-n transistor, a lateral p-n-p transistor, and a substrate p-n-p transistor.

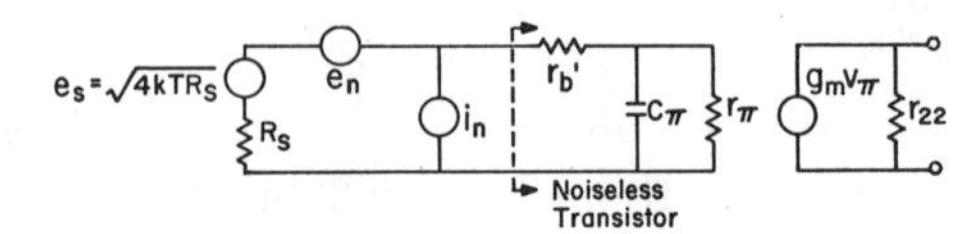

Fig. 10. Standard representation of the noise of a transistor. All noise is referred to the input leading to an input-noise voltage generator e_n, an input-noise current source i_n, and a noiseless transistor.

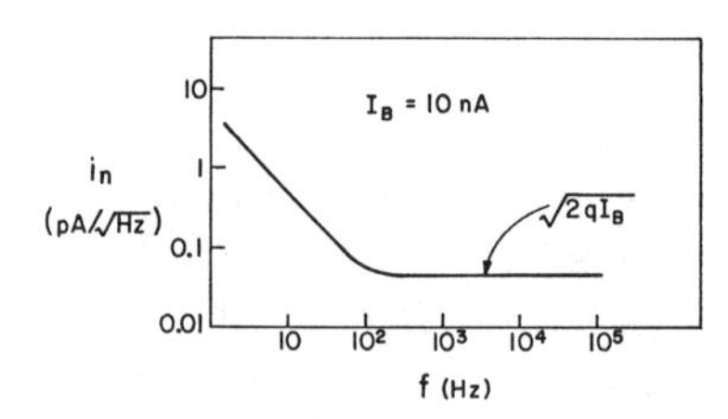

Fig. 11. Typical input noise characteristic for a micropower input transistor showing corner frequency of about 100 Hz.

All semiconductor devices show a component of noise with an f^{-n} dependence, where n is close to 1, in addition to the usual white (frequency-independent) thermal and shot noise components. This makes the low-frequency performance critical in meeting any noise specification. In bipolar transistors this excess noise component appears almost entirely in the input-referred current noise of the device. This is demonstrated by operating the device first with zero source resistance and then with very high source resistance and observing the spectrum of the output noise versus frequency for the two cases. The zero source case will show little or no excess noise at low frequencies whereas the high source resistance case will show a definite f^{-n} component.

Ideally the input current noise at all frequencies would be at its theoretical limit, i.e., the shot noise value,

$$i_n \big|_{\text{shot}} = \sqrt{2qI_B}.$$

In actual practice this limit is typically reached at frequencies of 10^2 to 10^3 Hz, as shown in Fig. 11.

In addition to its variation with frequency, the excess noise current component varies with dc base current. Experimentally this variation is observed to be pro-

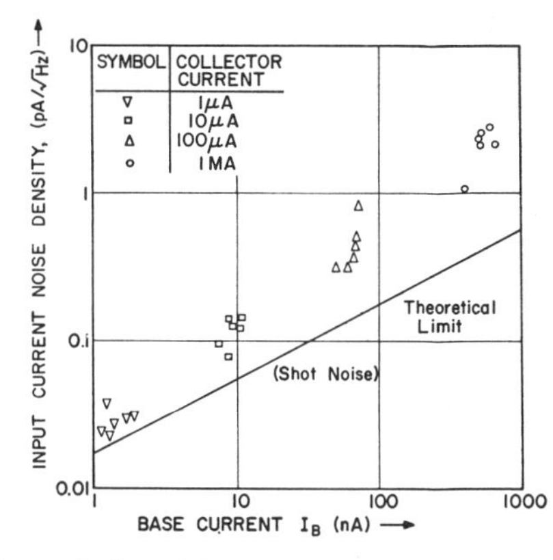

Fig. 12. The variation of input current noise with base current for high-beta low-noise transistors (F = 10 Hz).

portional to I_B^k where $\frac{1}{2} < k \leq 1$. This is illustrated in Fig. 12 where the total current noise density at 10 Hz is plotted versus I_B. These are measured results from six representative samples of discrete transistors fabricated by a high-beta low-noise IC process. It can be seen that for very low bias currents, the current noise density approaches the theoretical minimum, shot noise. When the data at, say, 1 kHz are plotted, they are found to lie along the shot noise line.

The measurements show that the lower the base current, the lower the input noise, and also that at very low input currents the input noise approaches the theoretical limit. *Thus, in considering any low noise circuit with a specified collector bias current, the input device should be made to operate at as low a base current as possible.* Thus, for low-noise performance, high current gains are required at low collector currents.

For a typical low-power amplifier, let us assume that the input devices are biased at $I_C = 1\ \mu$A. Assuming a typical $h_{FE} \simeq 200$ at room temperature, gives $I_B \simeq 5$ nA. Thus from Fig. 12

$$i_n = 0.1\ \text{pA}/\sqrt{\text{Hz}}$$

can be expected from each input device at 10 Hz.

The input-voltage noise has very little $1/f$ component. For a high-beta device, voltage noise is essentially determined by the collector-current shot noise. Referred to the input, this component is

$$e_n = \frac{kT}{q}\sqrt{\frac{2q}{I_C}}.$$

Since $I_C = 1\ \mu$A for the input stage of the amplifier,

$$e_n = 14\ \text{nV}/\sqrt{\text{Hz}}$$

from each input device.

For the amplifier, the input-referred noise is the square root of the sum of the squares of the independent noise sources of the two input devices. Hence, for the amplifier at 10 Hz, we expect

$$e_n = 20\ \text{nV}/\sqrt{\text{Hz}}$$
$$i_n = 0.15\ \text{pA}/\sqrt{\text{Hz}}.$$

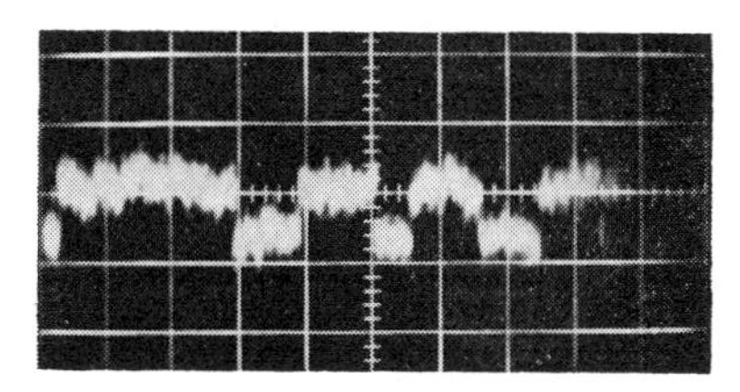

Fig. 13. Typical burst noise waveform as observed on an oscilloscope.

Since these two noise generators will be almost totally uncorrelated, we can determine the optimum source resistance R_0 and optimum noise figure F_0 expected from the amplifier at 10 Hz:

$$R_0 = 130\ \text{k}\Omega$$
$$F_0 = 1.5\ \text{dB}.$$

At midband frequencies, above approximately 100 Hz, the current-noise density will have dropped to the shot noise limit. Thus for the amplifier

$$e_n = 20\ \text{nV}/\sqrt{\text{Hz}}$$
$$i_n = 0.06\ \text{pA}/\sqrt{\text{Hz}}$$

and

$$R_0 = 330\ \text{k}\Omega$$
$$F_0 = 0.6\ \text{dB}.$$

At sufficiently high frequencies the input-referred noise of the amplifier will begin to increase again for two reasons. The input-referred noise of the input-stage devices begins to increase, and the input-stage gain drops low enough to make the noise contributions of the succeeding stages significant. But neither of these will occur in a typical micropower amplifier at frequencies below, say, 20 kHz.

Thus, achieving a 3-dB optimum noise figure from 10 Hz to 20 kHz appears feasible. This margin can accommodate the small additional noise contribution from base resistance thermal noise, leakage current noise, slight correlation between e_n and i_n, small contribution from the second stage, and process changes required for integrated-circuit fabrication rather than discrete device fabrication.

The experimental results presented so far are for devices that are free from burst (or popcorn) noise. Burst noise, when viewed on an oscilloscope across the output load, appears as shown in Fig. 13. When this type of bistable noise is present it completely dominates the noise performance of the device. Recently, Hsu and Whittier [12] have shown that such noise, when present in transistors, can be characterized by the equivalent circuit shown in Fig. 14. Burst noise is characterized by an intermittent current generator i_{BN}. The amplitude of the current generator (when on) is a function of V_{BE} as shown in the lower portion of the figure.

When burst noise is present, the input current noise at 10 Hz is typically increased by two to three orders of

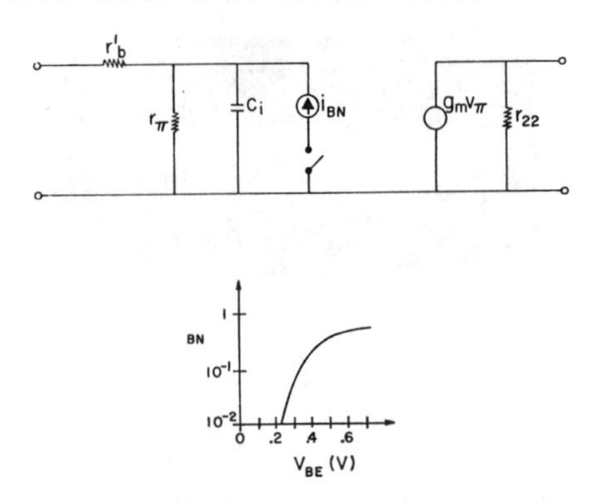

Fig. 14. Simple equivalent circuit for a transistor exhibiting burst noise. An additional current generator i_{BN}, in series with a switch, is placed in the emitter–base circuit.

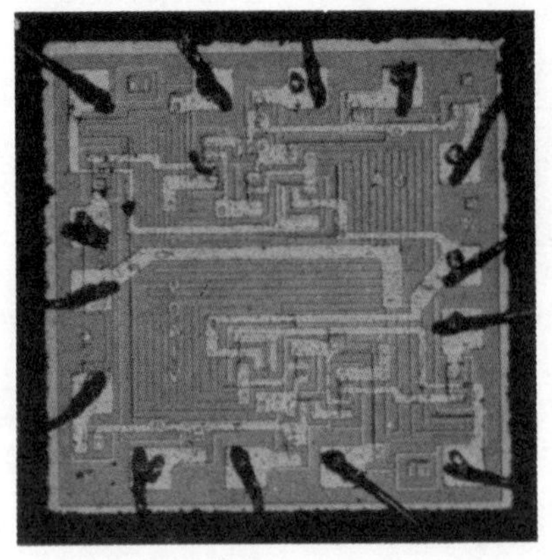

Fig. 15. An integrated circuit having an inordinate amount of resistor area.

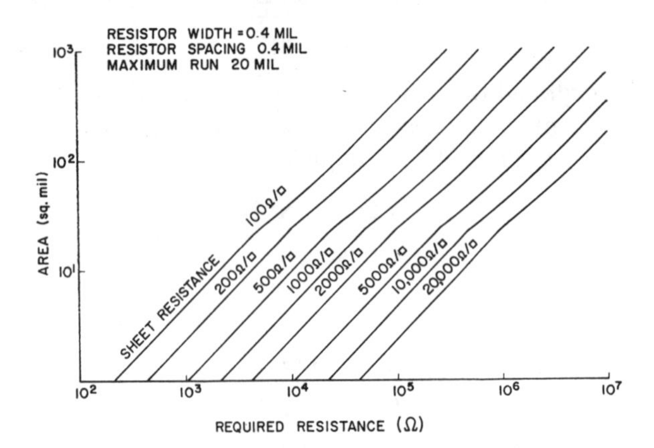

Fig. 16. Substrate area versus required resistance for sheet resistivities from 100 Ω/□ to 20 kΩ/□; resistor width 0.4 mil, resistor spacing 0.4 mil, maximum run 20 mils.

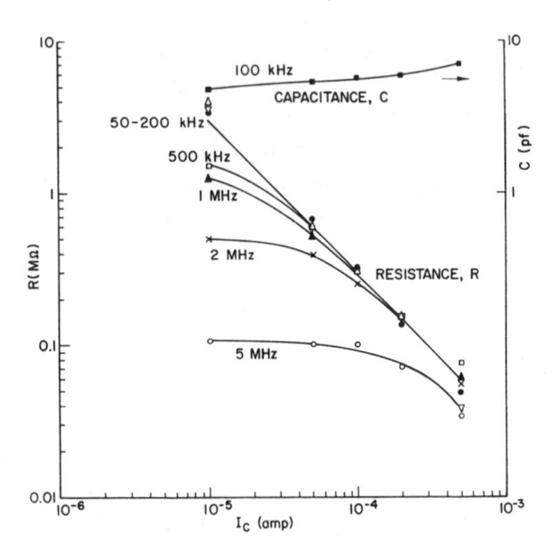

Fig. 17. Characteristics of an active load from an integrated circuit as a function of current and frequency.

magnitude. Thus, this type of noise must be eliminated entirely to meet any reasonable noise specification. It has been found that *most* of the processing steps leading to high-beta transistors also lead to a low incidence of popcorn noise.

IV. Resistors for Micropower Amplifiers

In order to get useful gain from a transistor operating at 1-μA collector current, the load resistance must be approximately 1 MΩ. The problem is to obtain this resistance within a reasonable space on an integrated circuit. If this is attempted using conventional diffused resistors, the circuit is dominated by the resistors and the active device area is almost insignificant. Such a circuit is shown in Fig. 15. In such cases, the cost in silicon real estate (the basic cost of integrated circuits) is prohibitive.

Fig. 16 shows the area required for a given total resistance for sheet resistivities from 100 Ω/□ to 20 kΩ/□. The resistor width and spacing are assumed to be 0.4 mil with a maximum run before doubling back of 20 mils. (This maximum run causes a slight inflection point in each curve.) To get 1 MΩ of resistance in an area of 200 mil^2 or less requires a sheet resistance of at least 20 kΩ/□.

This brings up the question as to the best way of achieving a high resistance. Two approaches are possible One is to use active devices for loads as pioneered by Widlar [13], and the other is to use high Ω/□ resistors. Active loads will, in some cases, take less space and be easier to fabricate than high-value resistors. On the other hand, they complicate circuit characteristics and can degrade the high-frequency performance of the device. For example, Fig. 17 shows the resistance and capacitance of an active load from an integrated circuit as a function of current and frequency. There is a fairly large amount of parallel capacitance (5 to 6 pF) that bypasses the load. In addition, there is a frequency-dependence of the real part of load resistance caused by the phase shift due to the emitter time constant. In spite of these complications, for many purposes an active load is still the easiest way to get a high load resistance. Widlar [14] has also used field-effect devices as active loads. These devices generally have undesirably high temperature coefficients; however, Widlar has cleverly used this effect to compensate another temperature coefficient elsewhere in his circuit.

The second alternative for high resistance is a high Ω/□ resistor. Such resistors can be diffused silicon, epitaxial silicon, or thin-film. Diffused and epitaxial silicon

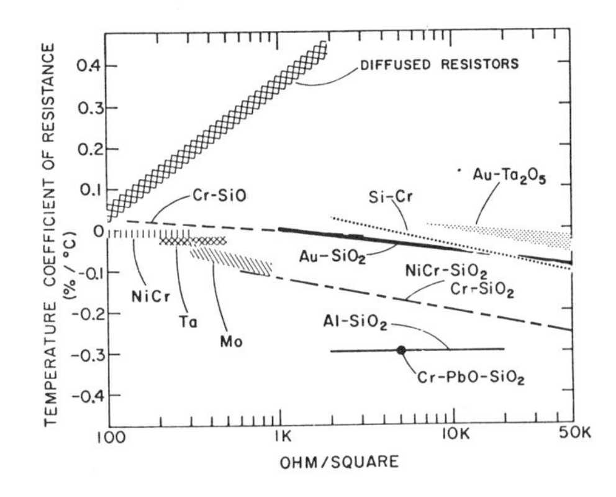

Fig. 18. Temperature coefficient versus sheet resistivity for diffused and thin-film resistors (see [15] for citations).

resistors have extremely high (>0.4 percent/°C) temperature coefficients of resistivity (TCR) at sheet resistivities over 1 to 2 kΩ/□. Thin-film polycrystalline silicon resistors are similar to diffused resistors in this regard. High sheet-resistivity thin-film resistors have recently been reviewed [15] and will be only briefly discussed here. TCR versus sheet resistivity for diffused resistors and several types of thin-film resistors are shown in Fig. 18. At sheet resistivities of 10 kΩ/□ or higher, gold-tantalum oxide [16], gold-silicon dioxide [17], [18], and silicon-chromium [19]–[21] films have reasonably low TCR's. The use of gold-oxide resistive films on integrated circuits has not been reported, and stability and reproducibility data are limited. Silicon-chromium films have been used successfully on integrated circuits (see [15]).

Fig. 19 shows the relation between resistivity and chromium content for silicon-chromium films annealed in nitrogen at 550 to 565°C for 5 to 10 minutes. For a 20 kΩ/□ film to be over 100 Å thick, the resistivity must be greater than 0.02 Ω·cm. This corresponds to about 20 atomic percent chromium. Fig. 20 shows the TCR versus resistivity curves for annealed silicon-chromium films. A 0.02 Ω·cm film can have a TCR of −400 to −1000 ppm/°C, depending on the deposition method. We have used approximately 10^{-2} Ω·cm silicon-chromium films deposited by electron-beam evaporation [20] and dc diode sputtering [21] as resistors in both linear and digital low-power integrated circuits. The highest sheet resistivity employed has been 20 kΩ/□. Stability on life test has been better than 1 percent for up to 5000 hours at 200°C, most of the drift occurring during the first few hundred hours.

V. Application to Micropower Operational Amplifier

Fig. 21 shows the circuit diagram of an amplifier in which the previously discussed technology has been applied to the design of a micropower operational amplifier. The amplifier employs an n-p-n input stage. By making use of the design principles heretofore stated, the stage operates with adequate gain at a collector current of 1.25 μA and has excellent balance between

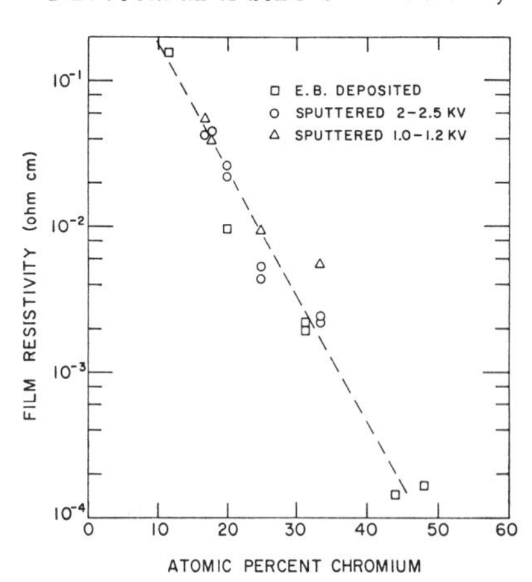

Fig. 19. Resistivity versus chromium content for annealed silicon-chromium films.

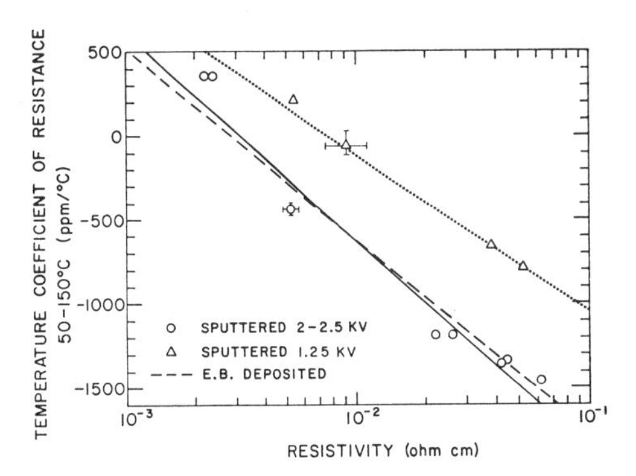

Fig. 20. Temperature coefficient versus resistivity for annealed silicon-chromium films.

the pair of input transistors as previously shown in Fig. 8. The second stage is a complex one involving both n-p-n and p-n-p transistors, which provides amplification, level shifting, and conversion from double-ended to single-ended output. The output stage is operated class AB and provides reasonably high efficiency and extremely low standby power. Note that in this design free use has been made of both p-n-p transistors and high Ω/□ chromium silicide resistors. The total resistance on the 56 × 56 mil chip is approximately 14 MΩ. The characteristics of this amplifier are given in Table I. Due to the low-current operation and the excellent characteristics of the input pair transistor, the amplifier has high input impedance, low input current, low input offset current, a reasonable input offset voltage, and extremely low input current noise. The common-mode rejection is adequate. The principal penalty paid for the micropower design is found in frequency response and slew rate.

VI. Conclusions

Transistors can be designed to have adequate current gain at collector currents below 1 μA. Design choices exist employing either high ohms/square resistor technology or active loads, which make possible the utilization of this current gain. Micropower operation, in addition to the obvious advantages of power saving, results fre-

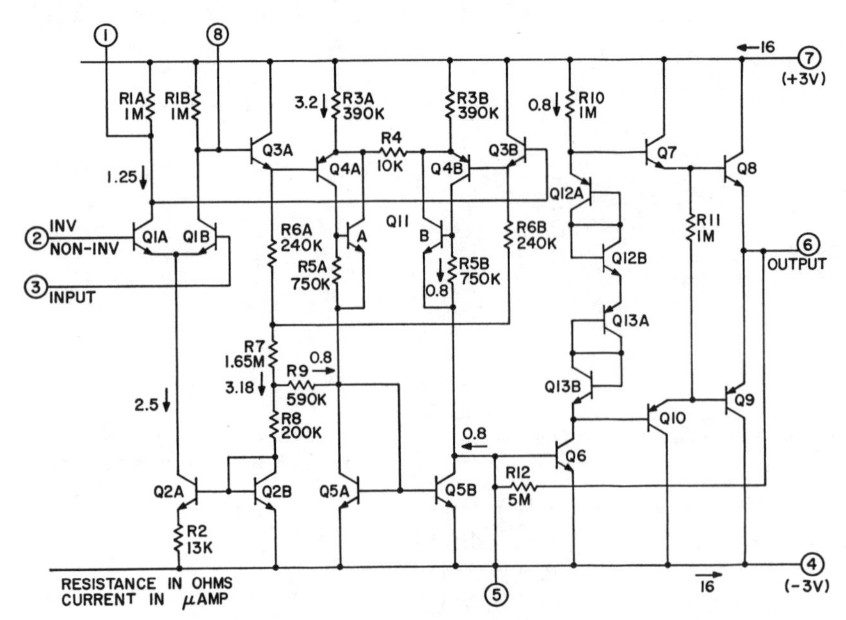

Fig. 21. Circuit diagram of micropower operational amplifier (fabricated using the technology discussed in this paper).

TABLE I
MICROPOWER AMPLIFIER TYPICAL CHARACTERISTICS

Dc supply voltage	±3 volts
Quiescent power drain	100 μW
Input impedance	10 MΩ
Input current (either input)	5 nA
Input offset current	2 nA
Input offset voltage	1 mV
Common-mode rejection ratio	10 μV/V
Power-supply rejection ratio	20 μV/V
Output impedance	180 Ω
Output voltage (R_L = 1 kΩ)	±1.5 volts
Output current (R_L = 1 kΩ)	±1.5 mA
Dc gain	25 000
Maximum frequency for full output	1 kHz
Maximum slewing rate	0.02 V/μs
Noise (referred to input):	
Voltage (rms) at 10 Hz	30 nV
Current (rms) at 10 Hz	0.1 pA

quently in improved circuit operation as exemplified by superior input characteristics in operational amplifiers and by low-noise circuits.

REFERENCES

[1] For a general discussion of micropower electronics, see J. D. Meindl, *Micropower Circuits*. New York: Wiley, 1969.

[2] For a general introduction to transistor design that emphasizes effects important in low-power operation, see A. S. Grove, *Physics and Technology of Semiconductor Devices*. New York: Wiley, 1967.

[3] R. L. Foglesong, "The effects of technology on digital-circuit design and performance at microwatt power levels," this issue, pp. 38–43.

[4] J. E. Iwersen, A. R. Bray, and J. J. Kleimack, "Low-current alpha in silicon transistors," *IRE Trans. Electron Devices*, vol. ED-9, pp. 474–478, November 1962.

[5] J. L. Moll and I. M. Ross, "Dependence of transistor parameters on distribution of base layer impurities," *Proc. IRE*, vol. 44, pp. 72–78, January 1956.

[6] C. T. Sah, "Effect of surface recombination and channel on p-n junction and transistor characteristics," *IEEE Trans. Electron Devices*, vol. ED-9, pp. 94–108, January 1962.

[7] V. G. K. Reddi, "Influence of surface conditions on silicon planar transistor current gain," *Solid-State Electron.* vol. 10, pp. 305–334, March 1967.

[8] A. S. Grove and D. J. Fitzgerald, "The origin of channel currents associated with p^+ regions in silicon," *IEEE Trans. Electron Devices*, vol. ED-12, pp. 619–626, December 1965.

[9] C. T. Sah, R. N. Noyce, and W. Shockley, "Carrier generation and recombination in *p-n* junctions and *p-n* junction characteristics," *Proc. IRE*, vol. 45, pp. 1228–1243, September 1957.

[10] R. J. Whittier and J. P. Downing, "Simple physical model for the injection efficiency of diffused p-n junctions," presented at the Internatl. Electron Devices Meeting, Washington, D.C., 1968, paper 12.4.

[11] See, for example, A. S. Grove and S. T. Hsu, "Don't just fight semiconductor noise," *Electron Design*, vol. 12, pp. 867–876, November 1969.

[12] S. T. Hsu and R. J. Whittier, "Characterization of burst noise in silicon devices," *Solid-State Electron.* (to be published).

[13] R. J. Widlar, "Some circuit design techniques for linear integrated circuits," *IEEE Trans. Circuit Theory*, vol. CT-12, pp. 586–590, December 1965.

[14] R. J. Widlar, "IC op amp with improved input-current characteristics," *EEE*, vol. 16, pp. 38–41, December 1968.

[15] R. K. Waits, "High-resistivity thin-film resistors for monolithic circuits," *Solid State Technol.*, vol. 12, pp. 64–68, June 1969.

[16] C. H. Lane and J. P. Farrell, "Cermet resistors by reactive sputtering," *Proc. 1966 IEEE Electronics Components Conf.*, pp. 213–224.

[17] N. C. Miller and G. A. Shirn, "Co-sputtered Au-SiO₂ Cermet films," *Appl. Phys. Letters*, vol. 10, pp. 86–88, February 1967; also, "Co-sputtered Cermet films," *Semiconductor Products*, vol. 10, pp. 28–31, September 1967.

[18] J. K. Pollard, R. L. Bell, and G. G. Bloodworth, "Structure and electrical properties of evaporated gold/silicon oxide Cermet films," presented at the 1969 Internatl. Conf. on Thin Films, Boston, Mass., paper M-5.

[19] A. P. Youmans, "Thin film resistor," U.S. Patent 3 381 255, 1968.

[20] R. K. Waits, "Silicon-chromium electron-beam-deposited resistive films," *Trans. Met. Soc. AIME*, vol. 242, pp. 490–497, March 1968.

[21] R. K. Waits, "Sputtered silicon-chromium resistive films," *J. Vac. Sci. Technol.*, vol. 6, pp. 308–315, March–April 1969.

Automotive and Industrial Electronic Building Blocks

RONALD W. RUSSELL AND THOMAS M. FREDERIKSEN

Abstract—**The advent of greater concern with the safety and environmental effects of automobiles has resulted in the generation of many new electronic systems (similar to those used in industrial control), which are directed at making improvements in these areas. The implementation of these systems requires a new generation of low-cost linear integrated circuits specifically designed for operation from the single power supply that is available. The operational amplifiers and comparators that are described in this paper are among the first of this generation. These circuits offer both cost and applications advantages in the emerging automotive electronic systems and also in existing industrial control applications.**

I. Introduction

MANY new automotive electronic systems are being developed today to satisfy the expected legislative requirements on passenger safety and exhaust emission. Monolithic integrated circuits (IC's) will be employed to realize these systems since they can provide maximum function and excellent system performance at the lowest possible cost. A major question in automotive electronics today is whether to implement these systems with standard IC's or to develop custom IC's for each specific application. If a system must be developed in a short period of time, due to either legislation or competition; or if the system concept is relatively new, and therefore subject to change during the development cycle, then the advantages of flexibility and short turn-around time possible through use of standard IC's makes this approach the most attractive. A proliferation of standard IC's is currently available, but when an attempt is made to use these in the control and warning systems that are being proposed for the automobile, problems arise. These circuits were not designed to operate from a single power supply (the car battery), which will vary from 4 to 24 V_{dc}. The existing linear circuits typically require both positive and negative power supply voltages. In addition, the circuit complexity associated with state-of-the-art performance undesirably increases costs and limits the number of circuit functions which can be placed on one die. Digital circuits, although offering multiple complex functions, require a low-voltage (5 V_{dc}) power supply and their fast responses cause poor noise immunity in the relatively slow automotive and industrial control systems. Therefore new types of "building blocks," are needed to provide in a single package, multiple independent circuits that will operate from a single power supply voltage.

Manuscript received April 14, 1972; revised July 6, 1972.
The authors are with the National Semiconductor Corporation, Santa Clara, Calif.

II. Input Stage for Single-Power-Supply Circuits

A basic consideration in the design of any standard circuit is the common-mode bias voltage levels that can be used at the inputs. Because ground is a convenient input biasing reference, it would be desirable for these new standard single-power-supply circuits to provide this capability. The downward dc level shift from base to emitter in the n-p-n transistors that are used in a conventional input stage (Fig. 1), prevents ground from being used as a reference on the inputs when operating from a single power supply. The lower end of the input common-mode voltage range is typically 2 V. Therefore common-mode biasing must be used if the dc input signal amplitude of concern is less than this value. This usually requires extra external components. In addition, the noise and pick-up contributed by the common-mode biasing components may exceed the desired input signal.

These problems are eliminated by using the input stage of Fig. 2, which employs p-n-p rather than n-p-n transistors in the input differential amplifier. The maximum voltage on the collectors of transistors Q_2 and Q_3 in the differential amplifier is one base–emitter voltage drop above ground due to the diode D_1 in the differential-to-single ended convertor and the V_{BE} of transistor Q_6. This allows the dc level shift from base to emitter of the substrate p-n-p followers Q_1 and Q_4 (which also provide lower input currents), to place the minimum collector-base potentials of all the p-n-p input transistors at zero for ground-level input voltages. The lower input common-mode voltage limit is reached when any of these devices saturate. For the low biasing current levels that are used, this generally occurs a few tenths of a volt below ground.

The upper limit on the input common-mode voltage of this stage is the supply voltage V_{CC} minus the sum of the base–emitter voltages of Q_1 and Q_2 and the saturation voltage of the current source I (which is a few tenths of a volt). This is the limit on the input common-mode voltage for which a small differential input voltage will be amplified. The maximum common-mode voltage that can be applied to the inputs without damage to the stage is limited by either the base-to-substrate breakdown or the reverse breakdown of the base–emitter junctions of the p-n-p transistors. The maximum differential voltage (without damage) is also limited by the reverse breakdown of the base–emitter junctions of the p-n-p transistors. These breakdowns and therefore the limits on the inputs are on the order of 90 V. In addition, the input voltages can exceed the power supply voltage.

Reprinted from *IEEE J. Solid-State Circuits*, vol. SC-7, pp. 446-454, Dec. 1972.

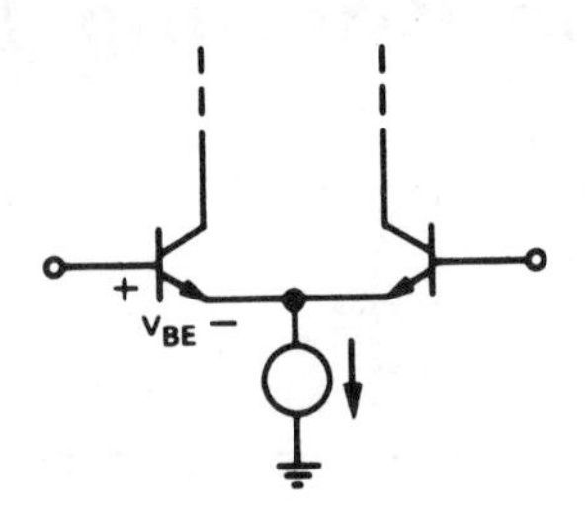

Fig. 1. Conventional input stage.

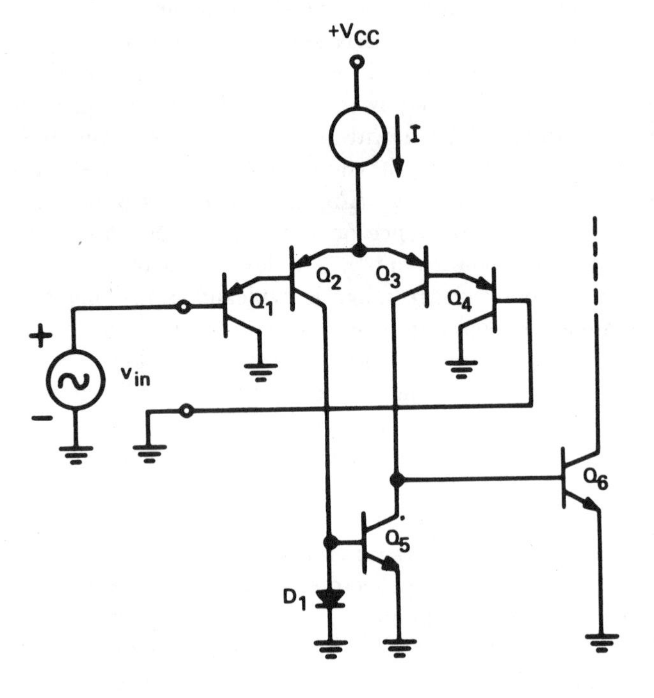

Fig. 2. Single-supply input stage.

The common-mode capabilities of this circuit have been achieved at the expense of a frequency response, which, due to the use of p-n-p transistors, is relatively poor in comparison to that of a conventional n-p-n input stage. It is, however, more than adequate in most automotive and industrial control applications. This input stage has been useful in many single power supply building blocks.

III. A Single-Supply Operational Amplifier

The most versatile linear building block is the operational amplifier (opamp). An opamp has therefore been realized by adding a second stage to the previously described input stage. To reduce the cost per circuit, a desirable goal is to have four independent amplifiers on a single die that can be accommodated in a 14-pin package. With only three pins allowed per amplifier, the frequency compensation for the amplifiers must be included on the chip. Because the available means for realizing the compensation capacitors on the die yield only approximately 0.1 pF/mil^2 of die area, the amplifiers should require a minimum amount of capacitance for compensation. A two-stage pole-split amplifier configuration was chosen for this reason. It has been shown [1] that this type of amplifier can be modeled as a transconductance stage driving a transimpedance amplifier, Fig. 3. The overall voltage gain, at high frequencies, is the product of the transconductance g_m and the transimpedance $1/sC$. Therefore the unity gain radian frequency ω_c is $\omega_c = g_m/C$.

A large dc voltage gain is desirable, but of more importance is the amount of capacitance required to make the amplifier stable in a noninverting unity gain configuration. In a typical internally compensated monolithic opamp a unity gain frequency of 1 MHz is provided by an input stage transconductance of 200 μmhos and roughly 30 pF of compensation capacitance. This capacitor occupies an area on the die that is roughly equivalent to that of 20 small-signal n-p-n transistors and requires approximately 15 percent of the total die. The unity gain frequency of an opamp that incorporates the input stage of Fig. 2 must be set at approximately 200 kHz to achieve an acceptable stability margin. In addition, the economics of small die size requires that, since there will be four amplifiers on the die, the required compensation capacitance for each must be small. Therefore a goal of 5 pF was established. In order to realize this goal some means had to be found to reduce the transconductance of the input stage to 6.6 μmhos without further degrading the frequency response of the amplifier.

A. Standard Techniques to Reduce Transconductance

There are two conventional techniques for reducing the transconductance of an emitter-coupled differential amplifier. One (Fig. 4) is to simply reduce the biasing current I to the emitters of the differential amplifier. The transconductance is directly proportional to this current. If an attempt is made to employ this technique, the bias current required for the desired transconductance is less than 1 μA. These small biasing currents in the p-n-p transistors cause the frequency response of the stage to degrade when compared to that obtained when higher biasing current levels can be used. The undesirable result is that a compensation capacitance of approximately 50 pF (not 5 pF) is required.

Another standard technique is the addition of emitter degeneration resistors R_E as shown in Fig. 5. The value of R_E can be made large enough to permit biasing the transistors (for improved frequency response) and also to obtain a small value of transconductance. An additional advantage of using this technique is the increase in the slew rate of the opamp [1]. Unfortunately, the amount of resistance required in the emitters of each of the transistors in the differential amplifier of the input stage of Fig. 2 is in excess of 100 kΩ. Diffused resistors of this magnitude would consume a large amount of die area, especially if these resistors are to be closely matched to prevent large input offset voltages from occurring. To use these large-valued area-consuming resistors would

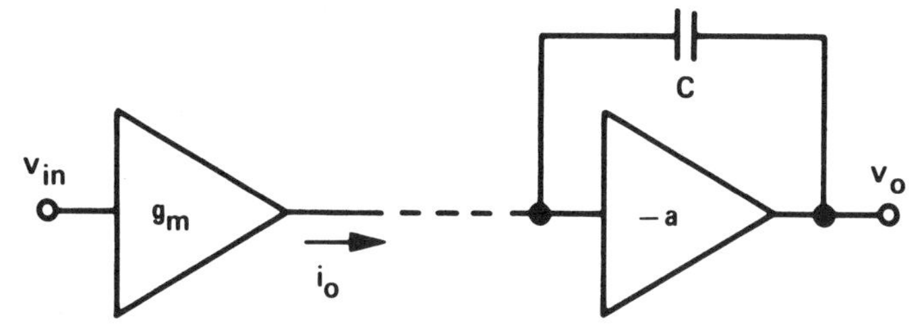

Fig. 3. Model for two-stage pole-split amplifier.

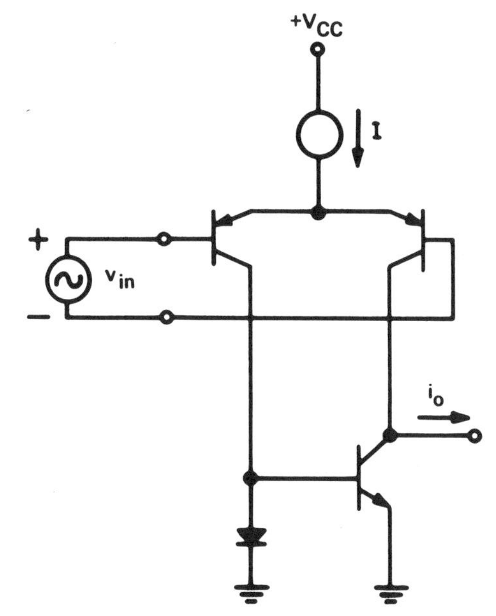

Fig. 4. Reducing transconductance by decreasing the bias current.

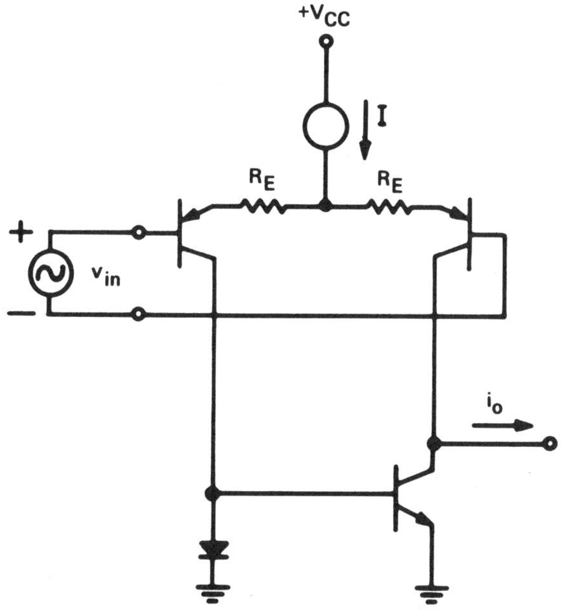

Fig. 5. Reducing transconductance by adding emitter degeneration resistors.

defeat the purpose of compensating the amplifier with a small value of capacitance.

Another technique for reducing transconductance is the addition of diodes in shunt with the base–emitter junctions of the transistors in the differential amplifier, Fig. 6. If these diodes match the transistors then the transconductance of the stage, assuming $\alpha = 1$, is reduced by a factor of 2 as a result of the decreased quiescent current in the transistors. A new technique that reduces the transconductance in a similar manner but without the disadvantages of additional devices and decreased quiescent current is presented in the following section.

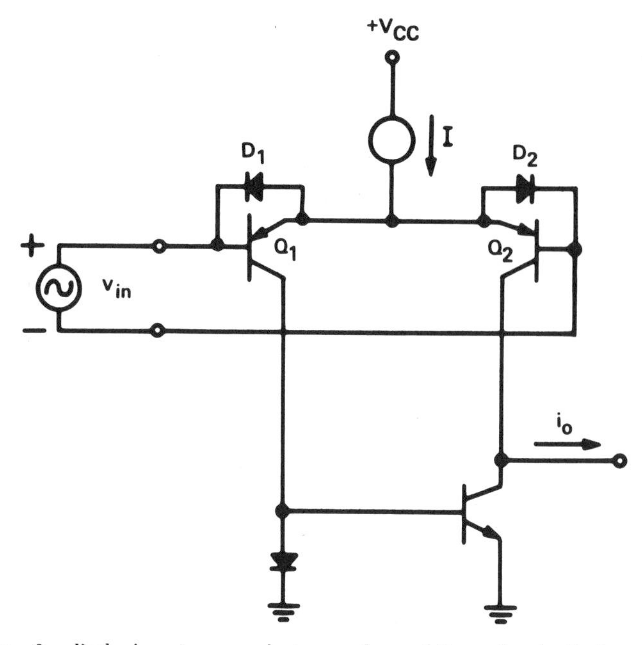

Fig. 6. Reducing transconductance by adding shunt diodes.

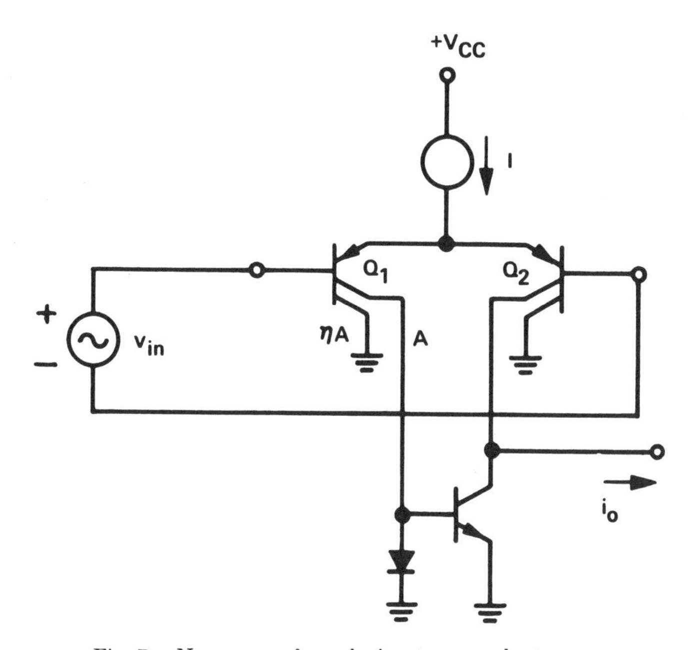

Fig. 7. New means for reducing transconductance.

B. A New Technique to Reduce Transconductance

A new means for reducing the transconductance of the input stage, which employs multiple rather than single collector lateral p-n-p transistors is shown in Fig. 7. By recovering the current in only one of the collectors,

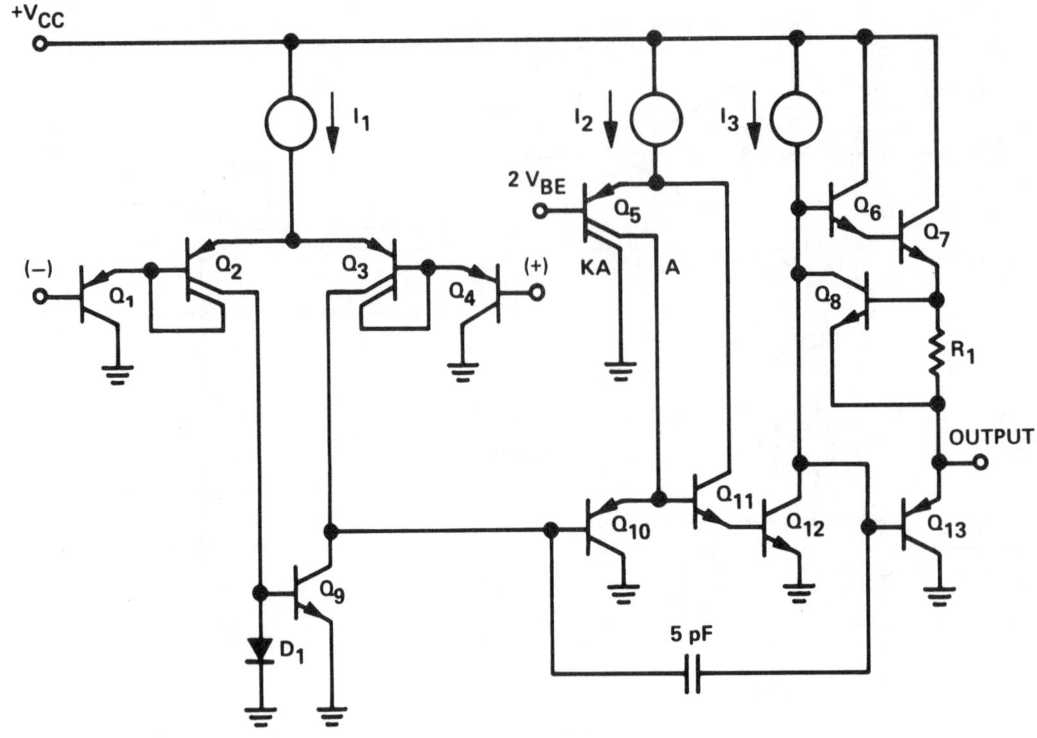

Fig. 8. Single-supply operational amplifier.

a reduction in the transconductance of the stage results. If the ratio of collector current recovered to that which is shunted to ground is $1:n$ then the transconductance is reduced by a factor of $(n + 1)$. This is easily seen by noting that, assuming $\alpha = 1$, a change in the current of the collector that is in the signal path is a factor of $(n + 1)$ less than the corresponding change in the emitter current. As a result, the emitter biasing current is a factor of $(n + 1)$ greater (as compared to that of Fig. 4) when the "split collector" p-n-p transistors are employed.

The advantages of this technique to reduce transconductance when compared to the alternatives are the following: 1) only one device is required, 2) the input parasitic capacitance to substrate C_{bs} is that of a single device, 3) the larger emitter current decreases emitter resistance r_e, which reduces detrimental effects of stray capacitance to ground from the emitters of Q_1 and Q_2, and 4) a reduced value of collector-to-base overlap diode capacitance C_{jc} as a result of a smaller collector area in the signal path. Consequently, the frequency response of the stage is improved. Another advantage is that the increased biasing current provides a proportional increase in the positive slew rate at the emitters of Q_1 and Q_2 (important in voltage follower applications). A minor disadvantage results since basewidth modulation resistance r_0 is smaller for the split-collector lateral p-n-p transistor due to the effects of edge (or sector size) modulation in addition to the normal base-width modulation. This causes a reduced dc gain.

In all of the transconductance reduction techniques that have been considered, a sacrifice in another area of performance has resulted: input offset voltage. The effect of a mismatch in the collector current ratios of transistors Q_1 and Q_2 is to force the emitter currents of these devices to be unbalanced for a balanced output (zero current out of the stage). The resulting input offset voltage can be modeled as a mismatch in the emitter areas of Q_1 and Q_2. A 10-percent mismatch in ratios for example, causes an offset of approximately 3 mV. In order to keep this offset contribution within a reasonable tolerance it appears that without increasing the area consumed by these devices, the limit on the maximum collector current ratio which can be used is approximately 5:1. In the single-supply input stage, a ratio of 2:1 is sufficient to obtain the desired transconductance reduction without a significant degradation in either frequency response or offset voltage.

C. The Complete Opamp Stage

The employment of this new stage in the operational amplifier shown in Fig. 8 has made it possible to achieve the objective of compensating the amplifier with a single 5-pF capacitor. To improve frequency response of the input transistors Q_1 and Q_4 the current that is available in the previously unused collectors of transistors Q_2 and Q_3 is used to bias these input emitter followers. This current also improves the slew rate at the emitters of these transistors. The differential-to-single-ended converter, diode D_1 and transistor Q_9, is used to eliminate the need for a common-mode loop to set the currents in the input stage. The output of this stage is at the collector of transistor Q_9 and is buffered from the relatively low input impedance of the second stage by the emitter followers Q_{10} and Q_{11}. These transistors are of

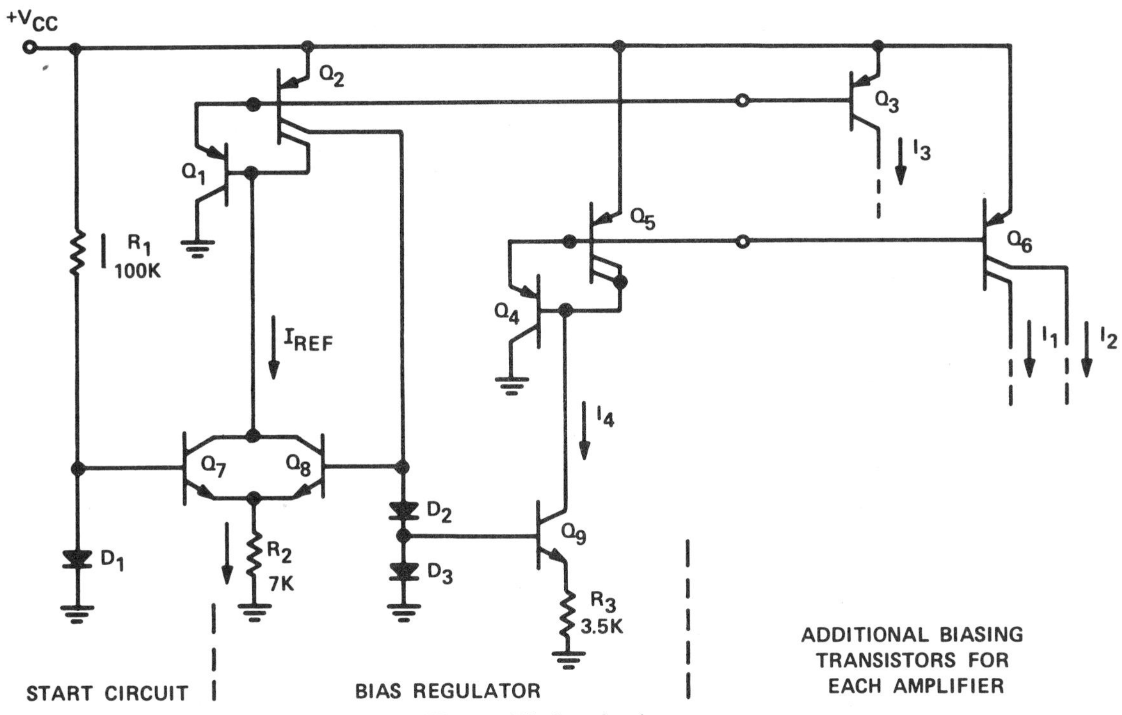

Fig. 9. Biasing circuitry.

opposite polarity such that the level shifts through them, to the first order, cancel. Therefore this opamp provides an input common-mode voltage range which, as in the circuit of Fig. 2, includes ground.

The drive to the second stage (Q_{12}) is limited to approximately the value of the current source I_2 due to the interconnection of Q_{11} and Q_5. In addition, one collector of Q_5 supplies the bias current for Q_{10}. The effective current gain of Q_{11} is reduced, as a result of this connection, to a value of approximately $K + 1$ (where K is the ratio of the collector currents of Q_5).

In normal operation the output will be returned to ground through an external load resistor. When used in this manner the output stage is biased class A and the emitter followers Q_6 and Q_7 provide increased output current capability and load isolation. Since the output of the second stage can easily turn these transistors OFF, the external load resistor can "pull" the output to ground. Under these conditions, both the input and the output voltage swing can include ground. Short circuit current limiting is provided by denying additional base drive to Q_6 when the current through resistor R_1 forward biases the base–emitter junction of transistor Q_8. The substrate p-n-p transistor Q_{13} provides increased "pull down" current under large signal conditions. This current is limited by the decreased h_{FE} of this transistor at large currents to approximately 20 mA.

D. Obtaining the Bias Currents for the Amplifiers

It is desirable that the three current sources that bias the opamp be independent of the magnitude of the power supply voltage V_{CC}. This has been accomplished by obtaining these currents from a regulator (which will operate from a 3-V power supply), Fig. 9. Start-up for the regulator is provided by the current flowing from V_{CC} through the epi resistor R_1 to the base of transistor Q_7 (and the diode D_1), which results in collector current in Q_7. Transistor Q_7 is biased OFF by the voltage at the common-emitter node once regulation is established and therefore the biasing currents are independent of the magnitude of V_{CC}. Fluctuations on the power supply are also rejected. The voltage across the resistor R_2, approximately one forward diode voltage, establishes the reference current I_{REF}. This current is independent of V_{CC} but it does have a negative temperature coefficient (TC). The collector current I_3 of transistor Q_3 has the same negative TC as I_{REF}, which approximately compensates for the positive TC of the current gain H_{FE} of output transistors Q_6 and Q_7 (Fig. 8).

In order to obtain a temperature-independent amplifier unity gain frequency ω_c the TC of the current I_1 (which biases the input stage of the opamp) must be inversely proportional to T. This results from the temperature stability of the compensation capacitor C and the temperature changes of g_m as $\omega_c = g_m/C$. The transconductance of a transistor is given by $g_m = qI_C/kT$, therefore temperature compensation can be obtained by designing I_C to be proportional to T. This has been achieved by canceling the negative TC of I_{REF} by using the voltage across diode D_3 as the input to the emitter-degenerated transistor Q_9. The resulting collector current of Q_9 (I_4) has the desired positive TC and is the reference for the current I_1. Another advantage is the small TC on the input currents to the opamp, which results from

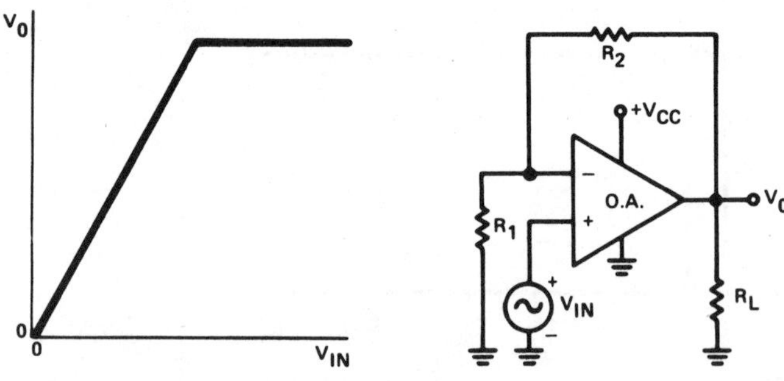

Fig. 10. Noninverting gain stage. (a) Circuit. (b) Transfer function.

the low TC on I_1 as the current gain H_{FE} of the p-n-p input transistors is relatively temperature insensitive.

The current sources for the other three amplifiers are obtained from the common biasing nodes shown. This bias regulator is also employed in the other building blocks presented in this paper.

E. Typical Performance and Applications

The typical performance characteristics at 25°C, V_{CC} = 15 V, of these single-supply operational amplifiers are given in the following list.

Open loop:
- voltage gain (R_L = 10 kΩ), 100 dB;
- unity gain bandwidth, 200 kHz;
- phase margin, 60°;
- input impedance, 500 kΩ;
- output impedance, 2 kΩ;

input common-mode voltage range, 0–13 V;
output voltage swing (R_L = 10 kΩ), 13 V peak to peak;
slew rate (unity gain, voltage follower), ±0.1 V/μs;
short circuit output current, ±20 mA;
input current, 50 nA;
input offset voltage, 5 mV.

The total bias current for all four amplifiers is 1 mA independent of the magnitude of V_{CC} and operation is possible over a power supply voltage range of 3 to 30 V_{dc}. Because of its unique characteristics, this single-supply opamp can be used to solve a major problem of the single-supply system designer amplification of low-level signals produced by transducers. This can be accomplished by using the amplifier as a noninverting gain stage, Fig. 10(a). Since both the input and output dynamic ranges of this stage include ground, the desired transfer function (extending to the origin) is produced, Fig. 10(b).

IV. Comparators

Although the previously described opamp can be used as a comparator, this function can be more optimally implemented. The typical application of comparators is in the processing of signals where a low power output stage is adequate. However, in some instances the comparator is used to control a high current warning device such as a light bulb, or a control relay. Thus two types of comparators, providing low and high output current, have been developed for use in single supply applications.

A. Comparator for Signal Processing

A simple comparator that exhibits high gain and low input current has been realized by employing the single-supply input stage of Fig. 2 and adding a current source I_2 (the load for the second stage) and a grounded emitter n-p-n transistor, Fig. 11. Flexibility is achieved by leaving the collector of this output transistor Q_7, uncommitted. As a result, the voltage to which the external load is returned is independent of the power supply and the outputs of more than one comparator can be connected to a common load to provide an output ORing function. The total biasing current for this comparator is only 150 μA, yet it achieves a transconductance of 5 mhos (the output will fully switch 1 mA of current for a change in the differential input voltage of 0.2 mV). The input currents and offset voltage are 50 nA and 5 mV, respectively. Because of their simplicity, four of these comparators are easily fabricated on one die. A photomicrograph of the quad comparator IC (which is 53 × 57 mils) is shown in Fig. 12.

The simultaneous realization of an input common-mode voltage range, which includes ground, high gain, and a flexible output circuit in these comparators makes them very useful for monitoring the outputs of several transducers and producing an output when any one of them exceeds or falls below (according to the input connections) its associated references. In Fig. 13, for example, the output voltage is high (+V) unless either V_{IN_1} exceeds V_{REF_1} or V_{IN_2} decreases below V_{REF_2}.

B. IC Comparator With a High Power Output Stage

The comparator shown in Fig. 14 will directly drive an automotive light bulb or a control relay. When the differential input signal causes transistor Q_2 to turn OFF, transistor Q_3 saturates at a current of approximately V_{BE}/R_1. This current is the base drive for the "on chip" Darlington n-p-n power transistor (Q_4 and Q_5) and is independent of V_{CC}.

The steady state current demanded by an automotive light bulb is approximately 300 mA but its cold temperature surge current is on the order of five times this value. For this reason, the output power Darlington was designed to withstand these current surges. The safe operating area is in excess of 20 V and 5 A. Safe area has been achieved by locating the metal contacts back from the edge of the emitters to provide a ballasting resistance from any point in the contact area to the edge of the emitter by simply making use of the resistance of the n^+ emitter diffusion, Fig. 15. In addition, separate ballast resistors are provided for each stripe of the multiple-stripe transistor geometry [2]. Approximately 2 Ω of resistance exist in the n^+ emitter diffusion from contact

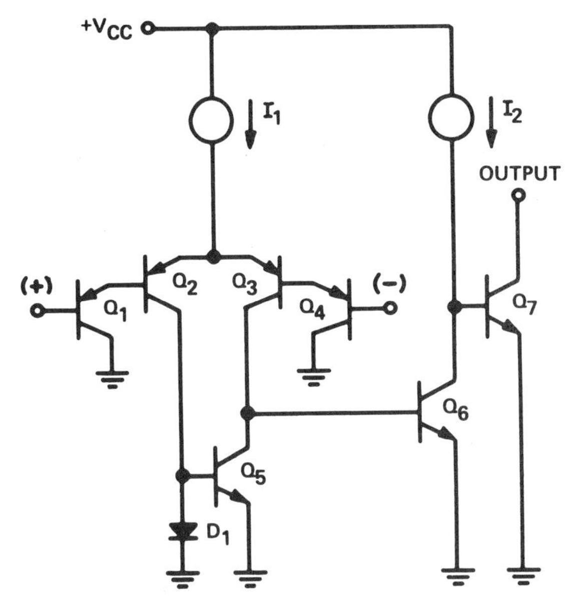

Fig. 11. Signal processing comparator.

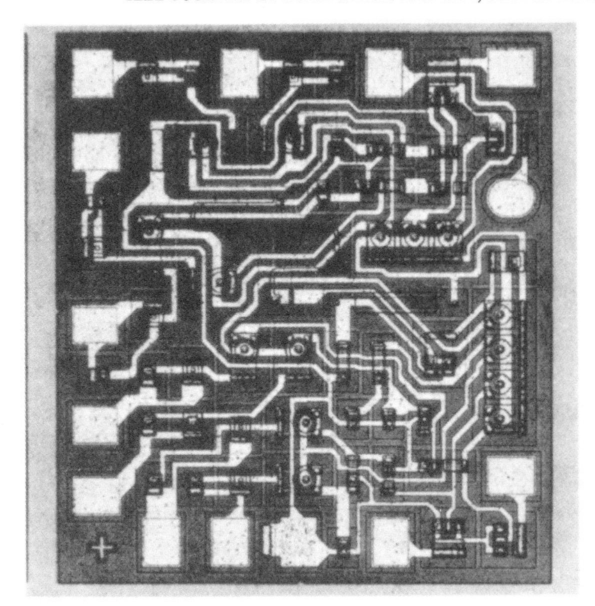

Fig. 12. Photomicrograph of quad comparator IC.

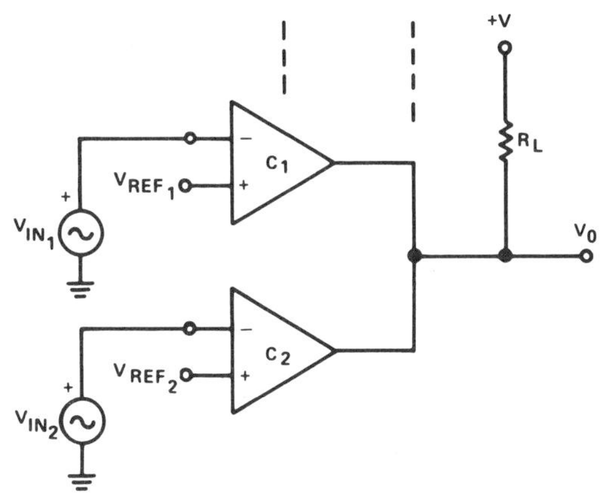

Fig. 13. Monitoring multiple transducers.

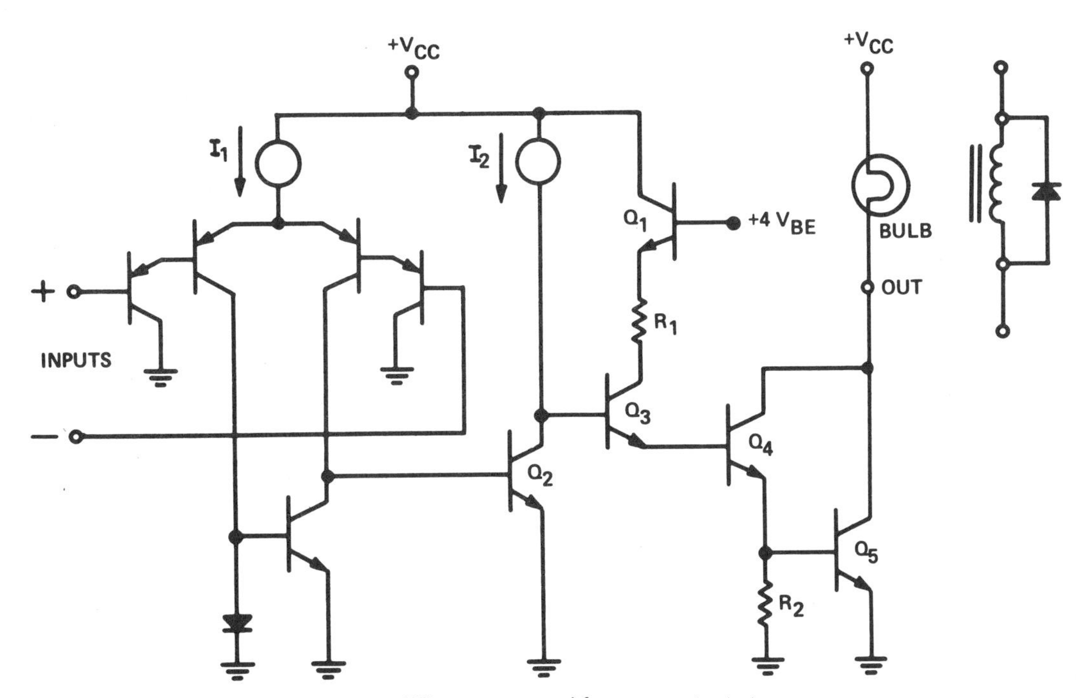

Fig. 14. IC comparator with power output stage.

C_2 to contact C_3 to provide this ballasting. Also this same region provides a larger valued resistor from base (contact C_1) to emitter (C_3), a means for base metal crossover and the top half of the n^+ ballast resistor is an active emitter.

A photomicrograph of a dual power comparator IC is shown in Fig. 16. For flexibility, either of two separate differential input stages can turn ON each power stage. The die size is 81 × 86 mils. The two power output stages (a single stripe for each driver and four stripes for each of the two power transistors) occupy approximately 30 percent of the die area. To reduce thermal coupling on the die, the input stages have been located across the die from the power transistors [3].

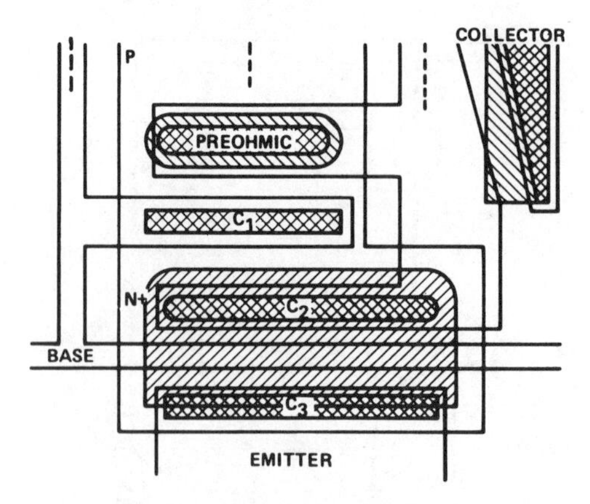

Fig. 15. Ballasting power transistor structure for safe operating area.

Fig. 16. Photomicrograph of power IC comparators.

C. IC Comparator that Drives an External n-p-n *Power Transistor*

In systems that require many comparators and in which each must handle a large load current the comparator of Fig. 17 is useful. By using the external resistor R_2, transistor Q_3 (in the IC) as well as the external n-p-n power transistor, Q_4, saturate when ON to limit the dissipation in these transistors. This approach reduces the chip size and the power dissipation of the IC circuit which allows IC costs to be minimized or additional circuit functions to be placed on the chip.

D. An IC Comparator that Drives an External p-n-p *Power Transistor*

The wiring convenience of a grounded load can be handled with a modification of the output stage as shown in Fig. 18. Once again the base drive to the output transistor Q_3 is limited by R_1 and both this transistor and the external power transistor Q_4, are allowed to fully

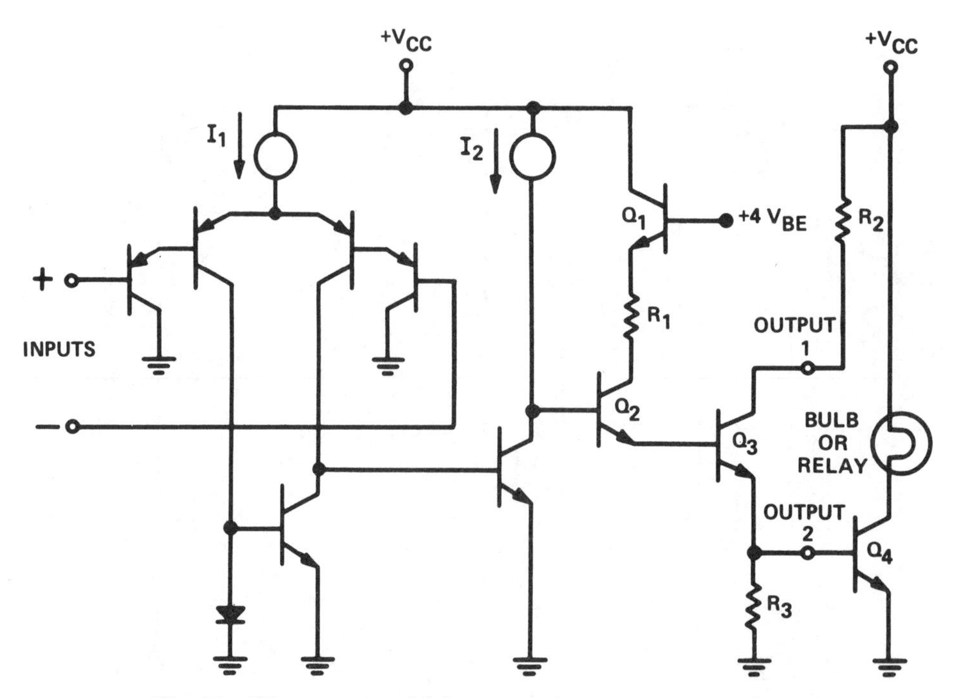

Fig. 17. IC comparator driving external n-p-n power transistor.

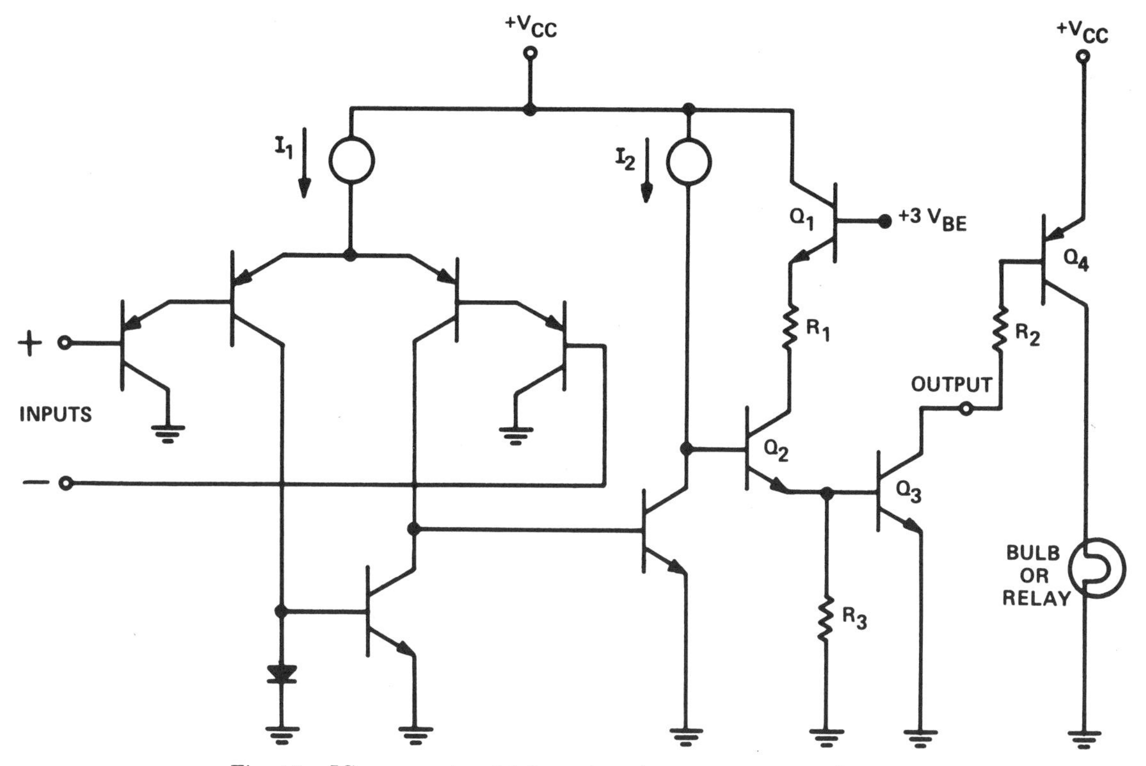

Fig. 18. IC comparator driving external p-n-p power transistor.

saturate to reduce dissipation. The external resistor R_2 limits the base drive to the p-n-p power transistor Q_4. As each of these comparators requires only three pins, four of these circuits can be supplied in a standard 14-pin package.

V. Conclusions

To meet the requirement for low-cost automotive electronics a new generation of relatively simple linear IC's has been designed to allow multiple independent circuits to be fabricated on one die and made available in a single package. A reduced speed of response compared to that of conventional IC's is not only permitted, but often desired. This allows additional flexibility in the IC design since lateral and substrate p-n-p transistors are not, for these circuits, slow devices. As a result, they have been incorporated in an input stage that provides a common-mode voltage range that includes ground (even though operated from a single power supply voltage). The development of a new technique for reducing transconductance has effected a single-supply opamp with a gain of 100 dB, which is internally compensated with only 5 pF of capacitance. All the application freedom of grounding either input is provided and the output voltage will also swing to ground so that small dc signals can be amplified. Many new comparator circuits have been developed, which range from the low-level signal processing type to a power IC that can sink 300 mA of current and also safely handle the surge current of an incandescent lamp. Because of the advantages offered by these new linear IC "building blocks," they are also useful in many existing linear and digital applications as well as in industrial control systems.

Acknowledgment

The authors wish to credit J. E. Solomon for providing the new g_m reduction technique, E. L. Long for contributing to the design of the IC n-p-n power transistor with excellent safe operating area, D. Culmer for assisting in circuit development and evaluation, and finally B. Owens and A. Smith for the IC layout design.

References

[1] J. E. Solomon, W. R. Davis, and P. L. Lee, "A self compensated monolithic operational amplifier with low input current and high slew rate," *ISSCC Dig. Tech. Papers*, pp. 14–15, Feb. 1969.

[2] E. L. Long and T. M. Frederiksen, "High-gain 15-W monolithic power amplifier with internal fault protection," *IEEE J. Solid-State Circuits*, vol. SC-6, pp. 35–44, Feb. 1971.

[3] W. F. Davis, "Thermal considerations in the design of a 500 mA monolithic negative voltage regulator," *NEREM Rec.*, pp. 84–85, 1969.

Low Voltage Techniques*

Robert J. Widlar

Consultant

Puerto Vallerta, Mexico

THE INTRINSIC OPERATING VOLTAGE limit of bipolar ICs is somewhat greater than the emitter-base voltage of the transistors. To date, this limit has been pushed only with digital circuitry. Techniques for constructing such devices as operational amplifiers, comparators, regulators and voltage references that will work from voltages as low as those supplied by a single nickel-cadmium cell, will be offered.

Obtaining high current gain per stage is one problem that becomes evident early in the design of low-voltage, micropower circuits where large output currents are required. The standard Darlington connection cannot be used as it raises input voltage unduly. Super-gain transistors severely restrict the maximum operating voltage, limiting the usefulness of an IC design using them as an output device.

Figure 1 shows a buffer that operates from little more voltage than an unbuffered transistor. A key to its usefulness is that the emitter-base voltage of Q2 can be made sufficiently lower than that of Q1 to insure proper operation. Both voltages are predictable functions of measured process variables, and the circuit can be reproduced without special manufacturing controls.

This circuit has obvious limitations that can be removed by including the refinements shown in Figure 2. A boost circuit is added to increase the bias current to Q3 with output current. This will not only increase the maximum output current for a given standby current, but also the input-voltage change required to produce it. Some linearization of the transfer function is also provided.

Considerable experience with such boost circuits has shown that they do not unfavorably alter the frequency response of the inverter, at least at frequencies below a few megahertz.

Figure 3 shows a class-B amplifier incorporating this design. At first glance, it would seem that the PNP output driver, consisting of three, common-emitter stages would be impossible to frequency compensate when included within a major loop. However, local feedback through Q6 and the inclusion of Q4 and R2 within the loop tends to control the frequency characteristic by making gain about equal to the NPN. Additional circuitry must be added to control the drive current to the output PNP in saturation. This can be accomplished with minimal increase in saturated operating current.

Integrated versions of this output stage have been built to supply ± 10mA at a total supply voltage of 1.1V with ± 200mV saturation voltage. Unloaded, saturation voltage is ± 10mV and operating current is 50μA. Maximum voltage is limited only by the BV_{CEO} of the transistors (60V – 80V).

Band-gap references can easily be designed to operate from a 1.3V supply. This is unnecessarily high in that it excludes the rechargable nickel-cadmium cell as a power source. The circuit in Figure 4 will provide a temperature compensated 200mV and will operate from less than 1V. It also has been corrected for the thermal nonlinearity of emitter-base voltage and the temperature drift of implanted resistors to provide observed drifts less than 0.1% over a -55°C to 125°C temperature range.

A revised equation for the emitter-base voltage of bipolar transistors is shown in Figure 5. This has been found to be more accurate than the original[1] in describing the observed nonlinearities in temperature drift. A new term, μ_B, has been introduced. This is the minority carrier mobility in the base which generally varies as some power of T giving a nonlinear component of the form Tlnt.

The collector current is generally a function of the resistors in the IC. Their temperature drift can likewise cause nonlinearities, usually of the form Tln $(T + T_o)$.

The reference in Figure 4 operates Q1 at a current proportional to absolute temperature. To a first approximation, the operating current of Q2 is independent of temperature. It can be shown that this gives rise to a Tlnt term that essentially cancels the nonlinearity introduced by emitter-base voltage and resistor drift.

Figure 5 also gives the correction term for the special case of double-diffused NPN transistors. The assumption is made that the effective base mobility is described by the behavior of pinch resistors and that operating current varies directly as temperature and inversely as resistor values.

This equation gives a fairly accurate representation of the nonlinearities observed. This implies that the extrapolated energy band-gap voltage of silicon is 1.183V. This is lower than the 1.205V suggested originally[2], but greater than the 1.153V obtained from other sources[3]. Without nonlinearity correction and with drift free resistors, the output voltage of the basic reference must be raised 52mV above the band-gap voltage for minimum drift over a -55°C to 125°C temperature range.

Linear ICs that operate only at very low voltages can be expected to have significantly higher yields than those that run at 30V – 40V. This means that a circuit that will operate at both will have a market for the low-voltage fallout. Further, if the IC can be designed so that the majority of the active area is biased at low voltage, even with high supply voltage, the yield to high voltage parts will be improved.

The foregoing could be important in the manufacture of complex linear circuits. Large die (approaching 100 x 100mils) are being made and sold in high volume at low cost. Linear voltage regulators are perhaps the best example of this. The power transistor, which occupies about half the active area, is forgiving of slight junction defects. Reducing bias voltage over a large area could provide an equivalent result for complex, low-current circuitry.

The high sheet resistivities obtainable with implanted resistors strongly recommend them for micropower devices, particularly

*Project performed under contract to National Semiconductor Corporation, Santa Clara, CA

[1]Widlar, R.J., "An Exact Expression for the Emitter-Base Voltage of BiPolar Tranistors", *Proc. IEEE* (Letters), Vol. 55, p. 96-97; Jan., 1967.

[2]Widlar, R.J., "New Developments in IC Voltage Regulators" *IEEE Journal of Solid-Circuits*, Vol. SC-6, No. 1, p. 2-7; Feb., 1971.

[3]Muller, R.S. and Kamins, T.J., "Device Electronics for Integrated Circuits", *John Wiley & Sons*, p. 32; 1977.

Reprinted from *1978 IEEE Internat. Solid-State Circuits Conf. Dig. Tech. Papers*, Feb. 15-17, 1978, pp. 238-239.

at high levels of complexity. But implanted resistors have a lower breakdown voltage than their diffused counterparts (40V vs 100V). This is caused by the reduced radius of curvature at the junction edges. With higher sheet resistances, a strong voltage coefficient of resistivity will be observed. These factors recommend that care be used when operating implant resistors at higher voltages or that they be operated at low voltages where possible.

Field-effect transistors have been considered for low-voltage applications because their operating voltage can be made less than V_{BE}. Although their transconductance equals that of bipolar devices at very low currents, it is considerably less even at moderate current densities. Thus, where currents vary from a moderate level to a high level (as in an output stage) the required change in gate voltage can easily exceed the turn-on voltage of bipolar devices. The use of FETs is not precluded here, but bipolar techniques are deemed necessary for the design of general-purpose ICs.

The examples suggest that many of the standard linear functions in use today can be made to work at low voltages. The application of these techniques to such functions as hearing aids, pacemakers, remote telemetry, etc., that must operate from batteries or solar cells is relatively obvious. Less obvious are the many general-purpose functions that can be performed by devices capable of operating well over such a wide supply-voltage range.

Acknowledgment

Special thanks are due M. Yamatake for his contribution in reducing these concepts to practice.

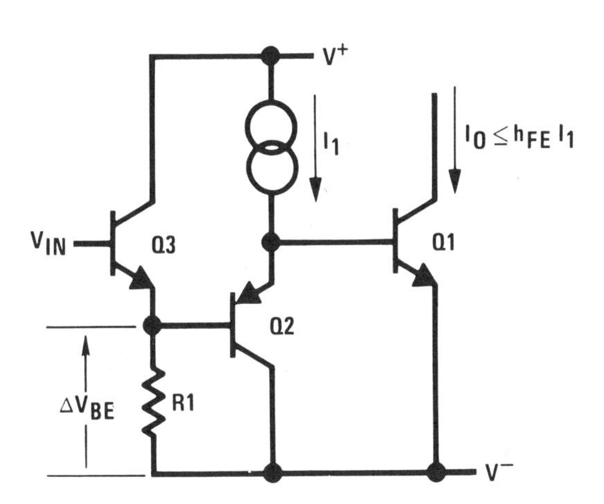

FIGURE 1—Inverter with compound buffer.

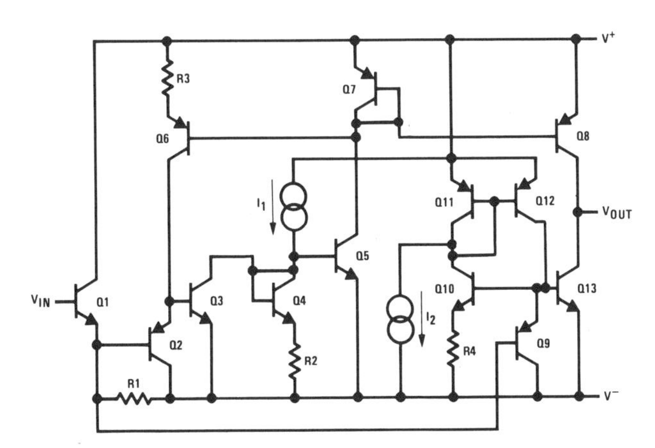

FIGURE 3—Complementary, class-B amplifier.

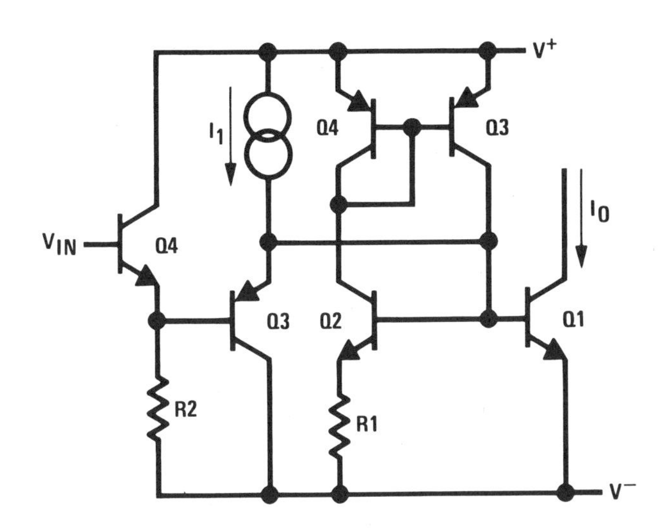

FIGURE 2—Buffered inverter with boost.

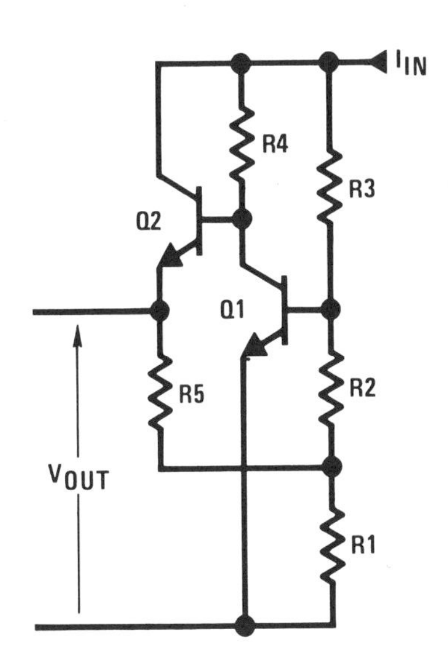

FIGURE 4—Curvature-corrected reference.

$$V_{BE} = V_{go}\left(1 - \frac{T}{T_o}\right) + V_{BEo}\left(\frac{T}{T_o}\right) + \epsilon$$

$$\epsilon = \frac{kT}{q}\left(4\,\ell n\,\frac{T_o}{T} + \ell n\,\frac{\mu_{Bo}}{\mu_B} + \ell n\,\frac{I_C}{I_{Co}}\right)$$

$$\mu_B \,\alpha\, T^{-1};\ I_C \,\alpha\, \frac{T}{R}$$

$$\epsilon = \frac{kT}{q}\left(2\,\ell n\,\frac{T_o}{T} + \ell n\,\frac{R_o}{R}\right)$$

FIGURE 5—The revised V_{BE} equation.

Part II
Operational Amplifiers

Because of its versatility and availability, the monolithic operational amplifier has been a universal building block of analog system design. Since their introduction in the mid 1960's, the performance characteristics of monolithic operational amplifiers have evolved and improved greatly. This part is made up of a number of selected papers which cover the basics and the evolution of the operational amplifier design techniques, over the recent years.

The first group of papers included in this part are tutorial in their nature. In the first paper, Solomon examines in detail the design trade-offs and performance limitations associated with the popular LM-101 and the μA741-type operational amplifiers. The second paper, by Widlar, presents some of the design techniques for monolithic operational amplifiers such as active-loads, controlled-gain lateral PNP and the super-beta NPN transistors. In the third paper, Gray and Meyer discuss the interrelations between key amplifier parameters such as offset-voltage and common-mode rejection, and define some of the commonly used specifications for op amps.

The fourth paper of this part, by Frederiksen *et al.*, discusses a new current-differencing operational amplifier technique which features a very simple circuit configuration that can operate with a single supply voltage. The fifth paper, by Boyle *et al.*,presents a macromodel for IC op amps which is well suited for efficient computer simulation. The next two papers deal with the use of different process technologies in op amp design. The ion-implanted junction FET technology, described in the sixth paper by Russell and Culmer, has led to new families of IC op amps which use matched JFET input devices for low input bias currents and high slew rates. The seventh paper, by Tsividis and Gray, describes how a basically digital device such as the NMOS transistor can be adapted to op any design. In the final paper of this part, Metzger gives a comparative survey on the characteristics of commercially available bipolar and FET-input op amps.

Because of its popularity, the monolithic operational amplifier has been the subject of many technical books and papers, in addition to the selected few included in this section. Some of these are listed at the end of this section, as recommended references. For a more detailed and in-depth coverage of IC op amps, the interested reader is also referred to an excellent companion book in this series, "Integrated Circuit Operational Amplifiers," edited by R.G. Meyer and published by the IEEE Press, in 1978.

ADDITIONAL REFERENCES

1. J.G. Graeme, G.E. Tobey, and L.P. Huelsman, *Operational Amplifiers* New York: McGraw-Hill, 1971.
2. J.K. Roberge, *Operational Amplifiers.* New York: Wiley, 1975.
3. J.V. Wait, L.P. Huelsman, and G.A. Kern, *Introduction to Operational Amplifier Theory and Applications.* New York: McGraw-Hill, 1975.
4. R.C. Jaeger, "Common-Mode Rejection Limitations of Differrential Amplifiers," *J. Sol. State Circuits,* vol. SC-11, pp. 411-417, June 1976.
5. D.L. Cave and W.R. Davis, "A Quad JFET Wide-Band Operational Amplifier Integrated Circuit Featuring Temperature-Compensated Bandwidth," *J. Sol. State Circuits,* vol. SC-12, pp. 382-388, August 1977.
6. R.J. Apfel and P.R. Gray, "A Fast-Settling Monolithic Operational Amplifier Using Doublet Compression Techniques," *J. Sol. State Circuits,* vol. SC-9, pp. 332-340, December 1974.
7. J.H. Huijsing and F. Tol, "Monolithic Operational Amplifier Design with Improved HF Behavior," *J. Sol. State Circuits,* vol. SC-11, pp. 323-328, April 1976.
8. M.A. Maidique, "A High-Precision Monolithic Super-Beta Operational Amplifier," *J. Sol. State Circuits,* vol. SC-7, pp. 480-487, December 1972.
9. P.R. Gray, "A 15-W Monolithic Power Operational Amplifier," *J. Sol. State Circuits,* vol. SC-7, pp. 474-480, December 1972.
10. O.H. Schade, Jr., "A New Generation of MOS/Bipolar Operational Amplifiers", *R.C.A. Review,* vol. 37, pp. 404-424, September 1976.
11. D. Jone and R.W. Webb, "Chopper-Stabilized Op Amp Combines MOS and Bipolar Elements on One Chip," *Electronics Magazine,* pp. 110-114, September 1973.
12. A.P. Brokaw and M.P. Timko, "An Improved Monolithic Instrumentation Amplifier," *J. Sol. State Circuits,* vol. SC-10, pp. 417-423, December 1975.

The Monolithic Op Amp: A Tutorial Study

JAMES E. SOLOMON, MEMBER, IEEE

Invited Paper

Abstract—**A study is made of the integrated circuit operational amplifier (IC op amp) to explain details of its behavior in a simplified and understandable manner. Included are analyses of thermal feedback effects on gain, basic relationships for bandwidth and slew rate, and a discussion of pole-splitting frequency compensation. Sources of second-order bandlimiting in the amplifier are also identified and some approaches to speed and bandwidth improvement are developed. Brief sections are included on new JFET-bipolar circuitry and die area reduction techniques using transconductance reduction.**

I. Introduction

THE integrated circuit operational amplifier (IC op amp) is the most widely used of all linear circuits in production today. Over one hundred million of the devices will be sold in 1974 alone, and production costs are falling low enough so that op amps find applications in virtually every analog area. Despite this wide usage, however, many of the basic performance characteristics of the op amp are poorly understood.

It is the intent of this study to develop an understanding for op amp behavior in as direct and intuitive a manner as possible. This is done by using a variety of simplified circuit models which can be analyzed in some cases by inspection, or in others by writing just a few equations. These simplified models are generally developed from the single representative op amp configuration shown in Figs. 1 and 2.

The rationale for starting with the particular circuit of Fig. 1 is based on the following: this circuit contains, in simplified form, all of the important elements of the most commonly used integrated op amps. It consists essentially of two voltage gain stages, an input differential amp and a common emitter second stage, followed by a class-AB output emitter follower which provides low impedance drive to the load. The two interstages are frequency compensated by a single small "pole-splitting" capacitor (see below) which is usually included on the op amp chip. In most respects this circuit is directly equivalent to the general purpose LM101 [1], μA 741 [2], and the newer dual and quad op amps [3], so the results of our study relate directly to these devices. Even for more exotic designs, such as wide-band amps using feedforward [4], [5], or the new FET input circuits [6], the basic analysis approaches still apply, and performance details can be accurately predicted. It has also been found that a good understanding of the limitations of the circuit in Fig. 1 provides a reasonable starting point from which higher performance amplifiers can be developed.

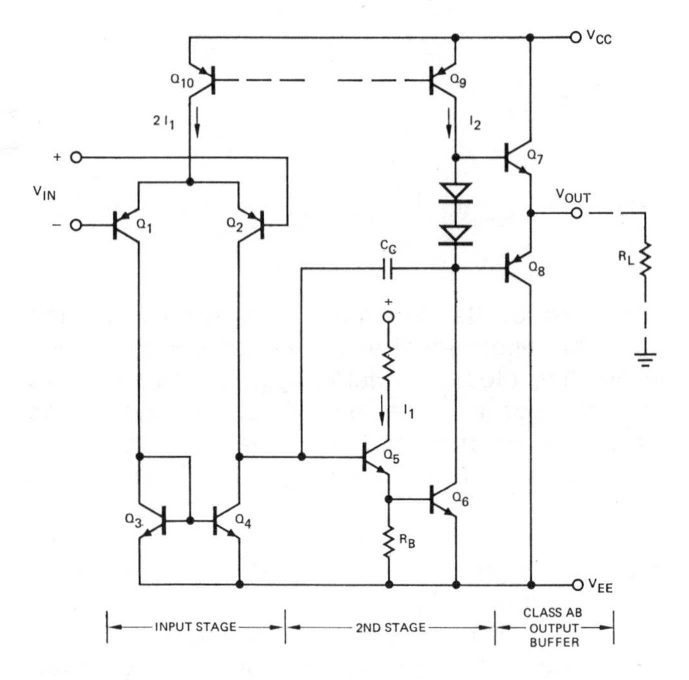

Fig. 1. Basic two-stage IC op amp used for study. Minimal modifications used in actual IC are shown in Fig. 2.

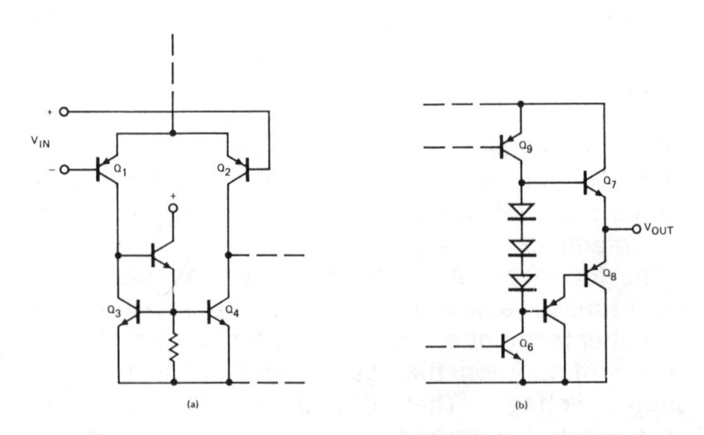

Fig. 2. (a) Modified current mirror used to reduce dc offset caused by base currents in $Q3$ and $Q4$ in Fig. 1. (b) Darlington p-n-p output stage needed to minimize gain fall-off when sinking large output currents. This is needed to offset the rapid β drop which occurs in IC p-n-p's.

The study begins in Section II, with an analysis of dc and low frequency gain. It is shown that the gain is typically limited by thermal feedback rather than elec-

Manuscript received July 1, 1974; revised August 8, 1974.
The author is with the National Semiconductor Corporation, Santa Clara, Calif. 95051.

Reprinted from *IEEE J. Solid-State Circuits*, vol. SC-9, pp. 314-332, Dec. 1974.

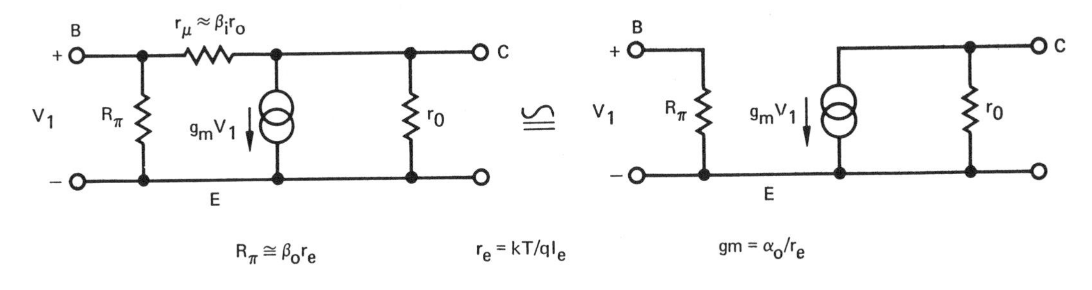

(a)

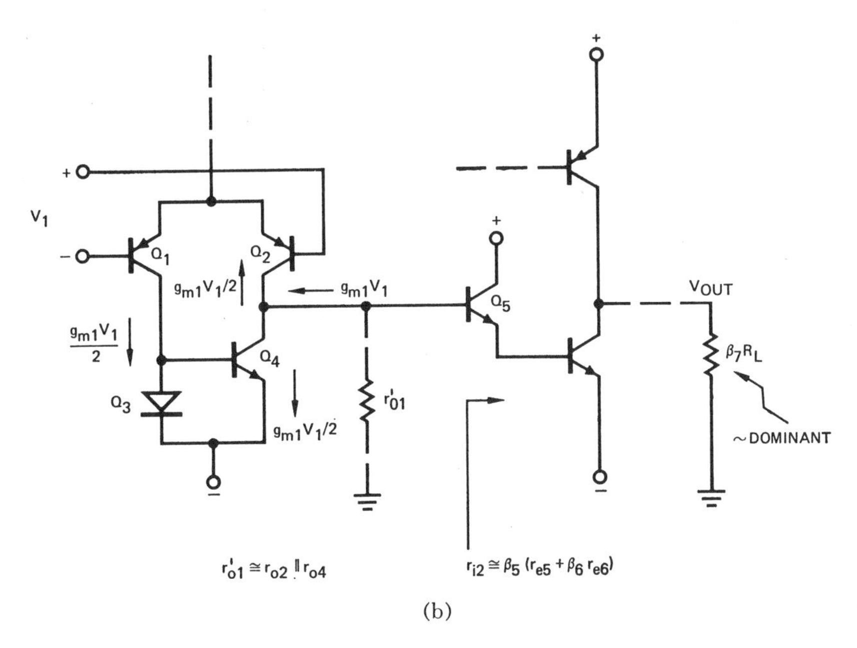

(b)

Fig. 3. (a) Approximate π model for CE transistor at dc. Feedback element $r_\mu \cong \beta_i r_o$ is ignored since this greatly simplifies hand calculations. The error caused is usually less than 10 percent because β_i, the intrinsic β under the emitter, is quite large. Base resistance r_x is also ignored for simplicity. (b) Circuit illustrating calculation of electronic gain for op amp of Fig. 1. Consideration is given only to the fully loaded condition ($R_L \cong 2$ kΩ) where β_7 is falling (to about 50) due to high current density. Under this condition, the output resistance of $Q6$ and $Q9$ are nondominant.

trical characteristics. A highly simplified thermal analysis is made, resulting in a gain equation containing only the maximum output current of the op amp and a thermal feedback constant.

The next three sections apply first-order models to the calculation of small-signal high frequency and large-signal slewing characteristics. Results obtained include an accurate equation for gain-bandwidth product, a general expression for slew rate, some important relationships between slew rate and bandwidth, and a solution for voltage follower behavior in a slewing mode. Due to the simplicity of the results in these sections, they are very useful to designers in the development of new amplifier circuits.

Section VI applies more accurate models to the calculation of important second-order effects. An effort is made in this section to isolate all of the major contributors to bandlimiting in the modern amp.

In the final section, some techniques for reduction of op amp die size are considered. Transconductance reduction and layout techniques are discussed which lead to fabrication of an extremely compact op amp cell. An example yielding 8000 possible op amps per 3-in wafer is given.

II. Gain at DC and Low Frequencies

A. The Electronic Gain

The electronic voltage gain will first be calculated at dc using the circuit of Fig. 1. This calculation becomes straightforward if we employ the simplified transistor model shown in Fig. 3(a). The resulting gain from Fig. 3(b) is

$$A_V(0) = \frac{v_{\text{out}}}{v_{\text{in}}} \cong \frac{g_{m1}\beta_5\beta_6\beta_7R_L}{1 + r_{i2}/r_{01}'} \tag{1}$$

where

$$r_{i2} \cong \beta_5(r_{e5} + \beta_6 r_{e6})$$

$$r_{01}' \cong r_{04}//r_{02}.$$

It has been assumed that

$$\beta_7 R_L < r_{06}//r_{09},\ g_{m1} = g_{m2},\ \beta_7 = \beta_8.$$

The numerical subscripts relate parameters to transistor Q numbers (i.e., r_{e5} is r_e of Q_5, β_6 is β_0 of Q_6, etc.). It has also been assumed that the current mirror transistors Q_3 and Q_4 have α's of unity, and the usually small loading of R_B has been ignored. Despite the several assumptions made in obtaining this simple form for (1), its accuracy is quite adequate for our needs.

An examination of (1) confirms the way in which the amplifier operates: the input pair and current mirror convert the input voltage to a current $g_{m1}v_{in}$ which drives the base of the second stage. Transistors Q_5, Q_6, and Q_7 simply multiply this current by β^3 and supply it to the load R_L. The finite output resistance of the first stage causes some loss when compared with second stage input resistance, as indicated by the term $1/(1 + r_{i2}/r_{01}')$. A numerical example will help our perspective: for the LM101A, $I_1 \cong 10\ \mu$A, $I_2 \cong 300\ \mu$A, $\beta_5 = \beta_6 \cong 150$, and $\beta_7 \cong 50$. From (1) and dc voltage gain with $R_L = 2$ kΩ is

$$A_V(0) \cong 625\,000. \tag{2}$$

The number predicted by (2) agrees well with that measured on a discrete breadboard of the LM101A, but is much higher than that observed on the integrated circuit. The reason for this is explained in the next section.

B. Thermal Feedback Effects on Gain

The typical IC op amp is capable of delivering powers of 50–100 mW to a load. In the process of delivering this power, the output stage of the amp internally dissipates similar power levels, which causes the temperature of the IC chip to rise in proportion to the output dissipated power. The silicon chip and the package to which it is bonded are good thermal conductors, so the whole chip tends to rise to the same temperature as the output stage. Despite this, small temperature gradients from a few tenths to a few degrees centigrade develop across the chip with the output section being hotter than the rest. As illustrated in Fig. 4, these temperature gradients appear across the input components of the op amp and induce an input voltage which is proportional to the output dissipated power.

To a first order, it can be assumed that the temperature difference (T_2-T_1) across a pair of matched and closely spaced components is given simply by

$$(T_2 - T_1) \cong \pm K_T P_d \quad °\text{C} \tag{3}$$

where

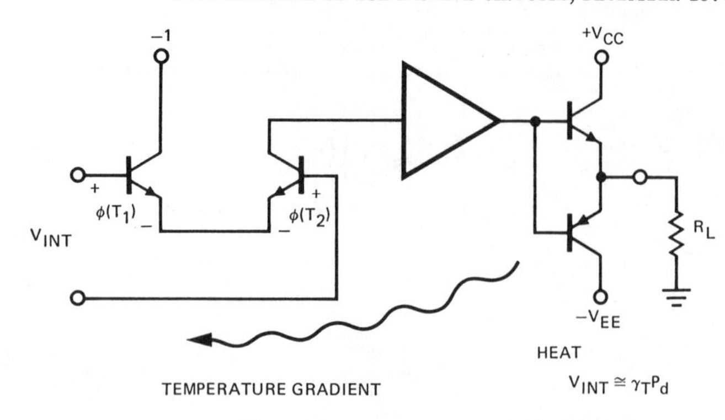

Fig. 4. Simple model illustrating thermal feedback in an IC op amp having a single dominant source of self-heat, the output stage. The constant $\gamma_T \cong 0.6$ mV/W and P_d is power dissipated in the output. For simplicity, we ignore input drift due to uniform heating of the package. This effect can be significant if the input stage drift is not low, see [7].

P_d power dissipated in the output circuit,
K_T a constant with dimensions of °C/W.

The plus/minus sign is needed because the direction of the thermal gradient is unknown. In fact, the sign may reverse polarity during the output swing as the dominant source of heat shifts from one transistor to another. If the dominant input components consist of the differential transistor pair of Fig. 4, the thermally induced input voltage V_{int} can be calculated as

$$V_{int} \cong \pm K_T P_d (2 \times 10^{-3})$$
$$\cong \pm \gamma_T P_d \tag{4}$$

where $\gamma_T = K_T(2 \times 10^{-3})$ V/W, since the transistor emitter-base drops change about -2 mV/°C.

For a thermally well designed IC op amp, in which the power output devices are made to approximate either a point or a line source and the input components are placed on the resulting isothermal lines (see below and Fig. 8), typical values measured for K_T are

$$K_T \approx 0.3\ °\text{C/W} \tag{5}$$

in a TO-5 package.

The dissipated power in the class-AB output stage P_d is written by inspection of Fig. 4:

$$P_d = \frac{V_0 V_s - V_0^2}{R_L} \tag{6}$$

where

$$V_s = +V_{cc} \quad \text{when } V_0 > 0$$
$$V_s = -V_{ee} \quad \text{when } V_0 < 0.$$

A plot of (6) in Fig. 5 resembles the well-known class-AB dissipation characteristics, with zero dissipation occurring for $V_0 = 0, +V_{cc}, -V_{ee}$. Dissipation peaks occur for $V_0 = +V_{cc}/2$ and $-V_{ee}/2$. Note also from (4) that the thermally induced input voltage V_{int} has this same double-humped shape since it is just equal to a constant times P_d at dc.

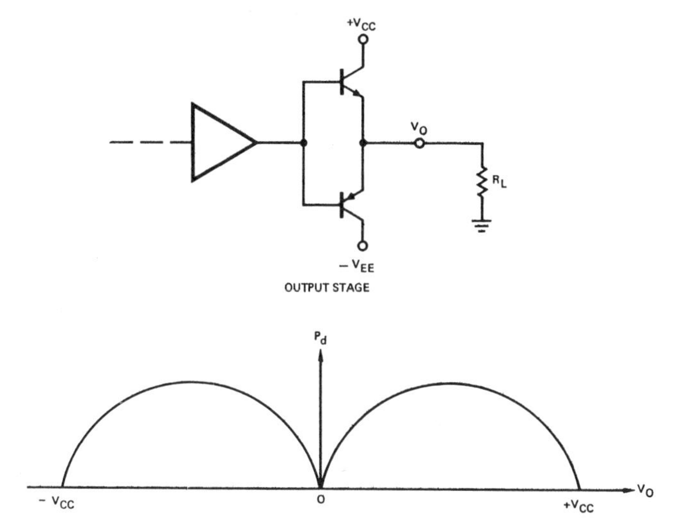

Fig. 5. Simple class-B output stage and plot of power dissipated in the stage, P_d, assuming device can swing to the power supplies. Equation (6) gives an expression for the plot.

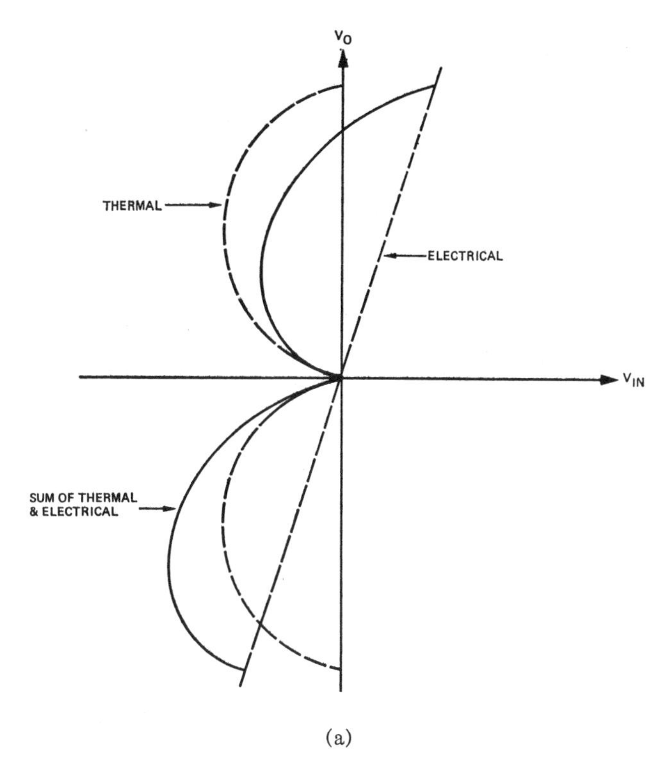

(a)

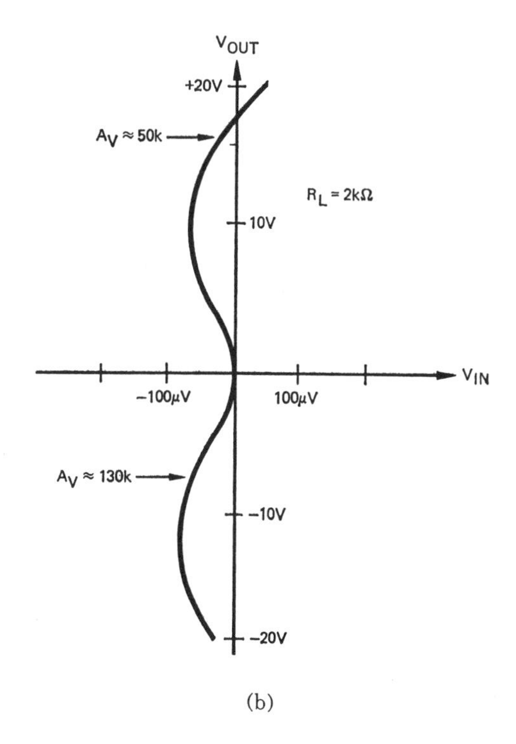

(b)

Fig. 6. (a) Idealized dc transfer curve for an IC op amp showing its electrical and thermal components. (b) Experimental open-loop transfer curve for a representative op amp (LM 101).

Now examine Figs. 6(a) and (b) which are curves of open-loop V_0 versus V_{in} for the IC op amp. Note first that the overall curve can be visualized to be made up of two components: a) a normal straight line electrical gain curve of the sort expected from (1) and b) a double-humped curve similar to that of Fig. 5. Further, note that the gain characteristic has either positive or negative slope depending on the value of output voltage. This means that the thermal feedback causes the open-loop gain of the feedback amplifier to change phase by 180°, apparently causing negative feedback to become positive feedback. If this is really true, the question arises: which input should be used as the inverting one for feedback? Further, is there any way to close the amplifier and be sure it will not find an unstable operating point and latch to one of the power supplies?

The answers to these questions can be found by studying a simple model of the op amp under closed-loop conditions, including the effects of thermal coupling. As shown in Fig. 7, the thermal coupling can be visualized as just an additional feedback path which acts in parallel with the normal electrical feedback. Noting that the electrical form of the thermal feedback factor is [see (4) and (6)]

$$\beta_T = \frac{\partial V_{\text{int}}}{\partial V_0} = \pm \frac{\gamma_T}{R_L}(V_S - 2V_0). \tag{7}$$

The closed-loop gain, including thermal feedback is

$$A_V(0) = \frac{\mu}{1 + \mu(\beta_e \pm \beta_T)} \tag{8}$$

where μ is the open-loop gain in the absence of thermal feedback [(1)] and β_e is the applied electrical feedback as in Fig. 7. Inspection of (8) confirms that as long as there is sufficient electrical feedback to swamp the thermal feedback (i.e., $\beta_e > \beta_T$), the amplifier will behave as a normal closed-loop device with characteristics determined principally by the electrical feedback (i.e., $A_V(0) \cong 1/\beta_e$). On the other hand, if β_e is small or nonexistant, the thermal term in (8) may dominate, giving an apparent open-loop gain characteristic determined by the thermal feedback factor β_T. Letting $\beta_e = 0$ and combining (7) and (8), $A_V(0)$ becomes

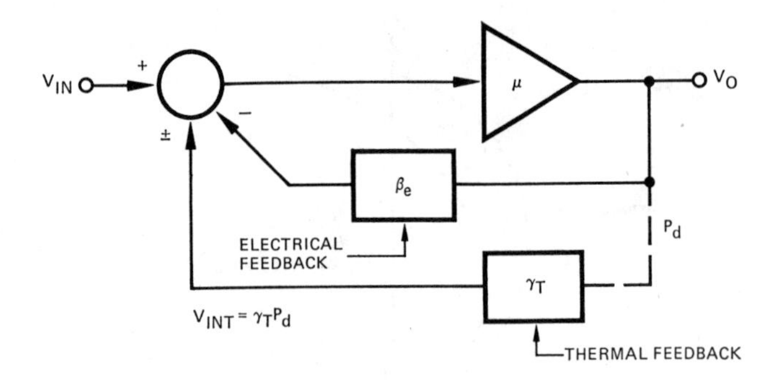

Fig. 7. Diagram used to calculate closed-loop gain with thermal feedback.

$$A_V(0) = \frac{\mu}{1 \pm \frac{\mu \gamma_T}{R_L}(V_S - 2V_0)}. \tag{9}$$

Recalling from (6) that V_0 ranges between 0 and V_S, we note that the incremental thermal feedback is greatest when $V_0 = 0$ or V_s, and it is at these points that the thermally limited gain is smallest. To use the amplifier in a predictable manner, one must always apply enough electrical feedback to reduce the gain below this minimum thermal gain. Thus, a *maximum usable gain* can be defined as that approximately equal to the value of (9) with $V_0 = 0$ or V_s which is

$$A_V(0)|_{\max} \cong \frac{R_L}{\gamma_T V_S} \tag{10}$$

or

$$A_V(0)|_{\max} \cong \frac{1}{\gamma_T I_{\max}}. \tag{11}$$

It was assumed in (10) and (11) that thermal feedback dominates over the open-loop electrical gain, μ. Finally, in (11) a maximum current was defined $I_{\max} = V_S/R_L$ as the maximum current which would flow if the amplifier output could swing all the way to the supplies.

Equation (11) is a strikingly simple and quite general result which can be used to predict the expected maximum usable gain for an amplifier if we know only the maximum output current and the thermal feedback constant γ_T.

Recall that typically $K_T \cong 0.3$°C/W and $\gamma_T = (2 \times 10^{-3})\ K_T \cong 0.6$ mV/W. Consider, as as example, the standard IC op amp operating with power supplies of $V_S = \pm 15$ V and a minimum load of 2 kΩ, which gives $I_{\max} = 15$ V/2 kΩ = 7.5 mA. Then, from (11), the maximum thermally limited gain is about:

$$A_V(0)|_{\max} \cong 1/(0.6 \times 10^{-3})(7.5 \times 10^{-3})$$
$$\cong 220\,000. \tag{12}$$

Comparing (2) and (12), it is apparent that the thermal characteristics dominate over the electrical ones if the minimum load resistor is used. For loads of 6 kΩ or more, the electrical characteristics should begin to dominate

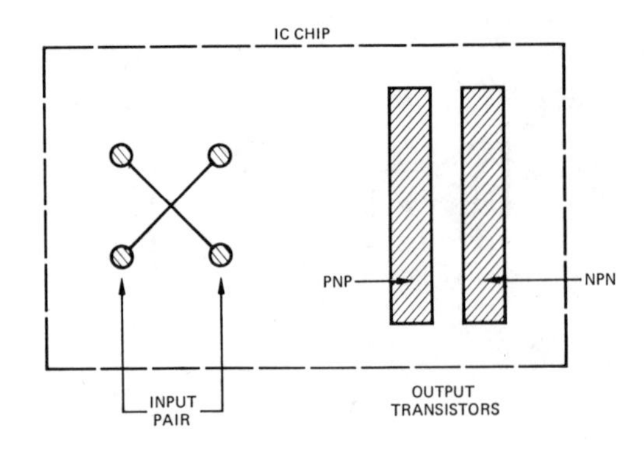

Fig. 8. One type layout in which a quad of input transistors is cross connected to reduce effect of nonuniform thermal gradients. The output transistors use distributed stripe geometries to generate predictable isothermal lines.

if thermal feedback from sources other than the output stage is negligible. It should be noted also that, in some high speed, high drain op amps, thermal feedback from the second stage dominates when there is no load.

As a second example, consider the so-called "power op amp" or high gain audio amp which suffers from the same thermal limitations just discussed. For a device which can deliver 1 W into a 16-Ω load, the peak output current and voltage are 350 mA and 5.7 V. Typically, a supply voltage of about 16 V is needed to allow for the swing loss in the IC output stage. $I_{\max}$ is then 8 V/16 Ω or 0.5 A. If the device is in a TO-5 package γ_T is approximately 0.6 mV/W, so from (11) the maximum usable dc gain is

$$A_V(0)|_{\max} \cong \frac{1}{(0.6 \times 10^{-3})(0.5)} \cong 3300. \tag{13}$$

This is quite low compared with electrical gains of, say, 100 000 which are easily obtainable. The situation can be improved considerably by using a large die to separate the power devices from the inputs and carefully placing the inputs on constant temperature (isothermal) lines as illustrated in Fig. 8. If one also uses a power package with a heavy copper base, γ_T's as low as 50 μV/W have been observed. For example, a well-designed 5-W amplifier driving an 8-Ω load and using a 24-V supply, would have a maximum gain of 13 000 in such a power package.

As a final comment, it should be pointed out that the most commonly observed effect of thermal feedback in high gain circuits is low frequency distortion due to the nonlinear transfer characteristic. Differential thermal coupling typically falls off at an initial rate of 6 dB/octave starting around 100–200 Hz, so higher frequencies are uneffected.

III. Small-Signal Frequency Response

At higher frequencies where thermal effects can be ignored, the behavior of the op amp is dependent on purely electronic phenomena. Most of the important small and large signal performance characteristics of the classical IC op amp can be accurately predicted from

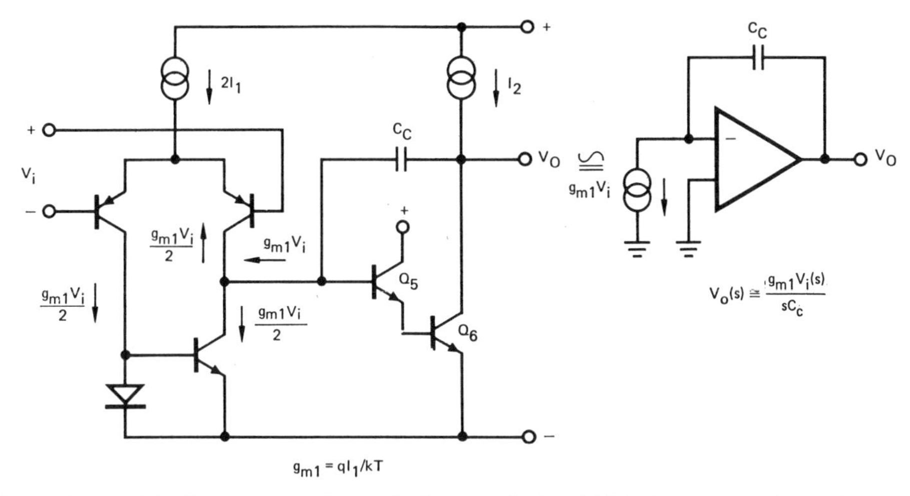

Fig. 9. First-order model of op amp used to calculate small signal high frequency gain. At frequencies of interest the input impedance of the second stage becomes low compared to first stage output impedance due to C_c feedback. Because of this, first stage output impedance can be assumed infinite, with no loss in accuracy.

very simple first-order models for the amplifier in Fig. 1[8]. The small-signal model that is used assumes that the input differential amplifier and current mirror can be replaced by a frequency independent voltage controlled current source, see Fig. 9. The second stage consisting essentially of transistors Q_5 and Q_6, and the current source load, is modeled as an ideal frequency independent amplifier block with a feedback or "integrating capacitor" identical to the compensation capacitor, C_c. The output stage is assumed to have unity voltage gain and is ignored in our calculations. From Fig. 9, the high frequency gain is calculated by inspection:

$$A_v(\omega) = \left|\frac{v_O}{v_i}(s)\right| = \left|\frac{g_{m1}}{sC_c}\right| = \frac{g_{m1}}{\omega C_c} \tag{14}$$

where dc and low frequency behavior have not been included since this was evaluated in the last section. Fig. 10 is a plot of the gain magnitude as predicted by (14). From this figure it is a simple matter to calculate the open-loop unity gain frequency ω_u, which is also the gain-bandwidth product for the op amp under closed-loop conditions:

$$\omega_u = \frac{g_{m1}}{C_c}. \tag{15}$$

In a practical amplifier, ω_u is set to a low enough frequency (by choosing a large C_c) so that negligible excess phase over the 90° due to C_c has built up. There are numerous contributors to excess phase including low f_t p-n-p's, stray capacitances, nondominant second stage poles, etc. These are discussed in more detail in a later section, but for now suffice it to say that, in the simple IC op amp, $\omega_u/2\pi$ is limited to about 1 MHz. As a simple test of (15), the LM101 or the μA741 has a first stage bias current I_1 of 10 μA per side, and a compensation capacitor for unity gain operation, C_c, of 30 pF. These amplifiers each have a first stage g_m which is half that

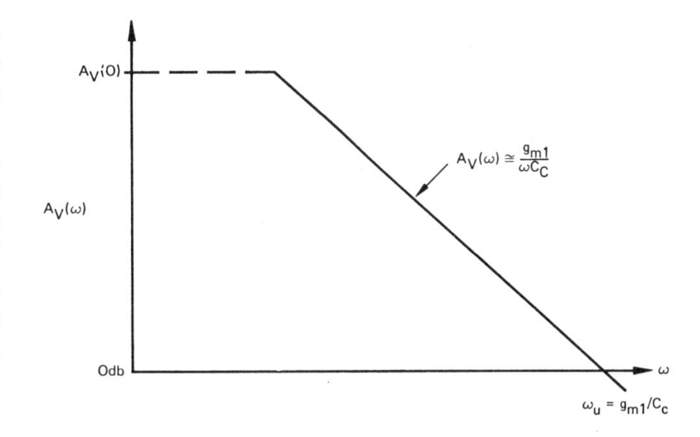

Fig. 10. Plot of open-loop gain calculated from model in Fig. 9. The dc and LF gain are given by (1), or (11) if thermal feedback dominates.

of the simple differential amplifier in Fig. 1 so $g_{m1} = qI_1/2kT$. Equation (15) then predicts a unity gain corner of

$$f_u = \frac{\omega_u}{2\pi} = \frac{g_{m1}}{2\pi C_c} = \frac{(0.192 \times 10^{-3})}{2\pi(30 \times 10^{-12})} = 1.02 \text{ MHz} \tag{16}$$

which agrees closely with the measured values.

IV. Slew Rate and Some Special Limits

A. A General Limit on Slew Rate

If an op amp is overdriven by a large-signal pulse or square wave having a fast enough rise time, the output does not follow the input immediately. Instead, it ramps or "slews" at some limiting rate determined by internal currents and capacitances, as illustrated in Fig. 11. The magnitude of input voltage required to make the amplifier reach its maximum slew rate varies, depending on the type of input stage used. For an op amp with a

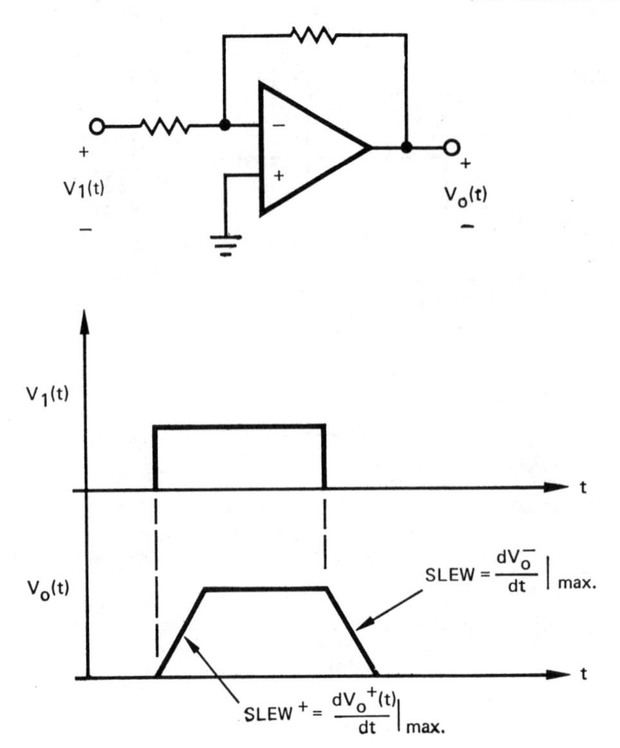

Fig. 11. Large signal "slewing" response observed if the input is overdriven.

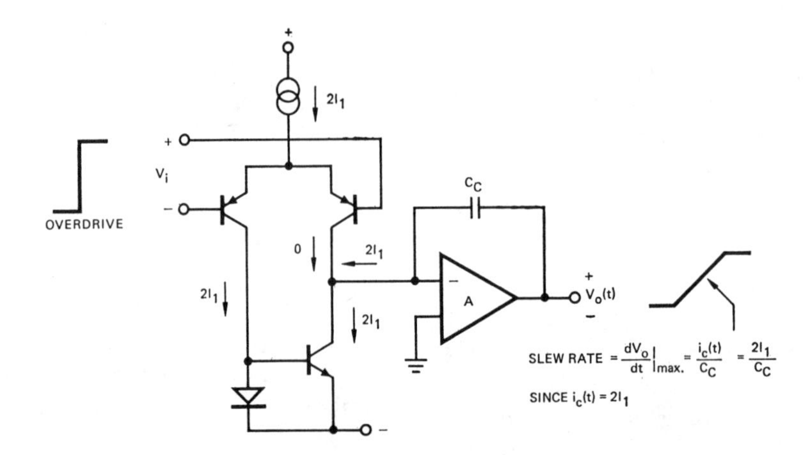

Fig. 12. Model used to calculate slew rate for the amp of Fig. 1 in the inverting mode. For simplicity, all transistor α's are assumed equal to unity, although results are essentially independent of α. An identical slew rate can be calculated for a negative-going output, obtained if the applied input polarity is reversed.

simple input differential amp, an input of about 60 mV will cause the output to slew at 90 percent of its maximum rate, while a μA741, which has half the input g_m, requires 120 mV. High speed amplifiers such as the LM 118 or a FET-input circuit require much greater overdrive, with 1–3 V being common. The reasons for these overdrive requirements will become clear below.

An adequate model to calculate slew limits for the representative op amp in the inverting mode is shown in Fig. 12, where the only important assumption made is that $I_2 \geq 2I_1$ in Fig. 1. This condition always holds in a well-designed op amp. (If one lets I_2 be less than $2I_1$, the slew is limited by I_2 rather than I_1, which results in lower speed than is otherwise possible.) Fig. 12 requires some modification for noninverting operation, and we will study this later.

The limiting slew rate is now calculated from Fig. 12. Letting the input voltage be large enough to fully switch the input differential amp, we see that all of the first stage tail current $2I_1$ is simply diverted into the integrator consisting of A and C_c. The resulting slew rate is then:

$$\text{slew rate} = \left.\frac{dv_0}{dt}\right|_{\max} = \frac{i_c(t)}{C_c}. \tag{17}$$

Noting that $i_c(t)$ is a constant $2I_1$, this becomes

$$\left.\frac{dv_0}{dt}\right|_{\max} = \frac{2I_1}{C_c}. \tag{18}$$

As a check of this result, recall that the μA741 has $I_1 = 10\ \mu$A and $C_1 = 30$ pF, so we calculate:

$$\left.\frac{dv_0}{dt}\right|_{\max} = \frac{2 \times 10^{-5}}{30 \times 10^{-12}} = 0.67\ \frac{\text{V}}{\mu\text{s}} \tag{19}$$

which agrees with measured values.

The large and small signal behavior of the op amp can be usefully related by combining (15) for ω_u with (18). The slew rate becomes

$$\left.\frac{dv_0}{dt}\right|_{\max} = \frac{2\omega_u I_1}{g_{m1}}. \tag{20}$$

Equation (20) is a general and very useful relationship. It shows that, for a given unity-gain frequency, ω_u, the slew rate is determined entirely by just the ratio of first stage operating current to first stage transconductance, I_1/g_{m1}. Recall that ω_u is set at the point where excess phase begins to build up, and this point is determined largely by technology rather than circuit limitations. Thus, the only effective means available to the circuit designer for increasing op amp slew rate is to *decrease* the ratio of first stage transconductance to operating current, g_{m1}/I_1.

B. Slew Limiting for Simple Bipolar Input Stage

The significance of (20) is best seen by considering the specific case of a simple differential bipolar input as in Fig. 1. For this circuit, the first stage transconductance (for $\alpha_1 = 1$) is[1]

$$g_{m1} = qI_1/kT \tag{21}$$

so that

$$\frac{g_{m1}}{I_1} = q/kT. \tag{22}$$

Using this in (20), the maximum bipolar slew rate is

$$\left.\frac{dv_0}{dt}\right|_{\max} = 2\omega_u \frac{kT}{q}. \tag{23}$$

This provides us with the general (and somewhat dismal) conclusion that slew rate in an op amp with a simple bipolar input stage is dependent only upon the unity gain corner and fundamental constants. Slew rate can be increased only by increasing the unity gain corner, which we have noted is generally difficult to do. As a demonstration of the severity of this limit, imagine an op amp using highly advanced technology and clever design, which might have a stable unity gain frequency of 100 MHz. Equation (23) predicts that the slew rate for this advanced device is only

$$\left.\frac{dv_0}{dt}\right|_{\max} = 33\ \frac{\text{V}}{\mu\text{s}} \tag{24}$$

which is good, but hardly impressive when compared with the difficulty of building a 100-MHz op amp.[2] But, there are some ways to get around this limit as we shall see shortly.

C. Power Bandwidth

Our intuition regarding slew rate will be enhanced somewhat if we relate it to a term called "power bandwidth." Power bandwidth is defined as the maximum frequency at which full output swing (usually 10 V peak) can be obtained without distortion. For a sinusoidal output voltage $v_0(t) = V_p \sin \omega t$, the rate of change of output, or slew rate, required to reproduce the output is

$$\frac{dv_0}{dt} = \omega V_p \cos \omega t. \tag{25}$$

This has a maximum when $\cos \omega t = 1$ giving

$$\left.\frac{dv_0}{dt}\right|_{\max} = \omega V_p, \tag{26}$$

so the highest frequency that can be reproduced without slew limiting, $\omega_{\max}$ (power bandwidth) is

$$\omega_{\max} = \frac{1}{V_p} \left.\frac{dv_0}{dt}\right|_{\max}. \tag{27}$$

Thus, power bandwidth and slew rate are directly related by the inverse of the peak of the sine wave V_p. Fig. 13 shows the severe distortion of the output sine wave which results if one attempts to amplify a sine wave of frequency $\omega > \omega_{\max}$.

Some numbers illustrate typical op amp limits. For a μA741 or LM101 having a maximum slew rate of 0.67 V/μs, (27) gives a maximum frequency for an undistorted 10-V peak output of

$$f_{\max} = \frac{\omega_{\max}}{2\pi} = 10.7\ \text{kHz}, \tag{28}$$

which is a quite modest frequency considering the much higher frequency small signal capabilities of these devices. Even the highly advanced 100-MHz amplifier considered above has a 10-V power bandwidth of only 0.5 MHz, so it is apparent that a need exists for finding ways to improve slew rate.

D. Techniques for Increasing Slew Rate

1) Resistive Enhancement of the Bipolar Stage: Equation (20) indicates that slew rate can be improved if we reduce first stage g_{m1}/I_1. One of the most effective ways

[1] Note that (21) applies only to the simple differential input stage of Fig. 12. For compound input stages as in the LM101 or μA741, g_{m1} is half that in (21), and the slew rate in (23) is doubled.

[2] We assume in all of these calculations that C_c is made large enough so that the amplifier has less than 180° phase lag at ω_u, thus making the amplifier stable for unity closed-loop gain. For higher gains one can of course reduce C_c (if the IC allows external compensation) and increase the slew rate according to (18).

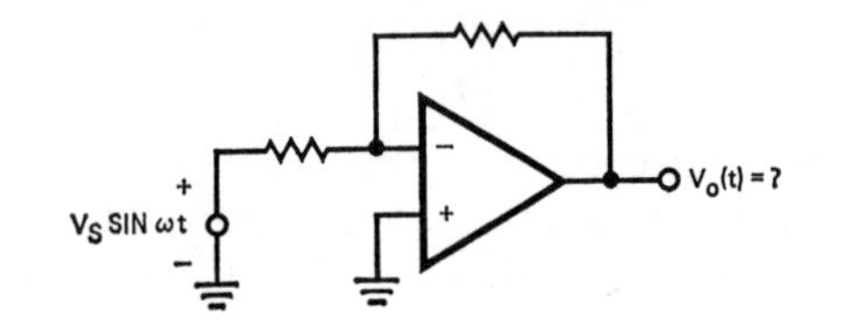

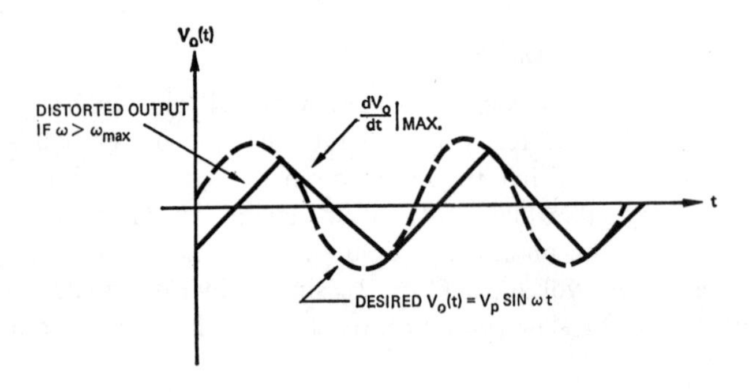

Fig. 13. Slew limiting effects on output sinewave that occur if frequency is greater than power bandwidth, ω_{max}. The onset of slew limiting occurs very suddenly as ω reaches ω_{max}. No distortion occurs below ω_{max}, while almost complete triangularization occurs at frequencies just slightly above ω_{max}.

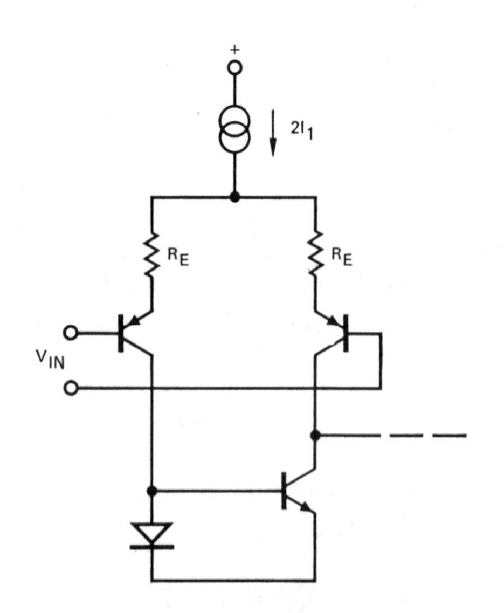

Fig. 14. Resistive degeneration used to provide slew rate enhancement according to (29).

of doing this is shown in Fig. 14, where simple resistive emitter degeneration has been added to the input differential amplifier [8]. With this change, the g_{m1}/I_1 drops to

$$\frac{g_{m1}}{I_1} = \frac{38.5}{1 + R_E I_1/26 \text{ mV}} \tag{29}$$

at 25°C.

The quantity g_{m1}/I_1 is seen to decrease rapidly with added R_E as soon as the voltage drop across R_E exceeds 26 mV. The LM118 is a good example of a bipolar amplifier which uses emitter degeneration to enhance slew rate [4]. This device uses emitter resistors to produce $R_E I_1 = 500$ mV, and has a unity gain corner of 16 MHz. Equations (20) and (29) then predict a maximum inverting slew rate of

$$\left.\frac{dv_0}{dt}\right|_{\max} = 2\omega_u \frac{I_1}{g_{m1}} = \omega_u = 100 \frac{\text{V}}{\mu\text{s}} \tag{30}$$

which is a twenty-fold improvement over a similar amplifier without emitter resistors.

A penalty is paid in using resistive slew enhancement, however. The two added emitter resistors must match extremely well or they add voltage offset and drift to the input. In the LM118, for example, the added emitter R's have values of 2.0 kΩ each and these contribute an input offset of 1 mV for each 4 Ω (0.2 percent) of mismatch. The thermal noise of the resistors also unavoidably degrades noise performance.

2) Slew Rate in the FET Input Op Amp: The FET (JFET or MOSFET) has a considerably lower transconductance than a bipolar device operating at the same current. While this is normally considered a drawback of the FET, we note that this "poor" behavior is in fact highly desirable in applications to fast amplifiers. To illustrate, the drain current for a JFET in the "current saturation" region can be approximated by

$$I_D \cong I_{DSS}(V_{GS}/V_T - 1)^2 \tag{31}$$

where

- I_{DSS} the drain current for $V_{GS} = 0$,
- V_{GS} the gate source voltage having positive polarity for forward gate-diode bias,
- V_T the threshold voltage having negative polarity for JFET's.

The small-signal transconductance is obtained from (31) as $g_m = \partial I_D/\partial V_G$. Dividing by I_D and simplifying, the ratio g_m/I_D for a JFET is

$$\frac{g_m}{I_D} \cong \frac{2}{(V_{GS} - V_T)} = \frac{2}{-V_T}\left[\frac{I_{DSS}}{I_D}\right]^{1/2}. \tag{32}$$

Maximum amplifier slew rate occurs for minimum g_m/I_D and, from (32), this occurs when I_D (or V_{GS}) is maximum. Normally it is impractical to forward bias the gate junction so a practical minimum occurs for (32) when $V_{GS} \cong 0$ V and $I_D \cong I_{DSS}$. Then

$$\left.\frac{g_m}{I_D}\right|_{\min} \cong -\frac{2}{V_T}. \tag{33}$$

Comparing (33) with the analogous bipolar expression, (22), we find from (20) that the JFET slew rate is greater than bipolar by the factor

$$\frac{\text{JFET slew}}{\text{bipolar slew}} \approx \frac{-V_T}{2kT/q} \frac{\omega_{uf}}{\omega_{ub}} \tag{34}$$

where ω_{uf} and ω_{ub} are unity-gain bandwidths for JFET and bipolar amps, respectively. Typical JFET thresholds are around 2 V ($V_T = -2$ V), so for equal bandwidths (34) tells us that a JFET-input op amp is about forty times faster than a simple bipolar input. Further, if

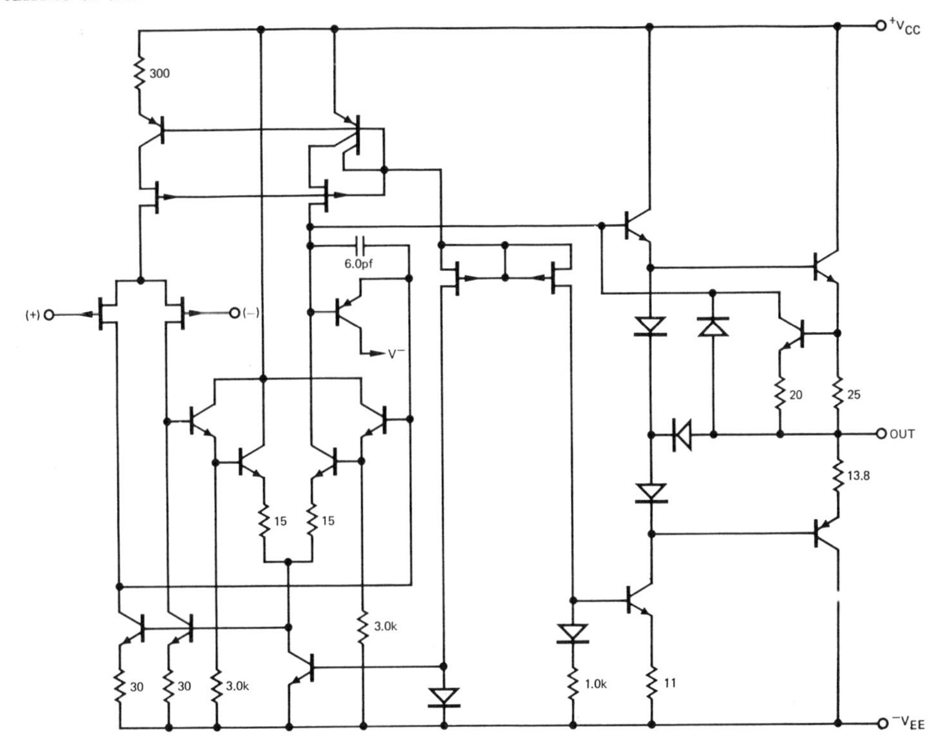

Fig. 15. Monolithic operational amplifier employing compatible p-channel JFET's on the same chip with normal bipolar components.

JFET's are properly substituted for the slow p-n-p's in a monolithic design, bandwidth improvements by at least a factor of ten are obtainable. JFET-input op amps, therefore, offer slew rate improvements by better than two orders of magnitude when compared with the conventional IC op amp. (Similar improvements are possible with MOSFET-input amplifiers.) This characteristic, coupled with picoamp input currents and reasonable offset and drift, make the JFET-input op amp a very desirable alternative to conventional bipolar designs.

As an example, Fig. 15, illustrates one design for an op amp employing compatible p-channel JFET's on the same chip with the normal bipolar components. This circuit exhibits a unity gain corner of 10 MHz, a 33 V/μs slew rate, an input current of 10 pA and an offset voltage and drift of 3 mV and 3 μV/°C [6]. Bandwidth and slew rate are thus improved over simple IC bipolar by factors of 10 and 100, respectively. At the same time input currents are smaller by about 10^3, and offset voltages and drifts are comparable to or better than slew enhanced bipolar circuits.

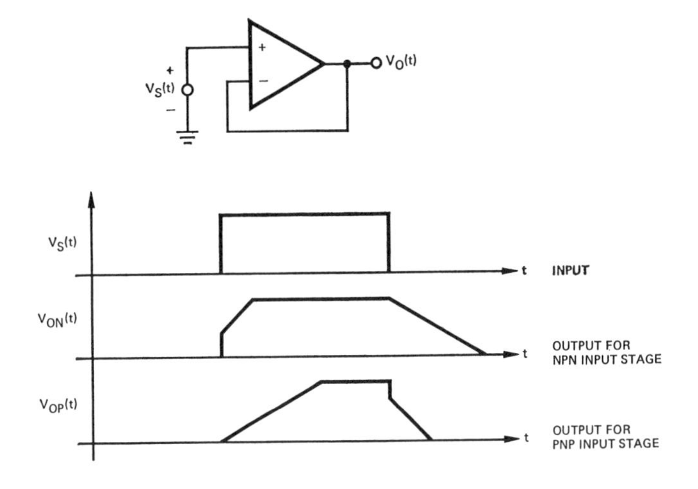

Fig. 16. Large signal response of the voltage follower. For an op amp with simple n-p-n input stage we get the waveform $v_{on}(t)$, which exhibits a step slew "enhancement" on the positive going output, and a slew "degradation" on the negative going output. For a p-n-p input stage, these effects are reversed as shown by $v_{op}(t)$.

V. Second-Order Effects: Voltage Follower Slew Behavior

If the op amp is operated in the noninverting mode and driven by a large fast rising input, the output exhibits the characteristic waveform in Fig. 16. As shown, this waveform does not have the simple symmetrical slew characteristic of the inverter. In one direction, the output has a fast step (slew "enhancement") followed by a "normal" inverter slewing response. In the other direction, it suffers a slew "degradation" or reduced slope when compared with the inverter slewing response.

We will first study slew degradation in the voltage follower connection, since this represents a worst case slewing condition for the op amp. A model which adequately represents the follower under large-signal conditions can be obtained from that in Fig. 12 by simply

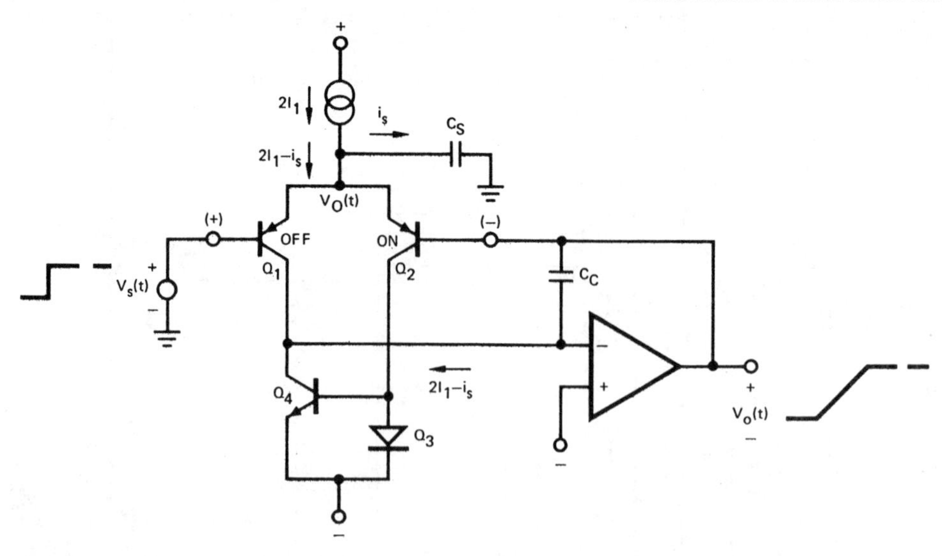

Fig. 17. Circuit used for calculation of slew "degradation" in the voltage follower. The degradation is caused by the capacitor C_s, which robs current from the tail, $2I_1$, thereby preventing the full $2I_1$ from slewing C_c.

tying the output to the inverting input, and including a capacitor C_s to account for the presence of any capacitance at the output of the first stage (tail) current source, see Fig. 17. This "input tail" capacitance is important in the voltage follower because the input stage undergoes rapid large-signal excursions in this connection, and the charging currents in C_s can be quite large.

Circuit behavior can be understood by analyzing Fig. 17 as follows. The large-signal input step causes Q_1 to turn OFF, leaving Q_2 to operate as an emitter follower with its emitter tracking the variational output voltage, $v_O(t)$. It is seen that $v_O(t)$ is essentially the voltage appearing across both C_s and C_c so we can write

$$\frac{dv_O}{dt} \cong \frac{i_c}{C_c} \cong \frac{i_s}{C_s}. \tag{35}$$

Noting that $i_c \cong 2I_1 - i_s$ (unity α's assumed), (35) can be solved for i_s:

$$i_s \cong \frac{2I_1}{1 + C_c/C_s} \tag{36}$$

which is seen to be constant with time. The degraded voltage follower slew rate is then obtained by substituting (36) into (35):

$$\left.\frac{dv_O}{dt}\right|_{\text{degr}} \cong \frac{i_s}{C_s} \cong \frac{2I_1}{C_c + C_s}. \tag{37}$$

Comparing (37) with the slew rate for the inverter, (18), it is seen that the slew rate is reduced by the simple factor $1/(1 + C_s/C_c)$. As long as the input tail capacitance C_s is small compared with the compensation capacitor C_c, little degradation occurs. In high speed amplifiers where C_c is small, degradation becomes quite noticeable, and one is encouraged to develop circuits with small C_s.

As an example, consider the relatively fast LM118 which has $C_c \cong 5$ pF, $C_s \cong 2$ pF, $2I_1 = 500\ \mu$A. The calculated inverter slew rate is $2I_1/C_c \cong 100$ V/μs, and the degraded voltage follower slew rate is found to be $2I_1/(C_c + C_s) \cong 70$ V/μs. The slew degradation is seen to be about 30 percent, which is very significant. By contrast, a μA741 has $C_c \cong 30$ pF and $C_s \cong 4$ pF which results in a degradation of less than 12 percent.

The slew "enhanced" waveform can be similarly predicted from a simplified model. By reversing the polarity of the input and initially assuming a finite slope on the input step, the enhanced follower is analyzed, as shown in Fig. 18. Noting that Q_1 is assumed to be turned ON by the step input and Q_2 is OFF, the output voltage becomes

$$v_O(t) \cong -\frac{1}{C_c}\int_0^t [2I_1 + i_S(t)]\, dt. \tag{38}$$

The voltage at the emitter of Q_1 is essentially the same as the input voltage, $v_i(t)$, so the current in the "tail" capacitance C_s is

$$i_s(t) \cong C_s \frac{dv_i}{dt} \cong \frac{C_s V_{ip}}{t_1} \qquad 0 < t < t_1. \tag{39}$$

Combining (38) and (39), $v_O(t)$ is

$$-v_O(t) \cong \frac{1}{C_c}\int_0^t 2I_1\, dt + \frac{1}{C_c}\int_0^{t_1} \frac{C_s V_{ip}}{t_1}\, dt \tag{40}$$

or

$$-v_O(t) \cong \frac{C_s}{C_c} V_{ip} + \frac{2I_1 t}{C_c}. \tag{41}$$

Equation (41) tells us that the output has an initial negative step which is the fraction C_s/C_c of the input voltage. This is followed by a normal slewing response, in which the slew rate is identical to that of the inverter, see (18). This response is illustrated in Fig. 18.

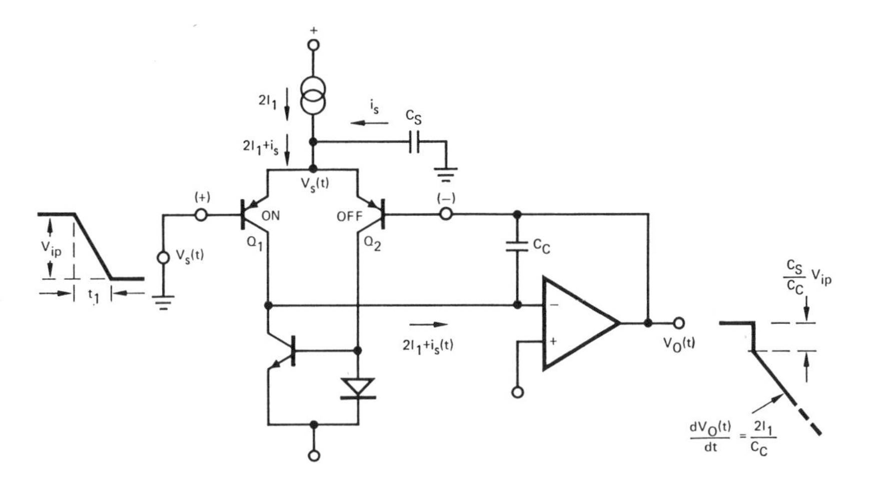

Fig. 18. Circuit used for calculation of slew "enhancement" in the voltage follower. The fast falling input causes a step output followed by a normal slew response as shown.

VI. Limitations on Bandwidth

In earlier sections, all bandlimiting effects were ignored except that of the compensation capacitor, C_c. The unity-gain frequency was set at a point sufficiently low so that negligible excess phase (over the 90° from the dominant pole) due to second-order (high frequency) poles had built up. In this section the major second-order poles which contribute to bandlimiting in the op amp are identified.

A. The Input Stage: p-n-p's, the Mirror Pole, and the Tail Pole

For many years it was popular to identify the lateral p-n-p's (which have f_t's $\cong$ 3 MHz) as the single dominant source of bandlimiting in the IC op amp. It is quite true that the p-n-p's do contribute significant excess phase to the amplifier, but it is not true that they are the sole contributor to excess phase [9]. In the input stage, alone, there is at least one other important pole, as illustrated in Fig. 19(a). For the simple differential input stage driving a differential-to-single ended converter ("mirror" circuit), it is seen that the inverting signal (which is the feedback signal) follows two paths, one of which passes through the capacitance C_s, and the other through C_m. These capacitances combine with the dynamic resistances at their nodes to form poles designated the mirror pole at

$$p_m \cong \frac{I_1}{C_m kT/q}, \tag{42}$$

and the tail pole at

$$p_t \cong \frac{2I_1}{C_s kT/q}. \tag{43}$$

It can be seen that if one attempts to operate the first stage at too low a current, these poles will bandlimit the amplifier. If, for example, we choose $I_1 = 1\ \mu$A, and assume $C_m \cong 7$ pF (consisting of 4-pF isolation ca-

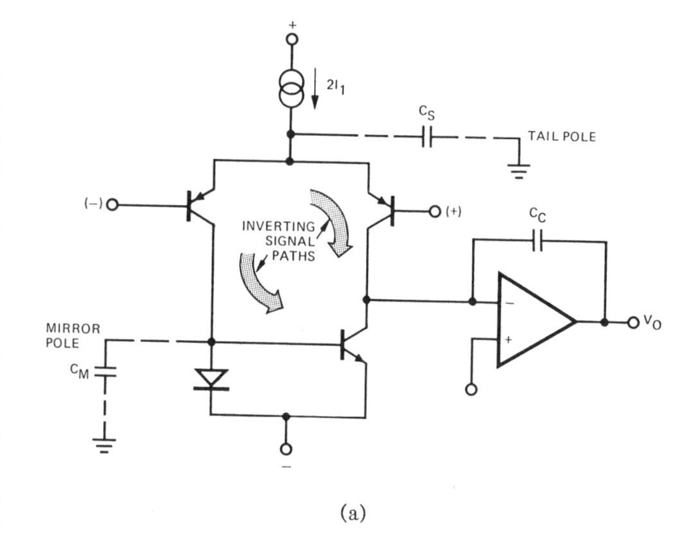

(a)

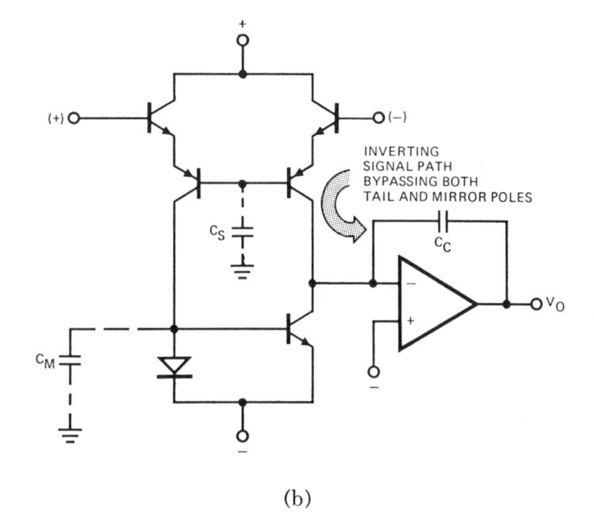

(b)

Fig. 19. (a) Circuit showing "mirror" pole due to C_m and "tail" pole due to C_s. One component of the signal due to an inverting input must pass through either the mirror or tail poles. (b) Alternate circuit to Fig. 19(a) (LM101, μA741) which has less excess phase. Reason is that half the inverting signal path need not pass through the mirror pole or the tail pole.

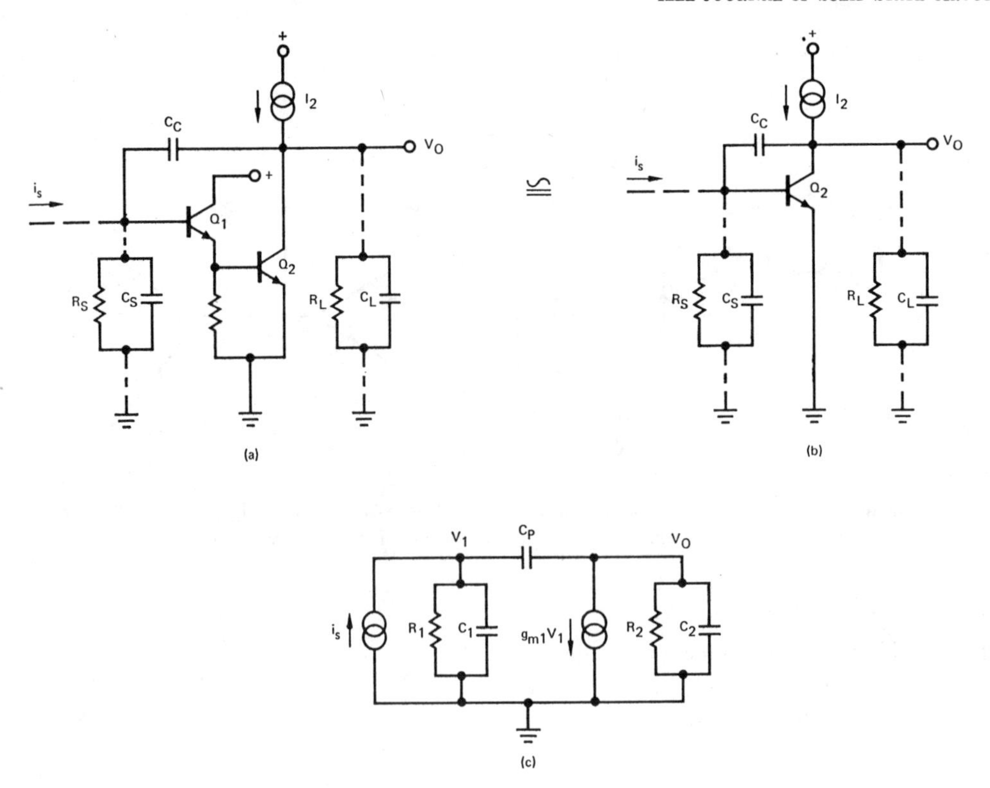

Fig. 20. Simplification of second stage used for pole-splitting analysis. (a) Complete second stage with input stage and output stage loading represented by R_s, C_s, and R_L, C_L, respectively. (b) Emitter follower ignored to simplify analysis. (c) Hybrid π model substituted for transistor in (b). Source and load impedances are absorbed into model with the total impedances represented by R_1, C_1, and R_2, C_2. Transistor base resistance is ignored and C_p includes both C_c and transistor collector-base capacitance.

pacitance and 3-pF emitter transition capacitance) and $C_s \cong 4$ pF,[3] $p_m/2\pi \cong 0.9$ MHz and $p_t/2\pi \cong 3$ MHz either of which would seriously degrade the phase margin of a 1-MHz amplifier.

If a design is chosen in which either the tail pole or the mirror pole is absent (or unimportant), the remaining pole rolls off only half the signal, so the overall response contains a pole-zero pair separated by one octave. Such a pair generally has a small effect on amplifier response unless it occurs near ω_u, where it can degrade phase margin by as much as 20°.

It is interesting to note that the compound input stage of the classical LM101 (and μA741) has a distinct advantage over the simple differential stage, as seen in Fig. 19(b). This circuit is noninverting across each half, thus it provides a path in which half the feedback signal bypasses both the mirror and tail poles.

[3] C_s can have a wide range of values depending on circuit configuration. It is largest for n-p-n input differential amps since the current source has a collector-substrate capacitance ($C_s \cong$ 3–4 pF) at its output. For p-n-p input stages it can be as small as 1–2 pF.

B. The Second Stage: Pole Splitting

The assumption was made in Section III that the second stage behaved as an ideal integrator having a single dominant pole response. In practice, one must take care in designing the second stage or second-order poles can cause significant deviation from the expected response. Considerable insight into the basic way in which the second stage operates can be obtained by performing a small-signal analysis on a simplified version of the circuit as shown in Fig. 20 [10]. A straightforward two-node analysis of Fig. 20(c) produces the following expression for v_{out}.

$$\frac{v_{out}}{i_s} = \frac{-g_m R_1 R_2 (1 - sC_p/g_m)}{1 + s[R_1(C_1 + C_p) + R_2(C_2 + C_p) + g_m R_1 R_2 C_p] + s^2 R_1 R_2 [C_1 C_2 + C_p(C_1 + C_2)]}. \tag{44}$$

The denominator of (44) can be approximately factored under conditions that its two poles are widely separated. Fortunately, the poles are, in fact, widely separated under most normal operating conditions. Therefore, one can assume that the denominator of (44) has the form

$$D(s) = (1 + s/p_1)(1 + s/p_2)$$
$$= 1 + s(1/p_1 + 1/p_2) + s^2/p_1 p_2. \tag{45}$$

With the assumption that p_1 is the dominant pole and

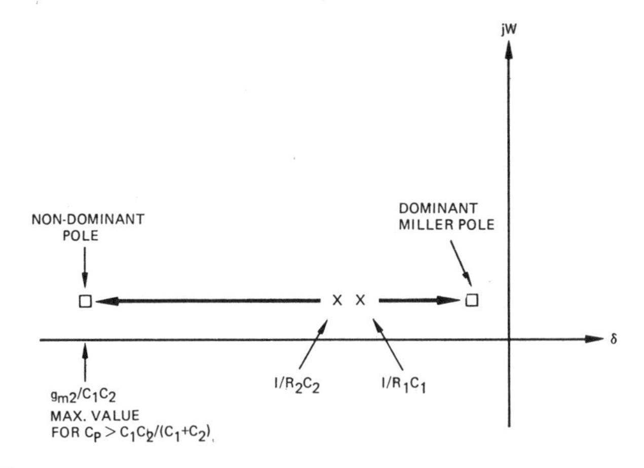

Fig. 21. Pole migration for second stage employing "pole-splitting" compensation. Plot is shown for increasing C_p and it is noted that the nondominant pole reaches a maximum value for large C_p.

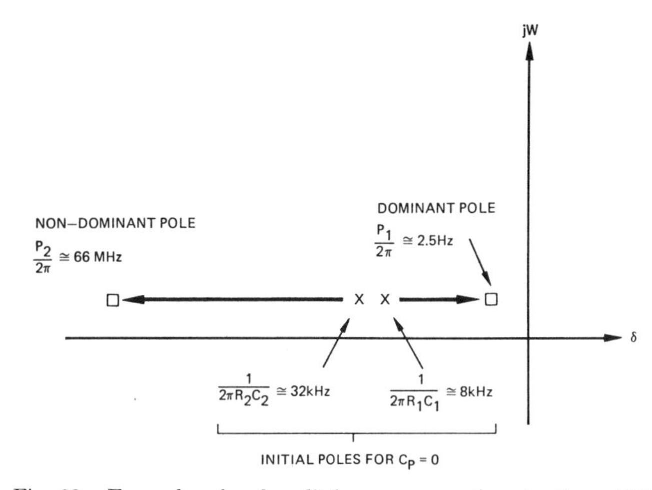

Fig. 22. Example of pole-splitting compensation in the μA741 op amp. Values used in (48) and (49) are: $g_{m2} = 1/87\ \Omega$, $C_p = 30$ pF, $C_1 \cong C_2 = 10$ pF, $R_1 = 1.7$ MΩ, $R_2 = 100$ kΩ.

p_2 is nondominant, i.e., $p_1 \ll p_2$, (45) becomes

$$D(s) \cong 1 + s/p_1 + s^2/p_1p_2. \tag{46}$$

Equating coefficients of s in (44) and (46), the dominant pole p_1 is found directly:

$$p_1 \cong \frac{1}{R_1(C_1 + C_p) + R_2(C_2 + C_p) + g_mR_1R_2C_p} \tag{47}$$

$$\cong \frac{1}{g_mR_1R_2C_p}. \tag{48}$$

The latter approximation, (48), normally introduces little error, because the g_m term is much larger than the other two. We note at this point that p_1, which represents the dominant pole of the amplifier, is due simply to the familiar Miller-multiplied feedback capacitance $g_mR_2C_p$ combined with input node resistance, R_1. The nondominant pole p_2 is found similarly by equating s^2 coefficients in (44) and (46) to get p_1p_2, and dividing by p_1 from (48). The result is

$$p_2 \cong \frac{g_mC_p}{C_1C_2 + C_p(C_1 + C_2)}. \tag{49}$$

Several interesting things can be seen in examining (48) and (49). First, we note that p_1 is inversely proportional to g_m (and C_p), while p_2 is directly dependent on g_m (and C_p). Thus, as either C_p or transistor gain are increased, the dominant pole decreases and the nondominant pole increases. The poles p_1 and p_2 are being "split-apart" by the increased coupling action in a kind of inverse root locus plot.

This *pole-splitting* action is shown in Fig. 21, where pole migration is plotted for C_p increasing from 0 to a large value. Fig. 22 further illustrates the action by giving specific pole positions for the μA741 op amp. It is seen that the initial poles (for $C_p = 0$) are both in the tens of kilohertz region and these are predicted to reach 2.5 Hz ($p_1/2\pi$) and 66 MHz ($p_2/2\pi$) after compensation is applied. This result is, of course, highly satisfactory since the second stage now has a single dominant pole effective over a wide frequency band.

C. *Failure of Pole Splitting*

There are several situations in which the application of pole-splitting compensation may not result in a single dominant pole response. One common case occurs in very wide-band op amps where the pole-splitting capacitor is small. In this situation the nondominant pole given by (49) may not become broadbanded sufficiently so that it can be ignored. To illustrate, suppose we attempt to minimize power dissipation by running the second stage of an LM118 (which has a small-signal bandwidth of 16 MHz) at 0.1 mA. For this op amp $C_p = 5$ pF, $C_1 \cong C_2 \cong 10$ pF. From (49), the nondominant pole is

$$\frac{p_2}{2\pi} \cong 16 \text{ MHz} \tag{50}$$

which lies right at the unity-gain frequency. This pole alone would degrade phase margin by 45°, so it is clear that we need to bias the second stage with a collector current greater than 0.1 mA to obtain adequate g_m. Insufficient pole-splitting can therefore occur; but the cure is usually a simple increase in second stage g_m.

A second type of pole-splitting failure can occur, and it is often much more difficult to cope with. If, for example, one gets over-zealous in his attempt to broadband the nondominant pole, he soon discovers that other poles exist within the second stage which can cause difficulties. Consider a more exact equivalent circuit for the second stage of Fig. 20(a) as shown in Fig. 23. If the follower is biased at low currents or if C_p, Q_2 g_m, and/or r_x are high, the circuit can contain at least four important poles rather than the two considered in simple pole splitting. Under these conditions, we no longer have a response with just negative real poles as in Fig. 21, but observe a root locus of the sort shown in Fig. 24. It is seen in this case that the circuit contains a pair of com-

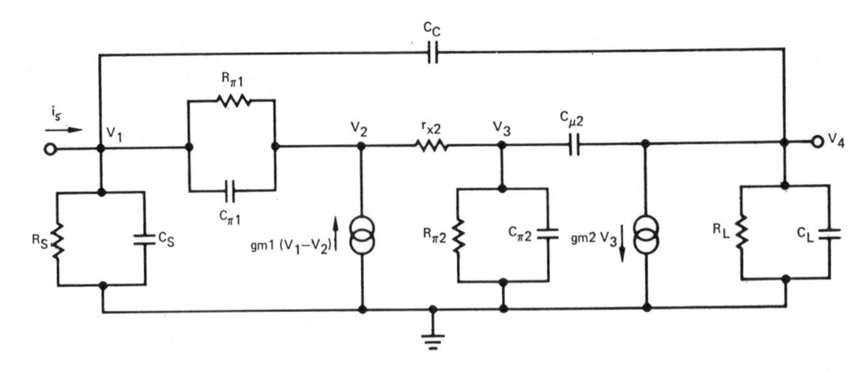

Fig. 23. More exact equivalent circuit for second stage of Fig. 20 (a) including a simplified π model for the emitter follower ($R\pi_1$, $C\pi_1$, g_{m1}) and a complete π for Q_2 (r_{x2}, $R\pi_2$, etc.).

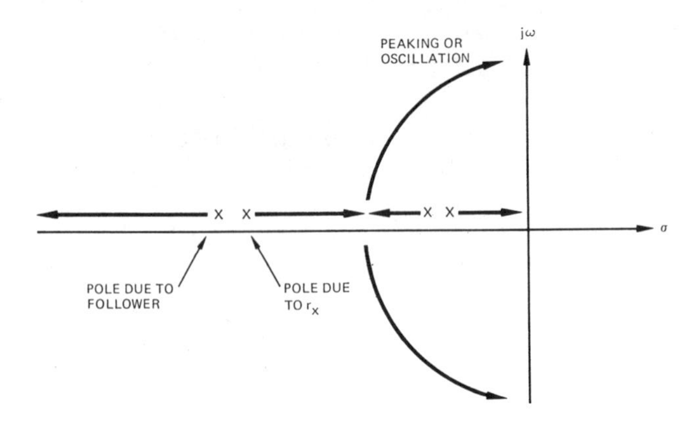

Fig. 24. Root locus for second stage illustrating failure of pole splitting due to high g_{m2}, r_{x2}, C_p, and/or low bias current in the emitter follower.

plex, possibly underdamped poles which, of course, can cause peaking or even oscillation. This effect occurs so commonly in the development of wide-band pole-split amplifiers that it has been (not fondly) dubbed "the second stage bump."

There are numerous ways to eliminate the "bump," but no single cure has been found which is effective in all situations. A direct hand analysis of Fig. 23 is possible, but the results are difficult to interpret. Computer analysis seems the best approach for this level of complexity, and numerous specific analyses have been made. The following is a list of circuit modifications that have been found effective in reducing the bump in the various studies: 1) reduce g_{m2}, r_{x2}, $C_{\mu 2}$, 2) add capacitance or a series RC network from the stage input to ground—this reduces the high frequency local feedback due to C_p, 3) pad capacitance at the output for similar reasons, 4) increase operating current of the follower, 5) reduce C_p, 6) use a higher f_t process.

D. Troubles in The Output Stage

Of all the circuitry in the modern IC op amp, the class-AB output stage probably remains the most troublesome. None of the stages in use today behave as well as one might desire when stressed under worst case conditions. To illustrate, one of the most commonly used output stages is shown in Fig. 2(b). The p-n-p's in this circuit are "substrate" p-n-p's having low current f_t's of around 20 MHz. Unfortunately, both β_0 and f_t begin to fall off rapidly at quite low current densities, so as one begins to sink just a few milliamps in the circuit, phase margin troubles can develop. The worst effect occurs when the amplifier is operated with a large capacitive load (>100 pF) while sinking high currents. As shown in Fig. 25, the load capacitance on the output follower causes it to have negative input conductance, while the driver follower can have an inductive output impedance. These elements combine with the capacitance at the interstage to generate the equivalent of a one-port oscillator. In a carefully designed circuit, oscillation is suppressed, but peaking (the "output bump") can occur in most amplifiers under appropriate conditions.

One new type of output circuit which does not use p-n-p's is shown in Fig. 26 [6]. This circuit employs compatible JFET's (or MOSFET's, see similar circuit in [11]) in a FET/bipolar quasi-complimentary output stage, which is insensitive to load capacitance. Unfortunately, this circuit is rather complex and employs extra process steps, so it does not appear to represent the cure for the very low cost op amps.

VII. The Gain Cell: Linear Large-Scale Integration

As the true limitations of the basic op amp are more fully understood, this knowledge can be applied to the development of more "optimum" amplifiers. There are, of course, many ways in which one might choose to optimize the device. We might, for example, attempt to maximize speed (bandwidth, slew rate, settling time) without sacrificing dc characteristics. The compatible JFET/bipolar amp of Fig. 15 represents such an effort. An alternate choice might be to design an amplifier having all of the performance features of the most widely used general purpose op amps (i.e., μA741, LM107, etc.), but having minimum possible die area. Such a pursuit is parallel to the efforts of digital large-scale integration (LSI) designers in their development of minimum area

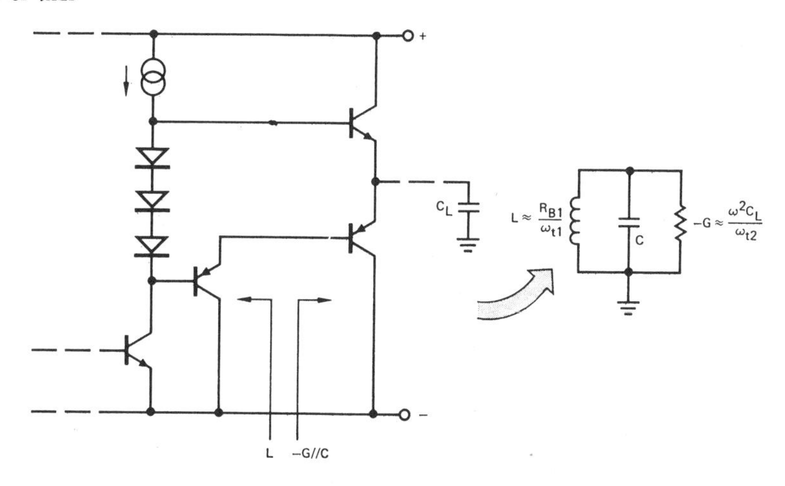

Fig. 25. Troubles in the conventional class-AB output stage of Fig. 2(b). The low f_t output p-n-p's interact with load capacitance to form the equivalent of a one-port oscillator.

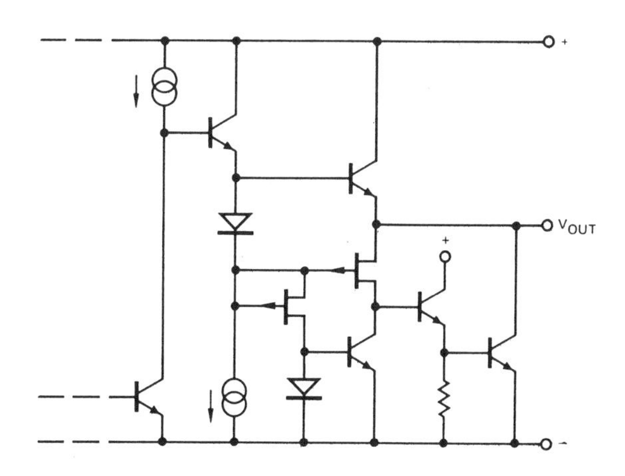

Fig. 26. The "BIFET" output stage employing JFET's and bipolar n-p-n's to eliminate sensitivity to load capacitance.

memory cells or gates. The object of such efforts, of course, is to develop lower cost devices which allow wide and highly economic usage.

In this section we briefly discuss certain aspects of the linear *gain cell,* a general purpose, internally compensated op amp having a die area which is significantly smaller than that of equivalent, present day, industry standard amplifiers.

A. Transconductance Reduction

The single largest area component in the internally compensated op amp is the compensation capacitor (about 30 pF, typically). A major interest in reducing amplifier die area, therefore, centers about finding ways in which this capacitor can be reduced in size. With this in mind, we find it useful to examine (15), which relates compensation capacitor size to two other parameters, unity gain corner frequency ω_u, and first stage transconductance g_{m1}. It is immediately apparent that for a fixed, predetermined unity gain corner (about $2\pi \times$ 1 MHz in our case), there is only one change that can be made to reduce the size of C_c: *the transconductance of the first stage must be reduced.* If we restrict our interest to simple bipolar input stages (for low cost), we recall the $g_{m1} = qI_1/kT$. Only by reducing I_1 can g_{m1} be reduced, and we earlier found in Section VI-A and Fig. 19(a) and (b) that I_1 cannot be reduced much without causing phase margin difficulties due to the mirror pole and the tail pole.

An alternate basic approach to g_m reduction is illustrated in Fig. 27 [12]. Here, a multiple collector p-n-p structure, which is easily fabricated in IC form, is used to split the collector current into two components, one component (the larger) of which is simply tied to ground, thereby "throwing away" a major portion of the transistor output current. The result is that the g_m of the transistor is reduced by the ratio of $1/(1 + n)$ (see Fig. 27), and the compensation capacitance can be reduced directly by the same factor. It might appear that the mirror pole would still cause difficulties since the current mirror becomes current starved in Fig. 27, but the effect is not as severe as might be expected. The

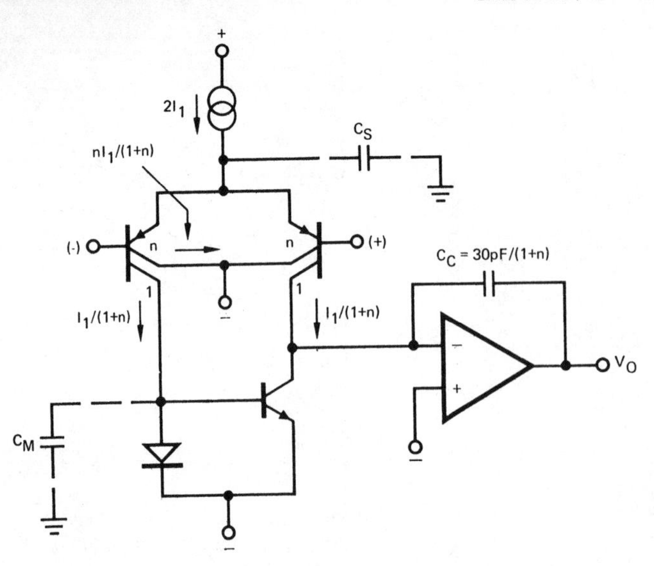

Fig. 27. Basic g_m reduction obtained by using split collector p-n-p's. C_c and area are reduced since $C_c = g_{m1}/\omega_u$.

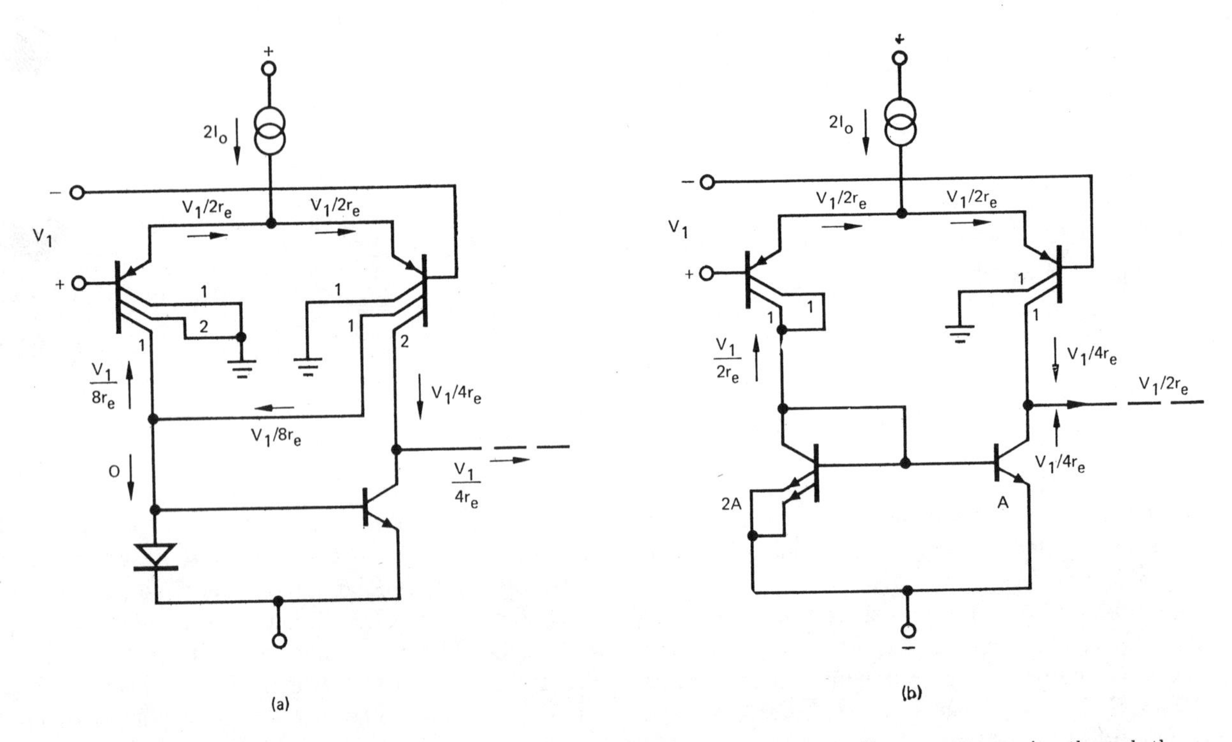

Fig. 28. Variations on g_m reduction. (a) Cross-coupled connection eliminates all ac current passing through the mirror, yet maintains dc balance. (b) This approach maintains high current on the diode side of the mirror, thereby broadbanding the mirror pole.

reason is that the inverting signal can now pass through the high current wide-band path, across the differential amp emitters and into the second stage, so at least half the signal current does not become bandlimited. This partial bandlimiting can be further reduced by using one of the circuits in Fig. 28(a) or (b).[4] In (a), the p-n-p collectors are cross coupled in such a way that the ac signal is cancelled in the mirror circuit, while dc remains completely balanced. Thus the mirror pole is virtually eliminated. The circuit does have a drawback, however, in that the uncorrellated noise currents coming from the two p-n-p's add rather than subtract at the input to the mirror, thereby degrading noise performance. The circuit in Fig. 28(b) does not have this defect, but requires care in matching p-n-p collector ratios to n-p-n

[4] The circuit in Fig. 28(a) is due to R. W. Russell and the variation in Fig. 28(b) was developed by D. W. Zobel.

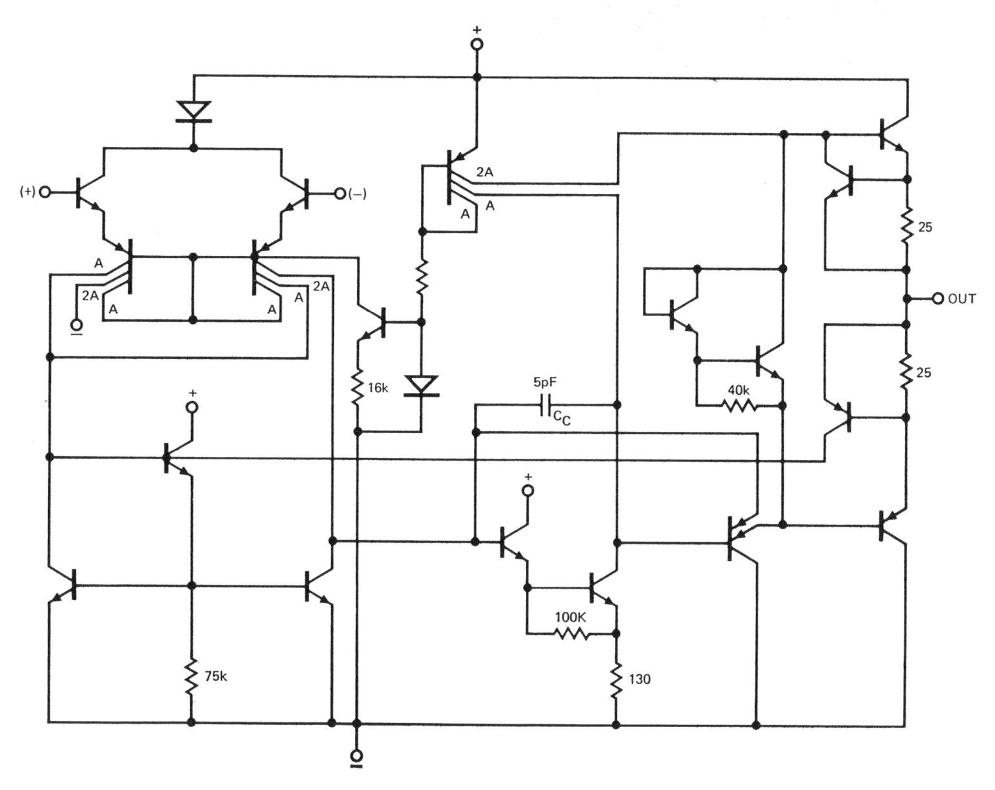

Fig. 29. Circuit for optimized gain cell which has been fabricated in one-fourth the die size of the equivalent μA741.

emitter areas. Otherwise offset and drift will degrade as one attempts to reduce g_m by large factors.

B. A Gain Cell Example

As one tries to make large reductions in die area for the gain cell, many factors must be considered in addition to novel circuit approaches. Of great importance are special layout/circuit techniques which combine a maximum number of components into minimum area.

In a good layout, for example, all resistors are combined into islands with transistors. If this is not possible initially, circuit and device changes are made to allow it. The resulting device geometries within the islands are further modified in shape to allow maximum "packing" of the islands. That is, when the layout is complete, the islands should have shapes which fit together as in a picture puzzle, with no waste of space. Further area reductions can be had by modifying the isolation process to one having minimum spacing between the isolation diffusion and adjacent p-regions.

An example of a gain cell which employs both circuit and layout optimization is shown in Fig. 29. This circuit uses the g_m reduction technique of Fig. 28(a) which results in a compensation capacitor size of only 5 pF rather than the normal 30 pF. The device achieves a full 1-MHz bandwidth, a 0.67-V/μs slew rate, a gain greater than 100 000, typical offset voltages less than 1 mV, and other characteristics normally associated with an LM107 or μA741. In quad form each amplifier requires an area of only 23 × 35 mils which is one-fourth the size of today's industry standard μA741 (typically 56 × 56 mils). This allows over 8000 possible gain cells to be fabricated on a single 3-inch wafer. Further, it appears quite feasible to fabricate larger arrays of gain cells, with six or eight on a single chip. Only packaging and applications questions need be resolved before pursuing such a step.

Acknowledgment

Many important contributions were made in the gain cell and FET/bipolar op amp areas by R. W. Russell. The author gratefully acknowledges his very competent efforts.

References

[1] R. J. Widlar, "Monolithic op amp with simplified frequency compensation," *EEE*, vol. 15, pp. 58–63, July 1967. (Note that the LM 101, designed in 1967, by R. J. Widlar was the first op amp to employ what has become the classical topology of Fig. 1.)

[2] D. Fullagar, "A new high performance monolithic operational amplifier," Fairchild Semiconductor Tech. Paper, 1968.

[3] R. W. Russell and T. M. Frederiksen, "Automotive and industrial electronic building blocks," *IEEE J. Solid-State Circuits*, vol. SC-7, pp. 446–454, Dec. 1972.

[4] R. C. Dobkin, "LM118 op amp slews 70 V/μs," *Linear Applications Handbook*, National Semiconductor, Santa Clara, Calif., 1974.

[5] R. J. Apfel and P. R. Gray, "A monolithic fast settling feed-forward op amp using doublet compression techniques," in *ISSCC Dig. Tech. Papers*, 1974, pp. 134–155.

[6] R. W. Russell and D. D. Culmer, "Ion implanted JFET-bipolar monolithic analog circuits," in *ISSCC Dig. Tech. Papers*, 1974, pp. 140–141.

[7] P. R. Gray, "A 15-W monolithic power operational amplifier," *IEEE J. Solid-State Circuits,* vol. SC-7, pp. 474-480, Dec. 1972.

[8] J. E. Solomon, W. R. Davis, and P. L. Lee, "A self compensated monolithic op amp with low input current and high slew rate," in *ISSCC Dig. Tech. Papers,* 1969, pp. 14-15.

[9] B. A. Wooley, S. Y. J. Wong, D. O. Pederson, "A computor-aided evaluation of the 741 amplifier," *IEEE J. Solid-State Circuits,* vol. SC-6, pp. 357-366, Dec. 1971.

[10] J. E. Solomon and G. R. Wilson, "A highly desensitized, wide-band monolithic amplifier," *IEEE J. Solid-State Circuits,* vol. SC-1, pp. 19-28, Sept. 1966.

[11] K. R. Stafford, R. A. Blanchard, and P. R. Gray, "A completely monolithic sample/hold amplifier using compatible bipolar and silicon gate FET devices," in *ISSCC Dig. Tech. Papers,* 1974, pp. 190-191.

[12] J. E. Solomon and R. W. Russell, "Transconductance reduction using multiple collector PNP transistors in an operational amplifier," U.S. Patent 3801923, Mar. 1974.

See also, as a general reference:

[13] P. R. Gray and R. G. Meyer, "Recent advances in monolithic operational amplifier design," *IEEE Trans. Circuits and Syst.,* vol. CAS-21, pp. 317-327, May 1974.

Design Techniques for Monolithic Operational Amplifiers

ROBERT J. WIDLAR

Abstract—**The characteristics of recently developed integrated-circuit components will be reviewed and some new devices will be described. Their impact on the design of monolithic operational amplifiers will also be discussed. Emphasis will be placed on realizing particularly good dc characteristics—especially low input current. However, techniques for obtaining higher operating speeds will also be covered.**

Introduction

NEARLY one million monolithic operational amplifiers per month are now being sold. Prices have dropped from $70 in 1965 to less than $2 today, even for relatively small quantities. In fact, the cost is low enough that these devices are used as simple components in applications where operational amplifiers would not even have been considered a couple of years ago.

The 709 [1] was the first monolithic circuit that approached discrete designs in general performance and usefulness, yet could be manufactured with high yields in volume production. It was also the first to be second sourced, and today there are eight major suppliers.

The LM101 [2],[1] announced in 1967, was the next major advance in monolithic amplifiers. It has essentially the same specifications as the 709, but it eliminates most of the application problems. The LM101 is not susceptible to latch-up when the common-mode range is exceeded, the inputs and output are protected from overloads, it operates over a wider range of supply voltages, and it is much less prone to oscillations. In addition, frequency compensation can be accomplished with a single 30-pF capacitor—a value that has been included on the silicon chip in the μA741 [3], the RM4101 and the LM107.[2] Hence, with this design it is practical to offer a fully compensated monolithic amplifier.

With the LM101A [4], brought out late in 1968, the input current specifications of discrete amplifiers were finally matched. This device offers input bias currents less than 100 nA and input offset currents less than 20 nA—guaranteed over a −55 to 125°C temperature range. This means that even FET-input amplifiers can be replaced with low-cost monolithic circuits, realizing improved performance in full-temperature-range applications.

The technology is available to make another order of magnitude improvement in input currents, making monolithics competitive with chopper stabilized amplifiers at high temperatures. This leaves only one major area where a significant performance improvement is needed: higher operating speeds. Present general-purpose amplifiers have slew rates of 0.5 V/μs and bandwidths of 1 MHz.[3] Many applications require more than ten times better performance than this.

Impact of New Components

Compared to discrete amplifiers of its time, even the 709 was a weird design, reflecting differences in the relative cost of components in an integrated circuit. This can be seen by comparing a typical discrete component design, shown in Fig. 1, with a 709 schematic in Fig. 2. The apparent complexity of the circuit was increased to minimize total circuit resistance, to make up for the poor gain characteristics of some transistors, and to reduce power consumption to a value that can be handled by integrated-circuit packages. For example, transistors were substituted for large-value resistors, where possible; buffer transistors were added to make the circuit operation insensitive to production variations in transistor current gains; and a class-B output stage was used to reduce quiescent current. In many cases, the integrated circuit gave significantly improved performance. And this was done without paying a price penalty, since the monolithic amplifier is now available at less than one-fifth the price of the discrete version.

Since the 709 was designed in the early days of linear circuits, it did not make use of components that were much different from those employed in digital circuits at that time. The singular exception was the lateral p-n-p [5]. But, the circuit was designed so that it could function properly with p-n-p current gains less than 0.1. Hence, the design was not too adventuresome, as should be the case when working with an emerging technology. The LM101A, on the other hand, makes extensive use of lateral p-n-p transistors, pinch resistors [6], and collector FET's.[4] Judging by the schematic diagram, it is a much more complicated circuit than the 709, as can be seen from Fig. 3. However, schematics are de-

Manuscript received April 2, 1969; revised June 11, 1969.
The author is with National Semiconductor Corporation, Santa Clara, Calif. 95051.

[1] The LM101 is similar to the LM101A described later in the text.

[2] The RM4101 and the LM107 are compensated versions of the LM101 and the LM101A, respectively.

[3] Slew rates around 10 V/μs and bandwidths of 10 MHz can be realized with certain amplifiers in specialized applications using appropriate compensation networks.

[4] To be described later.

Reprinted from *IEEE J. Solid-State Circuits,* vol. SC-4, pp. 184-191, Aug. 1969.

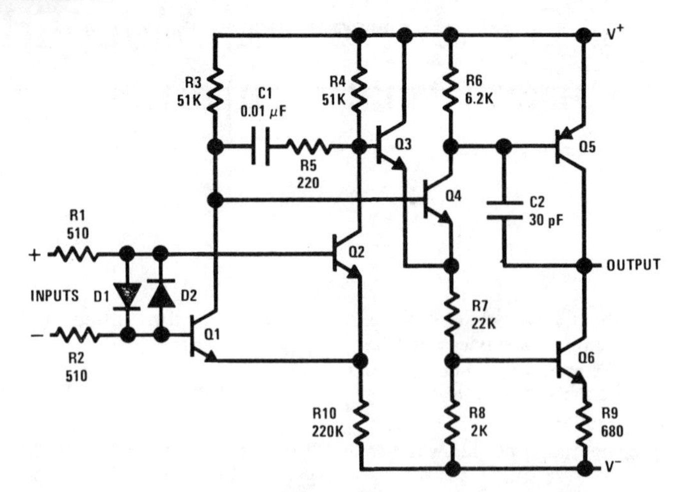

Fig. 1. Discrete component design for a low-cost operational amplifier.

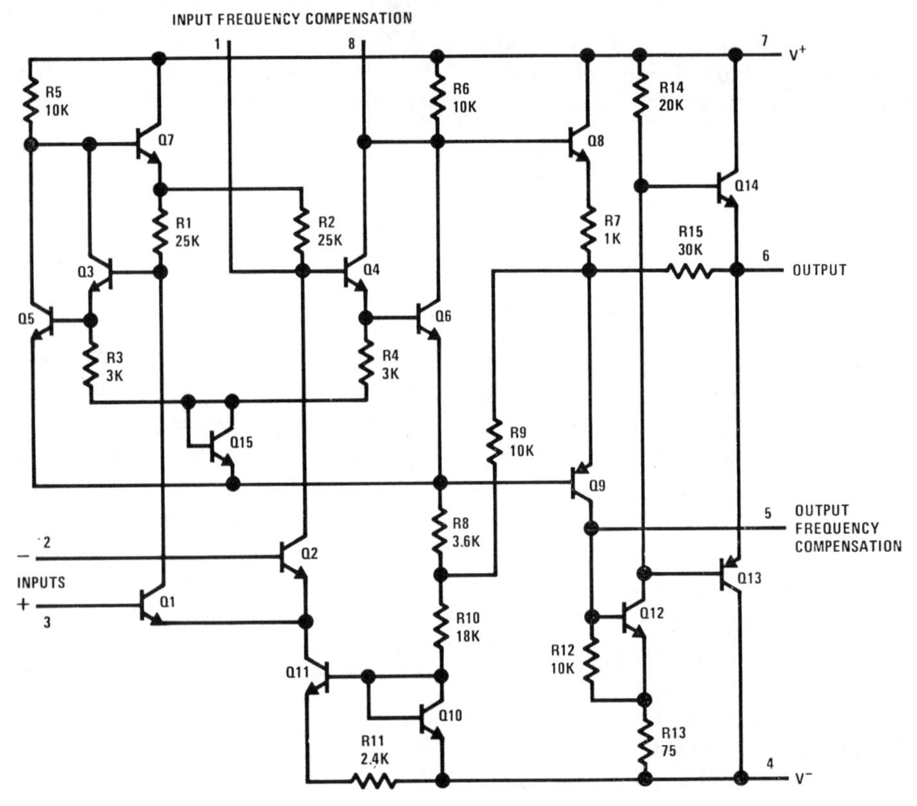

Fig. 2. Schematic diagram of the 709.

ceiving as it is fabricated on a 45-mil-square die, compared to 55 mil square for the 709. Hence, it is clear that new components have made possible much more sophisticated designs without making the circuit more difficult to manufacture.

One of the most significant departures from standard design on the LM101 and LM101A is the extensive use of active collector loads. Referring to Fig. 3, Q_5 and Q_6 serve as the collector loads for the input stage while Q_{17} is the load for the second-stage amplifier, Q_{10}. Active loads have several distinct advantages over the more common resistor loads. First, they permit low-current operation without large resistance values. This is important in reducing input bias currents and power consumption. Second, they do not require that much voltage be dropped for proper operation. This increases common-mode range, increases output swing, and permits the circuit to operate over a wider range of supply voltages. Last, they make possible much higher gain per stage, so fewer stages can be used. This simplifies frequency compensation immensely.

Another component first used in the LM101 design is the collector FET, Q_{18} in Fig. 3. This device, illustrated in Fig. 4(a), has proved invaluable for making low-power high-voltage circuits without large resistor values. The typical characteristics for an FET that is 3 mils wide between isolation cuts, and 20 mils long are shown in Fig. 4(b). Its usage is restricted somewhat by the fact

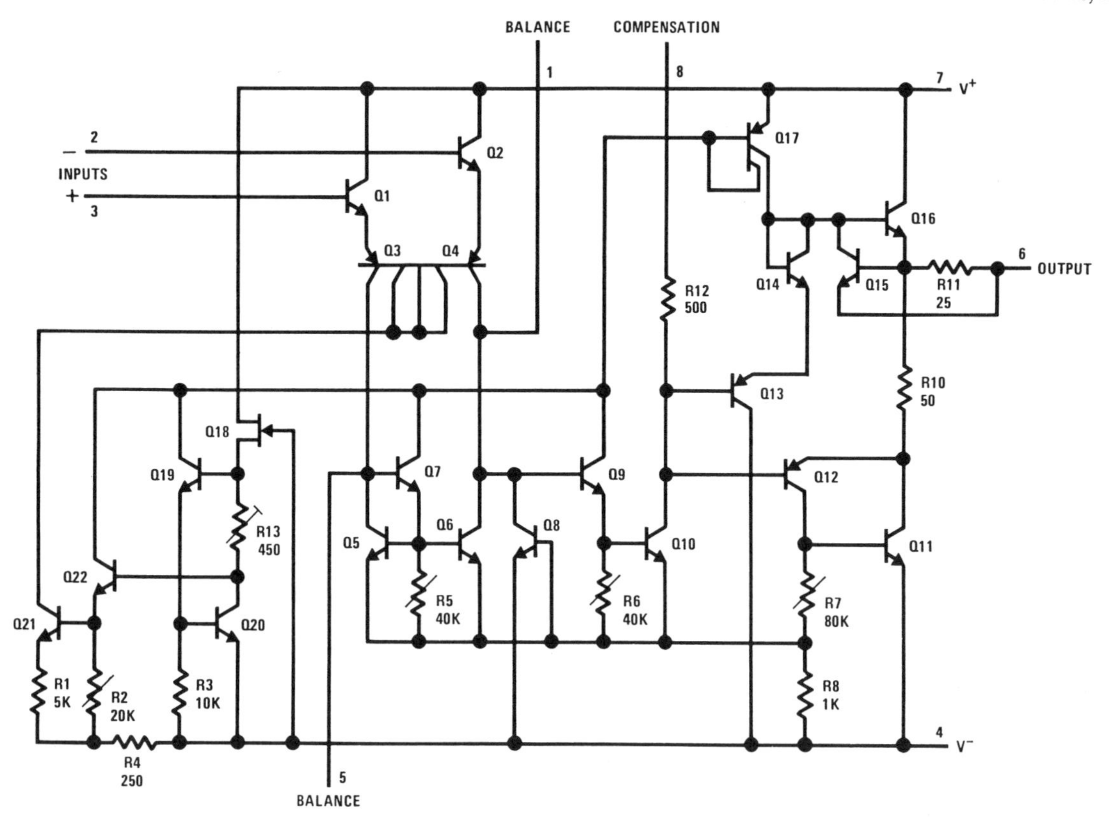

Fig. 3. Schematic diagram of the LM101A.

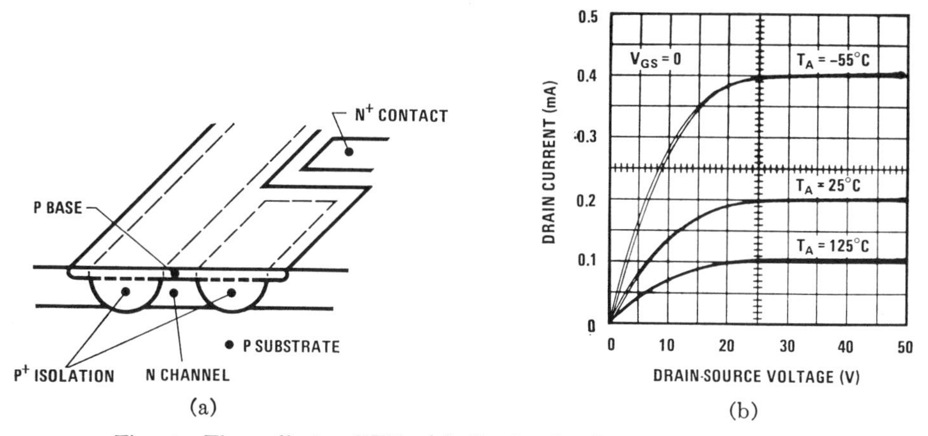

Fig. 4. The collector FET. (a) Sectional view. (b) Characteristics.

that the gate must be committed to the most negative point in the circuit.

The collector FET has proved most effective in circuits where the operating levels are determined by current sources. The FET provides turn-on, or bias, current for the current sources. All the currents can be determined by resistors in the current sources that have relatively small voltages dropped across them, and there are no resistors connected directly across the supplies. This approach not only minimizes chip area as well as current drain but also permits the circuit to operate over a wider range of supply voltages.

Pinched-base resistors [6] can give quite large resistance values in a small area, as they have a nominal sheet resistivity between 10 and 30 kΩ/□. However, it is difficult to maintain resistance tolerances within a factor of 2 or 3 of nominal, because of process variations in production. In addition, the resistors have a large positive temperature coefficient that causes them to vary by a factor of 4 over a −55 to 125°C temperature range. Pinch resistors also have an FET-like characteristic and a breakdown voltage around 6 volts, as shown in Fig. 5. Nonetheless, they have proved valuable for such uses as emitter–base bleed resistors (R_5, R_6,

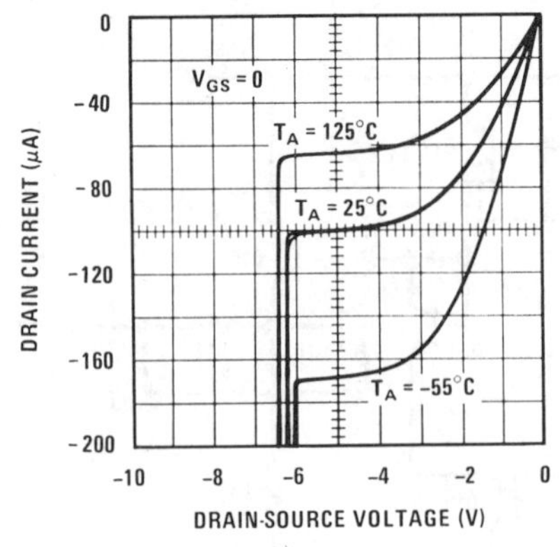

Fig. 5. Characteristics of pinched-base resistors.

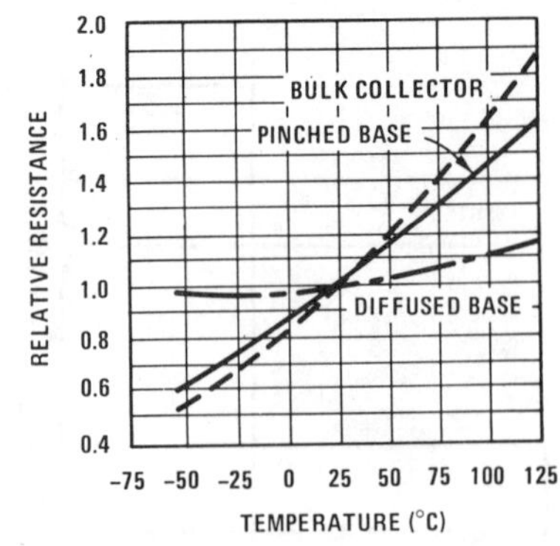

Fig. 6. Temperature characteristics of diffused-base resistors, pinched-base resistors, and bulk-collector resistors.

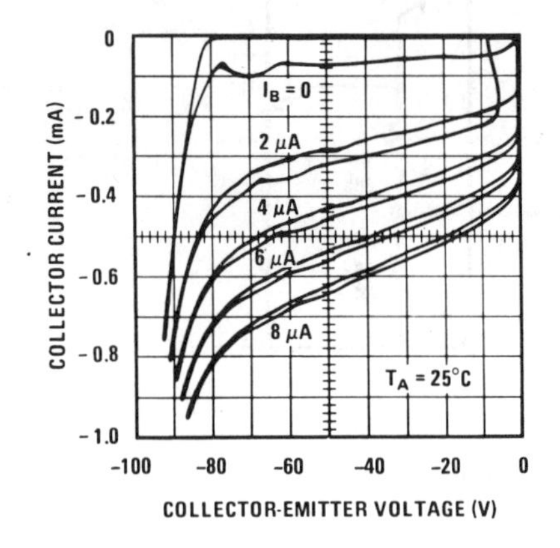

Fig. 7. Collector characteristiccs of a high-gain lateral p-n-p.

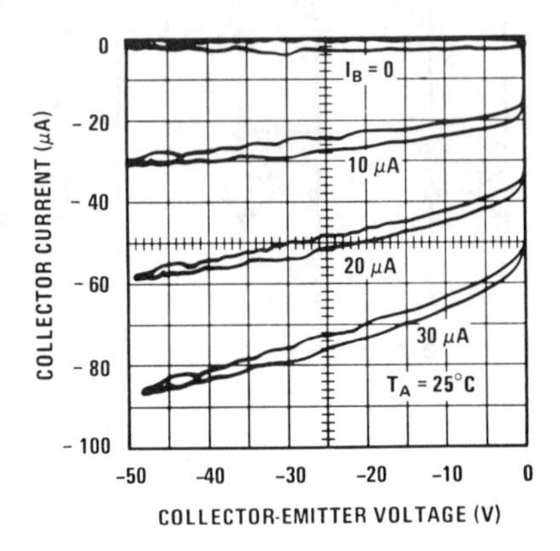

Fig. 8. Collector characteristics of a controlled-gain lateral p-n-p.

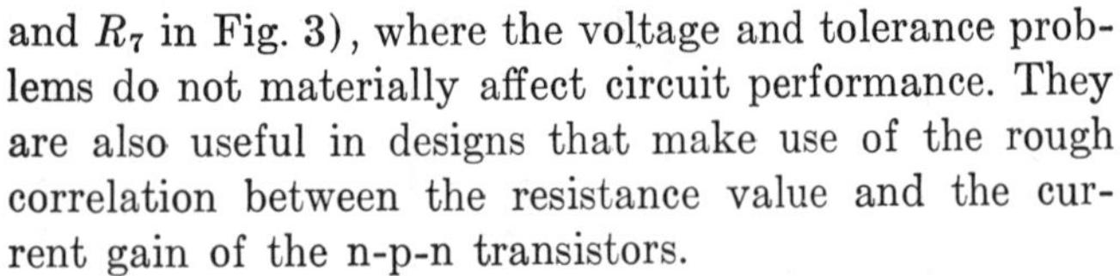
and R_7 in Fig. 3), where the voltage and tolerance problems do not materially affect circuit performance. They are also useful in designs that make use of the rough correlation between the resistance value and the current gain of the n-p-n transistors.

Collector resistors are made from the epitaxial collector material of the n-p-n transistors. The resistance tolerance is considerably better than pinch resistors, although it is not quite as good as base resistors. Resistance values from about 100 ohms to over 100 kΩ are practical. The interesting property of collector resistors is that they have temperature characteristics similar to Sensistors[5] and can be used for temperature compensation. One of these devices is used for R_{13} in Fig. 3. The temperature characteristics of collector resistors and pinched-base resistors are compared with standard diffused-base resistors in Fig. 6.

Lateral p-n-p transistors have long been known for their low dc gain. However, devices can now be fabricated with current gains greater than 100. This is done without significantly complicating processing, which is the major advantage of the lateral p-n-p over other complementary structures. High-gain p-n-p's open up new design possibilities. They can even be used for input stages, where their flat gain characteristics over a wide temperature range can be put to good use. The characteristics of a high-gain lateral p-n-p are shown in Fig. 7.

With higher current gain in the basic p-n-p structure, controlled-gain transistors can be made. This is done by breaking the collector into two segments and connecting one segment back to the base. The equivalent current gain is then determined by the relative size of the segments. Fig. 8 shows the collector characteristics of a device that has been designed for a current gain of 2. The current gain is not precisely fixed, varying between 2 and 3 as the collector–emitter voltage is increased to 50 volts. However, this tendency, caused by base-width modulation from the active collector, can be minimized by increasing the distance between the emitter and the collector.

The current gain of an n-p-n transistor depends, for one, on the length of the emitter diffusion cycle. Devices that are diffused for unusually long periods will exhibit increased current gain at the expense of breakdown voltage. Fig. 9 illustrates the characteristics of a transistor that has had the emitter driven in to the point where the device is nearly a collector–emitter short. Current gains in excess of 4000 can be obtained; however, the breakdown voltage is quite low. High-gain

[5] Registered trademark of Texas Instruments Incorporated.

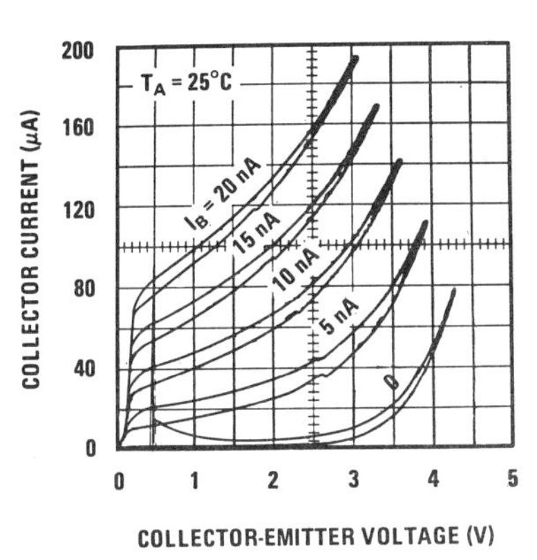

Fig. 9. Curve tracer display of high-gain transistors.

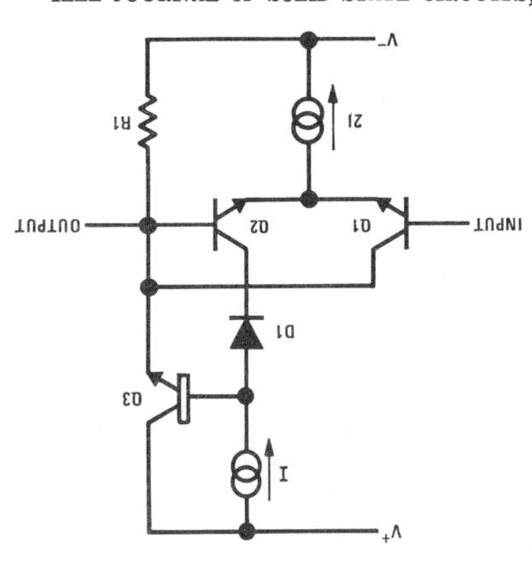

Fig. 10. Voltage follower using both high-gain and high-voltage transistors.

transistors, like these, can be built on the same chip with standard n-p-n transistors by using two separate emitter diffusions. With this technology, it is possible to design circuits that take advantage of the high current gain, yet can be operated at high voltages.

Reducing Input Currents

The low-voltage super-gain transistors can provide another order of magnitude improvement in the input current specifications of monolithic operational amplifiers like the LM101A. A circuit using them is the voltage-follower design shown in Fig. 10. High-gain primary transistors are used in the input stage to get very low input bias current. D_1 is included to operate Q_2 at near-zero collector–base voltage, and the collector of Q_1 is bootstrapped to the output for the same reason. Hence, low-voltage transistors can be used on the input. The only transistor that sees any voltage is Q_3, which buffers the output. Its current gain requirements are not stringent, so a moderate-gain secondary transistor can be used. In this particular design, the high-gain and high-voltage transistors are combined to take advantage of the best characteristics of both, without complicating the circuitry.

Because the input transistors are operated at zero collector–base voltage, high-temperature leakage currents do not show up on the input. Field-effect transistors, which, in the past, have been an obvious choice for the input stage of low-input-current operational amplifiers, suffer from leakage problems because there is no way to operate them with no voltage across the gate junction [7]. In applications covering a −55 to 125°C temperature range, super-gain transistors, which can give worst-case bias currents of 3 nA and worst-case offset currents of 400 pA, have a distinct advantage over FET's. With existing technology, they can equal FET's over a −25 to 85°C temperature range, and it is not difficult to foresee their superiority over a 0 to 70°C temperature range. Furthermore, matched pairs of super-gain transistors exhibit typical offset voltages of 0.5 to 1.0 mV with temperature drifts about 2 μV/°C, compared with 40 mV and 50 μV/°C for FET's. Certainly, discrete FET's can be compensated or selected for better offset or drift, but at a substantial increase in cost. MOS transistors, which do not have leakage problems, are no alternate solution [7] because they exhibit gross instabilities in offset voltage.

Standard bipolar transistors in a Darlington connection have been tried in a number of IC designs to get very low input currents. First-order calculations indicate that a Darlington should be competitive with super-gain transistors on input-current specifications. However, differential amplifiers, using Darlington-connected transistors, have problems that may not be immediately obvious.

The offset voltage depends not only on the inherent emitter–base voltage match of the transistors but also on the percentage match of current gains. A 10 percent mismatch in current gains give a 2.5-mV offset. Within a given process, bias currents drop faster than offset currents as the transistor current gains are raised. Hence, the better the transistors, the worse the offset voltage.

In addition to being a major contributor to offset voltage, this dependence on current-gain matching causes other problems. In a simple differential amplifier, the offset voltage drift can be correlated with offset voltage [8], [9] because of the predictable nature of emitter–base voltage. This is not so with Darlingtons, as bias current matching is not predictable over temperature. At high temperatures, this effect is aggravated further by leakage currents, so it is not possible to predict performance over a wide temperature range based on room-temperature tests.

Reducing the collector current of double diffused silicon transistors to very low currents (below 1 μA), as is done in a Darlington, does not improve input currents as much as might be expected. Lowering the collector current by a factor of 10 reduces the bias current by about a factor of 7 and the offset current by a factor of 3. These numbers are typical; the results obtained near the edges of a production distribution are significantly worse. In addition, the variation of input currents with

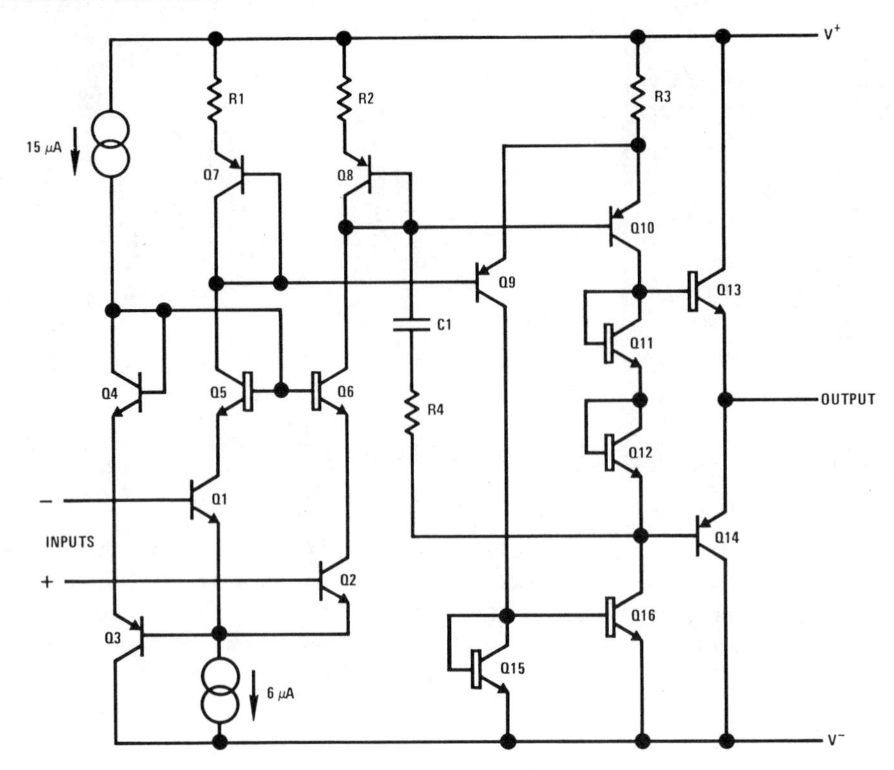

Fig. 11. General-purpose operational amplifier using bootstrapped high-gain transistors in the input stage.

temperature goes as the square of the current gain. Since the gain of integrated-circuit transistors falls off by a factor of 2 to 5 going from 25°C down to −55°C, the input currents obtained at the minimum operating temperature are considerably higher than at room temperature.

Other limitations of Darlingtons are that they have higher noise, lower common-mode rejection, and reduced common-mode slew rate. Further, they have one-half the transconductance of a simple differential stage; this doubles the effect of dc offset terms in their output circuitry.

To summarize, Darlingtons can give typical input currents competitive with super-gain transistors. However, if the full range of production variables is taken into account, along with −55 to 125°C operation, the performance is degraded considerably (or the yields reduced) both with respect to offset voltage and input current.[6] This is substantiated by the fact that a large number of IC operational amplifiers using Darlingtons have been marketed, but none have become industry standards. Nonetheless, the Darlington connection might become useful to compound super-gain transistors with standard transistors. This approach could give input currents less than 50 pA over a 0 to 70°C temperature range, which equals the best FET's. The offset voltage would not be as good as super-gain transistors alone, but it would be substantially better than monolithic FET's. It should be emphasized that this approach is not likely to work at temperatures much above 70°C.

Present data indicate that super-gain transistors are indeed the most effective and economical solution to the problem of making high performance integrated operational amplifiers. This approach cannot be considered to be speculation since an integrated voltage follower, the LM102 [10], which uses super-gain transistors to provide input bias currents less than 20 nA over a −55 to 125°C temperature range has been in volume production since early 1968.

Incorporating super-gain transistors in the input stage of a general-purpose operational amplifier is not as easy as in a voltage follower. Because the circuit must operate over a wide range of common-mode voltages, there is no simple way to operate the input transistors at zero collector–base voltage. However, this is certainly not impossible; and one circuit for doing it is shown in Fig. 11. The design uses more active components than a standard design, but this poses no problems in a monolithic circuit.

The input pair, Q_1 and Q_2, is operated in a cascode connection with Q_5 and Q_6, which stand off the common-mode voltage. The bases of Q_5 and Q_6 are bootstrapped to the common-mode voltage seen by the input transistors by Q_3 and Q_4. Hence, the input transistors are always operated with near-zero collector–base voltage.

[6] Modified Darlingtons, where the operating current of the input transistors is made large by comparison to the base current of the output transistors, do not suffer from problems caused by current-gain matching.

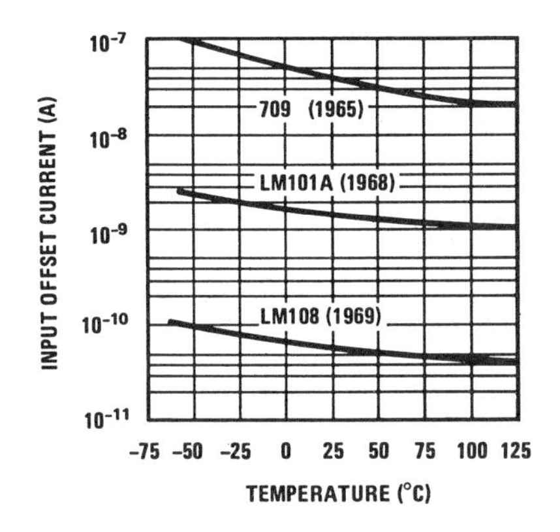

Fig. 12. Illustrating improvements in input current specifications.

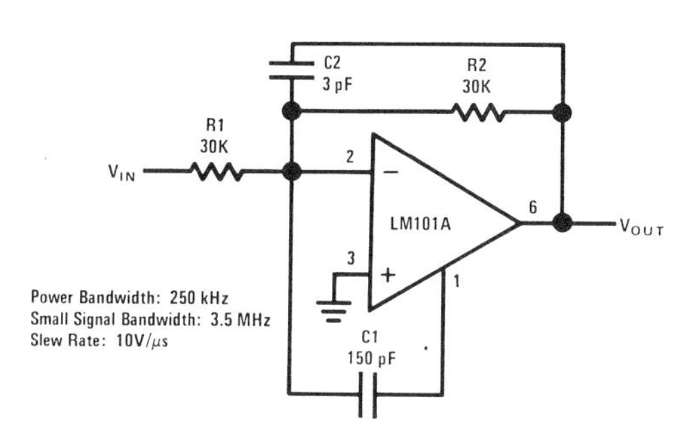

Fig. 13. Fast compensation for the LM101A in summing amplifier applications.

Fig. 12 shows the results of using this approach. The input currents are more than ten times better than the LM101A which, itself, is pretty close to the limits of what can be done with conventional transistors—FET or bipolar.

Fast Amplifiers

The ideal operational amplifier has been described as one with zero offset voltage, zero input current, infinite bandwidth, zero standby current drain, etc. No one can dispute that zero offset, no input current, infinite gain, and the like are desirable characteristics. However, a little practical experience quickly demonstrates that infinite bandwidth is not an unmixed blessing. High-speed amplifiers are definitely ore difficult to use. Capacitive loading, stray capacitances, improper supply bypassing, and poor physical layouts can all cause oscillation problems. Furthermore, fast amplifiers will, in general, have considerably higher power consumption; and it is harder to get effective protection of the input and output without adversely affecting the stability.

Most applications .require an operational amplifier with excellent dc characteristics but moderate high-frequency performance, so that a reasonable amount of slop can be tolerated in the physical layout before oscillation problems are encountered. However, there are a substantial number of applications where fast operation is definitely needed. It should be possible to make better high-frequency amplifiers with monolithic technology than with discretes. The basic reason is that stray and wiring capacitances can be virtually eliminated; an integrated circuit transistor has a collector–base capacitance of less than 0.1 pF. This value cannot be approached with discretes, especially if one considers the package and wiring capacitances. All that is needed for monolithic circuits to excel is a suitable circuit design.

There are two major parameters to consider in the design of a fast operational amplifier: small-signal bandwidth and slew rate. The major problem encountered in trying to improve small-signal frequency response has been the poor frequency characteristics of the lateral p-n-p [11], which has been used for level shifting. The response above 2 MHz, especially the excess phase shift, gets so bad that it cannot be used in a feedback amplifier at higher frequencies. A definite connection between the small-signal bandwidth, the unity gain slew rate, and the transconductance of the input transistors can be demonstrated [12] for amplifiers using 6 dB/octave rolloff.[7] With a simple, differential input stage using bipolar transistors, the p-n-p bandwidth forces the slew rate to be less than 1 V/μs. Higher slew rates can be obtained by reducing the input stage transconductance with emitter degeneration resistors, although this will reduce gain, decrease common-mode rejection and increase the offset voltage somewhat. It is possible to obtain any desired slew rate with degeneration, but slew rates much above 5 V/μs are not of much practical value. With faster slew rates, high-frequency gain error takes over as the biggest problem; and there is not much improvement in the end result. For example, with equal bandwidths and a 2-volt output step, an amplifier with a 50-V/μs slew would take about as long to settle within 1 percent of final value as an amplifier with a 5-V/μs slew. Hence, slew rates much above 5 V/μs will not be too useful without a proportionate increase in bandwidth.

Feedforward compensation [13] may be used to improve the high-frequency performance of standard amplifiers, in certain configurations. An example of this is shown in Fig. 13. An LM101A can be compensated by bypassing the lateral p-n-p to get slew rates of 10 V/μs, a power bandwidth of 250 kHz, and a small-signal bandwidth as high as 10 MHz. Further, high-frequency gain error is reduced (for example, by a factor of 50 at 100 kHz compared to standard compensation) so it is possible to take advantage of the faster slew. The feedforward compensation, which is no more complicated than standard compensation, provides more than an order of magnitude better performance in high-frequency applications. However, feedforward only works here when the device is used as a summing amplifier where the compensation capacitor does not have to be charged and discharged to swing the output.

[7] This relation does not apply if compensation is applied to the input terminals, but input compensation is not satisfactory in the majority of applications.

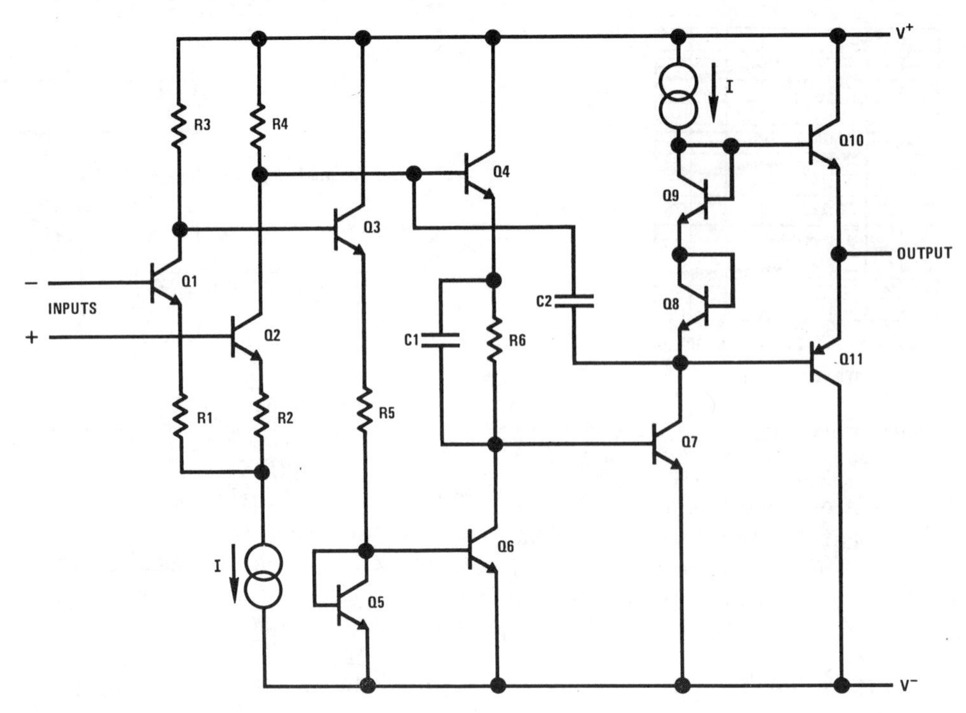

Fig. 14. Circuit configuration for a high-speed operational amplifier.

Even though good high-frequency performance can be obtained by using these tricks, in certain applications, it appears likely that a general-purpose high-speed amplifier design will somehow have to eliminate the lateral p-n-p's from the gain path. Fig. 14 shows one such amplifier. A standard, differential input stage is employed; and its output is buffered by a pair of emitter followers, Q_3 and Q_4. Level shifting is accomplished with resistors (R_5 and R_6) and a current inverter. The second-stage amplifier is Q_7, and it drives a complementary emitter-follower output stage. C_2 is the frequency compensation capacitor, while C_1 serves to minimize phase shift of the level shifting circuitry. Because the circuit can be made to look like a two-stage amplifier, using only n-p-n transistors, it should not be too difficult to frequency compensate. Bandwidths of 50 MHz with slew rates of 50 V/μs should be attainable with existing technology.

Conclusions

Monolithic operational amplifiers have been refined to the point where their dc performance and ease of use match amplifiers made with discrete and hybrid technology. Further improvements in dc characteristics can be expected soon. Monolithic amplifiers will provide better performance in full-temperature-range applications than present discretes, especially in obtaining low input currents.

This situation has come about because of improved processing and the development of new components useful only in monolithic circuits. In fact, the flexibility of designing the component parts of a monolithic device has become so great that it is impossible to draw a reasonably accurate schematic diagram of the circuit.

With very few exceptions, present-day monolithic amplifiers are lacking in high-frequency performance. This situation has come about because of the more obvious demand for moderate-frequency general-purpose devices. However, this can be expected to change. Dramatic improvements in the high-speed area will be made this year. Ultimately, monolithics can be expected to excel in this field because of lower stray and wiring capacitances.

References

[1] R. J. Widlar, "A unique circuit design for a high performance operational amplifier especially suited to monolithic construction," *Proc. 1965 NEC*, pp. 85–89.
[2] ——, "Monolithic op amp with simplified frequency compensation," *EEE*, vol. 15, pp. 58–63, July 1967.
[3] D. Fullagar, "A new high performance monolithic operational amplifier," Fairchild Semiconductor Appl. Brief, May 1968.
[4] R. J. Widlar, "I.C. op amp with improved input-current characteristics," *EEE*, pp. 38–41, December 1968.
[5] H. C. Lin, T. B. Tan, G. Y. Chang, B. VanDer Leest, and N. Formigoni, "Lateral complementary transistor structure for the simultaneous fabrication of functional blocks," *Proc. IEEE*, vol. 52, pp. 1491–1495, December 1964.
[6] G. E. Moore, "Semiconductor integrated circuits," in *Microelectronics*, E. Keonjian, Ed. New York: McGraw-Hill, 1963, ch. 5.
[7] R. J. Widlar, "Future trends in integrated operational amplifiers," *EDN*, vol. 13, pp. 24–29, June 1968.
[8] A. Tuszynski, "Correlation between the base emitter voltage and its temperature coefficient," *Solid-State Design*, pp. 32–35, July 1962.
[9] R. J. Widlar, "Linear IC's: Pt. 6; Compensating for drift," *Electronics*, vol. 41, pp. 90–93, February 1968.
[10] ——, "A fast integrated voltage follower with low input current," *Microelectronics*, vol. 1, June 1968.
[11] J. Lindmayer and W. Schneider, "Theory of lateral transistors," *Solid-State Electronics*, vol. 10, pp. 225–234, 1967.
[12] J. E. Solomon, W. R. Davis, and P. L. Lee, "A self-compensated monolithic operational amplifier with low input current and high slew rate," *1969 ISSCC Digest Tech. Papers*, vol. 12, pp. 14–15.
[13] R. C. Dobkin, "Feedforward compensation speeds op amp," National Semiconductor LB-2, March 1969.

Recent Advances in Monolithic Operational Amplifier Design

PAUL R. GRAY AND ROBERT G. MEYER, MEMBER, IEEE

Invited Paper

***Abstract*—The state of the art in monolithic operational amplifier design is surveyed. A set of large- and small-signal performance parameters are defined and discussed. Relationships between the important operational amplifier parameters of slew rate, offset voltage, and unity gain bandwidth are demonstrated. The dependence of settling time upon the fine structure of the open-loop frequency response is discussed, and the recent technological and circuit design approaches to the minimization of settling time are reviewed.**

I. Introduction

Since the introduction of the first monolothic operational amplifier (op amp) in 1963, their use in practical circuit and system design has grown steadily; in 1972, for example, \$50 million was spent in the United States for monolithic op amps. Because of the continually falling cost of monolithic op amps, their role as active analog circuit components might be compared to that of silicon transistors in the early 1960's. The ease of application of op amps in such fields as active filters, control systems, and instrumentation has made the op amp a much more widely used gain block than other alternatives such as gyrators or negative impedance converters.

The particular problems of op amp circuit design have not received wide attention from circuit theorists in the past, perhaps because in the spectrum of transistor circuits, op amps represent only a narrow, specialized segment. However, in view of the increasing importance of this segment, more attention seems justified. With this in mind, this paper is intended to define the parameters commonly used to describe performance of op amps, to show the interrelationship between these parameters and the fabrication technology used, and finally to assess the impact of upcoming new technology on performance levels to be expected in the next few years.

In Section II, the set of performance parameters commonly used to specify op amp behavior are given precise analytical definition, and the particular problems encountered in the measurement of each are described.

In Section III, several problems of current op amp circuit design are considered. The relationship between frequency response, transient response, and dc performance is discussed, and currently used techniques for improving amplifier settling time are described. Finally, the impact of fabrication technology on attainable amplifier performance levels is described.

Manuscript received October 18, 1973; revised January 8, 1974. This work was supported by the Joint Services Electronics Program under Contract F44620-71-C-0087.

The authors are with the Department of Electrical Engineering and Computer Sciences and the Electronics Research Laboratory, University of California, Berkeley, Calif. 94720.

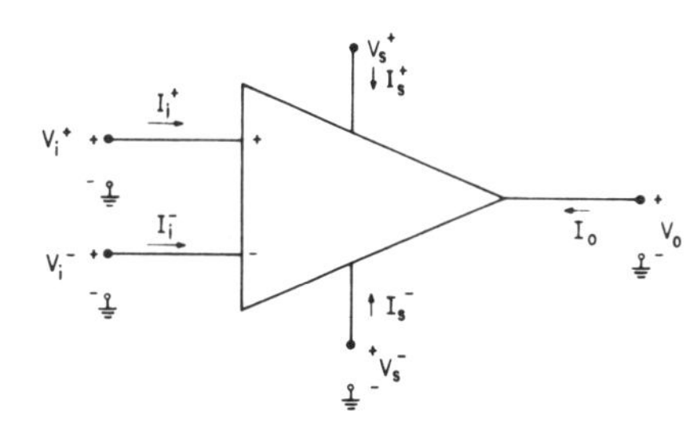

Fig. 1. The op amp as a six-terminal device.

II. Specification of Op Amp Behavior

The objective of this section is to define a set of basic parameters which characterize the small- and large-signal performance of an op amp, and which reflect the meaning most commonly attached to the parameters. In each case, the practical difficulties involved in measuring the particular parameter are discussed.

The differential input, single-ended output op amp can be characterized as a six-terminal nonlinear active circuit as shown in Fig. 1. While external components such as compensation capacitors and offset adjustment potentiometers are often required, these elements can be regarded as being contained within the circuit. It is convenient, from a measurement and application standpoint, to describe the behavior of this circuit with a set of equations of the form

$$I_i^+ = I_c^+(V_{id}, V_{ic}, V_s^+, V_s^-, I_o) \tag{1}$$

$$I_i^- = I_i^-(V_{id}, V_{ic}, V_s^+, V_s^-, I_o) \tag{2}$$

$$I_s^+ = I_s^+(V_{id}, V_{ic}, V_s^+, V_s^-, I_o) \tag{3}$$

$$I_s^- = I_s^-(V_{id}, V_{ic}, V_s^+, V_s^-, I_o) \tag{4}$$

$$V_o = V_o(V_{id}, V_{ic}, V_s^+, V_s^-, I_o) \tag{5}$$

where

$V_{id} \triangleq V_i^+ - V_i^-$ = differential mode input voltage;
$V_{ic} \triangleq (V_i^+ + V_i^-)/2$ = common mode input voltage;
$V_i^+ \triangleq$ voltage applied to the noninverting input;
$V_i^- \triangleq$ voltage applied to the inverting input;
$V_s^+ \triangleq$ positive power supply voltage;
$V_s^- \triangleq$ negative power supply voltage;
$I_o \triangleq$ output current;

and the dependent variables are

$I_s^+ \triangleq$ positive power supply current;
$I_s^- \triangleq$ negative power supply current;
$I_i^+ \triangleq$ current into noninverting input;
$I_i^- \triangleq$ current into inverting input;
$V_o \triangleq$ output voltage.

Reprinted from *IEEE Trans. Circuits and Syst.*, vol. CAS-21, pp. 317-327, May 1974.

These relationships can be used to define a set of dc and low-frequency incremental performance parameters which characterize the behavior of the amplifier in practical applications.

A. Behavior in the Quasi-Linear Region

In principle, (1)–(5) could be expanded in a Taylor series about an arbitrary operating point, and for sufficiently small signal excursions the behavior of the circuit represented by a set of linear equations. In practice, the small-signal behavior of the circuit is of interest only within the active, or quasi-linear, region in which the circuit behaves approximately as a linear amplifier. This mode of operation will take place under a set of constraints as follows:

$$V_{o\,\max}^- < V_o < V_{o\,\max}^+ \quad (6)$$

$$V_{ic\,\max}^- < V_{ic} < V_{ic\,\max}^+ \quad (7)$$

$$|V_{id} - V_{os}| < |V_{id\,\max}| \quad (8)$$

$$I_{o\,\max}^- < I_o < I_{o\,\max}^+ \quad (9)$$

$$V_s^+ > V_{s\,\min}^+ \qquad V_s^- < V_{s\,\min}^- \quad (10)$$

where

$V_{o\,\max}^-$, $V_{o\,\max}^+$	limits on the excursions of the output voltage for quasi-linear operation;
$V_{ic\,\max}^-$, $V_{ic\,\max}^+$	limits on the common mode input voltage for quasi-linear operation;
$V_{id\,\max}$	maximum differential input voltage prior to the onset of saturation or cutoff in the circuitry between the amplifier input and the particular node within the circuit whose capacitance limits the slew rate;
$I_{o\,\max}^+$, $I_{o\,\max}^-$	output current values beyond which current limit or overload occurs;
$V_{s\,\min}^+$, $V_{s\,\min}^-$	minimum power supply voltages;
V_{os}	input offset voltage, to be defined.

Under the restriction that the terminal voltages satisfy these restrictions, the behavior of the circuit for "small" excursions about a quiescent operating point can be described by hybrid matrix H as follows:

$$\begin{bmatrix} i_i^+ \\ i_i^- \\ i_s^+ \\ i_s^- \\ v_o \end{bmatrix} = [h_{ij}] \begin{bmatrix} v_{id} \\ v_{ic} \\ v_s^+ \\ v_s^- \\ i_o \end{bmatrix} \quad (11)$$

where the lower case designation indicates a small variation superimposed upon a quiescent value. It should be emphasized that (11) is a valid representation of an *actual* op amp circuit only for low frequencies and for sufficiently small signal variations about an operating point that the second-order terms in the Taylor expansion of (1)–(5) for that particular circuit are negligible. In most cases the hybrid parameters vary significantly as the operating point is moved around the quasi-linear region, and in particular h_{35} and h_{45} undergo drastic changes when a nonlinear class B output stage is used. The usefulness of (11) is as a first-order linear *model* for the op amp whose parameters are easily obtainable from data sheet specifications.

The relationship between the commonly used dc and incremental performance parameters and the parameters of (11) will now be considered.

1) Input Offset Voltage V_{os}: This is the differential voltage which must be applied at the input to drive the output voltage to zero:

$$0 = V_o(V_{os}, V_{ic}, V_s^+, V_s^-, 0). \quad (12)$$

This relation implicitly gives this offset voltage as

$$V_{os} = V_{os}(V_{ic}, V_s^+, V_s^-). \quad (13)$$

The offset typically specified is that for zero common mode voltage and nominal power supply voltages. While the nominal value of the offset voltage is usually zero, the actual value varies about zero from unit to unit due to random parameter variations which occur in production. A typical specification on offset would be ± 2 mV.

Input offset voltage is of particular importance in low-level instrumentation problems where small dc voltage must be detected. The offset voltage contributes a dc error which cannot be distinguished from the signal itself. Many amplifiers have a provision for nulling the offset, but this assures that it is zero at only one temperature. Some circuit considerations regarding offset voltage are discussed later.

A commonly used method to measure offset voltage is to simply connect the amplifier in a $\times$ 1000 inverting configuration, short the input to ground, and measure output voltage and divide by 1000. Sufficiently small gain-setting resistors must be used so that the error contributed by the input bias current is negligible.

2) Input Bias Current:

$$I_i^+ = I_i^+(V_{id}, V_{ic}, V_s^+, V_s^-, I_o) \quad (14)$$

$$I_i^- = I_i^-(V_{id}, V_{ic}, V_s^+, V_s^-, I_o). \quad (15)$$

The dc current which flows into the amplifier input terminals can cause dc errors because of the resulting voltage drops in the source or gain-setting resistors. A design objective is to make this current as small as possible, with 10–100 nA a typical value for bipolar input stages and 1–10 pA typical of field-effect transistor (FET) input stages. This parameter is often a function of power supply voltage. On amplifier data sheets, a single number is usually given which is a limit on the average of the two input currents, i.e., $(I_i^+ + I_i^-)/2$.

3) Input Offset Current:

$$I_{os} \triangleq I_i^+ - I_i^- \qquad V_{id} = V_{os},\ V_{ic} = 0. \quad (16)$$

While the effects of dc input currents I_i^+ and I_i^- can be made to cancel if the two currents are equal, random mismatches in components within the circuit cause them to be unequal. The offset current is a random parameter like offset voltage, and can take positive or negative values. It has a typical value of from 5 to 20 percent of the input bias current.

4) Differential Mode Voltage Gain:

$$A_{dm} = \frac{\partial V_o}{\partial V_{id}} (V_{id}, V_{ic}, V_s^+, V_s^-, I_o) = h_{51}. \tag{17}$$

The differential mode gain is quite large for practical circuits, ranging from 10^5 for general purpose circuits to 10^7 for circuits intended for instrumentation applications [1]. In many cases the electrical gain is so large that the actual measured value of dc open-loop gain is determined by thermal coupling on the integrated circuit (IC) die between the output stage of the circuit and the input [2]. Because the power dissipation on the die is a nonlinear function of the output voltage, this feedback mechanism tends to produce heavily distorted dc transfer characteristics. Further, it makes the open-loop measurement of gain very difficult, as the amplifier frequently behaves as a thermal multivibrator when operated open-loop.

A widely used technique for measurement of open-loop gain is illustrated in Fig. 2. A second op amp is used in a feedback configuration to force the *output* voltage of the device under test (DUT) to be equal to the input voltage V_a. The output of the second amplifier is applied to the DUT through a 1000:1 voltage divider, so that the voltage at the output of the second amplifier is 1000 times the voltage at the input of the DUT required to produce V_a at the output of the DUT. The gain is frequently measured as an average by setting V_a equal to, alternately, +10 and −10 V, and measuring the difference between the two values of V_b that result. The average gain $\bar{A}_{dw}$ is then

$$\bar{A}_{dm} = 1000 \left[\frac{\Delta V_a}{\Delta V_b}\right] \tag{18}$$

where ΔV_a in this case would be 20 V. This circuit can also be used to measure input offset voltage by setting $V_a = 0$.

5) Common-Mode Gain:

$$A_{cm} \triangleq \frac{\partial}{\partial V_{ic}} V_o(V_{id}, V_{ic}, V_s^+, V_s^-, I_o) = h_{52}. \tag{19}$$

Since an ideal op amp responds only to differential signals, A_{cm} is zero in the ideal case. However, limitations in practical circuit design and device mismatches produce some amount of common mode gain [3], [4]. In many instrumentation applications, a small difference voltage must be detected in the presence of large common mode signals, so that large common mode gain produces significant errors. This aspect of device performance is specified by the common mode rejection ratio (CMRR):

$$\text{CMRR} \triangleq \frac{A_{dm}}{A_{cm}} = \frac{h_{51}}{h_{52}}. \tag{20}$$

The practical effect of the common mode gain on device performance can be appreciated by considering the offset voltage to be a function of the common mode input voltage. We first differentiate (12), assuming V_s^+ and V_s^- to be constant, and I_c to be zero:

$$0 = \frac{d}{dV_{ic}} [V_o(V_{os}, V_{ic}, V_s^+, V_s^-, 0)] \tag{21}$$

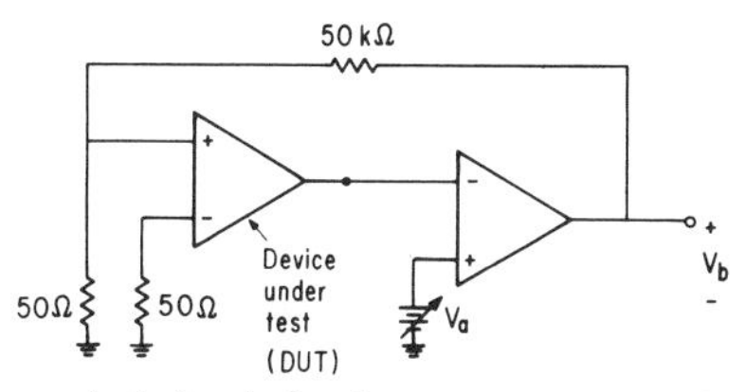

Fig. 2. A practical circuit for the measurement of loop gain.

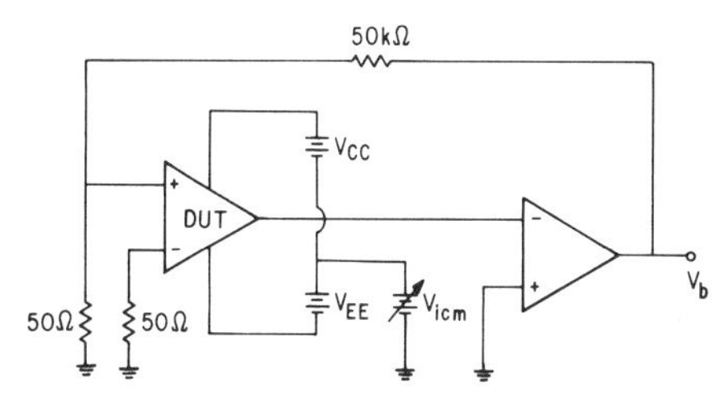

Fig. 3. A practical circuit for the measurement of CMRR. CMRR = $1000 V_{icm}/\Delta V_b$.

where $V_{os} = V_{os}(V_{ic}, V_s^+, V_s^-, 0)$. Using the chain rule,

$$0 = \left[\left.\frac{\partial V_o(V_{os}, V_{ic}, V_s^+, V_s^-, 0)}{\partial V_{os}}\right|_{V_{ic}}\right]\left[\frac{\partial V_{os}}{\partial V_{ic}}\right] + \left.\frac{\partial V_o(V_{os}, V_{ic}, V_s^+, V_i^-, 0)}{\partial V_{ic}}\right|_{V_{os}}. \tag{22}$$

Rearranging,

$$\frac{\partial V_{os}}{\partial V_{ic}} = \frac{-\left.\dfrac{\partial V_o(V_{os}, V_{ic}, V_s^+, V_s^-, 0)}{\partial V_{ic}}\right|_{V_{os}}}{\left.\dfrac{\partial V_o(V_{os}, V_{ic}, V_s^+, V_s^-, 0)}{\partial V_{os}}\right|_{V_{ic}}}. \tag{23a}$$

Clearly,

$$\left.\frac{\partial V_o(V_{os}, V_{ic}, V_s^+, V_s^-, 0)}{\partial V_{os}}\right|_{V_{ic}} = \left.\frac{\partial V_o(V_{id}, V_{ic}, V_s^+, V_s^-, 0)}{\partial V_{id}}\right|_{V_{ic}} = A_{dm}$$

and

$$\left.\frac{\partial V_o(V_{os}, V_{ic}, V_s^+, V_s^-, 0)}{\partial V_{ic}}\right|_{V_{os}} = A_{cm}.$$

Therefore,

$$\frac{\partial V_{os}}{\partial V_{ic}} = -\frac{A_{cm}}{A_{dm}} = -\frac{1}{\text{CMRR}}. \tag{23b}$$

Thus the reciprocal CMRR can be thought of as the change in offset voltage produced by a unit change in common mode input voltage. The contribution of offset voltage to dc errors has already been considered.

The CMRR is usually measured utilizing (23b). The change in offset voltage produced by a change in common mode input voltage is measured using the circuit of Fig. 3. As in Fig. 2, the second op amp forces the output of the DUT to equal V_a (in this case zero volts), and thus V_b in this case is just 1000 times the offset voltage of the DUT. Instead of applying a common mode signal to the input terminals, it is more convenient to leave these at constant potential and lower the voltages on the other device terminals by an amount ΔV_{icm} to simulate a positive common mode input voltage. The resulting change in V_b is just 1000 times the change in offset voltage, so that

$$\mathrm{CMRR} = 1000 \frac{\Delta V_{icm}}{\Delta V_b}. \tag{24}$$

The CMRR is often expressed in decibels, a typical value for this parameter being 80 dB.

6) *Power Supply Rejection Ratio (PSRR):*

$$\mathrm{PSRR}^+ \triangleq \frac{A_{dm}}{\dfrac{\partial}{\partial V_s^+} V_o(V_{id}, V_{ic}, V_s^+, V_s^-, I_o)} = \frac{h_{51}}{h_{53}} \tag{25}$$

$$\mathrm{PSRR}^- \triangleq \frac{A_{dm}}{\dfrac{\partial}{\partial V_s^-} V_o(V_{id}, V_{ic}, V_s^+, V_s^-, I_o)} = \frac{h_{51}}{h_{54}}. \tag{26}$$

These quantities reflect the gain observed between the power supply terminal and the output. Like CMRR, they can be related to the offset voltage:

$$\mathrm{PSRR}^+ = \left[\frac{\partial V_{os}}{\partial V_s^+}\right]^{-1} \tag{27}$$

$$\mathrm{PSRR}^- = \left[\frac{\partial V_{os}}{\partial V_s^-}\right]^{-1}. \tag{28}$$

These parameters are also often expressed in decibels, a typical value being 80 dB. They can be measured using a circuit similar to that shown in Fig. 3.

7) *Unity Gain Bandwidth* (ω_{co}): In the quasi-linear region, the differential voltage gain A_{dm} falls with increasing frequency. The nature of the falloff must be carefully shaped so that stability can be maintained at a wide variety of loop gains. The most common way of specifying the frequency response is to specify the unity gain bandwidth and phase margin. The former is the frequency at which $A_{dm} = 1$, and the latter is the phase shift of A_{dm} at that frequency, subtracted from 180°. The unity gain frequency is most easily measured using the circuit shown in Fig. 4. V_i is a small signal ac differential input voltage and V_{OUT} is the corresponding output voltage. The differential voltage gain is given by $A_{dm} = V_{\mathrm{OUT}}/V_i$. The 100-kΩ resistor gives unity feedback for dc and thus biases the amplifier with a dc output near zero volts. At the frequencies of interest, the 1-μF capacitor decouples the feedback so that the amplifier exhibits open loop gain characteristics.

8) *Other Incremental Parameters:*

a) *Differential input resistance:*

$$R_{inD} = \left[\frac{\partial(I_i^+ - I_i^-)}{\partial V_{id}}\right]^{-1} = \frac{1}{h_{11} - h_{21}}. \tag{29}$$

A typical value of R_{inD} is 1 MΩ.

b) *Common-mode input resistance:*

$$R_{inC} \triangleq \left[\frac{\partial(I_i^+ + I_i^-)}{2\,\partial V_{ic}}\right]^{-1} = \frac{2}{h_{12} + h_{22}}. \tag{30}$$

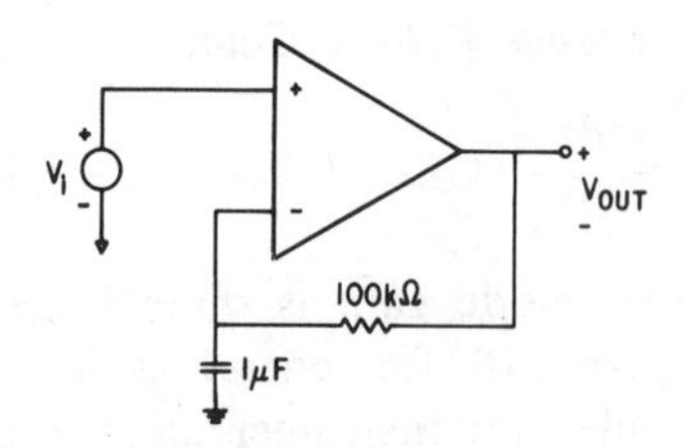

Fig. 4. A circuit for the measurement of unity gain bandwidth.

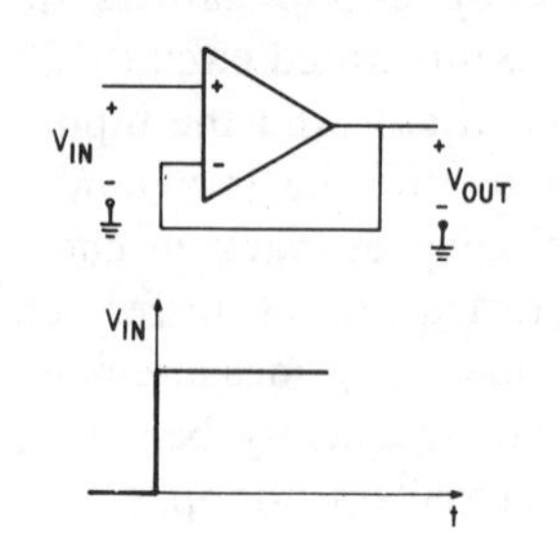

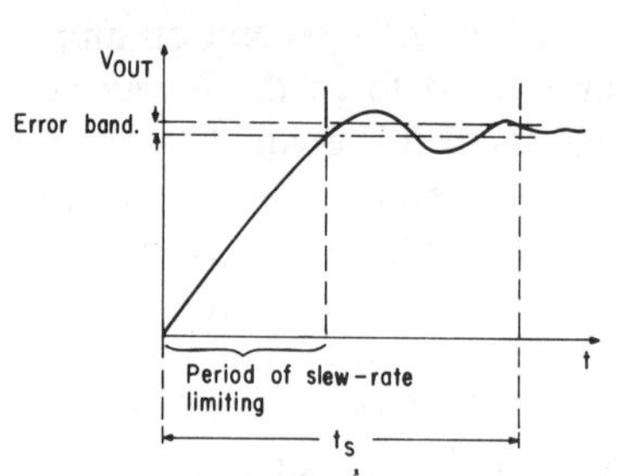

Fig. 5. Typical voltage follower step response.

Note that R_{inC} is defined as the resistance seen at *each* input terminal by a common mode signal. This is the quantity which is most important in practice. For example, in voltage follower circuits such as shown in Fig. 5, the differential input resistance is multiplied by the loop gain (which is about 10^5) and the circuit input resistance becomes approximately R_{inC}.

A typical value of R_{inC} is 10^9 Ω.

c) *Output resistance:*

$$R_{\mathrm{out}} \triangleq \left[\frac{\partial V_o}{\partial I_o}\right] = \frac{1}{h_{55}}. \tag{31}$$

A typical value of R_{out} is 50 Ω.

B. Behavior Outside the Quasi-Linear Region

While the behavior of op amps in the quasi-linear region can be characterized by means of the parameters so far defined, the behavior of such circuits when (6)–(10) are violated is of practical importance. In particular, when the differential input voltage is made to exceed a maximum value (i.e., when a large step input is applied) the response of the amplifier is no longer even approximately described by a quasi-linear model. While the negative feedback used with op amps causes the differential input voltage to be quite small in the steady state, it can become large during input transients. For the case of a large step input, for example, the linear model would predict a step response at the output whose initial slope was proportional to the input step. In most practical op amp circuits, large differential input voltages cause some of the transistors within the amplifier input stage to cutoff, so that no further drive can be delivered to later stages. This causes the step response to become independent of the differential step input magnitude once a certain threshold has been exceeded. Values of this input threshold for typical amplifiers range from about 50 mV to several hundred mV. When a step this large is applied to the input terminals, the response at the amplifier output is just the response to a step function equal in value to the input threshold value $V_{id\,max}$, which is a property of the amplifier and is independent of the step magnitude. This will be an approximate exponential waveform with a virtual final value of several thousand volts, so that the small initial portion of the waveform observed before the slew rate condition ceases to exist is very nearly a straight line. Performance in this mode is generally characterized by specifying this slope, called the slew rate, with no common mode input voltage and some standard loading and power supply conditions. The slew rates in the positive and negative directions can be different due to differing nonlinear behavior in the two directions. The presence of a transient common mode signal as is encountered in voltage follower configurations can enhance or degrade the slew rate. This is due to the presence of stray capacitances to ground in the earlier stages of the amplifier which charge and discharge with varying common mode signals. Practical values of slew rate range from 0.5 V/μs for general purpose amplifiers to 100 V/μs and more for high-speed amplifiers. This parameter is also sometimes specified by means of "full-power bandwidth." This is simply the maximum frequency at which the amplifier can reproduce a sinusoidal voltage from one specified output voltage limit to the other without slew rate limiting.

The practical significance of the slew rate parameter is that it limits the ability of the amplifier to follow step inputs. An important parameter in many analog systems is the total time required for a closed loop amplifier to settle to within a specified percentage of the new output voltage from the time a step is applied. This composite parameter is termed the settling time (t_s) and is dependent on both the slew rate and on the small-signal properties of the amplifier.

The settling time is composed of two intervals: an initial period dominated by slew rate limiting, followed by a period of settling in the quasi-linear region. The time periods are illustrated in Fig. 5. Usual error band specifications with a 10-V step are 0.1 percent (10 mV) and 0.01 percent (1 mV).

The settling time is particularly sensitive to the detailed shape of the open-loop frequency response and to the *phase margin*, since an underdamped response can grossly lengthen the settling time. The settling time is quite sensitive to circuit wiring details, stray capacitance, compensation, etc., so that in order for the specification to be definitive, the closed-loop configuration used, the type of load on the output, the nature of the input step and the details of the physical layout of circuitry used in the measurement must be specified.

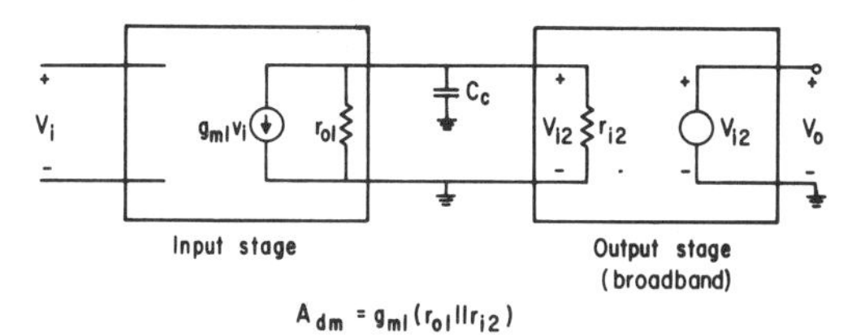

Fig. 6. Simplified small-signal equivalent circuit, single-stage op amp.

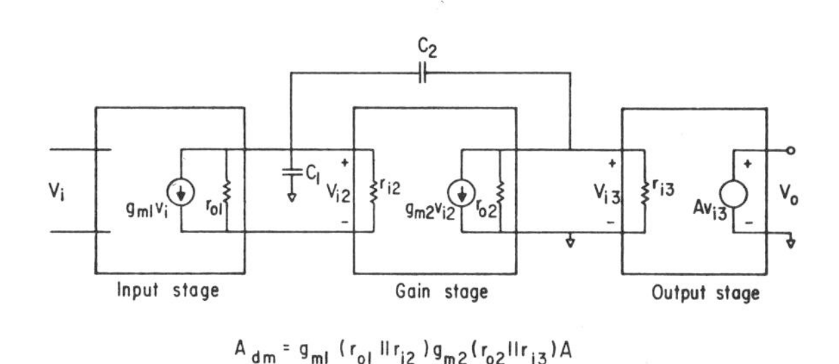

Fig. 7. Simplified small-signal equivalent circuit, two-stage op amp.

III. Op Amp Design Considerations

In this section, several practical aspects of op amp circuit design are considered. The various circuit approaches are first classified by the number of stages used. The relations between three important design parameters—bandwidth, slew rate, and offset voltage—are then described. The advantages and disadvantages of feedforward compensation for improving closed-loop bandwidth are described, and finally the limitations imposed by present technology, and the improvements expected from new technology, are described.

A. Practical Op Amp Configurations

Op amp circuits may be classified in many ways, but perhaps the most useful is by the number of stages within the amplifier providing voltage gain. This system of classification is useful because the voltage gain stages tend to contribute the dominant poles of the differential voltage gain function of the circuit, and hence dictate the amount and type of frequency compensation which must be used. For ease of compensation, the usual approach in op amp design is to provide one or two narrow-band, very high-gain stages, and to provide any impedance level conversion or buffering with broad-band, low-gain circuits (often unity gain emitter followers) [5]. It should be pointed out that this system of classification may break down for certain special purpose circuits, but is useful for the great majority of monolithic op amps. The most common configurations shown (Figs. 6 and 7) are the one- and two-stage circuits (referring to the number of high-gain stages).

The one-stage configuration is used in some low cost, low-performance, multiple op amp chips [6] and in certain very high-speed circuits [7]. While a single stage results in less overall phase shift through the amplifier, the gain stage

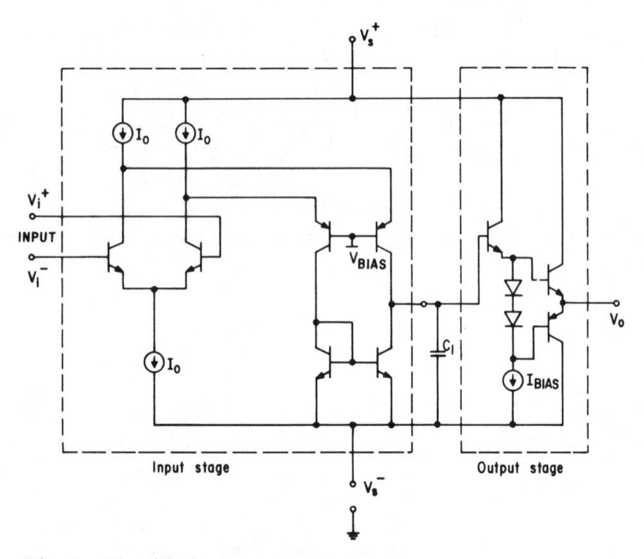

Fig. 8. Simplified schematic, Harris semiconductor 2500 op amp.

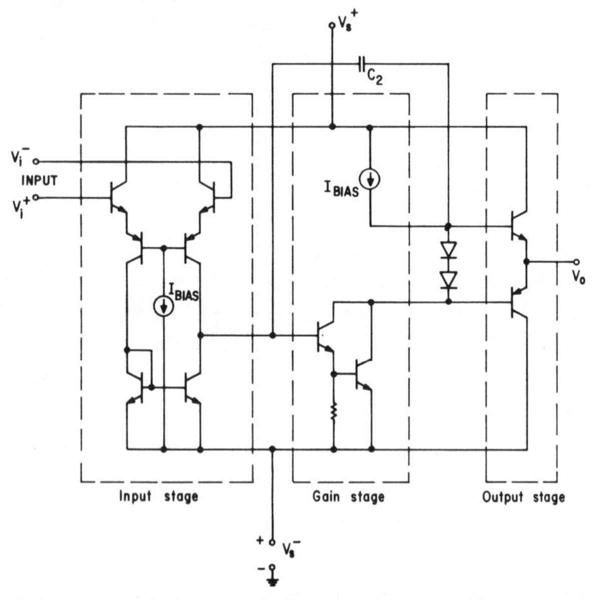

Fig. 9. Simplified schematic, Fairchild μA 741 op amp.

output impedance and the input impedance of the output buffer must be extremely high to obtain sufficient dc gain, usually at least 10 000 for general purpose amplifiers. A simplified example of such a circuit is shown in Fig. 8. Frequency compensation of this type of circuit can be accomplished by a single capacitor to ground.

The two-stage circuit of Fig. 7 is more commonly used, and consists of an input stage, intermediate gain stage, and broad-band output buffer. The μA 741 shown in Fig. 9 is an example of this type of circuit, where the broad-band output stage is a unity gain emitter follower. Large gain is easily achievable, and compensation can be accomplished with single capacitor to ground C_1 or with a feedback capacitor C_2 as in the μA 741, or both. It has been shown [8], [9] that one effect of C_2 is to broad-band the effect of any stray capacitance at the output of the second gain stage, so that a feedback capacitor C_2 is almost always used with this type of circuit. As a result capacitor C_2 is sometimes referred to in the literature as a pole-splitting capacitor [8], [9].

Amplifiers containing more than two stages are used when very large voltage gain is required [10]. The compensation of such circuits is more difficult since there are three high-impedance nodes in the signal path, although this problem can be eased using feedforward techniques. Since only one node can be broad-banded by pole-splitting, the crossover frequency is constrained to be fairly low.

B. Relationship Between Bandwidth, Slew Rate, and Offset Voltage

The representations of Figs. 6 and 7 can be used to obtain simple relationships between several of the performance parameters of any op amp which can be described by such a block diagram. A two-stage amplifier is assumed, but exactly the same results can be obtained for the one-stage circuit.

1) Unity Gain Bandwidth: In order to ensure stability of the circuit of Fig. 7 when used in a unity feedback configuration, the compensation capacitor must be chosen to provide a dominant pole. This dominant pole must produce a 6-dB/octave roll-off in gain so that unity gain is reached at some frequency (ω_{co}) lower than the next most important pole produced by the amplifier. Thus the value of ω_{co} and of compensation capacitance are determined by the frequency limitations of the devices in the amplifier.

A straightforward analysis [8] of the circuit of Fig. 7 shows that, neglecting all charge storage in the circuit other than the compensation capacitor, the unity gain frequency ω_{co} is given by

$$\omega_{co} \cong \frac{g_{m1}}{C_2} A, \qquad \text{for } C_2 \text{ only} \tag{32}$$

$$\omega_{co} \cong \frac{g_{m1}}{C_1} g_{m2}(r_{o2} \parallel r_{i3})A, \qquad \text{for } C_1 \text{ only.} \tag{33}$$

Here it has been assumed that

$$(r_{o1} \parallel r_{i2}) \gg \frac{1}{C_1 \omega_{co}} \tag{34}$$

and

$$(r_{o1} \parallel r_{i2}) \gg \frac{1}{C_2(g_{m2})(r_{o2} \parallel r_{i3})\omega_{co}}. \tag{35}$$

These assumptions are well satisfied in practical op amps.

2) Slew Rate: As mentioned earlier, the limitation on the maximum rate of change of the output voltage for a step input voltage results from the nonlinear behavior of the input stage. For example, in the typical input stage circuit of Fig. 10, the output current of the stage is a nonlinear function of the input voltage. In this case, the maximum current available from this input stage is I_o. Since the compensation capacitor C_1 or C_2 is connected to the output of this stage, the maximum charging current delivered to the compensation capacitor is I_o. As a result, the maximum rate of change of the voltage at the compensation point is I_o/C_{comp} and the maximum rate of change of the output

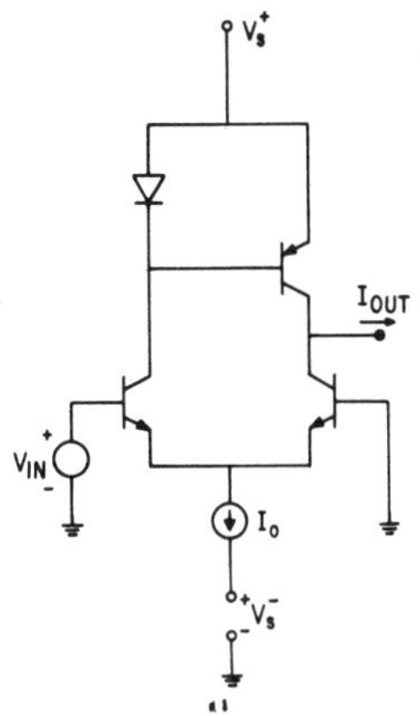

Fig. 10. Bipolar transistor differential input stage with active p-n-p load.

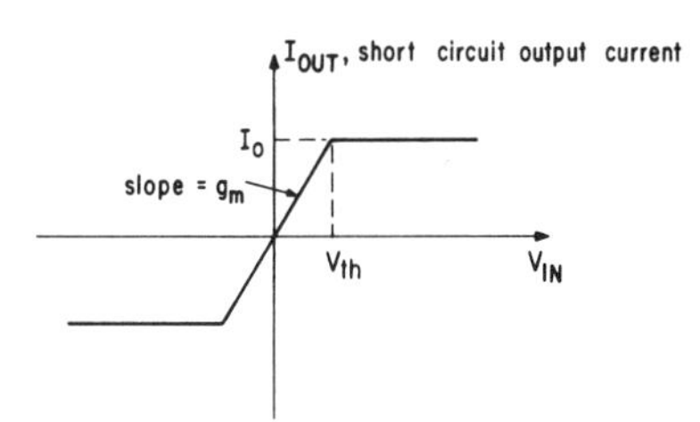

Fig. 11. Transfer characteristic of idealized piecewise linear input stage.

voltage (the slew rate) is

$$S_r = \frac{I_o}{C_2} A, \qquad C_2 \text{ only present} \tag{36}$$

$$S_r = \frac{I_o}{C_1} A g_{m2}(r_{02} \parallel r_{i3}), \qquad C_1 \text{ only present.} \tag{37}$$

Combining (36) and (37) with (32) and (33), we obtain

$$S_r = (\omega_{co}) \frac{I_o}{g_{m1}}, \qquad \text{both cases.} \tag{38}$$

Thus the attainable slew rate depends on the crossover frequency ω_{co} and the ratio I_o/g_{m1}. The former is dictated by stability considerations and is discussed in the next section. Consider the ratio of I_o/g_m for the simple bipolar differential input stage shown in Fig. 10:

$$\frac{I_o}{g_m} = \frac{I_o}{\left(\frac{1}{2}\frac{I_o}{V_T}\right)} = 2V_T = 52 \text{ mV}. \tag{39}$$

Further insight into the significance of (38) can be obtained by considering an idealized piecewise linear input stage as shown in Fig. 11. Here, the quantity I_o/g_m is equal to V_{th}, the voltage range of the input stage:

$$S_r = V_{th}\omega_{co}. \tag{40}$$

Thus slew rate can be increased by increasing input stage voltage range [8], [9], which can be achieved by the use of emitter degeneration resistors in the case of a bipolar circuit, or by the use of junction or metal oxide semiconductor (MOS) FET's. However, the input offset voltage tends to increase in proportion to V_{th} and is related to V_{th} by the accuracy of component matching that can be achieved. Consider, for example, the emitter degenerated bipolar input stage of Fig. 12:

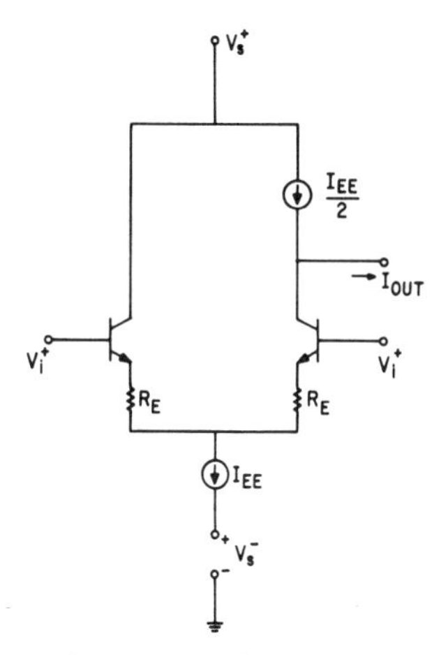

Fig. 12. Emitter degenerated bipolar input stage.

$$\frac{I_o}{g_m} = I_{EE}\frac{2V_T}{I_{EE}}\left(1 + \frac{I_{EE}R_E}{2V_T}\right) = 52\left(1 + \frac{I_{EE}R_E}{2V_T}\right) \text{ mV}. \tag{41}$$

Note that the $I_{EE}R_E/2$ is the dc voltage drop across each emitter resistor. As this is made significantly greater than $V_T = 26$ mV, the slew rate given by (38) can be significantly increased. The practical limit to this technique is the degradation of input offset voltage. If the matching between monolithic resistors is γ percent, then an input offset of $(\gamma/100)(I_{EE}R_E/2)$ is produced. If $\gamma = 1$ percent and a maximum increase in offset of 2 mV is specified, then $I_{EE}R_E/2$ must be less than 200 mV. Thus the maximum increase in slew rate from (41) and (34) is a factor of 9. In addition, offset in the active load is magnified by the gain loss caused by the emitter degeneration, causing even greater increases in offset.

In the above discussion, it has been assumed that slew rate as given by (38) can be increased by increasing the ratio I_o/g_{m1}. This assumes that ω_{co} is unaffected by any changes in g_{m1}. Also note from (32) and (33) that if g_{m1} is decreased and ω_{co} is constant, then the compensation capacitor must be decreased. For the op amp circuits considered in this paper, changes in g_{m1} and the compensation capacitor do not produce large changes in ω_{co} for a given phase margin in the amplifier.

The ratio I_o/g_m can also be increased by using special design techniques on the input stage [11]. Such techniques incorporate a nonlinear mode of operation in the input stage whereby large input voltages (as found in slew limiting) cause the input stage to deliver large charging currents to the compensation capacitor. The practical realization of such techniques has unfortunately required the use of multiple emitter follower inputs or lateral p-n-p's which again degrade the input offset voltage and frequency response.

Finally, the ratio I_o/g_m can be increased by the use of FET input stages (either junction FET's or insulated gate FET's). FET's have the basic property that their I_o/g_m ratio is much larger than that for a bipolar transistor. (Typically about 1 V compared to a bipolar transistor value of 52 mV.) However, these devices display a much larger offset voltage than bipolar transistors for a given percentage mismatch in the geometrical dimensions of the amplifier devices or of the load devices [12], so that the offset voltage still must be made poorer to achieve more slew rate at a given crossover frequency.

Future improvements in slew rate will require the development of techniques that allow large charging currents to be delivered to the compensation point, while minimizing contributions to input offset voltage and degradation of small-signal bandwidth due to the additional parasitic capacitance. This will probably be both technology dependent and circuit design dependent. Slew rate is also obviously increased by any techniques which increase the frequency of the next most dominant pole in the circuit. This problem is considered in the next section.

C. Limitations on Stable Bandwidth

In Section II, it was pointed out that the total time required for an op amp to settle to a new value of output voltage following a large step input is composed of two parts: a period during which the amplifier is outside the quasi-linear region and is slewing, and a second period during which the amplifier is in the quasi-linear region and settling to a final output voltage. The duration of the slewing period is inversely proportional to the unity gain bandwidth as shown in Section IIIA. The duration of the quasi-linear period is also inversely dependent on the unity gain bandwidth. Thus a principle aim in high-speed op amp design is the achievement of high crossover frequency, with sufficient phase margin to give an optimally damped response. The crossover frequency is thus determined by the location of the higher frequency poles, which contribute excess phase at frequencies at and above crossover. These higher frequency poles are contributed by the falloff of the gain of the various stages within the amplifier. For the purpose of this discussion, the mechanisms which result in these poles will be divided into three groups and discussed separately. One potential limitation on frequency response is the minority carrier base transit time of any lateral p-n-p transistors used in the signal path. Recent designs have overcome this limitation, and these techniques are discussed first. The more fundamental limiting factors are the parasitic capacitances associated with p-n junctions within the active devices and those p-n junctions used for isolation, and minority carrier charge storage in the base of the n-p-n transistors in the circuit. These two factors are discussed in the last section.

1) Effect of p-n-p Minority Carrier Transit Time on Crossover Frequency: Conventional integrated circuit technology inherently produces n-p-n transistors with peak f_T in the 600–800-MHz range. However, the p-n-p transistors produced by the same technology have peak f_T of about 2–5 MHz, and this has been a significant limitation in the past in the development of amplifiers with ω_{co} much beyond 1 MHz. This is so because of the requirement that a general purpose op amp have a common mode input range that extends to within a volt or two of each power supply. In order to achieve this, a dc level shift must be accomplished somewhere in the amplifier. An efficient way of accomplishing this without degrading offset voltage or PSRR is with a complementary device, such as a p-n-p transistor, in common base or common emitter configuration. p-channel MOS transistors can accomplish the same function. In the circuits shown in Figs. 8 and 9, common base p-n-p transistors are used.

It is interesting to note [13] that the primary limitation on the bandwidth of μA 741-type op amps ($\omega_{co} \approx 1$ MHz) is not the lateral p-n-p but rather an accumulation of effects due to parasitic junction capacitances. However, if one attempts to achieve broader bandwidth (for example, by raising currents and thus reducing impedance levels and parasitic capacitance effects) the frequency response of lateral p-n-p transistors quickly becomes dominant.

Two approaches have been taken to circumvent the p-n-p frequency response limitation. Several manufacturers have used a more complex technology to produce a high-speed double-diffused vertical p-n-p transistor on the same chip as the n-p-n transistor [7]. Very high performance levels (slew rate of 120 V/μs, $\omega_{co} = 25$ MHz) have been achieved, but at the expense of a complex and costly process. p-channel MOS devices have also been used, but again the process required is complex.

The second approach is to provide a second, ac coupled signal path in parallel with the path through the p-n-p transistor [14]–[16]. This approach has come to be called feedforward, and a typical example is illustrated in Fig. 13. The capacitor C_F provides a high-frequency feedforward path around the lateral p-n-p, as is illustrated schematically in Fig. 14. Using a simple single-lump hybrid transistor model, the short circuit output current of the circuit of Fig. 13 can be shown to be

$$i_{\text{out}} = g_{m1} v_i \left[\frac{1 + s\left(\dfrac{C_F}{g_{m2}}\right)}{1 + s\left(\dfrac{C_F + C_{\pi 2}}{g_{m2}}\right)} \right] \tag{42}$$

where charge storage in $Q1A$ and B has been neglected. This simple example illustrates both the advantages and drawbacks of feedforward compensation. A zero is introduced by C_F, which broad-bands the level shift stage, but in practice this zero does not precisely cancel the pole introduced by the lateral p-n-p. The resulting pole-zero pair, or doublet, can be shown to have a deleterious effect on the step response of the amplifier [17]. Although the doublet separation can be made nominally zero, component variations typically seen in IC's can cause a ± 30 percent mismatch in pole and zero locations.

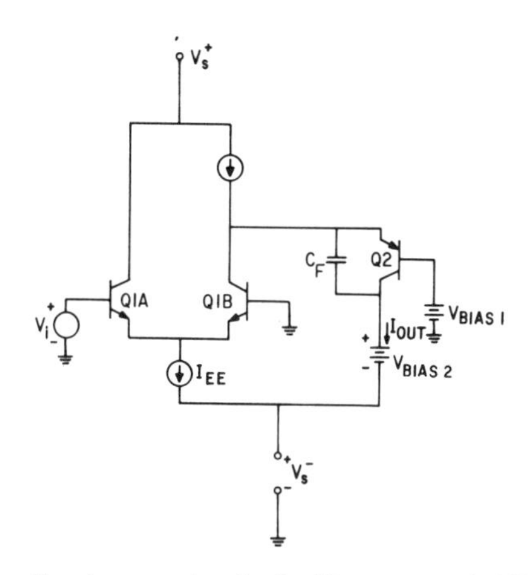

Fig. 13. Simple example of a feedforward level shift circuit.

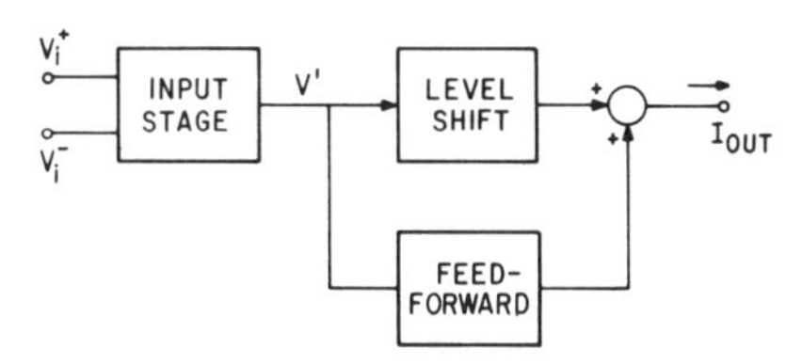

Fig. 14. Block diagram representation of feedforward level shift circuit.

This effect of this doublet can be illustrated by calculation of the step response of the amplifier in the quasi-linear region. For simplicity, we neglect all the higher order poles of the amplifier, so that the open-loop gain is given by

$$A_{oL}(s) = \frac{A_{dm}}{\left(1 + \dfrac{sA_{dm}}{\omega_{co}}\right)} \left[\frac{1 + \dfrac{s}{\omega_1}}{1 + \dfrac{s}{\omega_2}}\right] \tag{43}$$

where

ω_{co} unity gain bandwidth;
ω_1 doublet zero frequency;
ω_2 doublet pole frequency;
A_{dm} open-loop dc gain.

The unity gain closed-loop gain is then given by

$$G(s) = \frac{A_{oL}}{1 + A_{oL}}$$

$$= \frac{1 + \dfrac{s}{\omega_1}}{\left(1 + \dfrac{1}{A_{dm}}\right) + s\left[\dfrac{1}{\omega_{co}} + \dfrac{1}{\omega_1} + \dfrac{1}{A_{dm}\omega_2}\right] + \dfrac{s^2}{\omega_{co}\omega_2}}. \tag{44}$$

In practical amplifiers, $A_{dm} \gg 1$, so that

$$G(s) \cong \frac{1 + \dfrac{s}{\omega_1}}{1 + s\left(\dfrac{1}{\omega_{co}} + \dfrac{1}{\omega_1}\right) + \dfrac{s^2}{\omega_{co}\omega_2}} = \frac{1 + \dfrac{s}{\omega_1}}{\left(1 + \dfrac{s}{\omega_a}\right)\left(1 + \dfrac{s}{\omega_b}\right)} \tag{45}$$

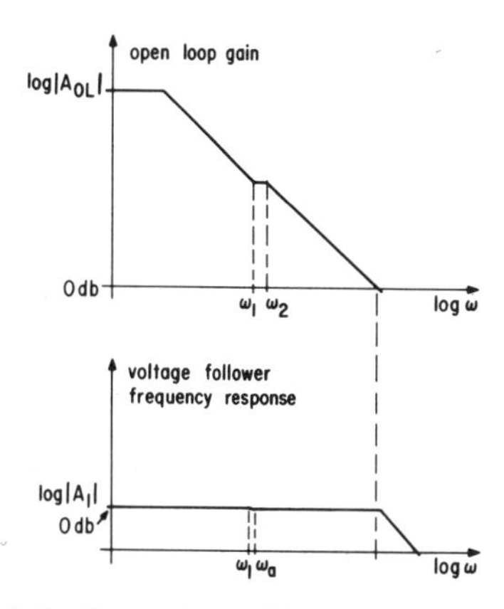

Fig. 15. Typical feedforward amplifier open-loop and closed-loop frequency response.

where

$$\omega_a \approx \omega_1 \left[1 - \frac{\left(1 - \dfrac{\omega_2}{\omega_1}\right)}{\left(\dfrac{\omega_{co}}{\omega_1} - 1\right)}\right] \tag{46}$$

$$\omega_b \approx \omega_{co}$$

if

$$|\omega_{co}| \gg |\omega_1 - \omega_2|. \tag{47}$$

Examination of (46) shows that in the closed-loop configurations, the doublet separation is reduced by an amount equal to the loop gain at the doublet frequency, which is ω_{co}/ω_1 under the assumption stated. A typical open-loop and closed-loop response is shown in Fig. 15.

Under the assumption of quasi-linear operation, the step response of the closed-loop circuit can be shown to be

$$V_{out}(t) = V_{in}[1 - k_1 e^{-t/\tau_1} - k_2 e^{-t/\tau_2}] \tag{48}$$

where

$$\tau_1 = \frac{1}{\omega_{co}} \tag{49}$$

$$\tau_2 = \frac{1}{\omega_a} \tag{50}$$

$$k_1 = 1 - k_2 \tag{51}$$

$$k_2 = \frac{\omega_1 - \omega_a}{\omega_1}\left[\frac{1}{1 - \dfrac{\omega_a}{\omega_{co}}}\right]. \tag{52}$$

The slow settling component k_2 has a magnitude which is proportional to the closed-loop doublet separation, which in turn is proportional to the open-loop doublet separation and inversely proportional to the loop gain at the doublet frequency. The resulting transient response in the quasi-linear region has a slow settling component in the output due to the doublet, with a typical shape shown in Fig. 16. This component is particularly troublesome when the settling time to high accuracy (1 mV, for example) is important, as is often the case in analog–digital conversion.

This problem has been alleviated in a recent design [18]

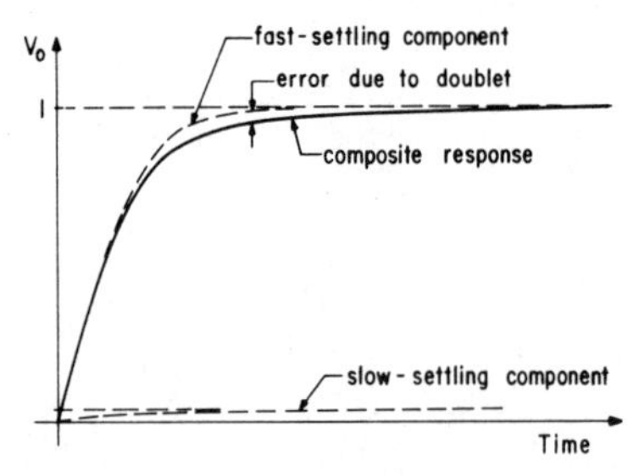

Fig. 16. Quasi-linear region step response corresponding to the amplifier of Fig. 15.

by including a feedback loop within the amplifier to compress the open-loop doublet, as illustrated in Figs. 17 and 18. This technique has been termed doublet compression. It can be shown that the doublet can be compressed by a factor roughly equal to the loop gain at the doublet frequency in the feedback loop. A complete circuit example, the Fairchild μA 772 amplifier, is shown in Fig. 19. By the use of these techniques, the frequency response limitations imposed by the lateral p-n-p level shift stage have been greatly eased.

2) Other Sources of Higher Frequency Poles: High frequency poles in the op amp frequency response are contributed in part by the depletion layer capacitance inherent in p-n junctions. In most op amp circuits, the excess phase shift in the transfer function which limits the stable bandwidth is not due to a single parasitic, but comes from a number of parasitics associated with different portions of the amplifier.

These parasitic capacitors have the property that they are only weakly dependent on the bias *currents* within the circuit. Thus if all the node bias voltages are kept the same, the natural frequencies of the circuit can be increased by simply lowering all the resistances in the circuit and increasing the bias currents. The input conductance, transconductance, and output conductance in the hybrid-pi model for bipolar transistors are proportional to collector bias current and will thus increase. An arbitrarily large bandwidth could be achieved at the expense of bias current, and hence power dissipation, if parasitic capacitance were the only factor limiting stable bandwidth. In fact, most high-speed op amps operate near the maximum power dissipation consistent with low-cost packaging. The use of large bias currents, within the amplifier circuit to achieve large bandwidth also results in a relatively large input bias current when bipolar transistors are used as the input devices. Another property of the parasitics is that they are proportional to area, so that the smaller the geometry of the devices used, the better the frequency response at a given power dissipation. Finally, the use of more sophisticated technology, such as dielectric isolation and thin film resistors, can reduce the parasitics associated with junction isolation and diffused resistors, respectively.

Unfortunately, the minority carrier storage within the n-p-n transistors ultimately limits the stable bandwidth to a value on the order of a fraction of the peak f_T of the transistor. This frequency is related to transistor base width and cannot be increased by increasing bias current levels. However, practical design constraints, such as those on input bias current and power dissipation, have prevented the approach to this fundamental limit. While current technology is capable of producing integrated circuit transistors with f_T of greater than 1 GHz, the highest stable bandwidth reported to date for a generally useful monolithic op amp has been only 50 MHz [19]. While clever circuit approaches will certainly result in continuing evolutionary performance improvements, the next step improvement in general purpose amplifier performance will probably result from technological reduction in the physical size of the circuit components, which will result in smaller parasitic capacitance and higher bandwidth for a given level of power dissipation and input bias current.

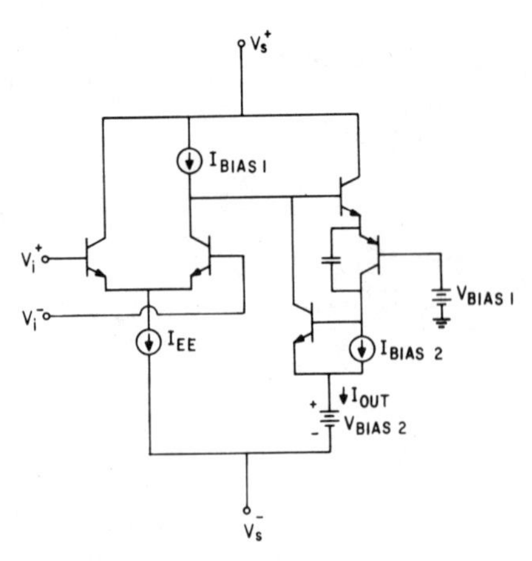

Fig. 17. Simplified doublet compression circuit.

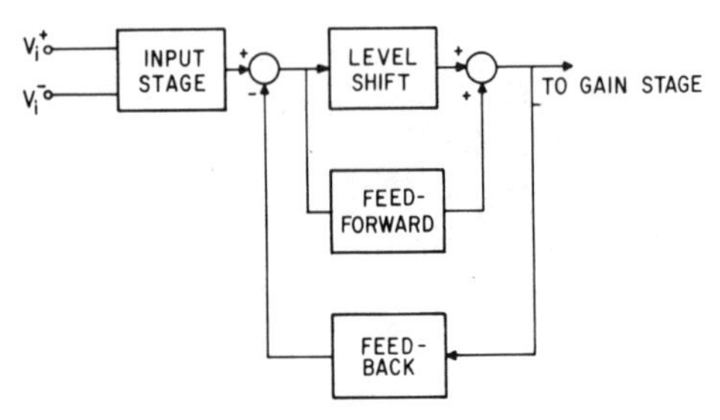

Fig. 18. Block diagram of doublet compression circuit.

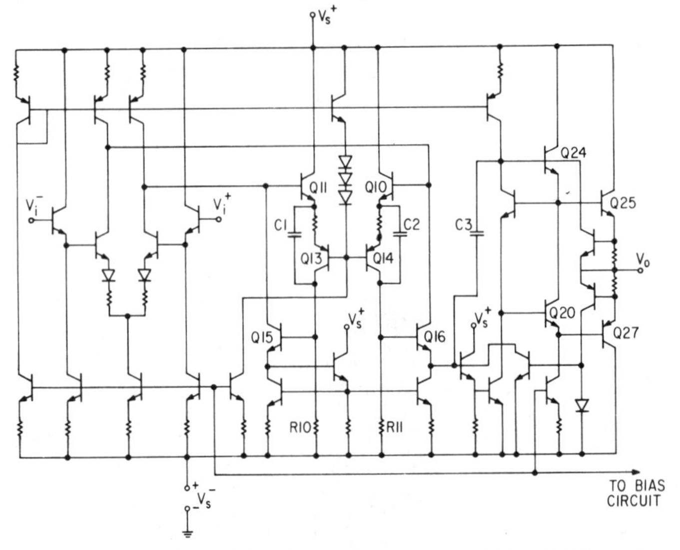

Fig. 19. Complete schematic of Fairchild μA 772 amplifier using doublet compression circuit.

References

[1] M. A. Maidique, "A high-precision monolithic super-beta operational amplifier," *IEEE J. Solid-State Circuits*, vol. SC-7, pp. 480–487, Dec. 1972.

[2] P. R. Gray, "A 15-W monolithic power operational amplifier," *IEEE J. Solid-State Circuits*, vol. SC-7, pp. 474–480, Dec. 1972.

[3] G. Meyer-Brotz and A. Kley, "The common-mode rejection of transistor differential amplifiers," *IEEE Trans. Circuit Theory*, vol. CT-13, pp. 171–175, June 1966.

[4] G. Erdi, "Common-mode rejection of monolithic operational amplifiers," *IEEE J. Solid-State Circuits* (Corresp.), vol. SC-5, pp. 365–367, Dec. 1970.

[5] R. J. Widlar, "Design techniques for monolithic operational amplifiers," *IEEE J. Solid-State Circuits*, vol. SC-4, pp. 184–191, Aug. 1969.

[6] T. M. Frederiksen, W. F. Davis, and D. W. Zobel, "A new current-differencing single-supply operational amplifier," *IEEE J. Solid-State Circuits*, vol. SC-6, pp. 340–347, Dec. 1971.

[7] HA-2600/2602/2605 Operational Amplifier Data Sheet, Harris Semiconductor Corp., June 1971.

[8] J. E. Solomon, W. R. Davis, and P. L. Lee, "A self compensated monolithic operational amplifier with low input current and high slew rate," *ISSCC Digest Tech. Papers*, pp. 14–15, 1969.

[9] R. W. Russell and J. E. Solomon, "A high-voltage monolithic operational amplifier," *IEEE J. Solid-State Circuits*, vol. SC-6, pp. 352–357, Dec. 1971.

[10] "A low drift, low noise monolithic operational amplifier for low level signal processing," Fairchild Semiconductor, Appl. Brief 136, July 1969.

[11] W. E. Hearn, "Fast slewing monolithic operational amplifier," *IEEE J. Solid-State Circuits*, vol. SC-6, pp. 20–24, Feb. 1971.

[12] H. C. Lin, "Comparison of input offset voltage of differential amplifiers using bipolar transistors and field-effect transistors," *IEEE J. Solid-State Circuits* (Corresp.), vol. SC-5, pp. 126–128, June 1970.

[13] B. A. Wooley, S. J. Wong, and D. O. Pederson, "A computer-aided evaluation of the 741 amplifier," *IEEE J. Solid-State Circuits*, vol. SC-6, pp. 357–366, Dec. 1971.

[14] T. J. van Kessel, "An integrated operational amplifier with a novel HF behaviour," *ISSCC Digest Tech. Papers*, pp. 22–23, 1968.

[15] H. Krabbe, "Stable monolithic inverting op amp with 130 V/μs slew rate," *ISSCC Digest Tech. Papers*, pp. 172–173, 1972.

[16] R. C. Dobkin, "LM 118 op amp slews 70 V/μs," National Semiconductor Linear Brief 17, Aug. 1971.

[17] F. D. Waldhauer, "Analog integrated circuits of large bandwidth," in *1963 IEEE Conv. Rec.*, Part 2, pp. 200–207.

[18] R. J. Apfel and P. R. Gray, "A monolithic fast-settling feed-forward operational amplifier using doublet compression techniques," *ISSCC Digest Tech. Papers*, pp. 134–155, 1974.

[19] P. C. Davis and V. R. Saari, "High slew rate monolithic operational amplifier using compatible complementary PNP's," *ISSCC Digest Tech. Papers*, pp. 132–133, 1974.

A New Current-Differencing Single-Supply Operational Amplifier

THOMAS M. FREDERIKSEN, MEMBER, IEEE, WILLIAM F. DAVIS, AND D. W. ZOBEL

Abstract—**The conventional integrated-circuit operational amplifier is not well suited to many system applications that operate from only a single power-supply voltage. To more optimally meet the requirements of industrial control systems, a new current-differencing opamp has been developed that uses a simple circuit to provide a gain element that out performs the 741 IC opamp. As a result of the circuit simplicity, multiple opamps are possible and six independent internally compensated amplifiers have been fabricated on a single 80 × 93-mil die. Many circuits are presented only not to show how this circuit can perform most of the application functions of a standard IC opamp, but also to indicate the increased usefulness of this new input current-differencing type of opamp circuit in single power-supply control system applications.**

I. Introduction

Many industrial electronic control systems are designed that operate off of only a single power-supply voltage. The conventional integrated-circuit operational amplifier (IC opamp) is typically designed for split power supplies (± 15 V_{dc}) and suffers from a poor output voltage swing and a rather large minimum common-mode input voltage range (approximately $+ 2$ V_{dc}) when used in a single power-supply application. In addition, some of the performance characteristics of these opamps could be sacrificed—especially in favor of reduced costs.

To meet the needs of the designers of low-cost single-power-supply control systems, a new internally compensated opamp has been designed that operates over a power-supply voltage range of $+3.5$ V_{dc} to $+35$ V_{dc} with small changes in performance characteristics and provides an output peak-to-peak voltage swing that is only 1 V less than the magnitude of the power-supply voltage. As many as 6 of these amplifiers have been fabricated on a single chip and are provided in the standard 14-pin dual-in-line package (inverting inputs only).

II. Basic Gain Stage

The simplest inverting amplifier is the common-emitter stage. If a current source is used in place of a load resistor, a large open-loop gain can be obtained, even at low power-supply voltages [1]. This basic stage (Fig. 1) is used for the single-supply opamp.

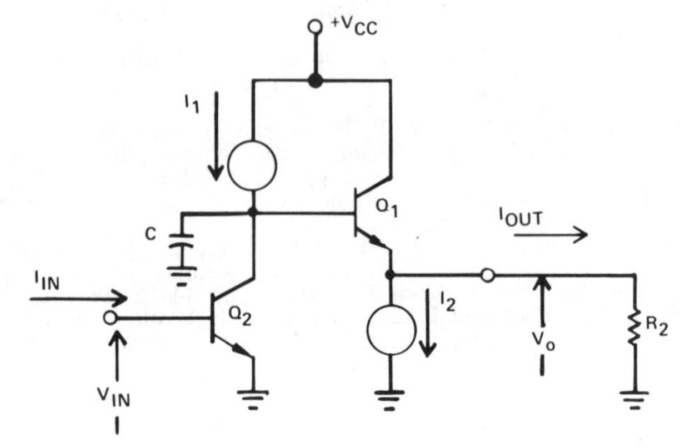

Fig. 1. Basic gain stage.

All of the voltage gain is provided by the gain transistor Q_2 and an output emitter–follower transistor Q_1 serves to isolate the load impedance R_2 from the high impedance that exists at the collector of the gain transistor Q_2. Closed-loop stability is guaranteed by the on-chip capacitor $C = 3$ pF, which provides the single dominant open-loop pole. The output emitter–follower is biased in class-A operation by the current source I_2 and will therefore source more current to the external load R_2 than it will absorb from this load. In single supply applications this is the typical requirement as the amplifiers are usually driving current into grounded load impedances.

This basic stage can provide an adequate open-loop voltage gain (70 dB) and has the desired large output voltage swing capability. A disadvantage of this circuit is that the dc input current I_{in} is large as it is essentially equal to the maximum output current I_{out} divided by β^2. For example, for an output current capability of 10 mA the input current would be at least 1 μA (assuming $\beta^2 = 10^4$). It would be desirable to further reduce this by adding an additional transistor to achieve an overall β^3 reduction. Unfortunately, if a transistor is added at the output (by making Q_1 a Darlington pair) the peak-to-peak output voltage swing would be somewhat reduced and if Q_2 were made a Darlington pair the dc input voltage level would be undesirably doubled.

To overcome these problems a p-n-p transistor has been added as shown in Fig. 2. This connection neither reduces the output voltage swing nor raises the dc input

Manuscript received May 12, 1971; revised August 6, 1971.
T. M. Frederiksen is with the Semiconductor Products Division, Motorola, Inc., Phoenix, Ariz. 85036.
D. W. Zobel was with Motorola, Inc., Phoenix, Ariz. He is now with National Semiconductor, Santa Clara, Calif. 95051.
W. F. Davis is with the Semiconductor Products Division, Motorola, Inc., Phoenix, Ariz. 85036.

Reprinted from *IEEE J. Solid-State Circuits*, vol. SC-6, pp. 340-347, Dec. 1971.

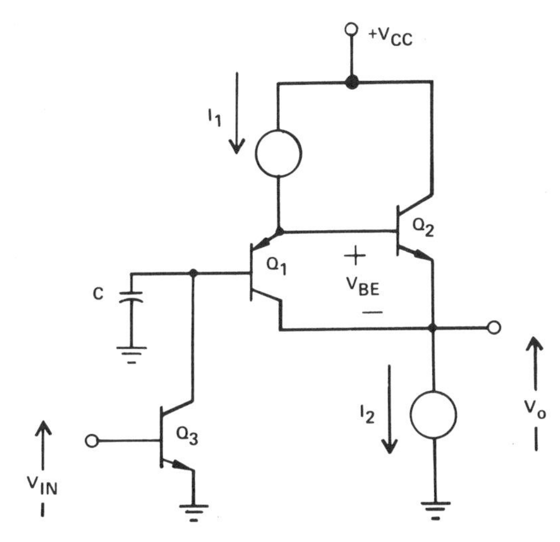

Fig. 2. Adding a p-n-p transistor to the basic gain stage.

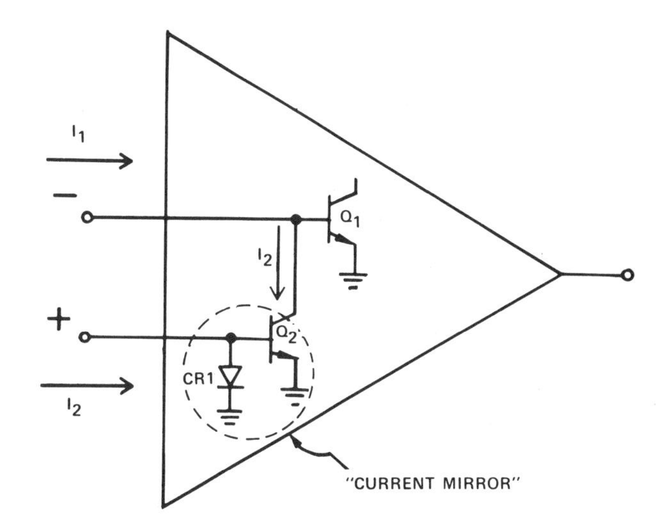

Fig. 3. Adding a current mirror to achieve a noninverting input.

voltage, but does provide the required additional gain that was needed to reduce the input current. Notice that the collector of this p-n-p transistor Q_1 is connected directly to the output terminal. This "bootstraps" the output impedance of Q_1 and therefore reduces the loading at the high-impedance collector of the gain transistor Q_3. In addition, the base-emitter junction of the output emitter-follower transistor Q_2 guarantees that the emitter-to-collector voltage of this added p-n-p transistor will always be clamped at one diode forward voltage (the V_{BE} of Q_2). This allows the design of a super-β lateral p-n-p transistor similar to the n-p-n input transistors that are used in the new IC opamp front-end designs. For the lateral p-n-p this is accomplished by providing a reduced lateral spacing between the emitter and collector regions in the device geometry and no additional steps are required in the processing.

A. Obtaining a Noninverting Input Option

The circuit of Fig. 2 has only the inverting input. A general purpose opamp requires two input terminals to obtain both an inverting and a noninverting input. In conventional opamp designs, an input differential amplifier provides these required inputs. The output voltage then depends upon the difference (or error) between the two input voltages. An input common-mode voltage range specification exists and, basically, input voltages are compared.

For circuit simplicity and ease of application (see Section IV) a noninverting input can be provided by adding a standard IC "current-mirror" circuit directly across the inverting input terminal as shown in Fig. 3. This operates in the current mode as now input currents are compared or differenced (this can be though of as a Norton differential amplifier). There is essentially no input common-mode voltage range directly at the input terminals—as both inputs will bias at one diode drop above ground—but if the input voltages are converted to currents (by use of input resistors), there is then no limit to the common-mode input voltage range. This is especially useful in high-voltage comparator applications. By making use of the input resistors, to convert input voltages to input currents, all of the standard opamp applications can be realized. Many additional applications are easily achieved, especially when operating with only a single power-supply voltage. This results from the built-in voltage biasing that exists at both inputs (each input biases at $+V_{BE}$) and additional resistors are not required to provide a suitable common-mode input dc biasing voltage level. Further, input summing can be performed at the relatively low impedance level of the input diode of the current-mirror circuit.

III. Complete Single-Supply Opamp

The circuit schematic for a single opamp stage is shown in Fig. 4. Due to the circuit simplicity, many of these amplifiers can be fabricated on a single chip. One common biasing circuit, which will be discussed in the next section, supplies a bias reference voltage to all of the individual amplifiers. This common bias voltage is equal to the forward voltage drop of three silicon diodes ($3\ V_{BE}$) and is used to bias both of the current sources of each of the amplifier stages.

Each amplifier uses an emitter-follower transistor Q_8 to isolate the stage from the common biasing voltage line. Therefore, even during saturated overdrive this common bias line is not disturbed. The magnitude of the biasing current is established by a resistor R in the emitter lead of this transistor and is equal to V_{BE}/R. This makes the magnitude of the biasing currents independent of the magnitude of the power-supply voltage and gives first-order compensation for β variations with temperature in the output emitter-follower, Q_5 (less drive current is needed at elevated temperatures).

The collector current of Q_8 (essentially also V_{BE}/R) is provided by the collector of the transistor Q_1 (an additional p-n-p transistor, Q_3 is used to reduce base cur-

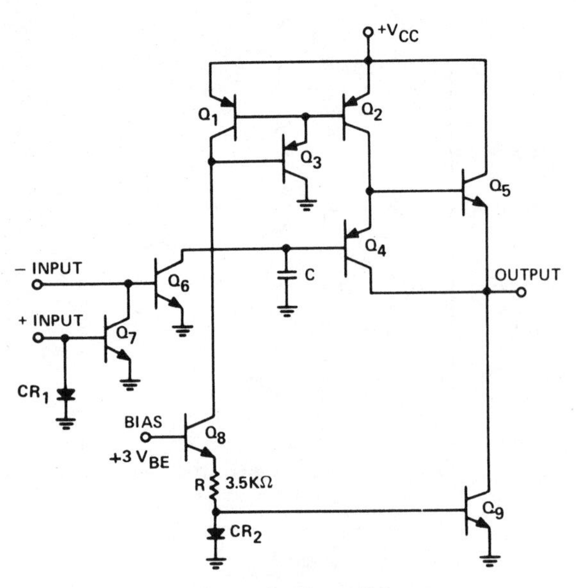

Fig. 4. Operational amplifier stage.

rent errors) and a matched lateral p-n-p transistor Q_2 provides a base drive to the output emitter–follower Q_5, which is also essentially equal to V_{BE}/R. Emitter area scaling is used between CR_2 and the output current source transistor Q_9 to give the 6:1 ratio that is needed to provide a 1-mA output current sink capability for the amplifier.

The performance characteristics of the opamp are given below.

Power-supply voltage range	3.5–35 V_{dc}.
Bias current drain per amplifier stage	1.4 mA dc.
Open loop:	
voltage gain (R_L = 10 k)	70 dB;
unity-gain bandwidth	4 MHz;
phase margin	70 degrees;
imput impedance	500 kΩ;
output impedance	3 kΩ.
Output voltage swing	$(V_{cc} - 1)V_{pp}$.
Input bias current	35 nA dc.
Slew rate	0.8 V/μs.

As the bias currents are all derived from V_{BE}/R, there is essentially no change in bias current magnitude as the power-supply voltage is varied. The open-loop gain varies less than ±2 dB over the temperature range of −55° to +125°C (as a result of only one kT/q term in the denominator of the gain expression) and is essentially independent of the magnitude of the power-supply voltage over the full 3.5–35 V_{dc} range. The open-loop frequency response is compared with the 741 opamp in Fig. 5 [1]. The higher unity-gain crossover frequency is seen to provide an additional 12-dB gain for all frequencies greater than 2 kHz.

A photomicrograph of the die is shown in Fig. 6. This is metalized for the six-amplifier option. The six on-chip MOS compensation capacitors can be seen, one associated with each amplifier cell. The die size is 80 × 93 mils.

A. Power-Supply Independent Biasing Reference

It is desirable to provide an amplifier stage in which the bias currents are independent of the magnitude of the power-supply voltage. Further, if it is also required that the amplifier operate from very small power-supply voltage levels (as 3–4 V_{dc}), the standard Zener-diode bias regulator circuits cannot be used. The forward voltage drop of three diodes was therefore used as the common bias voltage reference.

A schematic of the biasing circuit is shown in Fig. 7. To guarantee the initial bias setup, a START circuit is used. This consists of the resistor R_1, which provides current to the diodes CR_1 and CR_2. The transistor Q_3 of the differential amplifier Q_3 and Q_4 is thus initially brought into conduction (V_{BE}/R_2). This causes Q_2 to conduct; which, in turn, biases the multiple-collector lateral p-n-p transistor Q_1 into conduction. The three diodes CR_3, CR_4, and CR_5 now become forward biased and the differential ampilfier fully switches with Q_4 going ON and Q_3 turning OFF. This automatically disconnects the START circuit once the reference diode string is properly biased. The current-source transistor Q_5 is used to maintain the sustaining bias for the reference diode string.

B. DC Reference Voltage Generator

Many control systems require a reference voltage generator that provides a dc voltage that is constant in magnitude as the ambient temperature changes. This allows transducer output voltages to be monitored and compared with this dc reference voltage. Basically, if a Zener diode is added to the opamp as a shunt feedback element, a dc reference voltage generator can be obtained, see Fig. 8. As the temperature coefficient (TC) of the standard IC Zener diode is somewhat more positive than the negative TC of the V_{BE} of the inverting input transistor, resistors R_1 and R_2 have been incorporated to produce the desired zero TC reference voltage at the output. An input current I_0 is supplied to the noninverting input, which is equal to V_{BE}/R_1. The current-mirror circuit, which exists at this input, causes this same current I_0 to be drawn from the output terminal of the opamp through the Zener diode (which serves to bias the Zener) and through the resistor R_2. The voltage drop across R_2 is therefore nV_{BE} where n is adjusted by the ratio of R_2 to R_1. This introduces a variable negative TC into the network, which establishes the output voltage and allows the desired zero TC to be obtained. From Fig. 8 it can be seen that

$$V_{REF} = V_Z + V_{R2} + V_{BE} \tag{1}$$

and

$$V_{R2} = I_0 R_2 \tag{2}$$

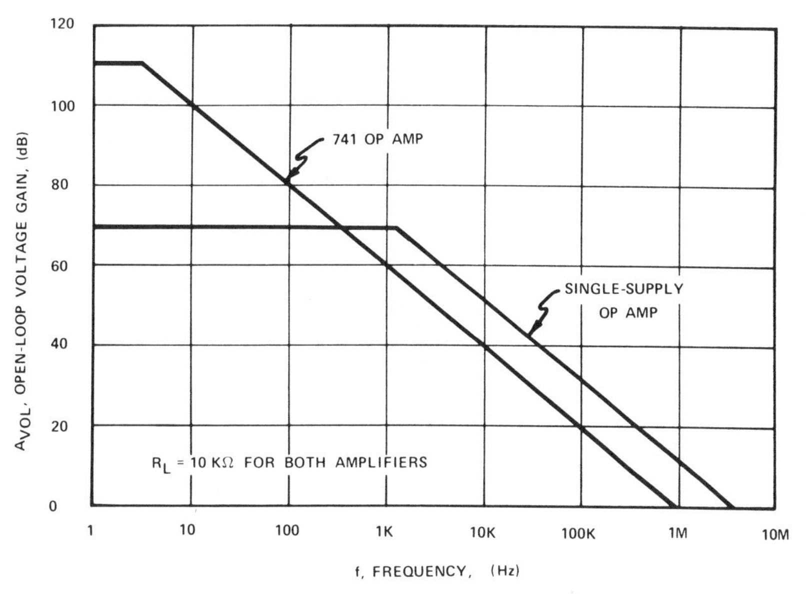

Fig. 5. Open-loop gain characteristics.

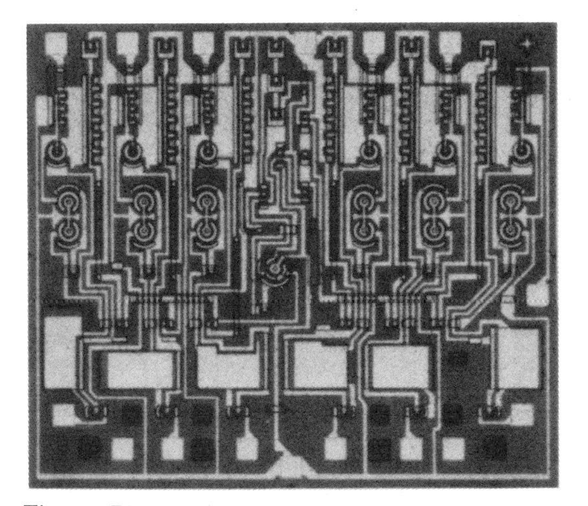

Fig. 6. Photomicrograph of the six-amplifier chip.

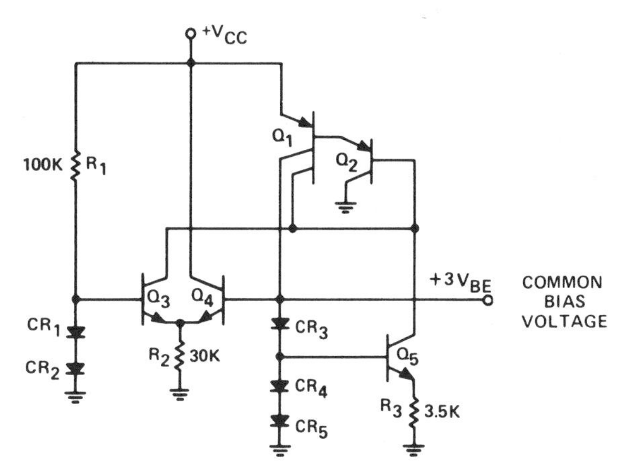

Fig. 7. Bias reference and START circuit.

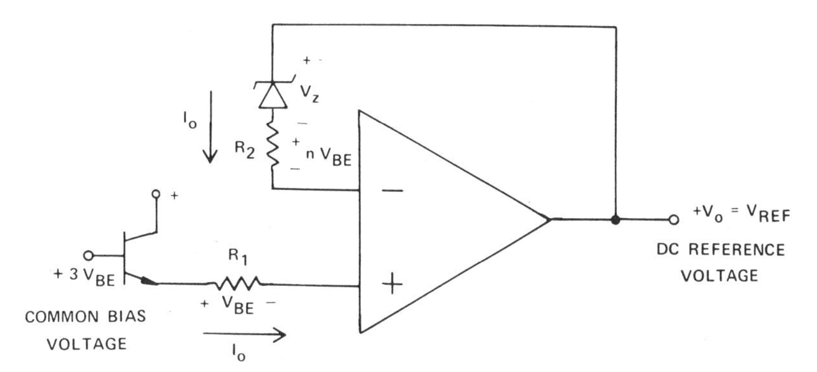

Fig. 8. Simplified dc reference voltage generator.

where

$$I_0 = V_{BE}/R_1 \tag{3}$$

combining (2) and (3) into (1) gives

$$V_{REF} = V_Z + (1 + R_2/R_1)V_{BE} \tag{4}$$

and the temperature change is given by

$$\frac{\partial V_{REF}}{\partial T} = \frac{\partial V_Z}{\partial T} + (1 + R_2/R_1)\frac{\partial V_{BE}}{\partial T}. \tag{5}$$

As a result of the opposite signs of $\partial V_Z/\partial T$ and $\partial V_{BE}/\partial T$, the proper choice of (R_2/R_1) can force

$$\frac{\partial V_{REF}}{\partial T} = 0,$$

which provides the desired temperature independent dc output reference voltage.

The complete circuitry for the dc reference voltage generator is shown in Fig. 9. An additional output emitter-follower transistor Q_6 has been added to provide a reduced output impedance. This circuit is a series voltage regulator that has the output voltage internally adjusted and is also internally frequency compensated. Only one pin in therefore needed to provide this dc reference generator option. The added 7-V Zenner diode does raise the minimum required supply voltage to 9.3 V_{dc}.

C. Functional Options

The standard 14-pin package allows four independent dual-input amplifiers to be provided in a single package—each amplifier requires three pins and $+V_{cc}$ and ground are common. Six independent single-input (inverting) amplifiers can also be obtained and other amplifier combinations that include the dc reference voltage generator are possible. Four of these options are shown in Fig. 10. The diode that is shown in series with the output terminal in Fig. 10 can be used to temperature compensate for the input V_{BE} of the amplifier.

IV. Typical Applications

A few typical applications will show the usefulness of this new type of input-current differencing opamp in single supply systems.

A. Biasing for AC Amplifiers

The circuits of Fig. 11 indicate how the dual-input amplifier can be setup in both an ac feedback mode and a dc biasing mode simultaneously by providing separate signals to the two input pins. The ac input signal is applied to the inverting input terminal and a separate dc input voltage is applied to the noninverting input terminal. The gains are individually adjusted to simultaneously give the desired closed-loop ac performance and the proper dc output level or quiescent point $(+V_{cc}/2)$.

A noninverting gain stage is shown in Fig. 12. In this

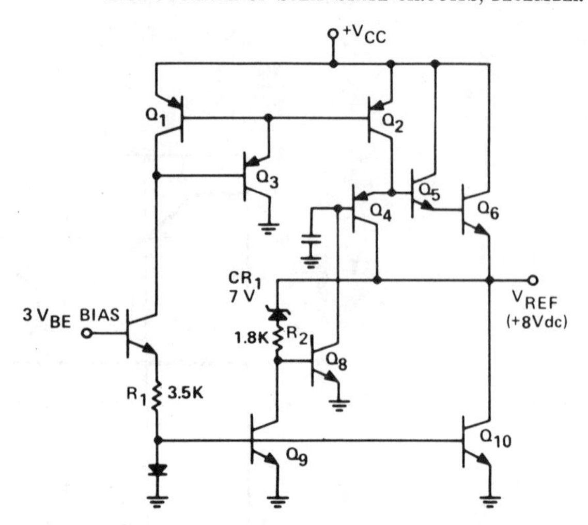

Fig. 9. Dc reference voltage generator.

circuit both the ac input signal and the dc input signal are summed at the noninverting input terminal.

B. Comparators

A comparator function can be achieved by simply comparing input currents as shown in Fig. 13(a), or hysteresis can be added as shown in Fig. 13(b). The reference input dc voltage can be supplied by the previously described dc reference voltage generator by providing for this option in the chip layout.

C. Averaging Circuits

Averaging circuits are useful in industrial tachometers. Either positive-going voltage pulses or current pulses are applied to the noninverting input terminal to provide an increasing dc output voltage for an increasing input frequency. A few averaging circuits are shown in Fig. 14.

D. Oscillators

A Wien-bridge sinewave oscillator is shown in Fig. 15(a) and a square wave oscillator is shown in Fig. 15(b). The basic Wien circuit has been slightly altered to allow biasing the output at $+V_{cc}/2$ and to add the required input resistors that are needed to convert the nodal voltages to input currents.

E. Biquadratic RC Active Filters

A useful application area for this multiple opamp is in the standard "biquad" *RC* active filter circuits [2]. These filters use additional opamps to reduce the sensitivity to the passive components. Two dc biasing techniques are indicated in Fig. 16. Fig. 16 (a) shows a simple connection that is suitable for small-signal levels (as all nodes bias at V_{BE}) and (b) indicates how all of the amplifier output nodes can be biased at $+V_{cc}/2$ for larger signal accommodation. The input coupling capacitor shown in Fig. 16(b) can be omitted if the value of the biasing resistor R, which is tied to the noninverting input of the first amplifier, is changed to compensate for the dc flow in the input lead.

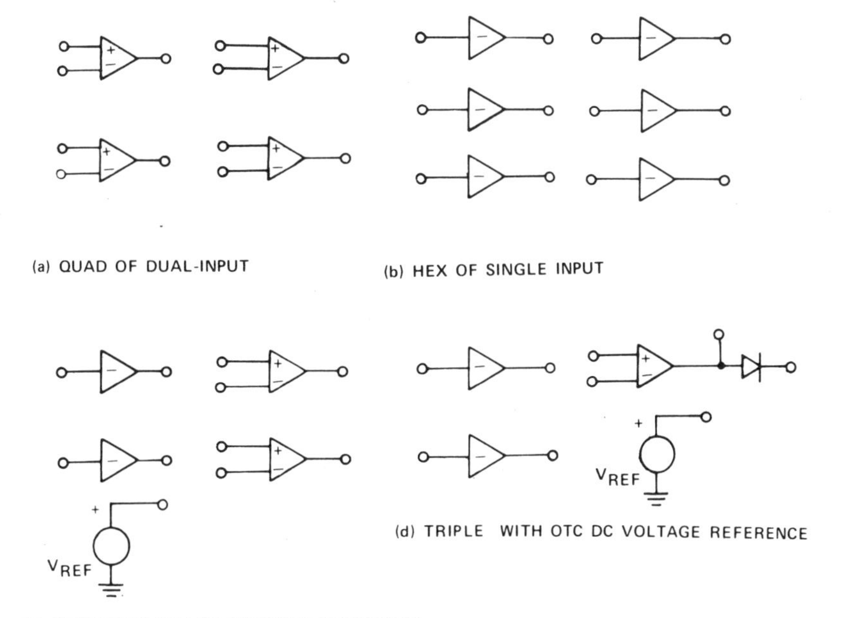

Fig. 10. Amplifier options.

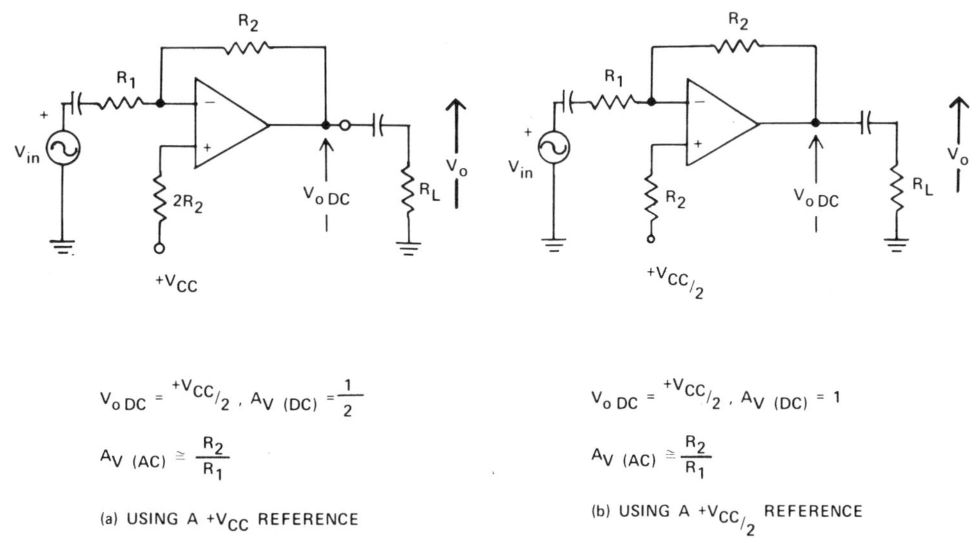

Fig. 11. Biasing circuits for ac amplifiers.

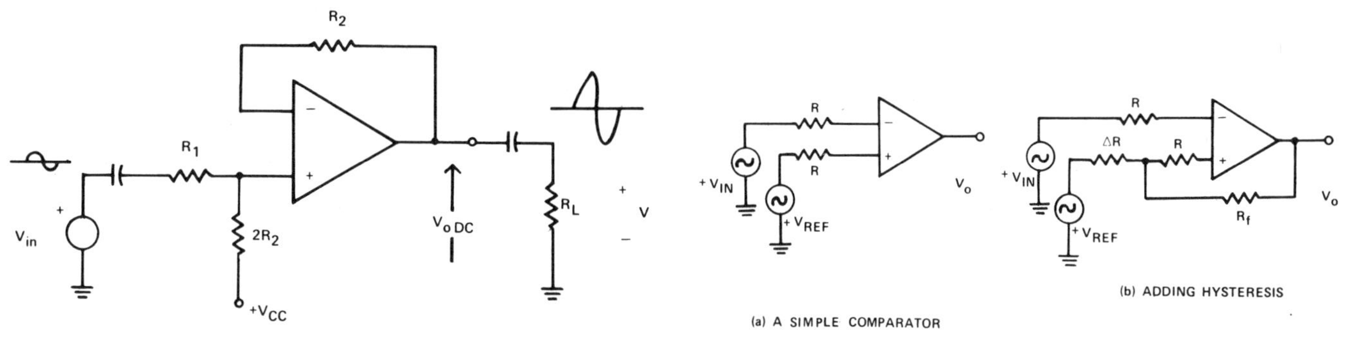

Fig. 12. Noninverting gain stage.

Fig. 13. Comparators.

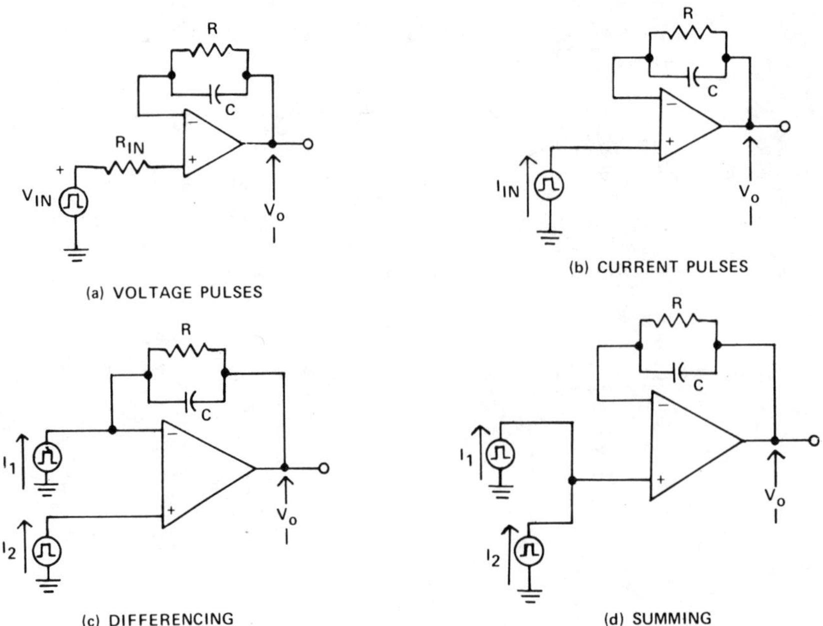

Fig. 14. Averaging circuits.

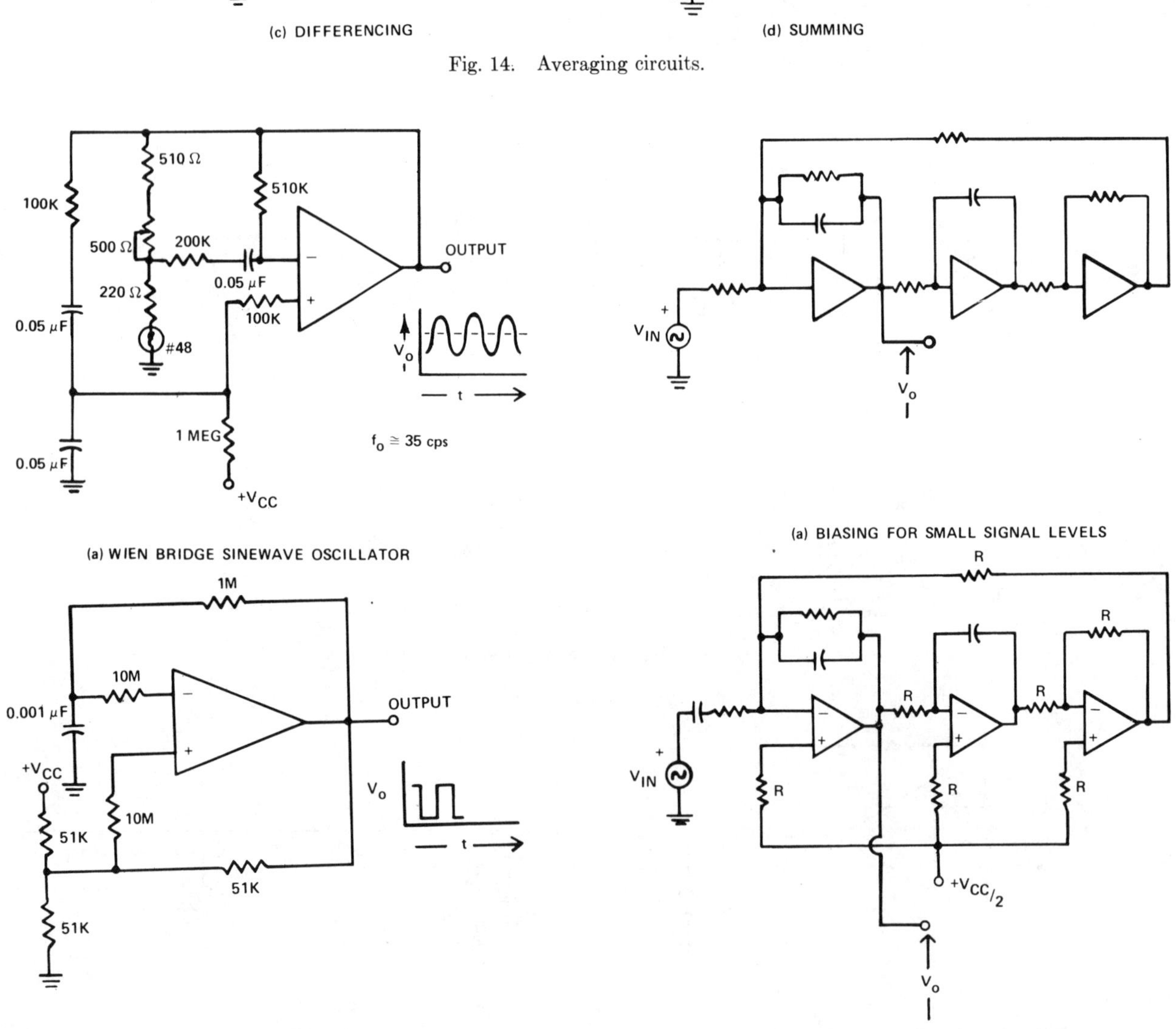

Fig. 15. Oscillator circuits.

Fig. 16. Biquadratic RC active filters.

V. Conclusions

A new type of opamp circuit has been presented that allows multiple amplifiers to be fabricated on a single chip. These amplifiers are well suited to applications in single-power-supply electronic systems and the performance characteristics are comparable to the 741 type of IC opamp. For single supply uses, these new circuits are actually superior in many respects to the standard IC opamps. The new Norton type of input-current differencing feature has also been shown to be actually more suitable than the conventional input differential amplifier in industrial control systems.

References

[1] E. L. Long and T. M. Frederiksen, "High-gain 15-W monolithic power amplifier with internal fault protection," *IEEE J. Solid-State Circuits,* vol. SC-6, Feb. 1971, pp. 35–44.

[2] J. Tow, "Design formulas for active RC filters using operational amplifier biquad," *Electron. Lett.,* July 24, 1969, pp. 339–341.

Macromodeling of Integrated Circuit Operational Amplifiers

GRAEME R. BOYLE, BARRY M. COHN, DONALD O. PEDERSON, FELLOW, IEEE, AND JAMES E. SOLOMON, MEMBER, IEEE

Abstract—**A macromodel has been developed for integrated circuit (IC) op amps which provides an excellent pin-for-pin representation. The model elements are those which are common to most circuit simulators. The macromodel is a factor of more than six times less complex than the original circuit, and provides simulated circuit responses that have run times which are an order of magnitude faster and less costly in comparison to modeling the op amp at the electronic device level.**

Expressions for the values of the elements of the macromodel are developed starting from values of typical response characteristics of the op amp. Examples are given for three representative op amps. In addition, the performance of the macromodel in linear and nonlinear systems is presented. For comparison, the simulated circuit performance when modeling at the device level is also demonstrated.

I. Introduction

INTEGRATED circuit (IC) simulators have proven to be a useful tool to the IC design engineer. Nonetheless, their widespread acceptance in the design of large-scale integrated circuits and IC subsystems has been impeded by excessive simulation costs and increasing convergence problems. Present simulators model semiconductor devices at the p-n junction and 2-terminal element level. Because of the large number of these devices in large-scale IC systems, the analysis can surpass the computer's memory capability, simulator circuit-size capability, or the inherent numerical accuracy of the computer. Even if an adequate simulator and computer are available, the required simulation time makes the analysis financially impractical. This paper describes one solution to this problem: macromodels which have been developed for IC's such as operational amplifiers and comparators.

The idea and use of macromodels in electronic circuit design is very common at the system level. For example, in developing an analog signal processor, one might utilize a number of ideal voltage amplifiers, integrators, and other subsystem blocks. In effect, a variety of zero-order circuit models are used. To determine the actual system performance, a prototype circuit is constructed and tested at the device level. The size and complexity of today's inexpensive IC's are large; therefore, the cost of using present simulators for design and evaluation can be very large. The cost for large IC's can only be justified if very large manufacture is anticipated. The costs and other problems can be relieved by the development of macromodels for IC's which provide an adequate pin-for-pin representation of the IC. For digital IC's, logic simulation and macromodels have been developed for digital logic blocks [1], [8]. For analog IC's, this paper describes a very effective macromodel that has been developed for IC op amps [2], [3], [9].

The aim of macromodeling is to obtain a circuit model of an IC or a portion of an IC which has a significantly reduced complexity to provide for smaller, less costly simulation time, or to permit the simulation of larger IC's or IC systems for the same time and cost. In the macromodel for IC op amps shown in Fig. 1, a reduction of approximately 6 in branch and node count has been achieved while providing a very close approximation to the actual performing op amp, i.e., accurate modeling of the input and output characteristics, differential- and common-mode gain versus frequency characteristics, quiescent dc characteristics, offset characteristics, and large-signal characteristics, such as slew rate, output voltage swing, and short-circuit current limiting. Further, since much of a simulation run is involved with iterative analysis to an equilibrium circuit solution, the reduction of 60 to 80 p-n junctions in an actual op amp to the 8 junctions in the macromodel of Fig. 1 indicates better how much faster and cheaper can be the simulation using the macromodels instead of device-level models. The results with amplifiers, timers, and filters that are cited in this paper show that a reduction in time of 6 to 10 is typical.

In many design or evaluation situations, it is not necessary to model an op amp in all of its performance characteristics. For example, maximum short-circuit current limiting may not be of interest. If the elements in the macromodel which provide this feature are eliminated, further simplification of the macromodel is obtained. As an example, the simulation time of the filter in Section IV is reduced by a factor of 1.4 if the current and voltage limiters are omitted.

Manuscript received August 2, 1974; revised August 16, 1974. This research was sponsored in part by the Joint Services Electronics Program under Contract F44620-71-C-0087 and by the National Science Foundation under Grant GK-17931. This paper was presented at the International Solid-State Circuits Conference, Philadelphia, Pa., February 1974.

G. R. Boyle and D. O. Pederson are with the Department of Electrical Engineering and Computer Sciences and the Electronics Research Laboratory, University of California, Berkeley, Calif.

B. M. Cohn is with Intel Corporation, Santa Clara, Calif.

J. E. Solomon is with National Semiconductor Corporation, Santa Clara, Calif.

Reprinted from *IEEE J. Solid-State Circuits*, vol. SC-9, pp. 353-363, Dec. 1974.

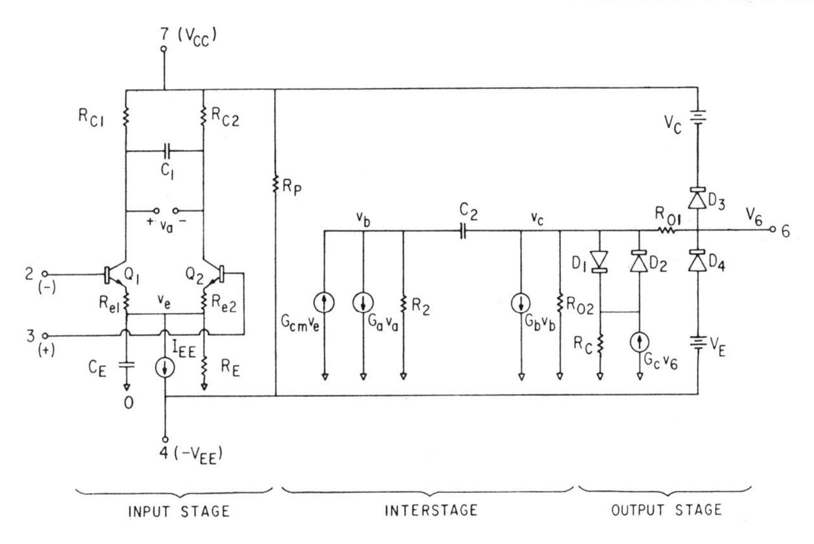

Fig. 1. Circuit diagram of the op amp macromodel.

II. Macromodel Development

The circuit model for an IC op amp which is developed in this paper is shown in Fig. 1. The configuration, with a suitable choice of parameters and elements, accurately models a broad class of IC op amps. For a given op amp, the model provides an essentially pin-for-pin correspondence with the op amp, and accurately represents the circuit behavior for nonlinear dc, ac, and large-signal transient responses.

The circuit of Fig. 1 is subdivided into three stages. The input stage consists of ideal transistors Q_1 and Q_2 and the associated sources and passive elements. This stage produces the necessary linear and nonlinear differential-mode (DM) and common-mode (CM) input characteristics. For convenience, the stage is designed for unity voltage gain. The stage can be designed to provide desired voltage and current offsets. As brought out in the next section, the capacitor C_E is used to introduce a second-order effect for the slew rate [4], and the capacitor C_1 introduces a second-order effect to the phase response.

The DM and CM voltage gains of the op amp are provided by the linear interstage and output stage elements consisting of G_{cm}, G_a, R_2, G_b, and R_{02}. The function of each element is presented in the next section. The dominant time constant of the op amp is produced with the internal feedback capacitor C_2. A feedback connection in the macromodel is used for C_2 in order to provide the necessary ac output resistance change with frequency. In addition, the two nodes of C_2 can be made available to the outside world in order that the circuit designer can introduce the same compensation modification as might be added to the actual op amp. Notice the complete isolation that exists between the input and the interior stages. This leads to a simplification of the frequency and the slew rate performances.

The output stage provides the proper dc and ac output resistance of the op amp. The elements D_1, D_2, R_C, and G_C produce the desired maximum short-circuit current. The elements D_3, V_C and D_4, V_E are voltage-clamp circuits to produce the desired maximum voltage excursion.

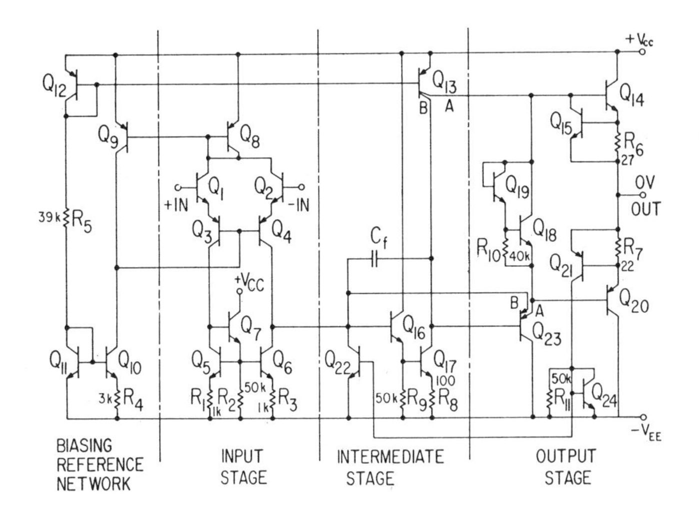

Fig. 2. Circuit diagram of the ICL8741 op amp.

The circuit model of Fig. 1 has been developed using two basic macromodeling techniques: simplification and build-up. In the simplification technique, representative portions of op amp circuitry are successively simplified by using simple ideal elements to replace numerous real elements. Thus, the final model using this approach bears a strong resemblance to the real circuit. In Fig. 1, the input stage design is an example of the simplification technique. In the build-up technique, a circuit configuration composed of ideal elements is proposed to meet certain external circuit specifications without necessarily resembling a portion of an actual op amp circuit configuration. The build-up technique is employed in the development of the output stage.

To illustrate these aspects further, consider the schematic diagram shown in Fig. 2 of the 741-type op amp

TABLE I
DESIGN EQUATIONS FOR THE OP AMP MACROMODEL

$$V_T = \frac{kT}{q} = 25.85 \text{ mV for } 300 \text{ K}$$

$$I_{S1} = I_{SD3} = I_{SD4} = 8 \cdot 10^{-16} \text{ A}$$

$$R_2 = 100 \text{ k}\Omega$$

$$I_{C1} = I_{C2} = \frac{C_2}{2} S_R^+$$

$$C_E = \frac{2I_{C1}}{S_R^-} - C_2$$

$$I_{B1} = I_B + \frac{I_{Bos}}{2}$$

$$I_{B2} = I_B - \frac{I_{Bos}}{2}$$

$$\beta_1 = I_{C1}/I_{B1}$$

$$\beta_2 = I_{C2}/I_{B2}$$

$$I_{EE} = \left(\frac{\beta_1 + 1}{\beta_1} + \frac{\beta_2 + 1}{\beta_2}\right) I_{C1}$$

$$R_E = 200/I_{EE}$$

$$I_{S2} = I_{S1}\left(1 + \frac{V_{os}}{V_T}\right)$$

$$\frac{1}{g_{m1}} = V_T/I_{C1}$$

$$R_{c1} = 1/2\pi f_{0\,\mathrm{dB}} C_2$$

$$R_{e1} = \left(\frac{\beta_1 + \beta_2}{\beta_1 + \beta_2 + 2}\right)\left(R_{C1} - \frac{1}{g_{m1}}\right)$$

$$C_1 = \frac{C_2}{2} \tan \Delta\phi$$

$$R_p = (V_{CC} + V_{EE})^2/(P_d - V_{CC}(2I_{C1}) - V_{EE} I_{EE})$$

$$G_a = 1/R_{c1}$$

$$G_{cm} = \frac{1}{R_{c1}\,(\mathrm{CMRR})}$$

$$R_{01} = R_{0-\mathrm{ac}}$$

$$R_{02} = R_{\mathrm{out}} - R_{01}$$

$$G_b = \frac{a_{VD} R_{c1}}{R_2 R_{02}}$$

$$I_X = (2I_{C1}) G_b R_2 - I_{SC}$$

$$I_{SD1} = I_{SD2} = I_X \exp - \frac{R_{01} I_{SC}}{V_T}$$

$$R_C = \frac{V_T}{100 I_X} \ln \frac{I_X}{I_{SD1}}$$

$$G_C = 1/R_C$$

$$V_C = V_{CC} - V_{\mathrm{out}}^+ + V_T \ln \frac{I_{SC}^+}{I_{SD3}}$$

$$V_E = V_{CC} + V_{\mathrm{out}}^- + V_T \ln \frac{I_{SC}^-}{I_{SD4}}$$

(Intersil ICL8741). This type is the most common, general purpose IC op amp. In developing a macromodel using the simplification technique, the circuitry employed for biasing can be replaced with ideal passive elements (pure current and voltage sources).

Similarly, the active load and balance-to-unbalance converter in the input stage can be replaced with ideal elements. Finally, it is not necessary to use composite transistors in the input stage. Thus, as shown in Fig. 1, a simple differential stage can be proposed to model accurately the nonlinear input characteristic of the op amp.

The op amp macromodel is developed keeping in mind existing IC simulators. Therefore, the model contains only elements which are common to most IC simulators (i.e., resistors, capacitors, inductors, dependent current sources, independent sources, diodes, and bipolar transistors). In addition, effort is made to minimize the number of p-n junctions. These nonlinear elements make necessary iterative analysis to obtain the equilibrium state of the circuit. A reduction of the number of nonlinear elements leads to smaller simulation time.

For the input stage, our investigations showed that at least four ideal junctions were necessary to provide the needed balanced, nonlinear behavior in the macromodel. It was determined that the simplest arrangement is that of Fig. 1 where the four ideal junctions were obtained with two ideal transistors, each modeled with the lowest order Ebers–Moll (E–M) transistor model which includes two ideal p-n junctions and two dependent current sources.

For the output stage, a simplified model of an actual op amp does not provide the best solution. A stripped-down class-AB stage with ideal transistors leads to a branch count of over 13 in comparison with 11 branches in the output stage of Fig. 1. In addition, the class-AB stage must be augmented with voltage limiters in the drive circuitry to limit the voltage excursion at the transistor bases to the supply potentials. It was found that the idealized built-up procedure provides an output stage which is considerably simpler.

III. PARAMETERS AND ELEMENT VALUES OF THE MACROMODEL

In this section, expressions are developed to relate the performance of the op amp and the macromodel to the parameters and elements of the macromodel. A summary of all design equations is presented in Table I. The determination of the element values of the macromodel proceeds from the input, transfer, and output characteristics of the op amp.

The Input Stage: I_{C1} and C_E

The value of the necessary collector current of the first stage is established by the slew rate of the op amp. If the op amp is connected as a voltage follower, the positive going slew rate S_R^+ is

$$S_R^+ = \frac{2I_{C1}}{C_2} \quad (1)$$

where an n-p-n stage has been assumed [4]. From a rearrangement of this expression

$$I_{C1} = \tfrac{1}{2}C_2 S_R^+. \quad (2)$$

For a quiescent situation, equal collector currents are used in the input stage $I_{C2} = I_{C1}$.

The negative going slew rate S_R^- is smaller because of the charge-storage effects in the input stage which is modeled by C_E [4].

$$S_R^- = \frac{2I_{C1}}{C_2 + C_E} \quad (3)$$

or

$$C_E = \frac{2I_{C1}}{S_R^-} - C_2. \quad (4)$$

If $S_R^+ < S_R^-$, the macromodel should be modified to use p-n-p transistors in the input stage. In the equations above S_R^+ and S_R^- should then be interchanged.

In addition to the transient slew rate effects, the element C_E also introduces a desirable modification to the ac response of the CM gain of the macromodel.

The Transistor Parameters

The values of β_1 and β_2 for the two ideal transistors are obtained from the specifications for the average input bias current I_B and the desired level of input current offset I_{Bos}.

$$I_{B1} = I_B + \frac{I_{Bos}}{2}, \qquad I_{B2} = I_B - \frac{I_{Bos}}{2}. \quad (5)$$

$$\beta_1 = \frac{I_{C1}}{I_{B1}}, \qquad \beta_2 = \frac{I_{C2}}{I_{B2}}. \quad (6)$$

The voltage offset V_{os} for the macromodel is produced by specifying different saturation currents I_S for the two transistors. Assume a given value for I_{S1} of Q_1

$$I_{C1} = I_{S1} \exp \frac{V_{BE1}}{V_T} \quad (7)$$

where $V_T = kT/q = 0.02585$ V at $T = 300$ K. A similar expression holds for $I_{C2} = I_{C1}$

$$I_{C2} = I_{S2} \exp \frac{V_{BE2}}{V_T}. \quad (8)$$

The offset voltage is

$$V_{os} = V_{BE1} - V_{BE2} = V_T \ln \frac{I_{S1}}{I_{S2}}. \quad (9)$$

This leads to

$$I_{S2} = I_{S1} \exp \frac{V_{os}}{V_T} \cong I_{S1}\left[1 + \frac{V_{os}}{V_T}\right]. \quad (10)$$

The Input Stage: R_{c1} and R_{e1}

Values for the resistors $R_{c1} = R_{c2}$ are derived from the required value of the 0 dB frequency $f_{0\,dB}$ of the fully compensated op amp. The 0 dB frequency is approximately the product of the DM voltage going a_{VD} and the -3 dB corner frequency $f_{3\,dB}$ of the gain function

$$f_{0\,dB} \simeq a_{VD} f_{3\,dB}. \quad (11)$$

The corner frequency can be estimated using a Miller-effect approximation in the interior stage.

$$f_{3\,dB} \simeq \frac{1}{2\pi R_2 C_2 (1 + G_b R_{02})} \simeq \frac{1}{2\pi R_2 C_2 G_b R_{02}}. \quad (12)$$

The DM voltage gain at very low frequencies is

$$a_{VD} = (G_a R_2)(G_b R_{02}). \quad (13)$$

G_a is chosen to be equal to $1/R_{c1}$ in order to obtain a convenient slew rate expression as in (1). The last three expressions lead to

$$f_{0\,dB} = \frac{1}{2\pi R_{c1} C_2} \quad (14)$$

or

$$R_{c1} = \frac{1}{2\pi f_{0\,dB} C_2}. \quad (15)$$

Alternately, a relationship between $f_{0\,dB}$ and S_R^+ can be written using (1).

$$f_{0\,dB} = \frac{S_R^+}{2\pi R_{c1}(2I_{C1})}. \quad (16)$$

The value of R_{c1} is usually small, of the order of $2/g_m$. R_{c1} and R_{c2} should be small in order that saturation of the input stage (and concomittent latchup of the op amp model) is avoided with maximum input. The resistances R_{e1} and R_{e2} in the input stage are introduced to provide a degree of freedom with respect to slew rate and 0 dB frequency, and to simulate better certain op amps which use emitter resistors for slew rate enhancement, e.g., the LM118. R_{e1} is found from the DM voltage gain of the first stage, which for convenience is taken to be unity.

$$\frac{v_a}{v_{in}} = \frac{\beta_1 R_{c1} + \beta_2 R_{c2}}{\dfrac{\beta_1}{g_{m1}} + (\beta_1 + 1)R_{e1} + \dfrac{\beta_2}{g_{m2}} + (\beta_2 + 1)R_{e2}} = 1. \quad (17)$$

If $I_{C1} = I_{C2}$, then $g_{m1} = g_{m2}$. If also $R_{c1} = R_{c2}$ and $R_{e1} = R_{e2}$,

$$R_{e1} = \frac{\beta_1 + \beta_2}{\beta_1 + \beta_2 + 2}\left[R_{c1} - \frac{1}{g_{m1}}\right]. \tag{18}$$

The Input Stage: I_{EE} and R_E

The value of the dc current source in the input stage for equal collector currents is

$$I_{EE} = \left(\frac{\beta_1 + 1}{\beta_1} + \frac{\beta_2 + 1}{\beta_2}\right) I_{C1}. \tag{19}$$

The resistor R_E is added to provide a finite CM input resistance. Because the current source I_{EE} is often realized with an n-p-n transistor, the resistance R_E is taken as its output resistance

$$R_E \simeq \frac{V_A}{I_C} = \frac{V_A}{I_{EE}} \tag{20}$$

where V_A is the early voltage of the device. V_A for a small n-p-n transistor is typically 200 V.

The Input Stage: C_1

To introduce excess phase effects in the DM amplifier response, another capacitor C_1 is added in the input stage. The second pole of the DM gain function is located at

$$p_2 = -1/2R_{c1}C_1. \tag{21}$$

Notice that there is no interaction amongst the three capacitors because of the use of unilateral devices and stages.

The excess phase at $f = f_{0\text{ dB}}$ due to the nondominant pole p_2 is

$$\Delta\phi = \tan^{-1}\frac{2\pi f_{0\text{ dB}}}{|p_2|} = \tan^{-1}(2\pi f_{0\text{ dB}})(2R_{c1}C_1) = \tan^{-1}\frac{2C_1}{C_2}. \tag{22}$$

The phase margin of the DM open-loop response is then

$$\phi_m = 90° - \Delta\phi. \tag{23}$$

The necessary value of C_1 to produce the excess phase is

$$C_1 = \frac{C_2}{2}\tan\Delta\phi. \tag{24}$$

DC Power Drain

To model the actual dc power dissipation of an op amp, a resistor R_p is introduced into the macromodel. For the circuit of Fig. 1, in a quiescent state, the power dissipation is

$$P_d = V_{CC}2I_{C1} + V_{EE}I_{EE} + \frac{(V_{CC} + V_{EE})^2}{R_p}. \tag{25}$$

The necessary value of R_p to produce this dissipation is

$$R_p = \frac{(V_{CC} + V_{EE})}{P_d - V_{CC}2I_{C1} - V_{EE}I_{EE}}. \tag{26}$$

In a typical op amp, most of the current drain from the voltage supplies is due to diode-resistor current defining paths. Therefore, as the supply voltages are changed, the current drain varies almost linearly and R_p will continue to model accurately the power dissipation.

The Interstage: G_a, R_2, and G_{cm}

As indicated earlier, the coefficient G_a of the voltage dependent current source $G_a v_a$ is chosen equal to $1/R_{c1}$ for convenience. Similarly, the value of R_2 or G_b can be arbitrarily chosen. Only the product is determined by the DM gain. For "active region" considerations, the choice of R_2 is not important. However, it must be kept in mind that the voltage response at node b is linear with R_2. If too large a value of v_b is developed during a transient excursion through the active region of the op amp, a considerable discharge or recovery time can be encountered after the active region excursion. To prevent these discharge delays in relation to actual op amp behavior, a small value of R_2 should be used. Empirically, a value of 100 kΩ is found to be appropriate.

If a second voltage-controlled current source is introduced across R_2, the CM voltage gain response can be introduced. The CM voltage gain in the input stage from v_{in} to v_e is approximately unity since R_E is large. The CM voltage gain from the input to v_b is then approximately

$$\frac{v_{b\text{CM}}}{v_{\text{inCM}}} \cong G_{cm}R_2. \tag{27}$$

The differential voltage gain from the input to v_b is

$$\frac{v_{b\text{DM}}}{v_{\text{inDM}}} = G_a R_2 = \frac{1}{R_{c1}}R_2. \tag{28}$$

The CM rejection ratio (CMRR) is the ratio of the two gains [5]

$$\text{CMRR} = \frac{a_{VD}}{a_{VC}} = \frac{1}{R_{c1}G_{cm}}. \tag{29}$$

Therefore,

$$G_{cm} = \frac{1}{(\text{CMRR})R_{c1}}. \tag{30}$$

The dominant behavior of the CM frequency response will be approximately the same as the DM frequency response except that the presence of the capacitor C_E in the input stage introduces a transmission zero in the CM gain function at $-1/R_E C_E$.

The Output Stage: R_{o1}, R_{o2}, and G_b

The output stage provides the desired dc and ac output resistances and the output current and voltage limitations. From Fig. 1, it is seen that the output resistance at very low frequencies for the quiescent state is

$$R_{\text{out}} = R_{o1} + R_{o2}. \tag{31}$$

At high frequencies, R_{o2} is shorted out by the (current) Miller-effect output capacitance across it due to C_2. The effective shunt capacitance is $C_{sh} \simeq C_2(1 + R_2 G_b)$. For the situation where a large load resistance is presented to the macromodel, the corner frequency of the output impedance is

$$f_c = \frac{1}{2\pi R_{02} C_2 (1 + R_2 G_b)}. \tag{32}$$

For frequencies well above this value, the output resistance is R_{01}. Therefore,

$$R_{01} = R_{0\text{-ac}}. \tag{33}$$

With this value established, R_{02} from (31) and G_b from (13) are

$$R_{02} = R_{\text{out}} - R_{01} \tag{34}$$

and

$$G_b = \frac{a_{VD} R_{c1}}{R_2 R_{02}}. \tag{35}$$

The Output Stage: Current Limiting

In the output stage of Fig. 1, the desired output-current limiting is provided by the elements $G_C V_6$, R_C, D_1, D_2, and R_{01}. The R_C, $G_C V_6$ combination is an equivalent to a voltage-controlled voltage source (which is not available in simulators such as program SPICE). Thus, $V_{\text{out}} = V_6$ also appears across R_C. If both of the voltage-clamp diodes D_3 and D_4 are off, the maximum current to the output is the ratio of the potential across D_1, D_2, and R_{01}

$$I_{SC} \simeq \frac{V_D}{R_{01}} \tag{36}$$

where

$$V_D = V_T \ln \frac{I_X}{I_{SD1}} \tag{37}$$

I_X Maximum current through D_1 or D_2.

I_{SD1} Saturation current of diodes D_1, D_2.

Since R_{01} is known, I_{SD1} can be established once I_X is determined. In Fig. 3, a reduced portion of the output stage is shown which applies for a positive output excursion, and where a very small load resistance is assumed. A Thévenin equivalent for $G_b v_b$ and R_{02} is used. The Thévenin open-circuit available voltage is $a_{VD} v_{\text{in}}$. An ideal voltage-controlled voltage source v_0 is also used in place of $G_C V_6$ and R_C. Assume first that the voltage $a_{VD} v_{\text{in}}$ is not large. The output current flowing through the resistor R_{01} then produces only a small voltage drop. The polarity of the voltage v_c is such as to forward bias diode D_1. If the voltage drop across R_{01} is small, the current through D_1 is very small and can be neglected.

As $a_{VD} v_{\text{in}}$ increases, so does I_L and the voltage drop across R_{01}. As the latter approaches the "ON" voltage of D_1, i.e., the voltage for appreciable current through the diode, the increasing current from the source I_{D1} flows through the diode. I_L is then approximately limited because of the exponential increase of I_{D1} with respect

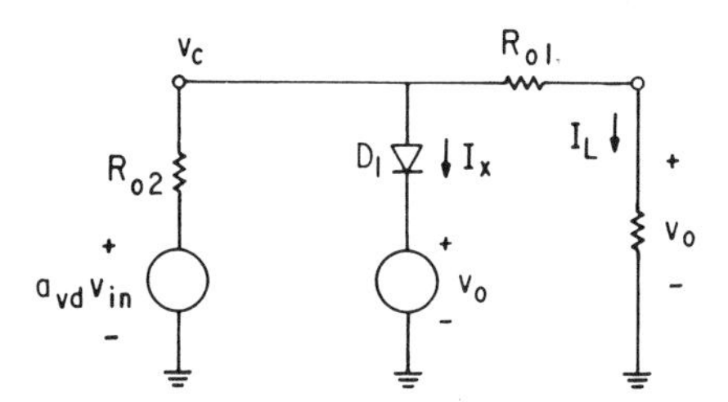

Fig. 3. Simplified circuit diagram of the output stage.

to $I_L R_{01}$. The approximate limiting condition is found from

$$I_X = I_{SD1} \exp \frac{I_{SC} R_{01}}{V_T}. \tag{38}$$

The limiting value of I_X is determined by an overdrive condition at the input. The short-circuit available current from $G_b v_b$ is $I_{\max}$

$$I_{\max} = I_X + I_{SC} = 2 I_{C1} R_2 G_b. \tag{39}$$

A typical value of $I_{\max}$ is 100 A.

From the equation above,

$$I_{SD1} = I_{SD2} = I_X \exp\left(-\frac{R_{01} I_{SC}}{V_T}\right). \tag{40}$$

For large required values of R_{01}, the value of I_{SD1} can be extremely small which may lead to numerical difficulty. In many applications when the output resistance is not critical, a smaller value of R_{01} can be used neglecting the exact realization of $R_{0\text{-ac}}$, e.g., if $I_{sD1} = I_{s1}$,

$$R_{01} \simeq \frac{V_T}{I_{SC}} \ln \frac{I_X}{I_{S1}}. \tag{41}$$

R_{02} is then increased to $R_{\text{out}} - R_{01}$, and G_b is decreased to maintain the same value of the $G_b R_{02}$ product.

In order to approximate well a voltage-controlled voltage source, R_C must be very small. If the voltage drop across R_C is to be only 1 percent of V_{D1} or V_{D2},

$$R_C = \frac{V_T}{100 I_X} \ln \frac{I_X}{I_{SD1}}. \tag{42}$$

The necessary value for the voltage-controlled current source $G_C V_6$ is

$$G_C = \frac{1}{R_C}. \tag{43}$$

The Output Stage: Voltage Limiting

The output voltage excursion is limited by the voltage source-diode clamp combinations V_C, D_3 and V_E, D_4, shown in Fig. 1. With a large, positive output voltage such as to forward bias D_3

$$\begin{aligned} V_{\text{out}}^{+} &= V_{CC} - V_C + V_{D3} \\ &= V_{CC} - V_C + V_T \ln \frac{I_{SC}^{+}}{I_{SD3}}. \end{aligned} \tag{44}$$

As indicated above, the diode current is limited to the

TABLE II

CIRCUIT DATA AND GUMMEL–POON TRANSISTOR PARAMETERS FOR THE ICL8741 OP AMP

Circuit Data			Gummel-Poon Transistor Parameters					
Element	Nodes	Value						
			.MODEL	BNP1	NGP	BFM=209.	BRM=2.5	RB= 670.
R1	02 17	1.0K	+		RC=300.	CCS=1.417P	TF=1.15N	TR= 405.N
R2	02 16	50.K	+		CJE=0.65P	CJC=0.36P	IS= 1.26E-15	VA=178.6
R3	02 18	1.0K	+		C2= 1653.	IK=1.611M	NE=2.0	PE=0.60
R4	02 08	3.0K	+		ME=3.	PC=0.45	MC=3.	
R5	04 05	39.K	.MODEL	BNP2	NGP	BFM=400.	BRM=6.1	RB= 185.
R6	12 26	27.	+		RC= 15.	CCS=3.455P	TF=0.76N	TR= 243.N
R7	12 25	22.	+		CJE=2.80P	CJC=1.55P	IS= 0.395E-15	VA=267.0
R8	02 23	100.	+		C2= 1543.	IK=10.00M	NE=2.0	PE=0.60
R9	02 21	50.K	+		ME=3.	PC=0.45	MC=3.	
R10	24 27	40.K	.MODEL	BPN1	PGP	BFM= 75.	BRM=3.8	RB= 500.
R11	02 22	50.K	+		RC=150.	CCS=2.259P	TF=27.4N	TR=2540.N
C	15 19	30.P	+		CJE=0.10P	CJC=1.05P	IS= 3.15E-15	VA=55.11
Q1	10 07 13	BNP1	+		C2= 1764.	IK=270.0U	NE=2.0	PE=0.45
Q2	10 06 11	BNP1	+		ME=3.	PC=0.45	MC=3.	
Q3	14 09 13	BPN1	.MODEL	BPN2	PGP	BFM=117.	BRM=4.8	RB= 80.
Q4	15 09 11	BPN1	+		RC=156.		TF=26.5N	TR=2430.N
Q5	14 16 17	BNP1	+		CJE=4.05P	CJC=2.80P	IS= 17.6E-15	VA=57.94
Q6	15 16 18	BNP1						
Q7	01 14 16	BNP1						
Q8	10 10 01	BPN1	+		C2= 478.4	IK=590.7U	NE=2.0	PE=0.60
Q9	09 10 01	BPN1	+		ME=4.	PC=0.60	MC=4.	
Q10	09 05 08	BNP1	.MODEL	BPN3	PGP	BFM=13.8	BRM=1.4	RB=100.
Q11	05 05 02	BNP1	+		RC= 80.	CCS=2.126P	TF=27.4N	TR= 55.N
Q12	04 04 01	BPN1	+		CJE=0.10P	CJC=0.30P	IS= 2.25E-15	VA=83.55
Q13A	20 04 01	BPN3	+		C2=84.37K	IK=5.000M	NE=2.0	PE=0.45
Q13B	19 04 01	BPN4	+		ME=3.	PC=0.45	MC=3.	
Q14	01 20 26	BNP2	.MODEL	BPN4	PGP	BFM=14.8	BRM=1.5	RB=160.
Q15	20 26 12	BNP1	+		RC=120.	CCS=2.126P	TF=27.4N	TR= 220.N
Q16	01 15 21	BNP1	+		CJE=0.10P	CJC=0.90P	IS= 2.25E-15	VA=83.55
Q17	19 21 23	BNP1	+		C2=84.37K	IK=171.8U	NE=2.0	PE=0.45
Q18	20 27 24	BNP1	+		ME=3.	PC=0.45	MC=3.	
Q19	20 20 27	BNP1	.MODEL	BPN5	PGP	BFM= 80.	BRM=1.5	RB=1100.
Q20	02 24 25	BPN2	+		RC=170.		TF=26.5N	TR=9550.N
Q21	22 25 12	BPN1	+		CJE=1.10P	CJC=2.40P	IS= 0.79E-15	VA=79.45
Q22	15 22 02	BNP1	+		C2= 1219.	IK=80.55U	NE=2.0	PE=0.60
Q23A	02 19 24	BPN5	+		ME=4.	PC=0.60	MC=4.	
Q23B	02 19 15	BPN6	.MODEL	BPN6	PGP	BFM= 19.	BRM=1.0	RB= 650.
Q24	22 22 02	BNP1	+		RC=100.		TF=26.5N	TR=2120.N
			+		CJE=1.90P	CJC=2.40P	IS=0.0063E-15	VA=167.1
			+		C2=57.49K	IK=80.55U	NE=2.0	PE=0.60
			+		ME=4.	PC=0.60	MC=4.	

short current I_{SC}^+. The necessary bias voltage is

$$V_C = V_{CC} - V_{out}^+ + V_T \ln \frac{I_{SC}^+}{I_{SD3}}. \quad (45)$$

Similarly,

$$V_E = V_{EE} + V_{out}^- + V_T \ln \frac{I_{SC}^-}{I_{SD4}}. \quad (46)$$

The Complete Model

A summary of the design equations for the parameters of the macromodel is given in Table I. An example of the use of these equations is given in the Appendix. The starting point is op amp performance data.

The particular IC used to illustrate the design procedure was the object of an earlier study [6]. The configuration of this IC was established to be that of Fig. 2 and the transistors were characterized by the Gummel–Poon (G–P) parameters of Table II.[1]

Program SPICE has been used to establish the performance and characteristics of the op amp [7]. The results are summarized in Table III, column 1. These values are used in the Appendix to develop the element and parameter values of the macromodel. As brought out in the Appendix, several parameters of the macromodel could be chosen arbitrarily:

[1] A slight modification of the G–P parameters of the output transistors has been made to produce a typical level of maximum short-circuit available current.

TABLE III

OP AMP PERFORMANCE CHARACTERISTICS

	8741 Device-Level Model	8741 Macromodel	LM741 Data Sheet	LM118 Data Sheet
C_2 (pF)	30	30	30	5
S_R^+ (V/μs)	0.9	0.899	0.67	100
S_R^- (V/μs)	0.72	0.718	0.62	71
I_B (nA)	256	255	80	120
I_{Bos} (nA)	0.7	<1	20	6
V_{os} (mV)	0.299	0.298	1	2
a_{VD}	$4.17 \cdot 10^5$	$4.16 \cdot 10^5$	$2 \cdot 10^5$	$2 \cdot 10^5$
a_{VD} (1 kHz)	$1.219 \cdot 10^3$	$1.217 \cdot 10^3$	10^3	$16 \cdot 10^3$
$\Delta\phi$ (°)	16.8	16.3	20	40
CMRR (dB)	106	106	90	100
R_{out} (Ω)	566	566	75	75
R_{o-ac} (Ω)	76.8	76.8	—	—
I_{SC}^+ (mA)	25.9	26.2	25	25
I_{SC}^- (mA)	25.9	26.2	25	25
V^+ (V)	14.2	14.2	14.0	13
V^- (V)	−12.7	−12.7	−13.5	−13
P_d (mW)	59.4	59.4	—	—

$$T = 300 \text{ K}(V_T = 25.85 \text{ mV}),$$

$$I_{S1} = I_{SD3} = 8 \cdot 10^{-16} \text{ A},$$

$$R_2 = 0.1 \text{ M}\Omega,$$

where I_{S1}, I_{SD3} are the saturation currents of the first transistor and the voltage-limiter diodes, respectively. In addition, the major compensation capacitor is fixed

TABLE IV
MACROMODEL PARAMETERS

	8741	LM 741	LM 118
T (K)	300	300	300
I_{S1} (A)	$8 \cdot 10^{-16}$	$8 \cdot 10^{-16}$	$8 \cdot 10^{-16}$
I_{SD3} (A)	$8 \cdot 10^{-16}$	$8 \cdot 10^{-16}$	$8 \cdot 10^{-16}$
R_2 (kΩ)	100	100	100
C_2 (pF)	30	30	5
C_E (pF)	7.5	2.41	2.042
β_1	52.6726	111.67	$2.033 \cdot 10^3$
β_2	52.7962	143.57	$2.137 \cdot 10^3$
I_{EE} (μA)	27.512	20.26	500
R_E (mΩ)	7.2696	9.872	0.40
I_{S2} (A)	$8.0925 \cdot 10^{-16}$	$8.309 \cdot 10^{-16}$	$8.619 \cdot 10^{-16}$
R_{c1} (Ω)	4352	5305	1989
R_{e1} (Ω)	2391.9	2712	1884
C_1 (pF)	4.5288	5.460	2.098
R_p (kΩ)	15.363	—	—
G_a (μmho)	229.774	188.6	502.765
G_{CM} (nmho)	1.1516	6.28	5.028
R_{01} (Ω)	76.8	32.13	32.13
R_{02} (Ω)	489.2	42.87	42.87
G_b (mho)	37.0978	247.49	92.792
I_X (A)	100.138	—	—
I_{SD1} (A)	$3.8218 \cdot 10^{-32}$	$8 \cdot 10^{-16}$	$8 \cdot 10^{-16}$
R_C (Ω)	$0.1986 \cdot 10^{-3}$	$0.02129 \cdot 10^{-3}$	$0.00279 \cdot 10^{-3}$
G_C (mho)	5034.3	46 964	358 000
V_C (V)	1.6042	1.803	2.803
V_E (V)	3.1042	2.303	2.803

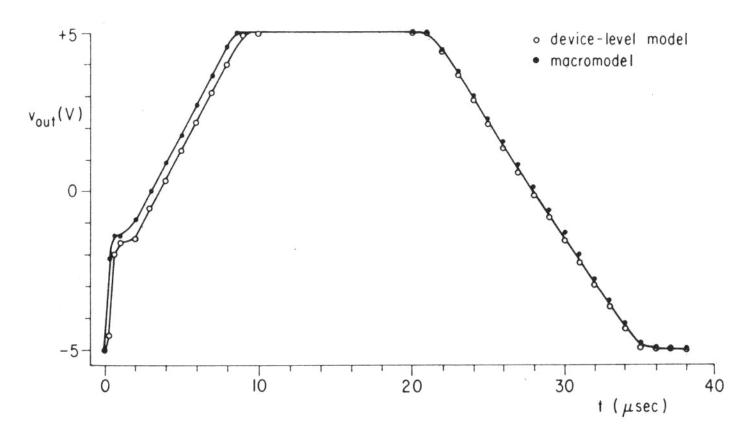

Fig. 4. Simulated voltage follower slew rate performance using both device-level models and macromodels.

by the type of op amp under study or is chosen appropriately. For the case at hand, $C_2 = 30$ pF.

The remaining values of the parameters of the macromodel are presented in Table IV, column 1.

IV. Comparison with Device-Level Models

Basic Macromodel Performance

The values for the macromodel of Table IV, column 1 were used to define an external model in program SPICE. The same set of computer runs was made as lead to the op amp performance results of Table III, column 1. The results for the macromodel are presented in column 2 of this table. It is seen that the comparison is excellent for both small-signal and large-signal experiments.

To provide a further comparison, the large-signal, slew rate performance for a voltage follower is shown Fig. 4 for both the device-level model and the macromodel. It is seen that the responses are very similar.

The presence of C_E produces a step in the initial response of the voltage follower. From simple theory [4], the jump should be approximately

$$\Delta V_{out} = \frac{C_E}{C_2} \Delta V_{in} .$$

For this example,

$$\Delta V_{in} = 10 \text{ V} \quad \text{and} \quad \Delta V_{out} = \frac{7.5}{30} (10) = 2.5 \text{ V}.$$

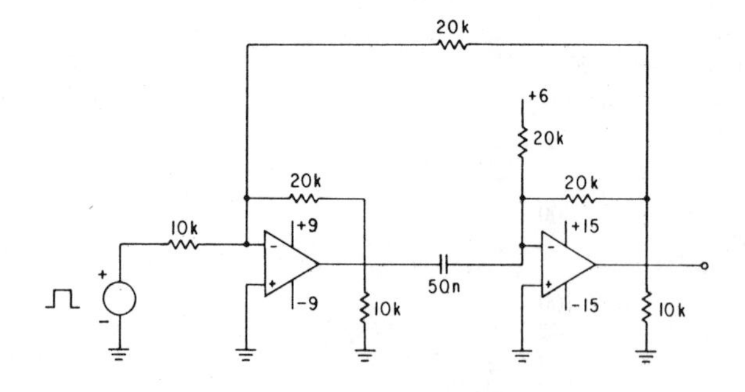

Fig. 5. A monostable time delay circuit.

From Fig. 4, the observed jump for the macromodel is 3.6 and 3.4 V for the device-level model.

A measure of the complexity of the two op amp models can be obtained by comparing the node and branch counts of each circuit. For the device-level model, where each G–P transistor model has 2 internal nodes and 7 branches, the totals are 81 nodes including the datum node, and 193 branches. For the macromodel, where each E–M transistor model has no internal nodes and 4 branches, the totals are 16 nodes and 28 branches. The ratios for the two models are 5.1 for the nodes and 6.9 for the branches. The number of p-n junctions in the device-level model is 52 and 8 in the macromodel, a ratio of 6.5.

The total computer central processing unit (CPU) time on a CDC 6400 to simulate the voltage follower slew rate performance is 39.2 s for the device-level model and 4.0 s for the macromodel, a ratio of 9.8. An alternate comparison is obtained if only the simulation times for the initial state and the transient analyses are used. The improvement ratio is then 12.0.

For the dc and ac simulations, the device-level model to macromodel CPU times, for the analyses only, have the ratios of 3.9 and 6.0, respectively. Note that the improvement ratio is less for the dc analyses.

A Regenerative Timer

The monostable time delay circuit of Fig. 5 provides a good test of the ability of the macromodel to perform as desired when demanding nonlinear performance is required. The voltage and timing levels of the circuit of Fig. 5 have been chosen to provide a maximum stress on the op amps with respect to voltage limits, critical voltage switching levels and speed of response, slew-rate limitations, etc. The output waveforms of the circuit as predicted by program SPICE using both the device-level model and the macromodel are shown in Fig. 6. It is seen that the responses compare closely. The leading and trailing edges of the output pulse differ in timing by less than 1 time step of the computer output, i.e., better than 3 percent of the overall pulsewidth. The total CPU times for the simulations using the two models are in the ratio of 8.9. If the common output time is deleted, the improvement ratio is 9.6.

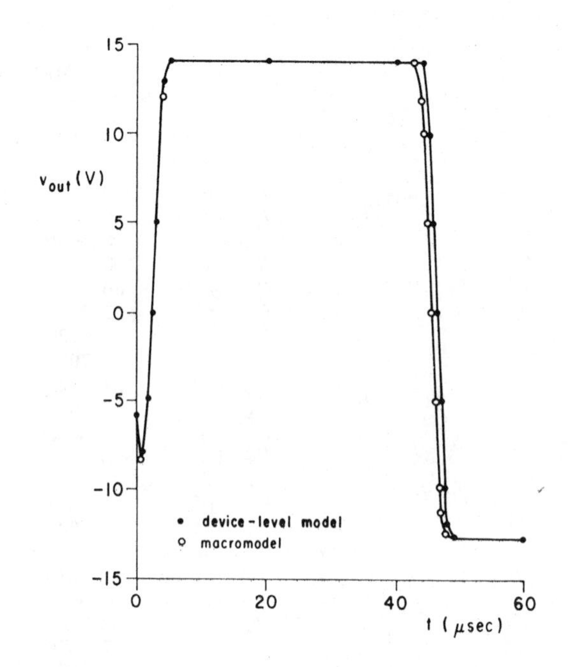

Fig. 6. Simulated output pulse response of time delay circuit using both device-level models and macromodels.

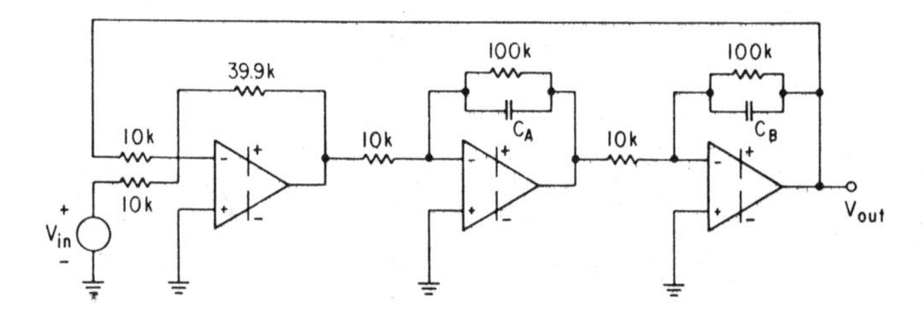

Fig. 7. "Ring of Three" bandpass filter.

TABLE V
RESPONSE DATA FOR THE "RING OF THREE" BANDPASS FILTER

Design Center Frequency f_{od} (kHz)	Actual Center Frequency f_{oa} (kHz)	Gain Magnitude at f_{oa}	Gain Magnitude at $0.9 f_{od}$	Gain Magnitude at $1.1 f_{od}$
1	0.998	11.01	4.806	4.311
	(0.998)	(11.00)	(4.806)	(4.311)
2	1.996	12.24	4.898	4.368
	(1.996)	(12.23)	(4.896)	(4.367)
10	9.934	112.1	5.547	4.398
	(9.934)	(107.1)	(5.542)	(4.401)

Numbers in parentheses refer to results with device-level model.

An Active RC Filter

To further check the second-order ac response of the macromodel, the simple "Ring of Three" op amp filter of Fig. 7 was designed for a center frequency of 1 kHz and a Q of 10. The frequency response from program SPICE for the two models is summarized in Table V. Again, it is seen that the comparison is very close.

At higher frequencies, the phase response of the (1 MHz) 741-type op amps comes into effect. A response comparison for the two models as shown in rows 2 and 3 indicates that the macromodel is providing the proper phase response.

The total CPU simulation times to determine the dc state and the frequency response using the two models have the ratio of 5.8. If the common output time is omitted, the ratio becomes 6.8.

In this application, the nonlinear performance of the op amp is not of major interest. In order to check on the improvement of computer run time for a reduced macromodel, the voltage and current limiting circuitry of Fig. 1 was omitted. The simulated response of the filter did not change, of course; however, the total CPU run time was reduced by a factor of 1.4.

V. Macromodel Parameters for Other IC Op Amps

The detailed, precise performance characteristics for an individual op amp as obtained from the use of the device-level models are usually not available. The precision used in the numerical example of this paper is employed in order to obtain an accurate estimate of the performance of a macromodel in relation to a known reference. Experimental results with actual op amps could include significant measurement inaccuracies.

Typically, one has a data sheet or averaged experimental data for a type of op amp which is to be included in a system. As an example of the choice of macromodel parameters in this situation, two further examples are given. In Table III, columns 3 and 4, measured typical op amp data are given for both the LM741 and the LM118. The macromodel parameters corresponding to these data are given in columns 2 and 3 of Table IV.

It is possible to introduce programming into a simulator to determine automatically the macromodel parameters. This has been done at one location for program SPICE. In this situation, all that is necessary to define a specific op amp model is to list its characteristics on an op amp "model card" in much the same way that one currently defines a transistor model by specifying its characteristics on a transistor model card. When an op amp characteristic is not inputed, a default value is used.

Appendix

The 8741 Macromodel

In this Appendix, a numerical example is used to illustrate the development of the parameters of the op amp macromodel. For the example, the response characteristics of the 8741 op amp are used as determined by several simulator runs using device-level modeling. The circuit of Fig. 2 together with the transistor parameters of Table II has been analyzed to obtain the characteristics which are summarized in column 1, Table III.

The development procedure follows the sequence of expressions of Table I. The final results are presented in column 1, Table IV.

From the slew rate performance of the 8741 as given in Table III

$$S_R^+ = 0.90\ \mathrm{V}/\mu\mathrm{s} \quad \text{and} \quad S_R^- = 0.72\ \mathrm{V}/\mu\mathrm{s}.$$

For the given compensation capacitor of $C_2 = 30$ pF, these values lead to

$$I_{C1} = \tfrac{1}{2} S_R^+ C_2 = 13.50\ \mu\mathrm{A}$$

$$C_E = \frac{2I_{C1}}{S_R^-} - C_2 = 7.50\ \mathrm{pF}.$$

The average base current is 256 nA and the desired base current offset is 0.7 nA.

$$I_{B1} = 256.3\ \mathrm{nA}, \qquad I_{B2} = 255.7\ \mathrm{nA}$$

$$\beta_1 = 52.6727, \qquad \beta_2 = 52.7962.$$

A high level of precision is used in this example in order to obtain an accurate comparison of macromodel performance in relation to that of the op amp modeled at the device level.

The necessary emitter current source for the input stage is $I_{EE} = 27.512\ \mu$A. The value of the CM emitter resistor is

$$R_E = 200/I_{EE} = 7.2696\ \mathrm{M}\Omega$$

where a value of $V_A = 200$ V has been used.

For Q_1, the assumed value of saturation current is $8 \cdot 10^{-16}$ A which is a typical value for a small n-p-n IC transistor. To produce the desired input offset voltage of 0.299 mV

$$I_{S2} = 8 \cdot 10^{-16}\left(1 + \frac{0.299}{25.85}\right) = 8.0925 \cdot 10^{-16}\ \mathrm{A}.$$

For a fully compensated op amp with a rolloff of -6 dB/octave, the 0 dB frequency can be calculated from the product of the gain and the value of the frequency at which it is measured providing that the frequency is well above the corner frequency of the gain characteristic. From the data of column 1, Table III,

$$f_{0\ \mathrm{dB}} = (1.219 \cdot 10^3)(10^3)\ \mathrm{Hz} = 1.219 \cdot 10^6\ \mathrm{MHz}.$$

The value of the collector resistors of the first stage is

$$R_{c1} = \frac{1}{2\pi f_{0\ \mathrm{dB}} C_2} = 4352\ \Omega.$$

The value of the reciprocal of g_m for the first stage is

$$\frac{1}{g_{m1}} = 1915\ \Omega.$$

The required value of the emitter resistor is

$$R_{e1} = 0.9814(4352 - 1915) = 2392\ \Omega.$$

The final element for the input stage is C_1, which produces the nondominant pole of the gain function. For $\Delta\phi = 16.80°$

$$C_1 = \frac{C_2}{2} \tan 16.80° = 4.529 \text{ pF}.$$

The value of the resistor R_p to simulate power dissipation for ±15 V supplies and a dissipation of 59.4 mW is

$$R_p = 15.363 \text{ k}\Omega.$$

For the interstage, R_2 is taken to be 100 kΩ and

$$G_a = \frac{1}{R_{c1}} = 229.774 \ \mu\text{mho}.$$

For a CMRR of 106 dB $(199.5 \cdot 10^3)$

$$G_{CM} = \frac{G_a}{\text{CMRR}} = 1.1516 \text{ nmho}.$$

In the output stage, the desired dc and ac output resistances are 566 Ω and 76.8 Ω, respectively. Therefore

$$R_{01} = 76.8 \ \Omega, \ R_{02} = 489.2 \ \Omega.$$

The value of G_b to provide the correct DM voltage gain of $417 \cdot 10^3$ is

$$G_b = \frac{a_{VD} R_{c1}}{R_2 R_{02}} = 37.097.$$

The maximum current through the diode D_1 or D_2 is

$$I_X = 2I_{C1} G_b R_2 - I_{SC} = 100.14 \text{ A}$$

where the desired value of I_{SC} is 25.9 mA. With these values, the saturation currents of D_1 and D_2 are

$$I_{SD1} = I_X \exp - \frac{R_{01} I_{SC}}{V_T} = 3.822 \cdot 10^{-32} \text{ A}.$$

The values for the approximate voltage-controlled voltage source are

$$R_C = \frac{V_T}{100 I_X} \ln \frac{I_X}{I_{SD1}} = 0.199 \text{ m}\Omega$$

and

$$G_C = \frac{1}{R_C} = 5034 \text{ mho}.$$

For the voltage-clamp circuits, the saturation currents of diodes D_3 and D_4 are chosen to be $8 \cdot 10^{-16}$ A. The voltage sources should be

$$V_C = 15 - 14.2 + 0.02585 \ln \frac{0.0259}{8 \cdot 10^{-16}} = 1.604 \text{ V}$$

and

$$V_E = 15 - 12.7 + 0.804 = 3.104 \text{ V}.$$

Acknowledgment

The authors are pleased to acknowledge the aid and discussion on this topic with C. Battjes, R. Bohlman, S. Taylor, R. Dutton, and I. Getreu.

The staffs at the Computer Centers at the University of California, Berkeley, and at Tektronix, Inc., Beaverton, Oreg., have been extremely generous and helpful in the support given for the numerous computer runs necessary for this project.

References

[1] J. R. Greenbaum, "Digital-IC models for computer-aided design," *Electronics*, vol. 46, 25, pp. 121–125, Dec. 6, 1973.

[2] B. M. Cohn, D. O. Pederson, and J. E. Solomon, "Macromodeling of operational amplifiers," in *ISSCC Dig. Tech. Papers*, Feb. 1974, pp. 42–43.

[3] D. O. Pederson and J. E. Solomon, "The need and use of macromodels in IC subsystem design," in *Proc. 1974 IEEE Symp. Circuits and Systems*, p. 488.

[4] J. E. Solomon, "The monolithic op amp: a tutorial study," *IEEE J. Solid-state Circuits*, this issue, pp. 314–332.

[5] P. R. Gray and R. G. Meyer, "Recent advances in monolithic operational amplifier design," *IEEE Trans. Circuits and Syst.*, vol. CAS-21, pp. 317–327, May 1974.

[6] B. A. Wooley, S.-Y. J. Wong, and D. O. Pederson, "A computer-aided evaluation of the 741 amplifier," *IEEE J. Solid-State Circuits*, vol. SC-6, pp. 357–366, Dec. 1971.

[7] L. W. Nagel and D. O. Pederson, "Simulation program with integrated circuit emphasis (SPICE)," Electron. Res. Lab., Univ. of California, Berkeley, Memo ERL-M382, and in *Proc. 16th Midwest Symp. Circuit Theory*, 1973.

[8] D. N. Pocock and M. G. Krebs, "Terminal modeling and photocompensation of complex microcircuits," *IEEE Trans. Nucl. Sci.*, vol. NS-19, pp. 86–93, Dec. 1972.

[9] D. H. Treleaven and F. N. Trofimenkoff, "Modeling operational amplifiers for computer-aided circuit analysis," *IEEE Trans. Circuit Theory* (Corresp.), vol. CT-18, pp. 205–207, Jan. 1971.

Ion-Implanted JFET-Bipolar Monolithic Analog Circuits

Ronald W. Russell and Daniel D. Culmer

National Semiconductor Corp.

Santa Clara, Cal.

IT HAS BEEN KNOWN for several years that operational amplifiers with FET input stages are significantly faster than amplifiers with bipolar inputs. This results from the inherently lower ratio of transconductance – to – operating current (g_m/I_1) in the FET[1]. Past efforts to build compatible JFET-bipolar monolithic circuits have met with limited success due to the difficulties of fabricating well matched, high voltage, wideband FETs[2]. A new process employing a shallow ion-implanted boron layer has produced P-channel JFETs with highly predictable pinch-off voltages, extremely good match between adjacent devices, high voltage breakdowns and complete compatibility with standard linear process. In addition to amplifiers, a wide variety of analog functions can be realized with higher performance and lower cost than was previously possible. Three representative analog ICs which take advantage of this new process will be described.

By employing a P-channel FET input stage in place of the conventional lateral PNP circuit a two-fold speed advantage can be realized: (1)–level shifting is accomplished without bandlimiting, and (2)–slew rate is enhanced by the low $g_{m1}/I_1 \cong 2$ as compared with the bipolar $g_{m1}/I_1 = q/kT \cong 40$.

Figure 1 shows a two-stage amplifier with FET inputs, bipolar current source loads and a wideband bipolar second stage. Bandlimiting in the second stage is minimized by using a differential circuit with asymmetric frequency compensation, rather than conventional differential-to-single ended conversion. The complete op-amp, Figure 2, achieves a bandwidth of 10 MHz, a slew rate of 40V/μs, settling to 0.01 per cent in 1/2 μs, along with input currents of 2pA and a 7-mV offset voltage. This op-amp is realized on a relatively small 48 mil x 50 mil die.

Although the common mode loop in Figure 1 has excellent wideband characteristics, the use of the bipolar current source loads for the FET input amplifier results in high input noise, offset and drift. This occurs because the noise and offset of the bipolar devices when referred to the input are amplified by the ratio of bipolar-to-FET transconductance.

By employing FET rather than bipolar current source loads in the two stage amplifier shown in Figure 3, low noise, offset voltage and drift are achieved without a reduction in dc gain. Asymmetric frequency compensation again allows wide bandwidth and doubles the effective input stage transconductance.

In the complete op-amp, Figure 4, FETs are employed for offset adjust to achieve a low offset drift contribution due to adjust, less than 1 μ V/°C per mV adjust, and low input noise contribution. In addition, a wideband FET-NPN composite solves the conventional problem of phase degradation with pull down loading in the output stage. Typical performance of the op-amp includes an input current of 2 pA, an offset voltage and drift of 3 mV and 2 μ V/°C, respectively, input voltage noise of 10 nV/Hz, a 5-MHz bandwidth and a slew rate of 15 V/μs.

For large swing applications, JFET switches must be mated with complex bipolar or MOS circuits to provide the needed logic interface and drive functions. Present-day JFET switches are, therefore, built using at least two chips. Competing MOS switch circuits can be built on a single chip, but generally suffer from high ON-resistance nonlinearity with swing. Using the implanted JFET-bipolar technology, both switch FETs and drive circuitry can be built on one die, providing high performance at low cost.

The analog switch discussed incorporates a unique high speed ON-resistance regulator; Figure 5. This circuit maintains a constant gate-to-source voltage to the switching FET, which keeps ON-resistance variation to less than 5 per cent over the analog range of ±10 V. In addition, the circuit produces virtually zero dc loading in the signal path. Operation is understood by noting that the current mirror (Figure 5) forces the current in J2 and diode D1 to be equal at half the saturation current of J1. J1 and J2 are designed such that the gate-source voltage of J2 is approximately a compensating diode drop in the level shift of the gate of J3. In the complete switch, Figure 6, FET, J3, can be turned ON by TTL compatible input circuitry to disable the regulator and turn the switch OFF.

A quad of these switches has been fabricated on a 72 mil x 72 mil die, with the following highly competitive performance per section: $r_{on} = 150\ \Omega$, $t_{on} = 90$ ns, $t_{off} = 250$ ns, and a maximum signal path slew rate without source loading of 50 V/μs.

Acknowledgments

The authors wish to acknowledge J. Solomon for his assistance in the projects and the paper, and J. Dunkley for the process development that made this work possible.

---

[1] Solomon, J., Davis, R., and Lee, P., "A Self-Compensated Monolithic Operational Amplifier with Low Input Current and High Slew Rate", *ISSCC DIGEST OF TECHNICAL PAPERS*, p. 14-15; Feb., 1969.

[2] Wilson, G., "A Monolithic Junction FET-NPN Operational Amplifier", *ISSCC DIGEST OF TECHNICAL PAPERS*, p. 20-21; Feb., 1968.

Reprinted from *1974 IEEE Internat. Solid-State Circuits Conf. Dig. of Tech. Papers*, Feb. 13-15, 1974, pp. 140-141, 243.

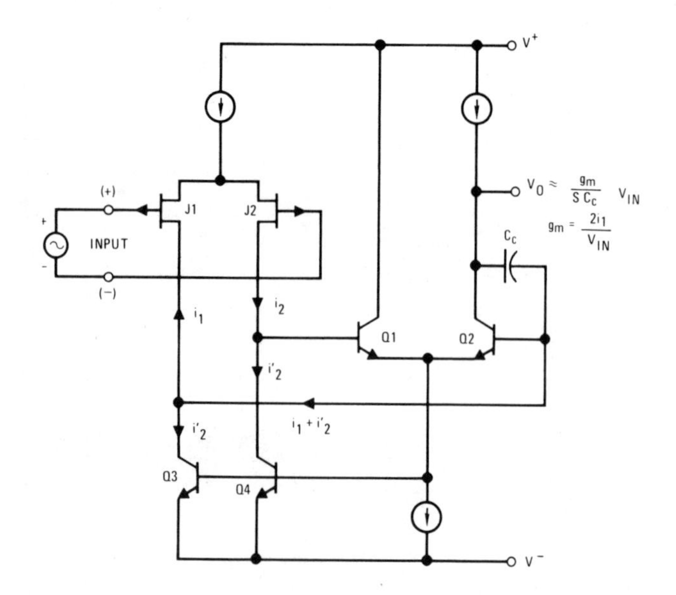

FIGURE 1—FET input op-amp which achieves wide bandwidth by using NPN current sources in the common mode loop and is compensated by a single capacitor.

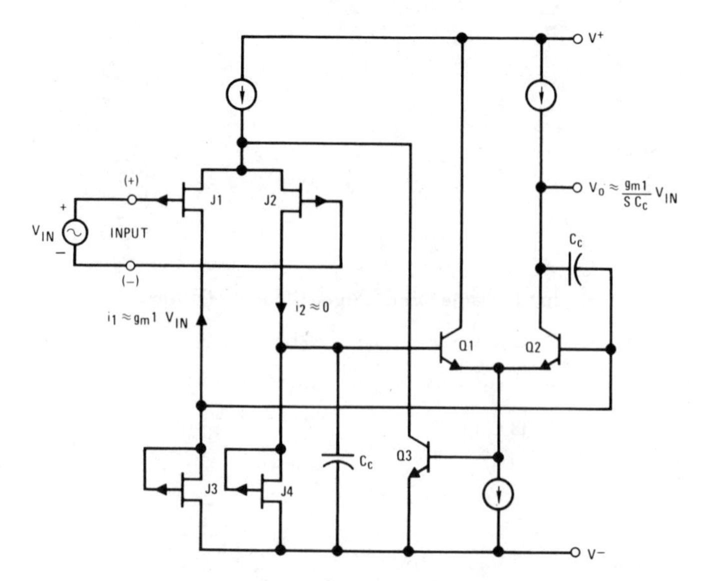

FIGURE 3—Low-noise, low-drift op-amp which employs FET current source loads with common mode feedback to the *tail* of the FET input differential amplifier.

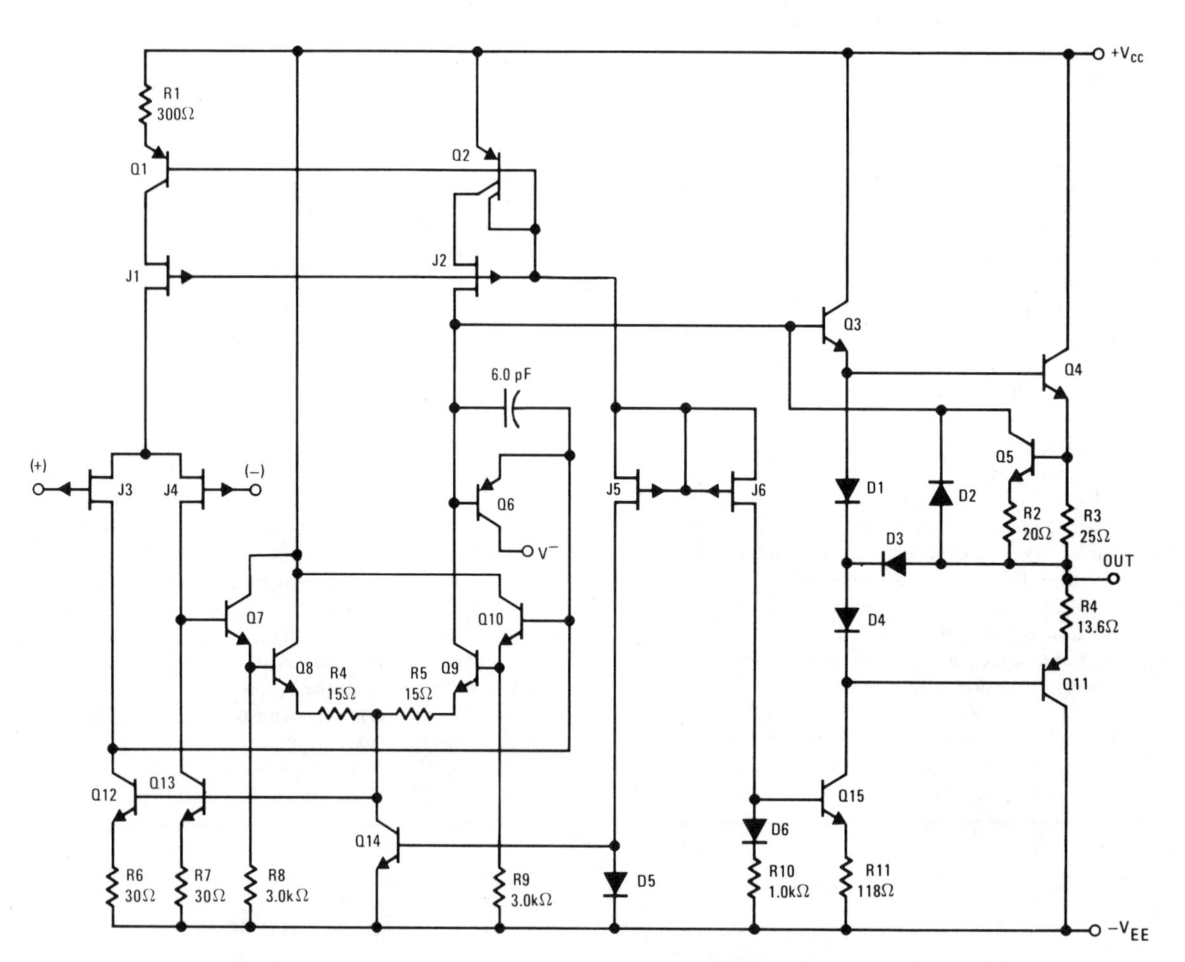

FIGURE 2—Schematic of the wideband, fast-settling op-amp including FET bias circuitry and a FET cascoded second stage current source load for high gain.

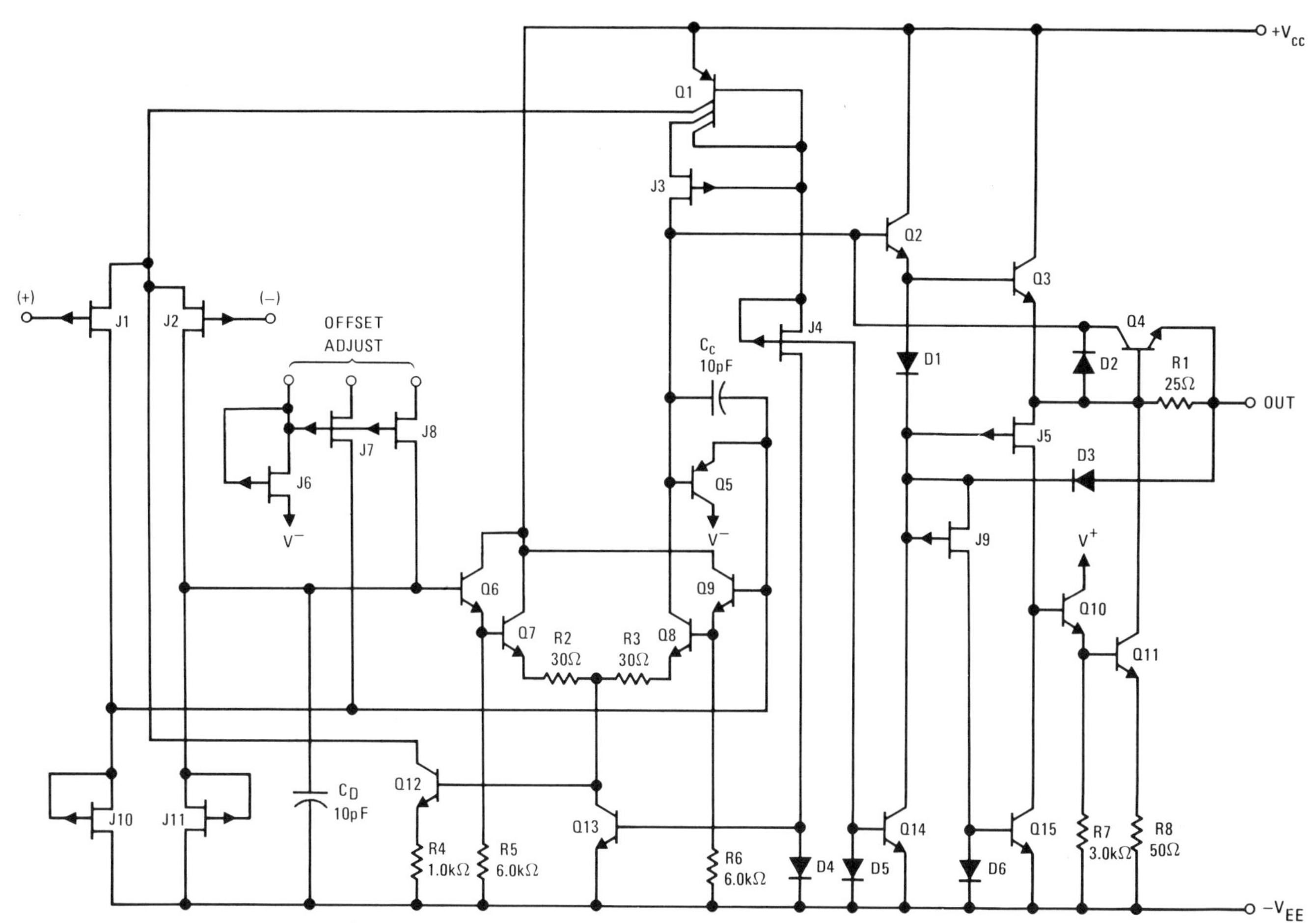

FIGURE 4—Schematic of the low-noise, low-drift FET op-amp including a low noise, drift-offset adjust and a wideband FET-NPN composite output stage.

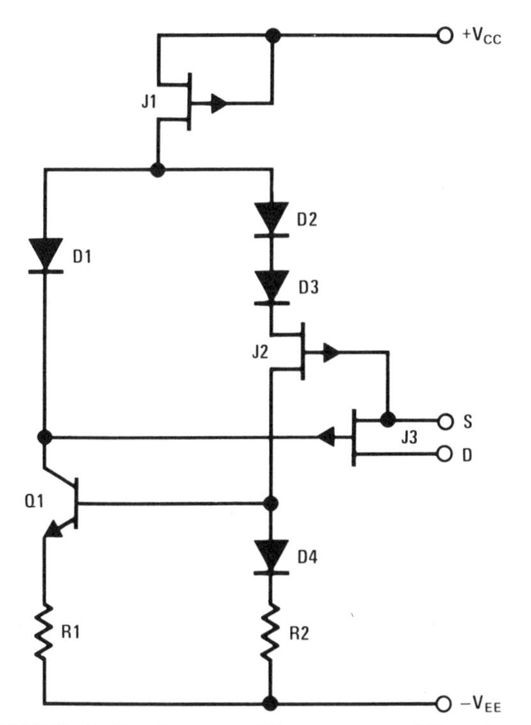

FIGURE 5—Regulator providing constant ON resistance during the analog switch ON condition. Since an FET senses the analog voltage, there is no signal loading.

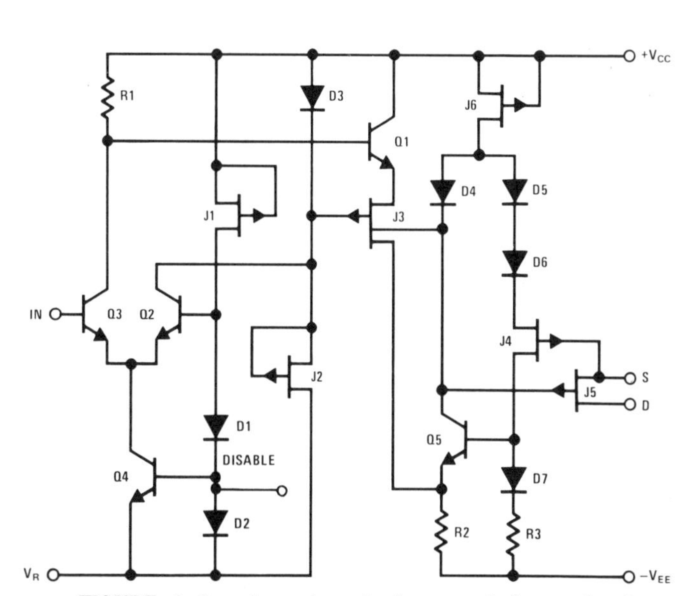

FIGURE 6—Complete schematic for one of the quads of analog switches including the disable input for independently turning all the switches OFF.

An Integrated NMOS Operational Amplifier with Internal Compensation

YANNIS P. TSIVIDIS, MEMBER, IEEE, AND PAUL R. GRAY, SENIOR MEMBER, IEEE

Abstract—**An internally compensated operational amplifier is described which has been fabricated using n-channel Al-gate MOS technology. Only enhancement mode devices are used, and the circuit has been designed so that its performance is insensitive to process parameters.**

I. Introduction

Although single-channel (NMOS or PMOS) technology has been used extensively in the realization of digital large-scale integrated (LSI) circuits, its use to perform analog functions has proved more difficult. Perhaps the most generally useful analog circuit function is that of an operational amplifier. For reasons to be discussed later, single-channel MOS technology is not as well suited as bipolar or CMOS to the realization of an operational amplifier-like function. Nevertheless, the ability to realize this function in single-channel MOS circuits is important for two reasons. First, as the level of circuit complexity which can be placed on a single chip increases, the partitioning of subsystems into separate analog and digital chips becomes cumbersome. Thus arises a need to fabricate both analog and digital circuit functions on the same chip. Examples of such subsystems are A/D and D/A converters, PCM encoders and decoders, analog sampled data filters, and so forth. In the realization of these functions, single-channel technology is only one of several alternatives, others being I^2L, mixed bipolar/MOS and CMOS. However, the single-channel technologies are the most cost-effective in terms of cost per unit of digital function on the chip because of their high yield, low fabrication cost, and high logic density. Thus, provided the design difficulties can be overcome, it is potentially less costly to realize these functions in single-channel technology.

The second motivation for the single-channel realization of the amplifier function is the application of such a circuit in on-chip peripheral circuitry for various types of charge-transfer device analog signal processors. Because of yield and other considerations, these types of circuits are most effectively realized in single-channel technology, and a strong need exists for on-chip charge-to-voltage and voltage-to-charge converters of reasonable performance.

Manuscript received June 11, 1976; revised July 28, 1976. This work was sponsored by the National Science Foundation under Grant ENG73-04184-A01.

Y. P. Tsividis was with the Department of Electrical Engineering and Computer Sciences, University of California, Berkeley, CA 94720. He is now with the Department of Electrical Engineering and Computer Science, Columbia University, New York, NY 10027.

P. R. Gray is with the Department of Electrical Engineering and Computer Sciences, University of California, Berkeley, CA 94720.

Previous attempts to design single-channel MOS amplifiers have resulted in single-ended input configurations [1], or differential amplifiers which require external frequency compensation with two capacitors [2]. This paper will describe an internally compensated differential amplifier which is applicable in many analog-digital MOS LSI circuits.

II. Circuit Objectives

The requirements imposed on an amplifier used in the applications mentioned in Section I are usually different than those for a general-purpose operational amplifier. Within a single-channel MOS LSI circuit, the amplifier typically must drive capacitive loads of 50 pF or more rather than resistive loads. In addition, low input offset voltage and offset voltage drift are not of primary importance since the offset voltage can be stored on a capacitor and then cancelled. In many of the applications, the amplifier would be used as a buffer in a unity gain feedback configuration. Only moderate open loop gain is required in most of these cases. However, such a buffer should be capable of fast settling when driving a capacitive load. In addition, it is desirable to devise a design whose performance is not sensitive to process parameters, especially the device threshold voltage. The latter can vary over a range of more than 1 V from wafer to wafer, and therefore a mechanism must be provided to render the circuit insensitive to such variations.

In view of the above considerations, we have concentrated our efforts in the design of a fast settling, low gain operational amplifier which is especially suited for use with capacitive loads.

III. Limitations in Single-Channel MOS Analog Circuit Design

Most of the problems encountered in single-channel MOS analog circuit design result from the small value of the ratio of transconductance to quiescent current inherent in the MOS device, and the lack of availability of complementary devices. The latter problem makes the realizations of large values of incremental load resistance within an amplifier impractical. As a result, the voltage gain attainable in a single stage is quite limited. For high gain in an inverter stage, high values of transconductance for the amplifying devices must be used. The transconductance g_m of an MOS transistor in the saturation region is given by:

$$g_m = 2\sqrt{k'\left(\frac{Z}{L}\right)I} \tag{1}$$

where k' is typically $5\ \mu A/V^2$ to $10\ \mu A/V^2$, Z is the width of

Reprinted from *IEEE J. Solid-State Circuits,* vol. SC-11, pp. 748-753, Dec. 1976.

the gate, L is its length, and I is the quiescent drain current. It is apparent that in order to obtain high transconductance, large values for one or both of the parameters Z and I must be used. A large value for Z will result in large area and large parasitic capacitances. A large value of I will require high power supply voltages and high power consumption. Also associated with the low value of g_m/I is the high input offset voltage expected from differential stages using MOS transistors [3].

The quiescent gate-to-source voltages required in MOS analog circuits are high, due to the low value of parameter k', the high value of the threshold voltage (in the case of enhancement mode devices), and the body effect. This makes the design of output stages difficult. The use of source followers at the output is undesirable in most cases, since the output swing would then be very limited. On the other hand, using simple inverters at the output stage results in a high output resistance unless excessively high quiescent currents are used.

IV. Circuit Description

A block diagram of the operational amplifier is shown in Fig. 1. The input stage is a source-coupled differential amplifier. The differential output signal of this stage is applied to a differential-to-single-ended converter, which develops a single-ended version of that signal. The converter drives a cascode stage, around which feedback is applied through a source follower and a capacitor, to implement a Miller compensation scheme. Another source follower, fed from the cascode stage, drives the output stage. The bias point interdependence of the stages is such that their quiescent voltages track one another, so that all stages except the last two remain in their active region independently of process parameter variations when the input voltage is zero.

The input stage is shown in Fig. 2. Only a limited voltage is dropped across $M6$ and $M9$ so as to ensure that $M7$ and $M10$ will not be driven out of their saturation region if a large common-mode input is applied. The differential gain of the stage is given by:

$$G_{dm1} = \frac{g_{m7}}{g_{m6}}. \tag{2}$$

The limited voltage drop across $M6$ and $M9$, along with power consumption, size and frequency response considerations, limit the differential gain to a nominal value of about 5.

The common-mode gain of the stage is given by:

$$G_{cm1} = \frac{1}{2 r_{08} g_{m6}} \tag{3}$$

where r_{08} is the incremental output resistance of $M8$. To increase the latter, and thus reduce G_{cm1}, $M8$ has been designed with a long channel. The resulting common-mode gain is nominally 0.017.

The differential output of the input stage is applied to the gates of $M4$ and $M11$ in the circuit shown in Fig. 3. The signals at the two gates are out of phase, one of them passing through an inverting path and the other through a noninverting path. The inverting path consists of source follower $M4$,

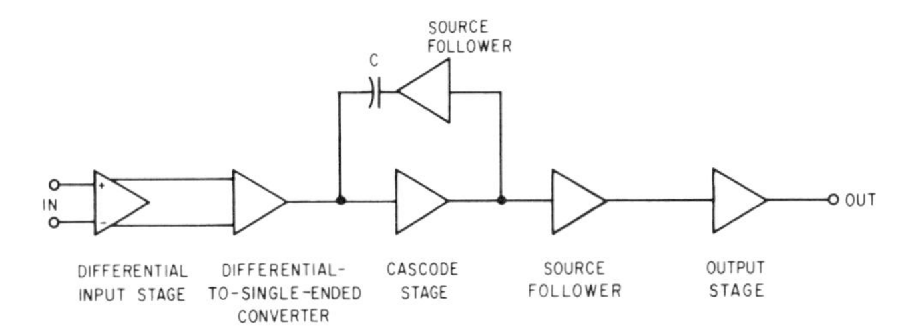

Fig. 1. Block diagram of the operational amplifier.

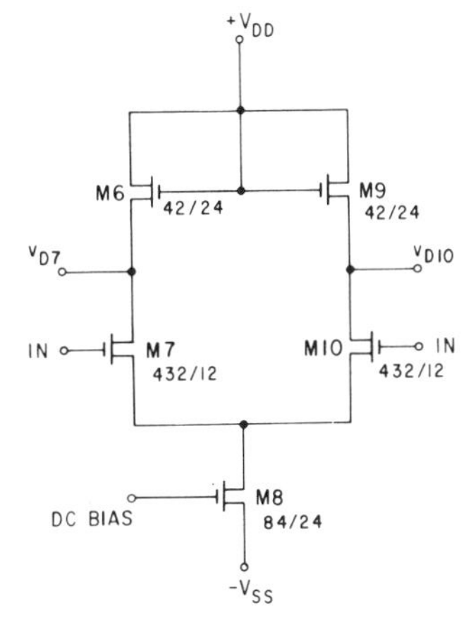

Fig. 2. Input stage. The channel width (Z) and length (L) is indicated in micrometers for each device (Z/L).

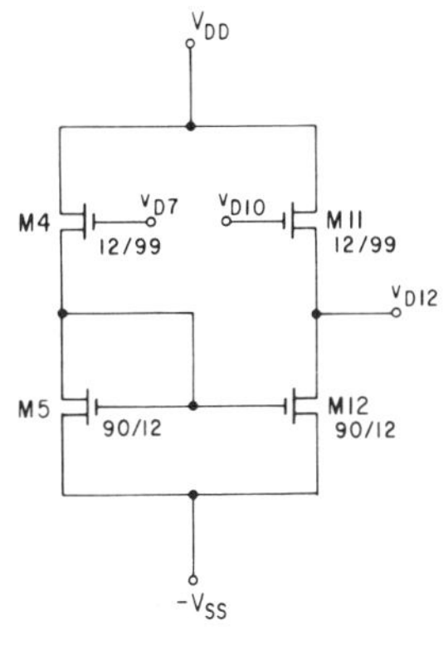

Fig. 3. Differential-to-single-ended converter.

its load $M5$, and the inverter $M12$. Transistor $M11$ acts as the load to $M12$, and the noninverting path consists of $M11$ acting as a source follower. Both signals appear in phase and are summed at the drain of $M12$. The differential-to-single-ended gain of this stage is given by [4]:

$$G_{dm2} = \frac{g_{m11} r_{012}}{2(1 + g_{m11} r_{012})} \left[1 + \frac{g_{m4} g_{m12}}{g_{m11}(g_{m4} + g_{m5})}\right]. \tag{4}$$

The value of this gain is 0.95.

The common-mode gain of the stage is given by [4]:

$$G_{cm2} = \frac{g_{m11} r_{012}}{1 + g_{m11} r_{012}} \left[1 - \frac{g_{m4} g_{m12}}{g_{m11}(g_{m4} + g_{m5})}\right] \tag{5}$$

and has a value of 0.09.

The differential-to-single-ended converter drives the stage

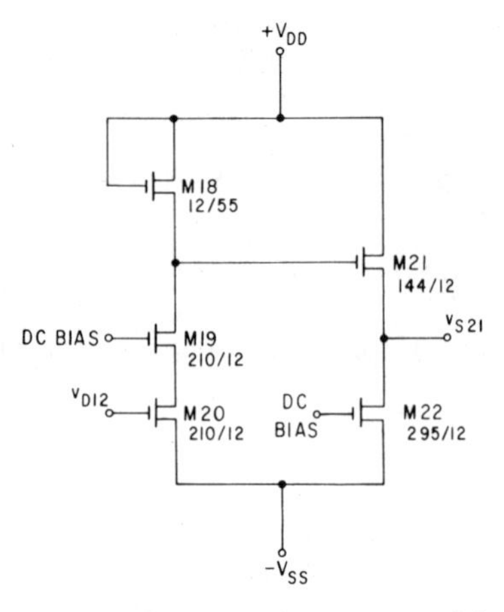

Fig. 4. Cascode stage and output stage driver.

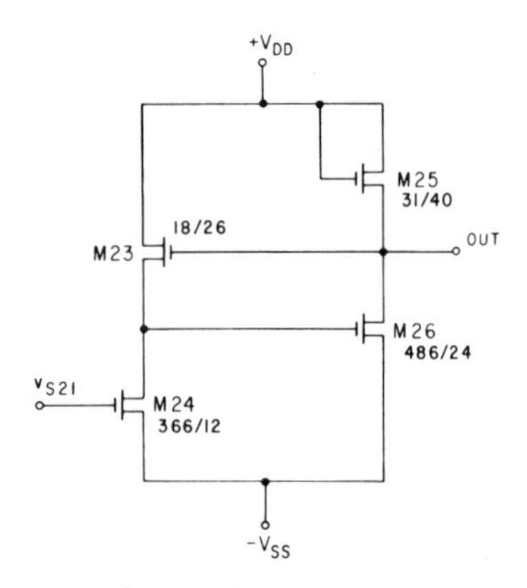

Fig. 5. Output stage.

shown in Fig. 4. A cascode configuration is used in order to reduce the Miller capacitance at the gate of $M20$, which would otherwise load the converter of Fig. 3 and degrade the frequency response. The gain of the stage is given by:

$$G_3 = \frac{g_{m20}}{g_{m18}} \tag{6}$$

and has a nominal value of 11. The output of this stage is fed to the source follower $M21$, also shown in Fig. 4, which is used both for level shifting purposes and to reduce the effect of the input capacitance of the output stage on the frequency response.

The output stage is shown in Fig. 5. The inverter $M25, M26$ has been designed with sufficiently large bias current so that capacitive loads up to 70 pF can be charged from 0 V to ±5 V in 1 μs. The output resistance of this inverter if no feedback is applied is $1/g_{m25} \simeq 6$ kΩ. This, in conjunction with a 70 pF capacitive load, would create a pole in the open loop frequency response at a frequency of 379 kHz, which would make internal frequency compensation virtually impossible. In order to lower the output resistance, shunt-shunt feedback is applied by the addition of $M23$. The gain of the combined stage $M23, M24, M25$, and $M26$ is given by:

$$G_4 = \frac{g_{m24}}{g_{m23}} \left[\frac{g_{m26}/g_{m25}}{1 + g_{m26}/g_{m25}} \right] \tag{7}$$

and is nominally equal to 6.6. The loop gain is approximately equal to the gain of the inverter $M25, M26$, and the output resistance of the stage is reduced to approximately 1.1 kΩ. Additional considerations in selecting the sizes for the four devices are given in [4].

Frequency compensation of the circuit is accomplished through the capacitance C, shown in Fig. 1, which is Miller-multiplied by the cascode stage and used to create a dominant pole in conjunction with the output resistance of the differential-to-single-ended converter, which is approximately equal to $1/g_{m11}$. The compensation capacitor C is not connected directly between the input and output of the cascode stage because for an inverting gain stage with simple pole-splitting compensation, the compensation capacitor, in addition to introducing a dominant pole, also introduces a right-half plane zero corresponding to $\omega_z = g_m/C$, where g_m is the transconductance of the stage and C the compensation capacitance. The right-half plane zero is clearly undesirable, since it degrades the phase of the complete amplifier by 90° at high frequencies, while at the same time eliminating the effect of the dominant pole on the gain magnitude frequency response by stopping the 20 dB/decade rolloff created by it. For the values of C required and for the values of g_m that can be achieved, the frequency of the zero is lower than that of the crossover frequency and thus frequency compensation becomes impossible. Although this zero often is present in internally compensated operational amplifiers made with bipolar technology, the transconductances there are large and the frequency of the zero extremely high.

Physically, the zero is caused by feedforward through the compensation capacitor [4]. To eliminate this feedforward, a buffer implemented by a source follower is placed in series with C, thus preventing the feedforward and therefore eliminating the right-half plane zero, while at the same time allowing feedback which gives rise to the desirable dominant pole.

Complete Amplifier

The complete amplifier schematic is shown in Fig. 6. It consists of the individual stages discussed above, in addition to two voltage dividers $M1, M2, M3$ and $M15, M16, M17$, which are used to bias the current sources $M8, M14$, and $M22$. Transistor $M13$ is the source follower used to prevent feedforward through the compensation capacitor, the value of which is 40 pF.

The circuit is designed so that its operation is largely independent of the threshold voltage V_T, which varies considerably from wafer to wafer. This is achieved by selecting the Z/L's of the devices so that the various quiescent voltages track one another in such a way that the devices up to $M21$ are maintained in the saturation region independently of the value of V_T. The voltage required at the gate of $M21$ to drive the output voltage to zero does depend on V_T. However, the range of values of that voltage for various values of V_T, when reflected to the input, corresponds to a small variation of the input offset precisely due to the fact that the devices are in the saturation region and therefore the gain between the input and the gate of $M21$ is high, independent of V_T.

The quiescent voltage tracking mentioned above will now be described. The symbol S will be used to denote the Z/L ratio

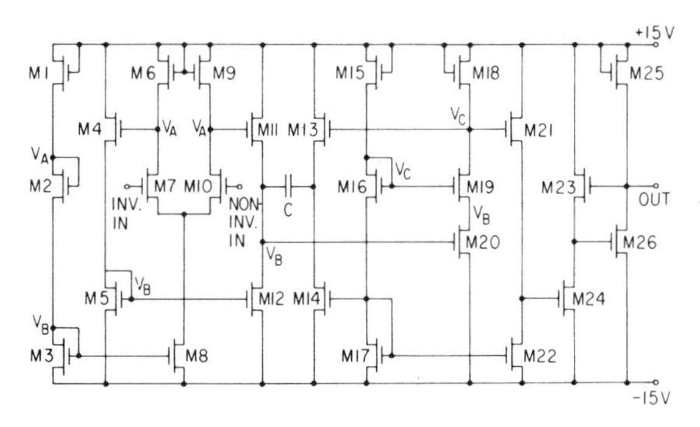

Fig. 6. Complete amplifier schematic diagram.

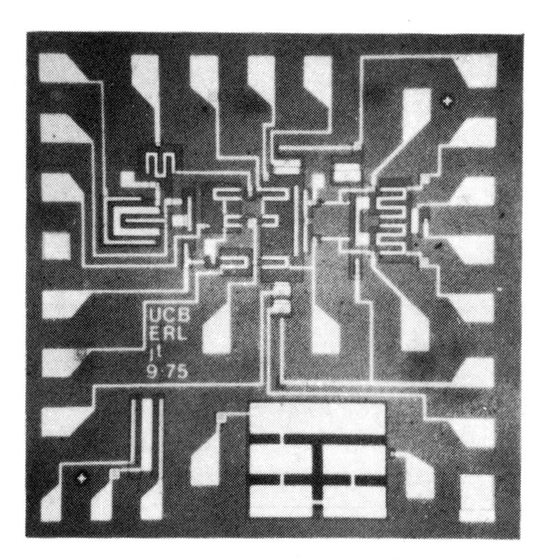

Fig. 7. Photograph of the integrated circuit.

after lateral diffusion has been taken into account. The input stage and the differential-to-single-ended converter are symmetric, so that for zero differential input, $V_{GS6} = V_{GS9}$, $V_{DG12} = V_{DG5} = 0$, and $V_{GS4} = V_{GS11}$, independent of V_T. We have chosen $S_8 = S_3$ so that $I_8 = I_3 = I_1$. Therefore, a current equal to $I_1/2$ flows through $M6$ and $M9$. Also, $S_6 = S_9 = S_1/2$, and therefore $V_{GS1} = V_{GS6} = V_{GS9}$. We therefore have $V_{GS2} + V_{GS3} = V_{GS4} + V_{GS5}$. By choosing $S_2/S_3 = S_4/S_5 = S_{11}/S_{12}$, we get $V_{GS2} = V_{GS4} = V_{GS11}$ and $V_{GS3} = V_{GS5} = V_{GS12}$. Therefore, since $V_{DG12} = 0$, we have $V_{DS12} = V_{GS12} = V_{GS3}$, and this has been shown to be the case independent of V_T. As long as $V_T > 0$, $M12$ will remain in the saturation region, where its transconductance is high.

The bias string $M15, M16, M17$ is used to bias the cascode stage and the source followers. There is no symmetry between this bias string and the string consisting of $M1, M2, M3$ due to the different biasing requirements of the cascode stage [4]. Specifically, in the string $M1$, $M2$, $M3$ we have $S_1 = S_3$, whereas in the string $M15$, $M16$, $M17$ we have $S_{16} = S_{17}$. However, the two strings have been designed so that $V_{GS17} = V_{GS3}$ for a typical value of V_T, and this relation is approximately true even if V_T deviates from the typical value. Therefore, since $V_{GS20} = V_{DS12} = V_{GS3}$, we have $V_{GS20} = V_{GS17}$. We have chosen $S_{19}/S_{16} = S_{20}/S_{17}$, and therefore $V_{GS16} = V_{GS19}$ and $V_{GS20} = V_{DS20}$, so that $M20$ is kept in the saturation region for any $V_T > 0$. Finally, we have $S_{15}/S_{16} = S_{18}/S_{19}$, and therefore $V_{GS18} = V_{GS15}$, so that $V_{DS19} = V_{GS19}$, which guarantees that $M19$ is also kept in the saturation region.

TABLE I
MASK DEVICE DIMENSIONS

Device	$Z(\mu m)$	$L(\mu m)$
$M1$	84	24
$M2$	12	276
$M3$	84	24
$M4$	12	99
$M5$	90	12
$M6$	42	24
$M7$	432	12
$M8$	84	24
$M9$	42	24
$M10$	432	12
$M11$	12	99
$M12$	90	12
$M13$	138	12
$M14$	132	12
$M15$	12	194
$M16$	54	12
$M17$	54	12
$M18$	12	55
$M19$	210	12
$M20$	210	12
$M21$	144	12
$M22$	295	12
$M23$	18	26
$M24$	366	12
$M25$	31	40
$M26$	486	24

V. EXPERIMENTAL RESULTS

The amplifier has been fabricated as an integrated circuit using n-channel Al-gate MOS technology. No p_+ isolation diffusion was used. The chip photograph is given in Fig. 7. As seen, numerous pads are used for experimentation. The active area of the amplifier is 0.77 mm^2. A 12 μm minimum feature size and a minimum alignment tolerance of 3 μm were used, and the devices $M6, M9, M7, M10, M4, M11, M5, M12$ have been laid out so that proper matching is retained in the presence of misalignment. The mask dimensions for the devices are given in Table I.

The device threshold measured was 0.2 V. To simulate higher threshold voltages typical of industrial processing, a substrate bias of 5 V was used so that the effective threshold of the devices whose source is connected to $-V_{SS}$ was 1.5 V. The range of threshold voltages encountered throughout the circuit will of course be different from that of a circuit where the devices have $V_T = 1.5$ V without substrate bias. However, the fact that the operation of the circuit is based on quiescent voltage tracking rather than exact threshold voltages justifies the use of substrate bias. Also, although the substrate bias used will decrease the junction capacitances, computer simulation shows that these capacitances are not the ones that dominate the circuit performance, and that the results obtained with substrate bias are representative of those expected with no substrate bias and a higher threshold.

The dc transfer characteristic of the amplifier is shown in

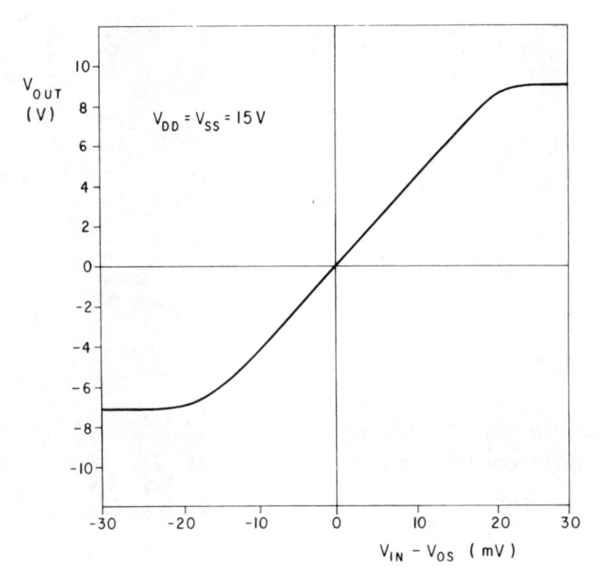

Fig. 8. DC transfer characteristic.

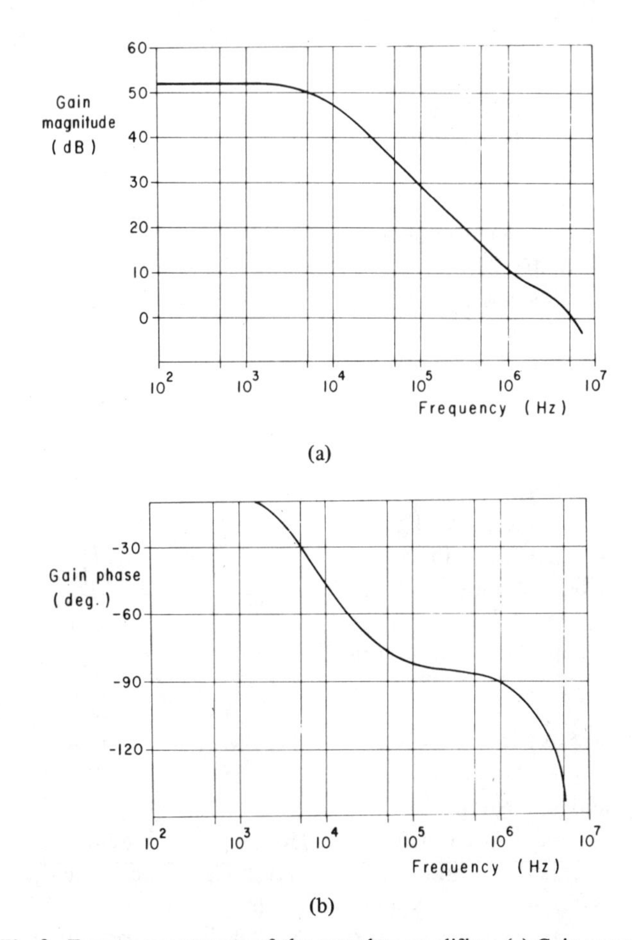

Fig. 9. Frequency response of the complete amplifier. (a) Gain magnitude. (b) Gain phase.

Fig. 8 and the frequency response is shown in Fig. 9. The step response of the amplifier when connected in a unity gain configuration is shown in Fig. 10 for a capacitive load of 70 pF connected to the output through a series device with $Z/L = 3$. This device is necessary, since direct connection of the load capacitance to the output stage would degrade the phase margin and cause oscillation. The acquisition to an accuracy of ±1 percent time is 2 μs, with an input step of 5 V.

The performance of the amplifier is summarized in Table II. The first two parameters describe the observed distribution of input offset voltage over a sample of three wafers. While these parameters are much worse than those observed for a bi-

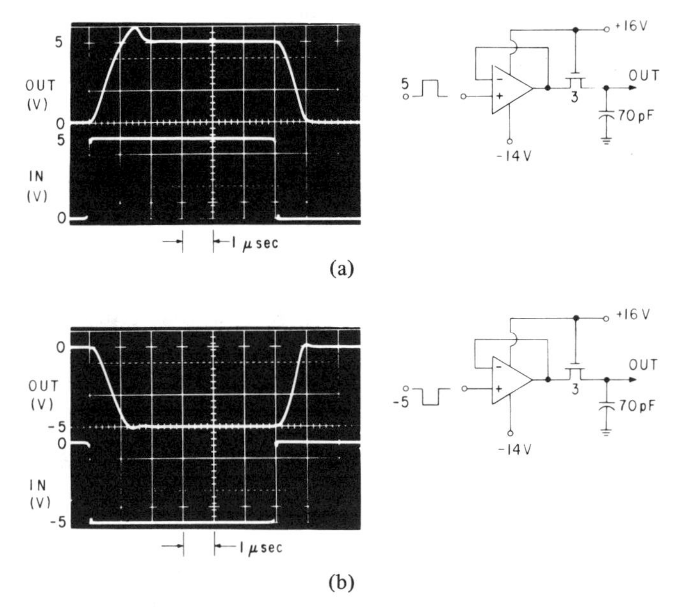

Fig. 10. Step response. (a) Positive input. (b) Negative input.

TABLE II
PERFORMANCE PARAMETERS FOR ±15 V POWER SUPPLIES

Input offset voltage mean	9 mV
Input offset voltage standard deviation	66 mV
Low frequency gain	51 dB
Common-mode rejection ratio	70 dB
Unity-gain frequency	5 MHz
Slew-rate (positive input)	7 V/μs
Slew-rate (negative input)	7 V/μs
Power supply rejection	4.5 mV/V
Input noise voltage (10 Hz to 10 kHz)	60 μV
Total harmonic distortion (1 kHz, output 10 V p-p)	1.5 percent
Power supply current	5 mA

polar circuit, the offset can be sampled and canceled in most applications because of the sample/hold capability inherent in MOS technology. The unity-gain frequency and slew rate are comparable with many bipolar amplifiers. The open loop gain of 51 dB restricts the circuit to low values of closed loop gain and moderate levels of closed loop gain accuracy.

VI. CONCLUSION

The realization of analog functions in single-channel MOS technology is both useful and possible. An operational amplifier using n-channel Al-gate technology has been realized which occupies 0.77 mm^2 of chip area, and has a power consumption of 150 mW. The performance of the amplifier is such that it can find use in A/D and D/A converter circuits, PCM encoders and decoders, and charge transfer device transversal and recursive filters.

ACKNOWLEDGMENT

The authors wish to thank Prof. D. A. Hodges and F. Hosticka for useful technical discussions and suggestions, K. P. Burns for trimming the mask and fabricating additional samples, and D. McDaniel for valuable lab assistance.

REFERENCES

[1] A. Boornard, E. Herrmann and S. T. Hsu, "Low-noise integrated silicon-gate FET amplifier," *IEEE J. Solid-State Circuits* (Corresp.), vol. SC-10, pp. 542–544, Dec. 1975.

[2] P. W. Fry, "A MOST integrated differential amplifier," *IEEE J. Solid-State Circuits* (Corresp.), vol. SC-4, pp. 166–168, June 1969.

[3] H. C. Lin, "Comparison of input offset voltage of differential amplifiers using bipolar transistors and field-effect transistors," *IEEE J. Solid-State Circuits* (Corresp.), vol. SC-5, pp. 126–128, June 1970.

[4] Y. P. Tsividis, "Nonuniform pulse code modulation encoding using integrated circuit techniques," Ph.D. dissertation, Univ. California, Berkeley, 1976.

[5] K. P. Burns, "Optimization of offset in an NMOS operational amplifier," Univ. California, Berkeley, M.S. Plan II Report, 1976.

Bipolar & FET op amps

by Jerry Metzger
Electronic Products Magazine

Sometimes quiet, sometimes widely heralded, changes are occurring in IC op amps. The changes lack the glamour of microcomputers and bubble memories. They aren't dynamic and flashy like the announcement of 64k RAM's and CCD devices. But op amps are pervasive, and the improvements in these universal blocks of gain — primarily in the FET input types — are important to engineers everywhere. So to find out what is happening now, to determine if the low cost FET units signal the demise of the 741 and other bipolars, **electronic products** surveyed a number of the major producers. Here's what we found.

The biggest change is in methods of construction: MOS FET inputs, the use of ion implant to place good performance JFET's on the same chip with bipolar transistors, the use of lasers or zener-zapping to trim the performance of the circuit, the ability to test and adjust circuit performance at the wafer level. These changes have evoloved over the past several years, but now a whole fist-full of manufacturers are putting the processes on-line. The results are new FET input, premium performance products, new wideband devices and, for the mass producers, a hastening down the learning curve.

Compared to the generally used bipolar units, the monolithic FET devices feature lower bias currents, faster slew rates and wider bandwidth. When it comes to offset voltage and drift characteristics, bipolars hold the edge. The Guide (page xx) presents the specs of a number of the newest and the most popular Bipolar-FET units. We've included only compensated single and quads, but duals and uncompensated singles as well as military versions are sometimes available. To assist in comparisons, the chart also presents various other devices. The μA741C is covered because it is the bipolar standard to which all inexpensive FET units are compared. Since these 741 specs were set almost a decade ago and manufacturers now build 741's much tighter, we've included the AD741L and OP-02E as examples of what can be selected on a commercial basis from the production mix. The bipolar LM308A is shown because it is a many sourced, super beta part with excellent input characteristics, and the AD515L two chip hybrid and Burr-Brown 3527BM four chip hybrid are listed to demonstrate what can be achieved with separate FET's, laser trimming and special testing. Similarly, we've listed four bipolar quads, the RC-136 and LM308 as widely second sourced devices plus the OP-09/11 and HA4605 for their offset and speed characteristics.

When FET op amps are mentioned, the first thing most users think about is low bias or leakage currents. They are apt to think of a 1000:1 or more improvement over bipolars. And something close to that is true at room temperature and over the temperature range where most devices are used; at elevated temperatures, however, the picture changes. Compared to the 741, the improvement is more like 100:1 at 70°C and virtually gone at 125°C (Fig. 1). In fact, most bipolars outshine the FET's at 125°C. Further-

Reprinted with permission from *Electron. Products Mag.*, pp. 51-58, June 1977.

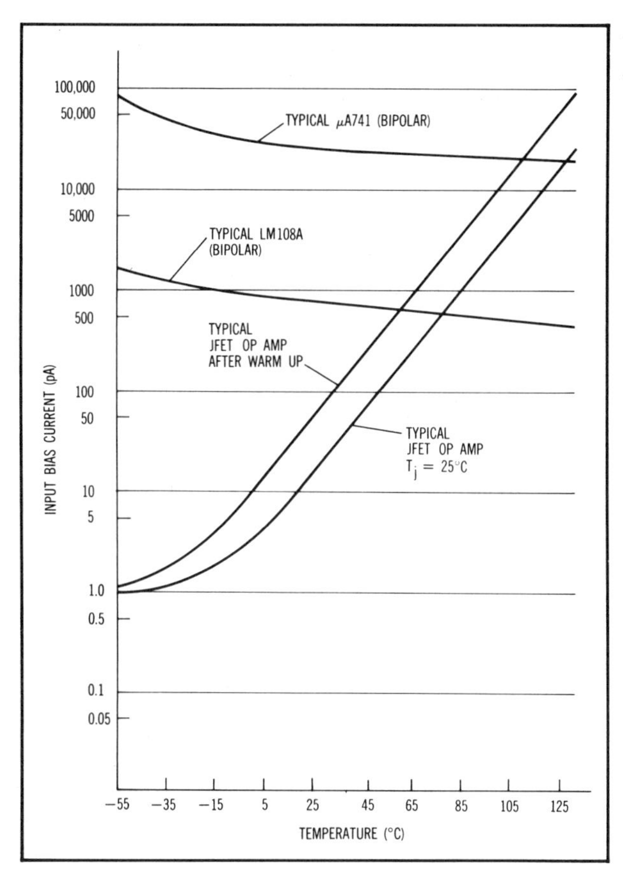

Fig. 1. Comparison of op amp input currents. The FET devices double in current every 10°C. Note the increase in the JFET current due to warm up (self heating when the power is turned on.)

more, it makes a difference whether the FET's are warmed up or not. That leads to the bias current story, a subject that every IC manufacturer we contacted said needs to be spelled out to all users — so it's told in detail at the end of this report.

The slew rate and bandwidth improvements of FET op amps are also significant. If reasonable input bias currents are maintained, bipolar devices will generally exhibit a slew rate of 0.5 V/μs and a unity gain bandwidth of 0.5 to 1 MHz. The ion implanted FET's of today permit 10 to 25 V/μs and 3 to 8 MHz bandwidth. This may stretch to 10 to 20 MHz as manufacturers tweak designs and processes. The improved bandwidth expands the use of op amps and is very important for audio applications. In audio equipment it means greater full power bandwidth, which could result in higher usage, and it means better TID (transient intermodulation distortion), which translates into more pleasing sound for the hi-fi enthusiast. Several manufacturers are investigating even wider bandwidth, but they believe that any big jump will require a new high frequency FET process.

Somewhat related to slew rate and bandwidth is settling time. Here again the FET units are superior, in fact many of the bipolars take forever to settle to 0.01%, if they settle at all. Settling time isn't generally a production test item, so guaranteed values are not stated. However to give perspective, a 356 typically settles to 0.01% in 1.5 microseconds.

The offset voltage of bipolar amplifiers is determined by a difference in V_{BE}'s, circuit design and mask layout. It's been suggested that "even with the world's worst processing, offsets will rarely exceed 10 mV." Today, manufacturers often hold bipolar offsets to less than 1 mV.

The offset of JFET's calls for the matching of pinch off voltages, that is, setting the difference of two larger numbers. Fortunately ion implant leads to fairly good matching and offsets of 10 to 15 mV max are standard. (Insight can be gained by looking at the manufacturer's poorest grade of a given device and assuming the spec's of that grade were initially set to cover most of the production distribution.) The manufacturers are working to reduce the offsets of low cost FET parts to 6 mV to compete with the 741's, and further to 1 or 2 mV with high yield to compete with most bipolar devices (Fig. 2). Various trimming methods are being investigated to achieve this improvement. Part of the trick is to trim the offset without degrading temperature drift. So when you evaluate a low offset device, check the drift characteristics. Admittedly these are usually given as "typicals," if stated at all, so you may have some trouble making this evaluation.

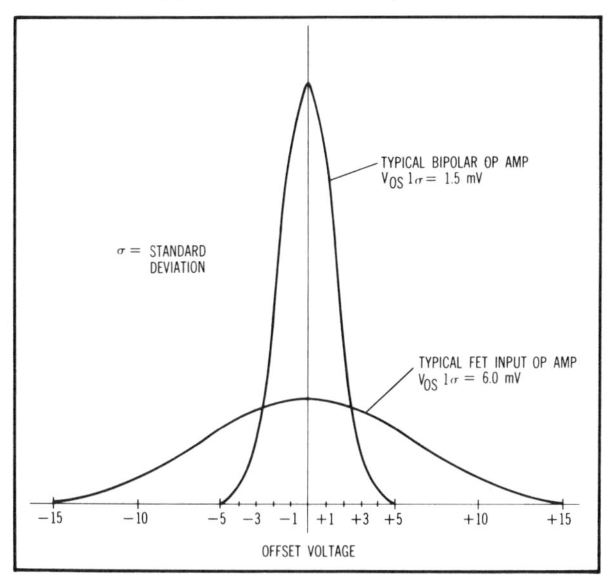

Fig. 2. Comparison of typical bipolar and FET input op amp offset voltage production distribution.

How does this stack up?

For many existing applications, any op amp — old or new — will do; the choice is just a matter of price, availability, familiarity, the cost of stocking several parts, etc. Today, 741's are selling for 20¢ to 25¢ in large volume quantities; it's hard to imagine them going lower. Some people look at the large die size and extra processing step of today's low cost FET

QUICK GUIDE TO BIPOLAR-FET OP AMPS

This guide does not list uncompensated versions, duals or military range versions.

Offset Voltage mV 25°C	Bias Current pA $T_j = 25°C$	Offset Current pA $T_j = 25°C$	Drift μV/°C	Bandwidth MHz	Slew Rate V/μs	Voltage Gain V/V	CMRR db	Supply Current per amp mA	Thermal Resistance °C/W	Model	Manufacturer
Single Units Compensated											
0.5	50	10	2T, 5M	6T, 4M	17T, 10M	100k	86	4.7	150	OP-15E	PMI
0.5	50	10	2T, 5M	8T, 6M	25T, 18M	100k	86	4.7	150	OP-16E	PMI
1	100	20	3T, 10M	5.7T, 3.5M	16T, 7.5M	75k	86	4.7	150	OP-15F	PMI
1	100	20	3T, 10M	7.6T, 5.5M	24T, 12M	75k	86	4.7	150	OP-16F	PMI
2	20	10	5T,15M	4T	10T	100k	86	15	135-145	CA3160B	RCA
2	30	10	5T	4.5T	9T	50k	86	6	135-145	CR3140B	RCA
2	50	10	3T, 5M	4.5T, 4M	12T, 10M	50k	85	7.	150	LF356A	National*
2	50	10	3T, 5M	2.5T	5T, 3M	50k	85	4	150	LF355A	National*
3	100	50	10T	5T	12T	50k	80	2.8	120-210	TL081BC	Texas Instrument
3	200	50	4T, 15M	5.4T, 3M	15T, 5M	50k	82	5.8	150	OP-15G	PMI
3	200	50	4T, 15M	7.2T, 5M	23T, 9M	50k	82	5.8	150	OP-16G	PMI
5	30	20	6T	4T	10T	50k	80	15	135-145	CA3160A	RCA
5	40	20	6T	4.5T	9T	20k	70	6	135-145	CA3140A	RCA
5	100	20	8T	4T	13T	50k	80	2.5	100	μAF771F	Fairchild
5	100	20	5T	5T	12T, 7.5M	50k	85	7	155	LF356B	National*
6	200	100	10T	3T	12T	50k	80	2.8	120-210	TL081AC	Texas Instrument
10	200	50	8T	4T	13T	25k	70	2.5	100	μAF771C	Fairchild
10	200	50	10T	5T	15T	25k	70	2.8	150	LF351	National
10	200	50	10T	3T	15T	25k	70	2.8	120-210	TL086C	Texas Instrument
10	200	50	5T	2.5T	5T	25k	80	4	150-175	LF355	National*
10	200	50	5T	5T	12T	25k	80	10	150-175	LF356	National*
15	50	30	8T	4.5T	9T	20k	70	6	135-145	CA3140	RCA
15	50	30	8T	4T	10T	50k	70	15	135-145	CA3160	RCA
15	200	50	10T	1T	0.5T	25k	70	4	150	LF13741	National
15	400	200	10T	3T	12T	25k	70	2.8	120-210	TL081C	Texas Instrument
References											
0.25	2	0.3T	5M	1T	0.9T, 0.6M	100k	76T	4	150	3527BM	Burr-Brown
0.5	7,000	1,000	2T, 5M	1T	0.3T	80k	96	0.8	150	LM308A	National**
0.5	30,000	2,000	2T, 8M	1.3T, 0.8M	0.5T, 0.25M	100k	90	2	—	OP-02E	PMI
0.5	50,000	5,000	5M	1T	0.5T	50k	90	2.8	—	AD741L	Analog Devices
1	0.075	—	25M	0.35T	1T	25k	66	1.5	—	AD515L	Analog Devices
6	500,000	200,000	—	1T	0.5T	20k	70	2.8	—	μA741C	Fairchild**
Quadruple Units											
3	100	50	10T	3T	12T	50k	80	2.8	85-97	TL084BC	Texas Instrument
5	100	20	8T	4T	13T	50k	80	2.5	70	μAF774E	Fairchild
6	200	100	10T	3T	12T	50k	80	2.8	85-97	TL084AC	Texas Instrument
10	200	50	8T	4T	13T	25k	70	2.5	70	μAF744C	Fairchild
10	200	50	10T	5T	15T	25k	70	2.8	100	LF347	National
15	200	50	—	10T, 7M	20M	25k	70	2.5	100	3471	Motorola
15	400	200	10T	3T	12T	25k	70	2.8	85-97	TL084C	Texas Instrument
References											
0.75	500,000	20,000	8M	2T	1T	100k	90	1.5	—	OP-09/11E	PMI
3.5	300,000	100,000	2T	8T	4T	75k	80	1.6	—	HA4605	Harris
6	200,000	50,000	—	1T	0.5T	25k	70	1.1	—	LM348	National**
6	500,000	50,000	—	3T	1T	20k	60	10	—	RC4136	Raytheon**

M — Maximum or Minimum.
T — Typical.

* Suppliers: AMD, Fairchild, Intersil, Motorola, National, PMI, Raytheon, and Texas Instruments.
** Suppliers: See IC Master.

amplifiers and think that prices can't go as low as 741's. That may be true, but in a couple of years, FET units will be optimized for cost, die size, and yield. Even now, the devices are being built on the same automated production lines as 741's. And, while the splicing and die processing might cost a few more mils, assembly and test will be the same. The costs will be close enough so that pricing will be based on other factors — enticing you to use FET's, competition, market share, which manufacturer needs the business. It's this writer's opinion that within a couple of years the general purpose class of FET op amps will sell for 10¢ more than 741's.

Obviously you will use the low cost FET units where needed. And there are performance advantages that might save a few cents elsewhere in your circuit.

Once you start using any low cost FET's it may make sense to standardize on one rather than to stock both a FET and a bipolar. There will be a concerted drive to get you to switch, and probably no one to champion 741's. So you may switch even when you don't have to.

The higher ticket FET amplifiers are only made available when the manufacturers see applications for the products and their performance exceeds existing devices. Therefore, the future for them looks good too. Let's take a look at what the manufacturers have introduced recently and what's almost here.

Advanced Micro Devices

AMD has come on stream with the 155/355 Series and is competing in price; in fact at the instant of writing this, the AMD prices are $0.05 to $0.25 less in 100 piece quantities than other nationally advertised prices.

Fairchild

Fairchild is shipping the 155/355 Series too. They are also introducing a 156-like family including a single, a dual and a quad; the μA771, μA772, and μA774, respectively. These units have a redesigned front end and feature typical slew rates of 13 V/μs, bandwidths of 4 MHz, and noise of 20 nV/$\sqrt{\text{Hz}}$ at 1 kHz. The current drain is 2.5 mA max per amplifier.

An important feature of all Fairchild DIP units is the use of copper lead frames to improve the heat transfer. This means lower chip temperatures for lower bias currents and higher reliability.

Harris

The Harris Semiconductor Group is developing particularly high speed op amps. They are doing this by adding FET inputs and then laser trimming their wideband, dielectrically isolated amplifiers. The dielectric isolation gives very high frequency performance and the laser trimming will give low osets. Like their bipolar counterparts, Harris plans both wideband and high slew rate families. Projected specications for the wideband series are: compensated versions unity gain bandwidth 25 MHz, decompensated versions (gain = 3) bandwidth 100 MHz; power bandwidth 200 kHz; 0.01% settling time 1.5 μs, offset 1 mV and drift 3 μV/°C. The bias currents will be on the order of 50 pA. The high slew rate family is designed to slew at 70 Vμs (gain = 1) to 120 V/μs (gain = 3). They will have gain bandwidth products of 20 to 30 MHz and settling times of 200 to 400 ns. Their front end characteristics will be similar to the wideband units.

Motorola

Motorola is developing the 155/355 Series to be offered in metal cans. They are also about to sample a pair of FET input quads; the commercial temperature range MC3471 and the military MC3571. Tentative specifications for the MC3471 are listed in the table. Its 7 MHz bandwidth is wider than other currently available quads. The MC3571 is expected to have similar performance with 8 MHz bandwidth, 6 mV offset and a supply current drain of 2 mA per amplifier.

National

National Semiconductor started the LF-155/355 Bipolar-FET Series several years ago, then introduced the tight specification LF-155A/355A versions. In this series there are three temperature prefixes, the 155 military, 255 industrial and 355 commercial versions. Within each temperature range, such as the 155 range, there is a lower power version (155); a faster, higher current version (156); and a decompensated unit (157). Recently National added the LF355B/356B/357B with specifications that fall between the 355 and 355A. This makes 18 versions of the same basic circuit, not to mention the different package styles, which include metal cans, plastic DIP's, and recently introduced ceramic DIP's. (The thermal resistance rating of the 155/355 Series plastic package units is being changed to 155°C/W.) National also offers a simple JFET-741 device called the LF13741.

At the end of the month the already announced LF351 single unit will become available. It will be followed by a quad version, the LF347. These units are designed to compete in the low cost, successor-to-the-741 applications. The specifications for both are similar to the 356, however the 351 has a different offset adjustment. The 351's adjustment can be connected to the negative supply just the same as the 741's, enhancing its use as a 741 substitute. Compared to the 356, the 351 is simpler and has a smaller die. While it may be hard to discern from the data sheets, the 356 has some fringe advantages in noise and drift performance as well as greater ability to drive capacitance loads.

Laser trimmed versions of the 156/356 are also in the works. With this trimming, offsets will be on the order of 1 mV and the temperature coefficients will be good too. However, since drift tests do not lend themselves to volume production, these units will simply be specified with a worst case offset over temperature. Such laser trimmed units can also provide an answer to the difficult to produce 155A Series. (The 155A requires individual temperature testing and low yields will keep the prices high well into the future.)

Another National product about to be launched is the LF353, a dual unit with special features for single supply operation. In the LF-353, the input will be able to swing to either supply voltage, which is not an easy chore for a JFET input device since the pinch off voltages must be maintained. Also, its output will swing within one diode drop of the rails. Thus this amplifier can be used with a single 5 V supply or with conventional ± supplies.

Intersil

Metal can versions of the 155/355 Series are in production and Intersil is considering becoming a second source for the CA3140. They are also investigating a very low leakage hybrid design.

Precision Monolithics

PMI concentrates on the premium performance products and while they offer the 155/355 Series with standard specifications their circuit design is different to give tighter control of the power supply currents, lower offset voltages (zener zap trimming), stabilized current in the output stage and increased speed. Also, PMI engineers have come up with a revolutionary circuit design (patent pending) that challenges one of the concepts of JFET amplifier circuits, that the input bias currents double every 10°C. With the ability to build a number of well matched FET's on the same monolithic device, Precision Monolithics has found that they can compensate the bias or leakage currents. In their new circuit, they develop a current from a separate FET transistor that leaks just the same as the input stage. Then they turn that current around into a current mirror and feed it into the input. Since the current is equal in magnitude but opposite in direction to the original, leakage is balanced out. In practice, the balancing out of picoamperes at room temperature is unimportant. The real advantage is the ability to balance the nanoamperes at temperatures over 50°C. There the circuit gives a 10 or 15:1 reduction in input currents. The first amplifiers to use this cancellation are the OP-15, 16, and 17 corresponding to the 155A/355A, 156A/356A and 157A/357A respectively. These units are also trimmed using zener zap techniques to reduce offsets to 0.5 mV maximum for the top mil and commercial grades. Moreover, they offer increased bandwidth with the OP-15 having the lower power drain of the 355A with the speed of the 356A. Specifications for the OP-16 and OP-17 are shown in Table 1. The OP-17 is not compensated. It has a minimum gain bandwidth product of 20 MHz and a slew rate of 45 V/μs (gain = 5).

Texas Instruments

TI is currently building the 155/355 Series and is refining the design, following the traditional semiconductor practice of getting something into production and then modifying it for high yields. A major thrust is TI's proprietary TL080 Series, which includes the externally compensated TL081 and TL 086 (singles), the TL082 and TL083 (duals) and TL084 (quad). The TL080, TL081, TL083 and TL086 permit 741 type offset null adjustments. This series is designed for low cost applications.

The next FET products from TI will be the TL071 (single), and the TL072 (dual), along with the TL074 and TL085 (quads). These units are similar to the TL081, et al, but feature a different design to give low harmonic distortion (output stage) and low noise (input stage). The typical noise performance is expected to be 10 nV/$\sqrt{}$ Hz at 1 kHz and 2 mV from dc to 10 kHz. The TL074 is pin out compatible with the TL084, while the TL085 has the RC4136 pin arrangement. All these amplifiers are suitable for hi-fi and stereo applications and the low noise may even permit use in phonograph pick-up preamplifiers.

RCA

The FET amplifiers from RCA all feature MOS FET inputs. The CA3130 uncompensated plus the CA3140 and CA3160 compensated units have been available for some time. The CA3140 directly competes with the 741 for low cost applications. The devices include special features for single supply operation; however, they can also be used with dual supplies. All these models are offered in 8 lead metal cans with the option of leads formed into a dual in-line configuration. RCA is now introducing plastic mini-DIP versions of all three products.

Raytheon

The first FET units from Raytheon, the 155/355 Series, are about ready for introduction.

Bias current — the FET hot button

All operational amplifiers have some temperature sensitivity, but in FET devices the change in bias current with temperature is significant. As mentioned before, at extremely high ambient temperatures, the current can approach and even exceed that of bipolar units. Since low bias current is a key parameter, a major reason for using FET devices, it is important for each user to know and understand what happens. Furthermore, one aspect of this phenomenon, the effect of self heating, is very obscure on the data sheets.

To present the story we'll go through a worst-worst case analysis and then indicate how you can make engineering judgments if you want to refine the numbers. Following that, we discuss some practical things you can do to minimize the bias current change in your circuits. The 355 and 356 were chosen as examples because they are popular and multiple sourced. Also, with the two we can point out the effect of changing one parameter while holding others constant. But while these were singled out, virtually every FET unit exhibits this performance.

In a JFET amplifier, the bias current is primarily due to the leakage through the reverse biased junctions of the input transistors. If the junction temperature is raised, the leakage current, I_B, approximately doubles every 10°C:

$$I_{B2} = I_{B1} \text{ x } 2 \exp \left[\frac{T_{j2} - T_{j1}}{10}\right]$$

Self heating of the whole chip when power is applied, additional dissipation when driving a load, and increases in the ambient temperatures around the unit all raise the temperature of the junction and thus increase the current.

Assume that a FET amplifier has an initial bias current of 50 pA with the ambient temperature at 25°C and with the unit just turned on so the junction temperatures are still 25°C. As the chip warms up, say in five seconds, the temperature can easily rise 20°C. So the bias current increases from 50 pA to:

$$I_B = 50 \text{ x } 2 \exp \left[\frac{20}{10}\right] = 50 \text{ x } 4 = 200 \text{ pA}$$

Now if the ambient temperature is raised to 70°C, this 200 pA is multiplied again:

$$I_B = 200 \text{ x } 2 \exp \left[\frac{70-25}{10}\right] = 200 \text{ x } 22.63 = 4526 \text{ pA}$$

For a military unit, if the ambient temperature is raised to 125°C, the 200 pA becomes:

$$I_B = 200 \text{ x } 2 \exp \left[\frac{125-25}{10}\right] = 200 \text{ x } 1024 = 204{,}800 \text{ pA} = 0.2\ \mu\text{A}$$

This 0.2 μA bias current could be compared to the military μA741 maximum specification: 1.5 μA at −55°C and 0.5 μA at 125°C. However most 741's now in production have a maximum bias current of 0.2 μA at −55°C and lower at 125°C. (Note that self heating tends to help bipolar devices.) This worst case calculation shows how some or all of the bias current advantages can disappear and most of them disappear right at the temperature extremes. At room temperature the FET's are much better than bipolar. But the bias current advantage may not be as great as it first appears from the data sheet.

What about this chip temperature rise due to self heating? Don't the published characteristics take care of this? No, and here is why: virtually all the manufacturers test op amps on automatic testers. They test them by the thousands, and they test each one in a few milliseconds. They can't afford to wait while each device heats up, and you generally wouldn't want to pay the extra cost if they did. So they measure the bias current of the amplifier right after turn on when, for all practical purposes, the

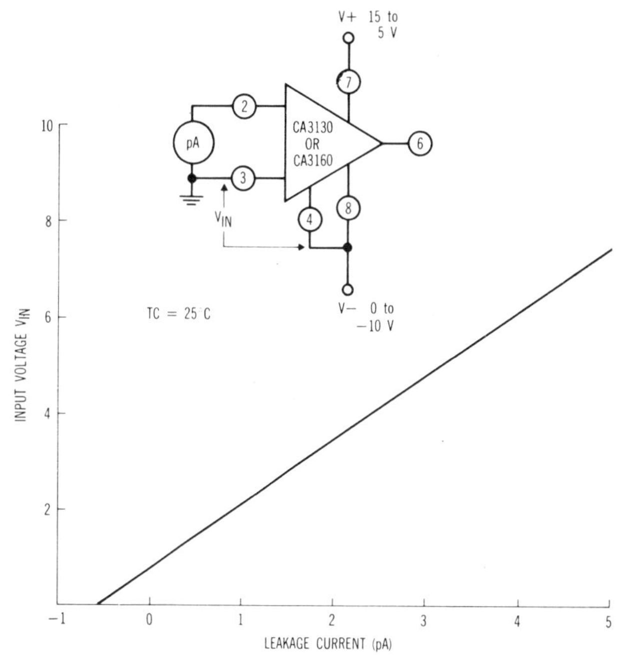

Fig. 4. With the input voltage close to the negative supply, this MOS FET amplifier measures zero leakage current.

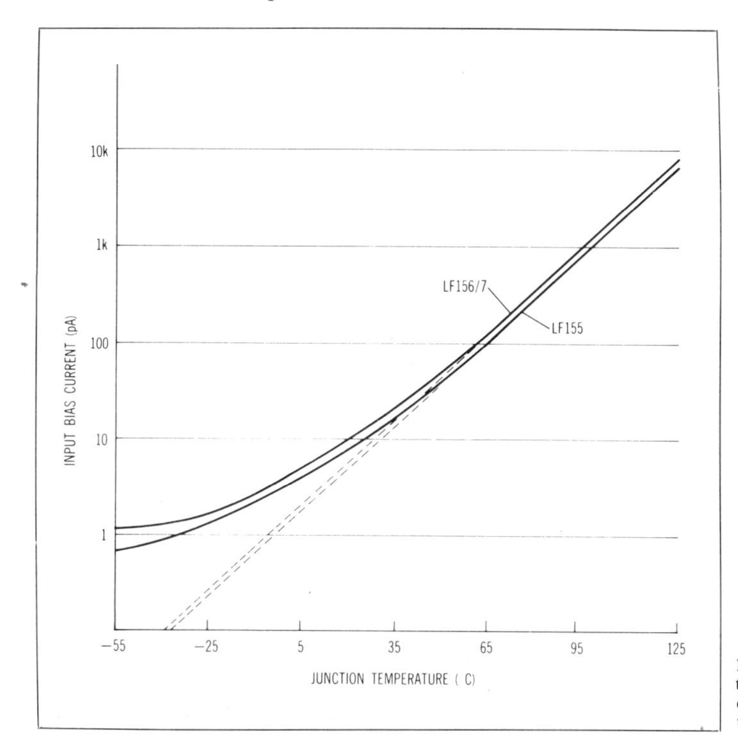

Fig. 3. Bias or leakage current versus temperature. At low temperatures the leakage currents of these JFET amplifiers deviate from the standard doubling every 10°C as shown by the deviation from the dashed lines.

junction temperature T_j is 25°C. Even if the data sheet says $T_A = 25°C$, interpret this as a junction temperature of 25°C and look for the footnote that explains the device isn't warmed up. There are a few exceptions, particularly hybrid IC's and these exceptions will have the bias current emblazoned with a statement showing that the specifications apply with the device warmed up.

Calculating heat rise

Well, how much does the chip warm up? We used the figure 20°C before, but is that reasonable? For a worst case calculation you simply multiply the junction-to-ambient thermal resistance, $R_{\theta jA}$, times the worst case power dissipation. (If you've tied the device into a infinite heat sink you can use the junction-to-case value, $R_{\theta jC}$.) The manufacturers tend to make it hard for you to determine $R_{\theta jA}$, probably because this is difficult for them to measure and it does vary depending upon chip size, how the die is mounted, lead material, etc. A few companies, including National, state values on their data sheets. That's great unless they've simply copied National's figures as some data sheets appear to have done. Other companies state a general value for each package style. This is better than no figure, but it may not take into account die size, etc. Texas Instruments takes a conservative approach by giving you a range of values to cover such variations and letting you decide how optimistic you want to be. We've listed some $R_{\theta jA}$ values for you in the table. The numbers came from the manufacturers and do give you an idea of what to expect. Readers who are highly involved in military calculations should review these with their favorite suppliers. Incidentally, the present Fairchild 355 Series data sheets give numbers for kovar lead frames, but Fairchild has converted to copper and the numbers stated here will appear on revised data sheets.

Taking the National $R_{\theta jA}$ values, neglecting any dissipation due to loading, and noting that the LF355 draws a maximum of 4 mA and the LF356, 10 mA, both at ±15 V, we can calculate a junction temperature rise:

LF 355 metal can (30 V)x(0.004 A)x(155° C/W) = 18°C

LF355 plastic DIP (30 V)x(0.004 A)x(155° C/W) = 18.6°C

LF356 metal can (30 V)x(0.010 A)x(150° C/W) = 45°C

LF356 plastic DIP (30 V)x(0.010 A)x(155° C/W) = 46.5°

The 20° figure used before doesn't seem unreasonable. Now, plugging the temperature rise into the exponential formula, we find the $T_j = 25°C$ bias current should be multiplied by the following factors.

LF355 metal can $2 \exp\left[\frac{18}{10}\right] = 3.48$

LF355 plastic DIP $2 \exp\left[\frac{18.6}{10}\right] = 3.63$

LF356 metal can $2 \exp\left[\frac{45}{10}\right] = 22.63$

LF356 plastic DIP $2 \exp\left[\frac{46.5}{10}\right] = 25.11$

Note the large effect of increased power dissipation (356 vs 355). Also observe the amount of change due to a small difference in thermal resistance (metal vs plastic). To complete the worst case calculation, you would multiply these numbers by the data sheet bias current limit, 200 pA. The manufacturers report that the data sheet bias current numbers are high to help take account of the difficulty in measuring low pA in a production environment, and that most 355/356 units are apt to fall in the 30 to 50 pA range. So if you want to estimate what happens in real life, you might want to start with 50 pA. Furthermore, the bias current measured at room temperature includes some other leakage in addition to the junction leakage. So the measured bias current may follow the curve shown in Fig. 3 and you could make allowances for this. (Fig. 3 shows more deviation than Fig. 1. The manufacturers we contacted agreed that there is deviation at low temperatures but disagree on how much would be found at room temperature.) The actual power consumption may also decrease at elevated chip temperature helping to reduce the temperature rise due to self heating. In evaluating devices for bias current rise due to self heating, calculate both the internal power dissipation and the extra dissipation with load. In low supply drain units, the load dissipation can become significant.

There are some things you can do to minimize heat rise. Locate FET amplifiers in the cooler portions of your circuit board. If you use metal can units, clip

THERMAL RESISTANCE

Company	Package	Thermal Resistance $R_{\theta jA}$°C/W
AMD	TO-99 Metal Package	150
Fairchild	8 Lead Plastic DIP (Copper Lead Frame)	150
	14 Lead Plastic DIP (Copper Lead Frame)	70
Motorola	14 Lead Ceramic DIP (MC3471)	100
National	TO-99 Metal Package (355 Series)	150
	8 Lead Plastic DIP (Kovar Lead Frame, 355 Series)	155
	14 Lead Plastic DIP (Kovar)	100
PMI	TO-99 Metal Package (Type J)	150
RCA	8 Lead Metal Package (3130/40/60)	145
	8 Lead Plastic DIP (3130/40/60)	135
Texas Instruments	8 Lead Metal Package (Type L)	120-210
	8 Lead Plastic DIP (Type P)	120-127
	14 Lead Plastic DIP (Type N)	85-97
	TO-99 Metal Package (Type J)	150

on heat sinks. These are not expensive and might save you from going to a higher cost amplifier. If you use plastic units, try to improve the heat sinking characteristics of the p-c board since much of the heat transfer is through the lead frame. Also consider the plastic units with copper lead frames; these leads are not as forgiving as kovar when bent too much, but the thermal resistances are significantly lower. Be careful about using quads, as their heat dissipation is four times that of a single, and the thermal resistance of a 14 lead DIP is not reduced to ¼ that of an 8 lead DIP. You might also consider lowering the supply voltages to reduce power dissipation. Last but not least, if you purchase parts to specification drawings, be sure to indicate that the bias current is to be measured at a junction temperature of 25°C. If you don't, the manufacturer should reject your order and/or your incoming inspection might start rejecting parts.

This discussion with the leakage current doubling every 10°C also applies for MOS input amplifiers such as the CA3130/40/60 Series. However, for these devices, most of the leakage current comes through the protection diodes connected to the input gates. RCA advises that by operating the circuit with very little voltage across the protection diodes, i.e., with the input voltage close to the negative supply, the leakage current becomes extremely low. In fact, a fortuitous combination of leakage effects brings the current down to about 1 pA (Fig. 4), and you could use this amplifier operating in this mode to measure the bias currents of the CA3130 Series and other amplifiers.

The effect of temperature on FET op amps is important but the solutions are easy. Just keep it cool — a drop of 10°C cuts the current in half. ⊕

Part III
Voltage Regulators and References

The temperature compensated voltage reference is one of the essential building blocks in analog design, particularly in the case of voltage regulators and digital-to-analog converters. This part of the book presents a number of selected papers which cover various approaches to obtaining precise voltage references, along with their application in the design of voltage regulators.

The first two papers of this part deal with the design of voltage regulator IC's. The first paper by Widlar covers the characteristics of the basic "band-gap" type voltage reference, and its application in high-current series regulator design. In the second paper by Davis, design of a high-voltage tracking regulator circuit is described, using the zener breakdown of the base-emitter junction of the conventional NPN transistor as the voltage reference. This paper also discusses the special geometry and bias considerations to optimize the zener breakdown characteristics.

The remaining papers in this part deal with the specifics of voltage reference design. The third paper by Brokaw describes an improved version of the band-gap reference. In the fourth paper, Dobkin descries a monolithic voltage reference which exhibits 1 ppm/°C temperature drift, which uses a special zener diode structure where the breakdown is confined to the subsurface area of the junction.

Until recently the performance requirements of an accurate voltage reference could only be met using the bipolar technology. However, the increased use of MOS technology in achieving analog functions has created a need for voltage references compatible with MOS process technology. The two short papers at the end of this section, by Tsividis and Ulmer, and by Blauschild *et al.,* address to this challenging problem and describe two separate approaches to obtaining stable reference sources using inherent characteristics of NMOS or CMOS transistors.

ADDITIONAL REFERENCES

1. L. Dixon and R. Patel, "Designers Guide to Switching Regulators," *EDN Magazine,* Part 1: pp. 53-59, October 20, 1974; Part 2: pp. 37-40, November 5, 1974; Part 3: pp. 76-82, November 20, 1974.
2. R. Dobkin, "Break Loose From Fixed IC Regulators," *Electronic Design,* pp. 118-122, April 12, 1977.
3. T.M. Frederiksen, "A Monolithic High Power Series Voltage Regulator," *J. Sol. State Circuits,* vol. SC-3, pp. 380-387, December 1968.
4. K.E. Kujik, "A Precision Reference Voltage Source,,"*J. Sol. State Circuits,* vol. SC-8, pp. 222-226, June 1973.
5. A. B. Grebene, "Analog Integrated Circuit Design," Chapter 6. New York: Van Nostrand Reinhold, 1972.

New Developments in IC Voltage Regulators

ROBERT J. WIDLAR

***Abstract*—A temperature-compensated voltage reference that provides numerous advantages over zener diodes is described along with the implementation of thermal overload protection for monolithic circuits. The application of these and other advanced design techniques to IC voltage regulators is covered, and an example of a practical design is given.**

Introduction

MONOLITHIC regulators have been receiving considerable attention during the last year or two. With the exception of operational amplifiers and voltage comparators, regulators lead other categories of linear devices in sales. Hence, considerable effort has gone into improving their performance and general utility.

Increasing the output-current capability, reducing the number of external components, and minimizing external trimming are obvious areas for improvement of older designs. Higher reliability under actual field conditions is less obvious, but probably more essential. Minimum input voltages less than the present 7–9 V would also increase the usefulness of these circuits.

Some techniques for realizing these objectives will be described. Further, a practical design for an on-card regulator that provides 5 V for logic circuits will be given. This circuit has three active leads, so it can be supplied in standard transistor power packages. It requires no external components and can reliably deliver output currents in excess of 1 A.

Design Concepts

A simplified schematic of a typical voltage regulator is shown in Fig. 1. An operational amplifier compares a reference voltage with a fraction of the output voltage and controls a series-pass transistor to regulate the output. Some form of overload protection is usually provided. Here, the output current is limited by Q_1 and R_1.

The low cost of IC regulators has created a considerable interest in on-card regulation, that is, to provide local regulation for each printed-circuit card in a system. Rough preregulation is used in the main power source, and the power is distributed without excessvie concern for line drops. The local regulators then smooth out the voltage variations caused by line drops and eliminate the transients on the main power bus.

A useful on-card regulator must include everything within one package, including the series-pass transistor. The author has previously advanced arguments against

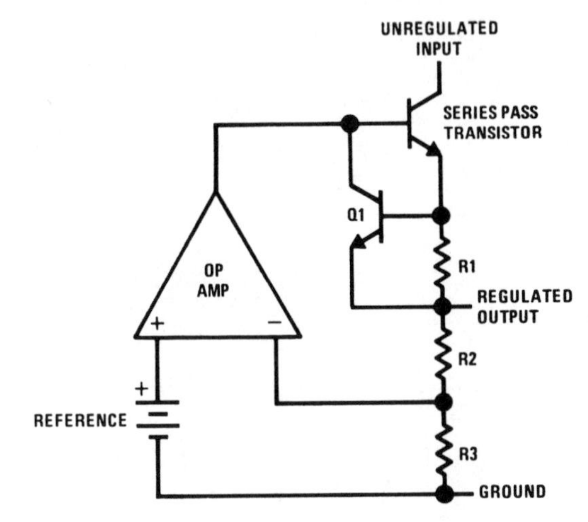

Fig. 1. Basic series regulator circuit.

including the pass transistor in an integrated-circuit regulator [1]. First, there are no standard multilead power packages. Second, integrated circuits necessarily have a lower maximum operating temperature because they contain low-level circuitry. This means that an IC regulator needs a more massive heat sink. Third, the gross variations in chip temperature due to dissipation in the pass transistors worsen load and line regulation. However, these problems can be largely overcome using the new techniques, especially for the kind of regulator required by digital systems.

It is conceptually possible to build a complete IC regulator that has only three external terminals. Hence, an ordinary transistor power package can be used, circumventing the package problem. The practicality of this approach depends on eliminating the adjustments usually required to set up the output voltage and limiting current for the particular application, as external adjustments require extra pins. A new solid-state reference, to be described later, has sufficiently tight manufacturing tolerances that output voltages do not always have to be individually trimmed. Further, thermal overload protection can protect an IC regulator for virtually any set of operating conditions, making adjustments unnecessary.

Thermal protection limits the maximum junction temperature, providing a constant-power limit that protects the regulator regardless of input voltage, type of overload, or degree of heat sinking. With an external pass transistor, there is no convenient way to sense junction temperature, so it is much more difficult to provide thermal limiting. Thermal protection is, in itself, a very good reason for putting the pass transistor on the chip.

Manuscript received March 23, 1970; revised July 10, 1970.
The author is with National Semiconductor, Santa Clara, Calif.

Reprinted from *IEEE J. Solid-State Circuits*, vol. SC-6, pp. 2-7, Feb. 1971.

When a regulator is protected by current limiting alone, it is necessary to limit the output current to a value substantially lower than is dictated by dissipation under normal operating conditions to prevent excessive heating when a fault occurs. Thermal limiting provides virtually absolute protection for any overload condition. Hence, the maximum output current under normal operating conditions can be increased. This tends to make up for the fact that an IC has a lower maximum junction temperature than discrete transistors.

Additionally, a 5-V regulator works with relatively low voltage across the integrated circuit. Because of the low voltage, the internal circuitry can be operated at comparatively high currents without causing excessive dissipation. Both the low voltage and the larger internal currents permit higher junction temperatures. This can also reduce the heat sinking required, especially for commercial-temperature-range parts.

Lastly, the variations in chip temperature caused by dissipation in the pass transistor do not cause serious problems for a logic-card regulator. The tolerance in output voltage is loose enough that it is relatively easy to design an internal reference that is much more stable than required, even for temperature variations as large as 150 °C.

Voltage Reference

Temperature-compensated zener diodes are normally used as the reference element in solid-state regulators. However, these devices have breakdown voltages greater than 6 V, which puts a lower limit on the input voltage to the regulator. Further, they require tight process control to maintain a given tolerance; and they are relatively noisy.

A new reference has been developed that does not use zener diodes. Instead, it uses the negative temperature coefficient of emitter-base voltage in conjunction with the positive temperature coefficient of emitter-base voltage differential of two transistors operating at different current densities to make a zero temperature coefficient reference. Practical references can be made at voltages as low as the extrapolated energy band-gap voltage of the semiconductor material, which is 1.205 V for silicon.[1]

A simplified version of this reference is shown in Fig. 2. In this circuit, Q_1 is operated at a relatively high current density. The current density of Q_2 is about 10 times lower and the emitter-base voltage differential ΔV_{BE} between the two devices appears across R_3. If the transistors have high current gains, the voltage across R_2 will also be proportional to ΔV_{BE}. Q_3 is a gain stage that will regulate the output at a voltage equal to its emitter-base voltage plus the drop across R_2.

Conditions for temperature compensation can be de-

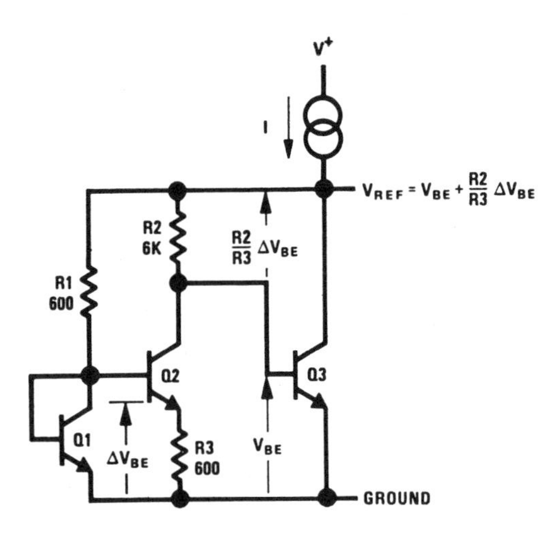

Fig. 2. The low voltage reference in one of its simpler forms.

rived starting with the equation for the emitter-base voltage of a transistor, which is [3]

$$V_{BE} = V_{g0}\left(1 - \frac{T}{T_0}\right) + V_{BEO}\left(\frac{T}{T_0}\right) + \frac{nkT}{q}\log_e\left(\frac{T_0}{T}\right) + \frac{kT}{q}\log_e\frac{I_C}{I_{CO}} \quad (1)$$

where V_{g0} is the extrapolated energy band-gap voltage for the semiconductor material at absolute zero, q is the charge of an electron, n is a constant that depends on how the transistor is made (approximately 1.5 for IC transistors), k is Boltzmann's constant, T is absolute temperature, I_C is collector current, and V_{BEO} is the emitter-base voltage at T_0 and I_{CO}.

Further, the emitter-base voltage differential between two transistors operated at different current densities is given by [4]

$$\Delta V_{BE} = \frac{kT}{q}\log_e\frac{J_1}{J_2} \quad (2)$$

where J is the current density.

Referring to (1), the last two terms are quite small and can be made even smaller by making I_C vary as absolute temperature. At any rate, the terms can be ignored because they are of the same order as errors caused by nontheoretical behavior of the transistors that must be determined empirically.

If the reference is composed of V_{BE} plus a voltage proportional to ΔV_{BE}, the output voltage is obtained by adding (1) in its simplified form to (2):

$$V_{ref} = V_{g0}\left(1 - \frac{T}{T_0}\right) + V_{BEO}\left(\frac{T}{T_0}\right) + \frac{kT}{q}\log_e\frac{J_1}{J_2}. \quad (3)$$

Differentiating with respect to temperature yields

$$\frac{\partial V_{ref}}{\partial T} = -\frac{V_{g0}}{T_0} + \frac{V_{BEO}}{T_0} + \frac{k}{q}\log_e\frac{J_1}{J_2}. \quad (4)$$

For zero temperature drift, this quantity should equal

[1] A similar reference based on emitter-base voltage was described in [2]; however, it did not operate at low voltages nor was it particularly suited to monolithic construction.

zero, giving

$$V_{g0} = V_{BE0} + \frac{kT_0}{q} \log_e \frac{J_1}{J_2}. \quad (5)$$

The first term on the right is the initial emitter–base voltage while the second is the component proportional to emitter–base voltage differential. Hence, if the sum of the two are equal to the energy band-gap voltage of the semiconductor, the reference will be temperature compensated.

In practice, for minimum drift, it is necessary to make the output voltage somewhat higher than was theoretically determined. This tends to compensate for various low-order terms that could not be included in the derivation. The typical performance of a circuit set for minimum temperature drift is shown in Fig. 3. It should be possible to get even lower drift as more experience is gained with the nontheoretical terms.

Some of the advantages of this reference that have not been pointed out are that the emitter–base voltage of n-p-n transistors is the most predictable and well understood parameter in an IC. And the generation of the ΔV_{BE} component depends only on matching, which is easily accomplished in a monolithic circuit. That is why the initial tolerance of the reference can be controlled easily, eliminating individual adjustments for many applications. In addition, the reference uses the same kind of parts that make up the error amplifier—transistors and resistors—and it employs them in essentially the same way. Therefore, it is no longer an outstanding source of noise, as is a zener diode; and large bypass capacitors are not needed to remove the noise. Further, long-term stability that is comparable to the error amplifier can also be expected.

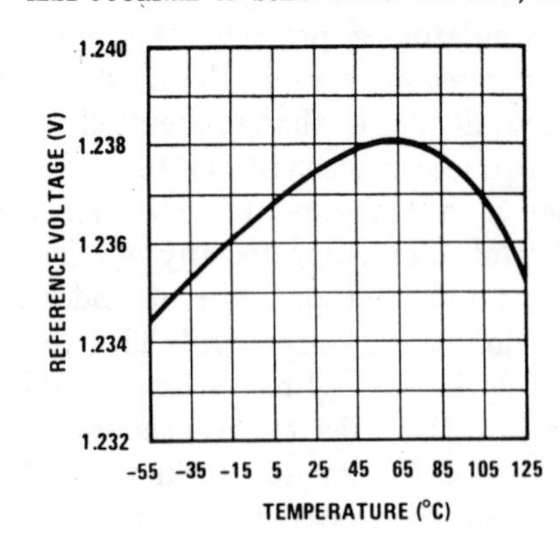

Fig. 3. Typical temperature characteristic of the low voltage reference.

Thermal Overload Protection

Older IC regulators relied on current limiting for overload protection. This required an external resistor to determine a safe limiting current that depended on the maximum input voltage, the highest operating ambient temperature, and the heat sinking available for the IC package.

Actually, the dominant failure mechanism of solid-state regulators is excessive heating of the semiconductors. Thermal protection attacks the problem directly by putting a temperature regulator on the IC chip. Normally, this regulator is biased below its activation threshold so it does not affect circuit operation. However, if the chip approaches its maximum operating temperature because of excessive internal dissipation, the temperature regulator turns on and reduces output current to prevent any further increase in chip temperature. Hence, if current limiting is included to ensure that the current-handling capability of the aluminum conductors and the secondary breakdown ratings of the pass transistor are not exceeded, it is virtually impossible to damage the circuit as long as the operating voltages are kept within ratings.

Although some form of thermal protection can be incorporated in a discrete regulator, ICs have a distinct advantage: the temperature sensing device detects increases in junction temperature within milliseconds. Schemes that sense case or heat-sink temperature take several seconds, or longer. With the longer response times, the pass transistor usually blows out before thermal limiting comes into effect.

The concept of thermal overload protection for ICs is not entirely new.[2] But the early designs did not have sufficient control on the limiting temperature to be practical. However, if the emitter–base turn-on voltage of a transistor is used to sense the chip temperature, a ±10 °C tolerance on limiting temperature can be maintained in a production environment without any external trimming.

Circuit Description

As mentioned earlier, a fixed 5-V regulator to be used as an on-card regulator for digital logic circuits was selected to evaluate the aforementioned techniques. A simplified schematic of the circuit used is shown in Fig. 4. The circuitry produces an output voltage that is approximately four times the basic reference voltage, or 5 V. The emitter–base voltage of Q_3, Q_4, Q_5, and Q_8 provide the negative-temperature-coefficient component of the output voltage. The voltage dropped across R_3 is the positive-temperature-coefficient component. Q_6 is operated at a considerably higher current density than Q_7, producing a voltage drop across R_4 that is proportional to the emitter–base voltage differential of the two transistors. Assuming large current gain in the transistors, the voltage drop across R_3 will be proportional to this differential, giving a temperature-compensated output voltage.

In this circuit, Q_8 is the gain stage providing regula-

[2] To the best of the author's knowledge, thermal protection was first used on a Continental Device regulator designed by G. Porter and announced in early 1967.

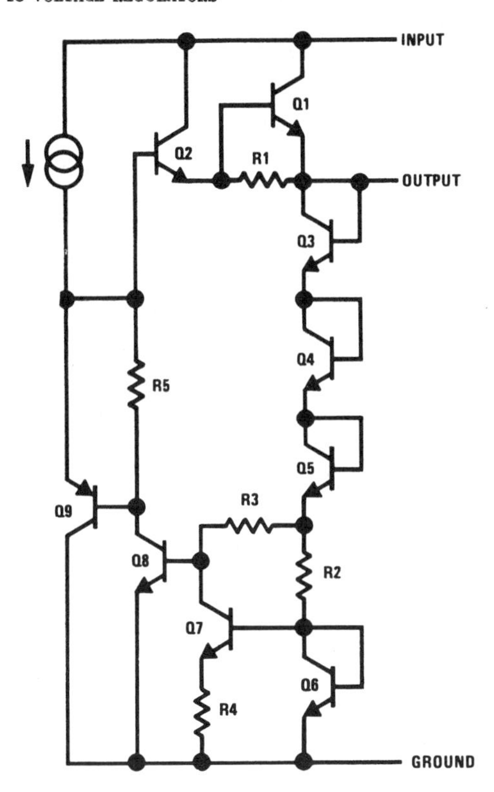

Fig. 4. Schematic showing essential details of the 5-V regulator.

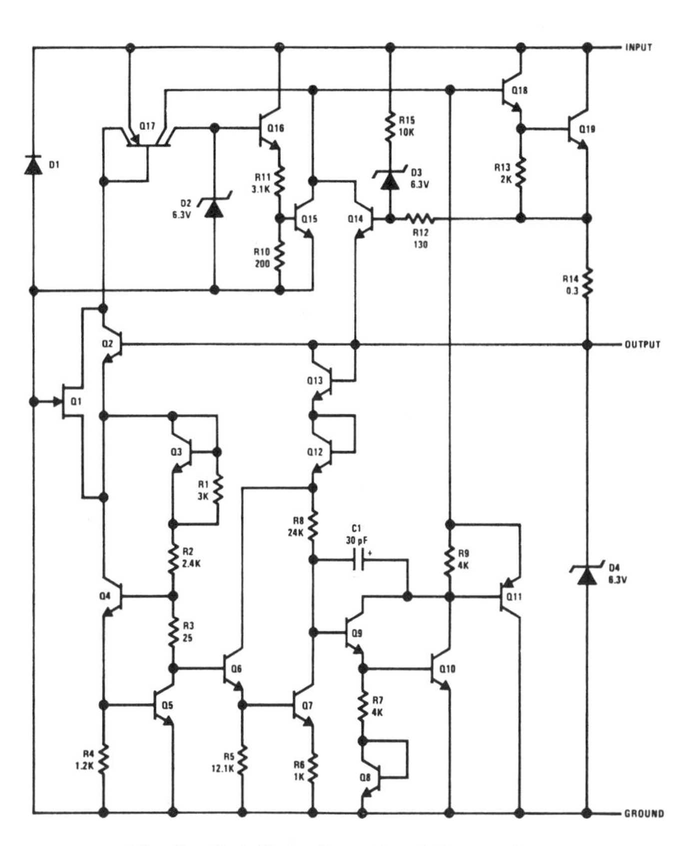

Fig. 5. Detailed schematic of the regulator.

tion. Its effective gain is increased by using a vertical p-n-p Q_9 that acts as a buffer driving the active collector load represented by the current source. Q_9 drives a modified Darlington output stage that acts as the series-pass element. With this circuit the minimum input voltage is not limited by the voltage needed to supply the reference. Instead, it is determined by the output voltage and the saturation voltage of the Darlington output stage.

Fig. 5 shows a complete schematic of the LM109 regulator. The reference is modified considerably over the simplified version to make it easier to manufacture. Q_4 and Q_6 are included both to develop a larger emitter–base voltage differential for Q_7 and to make the circuit less sensitive to production variations in current gain. Q_9 is also used to buffer the regulating transistor Q_{10}. The operating current of Q_9 is proportional to emitter–base voltage differential, as determined by R_7 and Q_8 [6]. R_3 serves to compensate for the transconductance of Q_5[6], so that the voltage delivered to Q_6 is relatively insensitive to changes in output voltage. An emitter–base junction capacitor C_1 frequency compensates the circuit so that it is stable even without a bypass capacitor on the output.

The active collector load for the output stage is Q_{17}. It is a controlled-gain p-n-p[6]. The output current is equal to the collector current of Q_2, with an auxiliary current being supplied to the zener diode controlling the thermal shutdown D_2. Q_1 is a collector FET [6] that along with R_1 ensures starting of the regulator under worst case conditions.

The output current of the regulator is limited when the voltage across R_{14} becomes large enough to turn on Q_{14}. This ensures that the output current cannot get high enough to cause the pass transistor to go into secondary breakdown or damage the aluminum conductors on the chip. Further, when the voltage across the pass transistor exceeds 7 V, current through R_{15} and D_3 reduces the limiting current, again to minimize the chance of secondary breakdown. The performance of this protection circuitry is illustrated in Fig. 6.

Even though the current is limited, excessive dissipation can cause the chip to overheat. However, if its temperature increases to about 175°C, Q_{15} will turn on and reduce the current further to maintain a constant temperature.

The thermal protection circuitry develops its reference voltage with a conventional zener diode D_2. Q_{16} is a buffer that feeds a voltage divider, delivering about 300 mV to the base of Q_{15} at 175°C. The emitter–base voltage of Q_{15} is the actual temperature sensor because with a constant voltage applied across the junction the collector current rises rapidly with increasing temperature.

Another protective feature of the regulator is the crowbar clamp on the output. If the output voltage tries to rise for some reason, D_4 will break down and limit the voltage to a safe value. If this is caused by failure of the pass transistor such that the current is not limited, the aluminum conductors on the chip will fuse, discon-

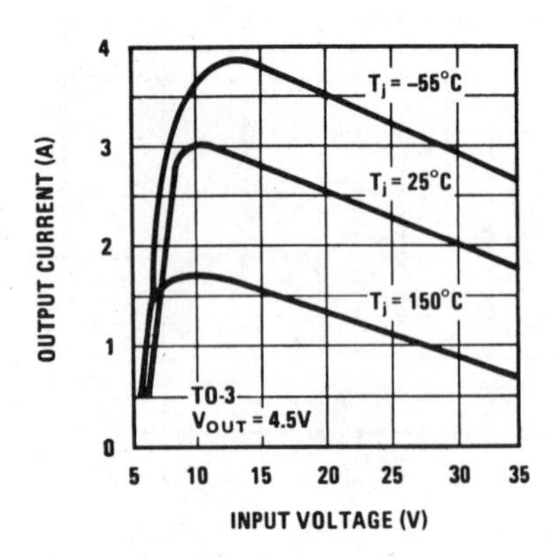

Fig. 6. Current-limiting characteristics.

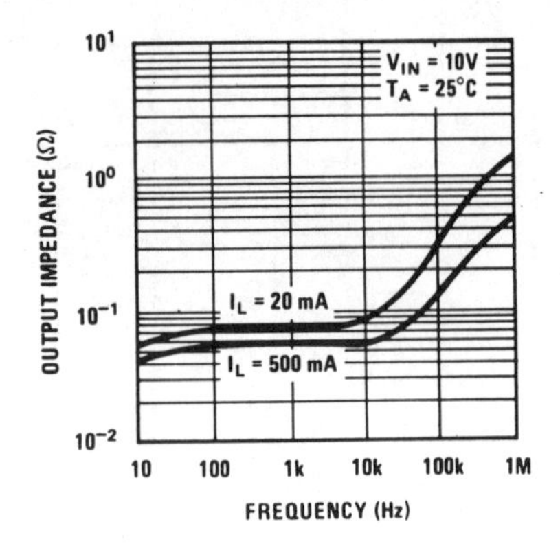

Fig. 7. Plot of output impedance as a function of frequency.

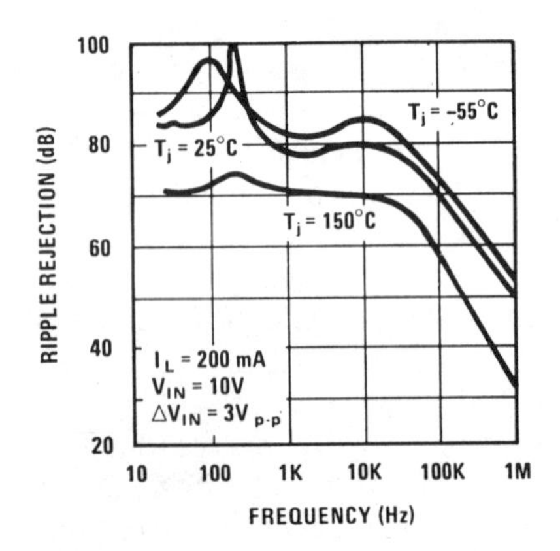

Fig. 8. Ripple rejection of the regulator.

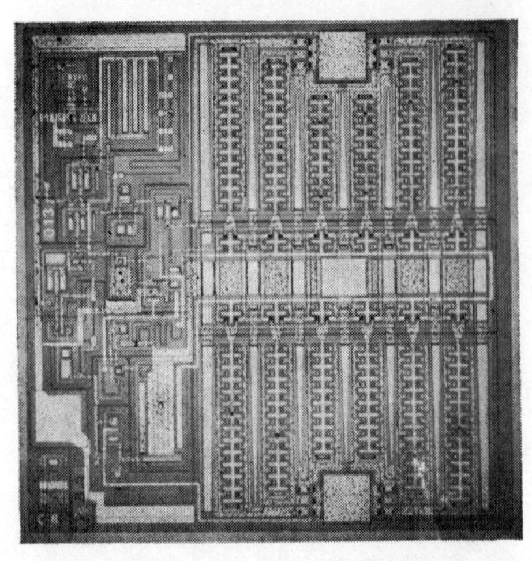

Fig. 9. Photomicrograph of the regulator shows that a high-current pass transistor takes more area than control circuitry.

TABLE I
TYPICAL CHARACTERISTICS OF THE LOGIC-CARD REGULATOR: T_A = 25 °C AND V_{IN} = 9 V

Output voltage		5.0 V
Output current		1.5 A
Output resistance		0.03 ohm
Line regulation	$7.0 \text{ V} \leq V_{IN} \leq 35 \text{ V}$	0.005 percent/V
Temperature drift	$-55 \text{ °C} \leq T_A \leq 125 \text{ °C}$	0.02 percent/°C
Minimum input voltage	$I_{OUT} = 1 \text{ A}$	6.5 V
Output noise voltage	$10 \text{ Hz} \leq f \leq 100 \text{ kHz}$	40 μV
Thermal resistance	LM109H (TO-5)	15 °C/W
junction to case	LM109K (TO-3)	3 °C/W

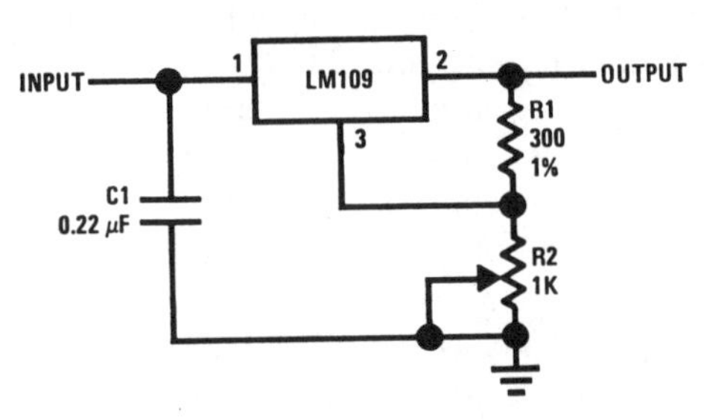

Fig. 10. Using the LM109 as an adjustable-output regulator.

necting the load. Although this destroys the regulator, it does protect the load from damage. The regulator is also designed so that it is not damaged in the event the unregulated input is shorted to ground when there is a large capacitor on the output. Further, if the input voltage tries to reverse, D_1 will clamp this for currents up to 1 A.

The internal frequency compensation of the regulator permits it to operate with or without a bypass capacitor on the output. However, an output capacitor does improve the transient response and reduce the high-frequency output impedance. A plot of the output impedance in Fig. 7 shows that it remains low out to 10 kHz without a capacitor. The ripple rejection also remains high out to 10 kHz, as shown in Fig. 8. The irregularities in this curve around 100 Hz are caused by thermal feedback from the pass transistor to the reference circuitry. Although an output capacitor is not required, it is necessary to bypass the input of the regulator with at least a 0.22-μF capacitor to prevent oscillations under all conditions.

Fig. 9 is a photomicrograph of the regulator chip. It can be seen that the pass transistors, which must handle more than 1 A, occupy most of the chip area. Q_{19} is actually broken into segments. Uniform current distribution is ensured by also breaking the current limit resistor into segments and using them to equalize the currents. The overall electrical performance of this IC is summarized in Table I.

Although the LM109 is designed as a fixed 5-V regulator, it is also possible to use it as an adjustable regulator for higher output voltages. One circuit for doing

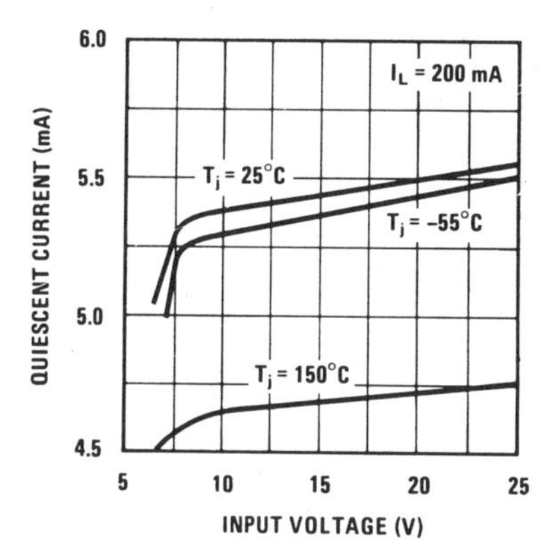

Fig. 11. Variation of quiescent current with input voltage at various temperatures.

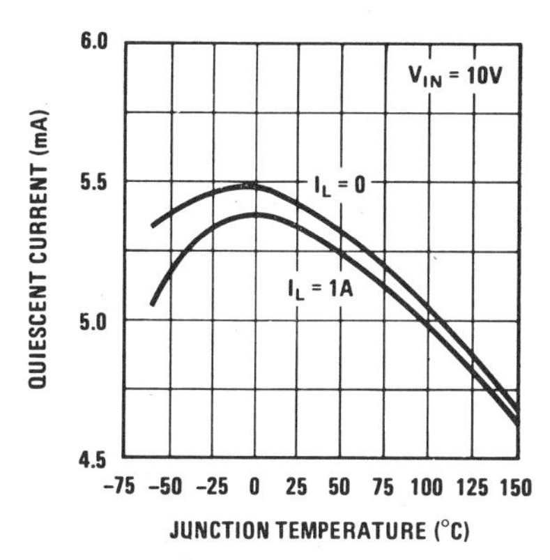

Fig. 12. Variation of quiescent current with temperature for various load currents.

this is shown in Fig. 10. The regulated output voltage is impressed across R_1, developing a reference current. The quiescent current of the regulator, coming out of the ground terminal, is added to this. These combined currents produce a voltage drop across R_2 that raises the output voltage. Hence, any voltage above 5 V can be obtained as long as the voltage across the integrated circuit is kept within ratings.

The LM109 was designed so that its quiescent current is not affected greatly by variations in input voltage, load, or temperature. However, it is not completely insensitive, as shown in Figs. 11 and 12 so the changes do affect regulation somewhat. This tendency is minimized by making the reference current through R_1 larger than the quiescent current. Even so, it is difficult to get the regulation tighter than two percent.

Conclusions

The techniques described significantly advance the state of the art in IC regulators and can be expected to make them even more attractive as a substitute for discrete designs. A 5-V regulator has already been made, and the validity of the approach established on mass-produced devices. In the future, fully adjustable regulators that operate on input voltages as low as 2 V can be expected. Furthermore, the reference shows definite possibilities for replacing zener diodes in applications requiring good long-term stability.

Thermal overload protection that directly senses and controls the junction temperature of the pass transistor should improve the reliability of regulators considerably. This is particularly obvious if the largely unpredictable electrical and environmental conditions to which a regulator may be exposed in field use are taken into account. Some degree of thermal protection can be incorporated into a discrete regulator. However, proper design is quite difficult and the protection is not completely effective because it is not possible to measure directly the junction temperature of the power transistor.

References

[1] R. J. Widlar, "Designing positive voltage regulators," *EEE*, vol. 17, pp. 90–97, June 1969.
[2] D. F. Hilbiber, "A new semiconductor voltage standard," *1964 ISSCC Digest Tech. Papers*, vol. 7, pp. 32–33.
[3] J. S. Brugler, "Silicon transistor biasing for linear collector current temperature dependence," *IEEE J. Solid-State Circuits*, vol. SC-2, pp. 57–58, June 1967.
[4] R. J. Widlar, "Some circuit design techniques for linear integrated circuits," *IEEE Trans. Circuit Theory*, vol. CT-12, pp. 586–590, December 1965.
[5] ——, "An exact expression for the thermal variation of the emitter base voltage of bi-polar transistors," *Proc. IEEE* (Letters), vol. 55, pp. 96–97, January 1967.
[6] ——, "Design of monolithic linear circuits," in *Handbook of Semiconductor Electronics*, L. P. Hunter, Ed. New York: McGraw-Hill, 1970, ch. 10, pp. 10.1–10.32.

A Five-Terminal ± 15-V Monolithic Voltage Regulator

WILLIAM F. DAVIS

***Abstract*—A five-terminal ±15-V monolithic voltage regulator has been developed that incorporates internal frequency compensation and internally provides a ±1 percent output voltage tolerance. In addition, a thermally symmetric layout design of the chip has been used to eliminate the detrimental effects of thermal feedback on the die and ensure that the complementary tracking output voltages will be independent of the power dissipation in the series-pass power transistors. Complete fault protection is accomplished by providing the power transistors with good dc safe operating area, internally limiting the short-circuit output currents, and accurately limiting the junction temperature to within 10°C of the specified maximum limit. Also, a new Zener-diode geometry is employed that significantly reduces the noise associated with the reference voltage. Other features include a maximum input voltage of ±40 V,**

Manuscript received May 26, 1971; revised August 9, 1971.
The author is with the Semiconductor Products Division, Motorola, Inc., Phoenix, Ariz. 85036.

Reprinted from *IEEE J. Solid-State Circuits,* vol. SC-6, pp. 366-376, Dec. 1971.

a maximum output current of ±200 mA, a current boost capability using external n-p-n power transistors for both outputs, and excellent load and line regulation.

I. Introduction

MONOLITHIC voltage regulators have been well received by the electronics industry and have become commercially successful. A large volume of these regulators are used to separately provide the complementary output voltages required to power operational amplifiers. To achieve this function more economically, a five-terminal monolithic dual ±15-V dc voltage regulator has been developed that operates from a maximum input voltage of ±40-V and supplies a maximum output current of ±200 mA.

In order to offer both excellent performance and extreme functional simplicity to the user, the basic design goal was to provide maximum function on the die without sacrificing regulator performance. By placing components such as the frequency compensation capacitors, the short-circuit current-limiting resistors, and the dc level shifting resistors on the die, the need to purchase and mount any external components to make the regulator function is eliminated. Also, with the output voltage adjusted internally, an external adjustment is not required. A five-terminal configuration can then be achieved. Although it is desirable for the performance of the regulator to include excellent load and line regulation, it is just as important that the complementary tracking output voltages are independent not only of the ambient temperature variations but also of the power dissipated by the series-pass power transistors that are located on the die. A thermally symmetric layout design technique is therefore employed to eliminate the detrimental effects of thermal feedback that occur from the power devices to the associated control circuitry. To ensure that the user cannot destroy or degrade the performance of the regulator by subjecting it to an output short-circuit condition, the output current and the junction temperature are both limited to a safe value and the power transistors are designed with a good dc safe operating area. Finally, since many operational amplifier applications require low noise voltages on the power-supply terminals, a new Zener-diode geometry was developed to significantly reduce the Zener noise since external capacitors, typically employed in the past for frequency compensation, are not available to sufficiently reduce this noise at the output.

II. Circuit Description

A. Basic Approach

As can be seen from Fig. 1, the basic regulator consists of two feedback amplifiers that are interconnected to provide the complementary tracking output voltages. The positive output voltage is established by multiplying an internally generated reference voltage by the closed-loop gain (determined by resistors R_1 and R_2) of a single-ended amplifier, A_1. The negative output voltage complementary tracks the positive output voltage, since the input of a unity-gain inverting dc amplifier (A_2 with resistors R_0) is connected to the positive output terminal. Since both amplifiers use negative feedback, the desired output impedances are determined by appropriately choosing the open-loop voltage gains A_1 and A_2.

B. Reference Voltage

The reference voltage is produced by the series connection of one Zener diode and two base–emitter junctions as shown in Fig. 2. By appropriately biasing transistors Q_1 and Q_2 with current sources I_1 and I_2, respectively, and by choosing an appropriate Zener breakdown voltage, the sum of the negative temperature coefficients associated with the base–emitter junctions can be made to cancel the positive temperature coefficient associated with the Zener diode to produce a voltage reference that is essentially independent of junction temperature. This can be accomplished since the base–emitter temperature coefficient is a function of the emitter current density and the temperature coefficient associated with the Zener diode is a function of the magnitude of the breakdown voltage.

In the past, the noise components of the reference voltage was not of particular concern since the large external capacitors used for frequency compensation would sufficiently band limit the amplifier and satisfactorily reduce the noise at the output terminal. Unfortunately, when internal frequency compensation is employed, the die area limitation allows the on-chip capacitance to be sufficient to provide an adequate unity-gain phase margin, but cannot satisfactorily band limit the reference noise. In this case, the noise must be reduced within the reference element.

Of the three basic series-connected elements that provide the reference voltage, the Zener diode contributes the most significant amount of noise. The geometry shown in Fig. 3(a) has typically been used as a Zener reference element but produces a substantial amount of undesirable noise (At 100-μA bias current, a noise voltage of 3 mV_{rms} is typical for a 10-MHz bandwidth.) In general, this Zener noise is inversely related to the current density in the device as indicated in Fig. 3(b). To realize optimum noise performance, the magnitude of the Zener current should be as large as practical while the area of the Zener diode should be as small as possible. Since the total quiescent bias current for the ±15-V regulator must be considerably smaller than the maximum load current for optimum power capability, the bias current is typically insufficient to bias the control circuitry and to satisfactorily reduce the noise voltage in this Zener geometry. A tolerable noise voltage can be achieved, however, if the area of the Zener breakdown is reduced to increase the current density. The minimum Zener junction area that can be realized with the geometry of Fig. 3(a) is limited by the minimum size allowed for the ohmic con-

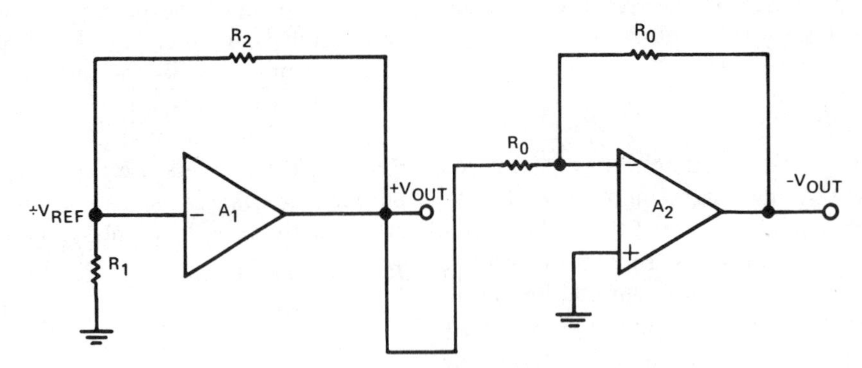

Fig. 1. Basic design approach for the ±15-V regulator.

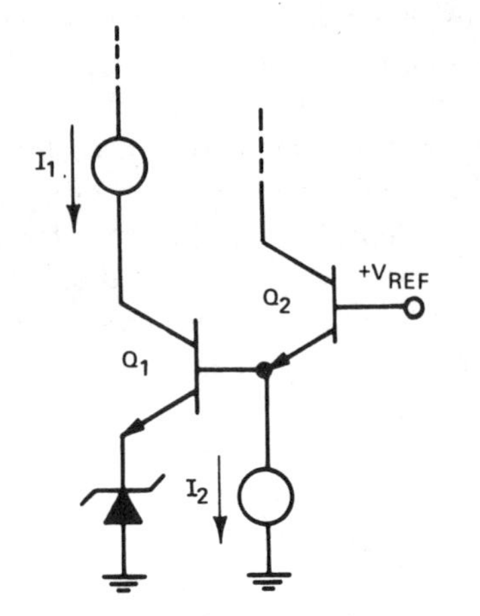

Fig. 2. Circuitry that provides the temperature independent reference voltage.

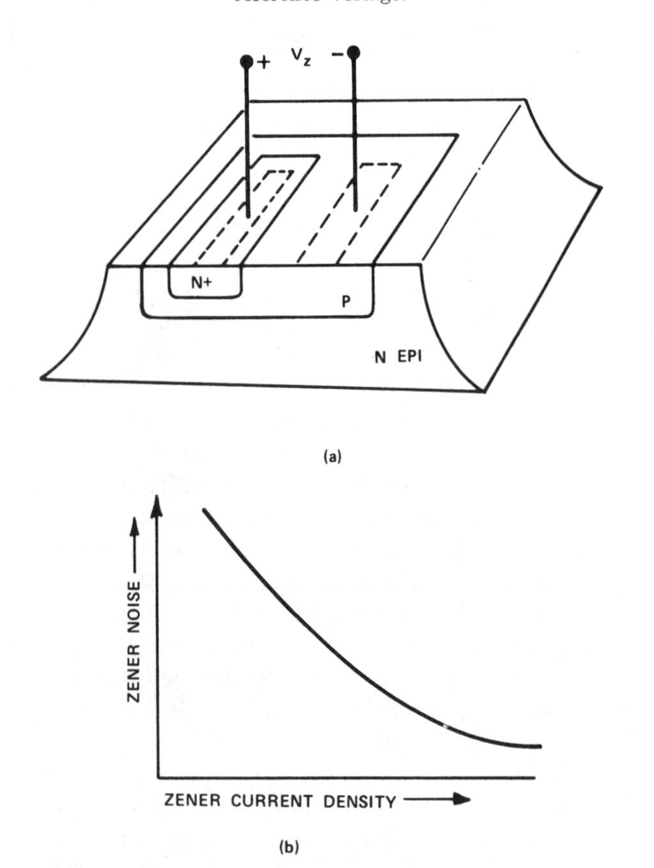

Fig. 3. (a) Zener-diode geometry that is typically used as a voltage reference element. (b) Noise associated with this geometry versus current density.

tacts which must be made to the n-type region. By using the Zener geometry shown if Fig. 4, the Zener junction area is limited only by the minimum width of the p-type "finger" that extends from the larger p-type region into the n-type region. Also, since the greatest doping concentration is at the narrow surface edge of the n-type p-type interface, the actual Zener breakdown area is confined to this small region. As a result, a noise voltage of less than 50 μV_{rms} at a biasing current of 100 μA (10-MHz bandwidth) has been accomplished with this geometry and represents a noise reduction of 35 dB when compared with the geometry of Fig. 3 (a).

C. Positive Voltage Regulator

The positive voltage regulator (Fig. 5) is a single-ended inverting amplifier that employs feedback to achieve the desired low output impedance and to level shift the reference voltage to the desired output voltage (+15-V dc). The input circuitry for this amplifier produces not only the reference voltage but also provides the total open-loop voltage gain determined by the collector and emitter impedances of the gain transistor Q_4. Since the closed-loop output impedance is equal to the open-loop output impedance divided by the feedback gain (loop transmission) and this feedback is proportional to the open-loop voltage gain (for a fixed resistive feedback network), the Zener impedance (determined by the length-to-width ratio of the p-type finger) was designed to provide the desired closed-loop output impedance. Transistors Q_1 and Q_2 are employed to handle the relatively large values of load current, to maintain a high open-loop voltage gain during loaded conditions and to achieve a low open-loop output impedance. The magnitude of the current source I_1 is made sufficiently large to drive the base of the Darlington connected series-pass power transistors Q_1 and Q_2 and to provide the necessary dc current density in the Zener diode to satisfactorily decrease the Zener noise. The current source I_2 minimizes the open-loop voltage gain losses by increasing the biasing current for transistors Q_3 and Q_1. This also improves the f_t of these devices to reduce the phase shift in the open-loop phase response. Closed-loop stability is maintained by creating a dominant pole at the input of the amplifier that is due to the relatively large impedance level at this node and

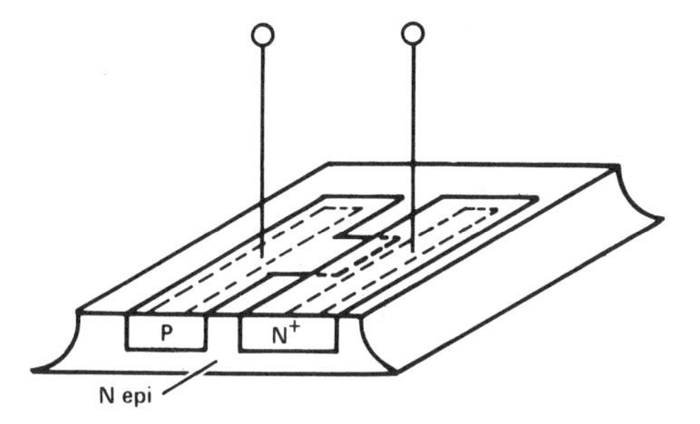

Fig. 4. Zener-diode geometry that maximizes the current density for improved noise performance.

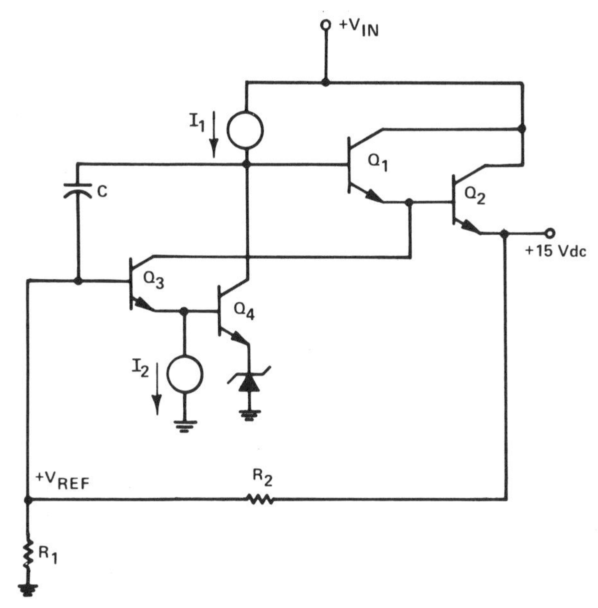

Fig. 5. Basic positive voltage regulator.

the Miller multiplication of the compensation capacitance C by the approximate voltage gain of transistor Q_4 [1].

D. Negative Voltage Regulator

The negative voltage regulator (Fig. 6) is a differential amplifier that incorporates feedback to the inverting input to provide a unity-gain inverting dc amplifier. This achieves the desired low output impedance and level shifts the positive output voltage to produce the complementary tracking negative output voltage. Three gain stages develop the overall open-loop voltage gain. The first stage consists of the differential transistors Q_1 and Q_2 operating into the relatively low emitter impedance of the p-n-p level shifting transistor Q_3. This p-n-p transistor operates as a common-base second stage and provides the major part of the open-loop voltage gain due to the large collector load impedance. It also incorporates a narrow basewidth and is appropriately biased to improve the f_t and allow an optimum open-loop phase response. The third stage consists of the common-emitter transistor Q_6 that operates essentially into the output load impedance. Emitter degeneration is used in the first stage (transistors Q_1 and Q_2) to establish the necessary open-loop voltage gain and achieve the desired closed-loop output impedance. The closed-loop stability is maintained by Miller multiplying the compensation capacitance C_1 by the voltage gain of transistor Q_6 to produce a dominent pole at the base of transistor Q_4. Transistor Q_4 is employed to raise this nodal impedance in order to obtain loop stability with a small value of on-chip capacitance and to provide sufficient base drive to the Darlington connected series-pass transistors Q_5 and Q_6. A zero is introduced in the feedback network (resistance R_0 and capacitance C_2) to further improve the stability.

E. DC Biasing

The biasing circuitry (Fig. 7) aids in providing the thermal performance required for this regulator by causing the temperature coefficient of the voltage across the resistors R_1 and R_2 to cancel with the temperature coefficient associated with these resistors to produce a reference current I that is essentially independent of junction temperature [2]. The collector currents of the transistor Q_1 and of the Darlington connected p-n-p transistors Q_3 and Q_4 are thus independent of junction temperature and are therefore employed are references to bias the current sources used in both the positive and the negative voltage regulators. Also, the sum of the base–emitter temperature coefficients of transistors Q_2 and Q_5 is cancelled with the sum of the base–emitter temperature coefficients of transistors Q_3 and Q_4 to establish a current through resistor R_3 that is proportional to the reference current and also independent of temperature. Since the collector and emitter biasing currents of the differential pair use a common reference, these currents will track forcing the ratio of emitter currents to remain constant independent of any variations that may occur in the common reference. A constant offset voltage is then realized across the differential amplifier, independent of any variations in the reference current or junction temperature.

III. Output Voltage Tolerance

The output voltage of present hybrid dual power supplies is generally specified to within a ± 1 percent tolerance limit that is typically achieved by a final adjustment in production. However, the output tolerance of monolithic voltage regulators depends on the tolerance of the reference voltage and the ratio tolerance of the level shifting resistors.

The Zener diode, which is commonly used as the reference element, essentially determines the reference voltage and thus controls the tolerance limit. This diode is produced by the diffusion of an n-type emitter material into a p-type base material. Although these diffusions are controlled in the standard production process, the tolerance of the breakdown voltage is only about ± 3 percent. The tolerance of other Zener structures, such as that produced by the p-type channel and the n-type buried layer, is even worse and thus these are undesirable reference elements. Tighter control of these diffusions could improve the reference voltage tolerance but this is difficult to realize and adds cost to the standard production process.

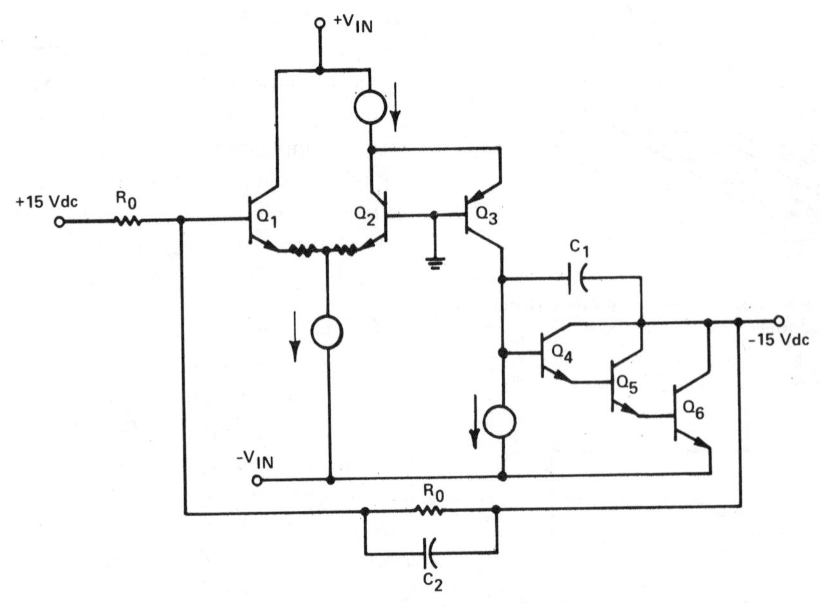

Fig. 6. Basic negative voltage regulator.

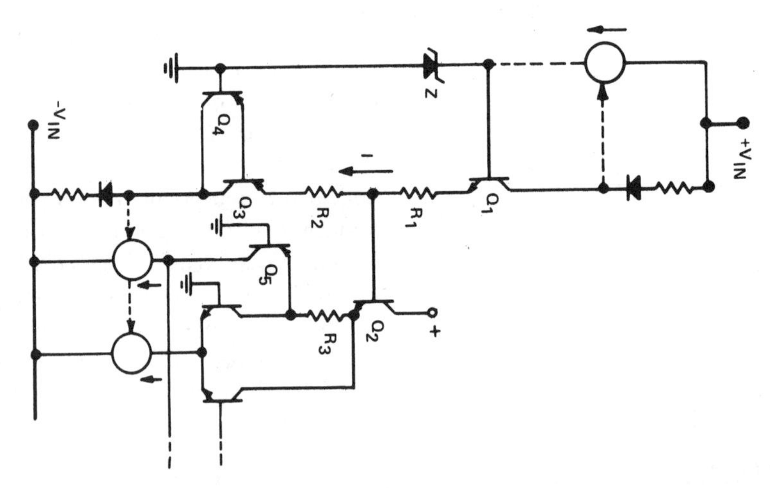

Fig. 7. Dc biasing technique to bias both the positive and negative voltage regulators with temperature-independent currents.

The ratio tolerance of the monolithic level shifting resistors is limited by the photographic delineation of the resistor edges on the photoresist mask plates. Although masking techniques are presently being developed to improve this delineation, the ratio tolerance can be improved in the standard process by employing wide resistor geometries so that the resistor value is not significantly affected by variations in the edge dimensions. The magnitude of the resistances, which are used to level shift the reference voltage to the positive and negative output voltages, should be as large as possible to minimize the total input bias current to the regulator. The area required for large-valued resistors (to allow a ±1 percent resistor ratio tolerance) is large, however, and thus economically undesirable. Also, the need to achieve a ±1 percent resistor ratio tolerance is questionable especially in view of the poor tolerance associated with the magnitude of the reference voltage.

It first appears that because of the requirement for larger die area and tighter control of the diffusion process, it would be uneconomical for monolithic voltage regulators to provide a ±1 percent output voltage tolerance to within a 3-sigma limit. A final test selection procedure could alternatively be used to grade the product, which employed standard processing, into two or three tolerance catagories eliminating the requirement for a yield to a ±1 percent tolerance. The additional tolerance categories should be close enough to the desired output voltage (±15-V dc in this case) to be meaningful to the user yet lax enough to sell all of the functional parts. This requirement makes the various categories difficult to define. In addition, the number of wafer starts required to produce a given amount of regulators with a ±1 percent output voltage tolerance is largely dependent on the yield to this tolerance and it is possible to acquire a substantial number of surplus regulators with a relaxed tolerance specification while trying to meet large volume ±1 percent tolerance demands. An alternate approach to this problem was selected that involves the adjustment of the monolithic level shifting resistors.

The area required to accomplish a ± 1 percent output voltage tolerance by resistor adjustment would be essentially the same as the area required by the level shifting resistors to satisfactorily accomplish the tolerance selection process. This seems like a reasonable trade since, in addition to providing a ± 1 percent output voltage tolerance with standard processing, the number of wafer starts required for a given amount of good die is reduced, and the need for a second and third tolerance category is eliminated—all of which are economically attractive.

The actual adjust procedure is accomplished during the wafer probe operation. Incremental resistance is added to the level shifting resistors by selectively disconnecting fuseable metal links, initially placed across the incremental resistances, until the desired output voltage tolerance is obtained. The number of incremental resistances required for the ± 1 percent voltage tolerance is determined by the reference voltage tolerance and the ratio tolerance of the level shifting resistors.

IV. Thermally Symmetric Layout Design

The performance of high-power monolithic voltage regulators can be considerably degraded by the effects of thermal coupling between the series-pass power device and the associated control circuitry. During changes in the dc load conditions, thermal feedback causes changes in the dc output voltage that must be small with respect to the changes caused by the output impedance of the amplifier. Standard circuit-design techniques have been developed to compensate for the uniform changes in surface temperatures that are caused by ambient temperature variations. However, changes in the power dissipation on an IC die create nonuniform changes in surface temperatures (thermal gradients) that require thermal compensation not only in the circuit design but also in the layout design where consideration must be given to the location of the circuit components with respect to the power devices.

Circuit-design techniques usually assume that all of the circuit components are at the same temperature. Most points on the surface of an IC die are not at the same temperature when power is being dissipated, and more important, the temperature difference between these points is a function of the power dissipation. Fortunately, it is not required that all of the components in an IC circuit design be at the same temperature. Many small circuit groups can be found that contain components that only require local thermal matching in order for the circuit to achieve temperature independent performance. Circuit components are defined to be thermally matched when they exhibit identical thermal properties and the temperature difference between them is always zero. (Strictly speaking, the temperature difference must only be held constant but, in practice, a constant other than zero is difficult to realize when changes in the power dissipation occur on the IC die.) As an example, local thermal matching of components shown in the shaded areas of Fig. 8 is the only requirement necessary for the negative output voltage to be independent of junction temperature. The temperature difference between each group of components is unimportant but the temperature difference among the components of each group must be held constant and preferably zero. The resistors R_3, R_4, and R_5 as well as the transistors Q_3, Q_4, and Q_5 must have identical temperature characteristics. The transistors, however, will not necessarily have the same geometry since the different emitter currents require that the emitter areas are ratioed to ensure identical current densities and thus equivalent temperature coefficients.

The groups of circuit components that must be thermally matched cannot be randomly located on the die but must be located on a locus of points where the temperature difference between these components is always zero and independent of the power dissipation (isothermal contours). The location of isothermal contours on the surface of a die for any random location of the power device can certainly be persued with mathematical rigor but thermal symmetry is an extremely powerful concept that can be used to easily predict the contour locations [3]. For example, with the power device distributed along one edge of the IC die as shown in Fig. 9, isothermal contours are produced that are essentially parallel to the power device except at the edges due to this boundary condition. Devices located on this die are thermally matched provided they are equidistant from both the axis of thermal symmetry and the power device. The actual curvature of the contours is seen to be unimportant. The resistance ratio of resistors located in a similiar manner is also independent of the power dissipation since the distributed thermal effect on each resistor will be identical. If three or more components exist in a thermally matched group, they are again located parallel to the power device but as close to the axis of symmetry as possible to take advantage of the linear portion of the isothermal contours. In general, if groups of thermally matched components are placed equidistant from the power device and about the axis of thermal symmetry, the IC performance will be essentially independent of the power dissipation and provides the basis for the layout design of the ± 15-V monolithic voltage regulator. The series-pass power devices have been located at the opposite edges of the die so they can share the same axis of thermal symmetry as shown in Fig. 10.

Since this layout-design procedure places constraints on the location of all of the thermally matched components, it becomes important to determine the relative placement priorities in order to facilitate the geometrical layout design. For this reason, it is convenient to define a thermal sensitivity factor S_T, which provides a quantitative measure to determine which thermally matched components are the most sensitive to thermal mismatch with respect to some output parameter. In general it can

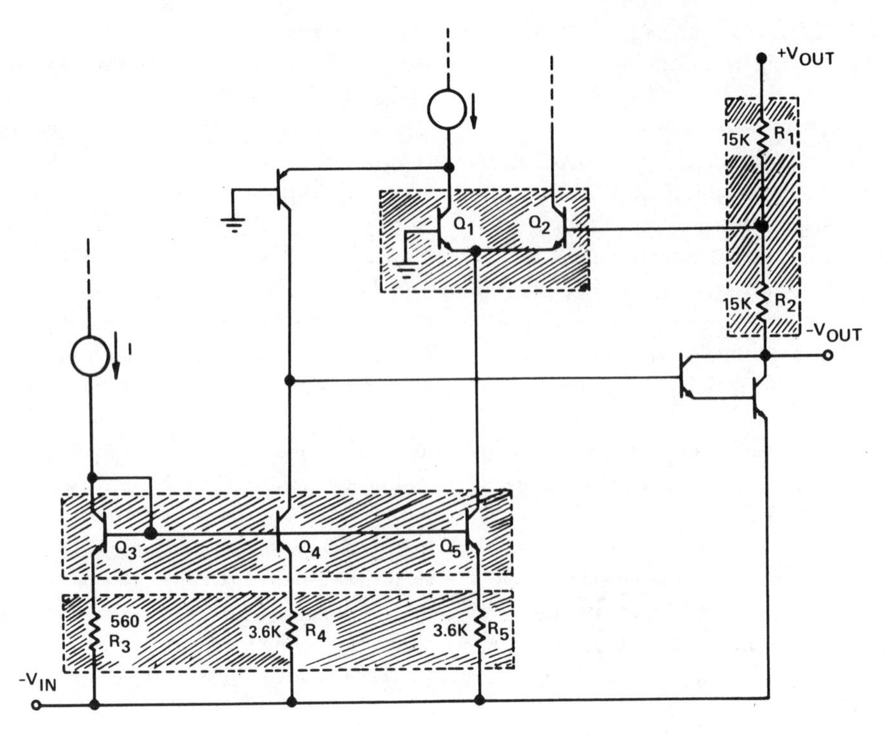

Fig. 8. Circuitry that illustrates the groups of components (shown in the shaded areas) that must be thermally matched for temperature-independent performance.

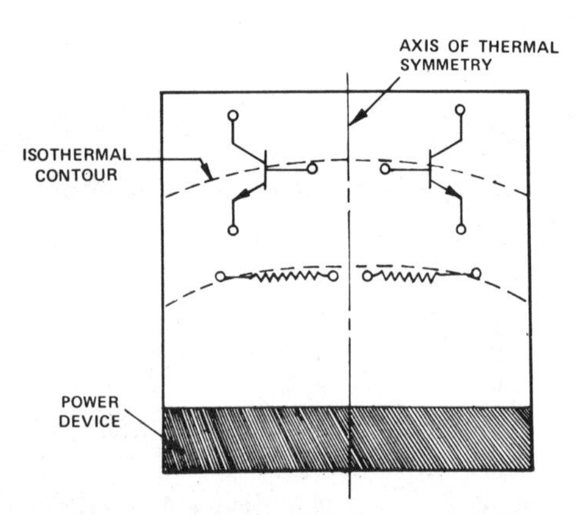

Fig. 9. Thermally symmetric die layout design. Components that are equidistant from the power device and symmetrical about the axis of symmetry are thermally matched.

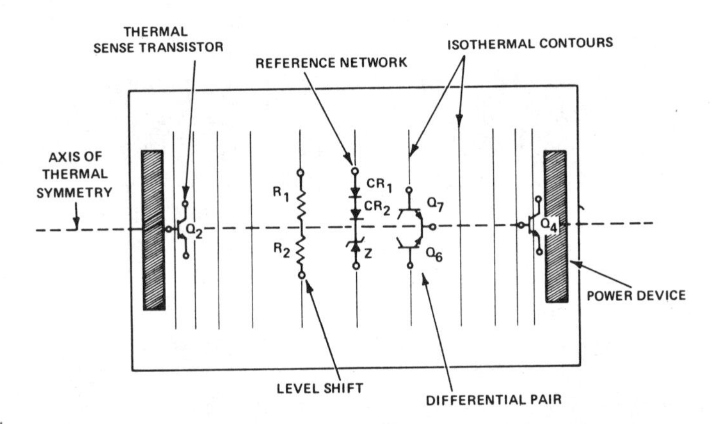

Fig. 10. Thermally symmetric layout design of the ±15-V regulator.

be defined as an output parameter change ΔP_0 per degree of temperature mismatch ΔT_m between one particular component and the rest of the components in a thermally matched grouping. All other junction temperatures T_n are assumed to be constant. The thermal sensitivity factor is then given by

$$S_T \equiv \left.\frac{\Delta P_0}{\Delta T_m}\right|_{\Delta T_n=0}.$$

The relative thermal sensitivity factors of Table I indicate, for example, the relative importance of achieving thermal matching between the various components in each of the shaded groupings shown in Fig. 8 where the negative output voltage is the output parameter ΔP_0. Those components with the largest sensitivity factors are placed in the center of the die as close to the axis of thermal symmetry as possible while components with smaller sensitivities can deviate from this ideal location without sacrificing overall thermal performance.

In general, the change in the dc output voltage during a dc load current change results from the output impedance of the amplifier, the thermal mismatch between IC components, and the change in the junction temperature. Since the changes due to the thermal feedback are negligible with respect to the output impedance as a result of this design technique, the regular performance is essentially independent of the power dissipation.

V. Fault Protection

Complete fault protection is achieved when the regulator output terminals can either be shorted to ground or shorted to each other indefinitely without failure or deg-

TABLE I
OUTPUT VOLTAGE THERMAL SENSITIVITY FACTORS FOR THE CIRCUITRY SHOWN IN FIG. 8

COMPONENTS	THERMAL SENSITIVITY FACTORS (mV/°C)
R_1	28
Q_1	3.88
Q_3	0.83
Q_4	0.55
Q_5	0.27
R_3	0.3
R_4	0.2
R_5	0.1

radation of reliability. This can be accomplished by internally limiting the short-circuit current, by accurately limiting the junction temperature, and by ensuring that the dc safe operating area of the series-pass power devices is never exceeded during a worst case operating condition. Methods to limit the output current during an output short circuit are well known but, unfortunately, offer little to protect the regulator unless the junction temperature is also limited to a safe value.

Fig. 11 demonstrates the basic junction temperature limiting technique. Basically, a thermal sense transistor Q_2 continuously monitors the junction temperature of the power transistor Q_1 and denies the base drive I_B at a particular junction temperature. This forces the output voltage to zero and causes the power dissipation on the die to decrease essentially to zero [2]. Since a relationship exists between the collector current I_T the junction temperature T_J and the base–emitter voltage of the sense transistor, the junction temperature for a specific value of collector current (the temperature at which the base drive is removed) is well defined for a given thermal reference voltage V_{TR} [4].

The performance of this junction temperature limiting technique is determined by the distance X between the thermal sense device and the power transistor on the die, and also by the gain of the collector current to junction temperature transfer function ($\Delta I_T / \Delta T_J$). Ideally, to eliminate serious degradation of performance and reliability, the regulator should function properly at the specified maximum junction temperature and yet reduce the output voltage to zero when the junction temperature increases above the specified limit, since the mean time to failure (MTF) is exponentially related to junction temperature [5]. To achieve this ideal limiting characteristic, the thermal sense transistor must be placed immediately adjacent to the power transistor (1–2 mils) to minimize incorrect temperature sensing caused by the differential temperature that exists between these devices when power is being dissipated. (Temperature gradients in excess of 100°C can easily exist within 15 mils of an IC power device during short-circuit conditions.) In addition, the transfer gain ($\Delta I_T / \Delta T_J$) must be sufficient to remove the total base drive from the power device within a few degrees of the specified junction temperature limit. For the circuit shown in Fig. 11, over 100°C is required for the sense transistor to totally remove the base drive (1 mA) from the power device. Unfortunately, the output voltage would not be reduced to zero, in this case, until the junction temperature was 100°C above the specified limit. This performance is quite unsatisfactory, especially from a reliability standpoint.

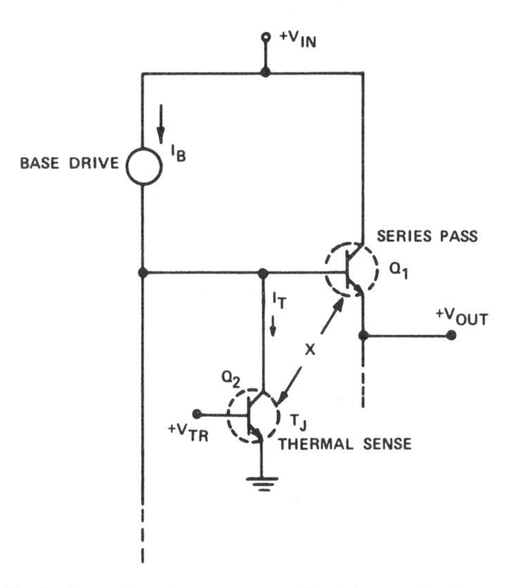

Fig. 11. Basic junction temperature limiting technique. The performance is a function of the distance X and the ($\Delta I_T / \Delta T_J$) transfer function.

In the ±15-V regulator design, transistors Q_1 and Q_3 were employed [Fig. 12(a)] as a cascade amplifier to increase the transfer gain as shown in Fig. 12(b). Two thermal sense devices were required to accurately monitor the junction temperatures of the widely separated power devices. Since all of the base drive to the power device associated with the positive voltage regulator is totally removed within 2°C, the actual junction temperature limit can be placed within 2°C above the specified operating limit of 175°C. Actually, the minimum difference between the specified and the actual junction temperature limit (10°C as shown) is primarily determined by the uncertainty in the thermal reference voltage. This reference voltage is designed to be independent of junction temperature and is derived from the START Zener since it remains biased ON even during the thermal shutdown period. Thus if the junction temperature exceeds 183°C in either of the power devices, both the positive output voltage and the complementary tracking negative output voltage are reduced to zero. Since the output load currents are also reduced to zero, the junction temperature decreases until the base drive is reestablished, providing the positive output voltage. This process alternately continues and results in a thermal relaxation oscillation typical of that shown in Fig. 13. If an attempt is made to increase the power dissipation in this device, the ratio of the ON–OFF time will decrease to maintain a constant average temperature of approximately 185°C.

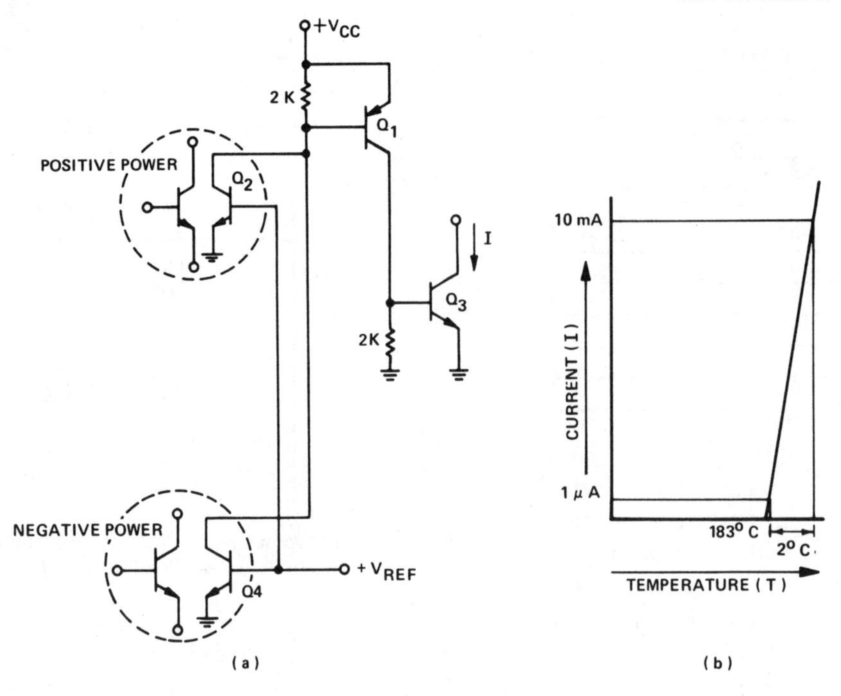

Fig. 12. (a) Improved junction temperature limiting technique for the ±15-V regulator. (b) Transfer characteristic.

Finally, the power devices were designed to exhibit a dc safe operating area characteristic that is sufficient to sustain worst case operating conditions as a final necessary requirement to achieve complete fault protection. A 40-V input–output differential with a 300-mA short circuit current represents the worst case condition that can occur, for example, during the turn ON transient when a large electrolytic output capacitor is connected to either of the output terminals or by a simple direct output short to ground.

VI. Complete ±15-V DC Monolithic Voltage Regulator

The schematic of the ±15-V regulator is shown in Fig. 14. Diodes are provided from each output terminal to ground to ensure that the output voltages will never change sign by more than one forward base–emitter voltage, independent of the input power-supply sequencing. Fig. 15 shows the typical connection required to provide the ±15-V dc output voltage and the typical performance characteristics are given below.

Input voltage (max)	40 V.
Output voltage	15 V dc ±1 percent.
Load current	200 mA.
Output impedance	50 mΩ.
Input regulation	0.004 percent/V.
Temperature coefficient	0.002 percent/°C.
Current limit	300 mA.
Junction temperature limit	185°C.
Noise voltage (10 Hz–10 MHz)	100 μV_{rms}.

Four additional terminals are provided to allow a current boost capability of up to ±10 A (see Fig. 16). Although the open-loop gains are increased with the addition of the n-p-n series-pass power transistors, the loop stability is still maintained with the internal frequency compensation capacitors since the poles also become proportionately more dominant in the amplifier. The ±300-mA internal current limit is also extended into the ampere range by shunting the internal short-circuit resistance with an appropriate external resistor.

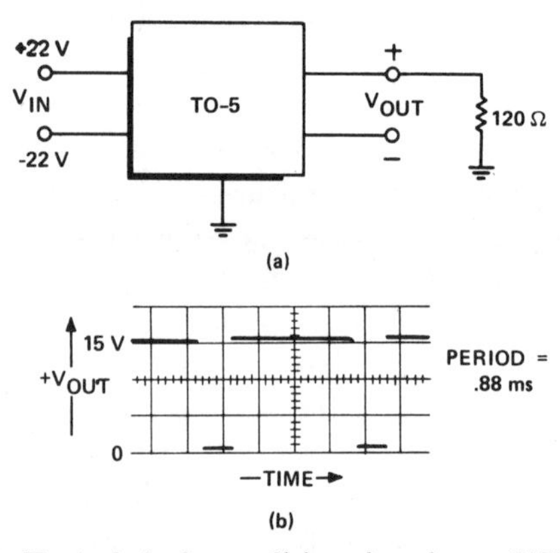

Fig. 13. (a) Typical fault condition for the ±15-V regulator. (b) Resulting thermal oscillation.

In the photomicrograph of Fig. 17, the series-pass power devices can be seen distributed along opposite edges of the die, immediately adjacent to the thermal sense transistors and the MOS frequency compensation capacitors. This arrangement allows the critical control circuitry to be located in the center of the die, far away from both of the power devices. The layout symmetry results from placing the sensitive components (which must be thermally matched) on the isothermal contours about the axis of thermal symmetry. The probe pads needed for the resistor adjust operation can be seen connected to the various taps on the ends of the level shift-

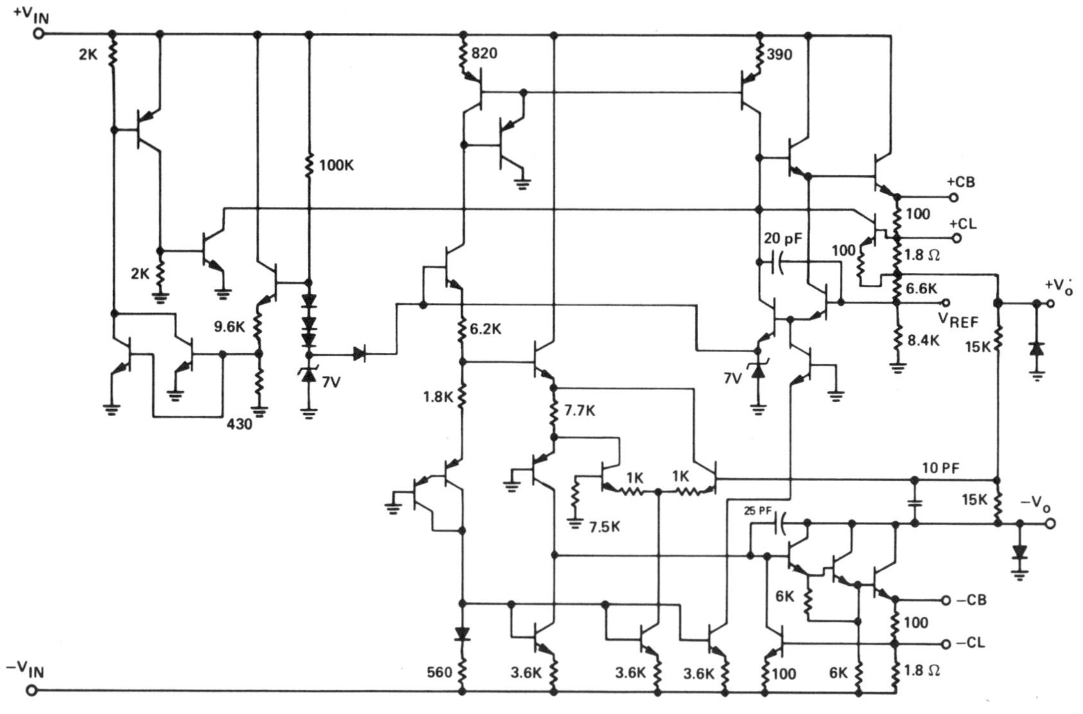

Fig. 14. Complete schematic of the ±15-V dc monolithic voltage regulator.

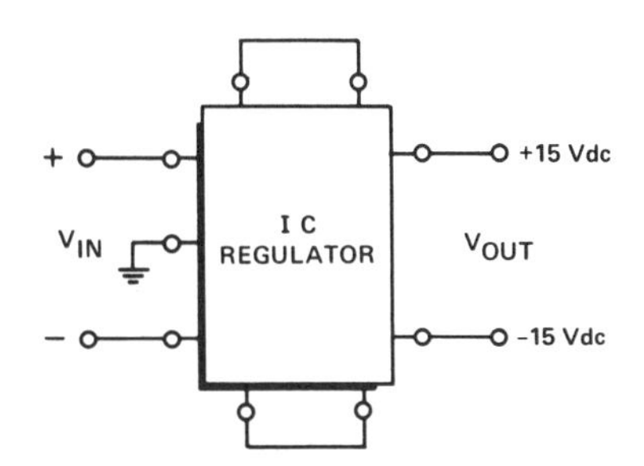

Fig. 15. Typical connection.

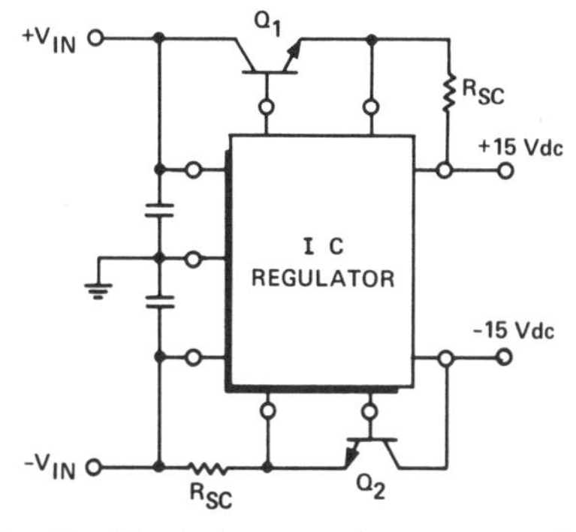

Fig. 16. Typical current boost connection.

ing resistors and also connected together by narrow metal links. A pad is also provided to access the thermal reference voltage for testing the thermal shutdown circuitry during wafer probe. Also, since the output impedance of the regulator is considerably smaller than the impedance of the aluminum bonding wire, it is necessary to supply two additional bonding pads in order to provide separate sense leads to sample the output voltage directly at the output wire bonding posts. The output impedance of the regulator can then be preserved at the terminals of the package since the impedance of the pins on the package is relatively small.

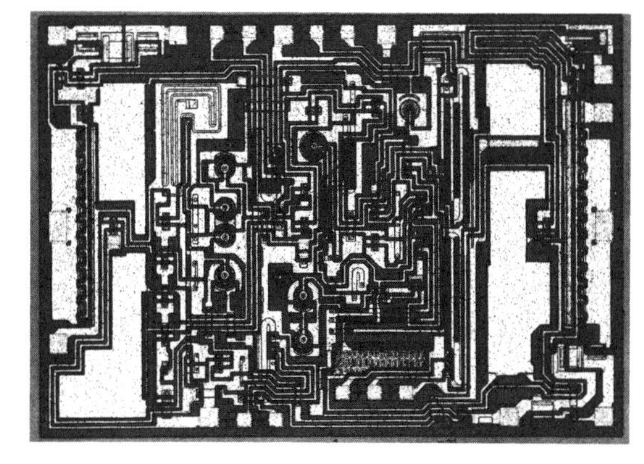

Fig. 17. Photomicrograph of the 80 × 114 mil regulator die.

VII. Summary

A dual complementary tracking monolithic voltage regulator has been built on a 80 × 114 mil die that requires no external components. Techniques have been used that eliminate the detrimental effects of thermal feedback on the performance characteristics of this ±15-V dc regulator. Complete fault protection has been provided by accurately limiting both the junction temperature and the short-circuit current and by providing excellent dc safe operating area in the design of the power transistors. In addition, excellent control over the ±1 percent output voltage tolerance is achieved without significantly adding to the manufacturing costs. A new Zener-diode geometry has also been developed that provides a significant reduction of noise associated with the avalanche breakdown. This, coupled with the overall regulator performance, demonstrates that the primary goal of providing maximum function while maintaining simplicity of use has been accomplished.

VIII. Acknowledgment

The author gratefully acknowledges the constructive criticism and encouragement of T. M. Frederiksen during the design of this regulator and the laboratory assistance of D. Culmer.

References

[1] C. L. Searle *et al., Elementary Circuit Properties of Transistors,* vol. 3. New York: Wiley, 1963, pp. 91–93.

[2] T. M. Frederiksen, "A monolithic high-power series voltage regulator," *IEEE J. Solid-State Circuits,* vol. SC-3, Dec. 1968, pp. 380–387.

[3] W. F. Davis, "Thermal considerations in the design of a 500 mA monolithic negative voltage regulator," *NEREM Rec.,* 1969, pp. 84–85.

[4] P. E. Gray *et al., Physical Electronics and Circuit Models of Transistors,* vol. 2. New York: Wiley, 1964, pp. 47–50.

[5] J. Black, "Metallization failures in integrated circuits," Semiconductor Div., Motorola, Inc., Phoenix, Ariz., Tech. Rep. RADC-TR-68, June, 1968.

A Simple Three-Terminal IC Bandgap Reference

A. PAUL BROKAW, MEMBER, IEEE

R₁, R₂, R₃ are resistors.

Abstract—**A new configuration for realization of a stabilized bandgap voltage is described. The new two-transistor circuit uses collector current sensing to eliminate errors due to base current. Because the stabilized voltage appears at a high impedance point, the application to circuits with higher output voltage is simplified. Incorporation of the new two-transistor cell in a three-terminal 2.5-V monolithic reference is described. The complete circuit is outlined in functional detail together with analytical methods used in the design. The analytical results include sensitivity coefficients, gain and frequency response parameters, and biasing for optimum temperature performance. The performance of the monolithic circuit, which includes temperature coefficients of 5 ppm/°C over the military temperature range, is reported.**

I. Introduction

The requirement for a stable reference voltage is almost universal in electronic design. The temperature-compensated avalanche breakdown diode fills many of the needs, but cannot be used with low voltage supplies and often suffers from long-term stability problems. Use of a transistor base emitter diode temperature compensated to the bandgap voltage of silicon is a technique which overcomes some of the avalanche diode limitations. Bandgap circuits can be operated from low voltage sources and depend mainly upon subsurface effects which tend to be more stable than the surface breakdowns generally obtained with avalanche diodes.

The conventional three-transistor bandgap cell works well for very low voltage two-terminal or "synthetic Zener diode" requirements. The three-transistor cell is less flexible in three-terminal applications and in circuits where the desired output is not an integral multiple of the bandgap voltage.

The two-transistor cell presented here is simpler, more flexible in three-terminal applications, and eliminates sources of error inherent in the three-transistor cell. The two-transistor cell offers separate control over output voltage and temperature coefficient in a circuit using only a single control loop.

The new bandgap circuit has been used as the basis of a monolithic three-terminal reference circuit supplying a stable 2.5-V output and operating down to 4-V input.

II. Conventional Circuit

Conventional bandgap circuits are based on the concept illustrated in Fig. 1. The transistors Q_1 and Q_2 are

Manuscript received May 6, 1974; revised July 25, 1974. This paper was presented at the International Solid-State Circuits Conference, Philadelphia, Pa., February 1974.

The author is with the Semiconductor Division, Analog Devices, Inc., Wilmington, Mass.

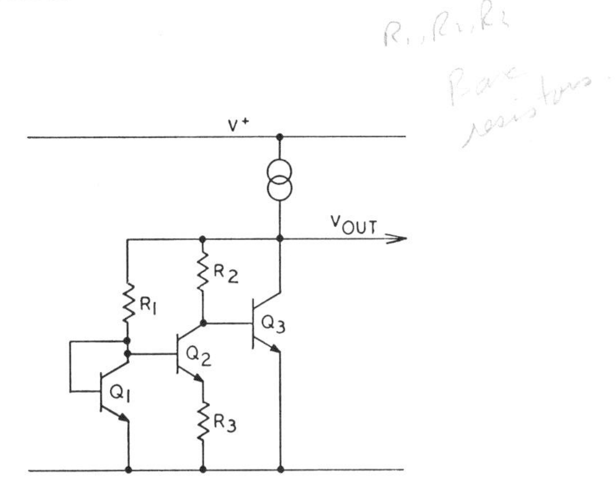

Fig. 1. Conventional bandgap circuit.

operated at different current densities to produce temperature proportional voltages across R_3 and R_2. A third transistor Q_3 is used to sense the output voltage through R_2. As a result, Q_3 drives the output to a voltage which is the sum of its V_{BE} and the temperature-dependent voltage across R_2. When the output voltage is set to approximate the bandgap voltage of silicon, the voltage across R_2 will compensate the temperature coefficient of V_{BE}, and the output voltage will be temperature invariant [1]. To minimize the output voltage temperature coefficient the collector current of Q_3 must be made proportional to temperature, as are the currents in Q_1 and Q_2. This large temperature-dependent current at the point where the stabilized voltage appears makes it inconvenient to produce an output greater than the bandgap voltage. Higher voltages can be generated by stacking several junctions to produce, in effect, several circuits like Fig. 1 in series [2].

The theory used to predict the temperature behavior of circuits like Fig. 1 neglects the effect of base current flowing in R_1 and R_2. The variability in this current due to processing and temperature effects on h_{FE} gives rise to an output voltage error and drift. This effect is particularly severe when the current in Q_2 is made much smaller than currents in Q_1 and Q_3 to produce the required current density difference. Use of "Super Beta" processing to reduce this problem results in low-voltage transistors not suitable for a three-terminal reference.

An additional temperature stability problem arises out of the nonlinearity and nonuniformity of the temperature characteristics of diffused resistors. The nonlinearity cannot easily be compensated, and the nonuniformity cannot be accommodated in the design.

The idealized circuit shown in Fig. 2 minimizes the difficulties of obtaining outputs above the bandgap voltage, reduces the problem of h_{FE} variability to one of α match, and can be implemented with thin-film resistors on the monolithic chip to virtually eliminate nonlinear

Reprinted from *IEEE J. Solid-State Circuits*, vol. SC-9, pp. 388-393, Dec. 1974.

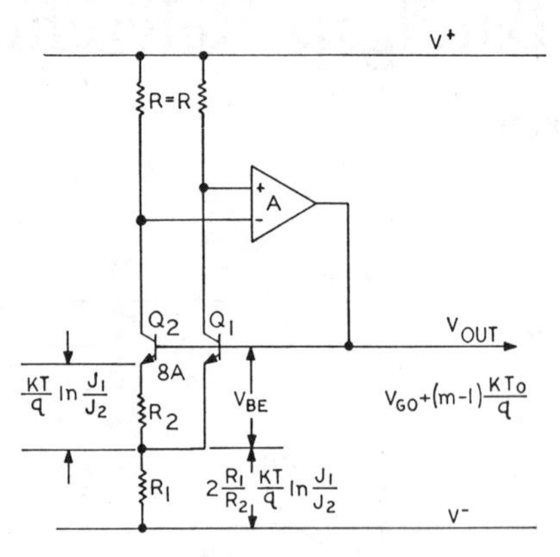

Fig. 2. Idealized circuit illustrating two-transistor bandgap cell.

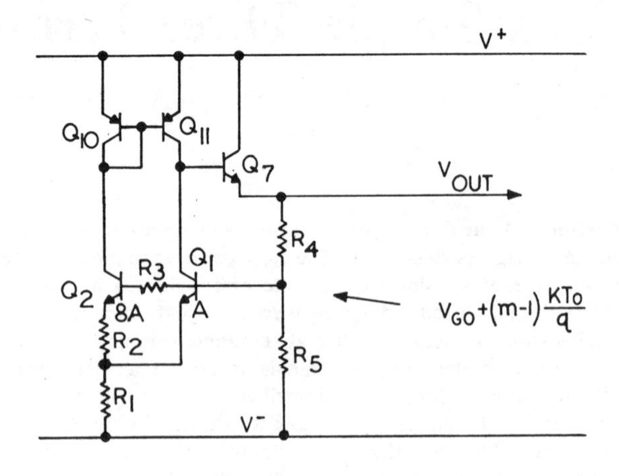

Eliminate error due to base current in R_4 by setting:

$$R_3 = \frac{R_2}{R_1}\,\frac{R_4R_5}{R_4+R_5}$$

Fig. 3. Simplified circuit for developing higher output voltages.

temperature coefficients of resistance (TCR) as an error factor. The circuit uses two transistors and collector-current sensing to establish the bandgap voltage. The voltage appears at active transistor (as opposed to diode-connected) bases, so that it is a straight forward and simple matter to obtain overall output voltages above the bandgap voltage.

III. Basis of the New Configuration

A. *Generating the Bandgap Voltage*

In the circuit of Fig. 2 the emitter area of Q_2 is made larger than that of Q_1 (by a ratio of 8-to-1 in the example given). When the voltage at their common base is small, so that the voltage drop across R_2 is small, the larger area of Q_2 causes it to conduct more of the total current available through R_1. The resulting imbalance in collector voltages drives the op amp so as to raise the base voltage. Alternatively, if the base voltage is high, forcing a large current through R_1, the voltage developed across R_2 will limit the current through Q_2 so that it will be less than the current in Q_1. The sense of the collector voltage imbalance will now be reversed, causing the op amp to reduce the base voltage. Between these two extreme conditions is a base voltage at which the two collector currents match, toward which the op amp drives from any other condition. Assuming equal α or common-base current transfer ratio for Q_1 and Q_2, this will occur when the emitter current densities are in the ratio 8-to-1, the emitter area ratio.

When this difference in current density has been produced by the op amp, there will be a difference in V_{BE}, between Q_1 and Q_2, which will appear across R_2. This difference will be given by the expression

$$\Delta V_{BE} = \frac{kT}{q}\ln\frac{J_1}{J_2}. \tag{1}$$

Since the current in Q_1 is equal to the current in Q_2, the current in R_1 is twice that in R_2 and the voltage across R_1 is given by

$$V_1 = 2\frac{R_1}{R_2}\frac{kT}{q}\ln\frac{J_1}{J_2}. \tag{2}$$

Assuming that the resistor ratio and current density ratio are invariant, this voltage varies directly with T, the absolute temperature. This is the voltage which is used to compensate the negative temperature coefficient of V_{BE}.

The voltage at the base of Q_1 is the sum of the V_{BE} of Q_1 and the temperature-dependent voltage across R_1. This is analogous to the output voltage of the conventional bandgap circuit and can be set, by adjustment of R_1/R_2, to a temperature stable value, as described in the Appendix.

B. *Increasing the Stabilized Output Voltage*

Assuming that the amplifier of Fig. 2 has sufficient gain, it will balance the collector currents of Q_1 and Q_2 despite an additional voltage drop added between its output and the common-base connection. This additional drop will not affect the base voltage which results in collector current balance. If the voltage is introduced by means of a resistive voltage divider, the op amp output voltage will be proportional to the common-base voltage.

The circuit of Fig. 3 uses an active load to sense the collector current of Q_1 and Q_2 more directly. The function of the op amp is replaced by Q_{10}, Q_{11}, and Q_7. The p-n-p transistors form a simple current mirror which takes the difference of the collector currents of Q_1 and Q_2. This difference current drives the base of Q_7 which supplies the circuit output voltage. This voltage is divided by R_4 and R_5 and applied to the base of Q_1. The sense of the signal to Q_7 drives Q_1 and Q_2 to minimize the collector current difference. By designing the circuit to stabilize the base voltage at the bandgap voltage the output will be stabilized at a higher voltage. Since the output voltage depends upon R_4 and R_5 it can be set to any convenient value and need not be an integral multiple of the bandgap voltage.

C. Base Current Error Correction

In the circuit of Fig. 3 the base current of Q_1 and Q_2 must flow through R_4. This current will require an increase above the nominal output voltage to bring the base of Q_1 to the proper level. This increase will be an h_{FE} dependent output voltage error which will vary from lot to lot and drift with temperature. The effect can be minimized by using relatively low values for R_4 and R_5, or R_3 can be added to compensate the effect. The proper value of R_3 is given by the following analysis.

To simplify the analysis, neglect the effects of finite h_{FE} and output conductance in Fig. 3 to idealize the performance of the Q_{10}, Q_{11}, Q_7 amplifier function.

If E is taken to be the circuit output voltage in the absence of base current for Q_1 and Q_2, then E' resulting from considering R_3 and the two base currents is given by

$$E' = E + R_4(i_{b1} + i_{b2}) - i_{b2}R_3(2R_1/R_2)(1 + R_4/R_5). \tag{3}$$

This relation contains a term due to the base currents through R_4 and an offsetting term due to reduction of ΔV_{BE} by base current through R_3. If E' is set equal to E, (3) can be reduced to a constraint on R_3. Expressing the relationship between the base currents in terms of a parameter $P_1 = i_{b1}/i_{b2}$ permits (3) to be reduced to

$$R_3 = (P_1 + 1)R_2R_4R_5/2R_1(R_4 + R_5). \tag{4}$$

In the case shown in Fig. 3 the collector currents and hence the base currents are assumed to match, making P_1 equal to 1 and resulting in the reduced expression shown in the figure. The general form of (4) is useful in circuits where the current density ratio is controlled by forcing unequal collector currents, rather than by emitter area ratios.

IV. Practical Realization of the Concept

Although the circuit of Fig. 3 can be used in some simple applications a number of factors limit its applicability. Base width modulation of Q_1 and Q_2, finite output impedance of Q_{11} and the finite h_{FE} of Q_7, Q_{10}, and Q_{11} all combine to raise the circuit's dynamic impedance and to degrade its input-voltage rejection. The configuration shown in Fig. 4 reduces these problems and has been built and tested in monolithic form.

The circuit elements shown in Fig. 4 correspond, roughly, to similarly numbered elements in the other figures. The basic two-transistor bandgap cell consisting of Q_1, Q_2, R_1, and R_2 is the same in all the figures. In Fig. 4 the current mirror transistor Q_{10} and Q_{11} are bootstrapped to the output voltage to improve the supply voltage rejection of the circuit. Degeneration resistors R_7 and R_8 have been added to raise the output impedance of Q_{11} and improve the emitter current match of the pair. To minimize the effect of p-n-p h_{FE} Q_8 drives the common bases of Q_{10} and Q_{11}. The output control voltage is picked off at the collector of Q_{11} by Q_9. A level translator consisting of Q_{12}, R_6, and Q_3 then applies the output

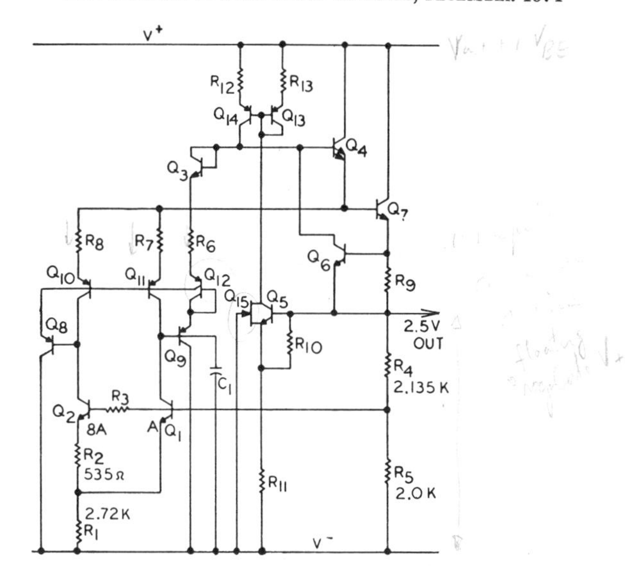

Fig. 4. Complete circuit of the monolithic three-terminal reference chip.

signal to the base of Q_4. This transistor forms a Darlington connection with Q_7, the output transistor, and provides the current-mirror bootstrap voltage as well.

The circuit, as shown in Fig. 3, has a stable "off" or no-current state. An epitaxial layer FET, Q_{15}, is incorporated into the circuit of Fig. 4 to provide starting. The FET insures that a minimum current flows into the current mirror Q_{13} and Q_{14} even when the base voltage of Q_1 and Q_2 is zero. This current is "reflected" by Q_{13} and Q_{14} to drive Q_4 and turn on the circuit. Once the circuit is on, the collector current of Q_{14} becomes nearly equal to the currents in Q_{10} and Q_{11}. As a result, the V_{BE} of Q_3 is equal to that of Q_4 which has twice the emitter area and supplies roughly twice the current. Moreover, the voltage drop across R_6 (which is equal in resistance to R_7 and R_8) is made equal to the voltage drops across R_7 and R_8. The emitter of Q_{12} is the same size as the emitters of Q_{10} and Q_{11} so that they all operate at the same current density and provide nearly equal emitter voltages for Q_8 and Q_9. The collector of Q_{12} is split to provide equal emitter currents for Q_8 and Q_9.

The operating bias level for Q_{14} is controlled by Q_5. This transistor is matched to Q_1 and has its emitter current forced by R_{10} and R_{11}. These resistors are in inverse ratio to R_4 and R_5 so that their open-circuit equivalent voltage with respect to the output is the same as the voltage across the base emitter of Q_1 in series with R_1. Their parallel resistance is twice the resistance of R_1 so that the current in Q_5 is matched to the current in Q_1. The fraction of the current bypassed by Q_{15} has a negligibly small effect on the V_{BE} of Q_5 and hence on the total current forced by R_{10} and R_{11}. The total current through Q_5 and Q_{15} drives Q_{13}. The resistors R_{12} and R_{13} raise the output impedance and improve the current matching of the simple current mirror Q_{13} and Q_{14}.

Output current limiting is provided by R_9 and Q_6.

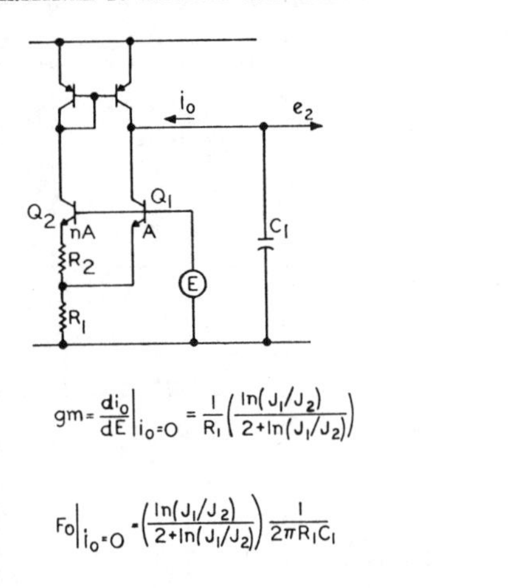

Fig. 5. Transconductance and frequency compensation model.

V. Frequency Compensation

The amplifier in this circuit operates in a closed loop to regulate the output voltage. A composite junction-MOS capacitor C_1 is used to control the open-loop crossover frequency and stabilize the closed-loop response. The analytical basis for this compensation is illustrated by Fig. 5. When the two-transistor bandgap cell is operated into a current mirror an output current is produced whenever the common-base voltage departs from the nominal voltage determined by the current density ratio and by R_1 and R_2. The change in this current as a function of the departure of base voltage from its nominal value has the dimensions of transconductance and can be used as such in design. The following incremental approximation gives a simple result which is more than adequate for most design procedures.

Incremental changes in the base voltage of Q_1 give rise to changes in collector current which can be approximated by the ratio of the voltage change to R_e, the incremental emitter resistance. This same voltage increment also drives R_2 and the incremental impedance of Q_2. If the transistors are operating at equal currents, the two R_e terms will be equal, making the total effective resistance in the Q_2 branch higher. This will result in a lower incremental current in Q_2. Equating the incremental base voltage changes gives

$$\Delta i_{e_1} R_{e_1} = \Delta i_{e_2} (R_{e_2} + R_2). \tag{5}$$

Substituting for R_e and for R_2 in terms of the voltage across it and current through it converts (5) to

$$\Delta i_{e_1} \frac{kT}{q i_{e_1}} = \Delta i_{e_2} \left(\frac{kT}{q i_{e_2}} + \frac{kT}{q i_{e_2}} \ln \frac{J_1}{J_2} \right). \tag{6}$$

A second approximation made is that the total incremental current is due to the voltage change across R_1 resulting from a voltage change ΔE at the common bases. That is

$$\Delta i_{e_1} + \Delta i_{e_2} = \frac{\Delta E}{R_1}, \tag{7}$$

which neglects R_e as compared with the value of R_1. After eliminating common factors and assuming $i_{e_1} = i_{e_2}$, the combination of (6) and (7) can be manipulated to form

$$\frac{\Delta i_{e_1} - \Delta i_{e_2}}{\Delta E} = \frac{1}{R_1} \left(\frac{\ln (J_1/J_2)}{2 + \ln (J_1/J_2)} \right). \tag{8}$$

The difference in the Q_1, Q_2 collector currents is the output current i_o and is approximately given by the difference of the emitter current increments. Taking (8) to the limit at $i_o = 0$ yields

$$\left. \frac{d_{io}}{dE} \right|_{i_o = 0} = \frac{1}{R_1} \left(\frac{\ln (J_1/J_2)}{2 + \ln (J_1/J_2)} \right). \tag{9}$$

With a current density ratio of 8:1, the term $\ln J_1/J_2$ is approximately 2, so that the "transconductance" of the entire circuit is approximately $1/(2R_1)$.

A capacitive load on the current output of the circuit in Fig. 5 will give a 6 dB/octave rolloff of voltage transfer ratio. The frequency at which the capacitive reactance equals the transconductance will be the unity-gain frequency of the simple circuit. This is given by the expression for F_0 in the figure. In the circuit of Fig. 4, the loop attenuation due to R_4 and R_5 reduces the overall unity-gain frequency by the ratio of the bandgap voltage to 2.5 V, which is approximately two.

The transconductance can also be used to estimate a low frequency "gain." In the simple circuit of Fig. 3, the gain is expressed as the ratio of the voltage at the base of Q_7 to a small-signal input applied to the base of Q_1 at balance. Using a value of 3 kΩ for R_1 and estimating the output impedance of Q_{11} at about 300 kΩ gives a gain of about 50. In the monolithic circuit, the effective open-loop gain is increased several orders of magnitude by the bootstrap connection to the current mirror.

VI. Monolithic Circuit Performance

The circuit of Fig. 4 is shown in Fig. 6 as it appears in monolithic form. Several diffusion lots have been made and measurements of these units indicate the typical properties given by the following table.

Typical Reference Circuit Parameters
(−55 to +125°C)

Output voltage	2.5 V ±2 percent
Minimum input voltage	4 V
Load regulation, 0 to 10 mA	3 mV
Supply rejection, 4.5 to 7 V	0.25 mV
Supply rejection, 7 to 30 V	0.25 mV
Standby current	1 mA
Output voltage temperature coefficient (γ) $\left(\gamma = \frac{(V_{\max} - V_{\min})}{(V_{\text{nominal}})(\Delta T)}\right)$	5 to 60 ppm/°C

The observed variation in temperature coefficients

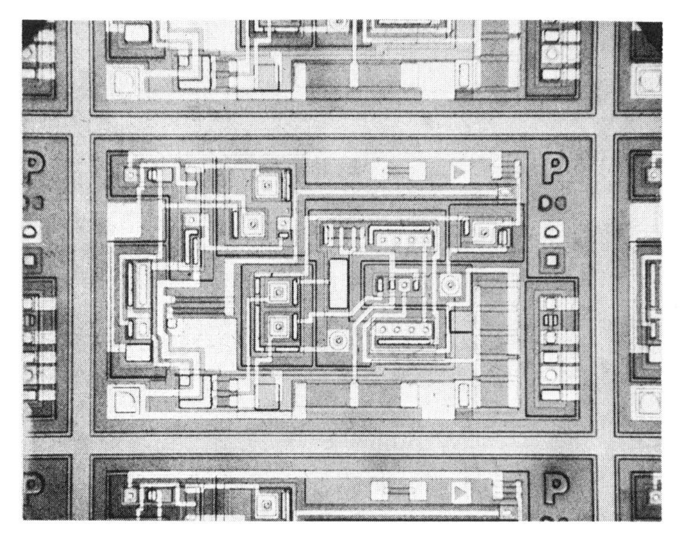

Fig. 6. Illustration of 37 × 62 mil monolithic die showing thin-film resistors and balanced thermal layout.

arises mainly from variability in absolute voltage at the base of Q_1 and variation in the coefficient m (see Appendix). Units showing a very constant temperature coefficient generally show a good correlation between absolute output voltage and drift.

Several units showing large but constant temperature coefficients have been adjusted to very small final temperature coefficients. This has been done by measuring the temperature coefficient calculating the ideal zero temperature coefficient voltage correction and laser-trimming R_1 or R_2 as required. Other units which are closer to the nominal output voltage exhibit very stable temperature characteristics initially. In extremely low-drift units, the performance of the basic cell appears to be masked by other drifts arising in the remainder of the circuit. These residual drifts are on the order of 2 to 4 ppm/°C over a temperature range of −55 to +125°C.

Some diffusion lots have shown a greater curvature in the temperature characteristic. These units exhibit the roughly parabolic temperature characteristic, which is implied by (14) of the Appendix. They show good temperature performance around a peak which occurs at a temperature related to initial output voltage. At temperature extremes the temperature coefficient may increase to 60 ppm/°C or more, if the peak is not centered in the −55 to +125°C range.

The elapsed time since obtaining the first completed units has not been sufficient to accumulate long-term drift results. Accelerated life tests have been made at high temperatures to uncover any gross drift problems. The temperature stability and monitoring equipment have not been adequate to determine the ultimate stability of the device. Examination of data taken over the course of 1000 hours at +125°C does not reveal any trends or systematic drifts at the 100 ppm level, which approximates the repeatability of the measurements.

Appendix

Theoretical Temperature Behavior

A. Optimum Cell Voltage

The optimum voltage at the base of Q_1 is approximately the bandgap voltage. A general analysis of the two-transistor bandgap cell yields a more-exact result and some insight into the residual temperature drift in an optimally adjusted circuit.

As a matter of convenience, the circuit description has involved the assumption that the required current density ratio is established by the use of emitter area ratio alone. The circuit can also be based on different collector currents and equal areas or a combination of unequal areas and currents.

The equation describing the voltage across R_1 can be generalized by including a parameter P_1 which is the ratio of emitter currents i_{e_1}/i_{e_2}, as follows:

$$V_1 = (P_1 + 1)\frac{R_1}{R_2}\frac{kT}{q}\ln\frac{J_1}{J_2}. \tag{10}$$

An expression given by Brugler [3] has been modified by the addition of a current-dependent term to give the V_{BE} of Q_1 as

$$V_{BE_1} = V_{go}\left(1 - \frac{T}{T_o}\right) + V_{BE_o}\frac{T}{T_o} + \frac{mkT}{q}\ln\frac{T_o}{T} + \frac{kT}{q}\ln\frac{J}{J_o} \tag{11}$$

where V_{go} is the bandgap voltage of silicon and V_{BE_0} is the value of V_{BE_1} at a reference temperature T_o.

Since the voltages across both R_1 and R_2 are proportional to temperature, it follows that the current and the current density in Q_1 are also proportonal to temperature. Therefore, the current density ratio in the last term of (11) can be equated to a temperature ratio as $J/J_o = T/T_o$. This relation can be used to reduce the sum of V_1 and V_{BE_1} to the form

$$E = V_{go} + \frac{T}{T_o}(V_{BE_o} - V_{go}) + (m - 1)\frac{kT}{q}\ln\frac{T_o}{T} + (P_1 + 1)\frac{R_1}{R_2}\frac{kT}{q}\ln\frac{J_1}{J_2}. \tag{12}$$

This represents the stable voltage established at the base of Q_1 in the circuit of Fig. 1. Differentiating this result twice with respect to temperature yields

$$\frac{dE}{dT} = \frac{1}{T_o}(V_{BE_o} - V_{go}) + (P_1 + 1)\frac{R_1}{R_2}\frac{k}{q}\ln\frac{J_1}{J_2} + (m - 1)\frac{k}{q}\ln\frac{T_o}{T} - 1 \tag{13}$$

and

$$\frac{d^2E}{dT^2} = -(m-1)\frac{k}{q}\frac{1}{T}. \tag{14}$$

Equating the first derivative to zero results in the equation

$$V_{BE_o} + (P_1+1)\frac{R_1}{R_2}\frac{kT_o}{q}\ln\frac{J_1}{J_2} = V_{go} + (m-1)\frac{kT_o}{q}. \tag{15}$$

The left side of (15) is the value of E at $T = T_o$. This means that, in principle, if the base voltage of Q_1 is set to $V_{go} + (m-1)\,kT_o/q$ at temperature T_o, the temperature coefficient of the output voltage will be zero. Assuming values of m greater than one in (14) implies, however, a nonzero temperature coefficient at temperatures other than T_o. Experimental data indicate that values of m as low as 1.2 have been achieved.

Examination of (13) and (15) shows that departures from the optimum output voltage will result in an approximately constant temperature coefficient. The magnitude of this coefficient will be the absolute error voltage divided by the absolute temperature. For example, a 3 percent absolute voltage error at 300 K will result in a 3 percent/300 K = 100 ppm/°C temperature coefficient.

B. Effects of Resistor Temperature Coefficients

Both the common and differential temperature coefficients of resistance of R_1 and R_2 enter into the overall output voltage stability. The differential TCR may be treated by assuming that the total change in resistor ratio is due to change in R_1. Since the remainder of the circuit forces a predetermined current in R_1, its temperature coefficient of resistance is translated to an error voltage in equal proportion. The effect on the overall output voltage is reduced by the fact that the voltage across R_1 accounts for only a fraction of the total output. This fraction is variable, but is about 1/2 at room temperature. Therefore, the differential TCR of R_1 and R_2 appears as about half the equivalent proportional change in the output voltage.

The common TCR affecting both resistors equally does not affect the voltage across R_1, which depends only upon their ratio. It does affect the total current in the transistors, however, and therefore affects the V_{BE} of Q_1.

This effect can be evaluated by expressing V_{BE_1} as

$$V_{BE_1} = \frac{kT}{q}\ln\left(\frac{i_{e_1}}{I_s}\right). \tag{16}$$

Differentiating

$$\frac{dV_{BE_1}}{dR_2} = \frac{kT}{q}\frac{I_s}{i_{e_1}}\left(\frac{-i_{e_1}}{R_2 I_s}\right) = \frac{-kT}{qR_2} \tag{17}$$

and

$$\frac{dV_{BE_1}}{dR_2}R_2 = \frac{-kT}{q} \tag{18}$$

which at room temperature is about −26 mV or −260 μV/percent. Dividing this coefficient by the cell's output voltage of about 1.22 V gives a relative coefficient of about -2.13×10^{-4}/percent of output voltage change as a function of relative resistance change. This is a reduction in sensitivity of about 47 times.

Despite this reduction, the TCR of diffused resistors is so large, nonlinear, and nonuniform that it still represents a serious temperature drift. The monolithic circuit described here uses thin-film-on-silicon resistors with a total TCR of about −60 ppm so that the 47 times reduction makes the uncorrectable component of TCR sensitivity less than 1 ppm/°C.

The effects of the TCR of other resistors, or of ratios such as R_4 and R_5, are readily evaluated by standard analytical techniques. The low differential TCR inherent in thin-film resistor pairs keeps these effects on the output voltage at a minimum.

References

[1] R. J. Widlar, "New developments in IC voltage regulators," *IEEE J. Solid-State Circuits*, vol. SC-6, pp. 2–7, Feb. 1971.

[2] K. E. Kuijk, "A precision reference voltage source," *IEEE J. Solid-State Circuits*, vol. SC-8, pp. 222–226, June 1973.

[3] J. S. Brugler, "Silicon transistor biasing for linear collector current temperature dependence," *IEEE J. Solid-State Circuits* (Corresp.), vol. SC-2, pp. 57–58, June 1967.

IC Voltage Reference has 1 PPM per Degree Drift

National Semiconductor
Application Note 161
Robert Dobkin
June 1977

A new linear IC now provides the ultimate in highly stable voltage references. Now, a new monolithic IC the LM199, out-performs zeners and can provide a 6.9V reference with a temperature drift of less than 1 ppm/° and excellent long term stability. This new IC, uses a unique subsurface zener to achieve low noise and a highly stable breakdown. Included is an on-chip temperature stabilizer which holds the chip temperature at 90°C, eliminating the effects of ambient temperature changes on reference voltage.

The planar monolithic IC offers superior performance compared to conventional reference diodes. For example, active circuitry buffers the reverse current to the zener giving a dynamic impedance of 0.5Ω and allows the LM199 to operate over a 0.5 mA to 10 mA current range with no change in performance. The low dynamic impedance, coupled with low operating current significantly simplifies the current drive circuitry needed for operation. Since the temperature coefficient is independent of operating current, usually a resistor is all that is needed.

Previously, the task of providing a stable, low temperature coefficient reference voltage was left to a discrete zener diode. However, these diodes often presented significant problems. For example, ordinary zeners can show many millivolts change if there is a temperature gradient across the package due to the zener and temperature compensation diode not being at the same temperature. A 1°C difference may cause a 2 mV shift in reference voltage. Because the on-chip temperature stabilizer maintains constant die temperature, the IC reference is free of voltage shifts due to temperature gradients. Further, the temperature stabilizer, as well as eliminating drift, allows exceptionally fast warm-up over conventional diodes. Also, the LM199 is insensitive to stress on the leads—another source of error with ordinary glass diodes. Finally, the LM199 shows virtually no hysteresis in reference voltage when subject to temperature cycling over a wide temperature range. Temperature cycling the LM199 between 25°C, 150°C and back to 25°C causes less than 50μV change in reference voltage. Standard reference diodes exhibit shifts of 1 mV to 5 mV under the same conditions.

SUB SURFACE ZENER IMPROVES STABILITY

Previously, breakdown references made in monolithic IC's usually used the emitter-base junction of an NPN transistor as a zener diode. Unfortunately, this junction breaks down at the surface of the silicon and is therefore susceptible to surface effects. The breakdown is noisy, and cannot give long-term stabilities much better than about 0.3%. Further, a surface zener is especially sensitive to contamination in the oxide or charge on the surface of the oxide which can cause short-term instability or turn-on drift.

The new zener moves the breakdown below the surface of the silicon into the bulk yielding a zener that is stable with time and exhibits very low noise. Because the new zener is made with well-controlled diffusions in a planar structure, it is extremely reproducible with an initial 2% tolerance on breakdown voltage.

A cut-away view of the new zener is shown in *Figure 1.* First a small deep P+ diffusion is made into the surface of the silicon. This is then covered by the standard base diffusion. The N+ emitter diffusion is then made completely covering the P+ diffusion. The diode then breaks down where the dopant concentration is greatest, that is, between the P+ and N+. Since the P+ is completely covered by N+ the breakdown is below the surface and at about 6.3V. One connection to the diode is to the N+ and the other is to the P base diffusion. The current flows laterally through the base to the P+ or cathode of the zener. Surface breakdown does not occur since the base P to N+ breakdown voltage is greater than the breakdown of the buried device. The buried zener has been in volume production since 1973 as the reference in the LX5600 temperature transducer.

CIRCUIT DESCRIPTION

The block diagram of the LM199 is shown in *Figure 2.* Two electrically independent circuits are included on the same chip—a temperature stabilizer and a floating active zener. The only electrical connection between the two

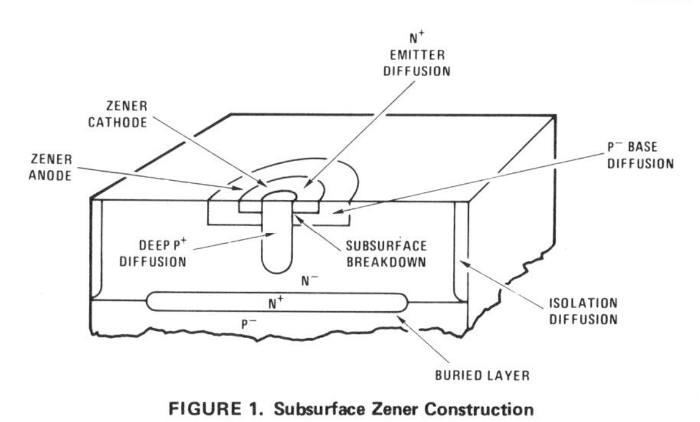

FIGURE 1. Subsurface Zener Construction

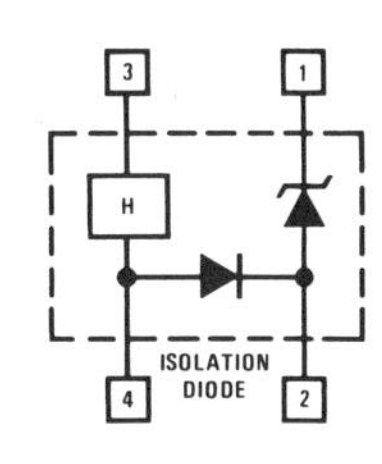

FIGURE 2. Functional Block Diagram

circuits is the isolation diode inherent in any junction-isolated integrated circuit. The zener may be used with or without the temperature stabilizer powered. The only operating restriction is that the isolation diode must never become forward biased and the zener must not be biased above the 40V breakdown of the isolation diode.

The actual circuit is shown in *Figure 3.* The temperature stabilizer is composed of Q1 through Q9. FET Q9 provides current to zener D2 and Q8. Current through Q8 turns a loop consisting of D1, Q5, Q6, Q7, R1 and R2. About 5V is applied to the top of R1 from the base of Q7. This causes 400μA to flow through the divider R1, R2. Transistor Q7 has a controlled gain of 0.3 giving Q7 a total emitter current of about 500μA. This flows through the emitter of Q6 and drives another controlled gain PNP transistor Q5. The gain of Q5 is about 0.4 so D1 is driven with about 200μA. Once current flows through Q5, Q8 is reverse biased and the loop is self-sustaining. This circuitry ensures start-up.

The resistor divider applies 400 mV to the base of Q4 while Q7 supplies 120μA to its collector. At temperatures below the stabilization point, 400 mV is insufficient to cause Q4 to conduct. Thus, all the collector current from Q7 is provided as base drive to a Darlington composed of Q1 and Q2. The Darlington is connected across the supply and initially draws 140 mA (set by current limit transistor Q3). As the chip heats, the turn on voltage for Q4 decreases and Q4 starts to conduct. At about 90°C the current through Q4 appreciably increases and less drive is applied to Q1 and Q2. Power dissipation decreases to whatever is necessary to hold the chip at the stabilization temperature. In this manner, the chip temperature is regulated to better than 2°C for a 100°C temperature range.

The zener section is relatively straight-forward. A buried zener D3 breaks down biasing the base of transistor Q13. Transistor Q13 drives two buffers Q12 and Q11. External current changes through the circuit are fully absorbed by the buffer transistors rather than D3. Current through D3 is held constant at 250μA by a 2k resistor across the emitter base of Q13 while the emitter-base voltage of Q13 nominally temperature compensates the reference voltage.

The other components, Q14, Q15 and Q16 set the operating current of Q13. Frequency compensation is accomplished with two junction capacitors.

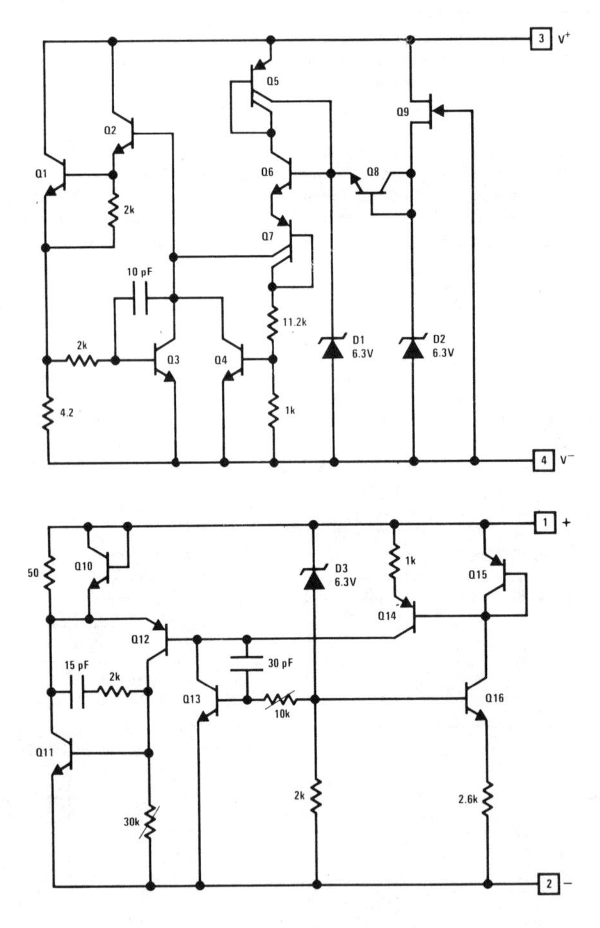

FIGURE 3. Schematic Diagram of LM199 Precision Reference

PERFORMANCE

A polysulfone thermal shield, shown in *Figure 4,* is supplied with the LM199 to minimize power dissipation and improve temperature regulation. Using a thermal shield as well as the small, high thermal resistance TO-46 package allows operation at low power levels without the problems of special IC packages with built-in thermal isolation. Since the LM199 is made on a standard IC assembly line with standard assembly techniques, cost is significantly lower than if special techniques were used. For temperature stabilization only 300 mW are required at 25°C and 660 mW at −55°C.

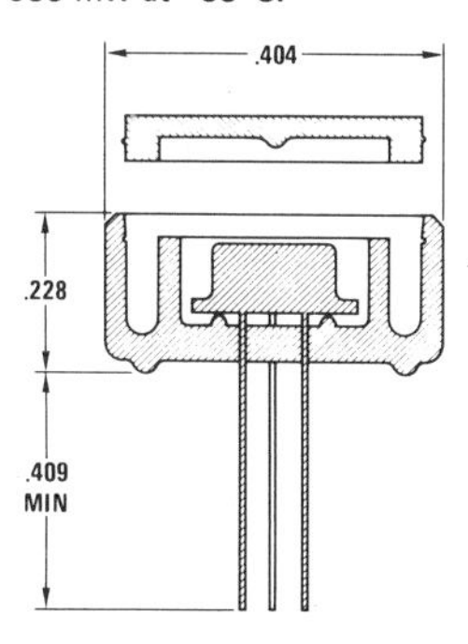

FIGURE 4. Polysulfone Thermal Shield

Temperature stabilizing the device at 90°C virtually eliminates temperature drift at ambient temperatures less than 90°C. The reference is nominally temperature compensated and the thermal regulator further decreases the temperature drift. Drift is typically only 0.3 ppm/°C. Stabilizing the temperature at 90°C rather than 125°C significantly reduces power dissipation but still provides very low drift over a major portion of the operating temperature range. Above 90°C ambient, the temperature coefficient is only 15 ppm/°C.

A low drift reference would be virtually useless without equivalent performance in long term stability and low noise. The subsurface breakdown technology yields both of these. Wideband and low frequency noise are both exceptionally low. Wideband noise is shown in *Figure 5* and low frequency noise is shown over a 10 minute period in the photograph of *Figure 6.* Peak to peak noise over a 0.01 Hz to 1 Hz bandwidth is only about 0.7μV.

Long term stability is perhaps one of the most difficult measurements to make. However, conditions for long-term stability measurements on the LM199 are considerably more realistic than for commercially available certified zeners. Standard zeners are measured in ±0.05°C temperature controlled both at an operating current of 7.5 mA ±0.05μA. Further, the standard devices must have stress-free contacts on the leads and the test must not be interrupted during the measurement interval. In contrast, the LM199 is measured in still air of 25°C to 28°C at a reverse current of 1 mA ±0.5%. This is more typical of actual operating conditions in instruments.

When a group of 10 devices were monitored for long-term stability, the variations all correlated, which indicates changes in the measurement system (limitation of 20 ppm) rather than the LM199.

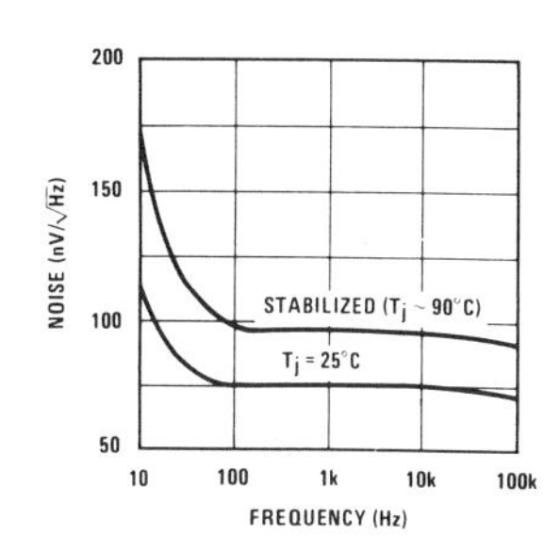

FIGURE 5. Wideband Noise of the LM199 Reference

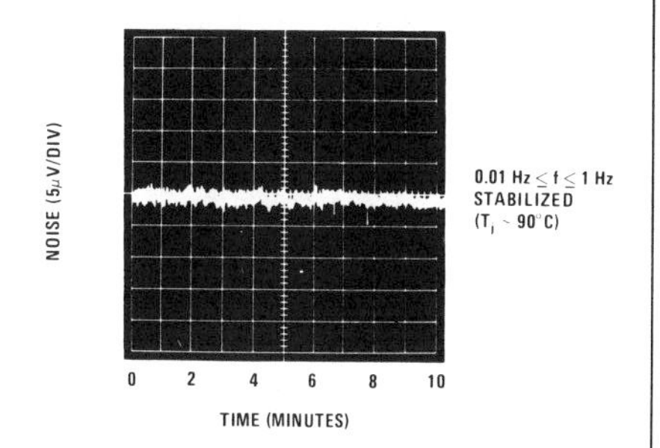

FIGURE 6. Low Frequency Noise Voltage

Because the planar structure does not exhibit hysteresis with temperature cycling, long-term stability is not impaired if the device is switched on and off.

The temperature stabilizer heats the small thermal mass of the LM199 to 90°C very quickly. Warm-up time at 25°C and −55°C is shown in *Figure 7.* This fast warm-up is significantly less than the several minutes needed by ordinary diodes to reach equilibrium. Typical specifications are shown in Table I.

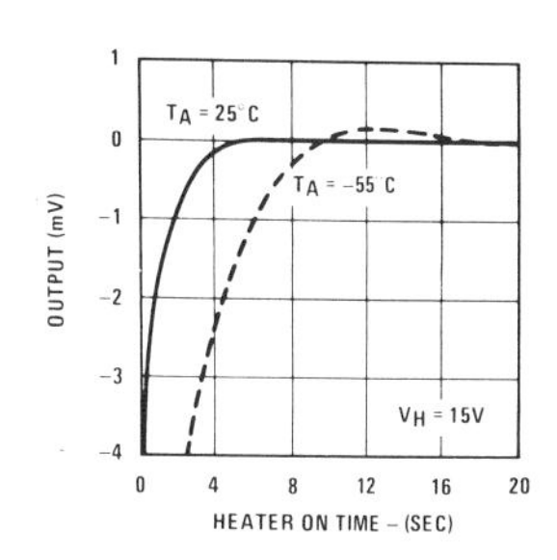

FIGURE 7. Fast Warmup Time of the LM199

Table I. Typical Specifications for the LM199

Reverse Breakdown Voltage	6.95V
Operating Current	0.5 mA to 10 mA
Temperature Coefficient	0.3 ppm/°C
Dynamic Impedance	0.5Ω
RMS Noise (10 Hz to 10 kHz)	7μV
Long-Term Stability	≤ 20 ppm
Temperature Stabilizer Operating Voltage	9V to 40V
Temperature Stabilizer Power Dissipation (25°C)	300 mW
Warm-up Time	3 Seconds

APPLICATIONS

The LM199 is easier to use than standard zeners, but the temperature stability is so good—even better than precision resistors—that care must be taken to prevent external circuitry from limiting performance. Basic operation only requires energizing the temperature stabilizer from a 9V to 40V power source and biasing the reference with between 0.5 mA to 10 mA of current. The low dynamic impedance minimizes the current regulation required compared to ordinary zeners.

The only restriction on biasing the zener is the bias applied to the isolation diode. Firstly, the isolation diode must not be forward biased. This restricts the voltage at either terminal of the zener to a voltage equal to or greater than the V^-.

A dc return is needed between the zener and heater to insure the voltage limitation on the isolation diodes are not exceeded. *Figure 8* shows the basic biasing of the LM199.

The active circuitry in the reference section of the LM199 reduces the dynamic impedance of the zener to about 0.5Ω. This is especially useful in biasing the reference. For example, a standard reference diode such as a 1N829 operates at 7.5 mA and has a dynamic impedance of 15Ω. A 1% change in current (75μA) changes the reference voltage by 1.1 mV. Operating the LM199 at 1 mA with the same 1% change in operating current (10μA) results in a reference change of only 5μV. *Figure 9* shows reverse voltage change with current.

Biasing current for the reference can be anywhere from 0.5 mA to 10 mA with little change in performance. This wide current range allows direct replacement of most zener types with no other circuit changes besides the temperature stabilizer connection. Since the dynamic impedance is constant with current changes regulation

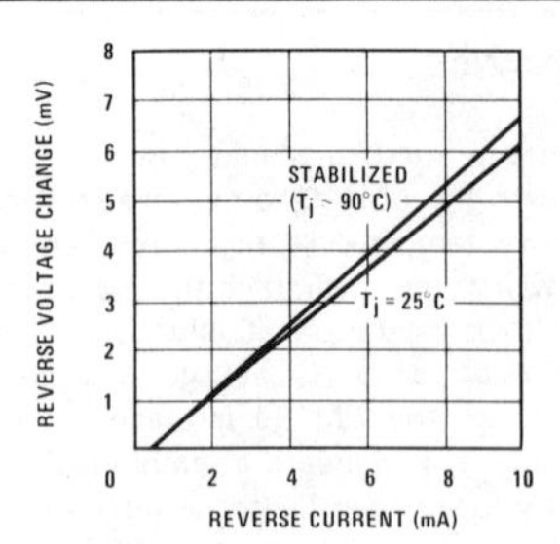

FIGURE 9. The LM199 Shows Excellent Regulation Against Current Changes

is better than discrete zeners. For optimum regulation, lower operating currents are preferred since the ratio of source resistance to zener impedance is higher, and the attenuation of input changes is greater. Further, at low currents, the voltage drop in the wiring is minimized.

Mounting is an important consideration for optimum performance. Although the thermal shield minimizes the heat low, the LM199 should not be exposed to a direct air flow such as from a cooling fan. This can cause as much as a 100% increase in power dissipation degrading the thermal regulation and increasing the drift. Normal conviction currents do not degrade performance.

Printed circuit board layout is also important. Firstly, four wire sensing should be used to eliminate ohmic drops in pc traces. Although the voltage drops are small the temperature coefficient of the voltage developed along a copper trace can add significantly to the drift. For example, a trace with 1Ω resistance and 2 mA current flow will develop 2 mV drop. The TC of copper is 0.004%/°C so the 2 mV drop will change at 8μV/°C, this is an additional 1 ppm drift error. Of course, the effects of voltage drops in the printed circuit traces are eliminated with 4-wire operation. The heater current also should not be allowed to flow through the voltage reference traces. Over a −55°C to +125°C temperature

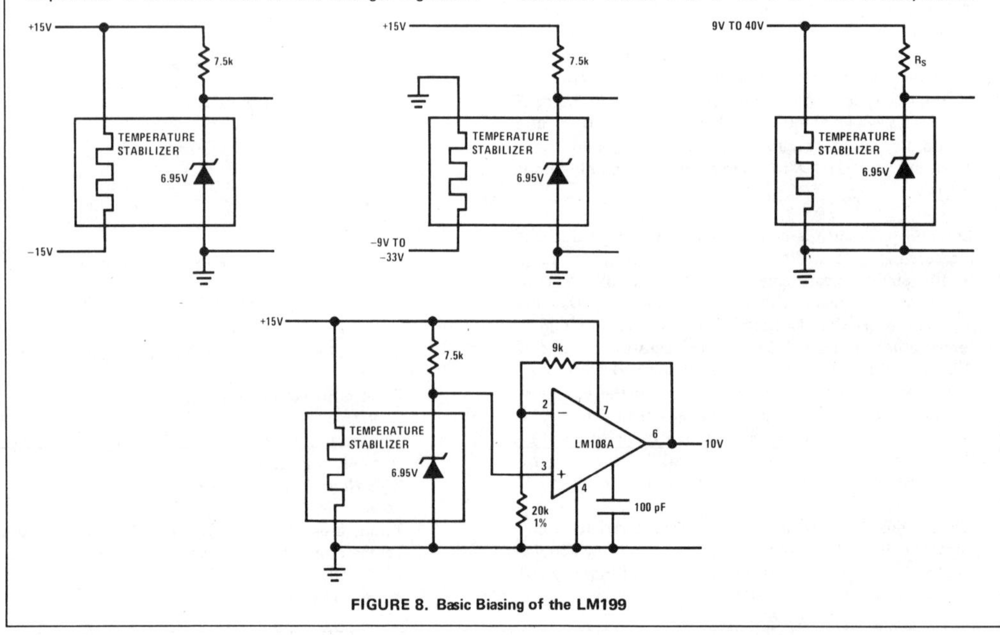

FIGURE 8. Basic Biasing of the LM199

range the heater current will change from about 1 mA to over 40 mA. These magnitudes of current flowing reference leads or reference ground can cause huge errors compared to the drift of the LM199.

Thermocouple effects can also cause errors. The kovar leads from the LM199 package form a thermocouple with copper printed circuit board traces. Since the package of the 199 is heated, there is a heat flow along the leads of the LM199 package. If the leads terminate into unequal sizes of copper on the p.c. board greater heat will be absorbed by the larger copper trace and a temperature difference will develop. A temperature *difference* of 1°C between the two leads of the reference will generate about 30μV. Therefore, the copper traces to the zener should be equal in size. This will generally keep the errors due to thermocouple effects under about 15μV.

The LM199 should be mounted flush on the p.c. board with a minimum of space between the thermal shield and the boards. This minimizes air flow across the kovar leads on the board surface which also can cause thermocouple voltages. Air currents across the leads usually appear as ultra-low frequency noise of about 10μV to 20μV amplitude.

It is usually necessary to scale and buffer the output of any reference to some calibrated voltage. *Figure 10* shows a simple buffered reference with a 10V output. The reference is applied to the non-inverting input of the LM108A. An RC rolloff can be inserted in series with the input to the LM108A to roll-off the high frequency noise. The zener heater and op amp are all powered from a single 15V supply. About 1% regulation on the input supply is adequate contributing less than 10μV of error to the output. Feedback resistors around the LM308 scale the output to 10V.

Although the absolute values of the resistors are not extremely important, tracking of temperature coefficients is vital. The 1 ppm/°C drift of the LM199 is easily exceeded by the temperature coefficient of most resistors. Tracking to better than 1 ppm is also not easy to obtain. Wirewound types made of Evenohm or Mangamin are good and also have low thermoelectric effects. Film types such as Vishay resistors are also good. Most potentiometers do not track fixed resistors so it is a good idea to minimize the adjustment range and therefore minimize their effects on the output TC. Overall temperature coefficient of the circuit shown in *Figure 10* is worst case 3 ppm/°C. About 1 ppm is due to the reference, 1 ppm due to the resistors and 1 ppm due to the op amp.

Figure 11 shows a standard cell replacement with a 1.01V output. A LM321 and LM308 are used to minimize op amp drift to less than 1μV/°C. Note the adjustment connection which minimizes the TC effects of the pot. Set-up for this circuit requires nulling the offset of the op amp first and then adjusting for proper output voltage.

The drift of the LM321 is very predictable and can be used to eliminate overall drift of the system. The drift changes at 3.6μV/°C per millivolt of offset so 1 mV to 2 mV of offset can be introduced to minimize the overall TC.

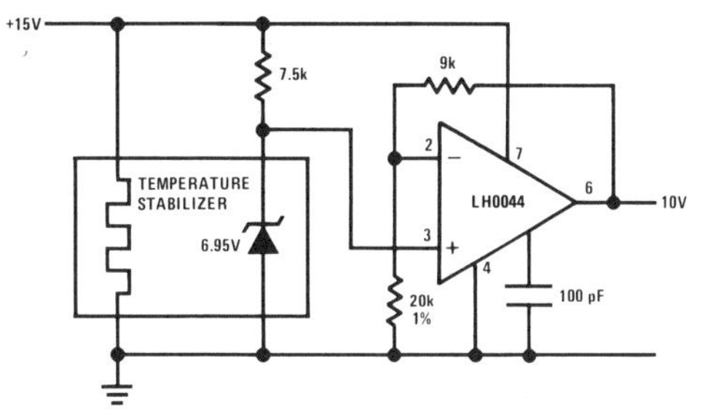

FIGURE 10. Buffered 10V Reference

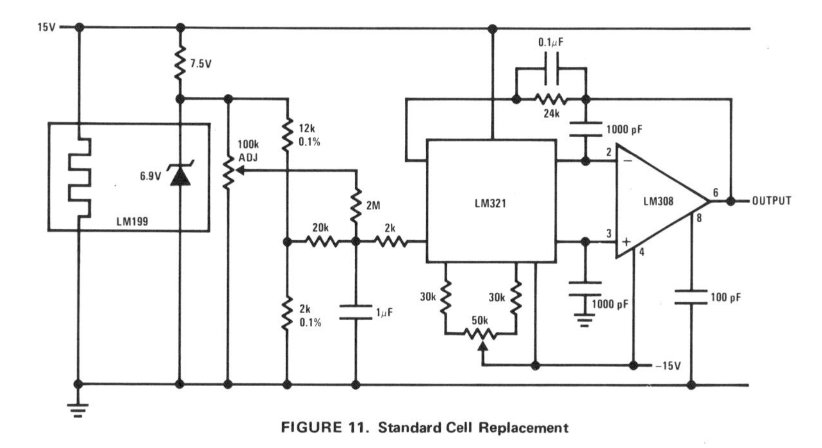

FIGURE 11. Standard Cell Replacement

For circuits with a wide input voltage range, the reference can be powered from the output of the buffer as is shown in *Figure 12.* The op amp supplies regulated voltage to the resistor biasing the reference minimizing changes due to input variation. There is some change due to variation of the temperature stabilizer voltage so extremely wide range operation is not recommended for highest precision. An additional resistor (shown 80 kΩ) is added to the unregulated input to insure the circuit starts up properly at the application of power.

A precision power supply is shown in *Figure 13.* The output of the op amp is buffered by an IC power transistor the LM395. The LM395 operates as an NPN power device but requires only 5μA base current. Full overload protection inherent in the LM395 includes current limit, safe-area protection, and thermal limit.

A reference which can supply either a positive or a negative continuously variable output is shown in *Figure 14.* The reference is biased from the ±15V input supplies as was shown earlier. A ten-turn pot will adjust the output from $+V_Z$ to $-V_Z$ continuously. For negative output the op amp operates as an inverter while for positive outputs it operates as a non-inverting connection.

Op amp choice is important for this circuit. A low drift device such as the LM108A or a LM108-LM121 combination will provide excellent performance. The pot should be a precision wire wound 10 turn type. It should be noted that the output of this circuit is not linear.

CONCLUSIONS

A new monolithic reference which exceeds the performance of conventional zeners has been developed. In fact, the LM199 performance is limited more by external components than by reference drift itself. Further, many of the problems associated with conventional zeners such as hysteresis, stress sensitivity and temperature gradient sensitivity have also been eliminated. Finally, long-term stability and noise are equal of the drift performance of the new device.

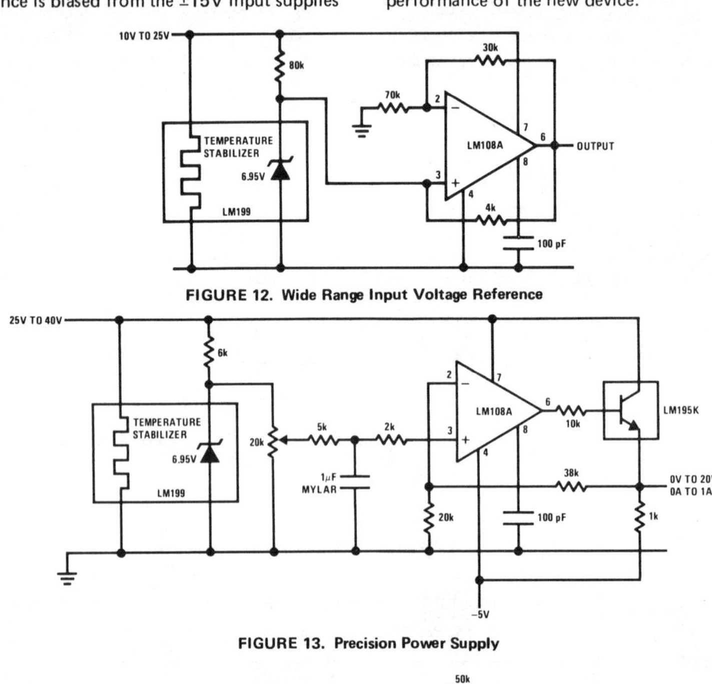

FIGURE 12. Wide Range Input Voltage Reference

FIGURE 13. Precision Power Supply

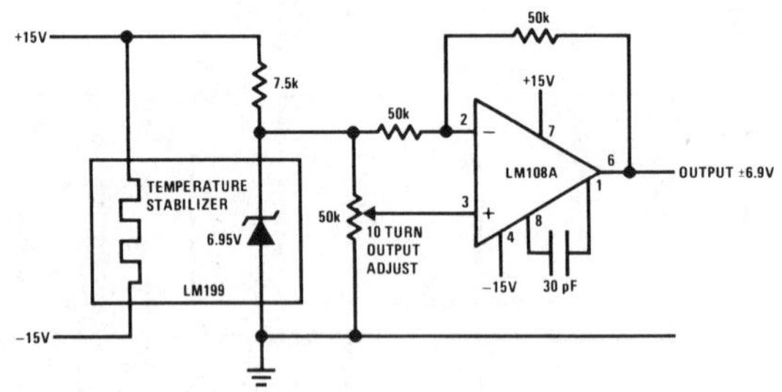

FIGURE 14. Bipolar Output Reference

National Semiconductor Corporation
2900 Semiconductor Drive, Santa Clara, California 95051, (408) 737-5000/TWX (910) 339-9240
National Semiconductor GmbH
808 Fuerstenfeldbruck, Industriestrasse 10, West Germany, Tele. (08141) 1371/Telex 05-27649

A CMOS Reference Voltage Source

Yannis P. Tsividis
Columbia University
New York, NY

Richard W. Ulmer
Motorola, Inc.
Austin, TX

THE IMPLEMENTATION OF MOS single chip systems for data acquisition and conversion has in the past been hindered by the lack of a reference voltage generated on-chip. This paper will present a reference voltage source implemented in CMOS technology. The circuit takes advantage of MOS devices operating in the weak inversion (subthreshold) region in two respects: (a)—the exponential nature of the I-V characteristic is used to make possible a predictable behavior over temperature, and (b)—the power consumption is made minimal due to the inherently low currents in that region.

The operation of an MOS transistor in weak inversion is described below;

$$I = \left(\frac{Z}{L}\right) \mu C_o \left(\frac{1}{m}\right)\left(\frac{nKT}{q}\right)^2 \exp\left[\frac{q}{nKT}\left(V_{GS}-V_T-\frac{nKT}{q}\right)\right] \left\{1-\exp\left[\frac{-mqV_{DS}}{nKT}\right]\right\} \qquad (1)$$

where I is the drain current, V_{GS} is the gate-to-source voltage, V_{DS} is the drain-to-source voltage, Z and L are the gate width and length, respectively, C_o is the oxide capacitance per unit area, μ is the effective mobility of carriers in the channel, V_T is the threshold voltage, K is the Boltzman constant, q is the electron charge, m and n are process-dependent parameters[1], and T is the absolute temperature.

It has previously been suggested[2] that it might be possible to somehow take advantage of the exponential nature of this equation to implement a reference voltage source. The idea was abandoned because most parameters in this equation are unreliable, notably C_o, μ, V_T, and the absolute value of Z/L (as opposed to ratios of (Z/L)s). Below is described an approach in which a voltage independent of these parameters is developed.

Consider two devices, M_1 and M_2, each with gate connected to drain and let the resulting two-terminal elements have voltages and currents V_1, I_1 and V_2, I_2, respectively. If $V_{DS} >> (mKT/nq)$, the last exponential in equation (*1*) can be neglected. Applying that equation for each device, and solving for $V_1 - V_2$, one obtains:

$$V_1 - V_2 = AT \qquad (2)$$

where:

$$A = \left(\frac{nK}{q}\right) \ln\left[\frac{I_1(Z/L)_2}{I_2(Z/L)_1}\right] \qquad (3)$$

It will be noted that the undesirable terms in equation (*1*) have been cancelled, and additionally, (V_1-V_2) is proportional to T. The constant of proportionality (equation (*3*)) depends on the physical constants K and q, *ratios* of currents and (Z/L)s, and n, which has proven to be a reliable parameter.

Voltage (V_1-V_2), with its positive temperature coefficient, can now be added to the voltage V_D of a forward-biased diode, the latter having a negative temperature coefficient, in such a way that the sum exhibits temperature independence. The circuit is, in principle, as shown in Figure 1, with the output voltage V_o, as per equation (*4*):

$$V_o = -(V_D + V_1 - V_2) \qquad (4)$$

In this circuit, I_1 and I_2 are chosen so that M_1 and M_2 operate in weak inversion, and are much less than I_B. The behavior of V_D with temperature is well documented. When higher order effects are included, (V_1-V_2) will not exhibit a completely linear behavior with respect to T, but A in equation (*3*) can still be chosen so that $dV_o/dT = 0$ at a desired temperature. The ratios I_1/I_2 and $(Z/L)_2/(Z/L)_1$ required in that equation, however, to yield the desired values of A, are exceedingly high. To avoid this problem, one can replace each one of devices M_1 and M_2 by strings of devices connected in series, so that a corresponding *multiple* of (V_1-V_2) is added to V_D, in which case the required ratios are easily realizable.

The foregoing concepts have been implemented as shown in Figure 2. The diode is realized by using a diode-connected bipolar transistor, the latter being formed by the N-substrate (collector), the P-tub (base) and the N^+ diffusion (emitter). The string $M_{1A}-M_{1B}-M_{1C}$ corresponds to M_1 of Figure 1, and $M_{2A}-M_{2B}-M_{2C}$ to M_2. The main biasing string of the circuit is R_1 and M_3. The bipolar transistor Q is biased through current source M_5. A fraction of V_{GS5} is developed across M_4 through the series resistor R_2, the latter being selected so that V_{GS4} guarantees operation in the subthreshold region. Current mirror M_6 is used to develop I_1+I_2 (see Figure 1) and M_7 is used to develop I_2, which is mirrored by M_8 and M_9 into string $M_{2A}-M_{2B}-M_{2C}$. Self-biasing and output buffering can be implemented in several ways.

A microphotograph of the chip is shown in Figure 3. The regular structure on the left is a binary weighted transistor array which has been used at the initial experimental stage to implement any desired value of I_1/I_2.

The resulting output voltage magnitude vs. temperature is shown in Figure 4. The upward trend of the curve at both sides

[1] Swanson, R.M. and Meindl, J.D., "Ion-implanted Complementary MOS Transistors in Low-Voltage Circuits", *IEEE J. Solid-State Circuits*, SC-7, 2, p. 146-153; April, 1972.

[2] Gray, P.R.; private communication.

Reprinted from *1978 IEEE Int. Solid-State Circuits Conf. Dig. of Tech. Papers,* Feb. 15-17, 1978, pp. 48-49.

of the first zero temperature coefficient point is due to higher order effects and the temperature dependence of the diode bias current. The second zero temperature coefficient point and the subsequent downward trend of the curve at high temperatures is due to the increased reverse current of the MOS device junctions at these temperatures. The output voltage exhibits an effective temperature coefficient of 90 ppm/°C from –55°C to 105°C. Coefficients of 45 ppm/°C were measured over a 100°C range. The total current drain is 16μA.

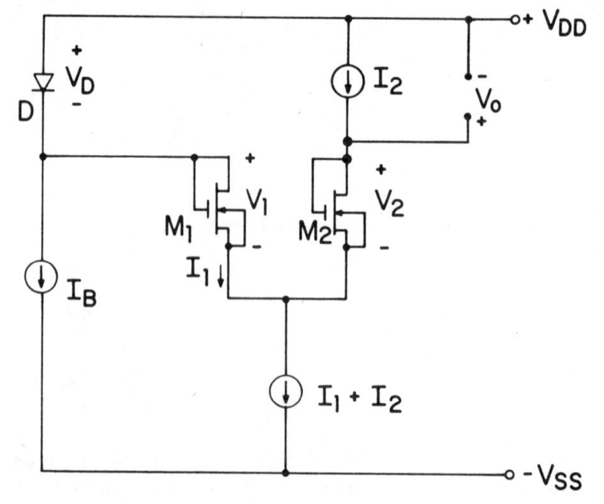

FIGURE 1—Simplified schematic illustrating the principle of operation.

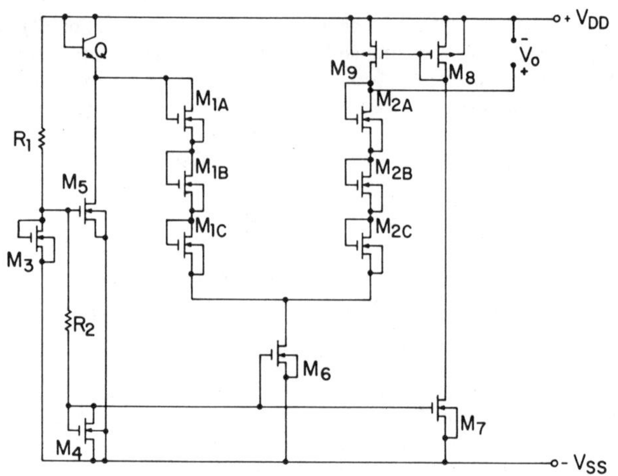

FIGURE 2—Circuit schematic.

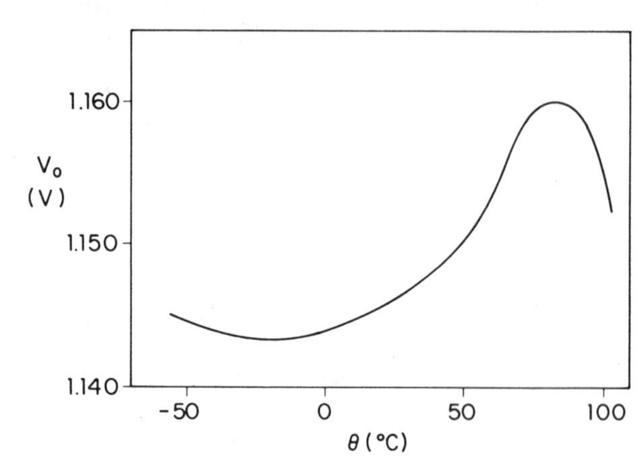

FIGURE 4—Output voltage vs temperature.

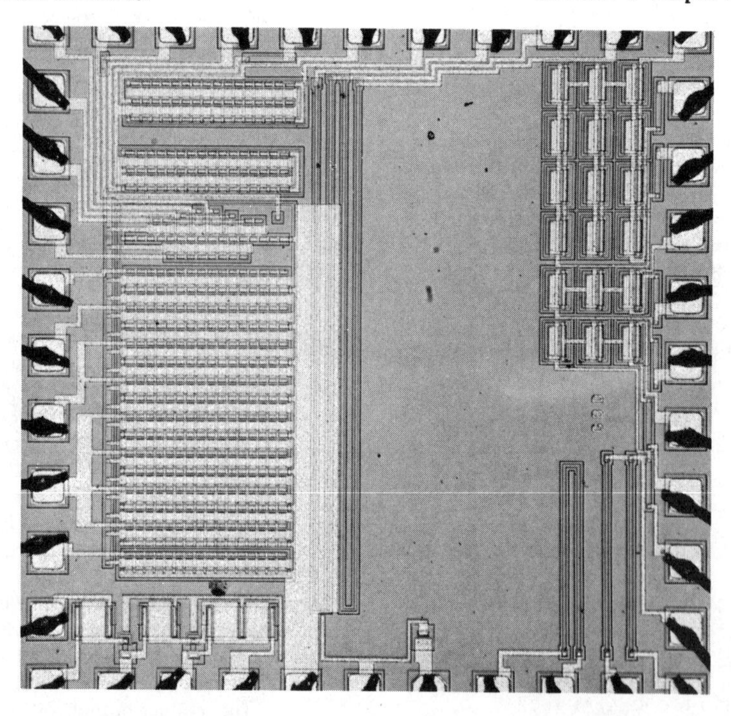

FIGURE 3—Test chip microphotograph.

An NMOS Voltage Reference

Robert A. Blauschild, Patrick Tucci, Richard S. Muller and Robert G. Meyer

Signetics Corp.

Sunnyvale, CA

A MOS temperature-stable voltage reference, using standard N-channel enhancement and depletion MOSFETs, that is process-compatible with dense NMOS logic, will be described.

With a >60dB supply rejection and a <6 ppm/°C (measured) temperature coefficient of output voltage, the reference can also be used as a discrete component in hybrid designs requiring moderate output currents (up to 50mA), or (with an external bypass transistor) for high-power operation.

The reference voltage is generated from the difference between the gate-source voltages of two N-channel MOSFETs. The transistors have had their thresholds adjusted by ion implantation so that one is an enhancement device and the other is a depletion device. Both MOSFETs operate under current-saturated conditions. The basic design principles of the circuit can be understood by considering the simplest formulation for the current-voltage behavior of the transistors under current-saturated conditions, where k is a prefactor which adjusts the charge-control equation (*1*) to fit an exact distributed analysis[1] and the other parameters have the usual meanings for MOS transistor analysis.

$$I_D = \frac{k\mu\, C_o\, W}{L}\, (V_{GS} - V_T)^2 \qquad (1)$$

If equation (*1*) is solved for the basic reference quantity V_{GS}, we can derive the temperature variation as:

$$\frac{d\, V_{GS}}{d\, T} = \frac{d\, V_T}{d\, T} + \sqrt{\frac{L}{kWC_o}}\, \frac{d}{dT}\left(\sqrt{\frac{I}{\mu}}\right) \qquad (2)$$

The temperature dependence of the terms in equation (*2*) can be formulated separately to derive an overall expression for the theoretical temperature sensitivity of the reference. This sensitivity is given by the difference in the variations predicted by equation (*2*), when that equation is applied first to the depletion and then to the enhancement MOSFET.

There are three terms in equation (*2*) that show appreciable sensitivity to temperature: the threshold voltage V_T, the device current I, and the channel mobility μ. The reference circuit is designed to maintain equality between the currents in the reference MOSFETs, so that the current variation does not lead to variation in the reference voltage. The temperature dependence of the mobility in both the depletion and enhancement-mode devices is proportional to $T^{-3/2}$ in the first order; thus, the sensitivity of the reference to this variation can be made negligible. The major variation in the basic reference is therefore caused by the dependence of the threshold voltage on temperature.

Analysis has shown that the variation in V_T over the useful temperature range can be expressed by a linear equation:

$$V_T = V_{TO} - \alpha T \qquad (3)$$

where V_{TO} and α are different for enhancement (V_{TOE}, α_E) and depletion (V_{TOD}, α_D) devices. The values of V_{TO} and α depend most strongly on the substrate dopant density and on the implant dosages used during processing. Both reference MOSFETs are in close proximity and go through the same process steps until the final implant. The only difference between them that affects V_{TO} and α, therefore, is the final implant which dopes the surface N-type in the channel of the depletion MOSFET. The stability and reproducibility of the reference thus depends on this well-controlled processing step.

A circuit realization for the basic reference is shown in Figure 1. The output voltage is seen to be the difference between V_{GSE} and V_{GSD}. By using an opamp in a negative feedback configuration, the output voltage is forced to a level that causes the enhancement and depletion devices (Q1 and Q2, respectively) to operate at equal V_{DS} and I_D values.

Although load resistors are shown in Figure 1, it is preferable to supply the current from matched depletion-mode devices having $V_{GS} = 0$. In this case, application of equations (*1*), (*2*), and (*3*) leads to a predicted temperature variation for the reference of

$$\frac{dV_{REF}}{d\, T} = -\alpha_E + \alpha_D \left[\sqrt{\frac{L_E}{W_E}\frac{W_L}{L_L}\frac{\mu_D}{\mu_E}} - \sqrt{\frac{L_D}{W_D}\frac{W_L}{L_L}} - 1\right] \qquad (4)$$

where the subscripts E and D denote the enhancement and depletion reference MOSFETs, the subscript L refers to the current source devices and μ_D/μ_E is the mobility ratio. The geometric ratios in equation (*4*) can be selected to compensate for unequal α values, although an exact cancellation cannot be achieved in practice because the mobilities do not track exactly over the temperature range. Computer calculations with SLIC predict a temperature coefficient for the reference voltage less than 2 ppm/°C for appropriately mismatched MOSFETs. The final reference circuit includes provisions for on-chip trimming to achieve this performance.

The complete voltage reference is shown in Figure 2. Transistors Q1 and Q2 are the matched differential enhancement-depletion pair, loaded by current sources Q3 and Q4. When

---

[1] Muller, R.S. and Kamins, T.I., "Device Electronics for Integrated Circuits", *John Wiley Inc.*, p. 395; 1977.

[2] Wang, P.P., "Double Boron Implant Short Channel MOSFET", *IEEE Trans. Electr. Dev.*, ED-24; March, 1977.

[3] Huang, J.S.T. and Taylor, G.W., "Modeling of An Ion-Implanted Silicon Gate Depletion Mode IGFET", *IEEE Trans. Electr. Dev.*, ED-22; Nov. 11, 1975.

Reprinted from *1978 IEEE Internat. Solid-State Circuits Conf. Dig. of Tech. Papers*, Feb. 15-17, 1978, pp. 50-51.

followed by a second gain stage (Q5–Q8), and an output source follower (Q9–Q10), loop gain is large enough (40dB) to yield proper load regulation. The equality of I_D and V_{DS} in Q1 and Q2 is controlled by matching depletion devices Q3 and Q4 and by the symmetry of the second gain stage.

Balance of the second gain stage is improved by use of Q7, which is also used to obtain a balanced to single ended conversion. If equal currents flow through Q7 and Q8 (which have equal aspect ratios), then V_{GS_7} must equal zero, since the gate and source of Q8 are shorted. Thus Q5 and Q6 always have equal V_{DS}, regardless of the value of V_{GS_9}, which increases with output current. Equal current through Q7 and Q8 is provided by designing Q16 to provide a current equal to $2\ I_{D_3} + 2\ I_{D_8}$.

Common-mode bias loops (including transistors Q11–Q15) provide feedback to insure that all depletion current sources operate in the saturated region. Excellent supply rejection is also achieved, since all voltages and currents are dependent only on current source Q16 and individual device geometries.

In the die layout (Figure 3), it can be seen that nearly half of the 10 x 12 mil active area contains the large output buffer device.

Acknowledgment

The authors would like to acknowledge the help of H. Sigg and S. Lai for device processing, and J. Mendoza for breadboard measurements.

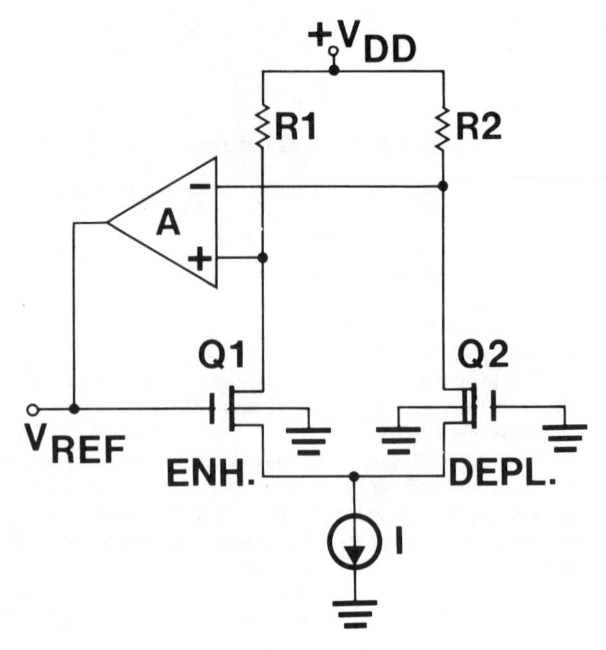

FIGURE 1–Basic reference.

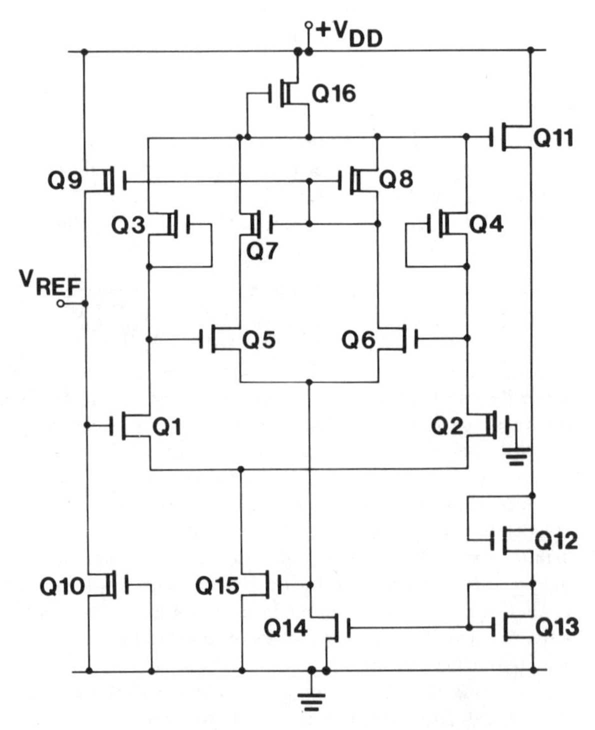

FIGURE 2–Complete NMOS reference circuit.

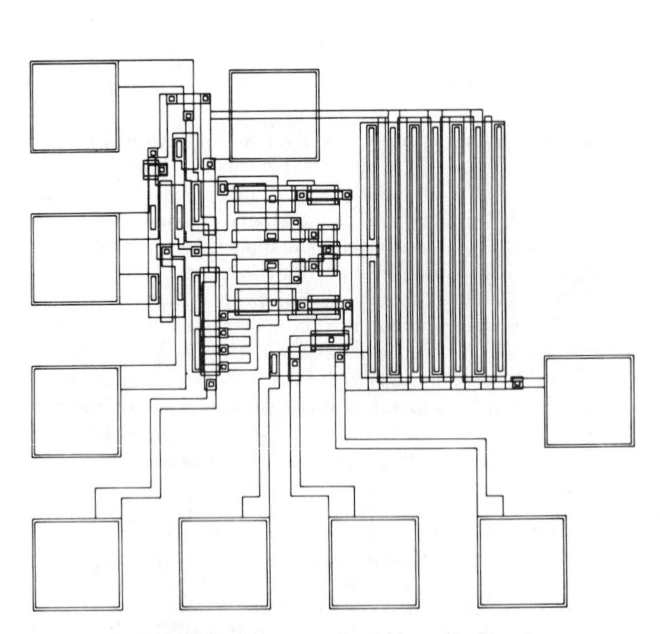

FIGURE 3–Layout of reference circuit.

Part IV
Wideband Amplifiers

In the design of broad-band high frequency amplifiers, the inherent limitations of monolithic devices impose severe restrictions on possible design approaches. The most significant design restriction is the lack of integrated inductors which make conventional broad-banding techniques such as shunt-peaking, unsuitable for monolithic design. In monolithic amplifiers, AC coupling between the stages is generally not possible; and complementary devices with comparable frequency capability are not readily available. Therefore, the AC design of the circuit can not be treated as a separate problem from DC bias considerations.

The papers collected in this part demonstrate various successful design approaches and techniques for wideband IC amplifiers. The first paper by Solomon and Wilson, gives a detailed analysis of a wideband amplifier configuration using the "series-series feedback triple" configuration. This paper also demonstrates the "pole-splitting" technique which is often used in IC design for broadbanding feedback amplifiers. In the second paper by Gilbert, a new wideband amplifier technique is demonstrated using current, rather than voltage, amplification. The third paper by Mataya *et al.* illustrates a novel design technique to cancel the effects of parasitic collector-base capacitance in monolithic transistors.

Electronic gain control is one of the desirable features in most wideband amplifiers. In the fourth paper, by Davis and Solomon, design of such a controlled-gain wideband amplifier is described. The subsequent paper by Wooley describes an approach to computer-aided analysis and optimization of DC-coupled monolithic wideband amplifiers. The last two papers of this section deal with distortion in wideband IC amplifiers. In the sixth paper by Sansen and Meyer, the sources of distortion in monolithic variable-gain amplifiers is analyzed. The final paper by Chan and Meyer describe an optimized broad-band amplifier design which provides 20-dB gain and 250-MHz bandwidth using 800-MHz f_T linear IC transistors.

ADDITIONAL REFERENCES

1. J.B. Couglin, R.J. Gelsing, P.J. Jochems and H.J.M. van der Laak, "A Monolithic Silicon Wide-band Amplifier from DC to 1 GHz," *J. Sol. State Circuits,* vol. SC-8, pp. 414-419, December 1973.
2. H.E. Abraham and R.G. Meyer, "Transistor Design for Low Distortion at High Frequencies," *IEEE Trans. Electron Devices,* vol. ED-22, pp. 1290-1297, December 1976.
3. C.R. Battjes, "A Wideband High-Voltage Monolithic Amplifier," *J. Sol. State Circuits,* vol. SC-8, pp. 408-413, December 1973.
4. K.E. Kujik, "A Fast Integrated Comparator," *J. Sol. State Circuits,* vol. SC-8, pp. 458-462, December 1973.
5. W.J. McCalla, "An Integrated IF Amplifier", *J. Sol. State Circuits,* vol. SC-8, pp. 440-447, December 1973
6. R.G. Meyer, R. Eschenbach, and W.M. Edgerley, "A Wideband Feedforward Amplifier," *J. Sol. State Circuits,* vol. SC-9, pp. 422-428, December 1974.
7. J. Choma, Jr., "A Model for the Computer-Aided Noise Analysis of Broad-Banded Bipolar Circuits," *J. Sol. State Circuits* vol. SC-9, pp. 429-435, December 1974.
8. R.J. van de Plassche, "A Wideband Monolithic Instrumentation Amplifier," *J. Sol. State Circuits*, vol. SC-10, pp. 424-431, December 1975.
9. R.I. Ollins and R.J. Ratner, "Computer-Aided Design and Optimization of a Broad-Band High Frequency Monolithic Amplifier," *J. Sol. State Circuits* vol. SC-7, pp. 487-492, December 1972.
10. W.M.C. Sansen and R.G. Meyer, "Distortion in Bipolar Transistor Variable-Gain Amplifiers," *J. Sol. State Circuits,* vol. SC-8, pp. 275-282, August 1973.
11. W.M.C. Sansen and R.G. Meyer, "An Integrated Wideband Variable-Gain Amplifier with Maximum Dynamic Range," *J. Sol. State Circuits,* vol. SC-9, pp. 159-166, August 1974.

A Highly Desensitized, Wide-Band Monolithic Amplifier

J. E. SOLOMON, MEMBER, IEEE, AND G. R. WILSON, MEMBER, IEEE

Abstract—**An all-diffused monolithic feedback triple with a maximum bandwidth of 50 MHz, 40 dB of feedback, and gain adjustable from 34 dB to 52 dB is described. Harmonic distortion is less than 0.15 percent at 0.5 MHz and gain variation is less than ±0.25 dB from −55°C to +125°C. Detailed analysis of the amplifier performance is carried out and an accurate design technique based upon transistor models is developed. A distributed base model is introduced which results in prediction of open loop gain and phase to within 10 percent of measured values at all frequencies below 200 Mc/s.**

I. Introduction

WIDE-BAND feedback amplifiers can be effectively fabricated using local feedback only [1], or by employing multistage amplifiers with overall feedback [2], [3]. The first method leads to amplifiers with large bandwidth but relatively poor desensitivity to device imperfections, while the second results in improved desensitivity at the expense of bandwidth and possible instability. In this paper an overall feedback amplifier is described, which achieves high desensitivity, reasonably wide bandwidth, and excellent stability.

Figure 1 shows the basic feedback pairs (a) and (b) and triples (c) and (d) which are most useful for obtaining large bandwidth and desensitivity. Amplifiers with more than three common emitter stages are not considered because their high-frequency phase shift is normally too great to permit stable wideband operation.

For high desensitivity, the feedback triples are more effective than the pairs because they have greater available loop gain. Of the two triples, the series-series triple of Fig. 1(d) exhibits better performance with regard to stability for the following reasons: 1) the shunt-shunt triple has at least one more important pole in the feedback loop, and 2) capacitive loading at either the input or the output degrades the stability of the shunt-shunt triple, but improves, or has no effect on, stability of the series-series triple. For these reasons, and because the configuration of Fig. 1(d) direct couples easily, the series-series triple is well suited for fabrication in integrated form.

In the discussion which follows, a practical configuration for a monolithic series-series feedback triple will be described. A large portion of the paper is devoted to analysis of performance and development of an accurate design technique. Care is taken to develop amplifier theory entirely from models which can be related to the physical structure of the integrated circuit. The advantage of this is that effects of either circuit or process variation on amplifier performance can be evaluated. Good agreement between theory and experiment is obtained by employing a distributed base model for the transistor which accurately accounts for high-frequency phase shift through the amplifier.

Manuscript received March 14, 1966.

J. E. Solomon is with the Semiconductor Division, Motorola Inc., Phoenix, Ariz.
G. R. Wilson is with Tektronix, Inc., Beaverton, Ore. He was formerly with the Semiconductor Division, Motorola Inc.

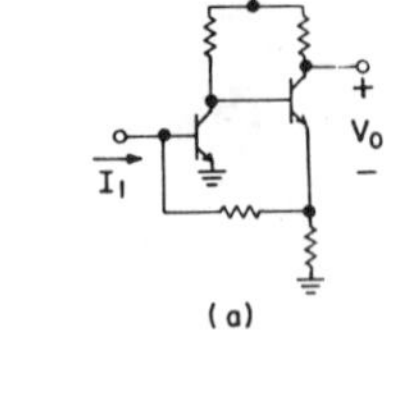

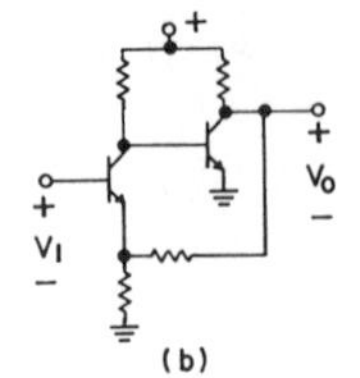

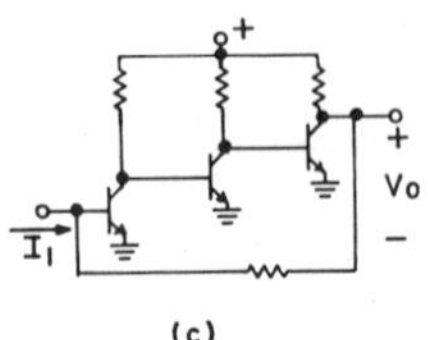

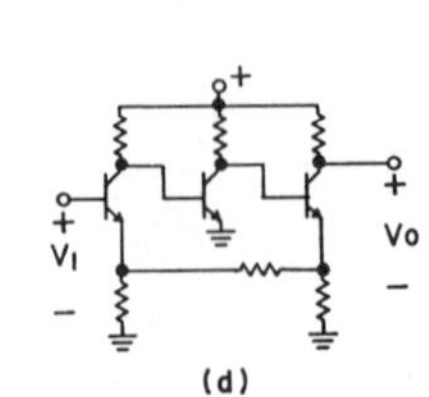

Fig. 1. Basic multistage feedback amplifiers. (a) Shunt-series pair. (b) Series-shunt pair. (c) Shunt-shunt triple. (d) Series-series triple.

II. Practical Feedback Triple

A practical configuration for the integrated series-series feedback triple is shown in Fig. 2. Major advantages of this configuration are that it can be direct coupled without level shifting and gain can be adjusted (by adjusting R_f) without upsetting the dc operating point. Low output impedance and large output swing are provided by an emitter follower at the output of the basic triple. The follower is biased with a current source rather than a resistor to improve the peak negative swing of the amplifier when capacitively coupled to a load. This current source is a multiple (n) emitter transistor biased by a diode-connected transistor with a single emitter as shown in Fig. 2. The junction currents are proportional to the emitter areas, so the bias transistor is supplied only $1/n$ times the current in the current source.

Stabilization of the output quiescent point is obtained by using an overall shunt-shunt dc feedback loop from the output emitter follower to the base of the input transistor. This loop is decoupled for ac signals by either driving the amplifier from a low impedance source or by using an external capacitor (C_B in Fig. 2). Changes in output quiescent point due to temperature variation are nearly eliminated by incorporating a temperature compensating current source in the dc loop. Operation of this loop can be understood by referring to Fig. 3 which shows that portion of the circuit which determines dc behavior.

Because the dc loop gain is high, the output dc level

Reprinted from *IEEE J. Solid-State Circuits,* vol. SC-1, pp. 19-28, Sept. 1966.

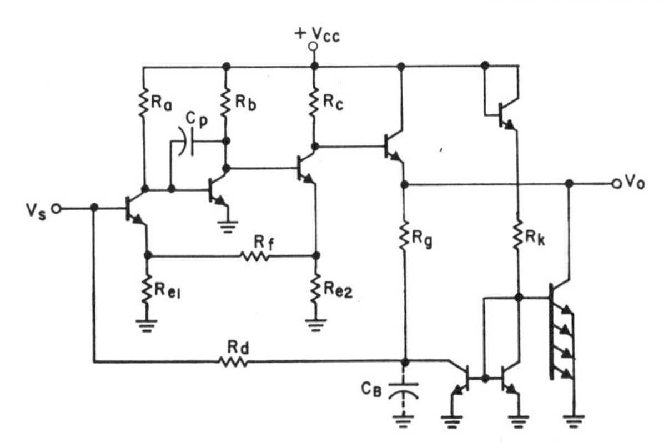

Fig. 2. Series-series feedback triple including bias circuitry and output emitter follower.

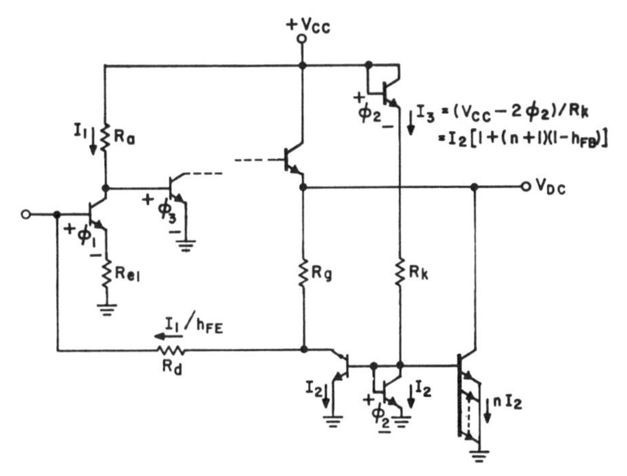

Fig. 3. Schematic for determining output quiescent point.

can be computed by summing voltage drops from the input emitter through the dc feedback network to the output

$$V_{DC} = I_1 R_{e1} + \phi_1 + I_1 \frac{R_d}{h_{FE}} + R_g\left(I_2 + \frac{I_1}{h_{FE}}\right). \quad (1)$$

Under conditions that $\phi_1 = \phi_2 = \phi_3 = \phi$ and $h_{FE} \gg (n+1)$, V_{DC} may be written as

$$V_{DC} \cong V_{CC}\left(\frac{R_g}{R_k} + \frac{R_{e1}}{R_a}\right) + \phi\left(1 - \frac{2R_g}{R_k} - \frac{R_{e1}}{R_a}\right)$$

$$+ \frac{(V_{CC} - \phi)}{h_{FE}}\left[\frac{R_g + R_d}{R_a} - \frac{(n+1)(V_{CC} - 2\phi)R_g}{(V_{CC} - \phi)R_k}\right] \quad (2)$$

where dependence of V_{DC} on ϕ and h_{FE} is shown explicitly. By properly choosing bias resistors, R_k, R_g, R_d, R_{e1}, and R_a, the second and third terms in (2) can be made approximately zero, thereby eliminating the dependence of V_{DC} on ϕ and h_{FE}. With $R_k = 6\text{k}\Omega$, $R_g = 3\text{k}\Omega$, $R_d = 12\ \text{k}\Omega$, $R_{e1} = 100\ \Omega$, $R_a = 9\ \text{k}\Omega$, and $n = 3$, (2) reduces to

$$V_{DC} \cong V_{CC}/2. \quad (3)$$

The remaining bias resistors are $R_b = 5\ \text{k}\Omega$, $R_c = 600\ \Omega$, and $R_{e2} = 100\ \Omega$ (R_f is determined by the closed loop gain desired).

The measured drift in output operating point (V_{DC}) for the monolithic amplifier was less than ±50 mV from −55°C to +125°C. By comparison, V_{DC} for the same amplifier employing no temperature compensation, i.e., purely resistive feedback, would drift about ±0.9 volt. Drift of this magnitude would reduce the useable output swing by about a factor of two.

III. Amplifier Theory

The amplifier theory is developed in several steps as follows. A general analysis of series-series connected two-ports is carried out first. This illustrates the separation of the feedback amplifier into forward and feedback networks and gives conditions under which the separation is valid. The feedback circuit resulting from this split consists only of a passive network which is easily analyzed. Evaluation of forward behavior requires greater effort since performance of the active portion of the circuit must be accurately determined at high frequencies.

The forward transfer function is first derived approximately using a simple transistor model. The gain magnitude and the most significant poles are accurately predicted from this analysis. A high-frequency analysis is then carried out using a more complex transistor model. This analysis accurately predicts phase shift through the amplifier at frequencies where the simple model becomes inaccurate. Using results from both analyses, a group of "equivalent" secondary poles is generated such that the gain and phase of the loop transmission function are accurately determined at all important frequencies. At this point, root locus techniques are employed to evaluate stability and optimize the design.

A. *Identification of Feedback Form* [3]

It is assumed initially that the amplifier can be modeled by series-series connected two-ports as shown in Fig. 4. An RC source impedance is included to consider the effects of source capacitance on stability.

The transadmittance for the complete amplifier in Fig. 4 is

$$A_y = \frac{I_2}{V_s} = \frac{-z_{21}^t Z_s/R_s(Z_c + z_{22}^t)(Z_s + z_{11}^t)}{1 - z_{21}^t z_{12}^t/(Z_c + z_{22}^t)(Z_s + z_{11}^t)} \quad (4)$$

where

$$z_{ij}^t = z_{ij}^a + z_{ij}^f$$

$$Z_s = R_s/(1 + sR_sC_s).$$

The superscript a pertains to the impedance parameters of the basic amplifier while the superscript f pertains to those of the feedback network. We now identify A_y with the ideal feedback equation

$$A_y = \frac{a_y}{1 + a_y f_z}. \quad (5)$$

A forward network (a_y) and feedback network (f_z) can be identified if:

1) the basic amplifier is approximately unilateral, i.e., $z_{12}^a \cong 0$, and
2) forward transmission through the feedback network is small compared with that through the basic amplifier, i.e., $z_{21}^a \gg z_{21}^f$.

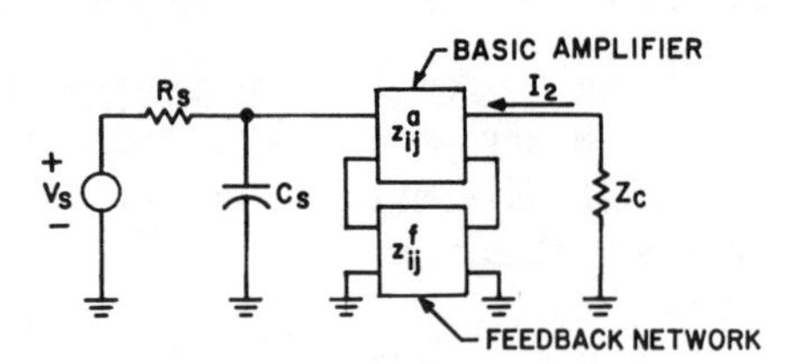

Fig. 4. Representation of amplifier as series-series connected two-ports.

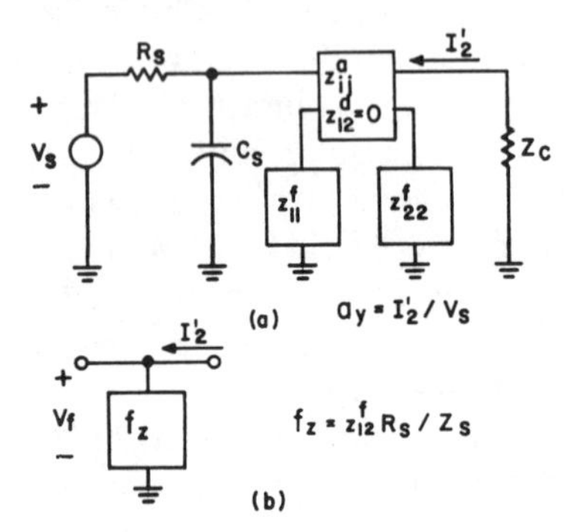

Fig. 5. Configuration of Fig. 4 split into (a) forward network and (b) feedback network.

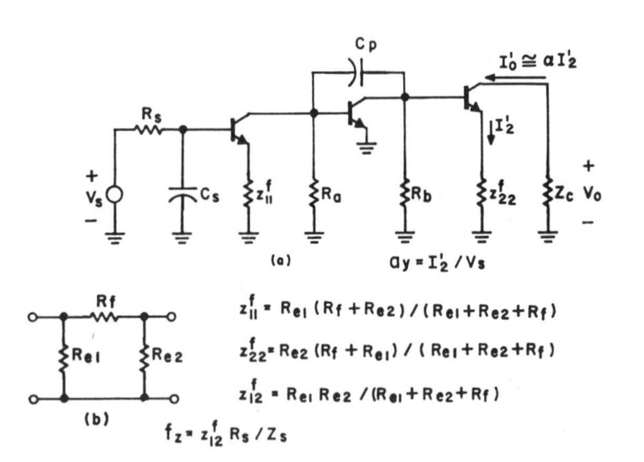

Fig. 6. Feedback triple of Fig. 2 split into (a) forward amplifier and (b) feedback network, for purposes of analysis. (The output emitter follower and bias circuitry are not shown).

With these assumptions

$$a_y = \frac{I_2'}{V_s}\bigg|_{\substack{z_{12}^a=0 \\ z_{12}^f=0,\ z_{21}^f=0}} = -z_{21}^a Z_s/R_s(Z_c + z_{22}^f)(Z_s + z_{11}^f) \tag{6}$$

$$f_z = \frac{V_f}{I_2'}\bigg|_{z_{11}^a=\infty} = z_{12}^f R_s/Z_s. \tag{7}$$

Forward and feedback networks defined by (6) and (7) are shown in Fig. 5.

By analogy with Fig. 5, the feedback triple can be split into the forward and feedback networks shown in Fig. 6. In reality, the basic amplifier is not properly represented by a two-port because the center stage emitter and bias resistors R_a and R_b return to an ac ground. A more detailed analysis shows that the split in Fig. 6 is indeed correct and that the assumptions required to make the feedback identification are not changed in any important way from those given.

B. Midband Gain

From Fig. 6 and (5), the closed loop voltage gain of the amplifier is

$$A_v = \frac{V_0}{V_s} \cong -\frac{\alpha a_y Z_c}{1 + a_y f_z}. \tag{8}$$

At low frequencies where the loop transmission, $a_y(0)f_z(0)$, is large compared with unity, the voltage gain becomes

$$A_v(0) \cong -\alpha_0 R_c/f_z(0) = -\alpha_0 R_c/z_{12}^f(0) \tag{9}$$

$$= \frac{-\alpha_0 R_c(R_{e1} + R_{e2} + R_f)}{R_{e1}R_{e2}}.$$

Equation (9) reflects an important benefit of the negative feedback; the amplifier gain is essentially dependent only upon ratios of resistors. These ratios are well controlled in circuit fabrication and remain nearly constant with temperature (although absolute values change considerably).

C. Effects of Large Source Resistance and Capacitance

Normally the amplifier is operated with small source resistance (R_s) and capacitance (C_s) to maintain maximum bandwidth. Under this condition the effects of C_s can be ignored, $Z_s \cong R_s$, and the feedback function f_z is equal to the open circuit transfer impedance of the feedback network, z_{12}^f [see Fig. 6(b)].

In some applications it is desirable to operate the amplifier with R_s large. In this case, effects of C_s become important and must be included in the analysis. From Fig. 6(b), the feedback function including C_s is

$$f_z = z_{12}^f R_s/Z_s = \frac{R_{e1}R_{e2}(1 + s/z_s)}{R_{e1} + R_{e2} + R_f} \tag{10}$$

where $z_s = 1/R_sC_s$.

The source capacitance is seen to cause a transmission zero z_s to appear in the feedback function. It will be shown later that this zero approximately cancels with the input pole of the forward amplifier under conditions that C_s is larger than the input capacitance of the forward amplifier. [See (19).] The result of the pole-zero cancellation is that detrimental effects of the amplifier input pole on stability can be eliminated for any value of R_s by appropriately padding source capacitance C_s.

D. Compensation with Feedback Zero

The maximum loop gain that can be achieved without causing a peak in the closed loop frequency response is determined largely by phase lag through the amplifier at high frequencies. This lag can be reduced by using a small compensation capacitor C_f in parallel with the feedback resistor R_f in Fig. 6(b). With this capacitor included (and $C_s = 0$) the feedback function becomes

$$f_z(s) = \frac{f_z(0)(1 + sR_fC_f)}{1 + sR_fC_f(R_{e1} + R_{e2})/(R_{e1} + R_{e2} + R_f)} \tag{11}$$

where

$$f_z(0) = R_{e1}R_{e2}/(R_{e1} + R_{e2} + R_f). \tag{12}$$

The capacitor C_f is seen to generate a feedback transmission zero at $-1/R_fC_f$.

E. Forward Transfer Computed from Simple Model

The complete forward transfer function for the three stage amplifier in Fig. 6(a) will now be derived using a simple model for the integrated transistor as shown in Fig. 7(a). This model ignores effects of lateral current flow in the transistor base and for this reason can be quite inaccurate at high frequencies. It does, however, predict with reasonable accuracy the gain magnitude and the most significant poles of the forward transfer function. In Section III-F, we again compute this forward transfer using a model which accurately accounts for transverse effects in the base.

The detailed development of the forward amplifier response (using the simple model) is given in the Appendix. From (52), the forward transadmittance of the amplifier is

$$a_y(s) = \frac{I_2'}{V_s} = \frac{a_y(0)(1 - s/z_1)}{(1 + s/p_s)(1 + s/p_1)(1 + s/p_2)} \tag{13}$$

where

$$a_y(0) = R_1R_2/K_sr_{e2}(r_{e1} + r_{11}^f)(r_{e3} + r_{22}^f) \tag{14}$$

$$p_1 \cong K_1(C_{c2} + C_p)/r_{e2}C_1C_2K_2 \tag{15}$$

$$p_2 \cong r_{e2}/R_1R_2K_1(C_{c2} + C_p) \tag{16}$$

$$p_s = K_s/R_s\left[C_s + C_{c1} + \frac{1}{\omega_t(r_{e1} + r_{11}^f)} + \frac{r_{e2}C_{c1}C_2}{(r_{e1} + r_{11}^f)(C_p + C_{c2})}\right]. \tag{17}$$

Constants K_1, K_2, and K_s, which are defined in (41) and (51), are real and have magnitudes in the vicinity of unity under usual circuit conditions. Other symbols are defined in Figs. 6 and 15, (39), (45), and below.

r_e = kT/qI_e

C_{ci}, C_{si} = collector-base and collector-substrate transition capacitances, respectively, for the ith transistor

β_0 = low-frequency common emitter short circuit current gain

ω_t = $|\beta(\omega)|$ ω, where ω is radian frequency at which $|\beta|$ has a 6 dB/octave slope.

C_{Ra}, C_{Rb}, C_{Rc} = junction capacitance associated with diffused resistors R_a, R_b, and R_c.

One pole p_s of (13) is associated with the input circuit of the amplifier and, with the usually small source resistance R_s and capacitance C_s, it is sufficiently broadbanded that it has minor effect on amplifier response. If R_s is large enough that p_s becomes important, sufficient C_s can

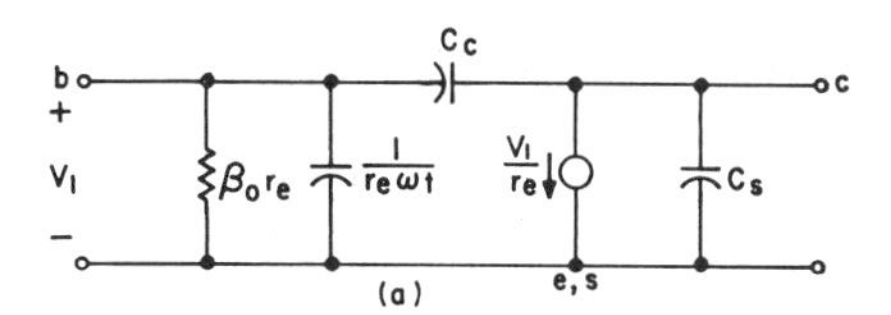

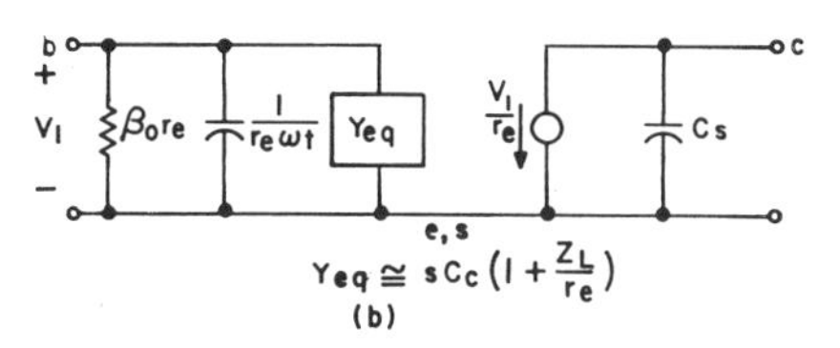

Fig. 7. (a) Simple model for integrated transistor. (b) Approximate unilateral equivalent to (a). (The transistor load impedance Z_L includes C_s.)

be padded so that[1]

$$C_s + C_{c1} > \frac{1}{\omega_t(r_{e1} + r_{11}^f)} + \frac{r_{e2}C_{c1}C_2}{(r_{e1} + r_{11}^f)(C_p + C_{c2})} \tag{18}$$

Then

$$p_s \cong K_s/R_s(C_s + C_{c1}) \cong z_s \tag{19}$$

where z_s was defined in (10) and R_s is not so large that K_s is significantly greater than unity. When (18) is satisfied, the pole-zero cancellation discussed in Section III-C occurs and the effects of source impedance on loop transmission, $a(s)f(s)$, can be ignored.

Pole Splitting

p_1 and p_2 are normally the most significant poles in the forward transfer function so they have a dominant effect on amplifier performance. Equations (15), (16), and (41) show that the compensation capacitor C_p can significantly alter the positions of both p_1 and p_2. More specifically,

for $(C_{c2} + C_p) < C_1, C_2$:

$$p_1 \propto (C_{c2} + C_p)$$
$$p_2 \propto 1/(C_{c2} + C_p). \tag{20}$$

The addition of C_p causes an interaction between the internal poles of the amplifier which, from (20), results in simultaneous broadbanding of p_1 and narrowbanding of p_2. It will be seen in Section III-H that this "pole splitting" is exactly the reverse of the "pole merging" which occurs when negative feedback is applied around the amplifier. Use of C_p, therefore, allows more negative feedback and greater desensitivity to be achieved than is possible without this compensation.

As with most compensation techniques of this sort, amplifier gain-bandwidth product is traded for the improved desensitivity. The trade is considerably more efficient with pole splitting, however, than with other techniques. For example, the conventional approach of

[1] C_{c1} is included with C_s because it essentially appears from the input base to ground and so cannot be distinguished from C_s. For this reason (19), with $K_s = 1$, defines Z_s more accurately than (10).

padding capacitance to ground at one of the interstages can result in a gain-bandwidth loss by as much as an order of magnitude when compared with C_p compensation.

F. Forward Transfer Computed From Distributed Model

A comparison of the measured forward amplifier response with that derived in Section III-E shows that the simple theory has considerable phase error at high frequencies. Most of the phase error can be attributed to transverse effects in the base and, to a lesser extent, feedforward in the collector-base capacitance.

In the planar transistor, the lateral sheet resistance in the active base region can be quite high (5 to 10 kΩ/□) due to the thin base width. As a result, significant ac voltage drops (ac crowding) may occur between the base contact and the center of the active base at high frequencies. A small signal equivalent circuit which accounts for the effects of lateral base current flow is shown in Fig. 8. The distributed r-y network represents the active base region under the emitter while r_b represents the spreading resistance between the base contact and the active base region. In the usual planar structure, most of the collector-base junction lies outside of the active base region, so the transition capacitance C_c can be lumped between the collector terminal and the active base input with little loss of accuracy.

The incremental current gain β is unchanged by ac crowding because it is not dependent upon the detailed charge distribution through the base. Therefore [4]

$$\beta = \frac{\beta_0 e^{-jm\omega/\omega_\alpha}}{1 + s\beta_0/\omega_t}, \qquad (21)$$

where $m\omega/\omega_\alpha$ represents the excess phase in β over that of a single pole approximation. For a typical planar diffused transistor $m \cong 1$ rad.

The distributed network in Fig. 8(a) can be solved exactly [5] as shown in Fig. 8(b), where

$$y_{be} \cong (1 - \alpha)/r_e \qquad (22)$$

$$\cong (1 - \alpha_0)/r_e + s/r_e\omega_t.$$

An important assumption in the model is that lateral voltage drops due to dc base current (dc crowding) are negligible, so that the emitter remains uniformly biased as implied in Fig. 8(a). Experiments carried out at this laboratory indicate that, at the bias currents used in this amplifier, dc crowding is indeed negligible. This, however, is in contradiction to calculations based on measured base current and sheet resistance. An explanation for this is that the measured base current results primarily from recombination in the inactive base region and at the surface, while bulk recombination and the associated base current under the emitter are relatively small.

The calculations that follow will be restricted to frequencies high enough that Y_1 in Fig. 8(b) simplifies to

$$Y_1 = \sqrt{s/R_b r_e \omega_t}\,\tanh\sqrt{sR_b/r_e\omega_t}. \qquad (23)$$

A simple approach to using (23) is to consider only the

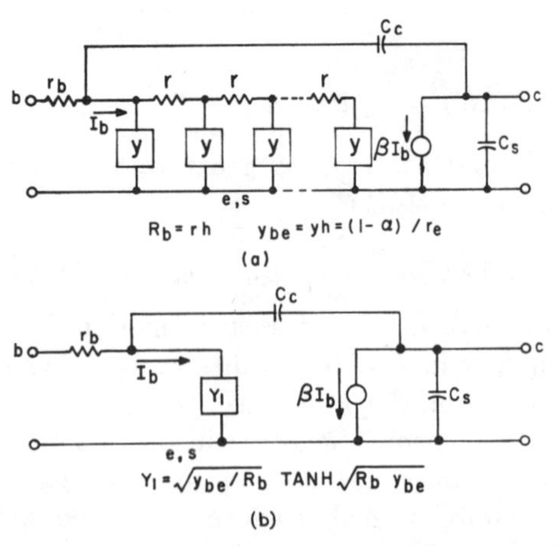

Fig. 8. (a) Distributed-base model for integrated transistor with transverse base dimension h. (b) Exact equivalent to (a), useful for circuit calculations.

j-axis response (i.e., let $s = j\omega$) and work with the exact $Y_1(j\omega)$. At high frequencies, for example, the real part of $\sqrt{j\omega R_b/r_e\omega_t}$ is large enough that the hyperbolic tangent is approximately unity. Equation (23) then becomes

$$Y_1(j\omega) \cong (1 + j)\sqrt{\omega/2R_b r_e\omega_t}. \qquad (24)$$

Equation (24) has negligible magnitude and phase error for

$$\sqrt{\omega R_b/2r_e\omega_t} \geq 2. \qquad (25)$$

At frequencies where (25) is satisfied, Y_1 is seen to have a constant 45° phase angle. This is considerably less than the 90° angle predicted by the simple model.

It is instructive to make a calculation demonstrating the manner in which the distributed model introduces phase shift in excess of that of the simple model. A voltage driven common emitter stage is considered in which r_b and C_c are ignored, and the transadmittance is computed using each model. (This calculation is similar to that made for the first and third amplifier stages.) From Fig. 7, the transconductance of the simple model is

$$I_{out}/V_{in} = 1/r_e. \qquad (26)$$

For the distributed model, using (24) and $\beta = (\omega_t/j\omega)$,

$$I_{out}/V_{in} = \beta Y_1 = \sqrt{\omega_t/r_e R_b\omega}\, e^{-j(\pi/4)}. \qquad (27)$$

Note that the distributed model predicts a phase lag of 45° while the simple model shows no phase shift.

Circuit Calculations

Because of the complexity of the distributed model, some results from the analysis in the Appendix are used to simplify the calculations. Specifically, the circuit is separated into parts as shown in Fig. 15 and the terminations for each transistor are assumed the same as those derived for the simple case.

Equivalent circuits used for each stage (analogous to Fig. 15) are shown in Fig. 9. The load impedance for the first stage [see (48)] has little influence on the stage gain, so a short circuit is assumed, as shown in Fig. 9(a). The center stage, Fig. 9(b), is assumed current driven by I_1, the short circuit output current of the first stage, and

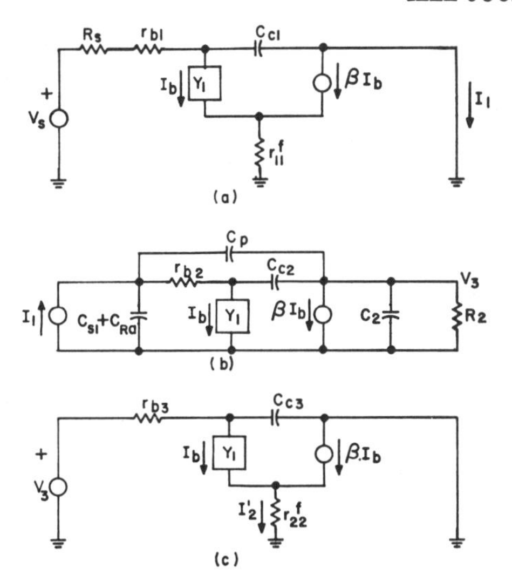

Fig. 9. Equivalent circuit for (a) input stage, (b) center stage, and (c) output stage of the forward amplifier using distributed-base transistor model.

has a parallel RC load defined in (39) and Fig. 15(b). In the last stage, Fig. 9(c), a short circuit load is again assumed because the important feedback effects of C_{c3} are included in the center stage equivalent circuit.

From the circuits in Fig. 9, the transfer function for each stage may be computed. For example, the first stage transfer is

$$\frac{I_1}{V_s}=\frac{sC_{c1}+r_{11}^f Y_1 sC_{c1}(1+\beta)-\beta Y_1}{1+(R_s+r_{b1})(sC_{c1}+Y_1)+r_{11}^f Y_1(1+\beta)[1+sC_{cL}(R_s+r_{b1})]}. \tag{28}$$

V_3/I_1 and I_2'/V_3 can be similarly obtained resulting in the complete forward transfer function

$$a_y = I_2'/V_s = (I_1/V_s)(V_3/I_1)(I_2'/V_3). \tag{29}$$

Using (23), $\beta \cong \omega_t/j\omega$, and letting $s = j\omega$ in (29), the high-frequency gain and phase can be numerically evaluated. For simplicity, the excess phase in β is ignored since it is small at frequencies of interest. Element values for the first transistor are

$$R_b \cong 250\Omega, \quad r_b \cong 30\Omega, \quad \omega_t \cong 2.8 \times 10^9 \text{ rad/s}$$

$$C_c \cong 1.5\text{pF}, \quad C_s \cong 4\text{pF} \quad r_e \cong 45\Omega.$$

The second and third transistors are biased at $I_e = 1$ mA and 4 mA, respectively, and have the following element values

$$R_b \cong 600\Omega, \quad r_b \cong 75\Omega, \quad \omega_t \cong 4 \times 10^9 \text{ rad/s}$$

$$C_c \cong 1\text{pF}, \quad C_s \cong 3\text{pF}.$$

Table I shows the calculated gain and phase for each stage at 100 and 200 MHz. Table II compares the measured transfer of the complete amplifier with that calculated from both the simple and distributed models. It is seen that the distributed model predicts magnitude within 1.3 dB and phase within 6 percent of the measured values, while the simple model has a phase error greater than 40 percent and a magnitude error as great as 10 dB.

TABLE I
CALCULATED FORWARD STAGE GAIN USING DISTRIBUTED MODEL

		100 MHz		200 MHz	
$T(\omega)$		$\lvert T\rvert$	arg $(-T)$	$\lvert T\rvert$	arg $(-T)$
I_1/V_s	Eq. 28	1/151 Ω	−38°	1/161 Ω	−52°
V_3/I_1	Fig. 9(b)	306 Ω	−185°	66 Ω	−239°
$-r_{22}^f(I_2'/V_3)$	Fig. 9(c)	0.76	−24°	0.64	−38°

TABLE II
CALCULATED AND MEASURED FORWARD TRANSFER, $r_{22}^f a_y$

		100 MHz		200 MHz	
	Equation	20 log $r_{22}^f a_y$	arg a_y	20 log $r_{22}^f a_y$	arg a_y
Simple Model	13	6 dB	−144°	−3 dB	−165°
Distributed Model	29	3.8 dB	−247°	−11.7 dB	−329°
Measured		4 dB	−240°	−13 dB	−310°

G. Generation of Secondary Poles

For root locus calculations it is useful to generate a group of "equivalent" poles that approximately account for the observed phase shift in excess of that of the simple solution of (13). Because of the distributed nature of the devices, this procedure is not exact, but little error is introduced at frequencies of interest.

A simple method for generating the poles is to assume the excess phase ϕ is due to an nth order negative real pole p_3, and match the observed phase to that calculated from

$$\phi(f) = -n \arctan(2\pi f/p_3). \tag{30}$$

Phase data at two frequencies are sufficient to determine n and p_3. Because of the approximate nature of the problem, some compromise in the solution to (30) may be necessary to obtain best magnitude agreement.

Using data from Table II, for example, the measured phase in excess of that predicted by the simple model at 100 MHz and 200 MHz is (−99°) and (−153°), respectively. These numbers do not include the phase of the zero ($z_1/2\pi = 1500$ MHz) in (13), so its phase contribution will be included in the "equivalent" poles. Solving (30), considering only integer values of n, either $n = 3$, $p_3/2\pi = -160$ MHz, or $n = 4$, $p_3/2\pi = -240$ MHz will provide good phase agreement. Best magnitude agreement occurs for $n = 4$, so this solution is chosen.

A comparison of measured gain and phase of the forward amplifier with that calculated using (13) and a fourth-order pole at $p_3/2\pi = -240$ MHz is shown in Fig. 10. The calculated and measured curves are seen to be very close. If the theoretical excess phase of the dis-

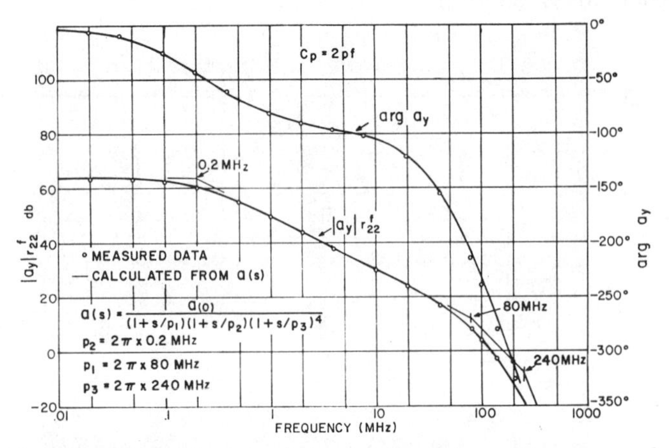

Fig. 10. Measured gain and phase of forward amplifier compared with that computed from (13) and (30).

tributed model is used, a fourth-order pole at -215 MHz is obtained and the agreement is still quite acceptable.

H. Root Locus Analysis

With the transfer function developed, root locus techniques can be used to optimize the feedback loop and predict closed loop response of the amplifier. The poles and zeros of the loop transmission function can all be related to the design of the device so effects of process modification, as well as circuit modification, can be evaluated.

In Fig. 11, root locus plots are shown for different types of capacitive compensation. The first plot, Fig. 11(a), shows the locus for no compensation in either the forward or the feedback network. Even with $C_p = 0$, considerable pole splitting occurs due to the presence of center stage collector-base capacitance C_{c2}. The maximum bandwidth that is achievable before the poles become underdamped is 17 MHz, and the corresponding loop gain, $a(0)f(0)$, is 22. The secondary poles at -2π (250 MHz) have little effect on the overall response. Figure 11(b) shows the plot obtained when a 2 pF pole splitting capacitor is added. Use of this capacitor has caused $p_1/2\pi$ to broadband from -30 MHz to -80 MHz and $p_2/2\pi$ to narrowband from -0.5 MHz to -0.2 MHz. These pole movements are in opposite direction to those caused by increased loop gain so $a(0)f(0)$ can be raised to 70 without causing the closed loop poles to become underdamped. The secondary poles at -2π (240 MHz) are now seen to have a significant effect on the response, limiting the maximum bandwidth to 32 MHz (as compared with 57 MHz if these poles were absent).

The last plot, Fig. 11(c) shows the effect of adding both the pole splitting capacitor C_p, and the feedback capacitor C_f. The positions of the feedback zero $1/R_fC_f$ at -2π (50 MHz) and its associated pole p_f, at -2π (350 MHz) (corresponding to a closed loop gain of 100), were selected to maximize loop gain and bandwidth. This compensation is seen to be quite successful, resulting in a maximum bandwidth improvement by a factor of three and a loop gain improvement by a factor of six, when compared with the uncompensated case.

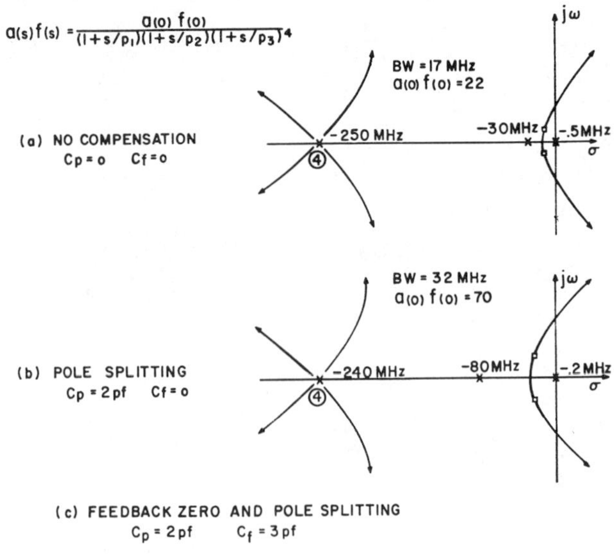

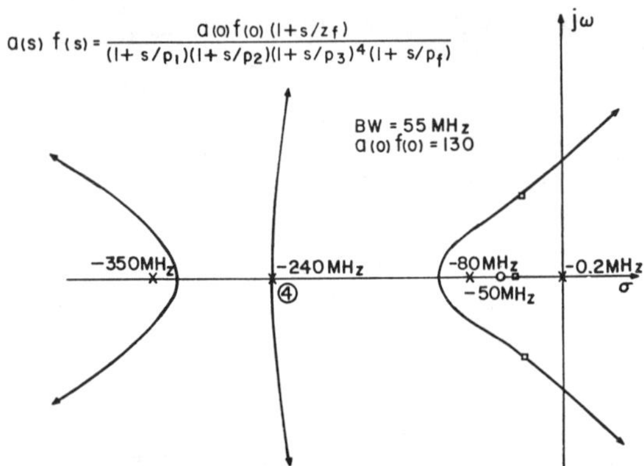

Fig. 11. Root locus plots for different types of compensation. The squares show positions of dominate closed loop poles corresponding to the maximum bandwidth (BW) and loop gain, $a(0)f(0)$, that can be achieved without passband peaking.

IV. Output Pole

From (8) the overall amplifier response is seen to depend not only on the closed loop feedback poles, but also on the third stage collector load impedance Z_c. Referring to Fig. 15(c) and (36), Z_c is seen to have a single pole:

$$p_0 = 1/R_{L3}(C_{s3} + C_{L3} + C_{RL3}) \tag{31}$$

where $R_{L3} \cong R_c$, C_{RL3} is the junction capacitance associated with R_c, and $C_{L3} \cong C_{c4}$ where C_{c4} is the collector-base capacitance of the emitter follower. If the overall bandwidth of the amplifier is to depend only upon the closed loop feedback poles and not on p_0, p_0 must be broadbanded beyond the amplifier bandedge (by making R_c small).

V. Diffused Resistors

The diffused resistor can be considered a uniform RC line in which an equipotential contact is made to the lower conductor of the capacitor. Transmission through the feedback network in Fig. 6(b) is approximately characterized by the short circuit transfer admittance y_{21} of the series resistor R_f if $R_f \gg R_{e1}, R_{e2}$. From [6], y_{21} of a uniform RC line is

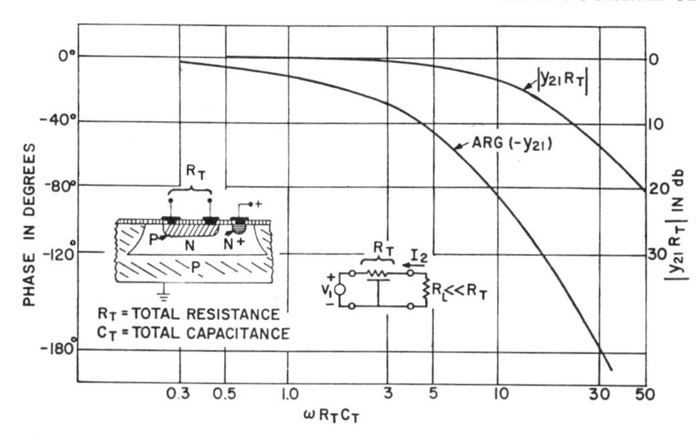

Fig. 12. Maximum attenuation and phase of y_{21} in a diffused resistor.

$$y_{21} = \theta / R_T \sinh \theta \tag{32}$$

where

$$\theta = b + jb \quad \text{and} \quad b = \sqrt{\omega R_T C_T / 2} \tag{33}$$

and R_T corresponds to the low-frequency value of the resistor R_f, while C_T corresponds to the total capacitance of the isolation junction.

The magnitude and phase of (32) can be written in the following form:

$$|y_{21}| = \frac{b}{R_T}\left[\frac{2}{\cosh^2 b - \cos^2 b}\right]^{1/2} \tag{34}$$

$$\arg y_{21} = \arc\tan\left[\frac{1 - \tan b \coth b}{1 + \tan b \coth b}\right]. \tag{35}$$

A plot of (34) and (35) as a function of frequency is given in Fig. 12. (It can be shown that nonzero resistivity in the resistor isolation island reduces both phase and attenuation of y_{21}, so Fig. 12 represents the maximum phase and attenuation expected.) Phase shift is minimized by minimizing the area of the diffused resistor, reducing its total RC product. Resistors in the feedback network are designed for $\omega R_T C_T < 1$ at 100 MHz which corresponds to a phase lag of less than 10° at that frequency.

VI. Monolithic Realization

The integrated amplifier die is shown in Fig. 13. Except for the input device and the current source, all transistors have single base and emitter stripes to minimize the capacitance of the collector-base and collector-substrate junctions. A special geometry is used for the input device to optimize operation at the low collector-base voltage (about 0 volts) and to improve noise figure. For reasons discussed in Section II, the output current source is essentially four parallel devices, each having base-emitter geometries identical to the reference transistor, and all fabricated in a common collector island.

Four gain options (V_0/V_s = 50, 100, 200, and 400) are obtained by using various configurations of feedback resistors. Two metal masks are used to obtain optimum performance at the different gains. In each version, a factor-of-two gain adjustment is made by shorting appropriate pins on the package.

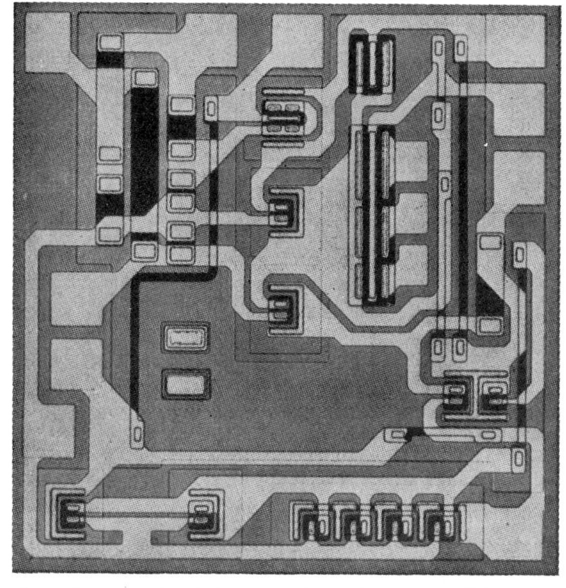

Fig. 13. Photograph of integrated amplifier die. Dimensions of the circuit are 45 mils per side.

VII. Experimental Performance

The measured frequency response of the amplifier operating at fixed voltage gains is shown in Fig. 14. In every case, the response agrees well with that predicted by the theory in Section III. The voltage gain-bandwidth product is seen to maximize at about 4.7 GHz for the gains of 100 and 200. A lower gain-bandwidth product is realized at the gain of 50 due to effects of the secondary poles and band limiting by the output pole p_0. At the gain of 400, some performance sacrifice is made in exchange for improved bandwidth at 200. Table III gives typical performance data for the amplifier operating from a 6 volt (11 mA) supply at a gain of 100. The total harmonic distortion is seen to be quite low out to 0.5 MHz, demonstrating excellent desensitivity to device nonlinearity. Gain variation with temperature is due primarily to variation in diffused resistor ratios which is quite small.

The optimum narrow-band noise figure increases from 1.9 dB at 1 MHz to 4 dB at 20 MHz. At frequencies above 20 MHz, the optimum noise source cannot be used without degrading bandwidth so the minimum noise figure can only be obtained by tuning. The input impedance is determined approximately by the 12 kΩ base resistor shunted by a capacitance of 3 pF. Output impedance is set entirely by the emitter follower with a 600 Ω base resistor (R_c).

As expected from the discussion in Section III, capacitive source impedances are observed to improve, rather than degrade, the amplifier stability. Capacitive loading on the emitter follower has no important effect (other than eventual bandwidth narrowing) because the follower is driven from a relatively low impedance.

The output current capability of the amplifier is sufficient to drive 1-volt peak pulses of either polarity into a 10 pF load without degrading rise or fall time. Larger capacitive loads can be driven by using an external bias resistor for the emitter follower.

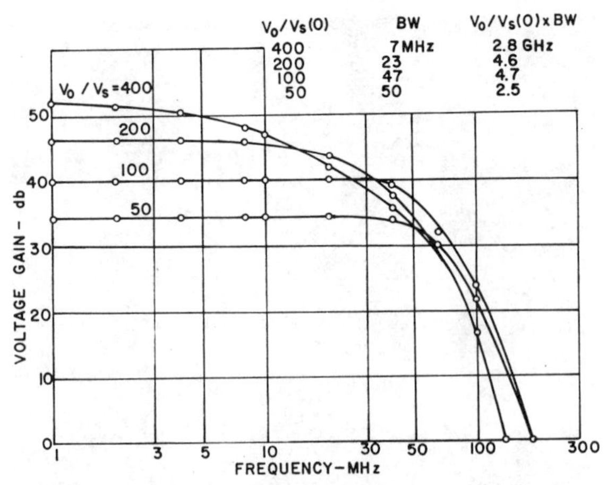

Fig. 14. Experimental frequency response of amplifier with R_s = 50 Ω and V_{CC} = 6V.

TABLE III
EXPERIMENTAL PERFORMANCE OF THE AMPLIFIER OPERATING AT A VOLTAGE GAIN OF 100 FROM A SUPPLY OF 6 VOLTS AND 11 mA

Total Harmonic Distortion (20 Hz–500 kHz, 2 V p-p)	0.15%
Gain Variation (−55°C to +125°C)	±0.25 dB
Loop Transmission, $a_y(0)f_z(0)$	36 dB
Noise Figure (R_s = 500 Ω, f = 1 MHz)	1.9 dB
Input Impedance	10 kΩ
Output Impedance	20 Ω
Maximum Output ($\|Z_L\|$ = 1 kΩ, V_{cc} = 6 V)	4 V p-p
Reverse Loss (R_s = 50 Ω, R_1 = 1 kΩ, $f \leq$ 50 Mc/s)	≥70 dB

APPENDIX

CALCULATION OF FORWARD TRANSFER FUNCTION USING SIMPLE MODEL

The transfer function I_2'/V_s for the forward amplifier of Fig. 6(a) is calculated using the simple transistor model in Fig. 7. By using the unilateral model [Fig. 7(b)] for the first and third transistors and the pi-model [Fig. 7(a)] for the center stage, the complete amplifier equivalent circuit shown in Fig. 15 is obtained.[2] For simplicity, a resistive feedback network is assumed, i.e., $z_{11}^f = r_{11}^f$ and $z_{22}^f = r_{22}^f$. Negligible error is introduced by this assumption as long as C_f is not too large [see Section III-D].

[2] The unilateral model of Fig. 7(b) may err considerably in predicting the effects of finite transistor output conductance on the circuit. For the first stage, no appreciable error is introduced because the stage is driven from a low impedance source, which causes the output conductance to be low enough that it can be ignored. Such is not the case for the center stage because it is driven from a high impedance source and has a low output resistance (on the order of $1/\omega_t(C_2 + C_p)$ at high frequencies. The complete pi-model is used, therefore, for this stage. Fortunately, the output impedance of the center stage is low enough so that the unilateral model can again be used for the third stage. The last approximation is further improved when feedback is applied because this raises the amplifier output impedance by a factor equal to $[1 + a(s)f(s)]$.

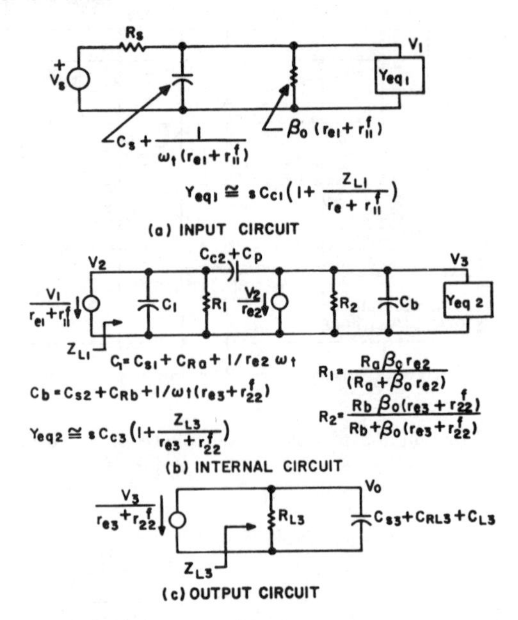

Fig. 15. Equivalent circuit for forward amplifier split into three separate parts by use of unilateral model for first and third stages.

Effects of reverse transmission in the transistors are accounted for by analyzing the output circuit first, then the interior circuitry, and finally the input circuit. The procedure is to compute the transfer function and input impedance for each stage so that the load for each preceding stage is known.

Output Circuit [Fig. 15(c)]

By inspection of Fig. 15(c) we see that the output circuit produces a single pole response

$$\frac{V_0}{V_3}(s) = -\frac{R_{L3}}{(r_{e3} + r_{22}^f)}\frac{1}{(1 + s/p_0)} \tag{36}$$

where $p_0 = 1/R_{L3}(C_{s3} + C_{L3} + C_{RL3})$. The output circuit is isolated from the feedback loop by the collector impedance of the third transistor so the pole p_0 is, to a good approximation, unaffected by the feedback. As a result, R_{L3} must be chosen small enough that p_0 lies beyond the upper band edge of the closed loop amplifier, or the bandwidth will be limited by p_0 rather than the closed loop poles of the feedback circuit.

Under the assumption that p_0 always lies beyond the amplifier band edge, Z_{L3} can be assumed real at all frequencies of interest, i.e., $Z_{L3} \cong R_{L3}$. Using this relationship and the expression for Y_{eq} in Fig. 7(b), the input admittance to the third transistor can be written

$$Y_{eq2} \cong sC_{c3}\left(1 + \frac{R_{L3}}{r_{e3} + r_{22}^f}\right). \tag{37}$$

Also, the transfer from input base to output emitter in the third transistor is

$$I_2'/V_3 \cong 1/(r_{e3} + r_{22}^f). \tag{38}$$

Internal Circuit [Fig. 15(b)]

Using (37), the load for the internal circuit in Fig. 15(b) is a parallel RC network with $R = R_1$ and C given by

$$C_2 = C_b + C_{e3}[1 + R_{L3}/(r_{e3} + r_{22}^f)]. \quad (39)$$

Calculation of the transfer function for the circuit in Fig. 15(b) yields

$$\frac{V_3}{V_1} = \frac{R_1R_2}{(r_{e1} + r_{11}^f)r_{e2}} \cdot \frac{[1 - sr_{e2}(C_{c2} + C_p)]}{[1 + sR_1R_2K_1(C_{c2} + C_p)/r_{e2} + s^2R_1R_2C_1C_2K_2]} \quad (40)$$

where the constants K_1 and K_2 are

$$K_1 = 1 + \frac{r_{e2}}{R_2}\left(1 + \frac{C_1}{C_{c2} + C_p}\right) + \frac{r_{e2}}{R_1}\left(1 + \frac{C_2}{C_{c2} + C_p}\right) \quad (41)$$

$$K_2 = 1 + (C_{c2} + C_p)(1/C_1 + 1/C_2).$$

Equation (40) has two negative real poles (p_1, p_2) which are usually separated widely due to the presence of C_p. If $p_1 \gg p_2$, then the denominator D of (40) can be approximately factored as follows

$$D = \left(1 + \frac{s}{p_1}\right)\left(1 + \frac{s}{p_2}\right) \cong 1 + \frac{s}{p_2} + \frac{s^2}{p_1p_2}. \quad (42)$$

Solving for p_1 and p_2 using (42) and the denominator of (40),

$$p_1 \cong K_1(C_{c2} + C_p)/r_{e2}C_1C_2K_2$$
$$p_2 \cong r_{e2}/R_1R_2K_1(C_{c2} + C_p). \quad (43)$$

These approximations are accurate to within 10 percent if p_1 and p_2 are separated by a factor of 10 or more, i.e., $p_1 \geq 10p_2$.

Using (42) and (43) the internal transfer (40) becomes

$$\frac{V_3}{V_1} \cong \frac{R_1R_2}{(r_{e1} + r_{11}^f)r_{e2}} \frac{(1 - s/z_1)}{(1 + s/p_1)(1 + s/p_2)} \quad (44)$$

where

$$z_1 = 1/r_{e2}(C_{c2} + C_p). \quad (45)$$

Using Y_{eq} as defined in Fig. 7(b), Z_{L1} can be written

$$Z_{L1} \cong 1/[SC_1 + 1/R_1 + s(C_{c2} + C_p)(1 + Z_{L2}/r_{e2})] \quad (46)$$

where, from Fig. 15(b) and (39)

$$Z_{L2} = 1/(SC_2 + 1/R_2). \quad (47)$$

Evaluation of (46) shows that the term involving Z_{L2} is dominant at all frequencies where Y_{eq1} is important. Consequently, (46) can be simplified to

$$Z_{L1} \cong \frac{r_{e2}C_2}{(C_{c2} + C_p)} + \frac{r_{e2}}{sR_2(C_{c2} + C_p)}. \quad (48)$$

Z_{L1} is seen to represent a series resistor and capacitor.

Input Circuit [Fig. 15(a)]

With Z_{L1} in the form given by (48), Y_{eq1} in Fig. 15(a) represents a simple parallel RC network. As a result, the input circuit is also a parallel RC network and exhibits a single pole response as follows

$$\frac{V_1}{V_s} = \frac{1/K_s}{1 + s/p_s} \quad (49)$$

where

$$p_s = K_s/R_s\left[C_s + C_{c1} + \frac{1}{\omega_t(r_{e1} + r_{11}^f)} + \frac{r_{e2}C_{c1}C_2}{(r_{e1} + r_{11}^f)(C_p + C_{c2})}\right] \quad (50)$$

$$K_s = 1 + \frac{R_s}{\beta_0(r_{e1} + r_{11}^f)} + \frac{R_sr_{e2}C_{c1}}{R_2(r_{e1} + r_{11}^f)(C_{c2} + C_p)}. \quad (51)$$

Complete Transfer Function

Equations (38), (44), and (49) can now be combined to give the complete forward transadmittance of the amplifier

$$a_y = \frac{I_2'}{V_s} = \left(\frac{I_2'}{V_3}\right)\left(\frac{V_3}{V_1}\right)\left(\frac{V_1}{V_s}\right) \cong \frac{a_y(0)(1 - s/z_1)}{(1 + s/p_s)(1 + s/p_1)(1 + s/p_2)}. \quad (52)$$

The low-frequency forward transconductance is

$$a_y(0) \cong \frac{R_1R_2}{K_sr_{e2}(r_{e1} + r_{11}^f)(r_{e3} + r_{22}^f)}. \quad (53)$$

ACKNOWLEDGMENT

The authors would like to acknowledge the helpful suggestions and criticisms made by T. M. Frederiksen and J. E. Thompson of this laboratory.

REFERENCES

[1] E. M. Cherry and D. E. Hooper, "The design of wideband transistor feedback amplifiers," *Proc. IEE (London)*, vol. 110, pp. 375–388, February 1963.
[2] M. S. Ghausi, "Optimum design of the shunt-series feedback pair with a maximally flat magnitude response," *IRE Trans. on Circuit Theory*, vol. CT-8, pp. 448–453, December 1961.
[3] M. S. Ghausi and D. O. Pederson, "A new design approach for feedback amplifiers," *IRE Trans. on Circuit Theory*, vol. CT-8, pp. 274–284, September 1961.
[4] D. E. Thomas and J. L. Moll, "Junction transistor short-circuit current gain and phase determination," *Proc. IRE*, vol. 46, pp. 1177–1184, June 1958.
[5] R. L. Pritchard, "Two dimensional current flow in junction transistors at high frequencies," *Proc. IRE*, vol. 46, pp. 1152–1160, June 1958.
[6] P. S. Castro, "Microsystem circuit analysis," *Elec. Engrg.*, vol. 80, pp. 535–542, July 1961.

A New Wide-Band Amplifier Technique

BARRIE GILBERT, MEMBER, IEEE

***Abstract*—Precision dc-coupled amplifiers having risetimes of less than a nanosecond have recently been fabricated using the monolithic planar process. The design is based on a simple technique that has a broad range of applications and is characterized by a stage gain accurately determined by the ratio of two currents, a stage-gain-bandwidth product essentially equal to that of the transistors, and a very linear transfer characteristic, free from temperature dependence.**

I. Introduction

AMPLIFIERS having wide bandwidths, in excess of 100 MHz, with low amplitude distortion and stable gain, are of increasing importance in modern electronics. By far the most prevalent circuit configuration is the emitter-degenerated amplifier, which suffers a familiar problem: the current- and temperature-dependent emitter impedance causes distortion [1], [2] and gain instability. Negative feedback techniques over more than two stages are rarely applicable at these bandwidths due to the excess phase shift around the loop.

A further characteristic of such amplifiers is that voltage swings at the signal frequencies are present throughout the chain, and, consequently, the parasitic capacitances have a part in limiting the bandwidth. This problem is especially severe in monolithic (junction-isolated) circuits. In circuits using discrete components, inductive elements are often employed [3] to improve the bandwidth, but no satisfactory way of incorporating inductive elements in a monolithic circuit is available. Another approach is to make these parasitic capacitances part of a lumped constant delay line (the distributed amplifier [4]), but this, too, is not applicable to microcircuits. Interestingly, however, some of the amplifiers described later behave very much like distributed amplifiers in that each stage works at an f_t-limited bandwidth, and all stages contribute to the total output capability.

Suggestions have been made from time to time [5]–[7] to include compensating diodes in the collector load circuit to mitigate the nonlinearities introduced by the emitter diode, or even eliminate the linear impedance at this point altogether [7]. This is a useful technique, but has several limitations, and the motivation is still to produce a stage having *voltage gain.*

The main objective of the work reported here was to develop a cascadable circuit form (a "gain cell") that could provide dc-coupled temperature-insensitive sub-nanosecond *current gain* with the virtual absence of voltage swings, and a theoretically perfect transfer characteristic, having constant slope between the upper and lower overload limits. Conversion between voltage and current can then be made only where needed—at the input and output terminals.

It became apparent that, in addition to meeting these goals, the principle developed had several other useful properties. For example, the gain of each stage can be electronically controlled with precision over a wide range at nanosecond speeds, a property exploited in a new four-quadrant multiplier [8], [9]. The analysis also shows that the stage gain of certain configurations is independent of beta, even at gains *close to beta* in magnitude. This is something that, to the author's knowledge, no other configuration permits.

Manuscript received June 28, 1968; revised September 16, 1968. This paper was presented at the 1968 ISSCC.
The author is with Tektronix, Inc., Beaverton, Ore.

II. Foundations of the Technique

Two very common circuits, shown in Fig. 1, were married to produce a new configuration. The first, (a), is the "differential pair," now widely used as a multiplier [10]–[13]. The second, (b), is the "current source" found in practically every linear IC. The married couple are shown in Fig. 3; this marriage has proven unexpectedly fruitful, and the prolific offspring will be described in this and later papers.

First, we will consider Fig. 1(a) in more detail. It can be used as a multiplier because the transconductance from base to collector is *proportional to the emitter tail current* I_E. However, it is far from being a precise element, because this transconductance is both nonlinear and temperature dependent.

In this paper, we will frequently use the junction-diode expressions [14].

$$I = I_s \exp \frac{qV}{mkT} \quad \text{or} \quad V = \frac{mkT}{q} \log \frac{I}{I_s} \tag{1}$$

where

I = forward conduction current, $\gg I_s$
I_s = reverse saturation current
V = voltage across the junction
q = charge on the electron
m = a constant near unity [14]
T = absolute temperature.

The quantity mkT/q is about 26 mV at 300°K. Applying (1) to Fig. 1(a), and solving for the variable

Reprinted from *IEEE J. Solid-State Circuits*, vol. SC-3, pp. 353-365, Dec. 1968.

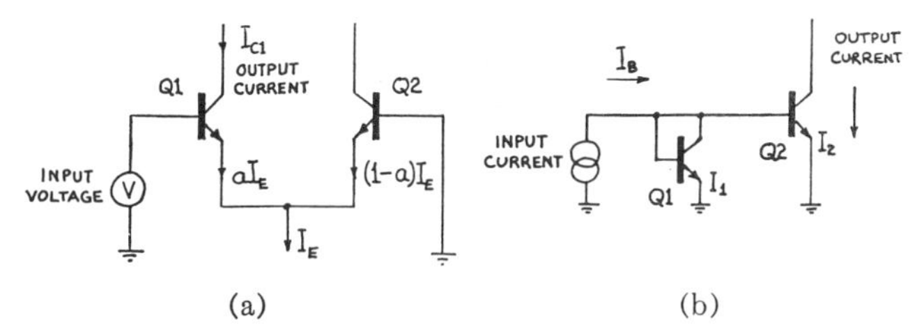

Fig. 1. Two common circuits. (a) The "differential amplifier." (b) The "current source."

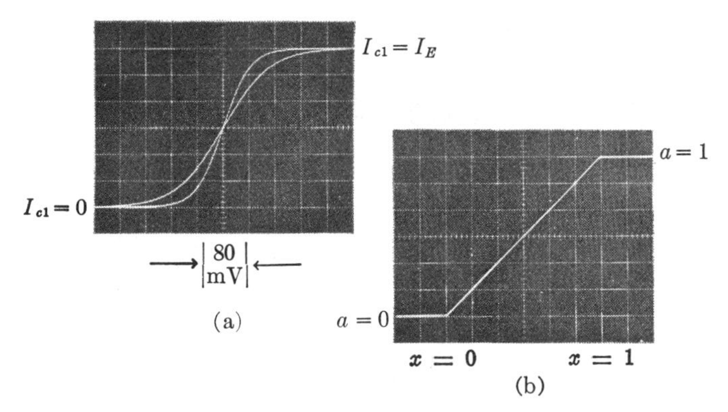

Fig. 2. Comparison of transfer curves for (a) Fig. 1(a) amplifier; (b) new circuit shown in Fig. 3. In each case, the curves for −50°C and +125°C are compared; in (b), they are virtually co-incident.

a we find

$$a = \frac{\exp \dfrac{qV}{mkT}}{\lambda + \exp \dfrac{qV}{mkT}}, \tag{2}$$

independent of I_E, where

$$\lambda = \frac{I_{s2}}{I_{s1}} = \frac{\text{area of } Q2 \text{ emitter}}{\text{area of } Q1 \text{ emitter}}, \tag{3}$$

and I_{s1}, and I_{s2} are the reverse saturation currents of the emitter diodes of $Q1$ and $Q2$. This area ratio is more usually expressed as an equivalent offset voltage,

$$V_0 = \frac{mkT}{q} \log \lambda, \quad \text{or} \quad \lambda = \exp \frac{qV_0}{mkT}. \tag{4}$$

Thus, in terms of V_0, we have

$$I_{c1} \simeq aI_E = \frac{I_E}{1 + \exp \dfrac{q}{mkT}(V_0 - V)}. \tag{5}$$

Characteristic transfer curves are shown in Fig. 2(a) for two temperatures. The transconductance is clearly a nonlinear, temperature-dependent quantity. Notice that an emitter-area mismatch merely shifts the transfer curve by an amount V_0, *without changing its shape.* This matter of area mismatch will be raised again in connection with the improved circuits, where the effects are very different.

The circuit of Fig. 1(b) can also be used as an amplifier, in which case the signal input is the *current* I_B. With reference to the figure it will be apparent that

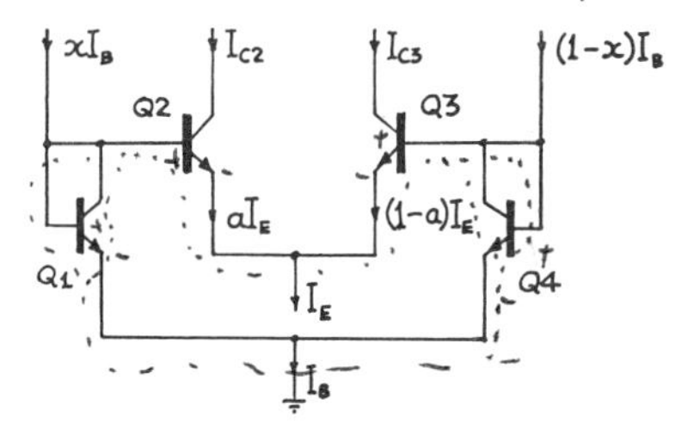

Fig. 3. The new circuit form. The text shows that the factors a and x are equal, for any I_E and at all temperatures.

$$\frac{mkT}{q} \log \frac{I_1}{I_{s1}} = \frac{mkT}{q} \log \frac{I_2}{I_{s2}} \tag{6}$$

or

$$I_2 = \frac{I_{s2}}{I_{s1}} \cdot I_1 = \lambda I_1. \tag{7}$$

This is a linear, temperature-insensitive relationship. By scaling the area of $Q2$ emitter relative to that of $Q1$, a stage gain can be realized; now *emitter-area mismatches cause a gain-error,* but do not jeopardize linearity.

Amplifiers built from directly cascaded stages like this have been successfully tested, and have the same high-speed properties as the circuits to be described—there is very little voltage swing, and, hence, f_t is the dominant factor in determining bandwidth. However, they lack the advantages of the differential configuration and have a fixed gain.

III. The Improved Circuit Principle

We can now examine Fig. 3, which can be considered as a differential amplifier in which the base-drive voltages are derived from a pair of junctions that, like the diode of Fig. 1(b), are current-driven. However, in this case there are two drive currents (the signal input), which are of the form

$$I_{B1} = xI_B$$
$$I_{B2} = (1 - x)I_B \tag{8}$$

where $0 < x < 1$, and will be termed the "modulation index" of the bias current I_B.

Temporarily ignoring the effects of junction area differences, finite beta and ohmic resistances, and summing the emitter voltages around the $Q1$–$Q4$ loop, we have

$$\frac{mkT}{q} \log \frac{xI_B}{I_s} - \frac{mkT}{q} \log \frac{aI_E}{I_s} - \frac{mkT}{q} \log \frac{(1 - x)I_B}{I_s} + \frac{mkT}{q} \log \frac{(1 - a)I_E}{I_s} = 0 \tag{9}$$

which collapses to

$$a = x. \tag{10}$$

Thus, the magnitudes of the output currents are simply

$$I_{C2} = xI_E$$
$$I_{C3} = (1 - x)I_E, \tag{11}$$

and the dc stage gain is

$$G_0 = \frac{I_{C2}}{I_{B1}} = \frac{I_E}{I_B}. \qquad (12)$$

Typical transfer curves for this circuit are shown in Fig. 2(b).

Notice the extreme insensitivity to temperature and the sharp overload points, making the whole dynamic range useful.[1] We will now discuss the effects of departures from the simple theory due to area mismatch, ohmic resistance, and beta.

A. Area Mismatches

Equation (9) assumed all the diodes had equal areas, hence, the same I_s. In practice, however, this will never be exactly the case. We will define the variable

$$\gamma = \frac{I_{s2} I_{s4}}{I_{s1} I_{s3}}. \qquad (13)$$

Reevaluating (9) to include γ, we find that

$$a = \frac{x\gamma}{1 + x(\gamma - 1)} \qquad (14)$$

which is *no longer linear* with respect to x, unless $\gamma = 1$, that is, unless the emitter areas of the inner and outer pairs of transistors are *mutually* equal.

Since we are concerned with achieving low distortion in the transfer function, the effects of even small departures from the ideal case must be examined. From (14),

$$\frac{da}{dx} = \frac{\gamma}{[1 + (\gamma - 1)x]^2}. \qquad (15)$$

At the extremes of the dynamic range, $x = 0$ and 1. The incremental slopes at these points are thus γ and $1/\gamma$, respectively. To a first approximation, then, small departures from $\gamma = 1$ cause the slope of the transfer curve to vary linearly over the dynamic range. For example, when $\gamma = 1.1$, corresponding to a total offset around the $Q1$–$Q4$ loop of 2.7 mV at 300°K, the gain will be 10 percent low at $x = 0$, rising to 10 percent high at $x = 1$.

The distortion caused by $\gamma = 2$ (for example, one of the four emitters has an area twice that of the others) is shown in Fig. 4(a). To confirm that area ratios are synonymous with offset voltages, this photograph also shows that linearity can be restored by introducing an error voltage of 18 mV (26 mV times log 2) into the loop.

It might seem that serious distortion would arise in this kind of amplifier due to the inevitable mismatches,

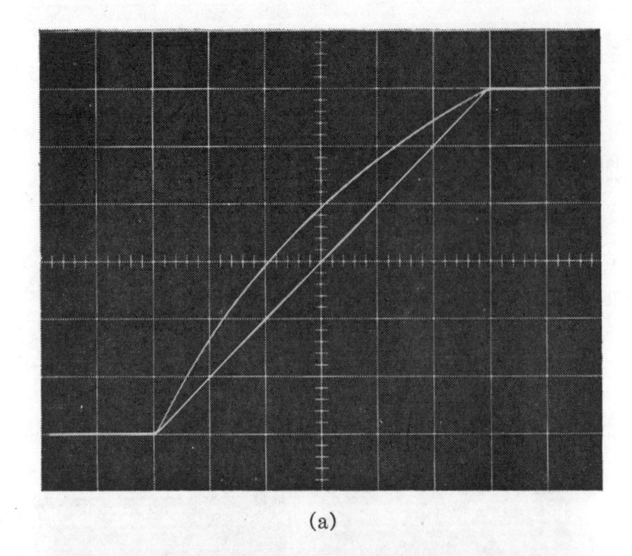

(a)

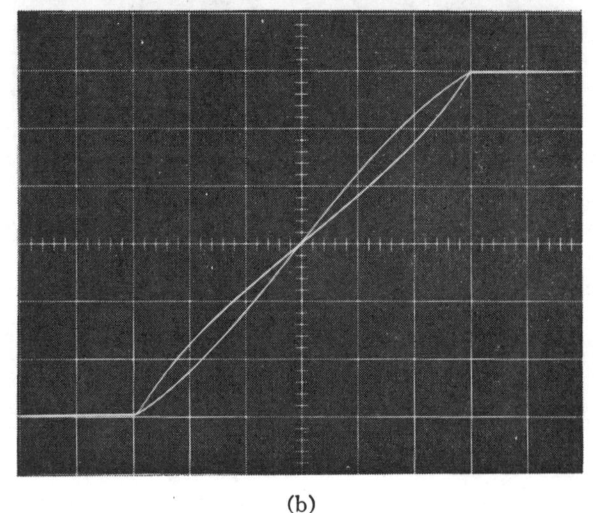

(b)

Fig. 4. Demonstration of distortions caused by (a) area mismatch ($\gamma = 2$), linearity is restored by a correction voltage; (b) ohmic resistances.

but measurements are surprisingly reassuring. Slope variations of less than ±2 percent are typical, and intermodulation products of −60 dB have been measured for two equal signals at 500 kHz having a peak-combined amplitude of 80 percent of the dynamic range. It should be pointed out that prevalent differential amplifiers of similar gain and bandwidth exhibit far greater distortion for *full-scale* signal swings.

B. Ohmic Resistances

In addition to the diode voltages described by (1) there will be components of voltage due to the ohmic (bulk) resistances of the base and emitter diffusions, metalization paths, and contact interfaces. These resistances are also current dependent due to crowding effects and to an extent dependent on the device geometry.

All of these resistive components can be referred to the emitter circuit as an *equivalent ohmic emitter resistance*. Also, we will assume that this resistance scales with emitter area (a safe assumption for many "stripe" geometries). With these resistances inserted into the

[1] These results were obtained using a single-sided input configuration (see Section III-G) driven from the oscilloscope sweep output of 100-volt amplitude via a large resistor. The input voltage variations over the dynamic range were thus negligible, and a true current-drive condition obtained.

circuit of Fig. 3 and (9) suitably modified, we find

$$\frac{mkT}{q} \log \left\{ \frac{x(1-a)}{(1-x)a} \right\} = R_E\{AI_B(1-2x) - I_E(1-2a)\} \quad (16)$$

where R_E = equivalent ohmic emitter resistance of each inner transistor

A = area ratio of inner to outer pairs of emitters.

When the right-hand side of this equation vanishes, $x = a$. This occurs when $A = G_0 = I_E/I_B$, that is, *the areas are scaled in proportion to the drive currents.* If $A \neq G_0$, distortion will arise. Equation (16) has no general explicit solution for a, but a good estimate of the magnitude of the distortion can be readily obtained by a small-signal analysis.[2]

At the quiescent point ($x = 0.5$), the mean normalized slope of the transfer curve is

$$\left.\frac{da}{dx}\right|_{0.5} = \frac{\frac{mkT}{q} + AR_E I_B}{\frac{mkT}{q} + R_E I_E}. \quad (17)$$

Also, putting $x = 0$, $x = 1$ in (16) yields $a = 0$, $a = 1$, respectively. Therefore, the *mean* normalized slope is still unity. Thus, when $A > G_0$ (area ratio too big) the transfer curve will have a *higher* slope at the center than at the extremes, and vice versa. Fig. 4(b) shows transfer curves for $I_B = 6$ mA, $I_E = 6$ mA $\pm$ 3 mA using four small-geometry transistors of the same emitter area.

C. Effects of Beta

For monolithic transistors, the match of the current gain between adjacent transistors is very good and will be assumed to be perfect. A further assumption is that the beta is not a function of I_E over the current range of interest. In the presence of finite beta, the emitter currents of $Q1$ and $Q4$ are modified so that

$$\begin{aligned} I'_{B1} &= xI_B - (1-\bar{\alpha})aI_E \\ I'_{B2} &= (1-x)I_B - (1-\bar{\alpha})(1-a)I_E \end{aligned} \quad (18)$$

where $\bar{\alpha}$ = large-signal common-base dc current gain.

Substituting these values into (9) and reducing the logarithmic terms to products and quotients as before, we have

$$\frac{\{xI_B - (1-\bar{\alpha})aI_E\}(1-a)I_E}{\{(1-x)I_B - (1-\bar{\alpha})(1-a)I_E\}aI_E} = 1 \quad (19)$$

which again reduces to $a = x$. This is an astonishing result, since it implies that no matter how low the beta is, the gain *to the emitter circuit* is still G_0. Of course,

[2] Another method of analyzing this kind of distortion is used in [9], where expressions for the form and amplitude of the nonlinearities are derived.

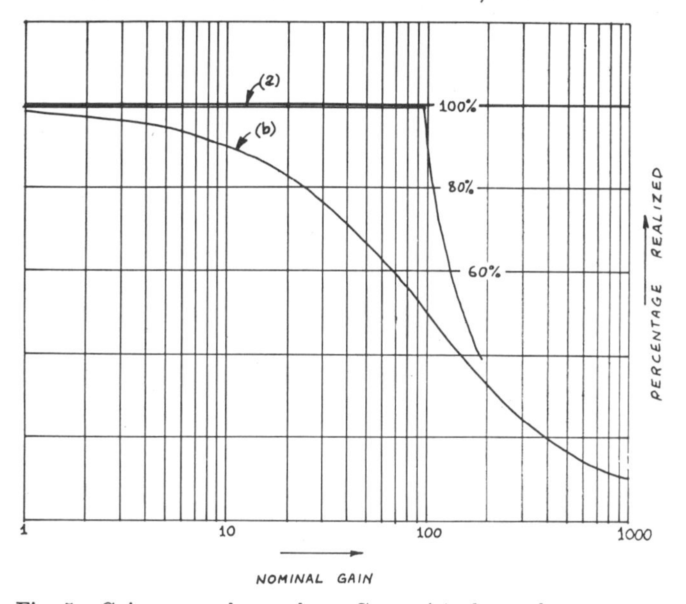

Fig. 5. Gain errors due to beta. Curve (a) shows that accurate gain up to beta is achieved with the new circuit. All prevalent transistor amplifiers have gain errors as shown in curve (b), calculated for $\beta = 100$.

I_B must still be sufficient to supply the base currents of $Q2$ and $Q3$; any excess flows into $Q1$ and $Q4$ to establish the correct drive voltages for linear operation. Since betas of 100 are typical, the gain reduction from emitter to collector is of the order of 1 percent and first order compensation can be readily made.

The significance of the above discussion is that stage gains close to beta in magnitude can be realized with very high accuracy. It will be appreciated that this is not possible with any of the standard circuit forms (including Fig. 1(b)) where the actual gain is given by an expression of the form

$$G'_0 = \frac{\beta}{\beta + (1 + G_0)} \cdot G_0, \quad (20)$$

where G_0 is now the nominal current gain (for example, the ratio of base to emitter impedances in an emitter-degenerated circuit, or the ratio λ in the circuit of Fig. 1(b)).

Measurements on an amplifier like Fig. 3 in which the betas of $Q2$ and $Q3$ were 95 showed a gain substantially equal to the ratio I_E/I_B right up to $I_E/I_B = 92$, with no degradation of linearity. The results are shown graphically in Fig. 5.

D. Collector Saturation Resistance

The input transistors are connected as diodes and thus operate with $V_{CE} = V_{BE} \approx 0.8$ volts. However, the internal collector bias is reduced to $V_{BE} - I_C R_{sat}$, and the transistor will not be a well-behaved diode when this voltage approaches zero. Since V_{BE} decreases with temperature but R_{sat} increases, the maximum usable current falls sharply with temperature. Circuits involving high-current operation must use a sufficiently large-geometry device with a buried-layer collector.

E. Thermal Effects

The analyses so far have assumed that all transistors operate at the same temperature, which need not be the case. For example, the inner pair may be at a higher temperature than the outer pair if they are in a separate package and dissipating considerable power. It can be shown by inserting the appropriate temperatures into (9) that the variable a becomes

$$a = \frac{x^n}{(1 - x)^n + x^n} \tag{21}$$

where

$$n = \frac{\text{temperature of } Q1 \text{ and } Q4 \text{ in °K}}{\text{temperature of } Q2 \text{ and } Q3 \text{ in °K}}. \tag{22}$$

Another potential source of nonlinearity is thus revealed. A temperature excess of 60°K will put $n = 1.2$ and introduce significant distortion, similar in form and magnitude to the effects of bulk resistances shown in Fig. 4(b).

In monolithic structures having close thermal coupling this problem does not arise, but *transient* thermal distortion can be a problem in critical applications. Following a step change of input current, the dissipation of each device in the quad changes, and some time is taken for the circuit to reacquire thermal equilibrium. During this time, the output signal will suffer a transient distortion whose magnitude, waveshape, and duration depend on the circuit-to-sink and transistor-to-transistor thermal impedances, and, of course, the power dissipation.

Measurements show that these effects are typically less than 1 percent of the step amplitude for the circuit being described. The gain cell described later, has less distortion than this.

F. DC Stability

Equation (14) showed that emitter-area mismatches introduce a shift in the output at the quiescent point ($x = 0.5$), but this shift is not temperature dependent. By adjusting the drive-current balance, the output can be zeroed, and remains this way over the temperature range. This is in contrast with the behavior of prevalent differential amplifiers, where the offset is corrected by a voltage, leaving a drift of 85 log λ μV/°K.

G. Unbalanced Drive Currents

The use of *complementary* drive currents is the key to linear large-signal operation of this type of circuit. However, a constant-current *offset* on one (or both) of the inputs will not impair linearity, but there will be a change in gain, and dc unbalance.

An amplitude *ratio* between the two inputs is more serious—for example, I_{B1} varies from 0 to 1 mA as I_{B2} varies from 1.5 mA to zero. This is equivalent to an area mismatch, as (14) will reveal, in which case λ represents the input ratio. If the ratio is known, linearity can be restored by an adjustment in the emitter area of one of the transistors.

The need for a pair of complementary currents can be eliminated altogether by connecting the $(1 - x)I_B$ input to ground and supplying a constant current of I_B to the junction of $Q1$ and $Q4$ emitters. Clamping diodes should also be added to the input point to control the input voltage during overload conditions ($x > 1$ or $x < 0$). This input point is a useful current-summing node close to ground potential since,

$$V_{\text{in}} = 26 \log \frac{x}{1 - x} \text{ mV at 300 °K} \tag{23}$$

thus

$$|V_{\text{in}}| < 36 \text{ mV} \quad \text{for} \quad 0.2 < x < 0.8.$$

This circuit no longer possesses the unique beta immunity of the original form and a small offset term ($a \neq 0.5$ at $x = 0.5$) arises. However, for typical betas, the errors in gain and dc balance are negligible.

IV. Alternative Configurations

Many variants of the basic circuit just discussed have been designed. Space permits only a few of these to be described here, but the generalized statement of the principle, given in the Appendix, will point the way to further forms. These examples will serve to illustrate the versatility of the principle.

A. Inverted Input Diodes

The input diodes $Q1$ and $Q4$ can be inverted and driven from current sinks (instead of current sources), as shown in Fig. 6. Neglecting the effects of beta, area mismatches, and bulk resistances, we can write[3]

$$xI_B a I_E = (1 - a)I_E(1 - x)I_B \tag{24}$$

or

$$a = 1 - x. \tag{25}$$

The circuit thus has a polarity reversal over the original form. It has the advantage that the input currents are *reusable* at the collectors of $Q1$ and $Q4$, a feature that is put to use in the gain cell described later.

This configuration lacks the beta immunity of the original form, because the base current of $Q2$ and $Q3$ *add* to the input currents. This, together with the fact that the input that is receiving the *smaller* fraction of the base-current drives the transistor that conducts the *larger* fraction of the emitter current, causes significant beta dependence.

By considering these currents, it is found that the output modulation index a is reduced to

$$a = \theta(1 - x) + \frac{G_0}{\beta} \tag{26}$$

where

$$\theta = \frac{\beta}{\beta + (1 + 2G_0)}. \tag{27}$$

[3] The Appendix shows how these quantities can be equated from an inspection of the circuit.

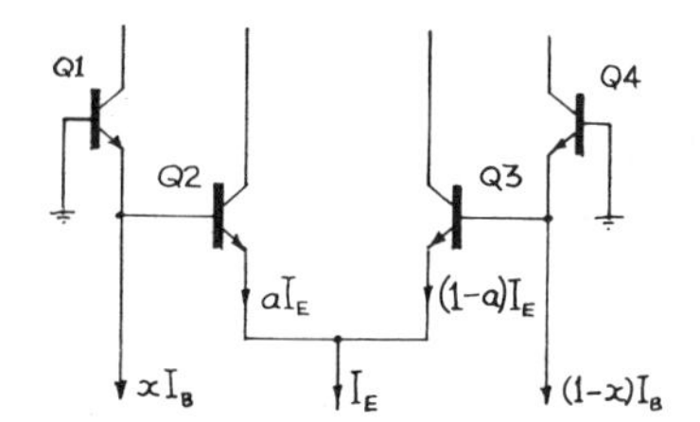

Fig. 6. Inverted form of circuit. Performance is similar, except for phase change in output ($a = 1 - x$).

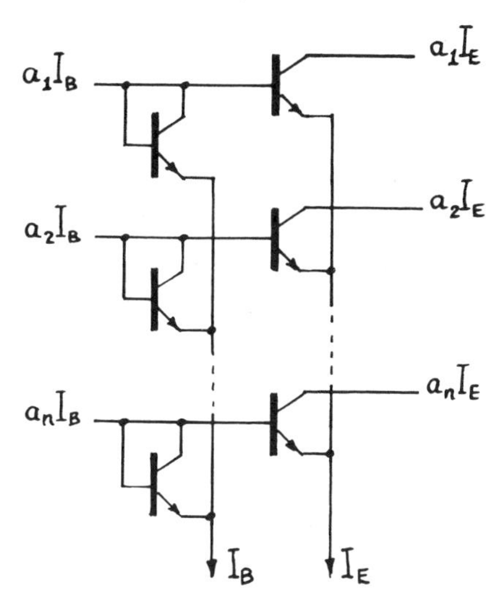

Fig. 7. Multiple-input version of Fig. 3. This circuit is useful to standardize the absolute magnitude of analog signals.

Linearity is not impaired, but the actual stage gain (to the collector circuit) is reduced to

$$G_0' = \theta G_0, \tag{28}$$

and the output swing is reduced.

For example, with $G_0 = 2$ and $\beta = 50$, G_0' is 1.8, 10 percent below the nominal gain, and the limiting values of a are 0.96 and 0.04.

The "inverted" circuit is identical in behavior to the earlier configuration as regards the effects of area mismatches, ohmic resistances, and temperature. It may also be driven by a single-sided signal, by a similar rearrangement to that suggested for Fig. 3. This time, the collectors and bases of $Q1$ and $Q4$ are joined and taken to a positive current source equal to I_B, and the emitter of $Q4$ and base of $Q3$ grounded. The input is then applied to the emitter of $Q1$, and must lie in the range 0 to I_B. Clamping diodes are needed to constrain the input voltage outside of this range.

B. Multiple-Input Configurations

It is not necessary to limit the input and output to a *pair* of complementary currents. Fig. 7 shows how n inputs may be accepted to produce n outputs. This is useful when the inputs are known to be in a certain *ratio* but have an absolute value which may vary widely, and it is desired to standardize the signals to a known amplitude. An example might be in connection with lateral *p-n-p* transistors as the drivers and level shifters.

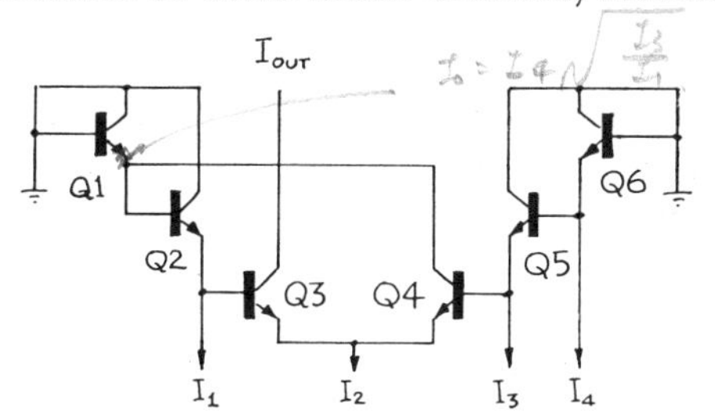

Fig. 8. Product-quotient circuit. By suitable imputs, various functions can be generated.

This circuit also possesses the beta immunity referred to earlier, and can be inverted to the Fig. 6 form.

C. Product-Quotient Configuration

Fig. 8 shows an interesting variant able to produce an output equal to the products or quotients of several inputs at nanosecond speeds. The significant feature of this circuit is the feedback connection of $Q4$'s output to the emitter of $Q1$. Neglecting second order effects, we have

$$(I_2 - I_{\text{out}})I_1 I_{\text{out}} = (I_2 - I_{\text{out}})I_3 I_4 \tag{29}$$

or

$$I_{\text{out}} = \frac{I_3 I_4}{I_1}.$$

The magnitude of the emitter current, I_2, although not present in the expression for I_{out}, determines the maximum value the output may assume.

The circuit may be expanded to accept any odd number of inputs to generate power terms such as x^2, x^3, x^3/yz, etc. Of course, there is a practical limit to the circuit complexity determined by the stacking of mismatch errors.

By putting extra diodes in series with the emitters of the center transistors, $Q3$ and $Q4$, and in $Q1$ and $Q6$, we get

$$I_{\text{out}} = I_4\sqrt{\frac{I_3}{I_1}}. \tag{30}$$

The number of circuits that can be devised to perform functions of this kind is legion.

V. The "Gain Cell"

We will now describe what is probably the most useful circuit to have arisen from this work, certainly as far as incorporation into cascaded amplifiers is concerned. It is shown in Fig. 9, and is similar to the "inverted" circuit of Fig. 6, except that the input currents that reappear at the collectors of $Q1$ and $Q4$ are added in phase with the outputs of $Q2$ and $Q3$. The gain is thus

$$G_0'' = 1 + G_0 = \frac{(I_B + I_E)}{I_B}. \tag{31}$$

This form is attractive for several reasons. Firstly, the inner stage can operate with a gain less than unity, yet still achieve a net stage gain greater than unity; we

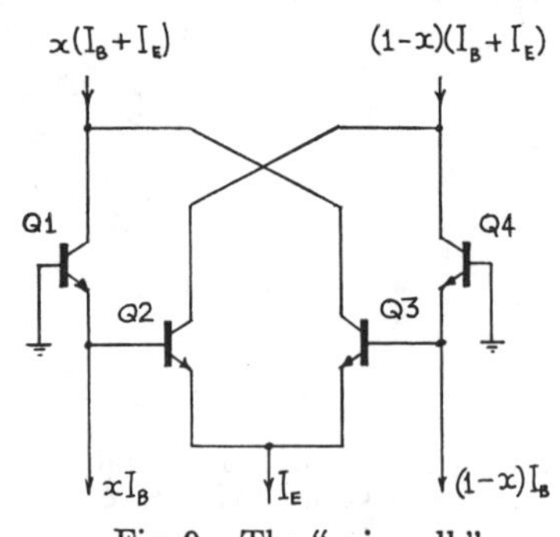

Fig. 9. The "gain cell."

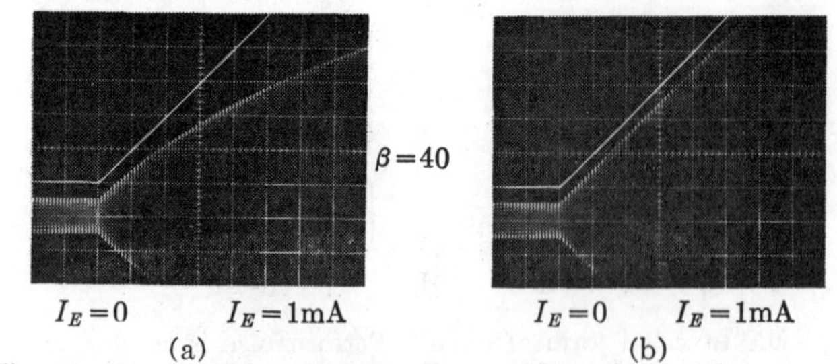

Fig. 10. By sweeping I_E, the effects of beta in limiting the accuracy of gain are demonstrated.

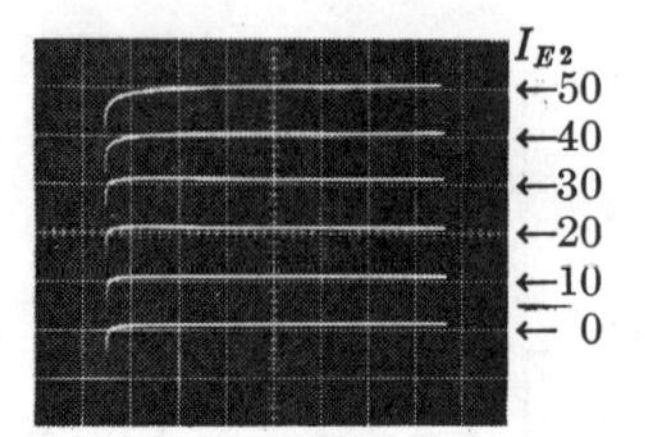

Fig. 11. Thermal distortion of gain cell. Vertical scale expanded to 1 percent/div. $I_B = 20$ mA.

would therefore expect to find this circuit faster for a given gain. Secondly, the cell is well suited to cascading, because the output of one stage can drive the next directly. This leads to a further advantage: all of the current injected at each emitter node contributes to the total output swing. Finally, the bias-voltage circuit for each stage has to supply only the base current for that stage, which is not signal dependent. Hence a low-power diode or resistor string can be used; only 0.5–1 volt per stage is needed.

A. More Exact Gain Expression

Equation (31) omits the effects of beta. It was previously shown that other parameters, such as ohmic resistances and area mismatches, did not affect the *mean* gain over the full dynamic range, but only introduced distortion. Thus, the accuracy limitations determined by beta are of most concern.

In a cascaded amplifier, the inputs of all stages, except the first, are simply the collector currents from the previous stage. Thus, starting with an input bias current, I_B, a modulation index, x, and an emitter supply current I_E, we can calculate the effective values of I_B' and x' for the input to the following stage, the stage gain G_0''', and the output swing ΔI_C.

The results are

$$\begin{aligned} I_B' &= \alpha I_B + I_E \\ \Delta I_C &= \alpha(I_B + \theta I_E) \\ G_0''' &= \alpha(1 + \theta G_0) \\ x' &= \frac{1}{2} + \frac{G_0'''}{\alpha + G_0}(x - \tfrac{1}{2}). \end{aligned} \tag{32}$$

The practical significance of these expressions is demonstrated by the photographs of Fig. 10, which show the swept gain of a gain cell using (a) transistors with a beta of 40, and (b) transistors with a beta of 200. For these measurements, $I_B = 200\ \mu$A and I_E was swept from 0 to over 1 mA. The input was modulated to a depth of 80 percent at 1 kHz. At more practical operating currents, the effects of bulk resistances must be taken into account, preferably by scaling the emitter areas.

B. Thermal Distortion

In many exacting applications, for example, oscilloscope vertical amplifiers, the transient thermal distortions mentioned earlier can be very disturbing. It can be shown that for the gain cell no thermal distortion arises if the total power dissipated in the outer and inner pairs of transistors is the same. Because the inner pair operates at a higher voltage (by one V_{BE}) than the outer pair, I_E must be less than I_B, so that to eliminate thermal distortion a gain of less than two must be used.

It can be simply shown that for no distortion

$$I_E = \frac{V_{CB}}{V_{CB} + V_{BE}} \cdot I_B, \tag{33}$$

where V_{CB} is the bias voltage supplied to each stage and appears across the collector-base junctions of $Q1$ and $Q4$. Typically, with $V_{CB} = 1.6$ volts, $V_{BE} = 0.8$ volts, we must use a stage gain of 1.67.

In practice, it proves very difficult to induce appreciable distortion even under high-power conditions. Fig. 11 shows the step response of a single integrated gain cell operated at very high currents with the vertical scale expanded to 1 percent per division.

VI. Transient Response

An accurate large-signal model for the transistor does not exist, and it is, therefore, not possible to determine the exact step response of this type of circuit. Various small-signal analyses have been made, however, and these indicate, not surprisingly, that the response time is dominated by the f_t of the devices. Further, if the geometries are scaled to the currents, the effects of r_b can be eliminated, and for fairly large stage-gains (greater than two or three), an adequate approximation for the stage response of the Fig. 3 circuit is obtained by treating it as a network with a single pole at f_t/G_0.

The gain-cell response is more complex, and can be approximated by a pole at $f_t/2$, flanked by a pole–zero pair. For $G_0 = 1$, ($G_0'' = 2$), the pole pair are co-incident and cancel, leaving a single-pole response with a gain–bandwidth product equal to f_t. For $G_0 < 1$, the zero moves toward the origin, while the second pole moves out,

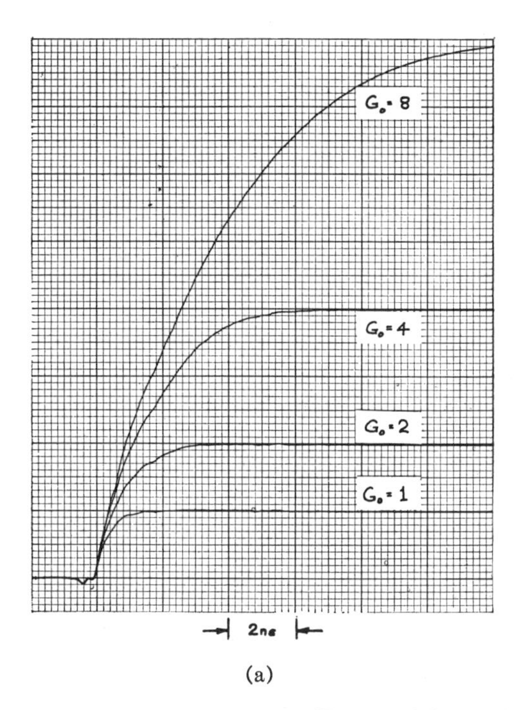

(a)

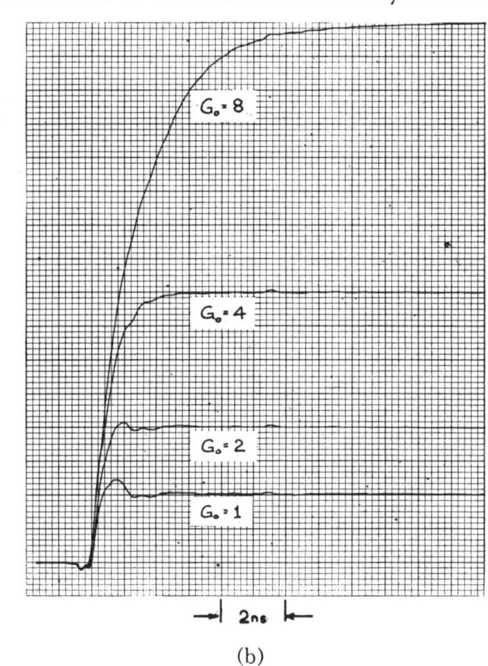

(b)

Fig. 12. Measured transient responses of Fig. 3 circuit.

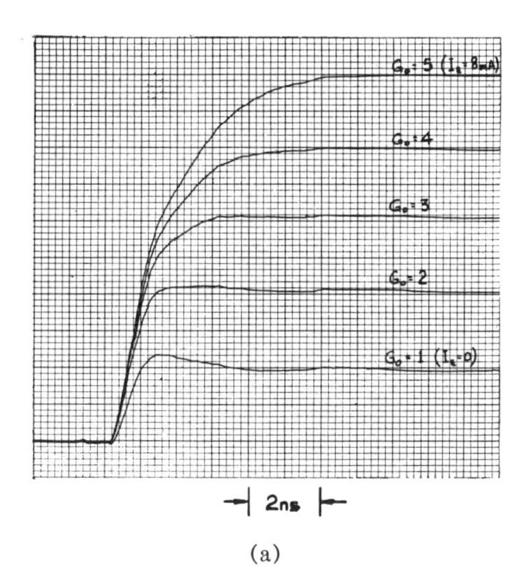

(a)

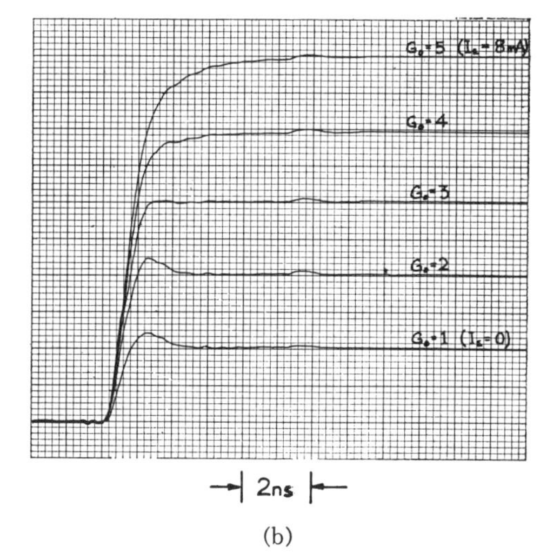

(b)

Fig. 13. Measured transient response of the gain cell.

causing an overshot response. For $G_0 > 1$, the reverse situation arises, and the response consists of an initial fast rise followed by a slower time constant. For $G_0 \gg 1$, the second pole dominates and is at approximately f_t/G_0.

Measurements on integrated forms of Fig. 3 and Fig. 9 have been made, using both the standard process (200 ohms-per-square base diffusion) and a shallower 450 ohms-per-square process. The device characteristics are summarized below.

Parameter	200-Ohm Process	450-Ohm Process	Test Conditions
f_t	600 MHz	1200 MHz	$V_{CE} = 2$ volts, $I_C = 10$ mA
C_{CS}	2 pF	2 pF	$V_{CS} = 10$ volts
C_{CB}	0.7 pF	0.7 pF	$V_{CB} = 0$ volts
β	40	200	$V_{CE} = 2$ volts, $I_C = 5$ mA
r_b	15 ohms	27 ohms	$I_{EB} = 10$ mA, $I_C = 0$

Response time was measured in a test system having an overall risetime of less than 75 ps. The waveforms were plotted on an X–Y plotter, which provided better resolution than photography for comparing several collated responses.

The results confirm the predictions of the small-signal analyses. In Fig. 12, the responses of the Fig. 3 circuits are shown, using (a) the 600-MHz devices, and (b) the 1200 MHz devices. The form of the response closely approximates a single time constant proportional to gain. Fig. 13 shows the more complex "gain cell" response for (a) the 600-MHz devices and (b) the 1200-MHz devices.

It should be mentioned that the test jig incorporated neutralizing capacitances and balun transforms at critical sites to eliminate preshoot and other aberrations in the responses, and those components markedly influenced the apparent risetimes at the lower gains. The improved performance obtained from the totally integrated amplifier (Section VII-D) supports this conclusion.

VII. Cascaded Amplifiers

A repetitive problem in dc amplifier design is that of cascading stages without "level shifting," that is, the buildup of supply–voltage requirements as each stage is added. Several techniques are used to overcome this problem. They include the use of

1) complementary *p-n-p–n-p-n* devices between two supply rails,
2) zener-diode or resistive level-shifting circuits,
3) very low bias voltages for each stage.

For monolithic designs, the third method is by far the most attractive, but is possible only where voltage swings are small, and hence is eminently suited to the circuits under discussion. Several cascaded forms will be presented that use this approach. In some cases, stages can be cascaded indefinitely without level shifting.

Other factors of interest in a cascaded amplifier are gain stability, drift, overload recovery, dinearity, noise level, and transient response. Typical results for these parameters will be given where appropriate.

A. Bandwidth of Cascaded Stages

It was shown earlier that each stage of an amplifier using the diode transistor circuit can be treated as a single pole at f_t/G_0, where G_0 is the low-frequency current gain of the stage.

The overall bandwidth, F, of N cascaded stages is thus

$$F = \frac{f_t}{G_0}(\sqrt[N]{2} - 1)^{1/2}. \tag{34}$$

Also, if the total current-gain, G, is shared equally over the N stages we can write

$$\frac{F}{f_t} = \frac{(\sqrt[N]{2} - 1)^{1/2}}{\sqrt[N]{G}}. \tag{35}$$

This "bandwidth-shrinkage" function is plotted in Fig. 14. It shows three things of interest. Firstly, there is an optimum number of stages required to maximize the bandwidth. Secondly, the maxima are well defined for low gains but vague at high gains. Finally, the bandwidth of a large number of stages is relatively insensitive to overall gain.

In discrete designs, it is frequently necessary to use less than the optimum number of stages for economic reasons. The possibility of integrating the entire amplifier removes this limitation.

B. A Practical Cascaded Amplifier

The first cascaded form we will discuss is made by taking the Fig. 3 circuit and connecting the collectors of one stage to the bases of the next, as shown in Fig. 15. The amplifying transistors operate with $V_{CB} = 0$, just

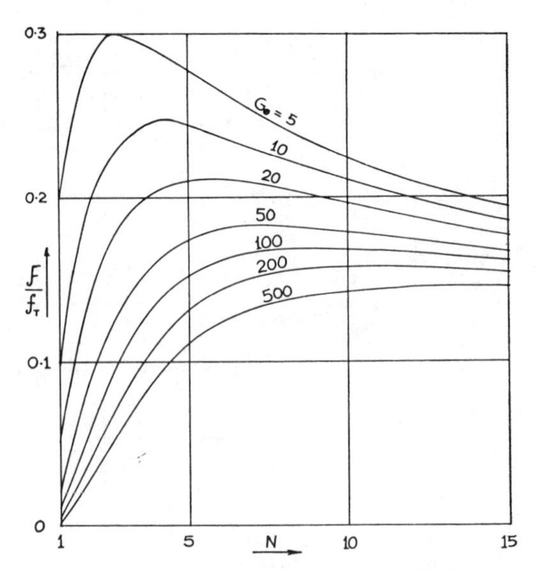

Fig. 14. Curves showing number of stages required to maximize bandwidth.

as the diode-connected transistors do. Currents must be supplied at each interface, and these, together with the emitter currents, determine the gain. The overall gain of N stages is

$$G = \frac{I_2}{I_1}\cdot\frac{I_4}{I_3 - I_2}\cdots\frac{I_{N+2}}{I_{N+1} - I_N}. \tag{36}$$

Consequently, the gain can be swept over a very wide range (zero to about β^N), but the gain is a sensitive function of the difference terms in the denominators of (36). Methods of overcoming this problem have been devised, using dependent current sources for the emitters. Diodes (not shown) prevent the transistors saturating at the extremes of the dynamic range.

A limited investigation of this form of amplifier has been made. Some results for a four-stage high-gain (about X1000) design are shown in Fig. 16. They demonstrate the linearity and overload recovery to a 3-μs ramp (Fig. 16(a)), and the transient response and noise level (Fig. 16(b)). In the second photograph, the responses of 600-MHz and 1200-MHz transistors are compared.

C. Cascaded Gain Cells

A series of three gain cells is shown in cascade in Fig. 17. The power is supplied as currents (to the emitter nodes) which set both the gain and the output swing capability. A low-power bias string, shown here as pairs of diodes, sets the operating voltage of each stage. The balanced nature of the amplifier results in very small signal currents in the base circuit, and permits a common undecoupled line to supply all the bases in a multistage amplifier. The overall gain, for sensible values of beta, is

$$G' = \frac{I_{E1} + I_{E2} + I_{E3} + I_{E4}}{I_{E1}}. \tag{37}$$

With this method of cascading, the correct static bias conditions to handle the steadily increasing signal ampli-

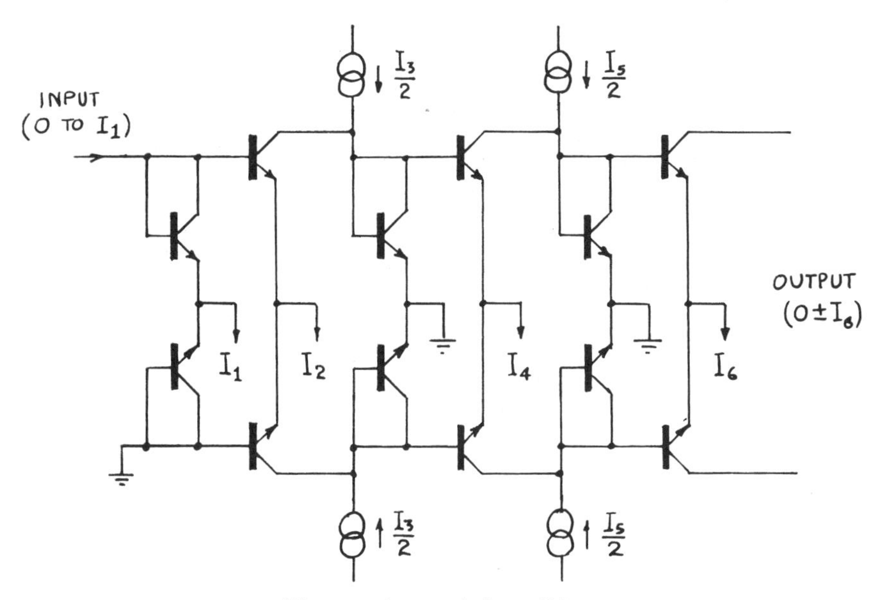

Fig. 15. A cascaded amplifier.

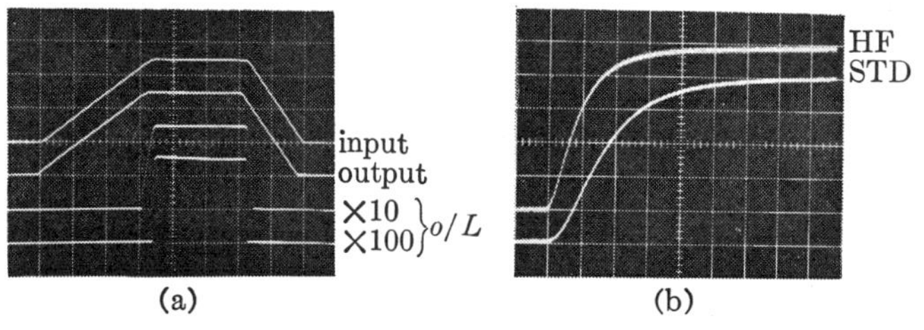

Fig. 16. Performance of amplifier similar to Fig. 15. In (a), a 2.5-μA input current (top trace) is accurately reproduced at × 1000 (second trace). When overloaded ten or a hundred times, (lower traces) recovery is rapid. The top portion of the waveform is within the linear range. Time scale is 1 μs/div. Waveforms (b) show transient response at × 1000, to a 1-μA input step using 600-MHz (top) and 1200-MHz (bottom) transistors. Time scale is 20 ns/div.

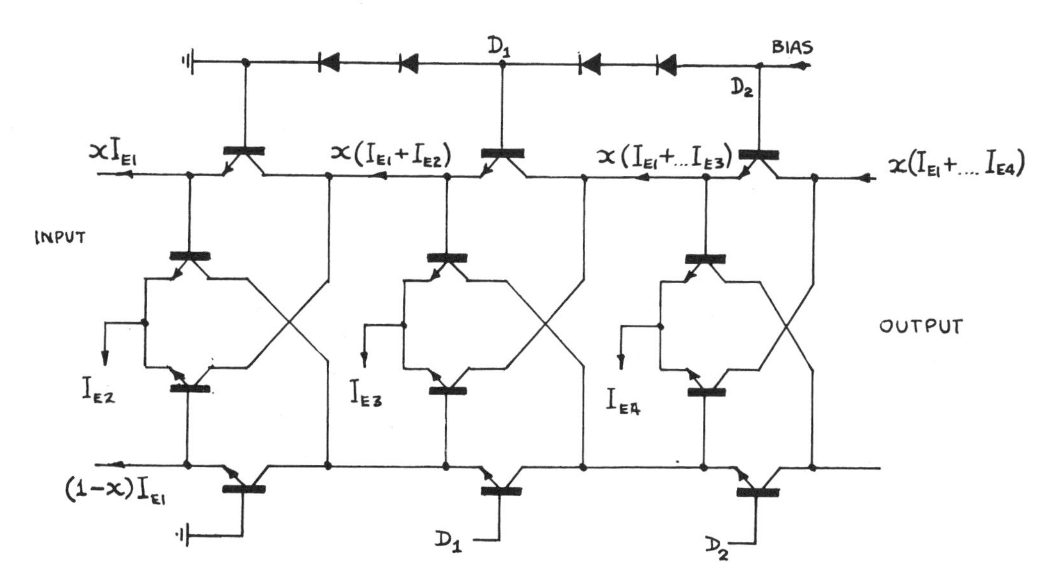

Fig. 17. Cascaded gain cells.

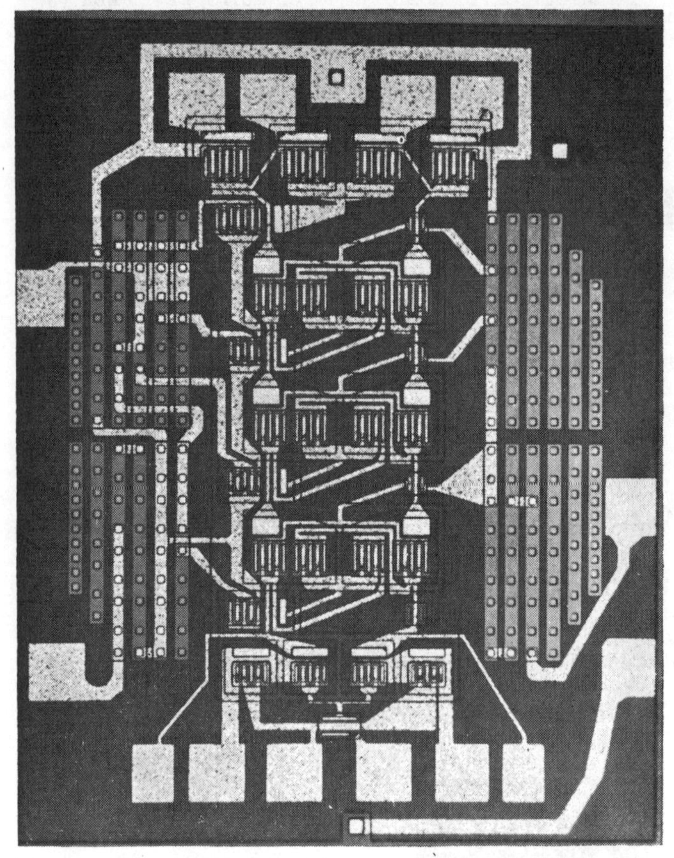

Fig. 18. Photomicrograph of integrated gain-cell amplifier. The emitter currents are supplied by the transistors just left of center, and the signal path is from bottom to top.

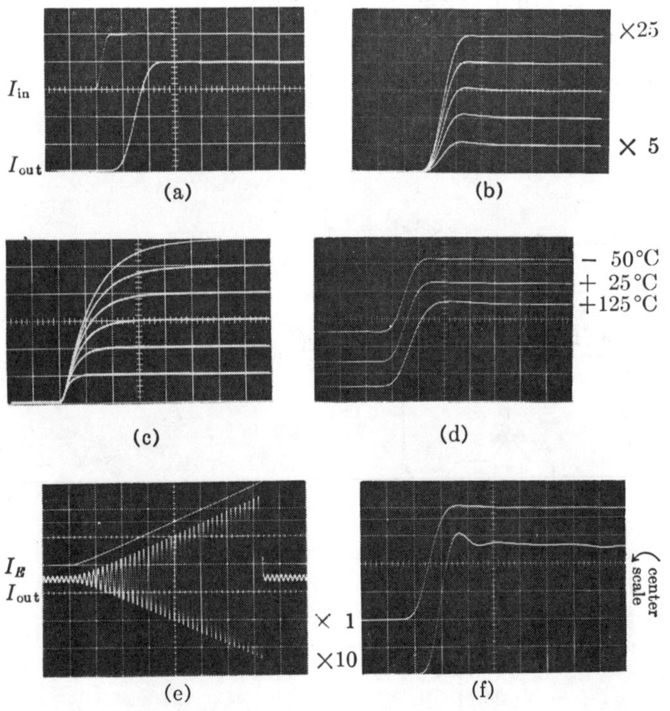

Fig. 19. Performance of the integrated amplifier. (a) Input/output delay at ×20, 1 ns/div. (b) Transient-response variations for gains of ×5 to ×25, 1 ns/div. (c) For gains ×50 to ×300, 5 ns/div. (d) Shows transient response for ×25 at −50°C, 25°C, and 125°C. (e) Demonstrates swept-gain linearity, ×1 to ×30. (f) Shows recovery from ×10 overload.

tude along the chain are automatically present, because *signal and bias currents increase at the same rate.* This is in contrast with conventional amplifiers where care must be taken to provide adequate range at each stage.

A corollary of this is that all stages overload at the same point on the overall transfer characteristic, when $x < 0$ or $x > 1$. Furthermore, at overload, no transistors saturate, and, because of the modulation reduction discussed above, the input to each successive stage never quite drops to zero. Consequently, the recovery characteristics are good.

D. An Integrated Gain-Cell Amplifier

Fig. 18 is a photomicrograph of a five-stage amplifier on a 50 × 60 mil die, using the 1200-MHz transistors. It comprises an input stage of the type shown in Fig. 3 followed by four cascaded gain-cell stages each with a current-source transistor, and a patchwork of resistors which can be used to set the base bias voltages and emitter currents.

The mask layout is of considerable importance in achieving optimum performance. For example, in order to maximize the f_t of each stage, the transistors must be increased in size from input to output, thus maintaining a constant current density. Also, to reduce collector-substrate capacitance, the gain-cell transistors are paired into two common-collector isolations, which also serve to provide a crossover region for the base circuit. Because of the low impedance level of the circuit, meticulous attention to balancing the metallization resistances is essential to achieve low distortion.

In this circuit, the inner and outer transistors have equal area. For minimum distortion due to bulk resistances a stage gain of two must be used, producing an overall gain of 16. (The input stage operates at a gain of unity.) However, quite large variations from this value can be tolerated without introducing significant distortion.

A summary of the performance of the integrated amplifier is given below, and in the waveforms of Fig. 19. The system risetime used for these measurements was 0.4 ns, and, accordingly, some corrections were necessary to the measured risetimes.

The input to the amplifier was supplied by a push-pull 50-ohm pulser driving $Q1$ and $Q4$ via 1-kΩ resistors, to ensure current-drive conditions. Also, both outputs were connected to the sampling oscilloscope via 50-ohm lines terminated at both ends. Under these conditions, no power gain was realized for current gains below 40. It is, however, entirely practical to match both input and output to 50 ohms.

By suitable choice of current ratios, a wide range of gain from zero to ×1000 could be obtained, although acceptable high-speed performance was possible only over a limited range. Some idea of the effect of gain on transient response is provided by Fig. 19(b) and (c). Considerable care had to be exercised to eliminate ringing and preshoot, particularly in respect to ground-plane currents. A bifilar transformer was required to balance the input drive. It is interesting to note the improvement that resulted from putting all the stages on one die. Results published earlier [8] showed a large amount of ringing. These waveforms were for an amplifier in which individually packaged gain cells were breadboarded together. Series damping resistors were required between stages to bring the ringing down to even this level, and these slowed the response.

Here is a summary of the performance of the fully integrated amplifier.

Current Gain	Corrected Risetime (ns)	Overshoot (percent)	Bandwidth (MHz to −3 dB)	Gain–Bandwidth Product (GHz)
5	0.58	12	>500	–
10	0.63	8	>500	–
15	0.69	3	500	7.5
20	0.81	–	420	8.4
25	0.92	–	370	9.2
50	2.0	–	170	8.5
100	4.0	–	92	9.2
200	8.5	–	45	9.0
300	12.0	–	31	9.3

The effect of temperature on transient response was also checked, and, as expected, the risetime decreases at low temperatures, due to increases in f_t. Also checked were the accuracy of gain in response to a linearly swept current (Fig. 19(e)) and overload recovery (Fig. 19(f)). The latter shows the response for an output equal to 80 percent of the dynamic range, and that for a ×10 increase in the input.

The CW response is shown in Fig. 20, for current gain from ×5 to ×50, and for a signal equal to 50 percent of the dynamic range. Other performance details, for a gain of ×25, are

Output current swing:	±60 mA into two 25-ohm loads
Output voltage swing:	3 volts peak–peak, differentially
Dissipation:	320 mW (64 mA at 5 volts)
Gain stability:	Within ±0.35 dB } over −35°C to +125°C
Zero drift:	<0.4 percent of full scale } over −35°C to +125°C

These results are very encouraging. Subsequent improvements in device technology will almost certainly permit faster gain-cell amplifiers to be fabricated.

VIII. Summary

This paper has described what is believed to be a new technique for the design of very linear transistor amplifiers especially suited to monolithic planar fabrication. It demonstrates that useful dc-coupled wide-band amplifiers can be made using transistors alone, operating in a strictly current-gain mode. A commercial integrated circuit giving a controllable power gain of 0 to 20 dB with a dc to 500-MHz response, and designed to be cascadable without additional components, is now fully practical. This building-block concept could be expanded

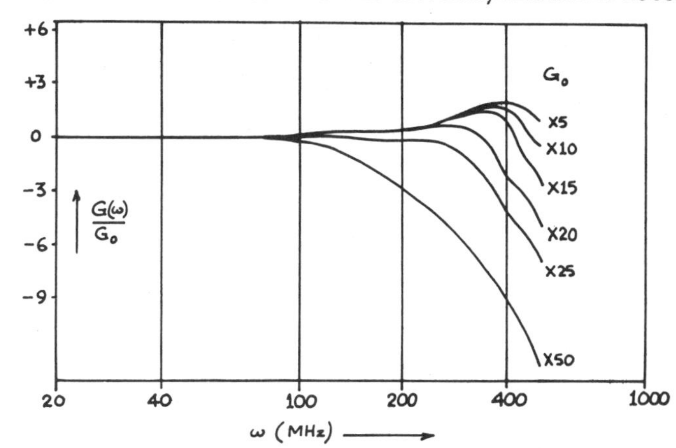

Fig. 20. CW response of integrated amplifier.

to include precision high-impedance input stages, power amplifiers, four-quadrant multipliers, etc., and would fill a long neglected need. Some of these topics will be taken up in subsequent papers.

Appendix

The Generalized Principle

The key equation for these circuits is of the form given in (9), an algebraic summation of an even number of logarithmic terms set equal to zero, having a common factor, mkT/q, which is, thus, of no consequence. The logarithmic arguments may be compounded into a single product-quotient term, and finally the antilogarithm of both sides is taken, leaving the product-quotient term equal to unity, for example

$$\frac{I_{D1}I_{D2}I_{S3}I_{S4}}{I_{S1}I_{S2}I_{D3}I_{D4}} = 1. \tag{38}$$

The saturation currents of the diodes in the loop can be extracted as the ratio

$$\gamma = \frac{\prod I_s(+)}{\prod I_s(-)} \tag{39}$$

where the plus sign indicates all diodes connected in the same direction around the loop, and the minus sign indicates those in the opposite direction. Since I_s is proportional to area, the γ ratio can also be calculated as the ratio of the *areas* of the emitter diodes in the loop. Also, every I_s varies with temperature in the same way, and thus temperature drops out of the analysis at this point, too, explaining the excellent freedom from temperature effects observed in these amplifiers. This reduces the key equation to the form.

$$\prod I(+) = \gamma \prod I(-). \tag{40}$$

Stated at length, in a closed loop containing an even number of perfect exponential diode voltages (not necessarily two-terminal devices) arranged in cancelling pairs, the currents are such that the product of the currents in diodes whose voltage polarities are positive with respect to a node in the loop is exactly proportional to the product of the currents in diodes whose voltages are negative with respect to that node, the constant of proportionality being the ratio of the product of the saturation currents of the former set of diodes to that of the latter set.

Acknowledgment

During the course of this work, the discussions held with L. Larson of the Advanced Instruments Group and G. Wilson of the Integrated Circuits Group proved stimulating and helpful. The experimental work of E. Traa is also gratefully acknowledged.

References

[1] A. E. Hilling and S. K. Salmon, "Intermodulation in common-emitter transistor amplifiers," *Electron. Engrg.*, pp. 360–364, July 1968.

[2] L. C. Thomas, "Eliminating distortion in broadband amplifiers," *Bell Sys. Tech. J.*, p. 315, March 1968.

[3] J. S. Brown, "Broadband amplifiers," in *Amplifier Handbook.* New York: McGraw-Hill, 1966, sec. 25, pp. 25–57.

[4] L. F. Roeshot, "U.H.F. broadband transistor amplifiers," *EDN Mag.*, January–March 1963.

[5] W. R. Davis and H. C. Lin, "Compound diode-transistor structure for temperature compensation," *Proc. IEEE (Letters)*, vol. 54, pp. 1201–1202, September 1966.

[6] A. Bilotti, "Gain stabilization of transistor voltage amplifiers," *Electron. Letters*, vol. 3, no. 12, pp. 535–537, 1967.

[7] A. M. VanOverbeek, "Tunable resonant circuits suitable for integration," *1965 ISSCC Digest of Tech. Papers*, pp. 92–93.

[8] B. Gilbert, "A dc-500 MHz amplifier multiplier principle," *1968 ISSCC Digest of Tech. Papers*, pp. 114–115.

[9] ——, "A precise four-quadrant multiplier with subnanosecond response," this issue, pp. 365–373.

[10] H. E. Jones, "Dual output synchronous detector utilizing transistorized differential amplifiers," U. S. Patent 3 241 078, June 18, 1963.

[11] A. Bilotti, "Applications of a monolithic analog multiplier," *1968 ISSCC Digest of Tech. Papers*, pp. 116–117.

[12] W. R. Davis and J. E. Solomon, "A high-performance monolithic IF amplifier incorporating electronic gain-control," *1968 ISSCC Digest of Tech. Papers*, pp. 118–119.

[13] G. W. Haines, J. A. Mataya, and S. B. Marshall, "IF amplifier using C-compensated transistors," *1968 ISSCC Digest of Tech. Papers*, pp. 120–121.

[14] C. T. Sah, "Effect of surface recombination and channel on *P-N* junction and transistor characteristics," *IRE Trans. Electron Devices*, vol. ED-9, pp. 94–108, January 1968.

IF Amplifier Using C_c Compensated Transistors

JOHN A. MATAYA, MEMBER, IEEE, GEORGE W. HAINES, MEMBER, IEEE, AND SUMNER B. MARSHALL

Abstract—**An integrated bridge network of four transistors is used as a self-neutralized active element in tuned RLC amplifier designs. The bridge network compensates for the transistor collector-base junction capacitance (C_c), yielding a 95-percent reduction in the common-emitter reverse transmission admittance. IF amplifier stages that achieve the maximum unilateral power gain of a common-emitter transistor while maintaining excellent alignability are realized using the C_c compensated transistor structure.**

Variations of the relative bias current levels of the transistors in the bridge network provides gain control by way of signal cancellation. This technique produces minimal frequency response variations of the amplifier stage being controlled. A noise analysis shows output signal to noise ratio at maximum attenuation can be a performance limitation.

A three-stage synchronously tuned 90-dB power gain amplifier with a center frequency at 44 MHz and a two-stage IF strip suitable for use in television receivers have been constructed as design examples.

INTRODUCTION

THE EFFECTS of the transistor collector-base depletion layer capacitance (C_c) on low-pass common-emitter amplifiers were discussed in previous work [1], [2]. In common-emitter tuned RLC bandpass amplifiers, the depletion layer capacitance couples the input and output tank circuits. This input–output interaction can cause the amplifier to be potentially unstable and thereby limit the usable power gain that can be achieved per stage. Further, this input–output coupling makes the tuning and alignment of a cascade of these tuned stages difficult, since the adjustment of one tuned network affects the adjustment of the other tuned networks in the cascade. If the collector depletion layer capacitance is large, the interaction is very strong, and it will be impossible to synchronously tune the cascade.

A measure of this coupling, or interaction, is the alignability factor k [3], which is defined by (1).

$$k = \frac{|y_{12}y_{21}|}{2(g_{11} + G_s)(g_{22} + G_L)} \tag{1}$$

where

$$y_{12} = g_{12} + jb_{12}$$

$$y_{21} = g_{21} + jb_{21}$$

where y_{12} and y_{21} are the reverse transmission and forward transmission admittance parameters of the transistor, respectively. g_{11} and g_{22} are the real parts of the input and output admittances of the device, respectively. G_s and G_L are the real part of the source admittance and the real part of the load admittance, respectively. As can be seen in (2), the alignability factor k is directly dependent on y_{12}, the reverse transmission admittance parameter of the active device. As k increases due to y_{12}, it becomes more difficult to align a cascade of tuned am-

Manuscript received June 5, 1968; revised August 26, 1968. This work was in part supported by the Air Force Avionics Laboratory, Research and Technology Division, Air Force System Command, Wright-Patterson AFB, Ohio. This paper was presented at the 1968 ISSCC.

The authors are with the Sprague Electric Company, Linear Integrated Circuits Development, North Adams, Mass.

Reprinted from *IEEE J. Solid-State Circuits*, vol. SC-3, pp. 401-407, Dec. 1968.

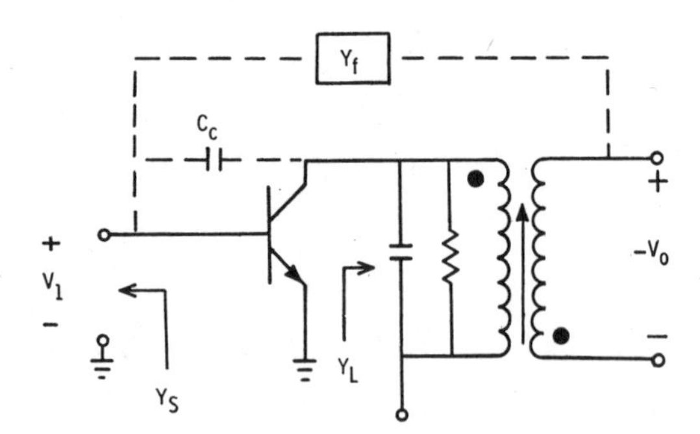

Fig. 1. Neutralized common-emitter tuned amplifier stage.

plifier stages and therefore the term alignability factor.

The power gain [4] of an RLC tuned amplifier at the center frequency of the amplifier is given by the equation

$$PG = \frac{P_0}{P_i} = \frac{|y_{21}|^2 G_L}{(g_{22} + G_L)^2(G_{11} + G_1)} \tag{2}$$

where

$$G_1 = \mathrm{Re}[Y_1] = \mathrm{Re}\left[\frac{y_{12}y_{21}}{y_{22} + Y_L}\right]. \tag{3}$$

Y_1 is the admittance reflected from output to input due to y_{12}. The presence of C_c yields Y_1, which decreases the power gain that can be obtained from a tuned RLC amplifier stage.

In designing a tuned RLC amplifier, k is typically adjusted to be less than or equal to 0.2 to assure reasonable ease of alignment.

There are a number of ways in which k can be decreased. One method is depicted in Fig. 1. This figure shows a simple RLC tuned amplifier with an external feedback network Y_f. If Y_f is adjusted to be equal to the negative of y_{12}, the transistor amplifier will be neutralized. In discrete amplifier designs, obtaining an exact match of Y_f to y_{12} is difficult, due to parameter changes of the active device with temperature and biasing conditions.

[1] The alignability factor k could also be called the stability factor. A tuned RLC amplifier will be stable if the following inequality is satisfied [3], [4].

$$1 - \frac{R_e[y_{12}y_{21}]}{2(g_{11} + G_s)(g_{22} + G_L)} - \frac{|y_{12}y_{21}|}{2(g_{11} + G_s)(g_{22} + G_L)}$$
$$= 1 - k\left[1 + \frac{R_e[y_{12}y_{21}]}{2(g_{11} + G_s)(g_{22} + G_L)}\right] > 0$$

For the case where the real part of Y_{12} is negligible (for the transistors considered, this approximation is valid beyond 100 MHz), this inequality can be rewritten as follows:

$$k\left[1 - \frac{I_m[y_{21}]}{|y_{21}|}\right] < 1.$$

Therefore the amplifier will be stable if

$$k < \frac{|y_{21}|}{|y_{21}| - I_m[y_{21}]}.$$

A second method of reducing k is to mismatch the load conductance (G_L) and the output conductance (g_{22}). If G_L is increased, we see from (1) that the alignability factor k will decrease, but also, as seen in (2), the power gain of the stage will be reduced. This second method has the advantage that the required adjustment of G_L to decrease k is not as critically dependent upon device parameters as is the neutralization method.

A third method to decrease k is to use a unilateral device, or device combination, in which y_{12} is zero or at least negligible.

C_c Compensated Transistor

One device combination for which y_{12} can be considered negligible is the C_c compensated transistor [1], [2], [5]. The effects of the collector-base depletion layer capacitance can be avoided in a differential amplifier stage by employing bridge neutralization, as shown in Fig. 2. If feedback capacitors C_{f1} and C_{f2} are adjusted to be identically equal to the collector-base depletion layer capacitance of transistors X_1 and X_2, a bridge network is formed. This bridge network removes the effects of C_c from the forward and reverse transfer functions of the stage, since forward and reverse transmission due to the transistor collector-base capacitance is compensated for by antiphase transmission through the feedback capacitor. As stated previously, this compensation is difficult to achieve in discrete form due to the tight tolerances involved, but has been shown to be feasible in monolithic integrated circuit form by the use of identical collector-base junctions to provide the feedback capacitors C_{f1} and C_{f2}.

Fig. 3(a) shows a C_c compensated transistor quadruplet. Transistors X_1 and X_2 are the differential pair; transistors X_{1c} and X_{2c} are the compensating transistors. Fig. 3(b) shows a photomicrograph of a monolithic integrated realization of the C_c compensated transistor quadruplet. Since these devices are fabricated in integrated form, good matching between devices is achieved, and very good C_c compensation is obtained. Measured data [1], [5] have indicated that the reverse transmission due to C_c in a compensated device is less than 5 percent of the value appropriate for a device fabricated in the same method, but without compensation. This compensation is independent of operating voltage and temperature within limits defined by junction breakdowns and maximum junction temperature. When used as a discrete device, the symmetrical layout of the devices also balances out any stray package or header capacitance.

Tuned Amplifiers

The use of C_c compensated transistors in a typical tuned interstage is shown in Fig. 4. When operated with a differential output, this tuned interstage will have the same gain as a common-emitter stage, but with the following advantages. First, since the amplifier is neutralized by C_c compensation, the maximum available power

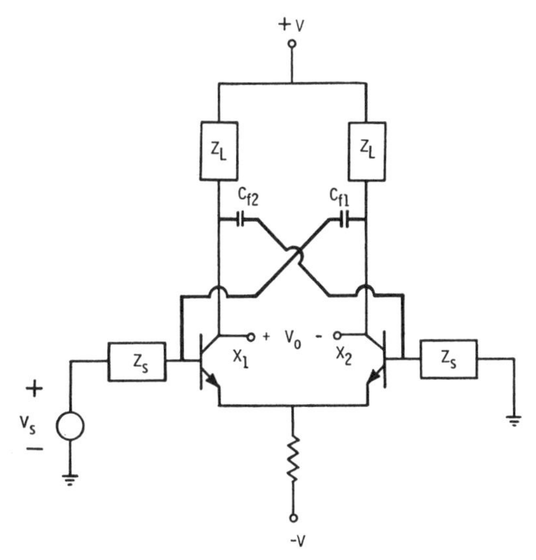

Fig. 2. Neutralized differential amplifier stage.

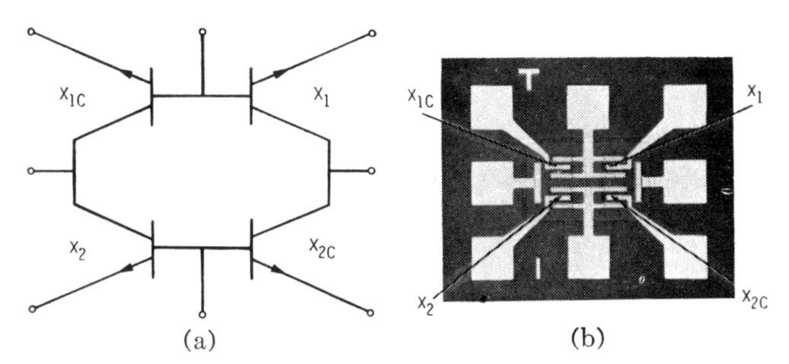

Fig. 3. C_c compensated transistor quadruplet. (a) Circuit diagram. (b) Photomicrograph of monolithic realization.

gain² of the stage can be utilized while maintaining a small alignability factor. Second, since the amplifier is a differential amplifier that is a balanced configuration, common-mode rejection of all signals on the power supply line is obtained. Therefore, no power supply decoupling networks are required between stages.

The power gain and alignability factor for a typical tuned interstage amplifier utilizing a C_c compensated device with 5-percent imbalance of C_c compensation is shown in Fig. 5. Similar curves for an uncompensated device of identical fabrication are also shown in Fig. 5. The curves are plotted as a function of various values of emitter current. The power gain of the uncompensated device as calculated for the case of using a mismatched load conductance (G_L) to reduce the alignability factor to 0.2. As can be seen in this figure, the

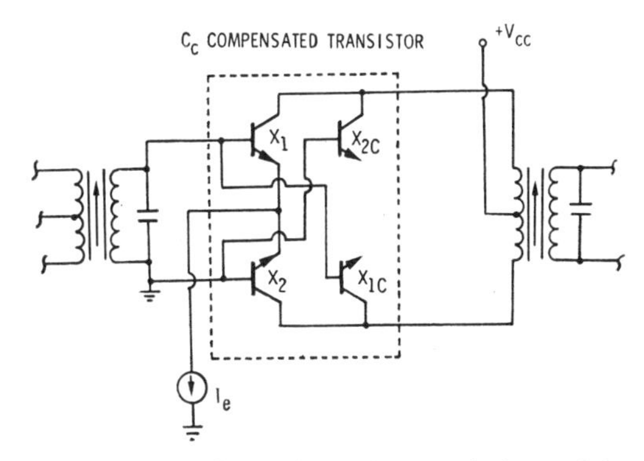

Fig. 4. C_c compensated transistors in a typical tuned interstage.

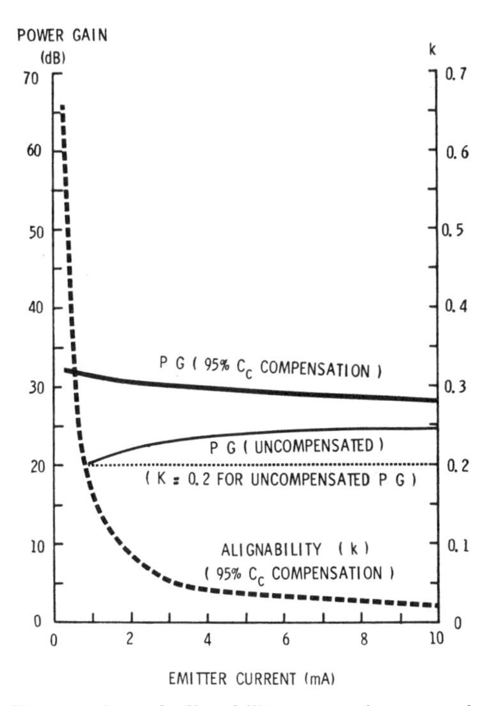

Fig. 5. Power gain and alignability comparison as a function of emitter current.

power gain of the 95-percent C_c compensated device is from 5 to 10 dB greater than the power gain available from the mismatched uncompensated device, and the alignability factor for the compensated device is as much as ten times less than that of the uncompensated device. Therefore, the amplifier using the compensated device will be much more stable.

Characteristics of Amplifiers Using C_c Compensated Transistors

An added feature obtained when using the C_c compensated device in IF amplifiers is that it provides a method of cancellation gain control. Fig. 6 again shows the typical tuned interstage using the C_c compensated transistors, but modified by the addition of a second current source I_{e2} connected to the emitters of the compensating devices X_{1c} and X_{2c}. The addition of the second current source modifies the voltage gain of the amplifier to be that given in (4).

² The maximum available power gain [6], when finite, occurs when the source and load admittances are chosen such that the following relationships of power (P) are obtained.

$$\frac{P_{\text{out}}}{P_{\text{in}}} = \frac{P_{\text{out}}}{P_{\text{available source}}} = \frac{P_{\text{available out}}}{P_{\text{available source}}}.$$

For a unilateral active device, the maximum available power gain will be obtained if the input and output admittances of the device are conjugately matched to the source and load admittances, respectively.

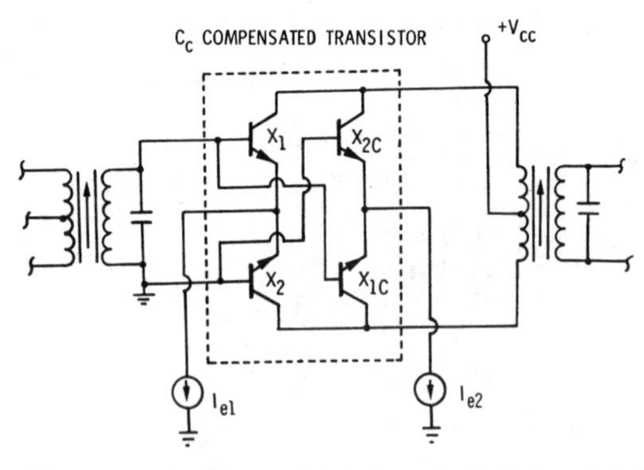

Fig. 6. C_c compensated transistors in a typical tuned interstage modified to provide gain control.

$$G_v = \frac{q}{KT}(I_{e1} - I_{e2})\frac{1}{G_L} \tag{4}$$

where G_L is the real part of the tuned load conductance, q equals the electronic charge, K is Boltzman's constant, and T is temperature in degrees Kelvin. If identically matched devices are used in the C_c quadruplet, when the emitter currents (I_{e1}, I_{e2}) are equal, theoretically the signal output will be reduced to zero as cancellation of the signal occurs. This form of gain control yields a number of advantages. If the sum of the emitter currents I_{e1} and I_{e2} is held constant while their ratio is varied to achieve gain control, the total input and output admittance parameters of the differential stage will have a minimum variation with gain reduction. The measured variation in the y parameters of a C_c compensated device quadruplet for a gain control range of -40 dB is

$$\Delta g_{11} < 10 \text{ percent}; \qquad \Delta b_{11} < 10 \text{ percent}$$

$$\Delta g_{22} < 10 \text{ percent}; \qquad \Delta b_{22} < 20 \text{ percent}$$

where $y_{11} = g_{11} + jb_{11}$ is the input admittance parameter (grounded emitter), and $y_{22} = g_{22} + jb_{22}$ is the output admittance parameter (grounded emitter).

A second advantage of using this form of gain control is the improvement in the cross-modulation characteristics of the amplifier. Fig. 7 presents the cross-modulation characteristics of a tuned amplifier interstage utilizing C_c compensated transistors. The data were measured for various total emitter currents ($I_{e1} + I_{e2}$ = constant = I_T) of 2, 5, and 10 mA.

For comparison purposes, Fig. 7 includes the cross-modulation characteristics of a differential amplifier ($I_{T0} = 5$ mA) utilizing reverse gain control. (Gain control is obtained by decreasing the g_m of a common emitter transistor by decreasing the emitter bias current.) As is indicated, the cancellation gain control method provides improved cross-modulation characteristics over the range of −40-dB gain reduction.

Fig. 8 shows the gain reduction of a differential compensated transistor amplifier as a function of emitter current ratio, I_{e2}/I_{e1} *for* $I_{e1} + I_{e2} = I_T$, a constant. A large amount of gain reduction is available with the can-

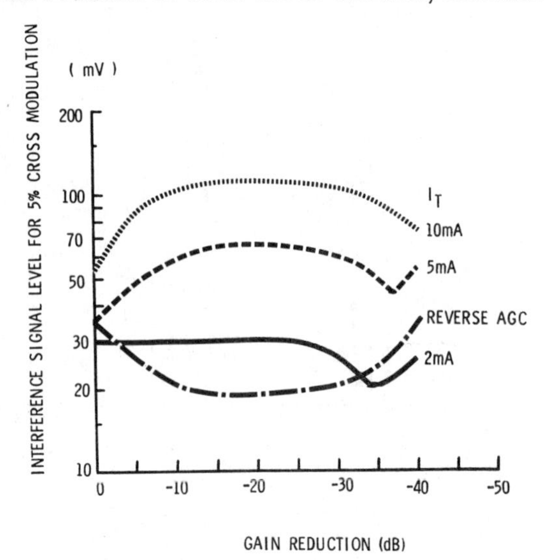

Fig. 7. Cross-modulation characteristics of typical C_c compensated transistor amplifier stage using cancellation gain control.

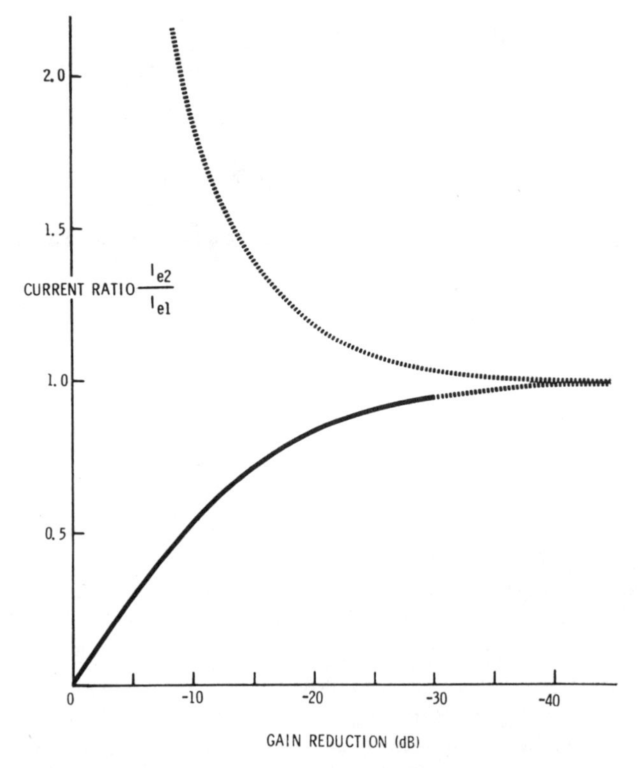

Fig. 8. Gain reduction as a function of bias current ratio of typical C_c compensated transistor amplifier stage.

cellation method. Since the change in I_{e2} over I_{e1} required to increase the gain reduction more than −40 dB is very small, achieving a gain reduction in excess of −40 dB is impractical. Further, if the I_{e2} to I_{e1} ratio should increase to greater than 1, the gain of the amplifier will increase, but with a reversed phase output signal. Because of this, a practical upper limit of −30 to −40 dB is set on the gain reduction to prevent reversal of the characteristics.

Additional gain reduction, as required in a cascade of tuned stages for IF amplifier use, is obtained by using successive stages. Since the cancellation form of gain reduction has a minimal effect on the input and output admittance parameters of the stage, little or no detuning will occur with attenuation.

Noise Analysis

Typically, the specifications given for an IF amplifier include noise figure, gain, attenuation range, etc. One parameter of interest, which is not always specified, is the signal-to-noise ratio at minimum gain. The noise figure and gain of a C_c compensated transistor tuned amplifier is the same as the noise figure and gain of a common emitter stage when operated at full gain. But due to the characteristics of the gain control method, signal-to-noise ratio at maximum attenuation can be different.

Fig. 9 shows the single-ended equivalent circuit of a C_c compensated transistor amplifier stage. Transistor X_1 is one half the differential stage pair; transistor X_{2c} is one half the compensating pair. The output current i_0 is given by (5),

$$i_0 = (i_{L1} - i_{L2}) \tag{5}$$

and is equal to the difference in load current 1 and load current 2. This circuit model is correct under the assumed conditions of a balanced input signal and a balanced output signal.

Fig. 10 is again the single-ended equivalent circuit, but with transistor 1 replaced by its equivalent noise circuit [7]. In this circuit, $\bar{v}_s^2$, $\bar{v}_{x1}^2$, $\bar{i}_{b1}^2$, $\bar{i}_{c1}^2$ are the source resistance, base resistance, base current, and collector current noise sources, respectively. This model includes the noise effects of the base-spreading resistance, but the base-spreading resistance itself is ignored in comparison to the source resistance.

Analysis of this circuit, under conditions of reduced gain, indicate that the output current due to the different noise generators are as follows.

$$i_0(v_s) = \frac{V_s G_s (G_{m1} - G_{m2})}{2(G_s + G_{i1} + G_{i2})}, \tag{6}$$

$$i_0(i_b) = \frac{(i_{b1} - i_{b2})(G_{m1} - G_{m2})}{2(G_s + G_{i1} + G_{i2})}, \tag{7}$$

$$i_0(i_c) = \frac{i_{c1}}{2} - \frac{i_{c2}}{2}, \tag{8}$$

$$i_0(v_x) = \frac{v_{x1}[(G_s + G_{i2})G_{m1} + G_{m2}G_{i1}] - v_{x2}[(G_s + G_{i1})G_{m2} + G_{m1}G_{i2}]}{2(G_{i1} + G_{i2} + G_s)}. \tag{9}$$

The output current, due to the source resistance noise is proportional to the difference in the G_m of transistor X_1 and the G_m of transistor X_{2c} (6). Therefore, this noise source is attenuated at minimum gain where $G_{m1} = G_{m2}$. Similarly, the output current due to the base current noise generator is also proportional to the difference in G_{m1} and G_{m2} (7). Again, this noise contribution will be attenuated at maximum gain reduction. The output current due to the collector current noise source is proportional to a constant (8) and will not be attenuated since the noise voltages are not correlated. The output noise, due to the base-spreading resistance $\bar{v}_x^2$ is proportional to the sum of G_{m1} and G_{m2} (9) and, therefore, is *not* attenuated as the gain is reduced.

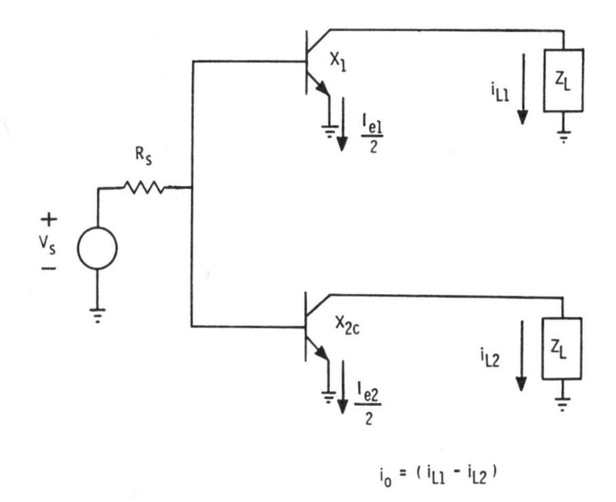

Fig. 9. Single-ended equivalent circuit of C_c compensated transistor amplifier stage.

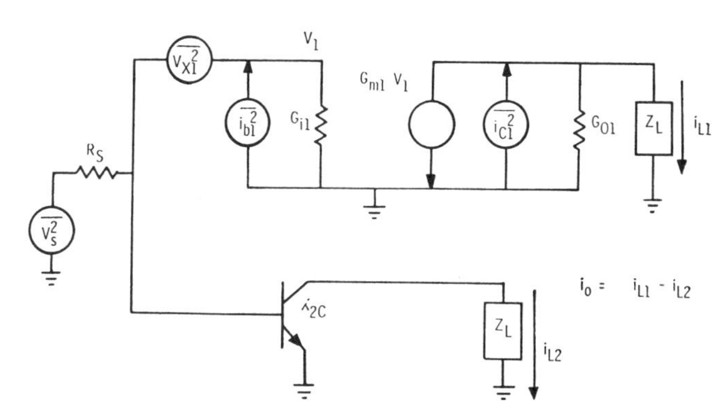

Fig. 10. Single-ended noise analysis circuit of C_c compensated transistor amplifier stage.

Thus the noise output due to $\bar{v}_s^2$ and $\bar{i}_b^2$ are reduced as the gain is reduced. Gain control methods not employing cancellation will also decrease the output noise due to $\bar{v}_s^2$ and $\bar{i}_b^2$ plus the noise contribution due to $\bar{v}_x^2$, and, therefore, may have a better signal-to-noise ratio at minimum gain.

Further improvement in the signal-to-noise ratio can be obtained by using more than one stage of attenuation,[3] since the second stage of gain control will reduce the output noise current due to the base-spreading resistance noise source of the first stage. This second stage of gain control is easily employed in the case of the cancellation gain control method. The second gain control stage using the cancellation method has little effect on the response of amplifier, whereas a second stage using conventional gain control methods will affect the bandpass response [8], [9].

[3] Another method of reducing the overall noise figure of a tuned RLC amplifier cascade would be to use a low noise input stage followed by stages using the C_c compensated transistors. The required gain control of the amplifier could be obtained using the cancellation gain control in the C_c compensated stages.

Fig. 11. A three-stage synchronously tuned 44-MHz 90-dB power gain amplifier using C_c compensated transistors.

Design Examples

A 44-MHz 90-dB power gain amplifier using the C_c compensated transistors in an IF amplifier is shown in Fig. 11. This amplifier is constructed of three identical RLC tuned stages under the conditions that the input and output admittances were conjugately matched to the source and load admittances. The voltage gain of this cascade from input to output was in excess of 100 dB at 44 MHz. The calculated maximum available power gain per stage of this amplifier is 32 dB, and the measured power gain per stage was 30 dB with a bandwidth of 1.9 MHz. A usable output signal of 1.8 V rms was obtained without appreciable distortion.

The first two stages of this amplifier were utilized for cancellation type gain control, with a control range in excess of -70 dB. One important fact should be pointed out about this amplifier. No power supply decoupling was required between stages, while the alignability of the cascade of stages was excellent. The primary concern in the construction of this amplifier was that the layout be symmetrical and that the interstage transformers used were shielded to prevent magnetic coupling.

TV IF Amplifier

A second design example of the use of the C_c compensated transistors in an IF amplifier is shown in Fig. 12. This amplifier was constructed using a single 14-lead flatpack containing two C_c transistor quadrupets for the two stages of gain. The characteristics of this amplifier are listed in Table I. While the observed 51-dB signal-to-noise ratio is satisfactory for low cost or small screen TV applications, a large-screen high-quality receiver may require as high as 56-dB signal-to-noise ratio at minimum IF gain. The voltage gain of 70 dB is measured for the condition of a 50-ohm source and output load resistance of 3000 ohms.

TABLE I

Voltage Gain, V_0/V_1	70 dB
Center Frequency, f_0	44 MHz
Bandwidth	3 MHz
AGC Range	> 50 dB
NF at Maximum Gain Without Traps	5 dB
Output S/N at -50-dB Gain Reduction	
$V_0 = 150$ mV rms	49 dB
$V_0 = 200$ mV rms	51 dB
Sync Tip Crushing	
$V_0 = 150$ mV rms	< 10 percent
$V_0 = 200$ mV rms	< 20 percent

The construction of this amplifier again required no power supply decoupling between stages, although RF filtering was required for each of the supplies used. As in the previous example, layout of the IF was symmetrical and magnetic shielding was used to prevent coupling between transformers. The attenuation range of -50 dB was obtained by controlling the gain of both stages with the cancellation gain control method.

Fig. 13 shows photographs of the bandpass characteristics of the TV IF amplifier without traps for various gain levels. The shift of center frequency is less than 1 percent, and the shift of Q is less than 5 percent.

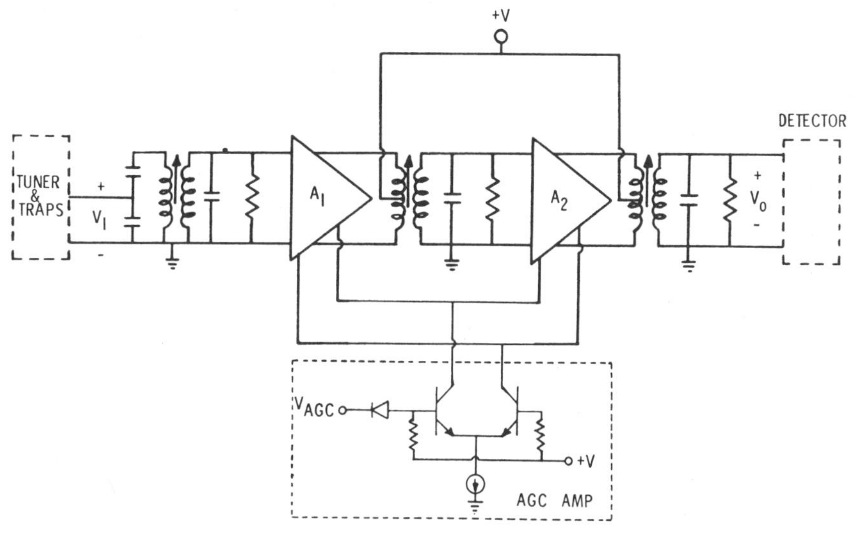

Fig. 12. A two-stage TV IF using C_c compensated transistors.

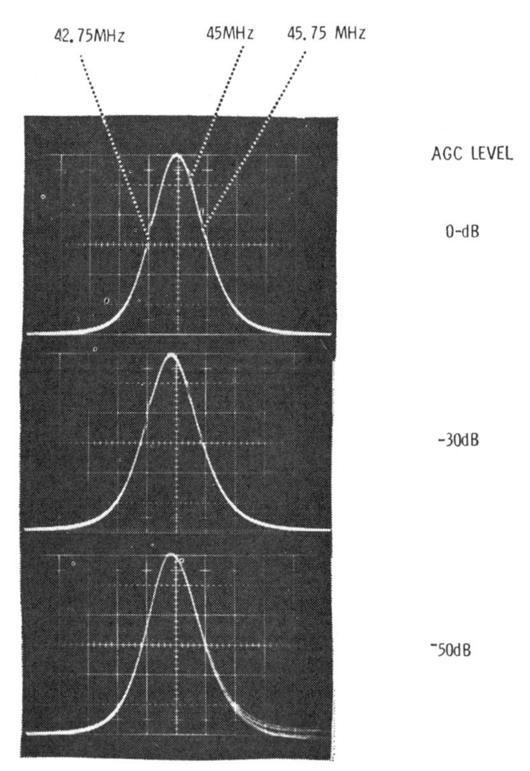

Fig. 13. Response characteristics for various levels of gain reduction of two-stage TV IF amplifier using C_c compensated transistors.

Conclusions

The use of C_c compensated transistor quadruplets in IF amplifier designs has been discussed. The balanced configuration of these devices and their inherent compensation for the reverse transmission parameter allows simple design of IF amplifiers. Designs obtaining the maximum available power gain with good alignability are possible while at the same time, power supply decoupling between amplifier stages is not required. It was shown that these devices provide a novel method of gain control which enables gain reduction with improved cross-modulation characteristics and with little detuning of the amplifier. However, a noise analysis of this signal cancellation technique shows that care must be taken to avoid degrading the signal-to-noise ratio of the amplifier under reduced gain conditions.

Specific design examples have been considered to illustrate the constructional ease and improved performance obtainable using the C_c compensated transistor quadruplet as a "unilateral" gain controllable device for general purpose IF amplifier designs.

References

[1] G. W. Haines, "C_c compensated transistors," presented at the 1966 IEEE Internat'l Electron Devices Meeting, Washington, D. C.
[2] G. W. Haines and B. E. Amazeen, "Amplifier designs using repetitive stabilized gain elements," *1967 IEEE Internat'l Conv. Rec.*
[3] C. L. Searle, R. D. Thorton, E. J. Angelo, D. O. Pederson, and J. Willis, *Multistage Transistor Circuits,* SEEC, vol. 5. New York: Wiley, 1965, ch. 7.
[4] C. L. Searle, A. R. Boothroyd, E. J. Angelo, P. E. Gray, and D. O. Pederson, *Elementary Circuit Properties of Transistors,* SEEC, vol. 3. New York: Wiley, 1964, ch. 8.
[5] R. S. Pepper, G. W. Haines, J. A. Mataya, and B. E. Amazeen, "Development of controllable selective amplification in silicon-based microcircuitry," Air Force Avionics Lab., Wright-Patterson AFB, Ohio. Tech. Rept. AFAL-TR-67-286, December 1967.
[6] J. G. Linville and J. F. Gibbon, *Transistors and Active Circuits.* New York: McGraw-Hill, 1961, chs. 10–11.
[7] D. O. Pederson, *Electronic Circuits,* preliminary ed. New York: McGraw-Hill, 1965, ch. 8.
[8] W. F. Chow and A. P. Stern, "Automatic gain control of transistor amplifiers," *Proc. IRE,* vol. 43, pp. 1119–1127, September 1955.
[9] M. D. Wood, "Gain controlled bandpass amplifiers—a new approach using integrated circuit video amplifiers," *Electron. Engrg.,* vol. 36, pt. 1, pp. 150–153, March 1964.

A High-Performance Monolithic IF Amplifier Incorporating Electronic Gain Control

W. RICHARD DAVIS, MEMBER, IEEE, AND JAMES E. SOLOMON, MEMBER, IEEE

Abstract—**A direct-coupled monolithic IF amplifier that incorporates an active gain control stage and exhibits a power gain of 50 dB and an AGC range of 60 dB at 50 MHz is described. This circuit has negligible change in either input or output admittance, and has excellent signal linearity over the full range of gain control. Experimental and theoretical analyses are made of the large signal reponse, stability, available gain, and noise behavior of the circuit. An application to color television is discussed in which the functions of the 45-MHz IF amplifier and the dc-AGC circuitry are fabricated on a single die.**

I. Introduction

ELECTRONIC gain control in solid-state amplifiers is most often obtained by shifting the operating point of a transistor amplifier, or by using attenuator networks that incorporate diodes, bipolar transistors, or FETs as electronically variable resistors [1]–[3]. These methods usually suffer from changes in input and output admittance, as gain is changed, and exhibit poor linearity when large signals are applied. Most approaches are difficult to realize in direct-coupled monolithic form, since dc shifts accompany the gain change.

A new circuit that overcomes these limitations and is particularly suitable for monolithic fabrication has been developed. This circuit consists of a high-gain wideband dc-coupled amplifier incorporating an active bipolar attenuator. Fixed gain stages are used at the input and output of the amplifier to eliminate input and output admittance variations as the gain is changed, and to provide amplification. Negligible distortion is produced by the control stage for a gain reduction greater than 100 dB at low frequencies and 60 dB at 50 MHz. The shape of the amplifier passband remains essentially constant from dc to 100 MHz over a gain control range in excess of 60 dB.

In the sections below, the ideal behavior and the performance limits of the basic gain-control circuitry are discussed. This is followed by a discussion of available power gain, stability, and noise performance of a complete monolithic IF amplifier incorporating the gain-control stage. In the final section, an application of the amplifier to a television receiver is considered. A circuit is described that incorporates the functions of the 45-MHz video IF amplifier, the AGC amplifiers, AGC gate, and RF AGC delay circuitry on a single monolithic die.

Manuscript received July 15, 1968; revised September 6, 1968. This paper was presented at the 1968 ISSCC.

The authors are with Motorola, Inc., Semiconductor Products Division, Phoenix, Ariz.

II. Transistor Gain Control

The basic all-transistor gain-control circuit consists of a double differential current divider driven by a differential input amplifier as shown in Fig. 1. Maximum gain occurs for negative V_{AGC}, which causes the emitter followers to be OFF. In this condition, the circuit functions as a differential cascode and exhibits the low reverse transfer and low noise figure normally associated with a cascode connection. As V_{AGC} is increased positively, the emitter followers gradually turn ON and divert current away from the output transistors, thereby reducing gain.

Details of the circuit operation are understood most easily by considering a single-ended equivalent to the differential circuit. Under the assumptions that the circuit is perfectly symmetrical, and small differential signals are applied, the differential-mode half circuit shown in Fig. 2 may be developed [4].[1]

The current transfer of the stage can be calculated by using the Ebers–Moll equations for the transistor [6]. Ignoring leakage currents and assuming identical transistors, the emitter currents of Q_1 and Q_2 are approximately

$$I_1 = a_{11}e^{q\phi_1/kT} \qquad I_2 = a_{11}e^{q\phi_2/kT}. \tag{1}$$

The current transfer can then be written

$$\frac{\alpha_F I_2}{\Delta I} = \frac{\alpha_F}{1 + I_1/I_2} = \frac{\alpha_F}{1 + \exp(qV_{AGC}/kT)} \tag{2}$$

where α_F is the low-frequency common-base short circuit current gain. Equation (2) indicates that, within the limits of the analysis, the current transfer exhibits nearly perfect linearity for arbitrary gain reduction or input current level. It is this excellent linearity that makes the approach of Fig. 2 attractive when compared with other gain-control techniques.

The differential response of the circuit in Fig. 1 is identical with (2) under the assumptions made. The differential circuit has several advantages over its single-ended analog however. AGC voltage variation in the single-ended circuit causes dc output variation, which

[1] The circuit in Fig. 2 has itself been used as a gain-controlled IF stage. See [5].

Reprinted from *IEEE J. Solid-State Circuits*, vol. SC-3, pp. 408-416, Dec. 1968.

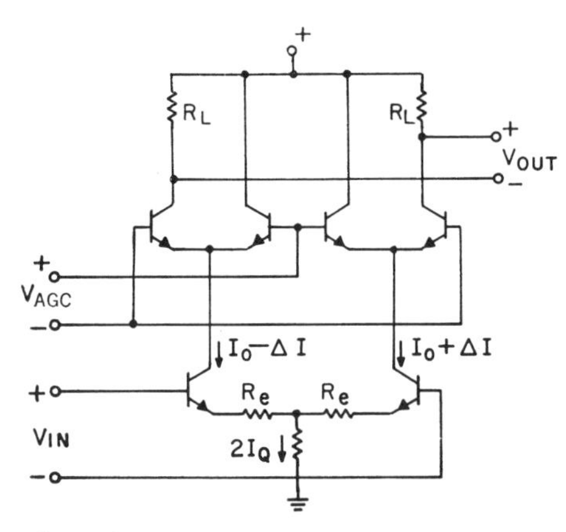

Fig. 1. Differential gain control circuit consisting of doubly differential transistor current divider.

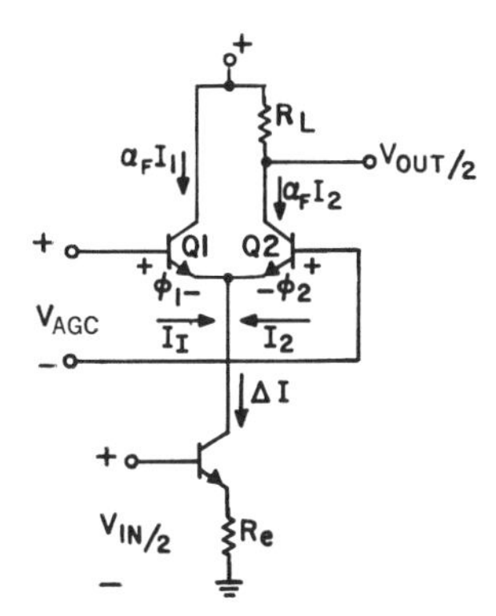

Fig. 2. Differential mode half circuit equivalent to Fig. 1.

prevents adding a direct-coupled second stage to the circuit. In the differential circuit, output due to V_{AGC} is a common-mode signal that can be eliminated in a following differential stage. Further, all of the nodes that require bypass capacitors in the circuit of Fig. 2 become virtual grounds in the differential circuit (under small signal conditions) and do not require bypassing. Thus, for monolithic fabrication, where capacitors are difficult to realize, the differential circuit is clearly more attractive.

It is of interest to derive the dependence of gain on AGC voltage. Taking the logarithm of (2) under the condition that exp $(qV_{AGC}/kT) > 1$, the circuit gain in decibels becomes

$$20 \log \frac{\alpha_F I_2}{\Delta I} \cong 20 \log \alpha_F \exp(-qV_{AGC}/kT)$$

$$= \frac{-8.68}{kT/q} \alpha_F V_{AGC} \text{ dB}. \quad (3)$$

Equation (3) shows that, after an initial gain reduction, the gain in dB is linearily dependent upon the control voltage V_{AGC}. Such dependence is highly desirable for most gain-control applications. If linear dependence of gain (not in decibels) on control voltage is desired, this can be obtained by deriving V_{AGC} from a pair of differentially driven diodes, as described in [7].

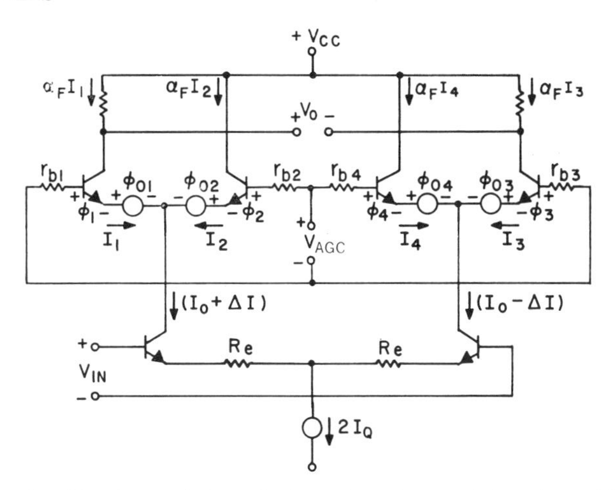

Fig. 3. Circuit used to calculate the effects of transistor V_{BE} mismatch and base resistance on gain-control stage.

III. Nonideal Behavior of Gain-Control Stage

There are two primary sources of nonlinearity which may arise in the circuit of Fig. 1. 1) Current-transfer nonlinearity can result from ohmic drops in the AGC transistor bases when ΔI becomes large, and 2) input voltage compression will occur when the maximum signal handling of the input differential amplifier is exceeded. It is shown below that the current nonlinearity produces signal expansion, so both compression and expansion of the signal are possible in the circuit, depending on signal conditions.

In addition to these effects, small mismatches between the AGC transistors produce a differential unbalance in the output as the gain is varied. This variation in output balance sets a limit on the dc gain that can be used following the gain-control stage.

In the sections below, the dc output unbalance is evaluated first. This is followed by a calculation of the current nonlinearity, and the voltage compression. In each case, simplifying assumptions are made that permit straightforward interpretation of the results. All calculations are made using the circuit of Fig. 3. In this circuit the transistors are assumed to follow ideal Ebers–Moll behavior, except for the inclusion of base resistance r_{bi}, as shown. The AGC transistors are further assumed to be identical in all respects, except for the presence of voltage offset generators, ϕ_{0i}, in series with each emitter. These offset generators represent the small base-emitter voltage mismatches (for a condition of equal collector currents), which occur in otherwise identical transistors fabricated on monolithic die.

DC Output Offset

It is shown in the Appendix, using Fig. 3, that the dc differential output current, $\Delta I_{out} = \alpha_F(I_3 - I_1)$, of the gain-control stage, with no signal applied, is

$$\Delta I_{out} = \frac{\alpha_F I_0(\phi_{01} + \phi_{04} - \phi_{02} - \phi_{03})}{[2kT/q](1 + \cosh qV_{AGC}/kT)}. \quad (4)$$

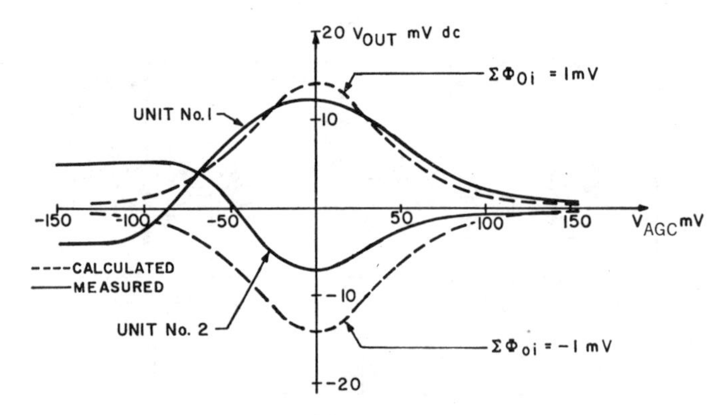

Fig. 4. Calculated and measured differential output voltage offset caused by mismatch in AGC transistors. $V_{out} = R_L \Delta I_{out}$ where $R_L = 470\ \Omega$ and ΔI_{out} is given in (4).

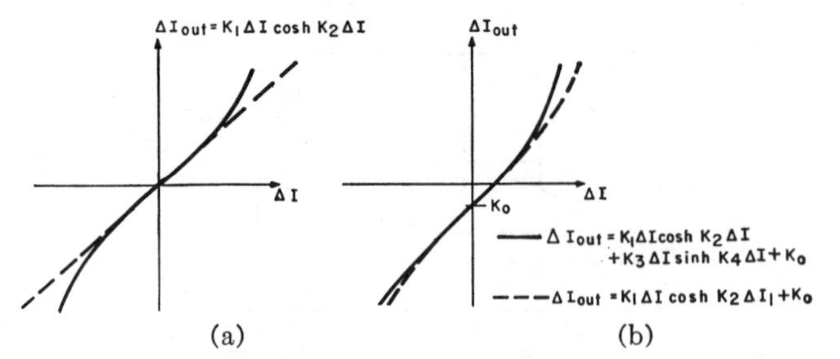

Fig. 5. (a) Plot of (5) showing nonlinear transfer arising from base resistance [$\phi_{oi} = 0$, and K_i defined in (5)]. (b) Plot of (5) showing nonlinear transfer due to voltage offsets and base resistance. K_i are defined by (5).

The hyperbolic cosine has a minimum (of unity) for zero argument, so the output offset predicted by (4) has a maximum for $V_{AGC} = 0$, and falls for V_{AGC} greater or less than zero. This behavior is understood by noting first that, when V_{AGC} is negative, the emitter followers are OFF, and the output offset must be the same as the input offset (ΔI), which was assumed zero. For V_{AGC} increasing positively, the magnitudes of the output currents decrease, which causes the magnitude of the output offset to decrease. Thus, a maximum output offset near zero V_{AGC} is expected.

Fig. 4 shows the output voltage offset, calculated using (4), and the measured output offset of two representative integrated circuits. Good correlation is seen to exist, with maximum offset occurring near $V_{AGC} = 0$ volts.

Nonlinear Current Transfer

The nonlinearity, which arises in the current transfer due to base resistance, is also derived in the Appendix. The expression obtained for the low-frequency differential output current, $\Delta I_{out} = \alpha_F(I_3 - I_1)$, valid for a gain reduction greater than 6 dB, is [see (49) and (47)]

$$\Delta I_{out} = \frac{-2\alpha_F}{\exp\left\{\frac{q}{kT}\left[V_{AGC} - r_b I_0(1-\alpha_F)\right]\right\}} \cdot \Big[\Delta I \cosh q r_b(1-\alpha_F)\,\Delta I/kT + \frac{q}{2kT}(\phi_{03} - \phi_{04} - \phi_{01} + \phi_{02}) \cdot (I_0 + \Delta I \sinh q r_b(1-\alpha_F)\,\Delta I/kT)\Big], \quad (5)$$

where $r_b = r_{b2} = r_{b4}$.

Signal distortion caused by the presence of base resistance is best seen from plots of (5), shown in Fig. 5(a) and (b). In Fig. 5(a), the voltage offsets are assumed to be zero, and the distortion expected under a condition of perfect balance is shown. This curve shows that the circuit produces expansion of the output, ΔI_{out}, for increasing input, ΔI. Odd symmetry exists, so only odd harmonic terms would be expected in the output. When the voltage offsets are included, as shown in Fig. 5(b), asymmetry is introduced by the sinh term in (5), which results in greater expansion for one output polarity than for the other. This produces even harmonics in addition to the odds.

It can be shown by expansion of (5) in a power series that the odd terms can be expected to dominate over the even terms as a result of the small voltage offsets in the monolithic circuit. Odd symmetry and signal expansion should, therefore, be the dominant characteristics of the large signal current response.

A practical result of the analysis leading to (5) is that signal distortion is shown to result primarily from the presence of the base resistors r_{b2} and r_{b4}, and not from the presence of r_{b1} and r_{b3}. As a result of this, the circuit linearity can be improved by using large transistor geometries for the emitter followers in Fig. 3, which reduces r_{b2} and r_{b4}. Little sacrifice is made in the high-frequency performance of the circuit, since the signal transistors can have small geometries which are optimized for high-frequency response.

The results of this section are valid only for low frequencies since the development of (5) was based on a low-frequency transistor model. However, an estimate of the high-frequency behavior can be obtained by letting $(1 - \alpha_F) \approx |1 - \alpha_F(\omega)| \approx \omega/\omega_t$ in (5). At high frequencies, the phase of the base current is no longer the same as that of the emitter current, so the calculation is in error, but not by a large factor.

The result of increasing the value of $(1 - \alpha_F)$ in (5) is to directly increase the arguments of the distortion terms. The signal expansion is, therefore, expected to increase with increasing frequency as the magnitude of $(1 - \alpha_F)$ increases. Experimentally, these predictions agree with the behavior of the integrated stage. For the circuit discussed below, the low-frequency linearity is good enough that negligible signal expansion is measured for a gain control range in excess of 100 dB. At higher frequencies, signal expansion is clearly observed, and restricts the gain reduction to smaller values (e.g., 60 dB at 50 MHz) if reasonable output linearity is to be maintained.

Input Overload

The signal compression arising from input overload is most conveniently evaluated by calculating the response ΔI due to a voltage input V_{in} as indicated in Fig. 3. Using the Ebers–Moll equations, ignoring base resistance, leakage currents, and voltage offsets in the input transistors, it is easily shown that

$$\Delta I = \alpha_F \left[\frac{2I_Q}{1 + e^{(q/kT)(\phi_2 - \phi_1)}} - I_Q \right] \tag{6}$$

where $(\phi_2 - \phi_1) = -V_{in} + 2R_e \, \Delta I/\alpha_F$.

An explicit expression for ΔI cannot be found, but (6) can be solved for V_{in} as follows:

$$V_{in} = \frac{2R_e \, \Delta I}{\alpha_F} - \frac{kT}{q} \ln \left[\frac{I_Q - \Delta I/\alpha_F}{I_Q + \Delta I/\alpha_F} \right]. \tag{7}$$

As $\Delta I/\alpha_F$ approaches $\pm I_Q$, the ln term in (7) blows up causing V_{in} to approach $\pm\infty$. Such behavior results in a typical saturation characteristic, with $\Delta I/\alpha_F$ compressing to a value I_Q as V_{in} increases. Graphical manipulation of (7) shows that the peak input voltage, which can be applied before the peak output, is compressed by a factor of $(1-\epsilon)$, i.e., $\Delta I = \alpha_F(1-\epsilon)V_{in}/2(R_e + kT/qI_Q)$ is (for $2R_eI_Q > kT/q\epsilon$):

$$V_{in_{max}} \approx 2R_eI_Q/(1 - \epsilon). \tag{8}$$

IV. Complete Amplifier Incorporating the Gain-Control Stage

Incorporation of the gain-control stage into a high-gain monolithic amplifier will now be considered. There are two primary requirements that must be satisfied by the added circuitry. 1) It must be dc compatible with the control stage, since capacitors are to be avoided, and 2) dynamic range must not be degraded.

Maximum dynamic range is assured only if amplification is added following the control stage. To be dc compatible, the added stage must 1) reject the common-mode output variation of the control stage, and 2) maintain a usable dc operating point, despite the variation in differential output offset expected from (4). These requirements are satisfied by a differential second stage using current source bias and resistive emitter degeneration. The current source reduces common-mode gain to a negligibly small value, while sufficient emitter resistance can be added to assure reasonable dc balance in the stage.

Fig. 6 shows the complete amplifier consisting of the balanced gain-control circuit at the input, followed by the differential second stage. The circuitry between the two stages is broadbanded by using small valued load resistors and emitter followers. This eliminates the need for interstage tuning, which would add additional pins and an external inductor. Two bias bleeders are used as shown to prevent signal feedthrough in the bias circuitry under conditions of high gain reduction.

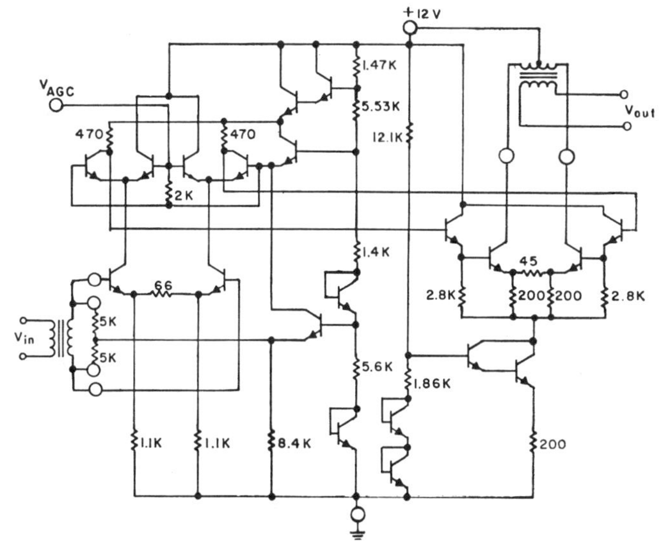

Fig. 6. Circuit diagram for complete IF amplifier. The circuit is fabricated on a 44 × 46 mil die using diffused technology.

IV. Performance of Complete Amplifier

The available power gain, stability, and noise performance of the complete amplifier with tuned terminations will now be evaluated. The short-circuit admittance parameters are first calculated and then used to study both gain and stability. For these calculations, the differential-mode half circuit and the transistor model shown in Fig. 7(a) and (b), are used.

Short-Circuit Admittance Parameters

Using Fig. 7, and ignoring a few small terms, the forward transadmittance, y_{21}, for a differential input and output is

$$y_{21}(s) \cong \frac{y_{21}(o)}{(1 + s/p_1)(1 + a_1s + a_2s^2)} \tag{9}$$

where

$$y_{21}(o) = \frac{\alpha_{0b}R_L/2}{(R_{e1} + r_{e1})(R_{e2} + r_{e2})(1 + r_{x1}/R_{\pi1})[1 + (r_{ef} + r_{x2})/R_{\pi2}]} \tag{10}$$

$$p_1 = (1 + r_{x1}/R_{\pi1})/r_{x1}C_{\pi1} \tag{11}$$

$$a_1 = C_{\pi2}(r_{ef} + r_{x2}) + R_LC_{sb} \tag{12}$$

$$a_2 = R_LC_{\pi2}[(1 + r_{x1}/R_L)/\omega_{tf} + C_{sb}(r_{ef} + r_{x2})]. \tag{13}$$

In the above, the subscripts 1, 2, b, or f refer to the input transistor, the output transistor, the common-base stage, or the emitter follower, respectively. Using the element values in Fig. 7(b), and $R_L = 470\ \Omega$, $R_{e1} = 32\ \Omega$, $R_{e2} = 20\ \Omega$, $I_{c1} = I_{cb} = I_{c2} = 3$ mA, $I_{cf} = 0.5$ mA, the low-frequency transconductance, $y_{21}(o) = 0.19$ mho. $y_{21}(o)$ is seen to be quite large when compared with conventional discrete devices, and if desired, could be increased by choosing a large R_L. The poles of $y_{21}(s)$ are

$$p_1/2\pi = -120 \text{ MHz}$$

$$p_2/2\pi, \quad \bar{p}_2/2\pi = -70 \text{ MHz} \pm j87 \text{ MHz}.$$

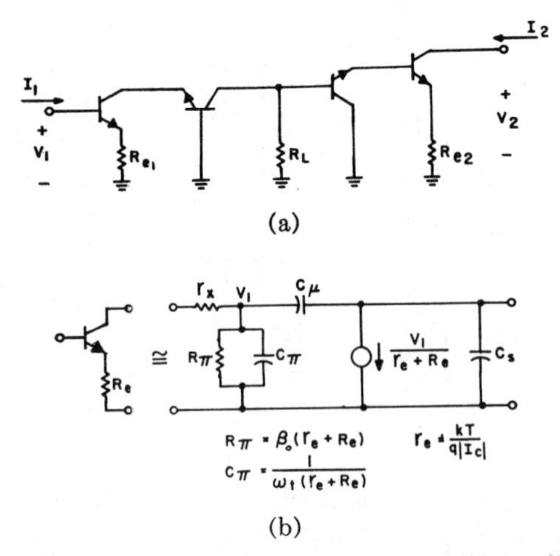

Fig. 7. (a) Equivalent half circuit for Fig. 6 valid for differential signals. (b) Transistor model used for calculations. Element values used are $\omega_t/2\pi = 500$ MHz, $C_\mu = 0.7$ pF, $r_x = 80$ Ω, $C_s = 1.2$ pF, $\beta_0 = 80$.

p_1 is seen from (11) to be set by the input impedance of the input transistor, while the complex pole pair, p_2 and $\bar{p}_2$, arises primarily from an interaction between the inductive output impedance of the emitter follower and the input capacitance of the output transistor. R_L has been chosen so that y_{21} has a 3-dB bandwidth of approximately 90 MHz, and a magnitude response that is reasonably flat. The calculated and measured $|y_{21}(j\omega)|$ for the integrated amplifier are shown in Fig. 8 and are seen to have good agreement. The measured phase shift agrees with that calculated from (9) for frequencies below 100 MHz. Variations in $|y_{21}(j\omega)|/y_{21}(o)$ with V_{AGC} for a gain reduction of 60 dB are also shown in Fig. 8 and are quite small.

The differential short-circuit input admittance y_{11} of the amplifier is, to a good approximation, equal to one half the input admittance for an input transistor. Using Fig. 7,

$$y_{11} \cong \frac{(1 + sR_{\pi 1}C_{\pi 1})}{2R_{\pi 1}(1 + s/p_1)} \tag{14}$$

where p_1 is defined in (11). The differential short-circuit output admittance, y_{22}, has a somewhat more complex form:

$$y_{22} = \frac{1}{2}\left[\frac{sC_{\mu 2}(y_1 + sC_{\pi 2} + 1/R_{\pi 2} + 1/r_{e2})}{y_1 + s(C_{\pi 2} + C_{\mu 2}) + 1/R_{\pi 2}}\right] \tag{15}$$

where

$$y_1 \cong 1/(R_1 + sL_1) \cong 1/[r_{e1} + s(R_L + r_{xf})/\omega_{tf}]. \tag{16}$$

y_1 is the predominantly inductive output admittance of the emitter follower and its presence causes the real part of y_{22} to rise with increasing frequency.

Curves showing the real parts of y_{11} and y_{22} (g_{11} and g_{22}), which are important for calculations of available power gain, are shown in Fig. 9. Experimental data are found to agree closely with those given by (14) and (15).

Calculation of the reverse transadmittance, y_{12}, from Fig. 7 indicates that reverse transfer in the integrated circuit is negligible under all conditions of interest. How-

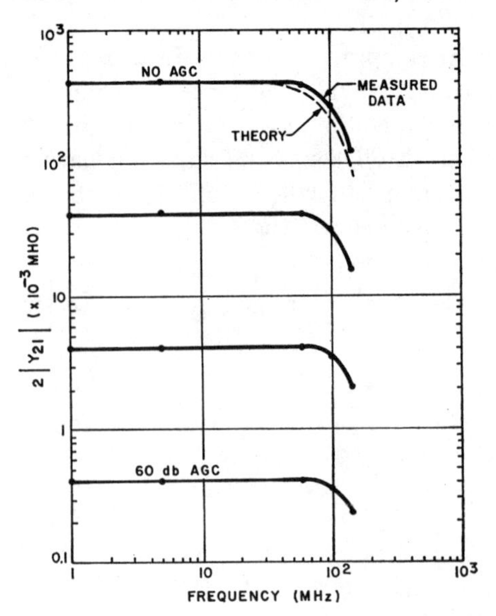

Fig. 8. Variation in $|y_{21}|$ with frequency and V_{AGC}.

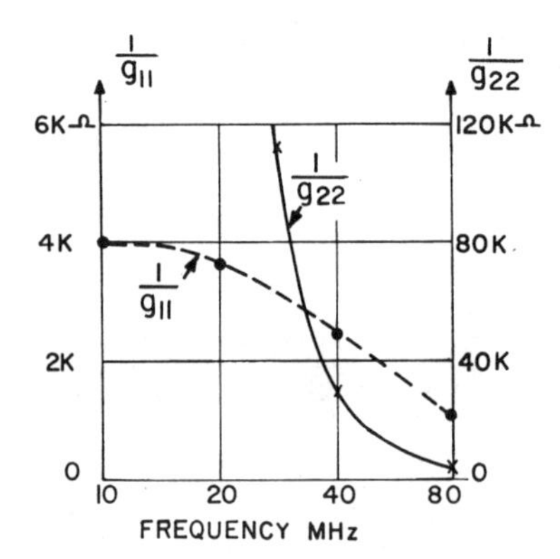

Fig. 9. Real parts of short circuit input and output admittance calculated using Fig. 7.

ever, pin-to-pin capacitance in the IC package and/or input–output coupling through external circuitry can make significant contributions to y_{12}, if care is not taken. The capacitance between pins of a grounded 10-pin TO-5 can is on the order of 0.001 pF. The pin-to-pin capacitance for a 14-pin dual-in-line plastic package varies from about 0.001 pF for diagonally opposite pins with intermediate pins grounded to 0.4 pF for adjacent pins. In the TO-5 package, Faraday shielding of the grounded metal can minimizes coupling, while in the latter, shielding other than intermediate grounded pins is not available. In any case, an input–output coupling of a few tenths of a picofarad can occur through the external circuitry if care is not taken. The significance of these capacitance values is considered in the next section.

Available Power Gain and Stability

The maximum available power gain, MAG, under conditions of negligible reverse transfer, is

$$\text{MAG} = |y_{21}|^2/4g_{11}g_{22}. \tag{17}$$

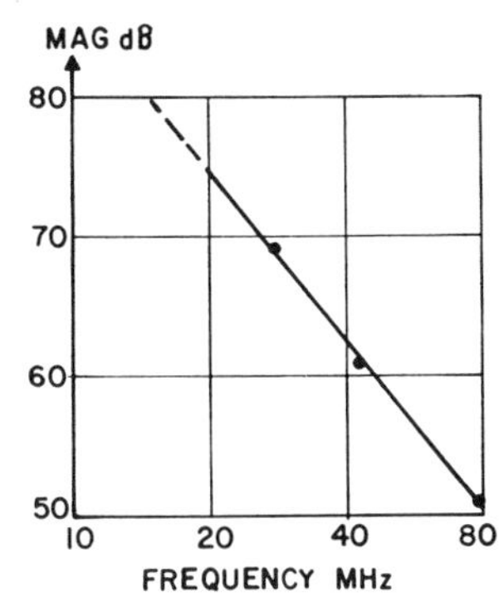

Fig. 10. Variation of MAG calculated using Figs. 8 and 9.

Using the y parameters given by (10), (14), and (15), the frequency dependence of MAG for the integrated amplifier can be calculated as shown in Fig. 10. It is seen that MAG falls from 75 dB at 20 MHz to 51 dB at 80 MHz. This fall has a 12 dB/octave slope, which results primarily from an increase in the real parts of the input and output admittance at high frequencies.

For the amplifier to be unconditionally stable, the following relationship must be satisfied [8]:

$$2(g_{11} + g_s)(g_{22} + g_L) > |y_{21}y_{12}| + \mathrm{Re}\,(y_{21}y_{12}) \tag{18}$$

where g_s and g_L are the real parts of the source and load admittance, respectively. It is reasonable to assume that the dominant reverse transfer is due to capacitive coupling C_f through the package or external circuitry. The short-circuit reverse transadmittance is, therefore,

$$y_{12}(j\omega) = \pm j\omega C_f. \tag{19}$$

The $\pm$ sign in (19) is possible, since transfer from the differential outputs to the differential inputs can have either polarity, depending on which path is dominant. Combining (18) and (19), the maximum C_f for stable operation is found to be

$$\omega C_f < \frac{2(g_{11} + g_s)(g_{22} + g_L)}{|y_{21}| \pm b_{21}} \tag{20}$$

where b_{21} = imaginary part of y_{21}. Considering the worst case in which there is no loading, i.e., $g_s = g_L = 0$, and $\pm b_{21} = |y_{21}|$, (20) becomes

$$\omega C_f < \frac{g_{11}g_{22}}{|y_{21}|}. \tag{21}$$

As an example, with the integrated circuit operating at 50 MHz (i.e., $|y_{21}| = 0.19$ mho, $g_{11} = 0.5 \times 10^{-3}$, $g_{22} = 0.8 \times 10^{-4}$), (21) shows that the circuit is potentially unstable unless

$$C_f < 0.66 \times 10^{-3} \text{ pF}.$$

Clearly the circuit can oscillate if care is not taken.

For a more realistic calculation, consider a loaded case with $g_s = 2g_{11}$ and $g_L = 2g_{22}$. (This corresponds to a 7-dB loss in power gain from the maximum available.) Also, by using care in packaging and layout, the balanced nature of the circuit can be exploited to partially cancel feedback effects and/or produce a favorable sign for the denominator of (20). It is reasonable to expect this to result in a reduction of the effective C_f by at least a factor of five. Under these conditions, (20) can be evaluated at 50 MHz to give

$$C_f < 0.03 \text{ pF}.$$

Precautions must still be taken to avoid feedback capacitances of 0.03 pF but coupling of this order can be avoided without great difficulty. Experimentally, the circuit produces a stable unilateral power gain of 50 dB at 50 MHz when packaged in either metal or plastic packages, and symmetry is used in package and circuit layout.

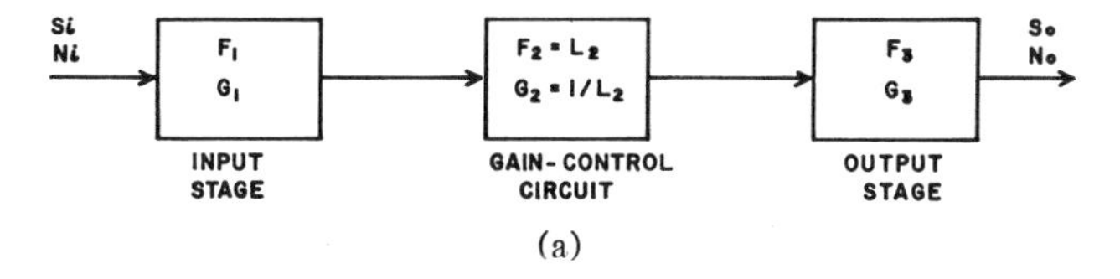

(a)

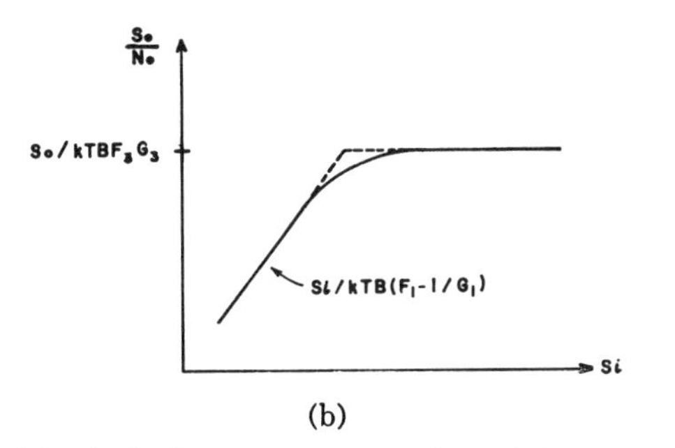

(b)

Fig. 11. (a) Block diagram representation of the circuit in Fig. 6 used for noise calculations. (b) Dependence of IF output signal-to-noise ratio S_o/N_o on input signal S_i for S_o held constant by AGC loop.

Noise

Two aspects of noise performance are of interest in a gain-controlled amplifier. One is the noise figure when gain is maximum, and the second is output signal-to-noise ratio when the input signal is large and gain is reduced.

It is shown in [9] that the noise figure of the gain-control stage at maximum gain can be as low as that of the best two-transistor cascade. In the particular example discussed here, high emitter currents and external emitter resistors have been used to improve input voltage handling, thus degrading the optimum noise figure to approximately 8 dB. In the usual application, where RF gains of 30 to 40 dB are used ahead of the IF amplifier, this number is low enough that the overall noise figure is not degraded.

The dependence of output signal-to-noise power ratio, S_o/N_o, on input signal power S_i can be determined by splitting the amplifier into three blocks as shown in Fig. 11(a). In this figure, F_1, F_2, F_3, and G_1, G_2, G_3 represent the noise figures and available power gains of the input amplifier, the gain control circuitry and the output amplifier, respectively. $N_i = kTB$ is the available power from the source, and it is noted that the noise figure of the gain-control circuit ($F_2 = L_2$) can be assumed equal to the reciprocal of the available power gain of this circuit ($G_2 = 1/L_2$).

Using the Friis formula for the noise figure of a cascade [10], the overall noise figure can be written directly from Fig. 11(a):

$$F_t = F_1 + \frac{(L_2 - 1)}{G_1} + \frac{(F_3 - 1)L_2}{G_1}$$

$$= (F_1 - 1/G_1) + F_3L_2/G_1. \quad (22)$$

From the definition of noise figure, the output signal-to-noise ratio is

$$S_o/N_o = (S_i/N_i)/F_t. \quad (23)$$

Noting that the available power gain of the complete amplifier is $S_o/S_i = G_1G_3/L_2$, (22) and (23) can be combined to give the output signal-to-noise ratio:

$$\frac{S_o}{N_o} = \frac{S_i}{kTB[(F_1 - 1/G_1) + F_3L_2/G_1]} \quad (24)$$

$$= 1/[(F_1 - 1/G_1)/S_i + F_3G_3/S_o]kTB. \quad (25)$$

In the usual application, an AGC loop adjusts L_2 to hold the output signal, S_o, constant. Thus the second term in the denominator of (25) is a constant, independent of input signal S_i. The first term in the denominator of (25) is dominant for small input signals, and is seen to produce an increase in the output signal-to-noise ratio as the input signal increases. For large enough inputs, the second term dominates and the output signal-to-noise ratio reaches a limiting value determined by the output stage noise figure and available gain, and S_o. This behavior is illustrated in Fig. 11(b).

The specific S_o/N_o limit for the IF amplifier in Fig. 6 is obtained most directly by expanding (24) in terms of circuit parameters. Considering only the second denominator term in (24) that dominates for large inputs, the maximum output signal-to-noise ratio is

$$[S_o/N_o]_{\max} = S_iG_1/kTBF_3L_2. \quad (26)$$

It is a straightforward matter to show that $S_i = V_s^2/4R_s$ and $G_1 = R_LR_s/4(r_{e1} + R_{e1})^2$ where R_s and V_s are the source resistance and voltage, respectively, and the other terms were previously defined. Equation (26) then becomes

$$[S_o/N_o]_{\max} = V_s^2R_L/16(r_{e1} + R_{e1})^2kTBF_3L_2. \quad (27)$$

The ratio V_s^2/L_2 is held constant by the AGC loop so it may be evaluated for any convenient input. Choosing V_s equal to the maximum source voltage before overload, $V_{s\max}$, which also corresponds to the maximum attenuation, $L_{2\max}$, we have $V_s^2/L_2 = V_{s\max}^2/L_{2\max}$.

As an example, in Fig. 6 let $V_{s\max} = 0.35$V rms, $L_{2\max} =$ 55 dB, $B = 3$ MHz, $F_3 = 6$ dB. Equation (27) reduces to

$$[S_o/N_o]_{\max} = 52 \text{ dB}.$$

This number agrees within 2 dB of the measured value. For many applications 52 dB is sufficient, but for some it is not, and the parameters in (27) must be modified to improve the S_o/N_o. A slightly modified design in which R_{e1} has been reduced and $I_Q \cong V_{s\max}/\sqrt{2}\,R_{e1}$ has been increased to improve S_o/N_o to 56 dB is considered in the next section.

V. An Application to Color Television

An interesting application for the monolithic amplifier is the 45-MHz video IF amplifier in the color television receiver. Here, in addition to its potentially low cost, it offers the important advantages of negligible detuning of the IF passband and excellent linearity for a wide range of input signals. Such characteristics are vital for good color reception, but are difficult to achieve in a conventional transistor IF amplifier.

The value of the integrated circuit can be increased without significantly increasing fabrication costs by adding additional circuit functions to the die containing the IF amplifier. As an example, the diagram for a monolithic circuit containing all of the dc AGC circuitry required in the TV receiver, in addition to the gain-controlled IF amplifier, is shown in Fig. 12. The AGC circuitry provides the functions of video gating and storage, IF AGC gain, and tuner AGC delay and gain. Inclusion of the AGC circuitry increases die area by about 40 percent (to 50 × 56 mils), but it approximately doubles the circuit value.

The integrated circuit is incorporated into the TV receiver as illustrated in Fig. 13 where a single output transistor is added to the IF circuit to provide an overall power gain of 75 dB, an AGC range of 57 dB, and a maximum output signal-to-noise ratio of 56 dB. Input signal handling is 400 mV p-p for less than 10 percent compression or expansion of the TV synchronizing pulse. The change of center frequency and bandwidth are both less than 1 percent over the full AGC range.

It is concluded that the integrated IF circuit can provide performance that is comparable with the most sophisticated discrete approach, yet it offers low cost and greater system simplicity than any available discrete approach.

Appendix

Voltage Offset and Base Resistance Effects on Gain-Control Stage

A low-frequency nonlinear transfer function will be derived in this Appendix, which shows the effects of base resistance and voltage offset on the gain-control stage. The circuit is modeled as in Fig. 3 where the transistors shown are assumed to be identical, and have ideal Ebers–Moll behavior, with negligible reverse leakage current. Base resistance and emitter-base voltage mismatch are represented by the base resistors r_{bi} and offset generators ϕ_{oi}, which are added to each AGC transistor as shown.

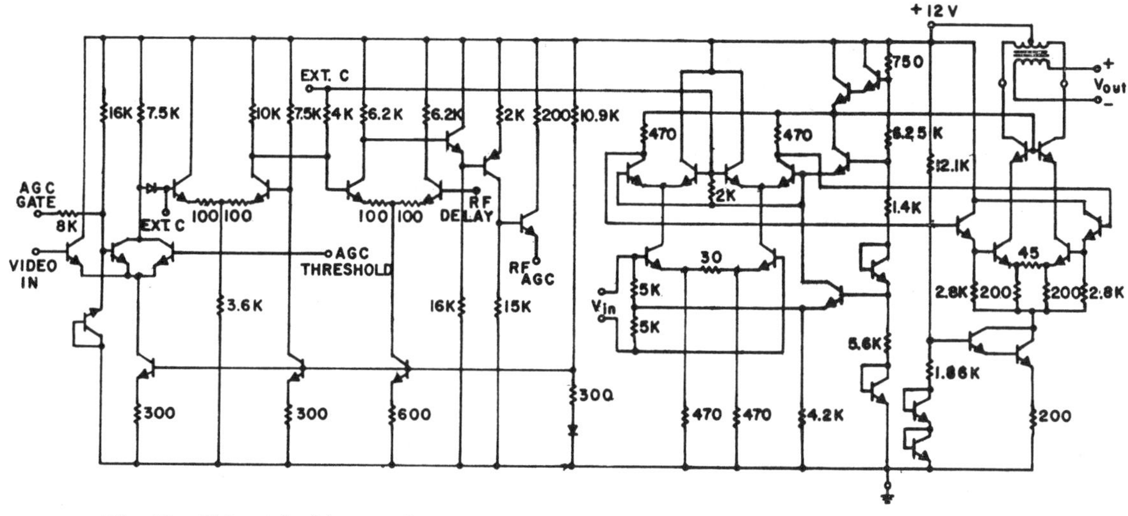

Fig. 12. Color television receiver circuit designed to perform the fucntions of the 45-MHz IF amplifier, AGC gate and store, IF AGC amplifier, tuner AGC delay and amplifier on a single chip.

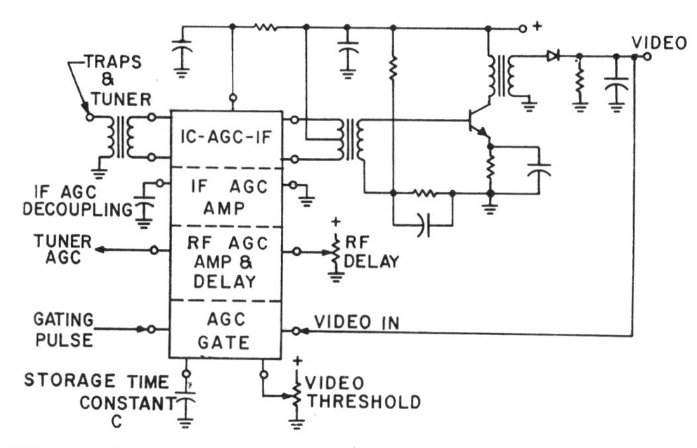

Fig. 13. Typical application of the chip in Fig. 12 to a television receiver.

The output currents, $\alpha_F I_1$ and $\alpha_F I_3$, are found in a fashion analogous to that used in deriving (2):

$$\alpha_F I_1 = \frac{\alpha_F(I_0 + \Delta I)}{(1 + I_2/I_1)} \tag{28}$$

$$\alpha_F I_3 = \frac{\alpha_F(I_0 - \Delta I)}{(1 + I_4/I_3)}. \tag{29}$$

The differential output current ΔI_{out} is given by

$$\Delta I_{\text{out}} = \alpha_F(I_3 - I_1) = \alpha_F\left[\frac{I_0 - \Delta I}{1 + I_4/I_3} - \frac{I_0 + \Delta I}{1 + I_2/I_1}\right]. \tag{30}$$

From the Ebers–Moll equations, the ratios of emitter currents are

$$I_2/I_1 = e^{q(\phi_2 - \phi_1)/kT} \tag{31}$$

$$I_4/I_3 = e^{q(\phi_4 - \phi_3)/kT}, \tag{32}$$

and the differential voltages across each AGC differential amplifier are found by summing voltage around their base-emitter loops:

$$(\phi_2 - \phi_1) = (\phi_{01} - \phi_{02}) + (1 - \alpha_F)(I_1 r_{b1} - I_2 r_{b2}) + V_a \tag{33}$$

$$(\phi_4 - \phi_3) = (\phi_{03} - \phi_{04}) + (1 - \alpha_F)(I_3 r_{b3} - I_4 r_{b4}) + V_a. \tag{34}$$

The object is to combine (30) through (34) for various conditions of interest.

Initially, the dc output offset, which results from the presence of the offset generators, ϕ_{oi}, with no signal applied (i.e., $\Delta I = 0$), is calculated. The r_b's can be ignored for this calculation, since they approximately match, and have small effect on balance. For $r_{bi} = 0$, and $(\phi_{01} - \phi_{02}) \ll kT/q$ [typical integrated transistors have $(\phi_{01} - \phi_{02}) \leq 1$ mV], (31) can be written as

$$1 + I_2/I_1 \cong 1 + \exp(qV_{\text{AGC}}/kT) \cdot [1 + q(\phi_{01} - \phi_{02})/kT]. \tag{35}$$

Using the relationships,

$$1 + \Delta \cong 1/(1 - \Delta) \qquad \Delta \ll 1 \tag{36}$$

and

$$1 + e^x(1 + \Delta) \equiv (1 + e^x)[1 + \Delta/(1 + e^{-x})] \tag{37}$$

where $\Delta = q(\phi_{01} - \phi_{02})/kT$ and $x = qV_{\text{AGC}}/kT$, (35) reduces to

$$1 + I_2/I_1 \cong \frac{1 + \exp(qV_{\text{AGC}}/kT)}{1 - q(\phi_{01} - \phi_{02})/kT[1 + \exp(-qV_{\text{AGC}}/kT)]}. \tag{38}$$

A similar expression for $(1 + I_4/I_3)$ can be developed from (32). Combining (38) and the expression for $(1 + I_4/I_3)$ in (30), and simplifying for $\Delta I = 0$,

$$\Delta I_{\text{out}} = \frac{\alpha_F I_0(\phi_{01} + \phi_{04} - \phi_{02} - \phi_{03})}{2(1 + \cosh qV_{\text{AGC}}/kT)(kT/q)}. \tag{39}$$

Equation (39) gives the output offset current resulting from AGC transistor voltage mismatch, ϕ_{oi}, as a function of AGC voltage, V_{AGC}.

For a second calculation, the nonlinear current transfer including both r_{bi} and ϕ_{oi} will be derived. At the outset, the analysis is simplified by assuming that sufficient V_{AGC} is applied to reduce the gain by a factor greater than 2. This can be justified by noting that nonlinear behavior is of concern only when large signals are present, and this normally corresponds to a condition of large gain reduction. Under the reduced gain, it is possible to assume that $I_2 > I_1$ and $I_4 > I_3$. It then follows that

$$1 + I_4/I_3 \cong I_4/I_3 \tag{40}$$

$$1 + I_2/I_1 \cong I_2/I_1. \tag{41}$$

Also, sufficient accuracy is retained in (33) and (34) if it is assumed that

$$I_2 \cong I_0 + \Delta I \tag{42}$$

$$I_4 \cong I_0 - \Delta I, \tag{43}$$

and ignore $r_{b3}I_3$ and $r_{b1}I_1$ compared with the larger I_2 and I_4 terms. Using (40) through (43), (33) can be combined with (31) to give

$$1 + I_2/I_1 \cong \exp\left\{\frac{q}{kT}\left[V_{AGC} - r_{b2}(1 - \alpha_F) \cdot (I_0 + \Delta I) + (\phi_{01} - \phi_{02})\right]\right\}. \tag{44}$$

An expression for $(1 + I_4/I_3)$ can be similarly found:

$$1 + I_4/I_3 = \exp\left\{\frac{q}{kT}\left[V_{AGC} - r_{b4}(1 - \alpha_F) \cdot (I_0 - \Delta I) + (\phi_{03} - \phi_{04})\right]\right\}. \tag{45}$$

The differential output current is found by combining (44), (45), and (30). In doing this, use the assumption that $r_{b4} = r_{b2}$, and $e^{\Delta} \cong 1 + \Delta \cong 1/(1 - \Delta)$, where $\Delta = q(\phi_{01} - \phi_{02})/kT$ or $q(\phi_{03} - \phi_{04})/kT$:

$$\frac{\Delta I_{out}}{K} = \Delta I\left[(1 - a)e^{x} + (1 - b)e^{-x}\right] + I_0\left[(1 - a)e^{x} - (1 - b)e^{-x}\right] \tag{46}$$

where

$$\begin{aligned} K &= -\alpha_F \exp -\left\{\frac{q}{kT}\left[V_{AGC} - r_{b2}I_0(1 - \alpha_F)\right]\right\} \\ x &= \frac{q}{kT} r_{b2}\, \Delta I\, (1 - \alpha_F) \\ a &= \frac{q}{kT}(\phi_{01} - \phi_{02}) \\ b &= \frac{q}{kT}(\phi_{03} - \phi_{04}). \end{aligned} \tag{47}$$

Equation (46) can be simplified by using the identities $e^{x} = \cosh x + \sinh x$ and $e^{-x} = \cosh x - \sinh x$:

$$\frac{\Delta I_{out}}{K} = (2 - a - b)(\Delta I \cosh x + I_0 \sinh x) + (b - a)(I_0 \cosh x + \Delta I \sinh x). \tag{48}$$

For $r_{b2}I_0(1-\alpha_F) < kT/q$, it can be shown by series expansion that all I_0 terms in (48) can be ignored except $(b - a)I_0$. Also $2 \gg (a + b)$, so (48) simplifies to

$$\frac{\Delta I_{out}}{K} \cong 2\,\Delta I \cosh x + (b - a)(I_0 + \Delta I \sinh x). \tag{49}$$

Acknowledgment

The authors are indebted to Dr. J. Narud and G. Lunn for important contributions to the work reported here.

References

[1] W. F. Bodtmann and C. L. Ruthroff, "A wideband transistor IF amplifier for space and terrestrial repeaters using grounded-base transformer-coupled stages," *Bell Sys. Tech. J.*, pp. 37–54, January 1963.

[2] J. A. Mataya and S. B. Marshall, "A monolithic integrated variable attenuator," *1968 ISSCC Digest of Tech. Papers*, p. 12.

[3] W. F. Chow and A. P. Stern, "Automatic gain control of transistor amplifiers," *Proc. IRE*, vol. 43, pp. 1119–1127, September 1955.

[4] R. D. Middlebrook, *Differential Amplifiers*. New York: Wiley, 1963, p. 16.

[5] J. E. Solomon, "An integrated 60 MC IF amplifier," *1965 Proc. Nat'l Electronics Conf.*, vol. 21, pp. 153–157.

[6] J. J. Ebers and J. L. Moll, "Large-signal behavior of junction transistors," *Proc. IRE*, vol. 43, pp. 1761–1772, December 1954.

[7] B. Gilbert, "A DC-500 MHz amplifier/multiplier principle," *1968 ISSCC Digest of Tech. Papers*, pp. 114–115.

[8] F. B. Llewellyn, "Some fundamental properties of transmission systems," *Proc. IRE*, vol. 40, pp. 271–283, March 1952.

[9] D. K. Lynn, C. S. Meyer, and D. J. Hamilton, Eds., *Analysis and Design of Integrated Circuits*. New York: McGraw-Hill, 1967, pp. 391–398.

[10] H. T. Friis, "Noise figures of radio receivers," *Proc. IRE*, vol. 32, pp. 419–422, July 1944.

Automated Design of DC-Coupled Monolithic Broad-Band Amplifiers

BRUCE A. WOOLEY, MEMBER, IEEE

Abstract—**A design automation program has been written to optimize the design of dc-coupled monolithic broad-band amplifiers. This program adjusts dc conditions, device geometry, and all passive elements to obtain the maximum small-signal bandwidth consistent with specified low-frequency gain and quiescent power dissipation. The principal features of the program are a frequency-response analysis subroutine based on a nodal admittance matrix formulation, a precise response sensitivity analysis using the adjoint network, and an optimization subroutine based on the Fletcher–Powell algorithm. Two complete design examples, based on the series–series feedback triple and series–shunt feedback pair, are presented. For typical integrated-circuit processing, a maximum bandwidth on the order of 100 MHz is obtained for these designs with a voltage gain of 34 dB and 96-mW power dissipation.**

INTRODUCTION

A DESIGN automation program has been implemented to optimize the design of dc-coupled monolithic broad-band amplifiers. The program significantly expands the degrees of freedom that can be utilized in the design of such circuits. Dc operating conditions, transistor geometry, and all passive elements are adjusted to achieve the maximum small-signal bandwidth consistent with specifications for low-frequency gain, power dissipation, gain sensitivity, dc output level, and processing capability.

Manuscript received May 5, 1970; revised June 1, 1970. This work was supported in part by Joint Services Electronics Program Grant AFOSR-68-1488 and by the U. S. Army Research Office-Durham under Contact DAHCO-4-67-C-0031.

The author was with the Department of Electrical Engineering and Computer Sciences, Electronics Research Laboratory, University of California, Berkeley, Calif. 94720. He is now with Bell Telephone Laboratories, Inc., Holmdel, N. J.

In this paper the important features of the design program are described and two complete amplifier designs are presented to illustrate the use of the program. These design examples are based on two configurations particularly suitable for broad-band voltage amplification in integrated circuits—the series–series feedback triple and the series–shunt feedback pair. On the basis of the design optimization results, the relative effectiveness of these configurations is established.

The class of circuits considered in this paper is defined by the general design requirements (GDR) below:

1) moderate voltage gain (34 dB);
2) low gain sensitivity to processing and environment;
3) standard IC processing;
4) dc coupling with 0-V output level;
5) maximum bandwidth for specified power dissipation.

The design program is, however, applicable to a broader class of problems. It can be used for optimization with respect to any criteria that can be related to the small-signal response of an amplifier.

The first requirement of the GDR is a moderate voltage gain, typically between 30 and 40 dB. The specification of *voltage* gain is meant to imply the assumption

Reprinted from *IEEE J. Solid-State Circuits,* vol. SC-6, pp. 24-34, Feb. 1971.

of a low source impedance (50 ohms) and the need for high-input and low-output amplifier impedance levels. Listed second is the requirement for low gain sensitivity to both environment (temperature) and processing. As indicated, a standard junction-isolated bipolar process is assumed. The fourth requirement given in the GDR is for dc coupling with 0-V input and output quiescent levels; dc level shifting must be incorporated within the amplifier. It is assumed that the dc output level should be relatively insensitive to temperature and processing. The final specification is for maximum −3-dB bandwidth consistent with the preceding requirements and a given quiescent power dissipation. As noted above, this final criterion is the objective of the automated design optimization.

Program Description

The essential requirements for a circuit-design optimization program are a scalar *performance index*, efficient automated circuit *analysis*, efficient evaluation of the performance index *gradient*, and an *optimization* algorithm. In this section, each of these requirements is described in relation to the design of monolithic broadband amplifiers.

Performance Index

The basic procedure of automated circuit design is an iterative numerical search for the minimum of a scalar performance index. This index, or objective, is a multivariable function of the design parameters and is formulated such that its minimum corresponds to an optimum design. A least-squared-error criterion has been used as a performance index for achieving a specified gain and maximum small-signal bandwidth:

$$E = \sum_{i=1}^{m} W(\omega_i)[|A_V(\omega_i)| - A]^2 \tag{1}$$

where ω_i, $i = 1, \cdots, m$ is a set of specified frequency points, $|A_V(\omega_i)|$ the magnitude of the small-signal voltage gain at ω_i, A the specified low-frequency voltage gain, and $W(\omega_i)$, $i = 1, \cdots, m$ a set of arbitrary nonnegative, scalar weighting parameters. The weighting parameters are used to permit the emphasis of various regions of the amplifier frequency response without analysis at a large number of frequency points. For example, to ensure adequate realization of the specified gain at dc, the point $\omega_1 = 0$ may be given a significantly larger weighting than points near the response bandedge.

The points ω_i define the frequency range over which response error is evaluated. The choice of this range is critical if the optimum design is to correspond to the maximum −3-dB bandwidth obtainable without peaking. If frequency points are selected too far beyond the amplifier bandedge, the minimization process may lead to heavy bandedge peaking in order to reduce the error beyond the bandedge. Conversely, if the frequency points are limited to a range well below the bandedge, the performance index is relatively insensitive to the response shape at the bandedge and convergence of the iterative optimization procedure may be drastically inhibited. The choice of frequency range thus requires an approximate knowledge of the amplifier response. This can usually be established from a preliminary circuit analysis; at worst, the frequency points can be rechosen after several design iterations.

Analysis

The performance index (1) is determined directly from the small-signal gain-frequency response $A_V(\omega)$. This response is evaluated at each frequency point ω_i through analysis of an appropriate linear small-signal circuit model. A nodal admittance matrix formulation is used for the analysis, and the admittance equations are solved using an LU factorization [1].

Because of the iterative nature of the automated design procedure, many analyses are needed to arrive at an optimum design. As a result, the principal factor governing the overall cost of the design process is the efficiency of the analysis subroutine, and much of the programming effort is directed toward improving this efficiency. For the design examples to be presented, the typical central processing unit (CPU) time required for evaluation of a performance index (1) with $m = 10$ is 0.2 s on a CDC 6400.[1]

Gradient Evaluation

The most effective algorithms for finding the minimum of a multivariable objective function by direct numerical search generally rely on repeated evaluation of the function's gradient. The classical technique for gradient evaluation is based on perturbations. Each of the variables is perturbed individually and the corresponding change in the objective function is determined. This information is then used to approximate the gradient.

For application to the automation of practical circuit design problems, the perturbation approach has two serious disadvantages. First, for n design variables, n additional analyses are needed to determine the gradient of the performance index at each trial point in the design procedure. Since most of the time required for a design run is consumed in the analysis subroutine, these additional analyses represent an unacceptable increase in computational effort. The second limitation of the perturbation technique is the approximate nature of the gradient information. This information is accumulated in the optimization algorithm, and the possible error build-up may seriously degrade the convergence capability of the algorithm.

An approach to gradient analysis that overcomes both of the disadvantages associated with perturbations is

[1] A preliminary investigation indicates that the implementation of sparse matrix techniques for the analysis may lead to as much as an order of magnitude reduction in this time.

based on the formulation of the adjoint relations to a set of network equations. Director and Rohrer have identified these relations in terms of the so-called linear adjoint network and have shown that the sensitivity of the response of a circuit to any of its branch elements may be expressed as a product of branch responses in the circuit and its adjoint network [2], [3]. Since the performance index (1) is related directly to the circuit response and the design variables are necessarily related to one or more of the branch elements in the circuit model, the evaluation of the response sensitivities from analyses of the circuit and its adjoint leads immediately to the gradient of the index.

The adjoint network is essential to the development of practical circuit-design automation. Through its use, the evaluation of the performance index gradient requires an additional computational effort less than that needed for analysis of the index itself. It can be shown that the nodal admittance matrix for the adjoint network is simply the transpose of the nodal admittance matrix for the circuit [4]. Hence, a substantial number of the computations performed in the original circuit analysis need not be repeated when solving the adjoint equations. In addition to its efficiency, the adjoint network approach also leads to a theoretically exact evaluation of the gradient.

Optimization Algorithm

The direct numerical search for a minimum of a performance function generally consists of repeating a specified sequence of operations for a number of iterations until convergence to the minimum is obtained. The first step at each iteration is to choose a direction for continuing the search. A one-dimensional search is then conducted to locate the minimum of the function in this direction. The directional minimum is taken as the starting point for the next iteration. Convergence is obtained when the directional minimum corresponds to a true minimum of the performance function.

When the gradient of the performance function is known, the classical search direction is that of steepest descent (the negative gradient direction). However, the method of steepest descent exhibits limited convergence ability if the search encounters a narrow valley in the shape of the performance function [5]. A number of algorithms have been developed to overcome this limitation. Among the most powerful of these is the method of Fletcher and Powell [5]–[8], which has been used with considerable success in the program described here. In the Fletcher–Powell algorithm, the search direction is chosen on the basis of the gradient and an approximation to the inverse of the matrix of second-order derivatives (Hessian matrix). The approximation is established from the gradient information of preceding iterations and, though necessarily poor at first, continues to improve throughout the search. For the initial iteration, the steepest descent direction is used.

Once a search direction has been chosen, steps are taken along it until a directional minimum is bounded. A cubic interpolation is then used to locate an approximate minimum. This procedure was suggested by Fletcher and Powell and has proven satisfactory.

Basic Amplifier Configurations

The program described above has been used to develop two complete amplifier designs that satisfy optimally the GDR. The first step is establishing each of these designs is the choice of a basic amplifier configuration. Possibilities for this choice will first be introduced by means of a brief description of commercially available amplifiers. Results of a preliminary investigation are then used to select the two most suitable configurations [9].

Commercial Amplifiers

A number of commercial integrated circuits more or less satisfy the GDR, with the exception of the dc-level shifting condition [10]–[14]. The only one of these circuits that does not rely on a feedback approach is the RCA CA3040. This amplifier is basically a differential cascode with emitter–follower input and output stages. The use of an emitter–follower output stage is virtually mandatory in any configuration meeting the GDR. The emitter–follower input is used in the CA3040 to achieve the required high input impedance and to buffer the input stage from the source conditions. The low-frequency voltage gain of this amplifier is given by

$$A_V \simeq (q/kT) I_C R_C \tag{2}$$

where I_C is the quiescent current in the cascode and R_C the diffused load resistor. The output emitter–follower buffers the cascode from the actual amplifier load. To achieve low gain sensitivity to temperature, the temperature dependence of I_C is established so as to cancel the sensitivities of R_C and q/kT. In doing this, however, a relatively sensitive quiescent voltage level results at the output of the cascode. This output dc level may be expressed as

$$V_{CL} = V_{CC} - I_C R_C \tag{3}$$

where V_{CC} is the positive supply voltage. In order to achieve a low sensitivity for (2), it is necessary that $I_C R_C \propto T$. The corresponding condition for V_{CL} is a direct dependence on $-T$.

The Fairchild μA733 is a differential amplifier configuration based on a series–shunt local feedback cascade. High input impedance is achieved with the series feedback in the input stage. An output emitter–follower is included within the shunt feedback loop of the second stage.

Two commercial monolithic broad-band amplifiers, the Sylvania SA-20 and the SL611C from Plessey Microelectronics of Great Britain, are based on a series–shunt overall feedback pair. Both amplifiers are single ended and include the output emitter–follower within the feedback loop.

The Motorola MC1553 is a single-ended amplifier based on a series–series feedback triple driving an emitter–follower output stage. The design of this amplifier has been described in detail by Solomon and Wilson [15]. This work is representative of the limits of complexity to which nonautomated broad-band amplifier design can be extended.

Feedback Configurations

As is evident from the above description of commercial amplifiers, a feedback approach is commonly used to satisfy specifications similar to those given in the GDR. Other approaches are possible but they generally entail severe disadvantages with respect to one or more of the specifications; for example, as cited above, low gain sensitivity is not compatible with an insensitive quiescent output voltage in the differential cascode circuit.

Shown in Fig. 1 are the basic local and overall feedback configurations suitable for meeting the GDR. Consideration is restricted to configurations with series feedback in the input stage to ensure realization of a high input impedance.

In previous work, a comparison of the configurations in Fig. 1 was made on the basis of extensive computer-aided analysis [9]. Equal voltage gain per common-emitter stage was specified for each configuration and the feedback elements were optimized on a trial-and-error basis. The results of the comparison indicate that for a corresponding number of stages, overall feedback leads to a substantially lower gain sensitivity than local feedback, while providing comparable gain and bandwidth. Nonetheless, a local feedback approach has been more commonly used in the design of discrete component feedback amplifiers [16]. This may be due, in part, to the stability problems often associated with discrete overall feedback amplifiers. To the extent that these problems are related to the physical circuit layout, they are virtually eliminated in monolithic realization. Therefore, overall feedback is regarded as superior for meeting the requirements under consideration here.

The relative merits of the overall feedback configurations of Fig. 1, the series–series triple and the series–shunt pair, can not be readily established from the results of the preliminary work referred to above [9]. While the triple provides substantially more loop gain than the pair, it also exhibits a more complex open-loop transfer function; hence, it can not be stated as a general conclusion that closed-loop bandwidths can be achieved for the triple comparable to those obtainable with the pair. For this reason, optimal amplifier designs based on both configurations will be presented.

The development of a complete amplifier design based on the series–series triple, illustrating the detailed design optimization procedure, is described first. A similar

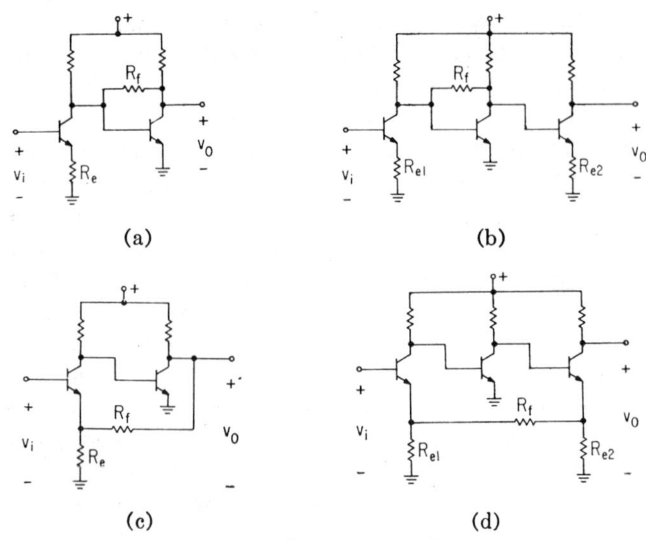

Fig. 1. Basic feedback configurations. (a) Series–shunt cascade. (b) Series–shunt–series cascade. (c) Series–shunt pair. (d) Series–series triple.

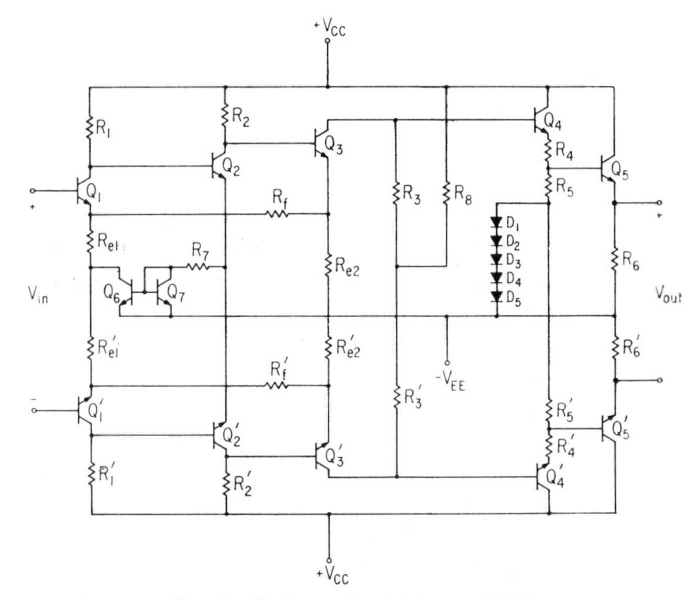

Fig. 2. Series–series triple amplifier.

design using the series–shunt pair is next presented, with emphasis on features that differ from the first design. Specific design optimization results are then given for both amplifiers and a comparison is made.

Series–Series Triple Design

Shown in Fig. 2 is the schematic of an amplifier based on the series–series triple and satisfying the GDR. The amplifier consists of a balanced, or differential, triple driving a level-shifting stage and emitter–follower output stage. A balanced configuration is used to achieve a dc-coupled response with temperature insensitive biasing and a differential input–output capability. The passive elements, the quiescent operating points, and the transistor geometry are to be adjusted for optimum performance. The principal specifications assumed are a differential voltage gain of 34 dB, ±6-V power supplies, and a quiescent power dissipation of 96 mW (corresponding to a total quiescent current of 8 mA).

DC Conditions

The quiescent emitter currents for the balanced triple in Fig. 2 are supplied by the common-mode current source Q_6–Q_7. The distribution of currents between the three stages is governed by R_f, R_7, and the ratio of currents in Q_6 and Q_7. As is shown in Appendix I, this form of biasing leads to a dc voltage at the collector of Q_3 that is relatively insensitive to both temperature and processing.

The dc voltage level is shifted across the resistor R_4 in the amplifier of Fig. 2. The emitter–follower Q_4 is used to isolate the basic triple from the level-shifting network. The dc current in the level-shifting stage is supplied through the resistor R_5 and the common-mode diode string D_1–D_5. Resistive biasing has been used, despite its associated attenuation, because it is the only level-shifting configuration that does not seriously limit the overall amplifier bandwidth. For example, if R_5 is replaced with a transistor current source, the attenuation is eliminated but the frequency response is severely degraded by the output capacitance of the current source [9].

The emitter–follower output stage provides the required low output impedance and also buffers the level-shifting network from the load. This buffer is necessary if the attenuation in the level-shifting network is to be relatively independent of the load.

The common-mode diode string is used in the level-shifting stage to cancel the effects of changes in the base–emitter voltages of Q_4 and Q_5 on the quiescent output voltage. This cancellation can be demonstrated through consideration of the generalized n-diode output stage shown in Fig. 3. In this figure, V_{C3} corresponds to the voltage at the collector of Q_3 in Fig. 2. If the transistors Q_4 and Q_5 and the diodes D_1–D_5 are fabricated such that $V_{BE}(Q_4) = V_{BE}(Q_5) = \phi$, then the dc output voltage may be expressed as

$$V_O = -V_{EE} + (V_{EE} + V_{C3})A_{LV} + \phi[n - 1 - (n + 1)A_{LV}] \quad (4)$$

where

$$A_{LV} = R_5/(R_4 + R_5) \quad (5)$$

is the small-signal differential voltage transmission of the level-shifting stage. The dependence of V_O on the junction voltages in the level-shifting and output stages can be eliminated by setting the coefficient of ϕ in (4) to zero, resulting in the requirement

$$A_{LV} = (n - 1)/(n + 1). \quad (6)$$

Under the condition (6), the specified 0-V quiescent

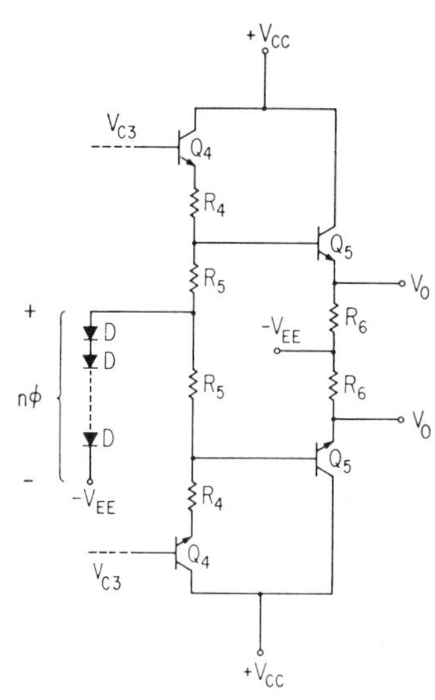

Fig. 3. Generalized level-shifting and output-stage configuration.

output level ($V_O = 0$) is obtained when

$$V_{C3} = V_{BE}((1/A_{LV}) - 1). \quad (7)$$

A 0-V dc output that is relatively insensitive to temperature and processing can be established with consistent choice of n and V_{C3} satisfying (6) and (7). This choice determines the transmission A_{LV} and, correspondingly, the ratio of resistors R_4 and R_5.

Several factors must be considered when choosing n and V_{C3}. First, n is obviously restricted to integer values. Second, the value of V_{C3} should correspond to a near-maximum available unclipped voltage swing at the amplifier output. Finally, the voltage transmission A_{LV} should be the maximum consistent with the first two considerations.

Values of V_{C3} and A_{LV} for several choices of n are given in Table I for ±6-V supplies. For the amplifier of Fig. 2, $n = 5$ has been chosen. This corresponds to a 3-V dc level at the collector of Q_3 and a voltage transmission of 0.67 for the level-shifting stage.

The values of the resistors R_4 and R_5 are determined from the dc current specification for the level-shifting stage. These values must be low enough so that the RC time constant at the base of Q_5 does not limit the amplifier bandwidth. For the total specified power dissipation of 96 mW, 48 mW is allotted to the level-shifting and output stages. This permits specification of 1-mA dc currents in each of the level-shifting and output-stage transistors. The corresponding resistor values are 1.6 kΩ for R_4 and 3.2 kΩ for R_5; the net differential-mode resistance from the base of Q_5 to ground is 1.07 kΩ.

Frequency Compensation

The level-shifting stage, the output emitter–follower, and the common-mode elements of the amplifier in Fig. 2 do not directly affect the overall differential frequency response of the amplifier. Therefore, this response can be determined from the differential-mode equivalent half

TABLE I
LEVEL SHIFTING AND OUTPUT STAGE DESIGN (FIG. 3) FOR ±6-V SUPPLIES

Number of Diodes n	Quiescent Input Voltage Level V_{C3}, V	Differential Voltage Transmission A_{LV}
4	4.0	0.60
5	3.0	0.67
6	2.4	0.715
7	2.0	0.75

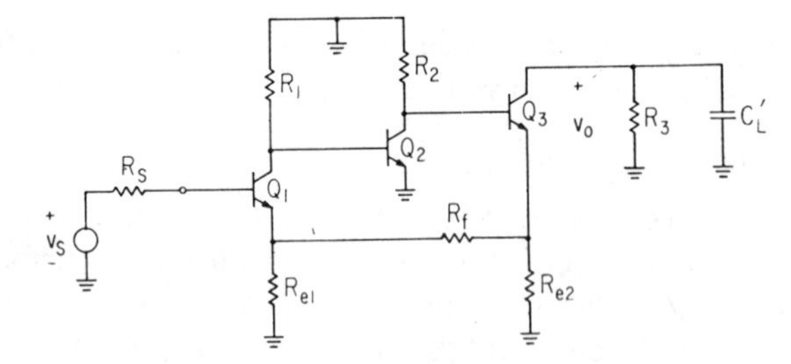

Fig. 4. Differential-mode equivalent half circuit for series–series triple of Fig. 2.

circuit for the basic triple, shown in Fig. 4 [17]. The loading on the triple in this circuit is represented by the resistor R_3 and the capacitor C_L', which models both the input capacitance of the level-shifting stage and the parasitic capacitance associated with R_3. The input resistance of the level-shifting network is much greater than R_3 and may be neglected.

A common approach to the frequency compensation of feedback configurations such as Fig. 4 is to introduce a zero into the feedback transmission (a phantom zero) with a shunt capacitor across the resistor R_f [18]. In the series–series triple, however, compensation can instead be established using an effective feedback zero that is inherent in the configuration. For a configuration such as Fig. 4, with a series feedback connection at the output, the feedback network samples the emitter current of the output transistor rather than the actual output signal (v_o in Fig. 4). This situation results in a zero of the feedback transmission that is given approximately by [9]

$$Z_o \approx -1/[R_3(C_{cb} + C_{cs} + C_L')] \tag{8}$$

where C_{cb} and C_{cs} are the collector–base and collector–substrate capacitance of the output transistor Q_3.

Optimal compensation of the series–series triple can be achieved through suitable choices for the resistor R_3 and capacitance C_L'. Introduction of a common-mode resistor, R_8 in Fig. 2, allows R_3 to be varied, below an upper bound, without disturbing the dc conditions for transistor Q_3. Small values for R_3 should, however, be avoided because they limit the available voltage swing at the output of the triple. If it is necessary, the capacitance C_L' may be increased by enlarging the area of the diffused resistor R_3. It should not be necessary to add separate capacitive elements for compensation of the triple.

Small-Signal Modeling

To determine the small-signal response of the circuit in Fig. 4, appropriate models must be introduced for the transistors Q_1–Q_3. Design variables that are not explicitly present as elements in the equivalent half-circuit of Fig. 4 are entered into the design process either through elements of the small-signal models or as constraints among the elements of the half circuit.

The general form assumed for the planar geometry of transistors Q_1, Q_2, and Q_3 is illustrated in Fig. 5. The figure is drawn to scale and the minimum allowed mask

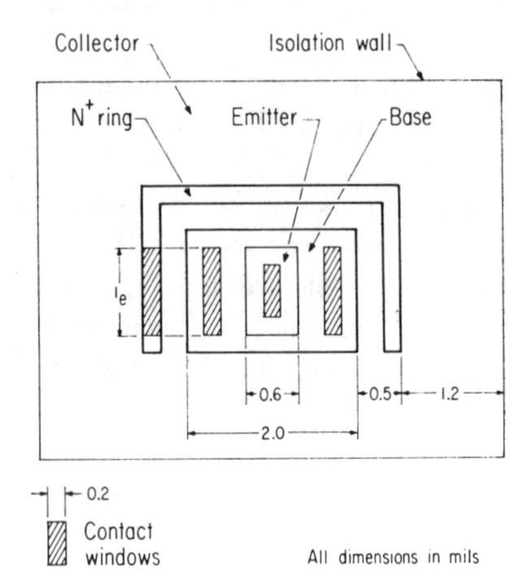

Fig. 5. Planar device geometry (to scale) with minimum mask dimensions except for l_e.

dimensions are used for all distances except the emitter stripe length l_e. These dimensions are well within present industrial capability. The device geometry design variables are the emitter stripe length l_e, the number of emitter stripes n_e, and the number of base contact stripes $n_b = n_e$ or $(n_e + 1)$. These variables govern the tradeoff between the ohmic resistances r_b' and r_c' and the junction depletion layer capacitances. This tradeoff is regarded as the principal effect of device geometry on the frequency response of a circuit such as that in Fig. 4.

The small-signal transistor model used in the design optimization program is shown in Fig. 6. The model is a basic hybrid-π configuration with RC π-sections used to represent the distributed base and collector structures. In a given design situation, the elements of the model must be characterized on the basis of the processing to be used. For the design examples presented here, the following relationships apply:

$$g_m = (q/kT)I_c \tag{9}$$

$$r_\pi = \beta_0/g_m, \qquad \beta_0 = 120 \tag{10}$$

$$r_0 = 1/\eta g_m, \qquad \eta = 0.000\,65 \tag{11}$$

$$C_\pi = C_{je} + g_m\tau_t, \qquad \tau_t = 0.22 \text{ ns} \tag{12}$$

$$C_{je}/\text{area} = 0.45/[(\psi_e - V_{BE})^{k_e}] \quad \text{pF/mil}^2 \tag{13}$$

$$C_{cb}/\text{area} = 0.11/[(\psi_c - V_{BC})^{k_c}] \quad \text{pF/mil}^2 \tag{14}$$

$$C_{cs}/\text{area} = 0.044/[(\psi_s - V_{SC})^{k_s}] \quad \text{pF/mil}^2 \tag{15}$$

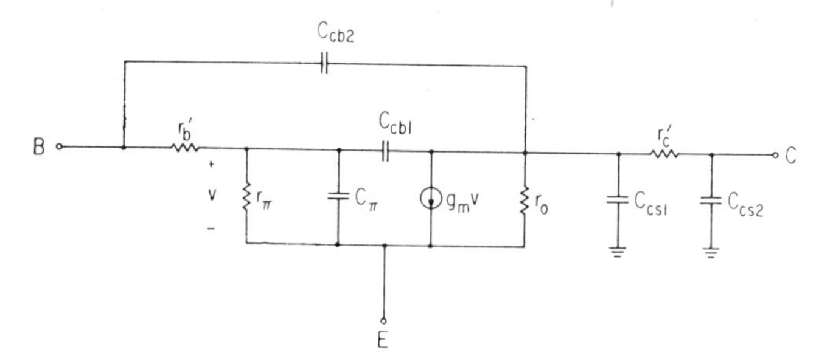

Fig. 6. Small-signal transistor model.

where

$$\psi_e = 0.75\text{ V} \qquad k_e = 0.35$$
$$\psi_c = 0.4\text{ V} \qquad k_c = 0.275$$
$$\psi_s = 0.4\text{ V} \qquad k_s = 0.34.$$

For an equal number of base contact and emitter stripes $n_b = n_e$,

$$r_b' = \frac{1}{n_e}\left[210 + \frac{200}{(I_c + 1.3)}\right]\frac{1}{(0.525l_e + 0.15)}\text{ ohms} \quad (16)$$

$$r_c' = 30\left[\frac{3.5}{2.3 + l_e + 1.4n_e} + \frac{0.2}{n_e}\right]\text{ohms} \quad (17)$$

where I_c is in milliamperes and l_e in mils. For the case $n_b = n_e + 1$,

$$r_b' = \frac{1}{(n_e + 1)}\left[210 + \frac{200}{(I_c + 1.3)}\right] \cdot \frac{1}{(0.525l_e + 0.15)}\text{ ohms} \quad (18)$$

$$r_c' = 30\left[\frac{3.5}{2.8 + l_e + 1.4n_e} + \frac{0.2}{n_e}\right]\text{ohms.} \quad (19)$$

These expressions correspond to a standard bipolar diffusion process with a 3.5-μ 120-$\Omega/\square$ base diffusion; their experimental basis is presented in Appendix II. A number of the expressions represent the best fit to experimental data and are not necessarily meaningful from a physical standpoint.

Design Variables

The design variables to be used in the optimization of the amplifier in Fig. 2 are the feedback resistors R_{e1}, R_{e2}, and R_f; the differential collector load resistors R_1, R_2, and R_3; possible common-mode collector resistors, such as R_8; the dc current in the triple as governed by the resistor R_7 and the ratio of currents in the transistors Q_6 and Q_7; the collector–emitter voltages of the transistors Q_1, Q_2, and Q_3; and the device geometry variables, as described above, for Q_1, Q_2, and Q_3. Design constraints among these variables are 1) all passive elements must be nonnegative, 2) a minimum collector–emitter voltage of 1.4 V for transistors Q_1 and Q_2, 3) a collector voltage of 3 V for Q_3, as required by the output stage design, 4) an upper bound on the differential collector load resistance of each stage, determined by dc conditions, and 5) a minimum emitter-stripe length of 0.6 mil.

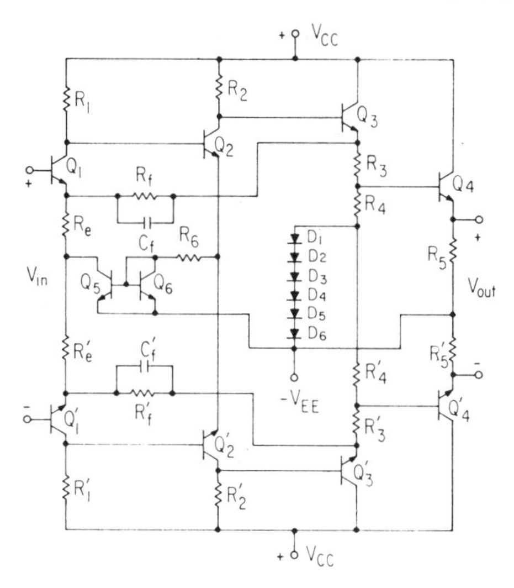

Fig. 7. Series–shunt pair amplifier.

Series–Shunt Pair Design

An amplifier configuration very similar to that of Fig. 2, but based on the series–shunt feedback pair, is shown in Fig. 7. As for the design based on the series–series triple, a balanced configuration is used and the principal specifications are a voltage gain of 34 dB, ±6-V supplies, and 96-mW power dissipation. The emitter–follower for the level-shifting stage in Fig. 7 is included within the basic amplifier feedback loop. This increases substantially the loop gain of the feedback pair by reducing the loading at the collector of Q_2; also, the dc drop across R_f is decreased, correspondingly lowering the quiescent current that must be drained through this resistor.

The biasing approach used for the amplifier of Fig. 7 is identical to that of the series–series triple design. The common-mode current source supplies the currents for the basic pair and the drain through R_f. As is demonstrated in Appendix I, this configuration desensitizes the quiescent voltage at the collector of Q_2 (not the voltage at the emitter of Q_3) to processing and environment. Hence, the same general level-shifting and output-stage configuration used for the previous design can be employed to realize an insensitive zero-voltage quiescent output level. Because the voltage at the emitter of Q_2 in the amplifier of Fig. 7 is lower than that at the emitter of Q_3 in Fig. 2, a comparable available voltage swing can be achieved with a lower quiescent output voltage for the basic amplifier. For this reason, a six-diode string has been used to bias the level-shifting network in Fig. 7. This results in an increased voltage transmission in the level-shifting stage, relative to the five-diode configuration. From the data in Table I, for $n = 6$, $V_{C2} = 2.4$ V and $A_{LV} = 0.715$. The use of six diodes also reduces the current needed in the level-shifting resistors to maintain the same resistance level at the base of the output emitter–follower. To realize a 1.07-kΩ equivalent resistance from this point to ground, the required current is 0.667 mA and the resistor values are $R_3 = 1.5$ kΩ and $R_4 = 3.75$ kΩ.

Frequency compensation is most effectively achieved in the series–shunt pair with a capacitance in shunt with the feedback resistor R_f. Unlike the series–series triple, the pair can not generally be compensated without the addition of an actual capacitance element.

Design Results

Series–Series Triple

The design optimization results for the series–series triple amplifier of Fig. 2 are summarized in Fig. 8 and Tables II and III. The dashed line responses in Fig. 8 correspond to several designs used as starting points for the optimization procedure. All design runs converged to the same optimum, represented by the solid-line response. The optimum design achieves the specified voltage gain of 34 dB with a −3-dB bandwidth of 85 MHz.

Resistor values for the optimum design are given in Table II. The constraint $R_{e1} = R_{e2}$ has been imposed to ensure good matching between the relatively small emitter resistors. In Table III, the optimum geometry and dc level is given for each of the transistors Q_1, Q_2, and Q_3; included in the table are the values of transistor f_t for the geometry and current level shown. A total dc of 2 mA is available for each side of the balanced triple for the specifications (96-mW total power dissipation and ±6-V supplies) and output stage design (48-mW power dissipation) presented earlier. To establish the current levels given in Table III, the ratio of currents in Q_6 and Q_7 is $I_{C6}/I_{C7} = 1.6$.

The optimum value of collector–emitter voltage for both Q_1 and Q_2 is the minimum allowed, 1.4 V. The values of resistance given in Table II for R_1 and R_2 correspond to the maximum allowed by dc conditions.

The temperature dependence of the low-frequency voltage gain has been determined from dc and ac analysis of the complete configuration of Fig. 2, using the program BIAS-3 [19]. If typical first-order temperature sensitivities are assumed for all elements in the amplifier, the total gain variation is 2.6 percent over the full temperature range −55 to 125°C. A feedback analysis of the optimum design indicates a low-frequency loop gain of 175 for the triple.

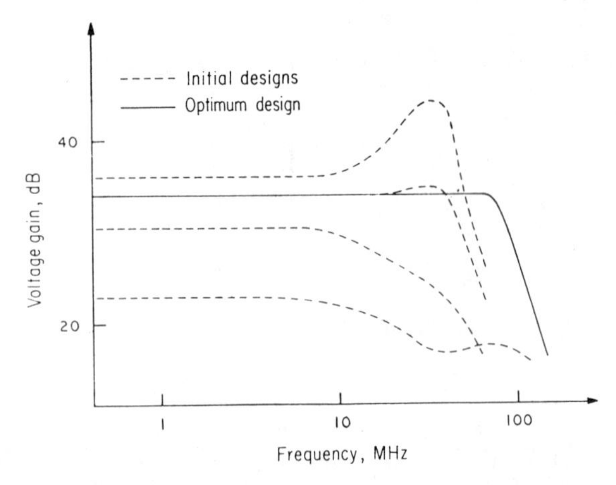

Fig. 8. Design optimization results for series–series triple amplifier of Fig. 2.

TABLE II
RESISTOR VALUES FOR OPTIMUM SERIES–SERIES TRIPLE DESIGN (FIG. 2)

$R_{e1} = R_{e2} = 290$ ohms	
$R_f = 1.9$ kΩ	$R_3 = 2.5$ kΩ
$R_1 = 10.6$ kΩ	$R_8 = 810$ ohms
$R_2 = 6.0$ kΩ	$R_7 = 3.4$ kΩ

TABLE III
DEVICE SPECIFICATIONS FOR OPTIMUM SERIES–SERIES TRIPLE (FIG. 2)

Device	n_e	n_b	l_e, mils	I_C, mA	f_t, MHz
Q_1	1	2	0.90	0.50	470
Q_2	1	2	0.71	0.77	560
Q_3	1	1	0.95	0.73	550

Series–Shunt Pair

The design results for the series–shunt pair amplifier of Fig. 7 are given in Fig. 9 and Tables IV and V. The frequency response for the optimum design is shown in Fig. 9. The 34-dB gain is obtained over a bandwidth of 112 MHz. Optimum element values are given in Table IV and the device data are summarized in Table V. The level-shifting emitter–follower, Q_3 in Fig. 7, was included in the optimization process and a minimum area structure was specified.

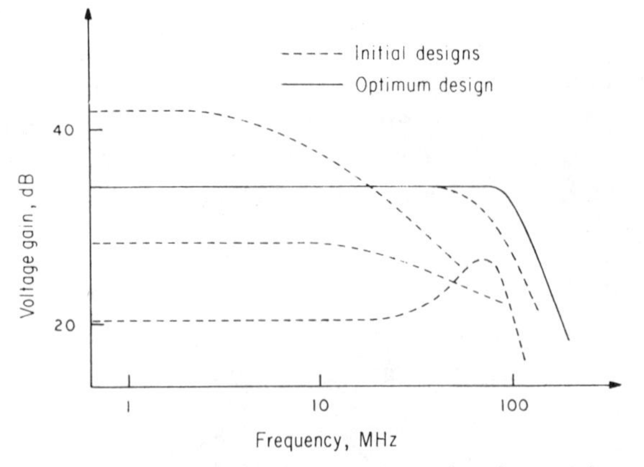

Fig. 9. Design optimization results for series–shunt pair amplifier of Fig. 7.

A total dc current of 3 mA is available for each side of the combined feedback pair and level-shifting stage. The optimum collector–emitter voltage in the first stage of the series–shunt pair is the minimum 1.4 V. As for the series–series triple design, the maximum values allowed were optimum for the differential load resistors R_1 and R_2.

TABLE IV
PASSIVE ELEMENT VALUES FOR OPTIMUM SERIES–SHUNT PAIR DESIGN (FIG. 7)

R_e = 31 ohms	R_6 = 6.7 kΩ
R_f = 2.2 kΩ	
R_1 = 6.4 kΩ	C_f = 0.83 pF
R_2 = 9.1 kΩ	

TABLE V
DEVICE SPECIFICATIONS FOR OPTIMUM SERIES–SHUNT PAIR (FIG. 7)

Device	n_e	n_b	l_e, mils	I_C, mA	f_t, MHz
Q_1	1	2	1.3	0.83	520
Q_2	1	1	0.6	0.39	500
Q_3	1	1	0.6	1.78	640

The value of compensation capacitance C_f needed for the optimum series–shunt pair design, 0.83 pF, can be realized with base–collector junction 14 mil^2 in area under a reverse bias greater than 5 V.

A low-frequency analysis of the amplifier in Fig. 7, carried out with the BIAS-3 program, indicates a total gain variation with temperature of 0.4 percent between −55 and 125°C. As for the amplifier of Fig. 2, only first-order temperature sensitivities were assumed. The low-frequency loop gain of the optimum series–shunt pair design is 105.

Comparison

The design optimization results presented above provide a basis for comparison of the series–series triple and series–shunt pair. As used in the amplifiers of Figs. 2 and 7, these configurations appear very similar because of the inclusion of the emitter–follower within the feedback loop of the pair. This emitter–follower is essential if a high loop gain is to be maintained for the pair.

For the design specifications and constraints assumed here, the maximum unpeaked −3-dB bandwidth for the series–shunt pair is 32 percent greater than the maximum for the series–series triple. In addition, on the basis of a first-order analysis, the dc gain of the pair is much less sensitive to variations in temperature. This latter result is somewhat surprising in view of the larger loop gain of the triple. It is due to a fortuitously low open-loop response sensitivity for the configuration of Fig. 7. Coincidentally, a first-order zero in gain sensitivity to temperature exists for this amplifier near room temperature.

For practical applications, the very low gain sensitivity found for the pair may be somewhat misleading because, in the first-order analysis, ideal tracking has been assumed between all diffused resistors. This may be difficult to achieve for the optimum series–shunt pair design because of the very low value (31 ohms) for the series emitter resistor R_e. This small value represents a significant disadvantage for the amplifier of Fig. 7. In contrast, the smallest resistor needed for the series–series triple design of Fig. 2 is 290 ohms. A minimum value, such as 100 ohms, could be specified for R_e in the pair. This, however, would degrade both bandwidth and sensitivity performance.

A second processing limitation of the pair is the need to realize a capacitive element in the feedback network for frequency compensation. For the triple, compensation is established using parasitic and internal device capacitances.

EXPERIMENTAL RESULTS

In order to verify the feasibility of realizing configurations such as that shown in Fig. 2, the balanced series–series triple portion of this amplifier has been fabricated. The diffusion processing used is the same as that employed in the device characterization. However, the minimum allowable mask dimensions are somewhat larger than those indicated in Fig. 5. The design optimization process was repeated for the larger geometry and the resulting optimum design was realized. The measured frequency response for a typical unit is given in Fig. 10, along with the thoretical response predicted by the design program. The −3-dB bandwidth of the experimental response is 34 MHz. The transistors in the realization exhibited an f_t of 200 MHz at a collector current of 1 mA.

CONCLUSION

The implementation of an effective circuit design optimization program has been described. Design examples based on the series–series triple and the series–shunt pair were optimized and the results used to compare these two basic feedback configurations. The pair was found to provide significantly better performance at a cost of increased processing difficulty.

Design results such as those given in Figs. 8 and 9 indicate that automated design optimization may lead to as much as a 50 percent increase in bandwidth over what might be considered a good design. Of course, the improvement obtainable in specific design situation, relative to nonautomated design results, depends in large part on 1) the influence on the response of parameters, such as dc conditions, that are chosen somewhat arbitrarily in a nonautomated procedure, and 2) whether or not a fortunate choice has been made for such parameters on a nonautomated basis.

Perhaps more important than the improvement obtainable with automated design in a specific situation is the ability to document the existence of an optimum design. Once the design procedure is completed for a given amplifier, no further design effort is necessary. Additionally, comparison of various design approaches can be made on the basis of the best performance obtainable with each approach.

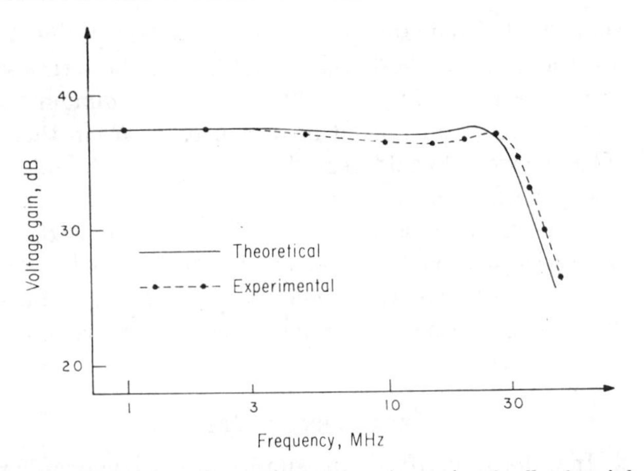

Fig. 10. Response of experimental series–series feedback triple.

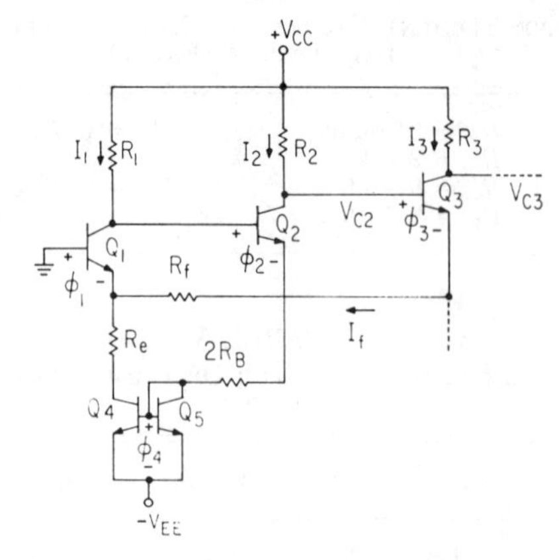

Fig. 11. Common-mode equivalent half-circuit for basic amplifiers of Figs. 2 and 7.

Although this paper deals only with a specific class of analog integrated circuits, the design approach is readily extended to a much broader class of circuit-design problems. The techniques employed have proven to be highly effective and the implementation of a general-purpose program for the automated design of linear integrated circuits is a realistic objective for the immediate future.

Appendix I

The common-mode equivalent half circuit of Fig. 11 can be used to show that the current source biasing in the amplifiers of Figs. 2 and 7 desensitizes the quiescent output collector voltage for the basic feedback amplifiers. For the series–series triple of Fig. 2, $I_3 = I_f$, and, neglecting base currents, the dc voltage V_{C3} may be expressed as

$$V_{C3} = V_{CC} + \frac{V_{CC} + V_{EE} - (\phi_2 + \phi_4) - [\gamma(R_1/R_2) + 2(R_B/R_2)](V_{CC} + \phi_1 - \phi_3)}{(R_f/R_3)[\gamma(R_1/R_2) + 2(R_B/R_2)] + (R_1/R_3)} \tag{20}$$

where $\gamma = I_C(\phi_4)/I_C(\phi_5)$. If the supply voltages and resistor ratios in (20) are assumed to be temperature insensitive, and if $\phi \triangleq \phi_1 \approx \phi_2 \approx \phi_3 \approx \phi_4$, then the temperature dependence of V_{C3} is given approximately by

$$\partial V_{C3}/\partial T \approx -2 \bigg/ \left(\frac{R_f}{T}\left(\gamma \frac{R_1}{R_2} + 2\frac{R_B}{R_2}\right) + \frac{R_1}{R_3}\right) \partial\phi/\partial T. \tag{21}$$

For the data of Tables II and III,

$$\partial V_{C3}/\partial T \approx -0.25\ \partial\phi/\partial T. \tag{22}$$

Thus, the change in V_{C3} with temperature corresponds to approximately 25 percent of the change of a single base–emitter drop.

The voltage V_{C2} in Fig. 11 corresponds to the quiescent voltage at the collector of the second stage of the feedback pair in Fig. 7. Independent of the current in Q_3, this voltage is given by

$$V_{C2} = V_{CC} - \frac{V_{CC} + V_{EE} - (\phi_2 + \phi_4) + (R_1/R_f)(V_{CC} + \phi_1 - \phi_3)}{2R_B/R_2 + [(R_2/R_f) + \gamma]}. \tag{23}$$

Under the same assumptions as used for (21),

$$\frac{\partial V_{C2}}{\partial T} \approx -2 \bigg/ \left[2\frac{R_B}{R_2} + \left(\frac{R_2}{R_f} + \gamma\right)\right] \frac{\partial\phi}{\partial T}. \tag{24}$$

For the data of Tables III and V

$$\partial V_{C2}/\partial T \approx -0.14\ \partial\phi/\partial T. \tag{25}$$

Appendix II

The basis for device characterization in terms of the structure shown in Fig. 5 and the model of Fig. 6 is described briefly below. The diffusion process assumed, and used for the experimental data, results in a 120 Ω/□, 3.5-μ base diffusion. The basewidth is 0.8 μ.

Current Gain β_0

As indicated in (10), a constant value has been assumed for the transistor β_0. Dependence of this parameter on current and geometry, over the ranges of interest, are generally masked by run-to-run and slice-to-slice processing variations. In addition, the amplifier response has been found to be relatively insensitive to β_0, as determined using the sensitivity analysis capability of the design program.

Basewidth Modulation η

The basewidth modulation was determined from measurements of the small-signal transistor output conductance. For the value $\eta = 0.00065$, the basewidth modulation contribution to the collector–base capacitance is completely negligible relative to the junction depletion layer capacitance.

Base Resistance r_b'

The expressions (16) and (18) for r_b' are based on both experimental evidence and geometrical estimates corresponding to the structure of Fig. 5. Experimental values of r_b' were obtained from high-frequency input impedance measurements. This method leads to a value appropriate for frequencies near the bandedge of amplifiers such as shown in Figs. 2 and 7.

Series Collector Resistance r_c'

The expressions (17) and (19) for r_c' were established from dc measurements and estimates based on the device geometry. An n^+ buried layer structure is assumed. For geometries such as shown in Fig. 5, the principal component of r_c' is the vertical collector resistance between the n^+ collector ring and the buried layer.

Base Transit Time τ_t

The typical value of 0.22 ns for the base transit time was determined from measurements of transistor f_t as a function of dc current and voltage.

Junction Capacitances C_{je}, C_{cb}, and C_{cs}

The small-signal junction capacitances were established from geometric considerations and measurements of capacitance per area. The expression (13) represents the best fit to experimental data for the base–emitter junction over the bias range of -4 to $+0.5$ V. Equations (14) and (15) are best fit curves over the range -12 to $+0.3$ V for the base–collector and collector–substrate junctions. The divisions of collector–base and collector–substrate capacitance in the model of Fig. 6 are based on geometrical considerations. In establishing junction area from mask dimensions, such as given in Fig. 5, both lateral diffusion and sidewall area must be considered.

Acknowledgment

The author gratefully acknowledges numerous helpful discussions with Prof. D. O. Pederson and Prof. R. A. Rohrer of the University of California, Berkeley. He also wishes to thank Mrs. Dorothy McDaniel for her assistance in fabricating the numerous experimental devices.

References

[1] A. Ralston, *A First Course in Numerical Analysis.* New York: McGraw-Hill, 1965.
[2] S. W. Director and R. A. Rohrer, "The generalized adjoint network and network sensitivities," *IEEE Trans. Circuit Theory,* vol. CT-16, pp. 318–323, August 1969.
[3] ——, "Automated network design—The frequency domain case," *IEEE Trans. Circuit Theory,* vol. CT-16, pp. 330–337, August 1969.
[4] S. W. Director, "Increased efficiency of network sensitivity computations by means of the LU factorization," *12th Midwest Symp. Circuit Theory Conf. Rec.,* pp. III.3.1–III.3.8, 1969.
[5] D. J. Wilde and C. S. Beightler, *Foundations of Optimization.* Englewood Cliffs, N. J.: Prentice-Hall, 1967.
[6] R. Fletcher and M. J. D. Powell, "A rapidly convergent descent method for minimization," *Computer J.,* vol. 6, pp. 163–168, 1963.
[7] G. C. Temes and D. A. Calahan, "Computer-aided network optimization—The state-of-the-art," *Proc. IEEE,* vol. 55, pp. 1832–1863, November 1967.
[8] "System/360 scientific subroutine package," IBM Corp., White Plains, N. Y., Appl. Prog. 360-CM-03X, version III, 1968, pp. 221–225.
[9] B. A. Wooley and D. O. Pederson, "An optimized design for a dc-coupled monolithic wideband amplifier," *Proc. NEC,* vol. 24, pp. 742–747, 1968.
[10] "Video and wide-band amplifier, CA3040," RCA Linear Integrated Circuits, Data Sheet, File 363, 1968.
[11] "Differential video amplifier μA733, " Fairchild Linear Integrated Circuits, Data Sheet, 1968.
[12] L. Pope, "SA-20 series wideband amplifier," Sylvania Integrated Circuits, Appl. Note 13, 1966.
[13] "SL610 and SL611 r.f., amplifiers," Plessey Microelectronics, England, Provisional Data SL600 Series, 1968.
[14] "A wide band monolithic video amplifier," Motorola Semiconductor Products, Inc., Phoenix, Ariz., Integrated Circuit, Appl. Note AN-404, 1968.
[15] J. E. Solomon and G. R. Wilson, "A highly desensitized, wide-band monolithic amplifier," *IEEE J. Solid-State Circuits,* vol. SC-1, pp. 19–28, September 1966.
[16] E. M. Cherry and D. E. Hooper, "The design of wideband transistor feedback amplifiers," *Proc. Inst. Elec. Eng.* (London), vol. 110, pp. 375–389, February 1963.
[17] R. D. Middlebrook, *Differential Amplifiers.* New York: Wiley, 1964.
[18] M. S. Ghausi and D. O. Pederson, "A new design approach for feedback amplifiers," *IRE Trans. Circuit Theory,* vol. CT-8, pp. 274–284, September 1961.
[19] W. J. McCalla and W. G. Howard, "BIAS-3—A program for the nonlinear dc analysis of bipolar transistor circuits," *ISSCC Digest Tech. Papers,* pp. 82–83, 1970.

Distortion in Bipolar Transistor Variable-Gain Amplifiers

WILLY M. C. SANSEN AND ROBERT G. MEYER

Abstract—**Wide-band variable-gain amplifiers consisting of bipolar junction transistors and exhibiting maximum gain larger than unity are considered. The mechanisms of distortion are analyzed at low and high frequencies. Approximate expressions for distortion are derived and give good agreement with computational results and measurements. The most common high-performance variable-gain circuit realizations are discussed and compared for distortion performance.**

Manuscript received September 6, 1972; revised April 25, 1973. This research was supported by the U. S. Army Research Office, Durham, N. C., under Grant DA-ARO-D31-124-72-G52.

W. M. C. Sansen was with the Department of Electrical Engineering and Computer Sciences and the Electronics Research Laboratory, University of California, Berkeley, Calif. 94720. He is now with the Laboratorium Fysica en Electronica van de Halfgeleiders, Katholieke Universiteit, Leuven, Belgium.

R. G. Meyer is with the Department of Electrical Engineering and Computer Sciences and the Electronics Research Laboratory, University of California, Berkeley, Calif. 94720.

I. Introduction

VARIABLE-GAIN amplifiers are widely used in automatic-gain-control (AGC) circuits and programmable attenuators. An important parameter of a variable-gain circuit is its dynamic range. This is

Reprinted from *IEEE J. Solid-State Circuits,* vol. SC-8, pp. 275-282, Aug. 1973.

the maximum range of input signal amplitudes that the circuit can handle for given limits on noise for small signals and distortion for large signals. In this paper the subject of distortion in variable-gain circuits is considered at both low and high frequencies. Considerations are restricted to wide-band dc-coupled variable-gain amplifiers using bipolar transistors. Such circuits are suitable for realization in monolithic integrated form.

Distortion in single transistors and differential pairs is first considered. Using these results three high-performance integrated gain-control circuits incorporated transistor quads are analyzed and compared for distortion performance. This work allows the selection of the appropriate configuration for best distortion performance in any given application and the optimization of the performance of the chosen circuit.

II. Gain Control in a Single Transistor and a Base-Driven Pair

In a bipolar transistor the ac-collector current i_C depends on the ac base–emitter voltage v_{BE} by an exponential relationship of the form

$$I_Q + i_C = I_{CS} \exp\left(\frac{V_{BE} + v_{BE}}{V_T}\right) \quad (1)$$

in which

I_Q the dc collector current,
I_{CS} the collector-saturation current,
V_{BE} the dc base–emitter voltage, and
$V_T = kT/q \simeq 26$ mV at 302 K.

For small v_{BE} the exponential can be expanded in v_{BE} and after rearrangement (1) becomes

$$i_C = g_m v_{BE}\left[1 + \frac{1}{2}\left(\frac{v_{BE}}{V_T}\right) + \frac{1}{6}\left(\frac{v_{BE}}{V_T}\right)^2 + \cdots\right]. \quad (2)$$

The first-order coefficient g_m is given by

$$g_m = I_Q/V_T, \quad (3)$$

where g_m is the small-signal transconductance of the transistor. Consequently, gain can be varied by changing I_Q.

The higher order terms in (2) are due to the exponential nonlinearity and give rise to signal distortion. Common measures of this distortion are intermodulation (IM) and harmonic distortion (HD). IM distortion is defined for two-sinusoidal input signals at frequencies ω_1 and ω_2 applied to a circuit. Second-order intermodulation (IM_2) is defined as the ratio of the amplitude of the output distortion product at frequency $\omega_1 \pm \omega_2$ to the fundamental signal. This arises from the square term in the power series expansion. Third-order intermodulation (IM_3) is similarly defined for distortion at $2\omega_2 \pm \omega_1$ and $2\omega_1 \pm \omega_2$ and arises from the cubic term in the power series. At low frequencies, IM_2 and IM_3 are related to second-harmonic distortion (HD_2) and third-harmonic distortion (HD_3) by [4]

$$IM_2 = 2 \cdot HD_2 \quad (4)$$
$$IM_3 = 3 \cdot HD_3.$$

The specification of one form of distortion thus determines the other. Under low-distortion conditions the power series, given by (2), converges rapidly so that only the second- and third-order terms need to be considered. The distortion is then given by

$$IM_2 = \tfrac{1}{2}(v_{BEp}/V_T) \quad (5)$$

and

$$IM_3 = \tfrac{1}{8}(v_{BEp}/V_T)^2$$

in which v_{BEp} is the peak value of v_{BE}. As an example, 1-percent IM_3 is reached for a peak input voltage of 5.2 mV_{rms}.

Series base and emitter resistances in the transistor linearize the exponential $I_C - V_{BE}$ relationship and thus reduce the distortion [5], [6]. This corresponds however with a reduction in available gain variation. In the limit of very high series resistance, the transistor is current driven and gain variation is possible only as far as current gain β depends on collector current I_C. The distortion given by (5) is then negligible with respect to distortion caused by this I_C dependence of β.

The large values of even-order distortion in the single transistor can be balanced out by applying the input signal v_i between the bases of a matched pair [Fig. 1(a)]. Varying the common-emitter current I_E changes the gain. Signal output current i_{C2} is given by

$$i_{C2} = I_E\left(\frac{1}{1 + \exp(v_i/V_T)} - \frac{1}{2}\right) \quad (6)$$

if $\beta \gg 1$. Under low-distortion conditions (6) can be expanded in v_i and becomes

$$i_{C2} = -\tfrac{1}{4} g_m v_i [1 - \tfrac{1}{12}(v_i/V_T)^2 + \cdots] \quad (7)$$

in which $g_m = I_E/V_T$. The second-order term is zero. The third-order term yields

$$IM_3 = \tfrac{1}{16}(v_{ip}/V_T)^2 \quad (8)$$

in which v_{ip} is the peak value of v_i. In this case 1-percent IM_3 is reached at 7.4-mV_{rms} input signal that is $\sqrt{2}$ times higher than for a single transistor. In practice, small mismatches between transistors cause some second-order distortion. However, since collector currents i_{C1} and i_{C2} have an opposite phase, a differential output [Fig. 1(a)] can be taken that significantly improves the rejection of even-order distortion.

As for a single transistor, the presence of base and emitter resistance linearizes the transfer characteristic so that the available gain variation is exchanged for distortion reduction.

The base-driven pair is thus able to suppress even-order distortion but decreases the odd-order distortion by a factor of only 1.4 compared with a single transistor. A significant improvement in distortion performance can be obtained by applying the signal input as a current via the current source of the differential pair.

III. The Emitter-Driven Pair

Interchanging the positions of input signal and gain control in the pair of Fig. 1(a) results in the emitter-

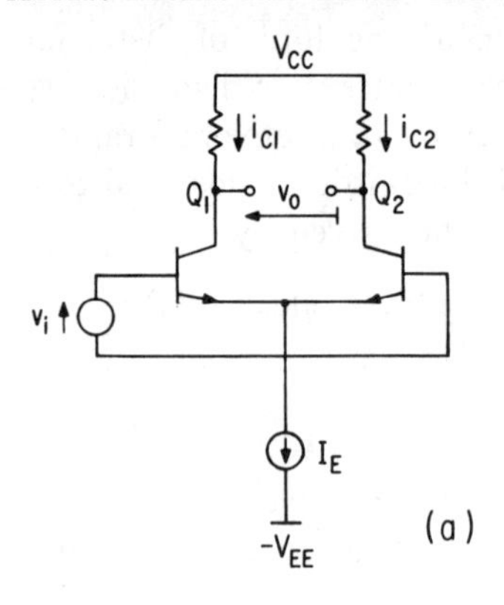

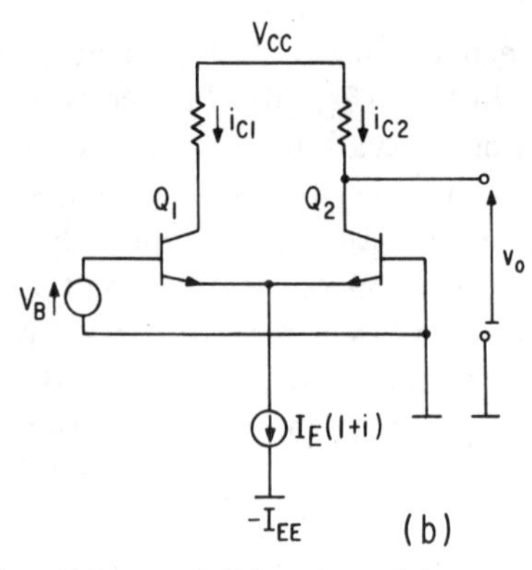

Fig. 1. (a) Base-driven and (b) emitter-driven variable-gain pair.

driven pair shown in Fig. 1(b). Input current $I_E(1+i)$, in which i represents the fractional signal level, feeds both transistors in parallel. DC voltage V_B determines the fraction of the input current that flows in the second transistor and develops output voltage v_0. At low frequencies the signal output current i_{C2} is readily found from (6) for $\beta \gg 1$ and no series emitter resistance and is given by

$$i_{C2} = I_E \frac{i}{1 + \exp(b)} \tag{9}$$

in which $b = V_B/V_T$. Since (9) is linear in i, no distortion occurs so that the emitter-driven pair is far superior to the base-driven pair. However the input signal must be available as a current. The required voltage-current conversion implies a tradeoff between circuit gain and distortion as will be illustrated in the circuit realizations of Section IV. Thus the distortion actually present in i_{C2} is never zero but can be made very low at the expense of circuit gain.

Whereas the presence of base and emitter resistance improves the distortion performance of the base-driven pair, it has a deteriorating effect for the emitter-driven pair. Inclusion of base resistances r_{B1} and r_{B2} in the transistors of Fig. 1(b) gives the circuit equations

$$I_{C1}\left(1 + \frac{1}{\beta_1}\right) + I_{C2}\left(1 + \frac{1}{\beta_2}\right) = I_E(1 + i)$$

$$\frac{I_{C1}}{I_{C2}} = \exp\left[b - \frac{1}{V_T}\left(\frac{r_{B1}I_{C1}}{\beta_1} - \frac{r_{B2}I_{C2}}{\beta_2}\right)\right] \tag{10}$$

in which I_{C1} and I_{C2} represent total collector currents. The second equation of (10) is nonlinear if the voltage drops across the base resistances are different. Elimination of I_{C1} in (10) gives an expression for I_{C2}. This nonlinear equation is solved point by point for a sinusoidal input and a Fourier analysis is taken of the output waveform. The IM_2 and IM_3 products are plotted versus relative signal attenuation in Fig. 2. Note that the second-order distortion is again present due to the unbalancing of the differential pair. Maximum current gain (0-dB attenuation) is achieved for high negative V_B. Almost all the current then flows in transistor Q_2 [Fig. 1(b)], which thus acts as a current-driven common-base stage. As a result there is no distortion (it is assumed throughout that β is approximately independent of current). This is also seen from (10). I_{C1} is then negligible in the first equation and thus I_{C2} is linear in i. When both transistors carry the same current the exponential in (10) equals unity. Voltage V_B is zero. Currents I_{C1} and I_{C2} are equal and again linear in i. There is thus no distortion at an attenuation of 6 dB. For high attenuation nearly all the current flows in transistor Q_1. Voltage V_B is both positive and large compared with V_T. Since current I_{C2} is small, elimination of I_{C1} in (10) gives

$$I_{C2} = \alpha_1 I_E(1 + i) \exp(-b + r_1) \exp(r_1 i) \tag{11}$$

in which

$$r_1 = \frac{r_{B1}I_E}{(1 + \beta_1)V_T} \quad \text{or} \quad \frac{r_{B1}}{r_{\pi 1}}$$

$$\alpha_1 = \frac{\beta_1}{1 + \beta_1}.$$

The ac voltage drop across the base resistance of the current-carrying transistor represented by $\exp(r_1 i)$ in (11), thus causes distortion in the attenuated output current. This distortion is found by expanding $\exp(r_1 i)$ in a power series in $(r_1 i)$ and is given by

$$IM_2 = r_1 i_p \frac{1 + r_1/2}{1 + r_1} \tag{12a}$$

$$IM_3 = \tfrac{3}{8}(r_1 i_p)^2 \frac{1 + r_1/3}{1 + r_1} \tag{12b}$$

in which i_p is the peak value of i. The values given by (12a) and (12b) constitute an upper limit for the distortion. They agree very well with the ones obtained by exact calculations (Fig. 2).

Experiments have been performed on a matched pair of transistors (CA 3018). Base resistance was determined by means of the circle diagram and phase cancellation methods [7]. A single transistor with large emitter resistance was used as a current source. The distortion associated with the current source was thus negligible compared with the distortion generated in the pair itself. HD and IM distortion have been measured at low frequencies. Their ratio was in very close agreement with the relations given in (4) provided the distortion was low. The experimental results are given in Fig. 2.

IV. Practical Realizations

The excellent properties of an emitter-driven pair are realized in two commonly known transistor quads: the

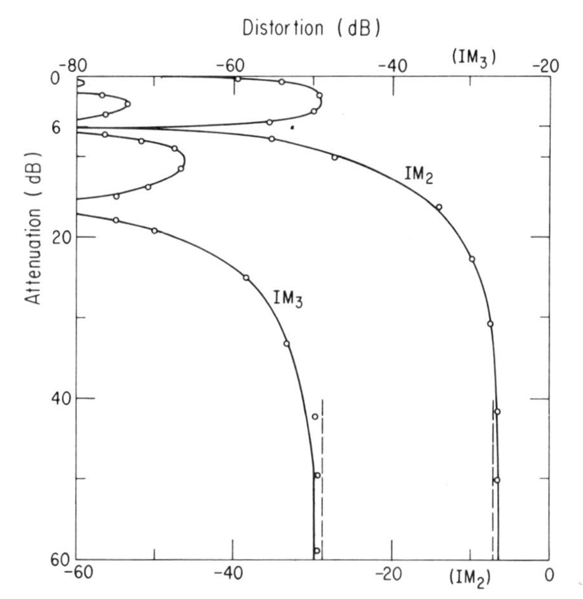

Fig. 2. Low-frequency IM_2 and IM_3 distortion for the emitter-driven pair for CA 3018 with $r_1 = 0.385$ and $i_p = 0.25$;—computed; --- approximated (12); ○ measured.

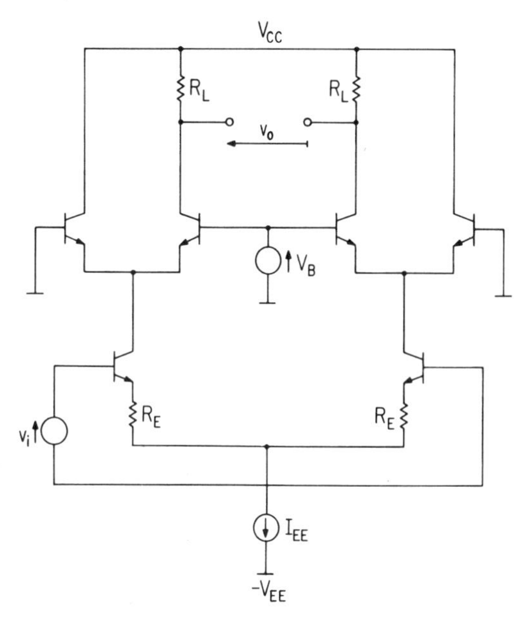

Fig. 3. Variable-gain quad based on signal summation (AGC amplifier).

AGC amplifier [1] represented in Fig. 3 and the multiplier [3] shown in Fig. 4. The use of a balanced quad arrangement results in cancellation of the second-order distortion of the emitter-driven differential pair discussed previously.

Both circuits employ a base-driven pair with emitter degeneration as a differential current source. The distortion of the input current with no feedback is thus given by (8). At low-frequencies feedback caused by emitter resistance R_E however reduces this to

$$IM_{3F} = \frac{IM_3}{1 + (R_E I_{EE}/2V_T)}, \tag{13}$$

if the signal output current has the same amplitude as without feedback [8]. Arbitrarily increasing R_E thus decreases the distortion but also the maximum current gain, which is given by

$$A_V = \frac{R_L}{R_E + (2V_T/I_{EE})}. \tag{14}$$

Consequently, a tradeoff has to be made in the choice of emitter resistance R_E.

The AGC quad shown in Fig. 3 is formed by balancing two emitter-driven pairs. Even-order distortion is thus absent for perfectly matched transistors. Odd-order components however add. The IM_3 distortion versus attenuation is thus as given in Fig. 2 and the maximum value is given again by (12b).

The AGC quad using the multiplier configuration is shown in Fig. 4. For zero input voltage V_B, all collector currents are equal and the signal output v_O is zero. For negative V_B the attenuated ac collector currents i_{C2} and i_{C3} become highly distorted but are negligible in magnitude with respect to the increasing ac collector currents i_{C1} and i_{C4}. Thus the amount of third-order distortion at low frequencies is similar to that in Fig. 2 for attenuation between 6 and 0 dB, but now expanded over the whole attenuation range of the multiplier. However, as

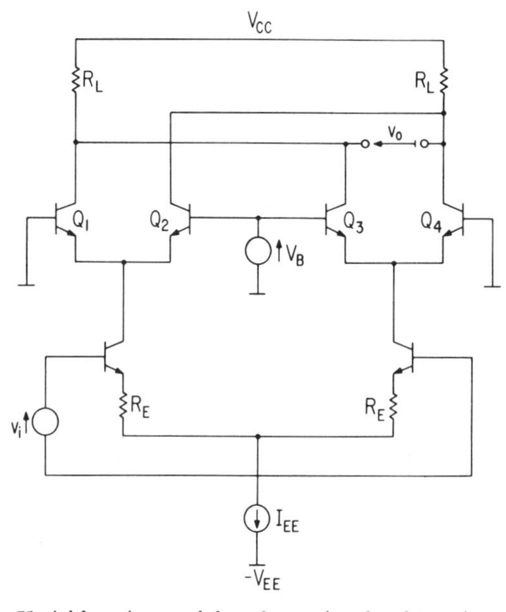

Fig. 4. Variable-gain quad based on signal subtraction (multiplier).

attenuation increases, the fractional third-order distortion approaches a constant value because the fundamental and the third-order component itself both decrease at the same rate. This is true only if the transistors are perfectly matched. The IM_3 distortion is computed from (10) and plotted versus attenuation in Fig. 5. The constant amount of distortion at high attenuation is obtained by expanding (10) for very small b and is found to be about 10 dB lower than the maximum distortion of the AGC amplifier (Fig. 2) given by (12b). Even-order distortion is canceled if perfect matching is achieved. The multiplier has thus actually a better distortion performance than the so-called AGC amplifier itself.

However, if the multiplier is not fully compensated [3] for mismatches, the third-order component cancels at a different value of V_B than does the fundamental. This results in an infinite value for IM_3 at an attenuation level

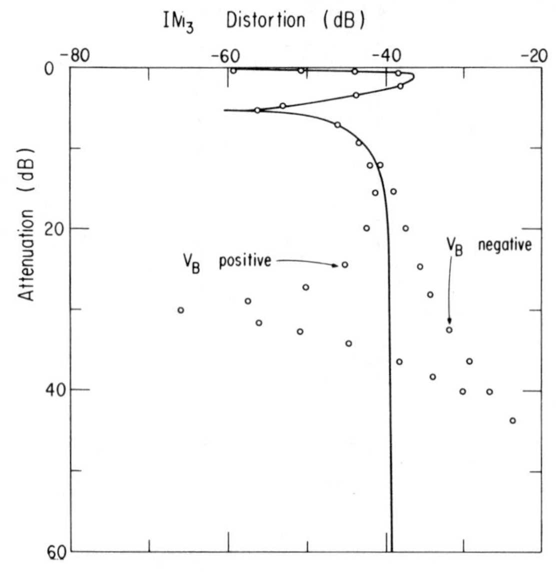

Fig. 5. Low-frequency IM_3 distortion for multiplier using CA 3045 with $r_1 = 0.35$ and $i_p = 0.5$; —— computed for matched transistors; ○ measured.

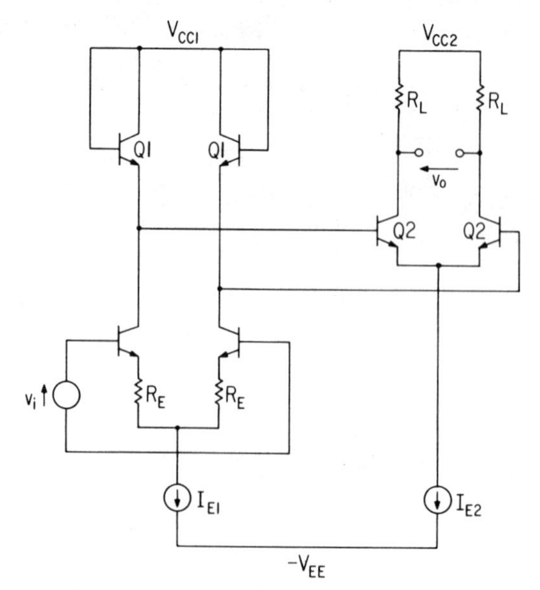

Fig. 6. Variable-gain quad based on predistortion (Gilbert's quad).

determined by the degree of mismatch. This is illustrated in Fig. 5 for a multiplier built with four-matched transistors (CA 3045). Base resistance is added to make distortion measurement easier. Mismatch is such that for positive V_B the third-order component is zero but not the fundamental. The waveform is thus distortionless at an attenuation of about 30 dB (Fig. 5). For negative V_B however, the fundamental cancels but not the third-order component. The fractional distortion thus increases with a slope of 20 dB/decade. It can thus be concluded that a multiplier has superior distortion performance compared with the AGC amplifier provided full four-potentiometer compensation [3] is applied.

V. Variable-Gain Quad Based on Predistortion

The gain variation in the AGC amplifier and the multiplier are based on unbalancing a differential pair. Even-order distortion is canceled by taking a differential output of two pairs. A different variable-gain quad is based on Gilbert's wide-band amplifier technique [2]. In this case, a single-ended output ideally does not contain second-order distortion. The quad is shown in Fig. 6.

The input signal is converted into a current and predistorted by the transistors $Q1$ in order to cancel the distortion generated in the base-driven pair of transistors $Q2$. The same predistortion concept is applied in high-performance multipliers [3].

The gain is controlled by the ratio of pair currents I_{E1} and I_{E2}. However, distortion is absent only if [2]

$$r_1 - r_2 = 0 \tag{15}$$

in which r_1 and r_2 are defined as in (11) for the first and second pair, respectively. Condition (15) is fulfilled for only one gain value, and distortion occurs for lower and higher gains. At low frequencies this distortion is found from the nonlinear equation describing the quad [3] using the methods described in Section III, and is plotted versus attenuation in Fig. 7. Using a power series ex-

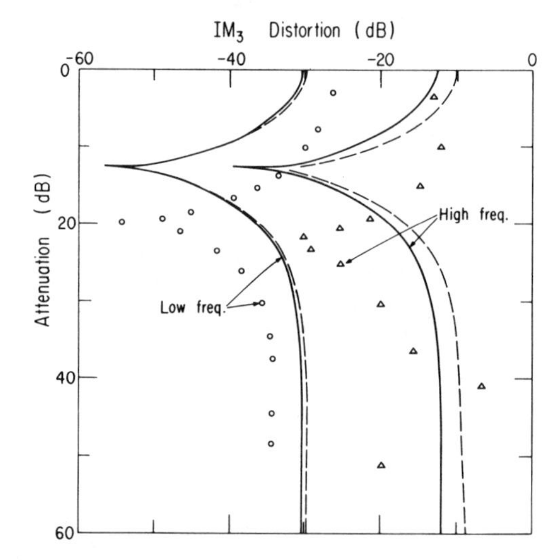

Fig. 7. IM_3 distortion versus attenuation for Gilbert's quad (Fig. 6) using CA 3045; $r_{1,2\,\max} = 0.35$ and $i_p = 0.5$; —— computed; --- approximated at low (16) and high (25) frequencies; ○ measured at low frequencies; Δ measured at high frequencies for $B_{1,2\,\max} = 4$; $B = 4$ corresponds to a frequency of 48 MHz.

pansion, the third-order distortion can be approximated by

$$IM_3 = \tfrac{3}{8}(r_1 - r_2)i_p^{\,2} \tag{16}$$

in which i_p is the fractional current swing in transistors Q_2. Ideally, no second-order distortion is present, even from a single-ended output. Residual second-order distortion, due to device imbalances can be further reduced by taking a differential output (Fig. 6). The results given by (16) and the measured data are plotted also in Fig. 7. The null in distortion is measured at a lower gain than predicted. This is caused by some residual distortion present in the current source that adds on one side of the null and subtracts on the other side [4]. For a given operating current, the maximum amount of distortion given by (16) is larger than the maximum distortion in the AGC amplifier, which is given in (12b). The distortion null however, is less sharp and thus Gilbert's variable-gain quad is quite attractive for a limited attenuation range around the null in distortion.

VI. High-Frequency Analysis

The previous analysis was restricted to low frequencies where charge storage is unimportant. Consider now the situation at high frequencies. The distortion in the emitter-driven pair [Fig. 1(b)] is analyzed first. Then a comparison is made of the distortion performance of the AGC amplifier (Fig. 3), the multiplier (Fig. 4), and Gilbert's quad (Fig. 6).

For the emitter-driven pair, initially, assume that the device f_T is constant (i.e., I_C is not low) and that the frequencies are high enough that recombination current is negligible. Then

$$I_B = \tau(dI_C/dt) \tag{17a}$$

$$I_C = I_{CS} \exp(V_{BE}/V_T), \tag{17b}$$

where I_B, I_C, and V_{BE} are total values for base current, collector current, and base–emitter voltage, respectively; τ is the base transit time. The circuit equations are then given by

$$\left(1 + \tau_1 \cdot \frac{d}{dt}\right) I_{C1} + \left(1 + \tau_2 \cdot \frac{d}{dt}\right) I_{C2} = I_E(1 + i) \tag{18a}$$

$$\frac{I_{C1}}{I_{C2}} = \exp\left[b - \frac{1}{V_T}\left(r_{B1} \cdot \tau_1 \frac{dI_{C1}}{dt} - r_{B2} \cdot \tau_2 \cdot \frac{dI_{C2}}{dt}\right)\right]. \tag{18b}$$

Equation (18a) shows that both collector currents roll off at their common base cutoff frequency, which is about

$$f_T = 1/2\pi\tau. \tag{19}$$

Equation (18b) describes the nonlinearity. As for low frequencies, this is caused by the difference in voltage drop across the base resistances. These voltage drops are now time differentials so that the distortion is frequency dependent. Eliminating I_{C1} in (18a) and (18b) gives a nonlinear differential equation of the first order. This equation has been solved directly by computer aids and a Fourier analysis has been taken of the output waveform. The amount of distortion at one specific frequency is plotted in Fig. 8. Frequency is normalized as given by

$$B = \omega\tau_1 \frac{r_{B1}I_E}{V_T} \quad \text{or} \quad r_1 \frac{f}{f_\beta} \quad \text{with} \quad f_\beta = \frac{f_T}{\beta}. \tag{20}$$

As for low frequencies, distortion due to the presence of base resistance is zero at full gain and at half that gain. For high attenuation, distortion becomes maximum. This maximum value can be predicted by an analytical expression derived as follows.

In (18a) and (18b) I_{C2} is neglected with respect to I_{C1}. The ac part of I_{C1} is then found from (18a) and given in the frequency domain by

$$i_{C1} = I_E \cdot \frac{i}{1 + j\gamma B} \tag{21}$$

in which $\gamma = V_T/r_{B1}I_E$ and thus $\gamma B = \omega\tau_1$ or f/f_T. The ac part of collector current i_{C2} can then be represented by a Volterra series of the form [9]

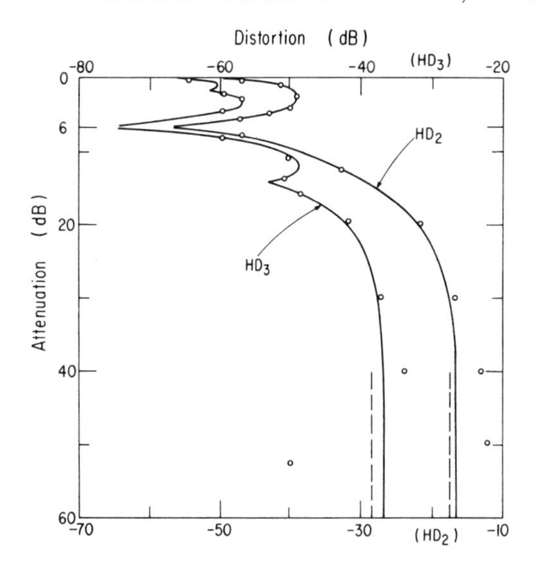

Fig. 8. High-frequency HD_2 and HD_3 distortion for the emitter-driven pair using CA 3018; $B = 4$ (60 MHz) and $i_p = 0.25$; —— computed; --- approximated (24; ○ measured.

$$i_{C2} = [I_E \exp(-b)][H_1(B_1) \circ i + H_2(B_1, B_2) \circ i^2 + H_3(B_1, B_2B_3) \circ i^3 + \cdots] \tag{22}$$

in which H_i $(B_1, \cdots B_i)$ are the ith order Volterra kernels operating on input signal i. They are found by substituting (21) and (22) in (18b) and they are given by

$$H_1(B_1) = \frac{1 + jB_1}{1 + j\gamma B_1}$$

$$H_2(B_1, B_2) = \frac{jB_1\left(1 + \frac{jB_2}{2}\right)}{(1 + j\gamma B_1)(1 + j\gamma B_2)}$$

$$H_3(B_1, B_2, B_3) = \frac{jB_1 jB_2\left(1 + \frac{jB_3}{3}\right)}{2(1 + j\gamma B_1)(1 + j\gamma B_2)(1 + j\gamma B_3)}. \tag{23}$$

The HD components [which at high frequencies are not necessarily related to the IM products by a constant ratio as given by (4)] are derived from (23) and given by

$$HD_2 = \frac{i_p}{2} \cdot \frac{B}{(1 + \gamma^2 B^2)^{1/2}} \left(\frac{1 + (B/2)^2}{1 + B^2}\right)^{1/2} \tag{24a}$$

$$HD_3 = \frac{i_p^2}{8} \cdot \frac{B^2}{1 + \gamma^2 B^2} \left(\frac{1 + (B/3)^2}{1 + B^2}\right)^{1/2}. \tag{24b}$$

HD as predicted by the previous equation is plotted versus frequency in Fig. 9 and as asymptotes in Fig. 8. Experimental results are also shown in Figs. 8 and 9. Agreement is satisfactory. However, at high-attenuation levels, deviations occur as can be seen in Fig. 8. Also the maximum available attenuation is limited. Both effects are caused by direct signal feedthrough via the junction capacitances C_{jE2} and C_{jC2} of output transistor $Q2$ [Fig. 1(b)]. For high attenuation, emitter voltage v_i is determined entirely by the common-base input impedance of transistor $Q1$. This voltage is highly distorted because of the exponential $I_C - V_{BE}$ relationship. Also the magnitude of v_i increases with frequency due to the presence of the base resistance [7]. The output

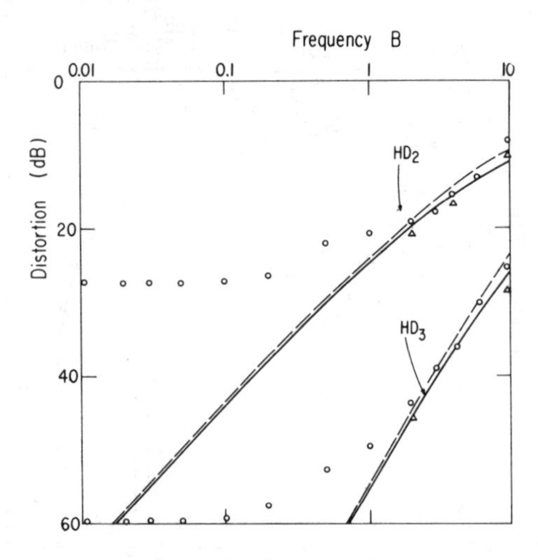

Fig. 9. Maximum HD_2 and HD_3 distortion versus frequency for the emitter-driven pair; $i_p = 0.25$ and $\gamma = 0.06$; —— computed; --- approximated (24); 0 measured.

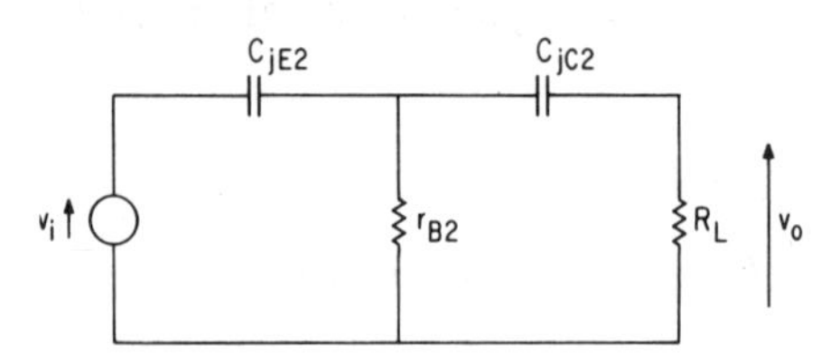

Fig. 10. Model of output transistor Q_2 with collector load R_L at high attenuation; v_i is the emitter voltage and v_0 the collector voltage.

transistor with collector load R_L is nearly cutoff and can thus be simulated by a second-order high-pass filter shown in Fig. 10. Output signal v_0 is thus even more distorted than v_i. Also its amplitude increases with frequency. Although other parasitic capacitances such as collector substrate capacitance are present the model of Fig. 10 has yielded values of maximum available attenuation that are in good agreement with experimental data.

As for low frequencies the HD_3 curve shown in Fig. 8 for the emitter-driven pair applies directly for the AGC amplifier (Fig. 4). The HD_2 is reduced by taking a differential output and has usually become smaller than HD_3. In the multiplier (Fig. 4) full compensation at high frequencies is not feasible and the attenuation range for a given distortion level is quite limited. The multiplier is thus not very attractive as an AGC circuit at high frequencies.

The distortion at high frequencies in Gilbert's variable-gain quad is obtained by means of a Volterra series expansion using the model described by (17). The distortion is given approximately by

$$HD_3 = (i_p^2/8)(B_1 - B_2) \tag{25}$$

in which B_1 and B_2 are defined as in (20) for transistors Q_1 and Q_2, respectively. Computed, predicted, and measured results are represented in Fig. 7. Agreement is satisfactory.

The maximum distortion at high frequencies in Gilbert's variable-gain quad increases only linearly with frequency whereas in the AGC amplifier (Fig. 4) the maximum distortion, given by (24b) increases twice as fast. For high frequencies ($B_{1,2\,\max} > 1$) Gilbert's quad has a better distortion performance than the AGC amplifier. However, direct feedthrough in Gilbert's quad is worse than in the AGC amplifier. The signal voltage developed at the emitters of transistors Q_1 (Fig. 6) is applied directly to the junction capacitances C_{jC2} of transistors Q_2.

VII. The Influence of Nonideal Current Sources

A current source is nonideal when it is shunted by a finite resistance or by capacitance.

Presence of shunt resistance R_S at the common-emitter points of the AGC circuits in Figs. 3 and 4 or at the emitters of transistors Q_1 in Gilbert's quad in Fig. 6 introduces distortion that subtracts from the distortion obtained for high attenuation. At low frequencies and for high attenuation it is found by power series expansion and given by

$$IM_3 = \frac{1}{3}\cdot\frac{2V_T}{R_S I_{EE}}\cdot i_p^2 \tag{26}$$

(for Gilbert's quad take $I_{EE} = 2\,I_{E1}$). This amount of distortion is usually small enough to be neglected unless the current sources are replaced by resistors. At high frequencies, constant capacitances at the emitters of the quad, such as collector-substrate capacitance C_{CS} and emitter–base junction capacitance C_{jE}, become important. They can be modeled by assuming a current dependent base-transit time τ for the quad transistors. Expression (17a) becomes then

$$I_B = \left(\tau + \frac{C_C}{I_C}\cdot\frac{dI_C}{dt}\right). \tag{27}$$

The ac component of base current i_B can be expanded in i_C, the ac part of I_C, as given by

$$i_B = \tau_0\cdot\frac{di_C}{dt} + \frac{\tau_1}{2}\cdot\frac{di_C^2}{dt} + \frac{\tau_2}{3}\cdot\frac{di_C^3}{dt} + \cdots \tag{28}$$

with

$$\tau_0 = \tau + \frac{2C_C\cdot V_T}{I_{EE}}$$

$$\tau_1 = -\frac{4C_C\cdot V_T}{I_{EE}^2}$$

$$\tau_2 = \frac{8C_C\cdot V_T^2}{I_{EE}^3}$$

and $C_C = 2C_{jE} + C_{CS}$ for one pair. In a first-order analysis C_{jE} is voltage independent and thus the same for both transistors of the pair under any attenuation level. Base resistance is neglected also. The distortion is then found in the collector current of one emitter-driven transistor. Using a Volterra series approximation the distortion is given by

$$HD_3 = \frac{i_p^2}{4}\cdot\frac{V_T C_C \omega}{2I_{EE}} \tag{29}$$

for small C_C and frequencies below $f_T/3$. As for low frequencies this distortion subtracts from the distortion level obtained for high attenuation. Omission of (26) and (29) leads thus to worst case analysis.

VIII. Conclusion

Distortion in wide-band transistor variable-gain amplifiers has been analyzed at both low and high frequencies. At high-attenuation levels all circuits considered here approach constant distortion values that are given by simple analytic expressions. Experimental results show good agreement with these calculations.

Three high-performance variable-gain amplifier configurations are identified and compared. The AGC amplifier covers the widest attenuation range at both low and high frequencies. However, at high attenuation its distortion is relatively high. A multiplier used as a variable-gain amplifier exhibits less distortion but needs full compensation for offset in order to obtain the same attenuation range as in an AGC amplifier. This is not feasible however at high frequencies. Gilbert's quad gives higher distortion than both the previous circuits at both high- and low-attenuation levels. At high frequencies the distortion is lower than in both other amplifiers but the attenuation level is severely limited by feedthrough.

References

[1] W. R. Davis and J. E. Solomon, "A high-performance monolithic IF amplifier incorporating electronic gain control," *IEEE J. Solid-State Circuits,* vol. SC-3, pp. 408–416, Dec. 1968.

[2] B. Gilbert, "A new wide-band amplifier technique," *IEEE J. Solid-State Circuits,* vol. SC-3, pp. 353–365, Dec. 1968.

[3] ——, "A precise four-quadrant multiplier with subnanosecond response," *IEEE J. Solid-State Circuits,* vol. SC-3, pp. 365–373, Dec. 1968.

[4] K. A. Simons, "The decibel relationships between amplifier distortion products," *Proc. IEEE,* vol. 58, pp. 1071–1086, July 1970.

[5] H. K. V. Lotsch, "Theory of nonlinear distortion produced in a semiconductor diode," *IEEE Trans. Electron Devices,* vol. ED-15, pp. 294–307, May 1968.

[6] L. C. Thomas, "Eliminating broadband distortion in transistor amplifiers," *Bell Syst. Tech. J.,* vol. 47, pp. 315–342, Mar. 1968.

[7] W. M. C. Sansen and R. G. Meyer, "Characterization and measurement of the base and emitter resistances of bipolar transistors," *IEEE J. Solid-State Circuits,* vol. SC-7, pp. 492–498, Dec. 1972.

[8] A. T. Starr, *Radio and Radar Technique.* London: Pitman, 1953, p. 416.

[9] S. Narayanan, "Transistor distortion analysis using Volterra series representation," *Bell. Syst. Tech. J.,* vol. 46, pp. 991–1024, May/June 1967.

A Low-Distortion Monolithic Wide-Band Amplifier

KAM H. CHAN AND ROBERT G. MEYER, SENIOR MEMBER, IEEE

***Abstract*—A wide-band monolithic amplifier is described which realizes 20-dB gain and 250-MHz bandwidth using an 800-MHz integrated-circuit process. At 0-dBm signal levels, second- and third-order intermodulation distortion levels are below -55 and -42 dB, respectively, across the band. Terminal impedances are matched to 75 Ω with VSWR better than 1.7. Both circuit and process design are described as well as computer optimization of circuit performance.**

I. Introduction

WIDE-BAND low-distortion amplifiers are widely used in communication systems and instrumentation applications. In communication systems, for example, amplifiers are often required with bandwidths of several hundred megahertz and very low distortion levels. At present, amplifiers of this type are generally realized as thin-film hybrids [1].

In this paper, a wide-band monolithic amplifier with characteristics comparable to many hybrid units, but with much lower manufacturing cost, is described. The circuit requirements are first described, followed by a discussion of the circuit configuration. Next the monolithic process is described and finally the measured circuit data are presented.

II. Amplifier Requirements

The choice of amplifier specifications involves compromises, and this is reflected in the amplifier requirements listed below. A gain of 20 dB was chosen as a compromise value that could be realized in a single package. It is high enough to overcome the noise of the following stages and also high enough that the input signal requirements to the amplifier are moderate. A bandwidth greater than 200 MHz was specified, and this enables the amplifier to be used in many common applications.

Most applications of these amplifiers are in systems designed around matched impedances of 50 or 75 Ω. The amplifier was designed for input and output impedance matches to 75 Ω, although 50 Ω could equally well be used.

Three important specifications that are related are amplifier distortion, noise figure, and power requirements. Since many applications of these amplifiers require very low distortion levels, a design aim was to achieve second- and third-order intermodulation distortion below -50 dB at signal levels of 0 dBm across the frequency band [2]. As will be shown later, low distortion generally requires use of high bias currents in the active devices with consequent high-power requirements. A limit of 2 W was set for power dissipation in this design, although later versions of the circuit could use higher values, and achieve even lower distortion values.

Manuscript received May 20, 1977. This work was supported by the U.S. Army Research Office under Grant DAHCO4-74-G-151.

The authors are with the Department of Electrical Engineering and Computer Sciences, University of California, Berkeley, CA 94720.

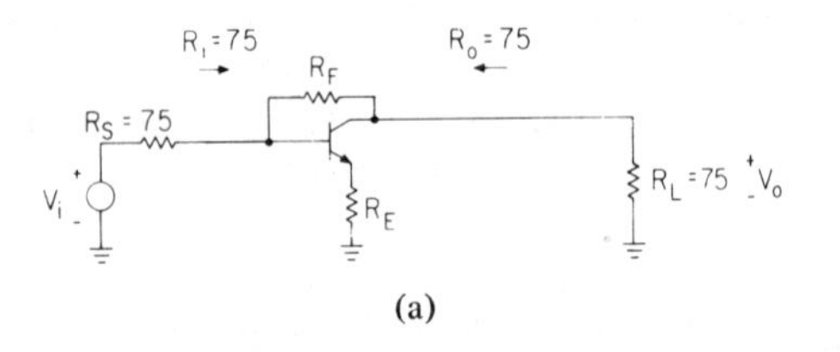

(a)

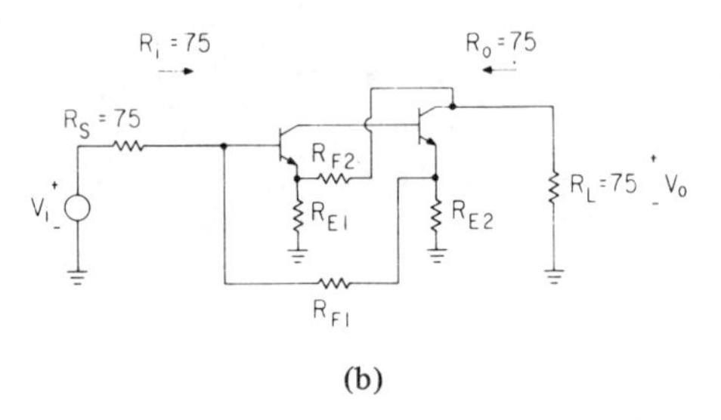

(b)

Fig. 1. Shunt and series feedback circuits. (a) Single stage. (b) Two stage.

The noise performance of the amplifier is important for several reasons. First, the amplifier itself can be used as a preamplifier in many applications if its noise performance is adequate. Second, if very low noise is required, a low-noise preamplifier can be added and the noise of the main amplifier should then be low enough to make a negligible contribution. The compromise here is that the high bias currents needed for low distortion generally lead to a degradation in circuit-noise performance. A specification of 9 dB was set for the circuit-noise figure.

Finally, in order to reduce the cost of fabrication, the circuit was designed for realization using fairly standard bipolar monolithic processing. The process, to be described later, incorporates shallow diffusion but uses standard diffused isolation techniques and regular single-level aluminum metallization.

III. Circuit Configuration

The realization of the amplifier specifications previously described requires careful choice of the amplifier-circuit configuration. The simultaneous requirements of wide bandwidth, matched terminal impedances, low noise, and low distortion are difficult to achieve, particularly when realized in a monolithic integrated circuit (IC). One possible configuration is shown in Fig. 1(a). This incorporates simultaneous shunt and series feedback via R_F and R_E [1], [3]. It can be shown that the terminal impedances of this circuit are resistive and broadband and can be designed for 75-Ω operation by a suitable choice of R_F and R_E. The gain characteristic is broad-band,

Reprinted from *IEEE J. Solid-State Circuits*, vol. SC-12, pp. 685-690, Dec. 1977.

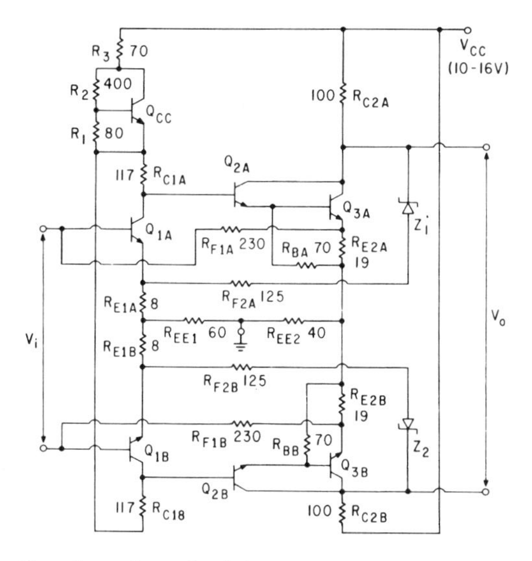

Fig. 2. Complete schematic of the wide-band monolithic amplifier.

but computer simulation showed that the gain and bandwidth requirements for this design could not be met in a single-stage circuit of this type using monolithic technology.

The advantages of the circuit of Fig. 1(a) can be retained while the gain-bandwidth product is increased if a two-stage version of the circuit [1] is used. This is shown schematically in Fig. 1(b), where simultaneous shunt and series feedback is applied over two stages to realize resistive terminal impedances of 75 Ω, together with a much larger gain-bandwidth product. Since feedback is applied over two stages, substantial feedback-loop gain is realized for the minimization of distortion.

In the final circuit a balanced version of this configuration was used, and input and output signals were taken differentially. This eases the biasing problem in the IC and also gives a large amount (20-30 dB) of second-order distortion cancellation.

The complete circuit of the monolithic amplifier is shown in Fig. 2. The input and output signals appear differentially at V_i and V_o, respectively. The differential terminal impedances are 75 Ω and the terminal impedances of the half-circuits are 37.5 Ω. The component values in the circuit were optimized by computer to yield 20-dB gain with maximum bandwidth and adequate input and output impedance matches. This is achieved by minimizing a figure of merit M where

$$M = \sum_{i=1}^{n} \{A_i[G(\omega_i) - G_0]^2 + B_i[\rho_1(\omega_i) - 1]^2 + B_i[\rho_2(\omega_i) - 1]^2\}. \qquad (1)$$

Coefficients A_i and B_i are weighting functions at different frequency points, G is the circuit gain, ρ_1 and ρ_2 are input and output voltage standing-wave ratios (VSWR), and ω_i is the frequency.

The circuit is biased from a single supply V_{CC} which can have values from 10 to 16 V. Resistors R_{F1} and R_{E2} provide shunt-series feedback and R_{F1} also provides dc bias to the input base. Resistors R_{E1} and R_{F2} provide series-shunt feedback. A Darlington output connection is used to improve the circuit distortion and high-frequency gain. Resistors R_{EE} are bias resistors common to both half-circuits. Transistor Q_{CC} is connected as a V_{BE}-multiplier and sets the bias voltage at the collectors of Q_{1A} and Q_{1B}. Other important components in the bias circuit are Zener diodes Z_1 and Z_2 which have a 5.5-V dc voltage drop and allow direct connection of R_{F2} to the output.

Although the Zener diodes greatly facilitate biasing in the amplifier, they must be carefully designed to minimize their distortion and noise contributions. The noise is minimized by designing the devices to operate at high-current densities, as described in the next section. Distortion is minimized by operating the Zeners well beyond the knee of the Zener characteristic. In addition, series resistance in the Zeners must be minimized since this causes errors in the values of R_{F2A} and R_{F2B}. In the final circuit, the Zeners made no measureable contribution to the distortion performance and added only about 1 dB to the circuit-noise figure.

The distortion performance of the circuit of Fig. 2 was also simulated by computer using the program SPICE [4], and the bias currents in the various stages were set largely on the basis of computer prediction of best distortion performance. At low frequencies, the distortion is due almost entirely to nonlinearity in Q_{3A} and Q_{3B}. At the upper end of the frequency range, the output devices are still the most important, but distortion due to the input stage is significant because of the decrease in gain of the output stage.

IV. Monolithic Process and Device Structure

The distortion specification of the amplifier is one of the most important characteristics and is heavily dependent on the characteristics of the monolithic transistors used. In addition, realization of the gain and bandwidth requirements of the amplifier requires transistors with f_T in the region of 1 GHz.

The design of low-distortion transistors requires optimization of device geometry and epitaxial (epi) layer characteristics. The effects of device geometry can be seen by an approximate analysis of high-frequency distortion in the transistor. It can be shown [5] that third-order intermodulation distortion in a bipolar transistor at high frequencies is given approximately by

$$\mathrm{IM}_3 = \frac{\dfrac{\hat{V}_0^2}{4}\dfrac{C_{jEQ}V_T}{R_L^2 I_Q^3} + C_{jCQ}R_L\dfrac{1}{n}\left(\dfrac{1}{n}+1\right)\dfrac{1}{V_Q^2}}{1 + \dfrac{C_{jEQ}V_T}{I_Q} + C_{jCQ}R_L} \qquad (2)$$

where

$$C_{jC} = \frac{K}{(\phi_B + V_{CB})^{1/n}} \qquad (3)$$

and

$$V_T = \frac{kT}{q}. \qquad (4)$$

In these equations I_Q is the collector current, V_Q the collector-

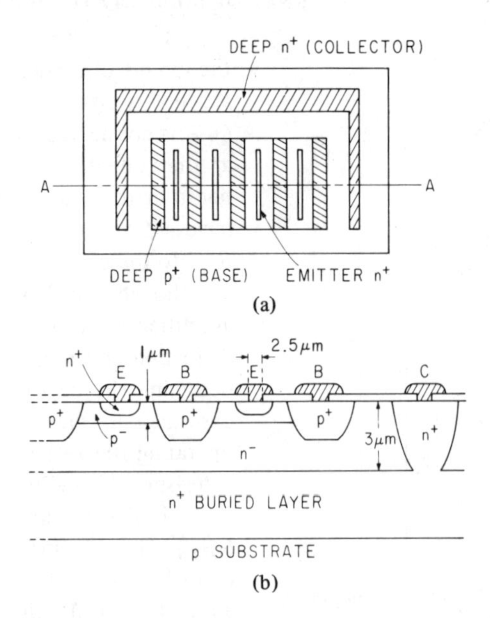

Fig. 3. Monolithic transistor structure. (a) Plan. (b) Section AA.

base voltage, R_L the load resistance, $\hat{V}_0$ the peak output voltage per signal, C_{jEQ} the emitter depletion capacitance, and C_{jCQ} the collector depletion capacitance. If two sinusoidal signals of frequencies ω_1 and ω_2 are applied to the transistor to produce equal output voltages V_0, then IM_3 is the ratio of the distortion component at $(2\omega_2 - \omega_1)$ to the fundamental signal at ω_1 or ω_2. Second-order intermodulation distortion IM_2 is the ratio of the distortion component at $(\omega_1 \pm \omega_2)$ to the fundamental signal.

Equation (2) shows that IM_3 increases as I_Q is reduced because of the presence of C_{jEQ} in the transistor. The presence of C_{jEQ} also causes the device f_T to fall at low I_Q. In order to minimize this source of distortion, the device must be designed for minimum C_{jEQ} and operation at moderately high I_Q is essential. This also minimizes second-order intermodulation.

An additional source of distortion not considered in (2) is high-current f_T falloff due to the Kirk effect [5]. This can be minimized by the choice of epi layer characteristics. Because of difficulty in controlling out-diffusion of the buried layer, no attempt was made to operate with the epi region depleted, and epi thickness and doping of 3 μm and 1 Ω · cm were used to minimize the Kirk effect.

The monolithic device structure used is shown in Fig. 3. An interdigitated base and emitter structure is employed to maximize the emitter periphery which is the active region of the emitter at high-current levels. In order to minimize the total emitter area (and thus C_{jEQ}), the minimum possible emitter contact opening is used. In this process, an emitter contact opening of 2.5 μm was used with an emitter-wash technique. Developments in lithography allowing submicron dimensions should allow significant further improvements in distortion in devices of this type.

A cross section of the device is shown in Fig. 3(b). A deep n^+-diffusion is used to contact the buried layer and reduce the collector series resistance. A deep p^+-diffusion is also used to reduce the base resistance and improve the high-frequency gain and noise figure. The base diffusion is about 1-μm deep and the active basewidth about 0.4 μm. The process uses eight masks and standard aluminum metallization.

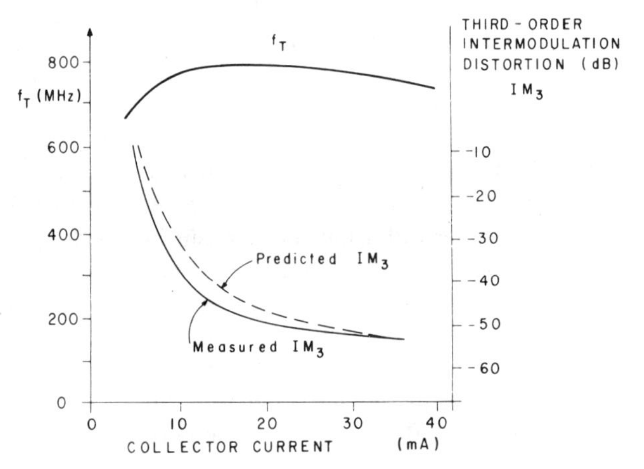

Fig. 4. Transistor distortion and f_T versus collector current.

In Fig. 4 measured curves of f_T and IM_3 (at 200 MHz) versus collector current are shown for a typical device. Distortion predicted for the device by SPICE is also shown and is close to the measured value. The behavior predicted earlier is observed in that distortion increases rapidly as I_Q is reduced and f_T becomes dependent on I_Q. Thus typical bias currents in the output devices of the circuit are in excess of 20 mA where the device f_T is approximately constant and equal to 800 MHz.

The amplifier circuit of Fig. 2 utilizes Zener diodes Z_1 and Z_2 as part of the bias circuit. In order to reduce the noise contribution of the Zener diodes, the devices are designed for small geometries [6] and high-current density. An additional requirement in this application is that Z_1 and Z_2 should be reasonably well matched and this precludes the use of minimum-area devices. The Zener is fabricated using the n^+- and

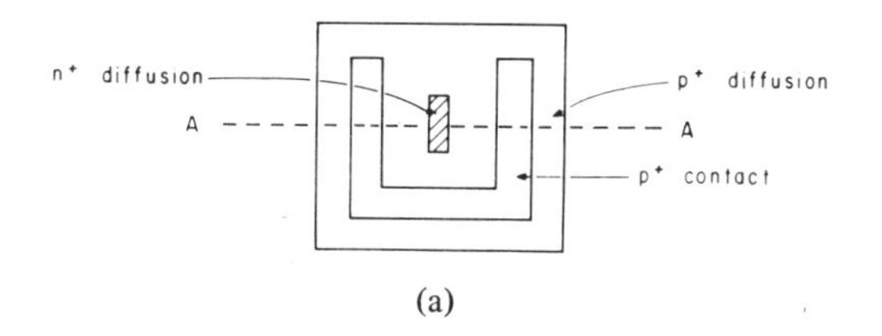

(a)

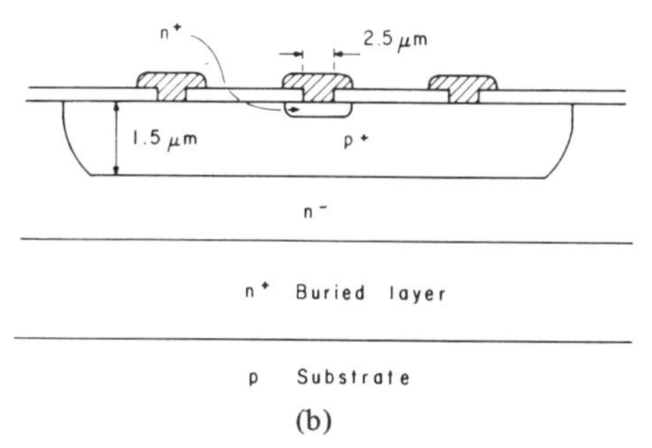

(b)

Fig. 5. Low-noise Zener-diode structure. (a) Plan. (b) Section *AA*.

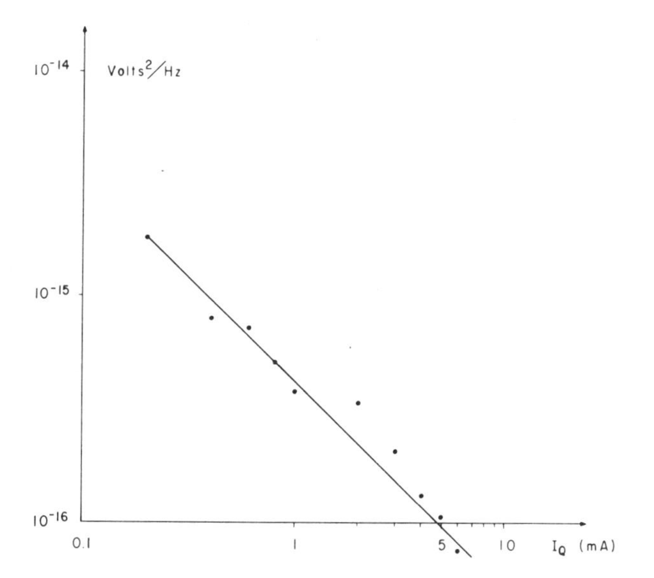

Fig. 6. Measured noise-voltage spectral density versus bias current for the Zener diodes.

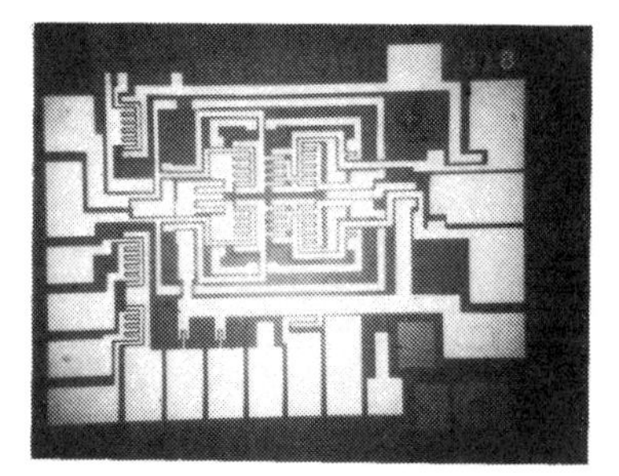
Fig. 7. Die photograph.

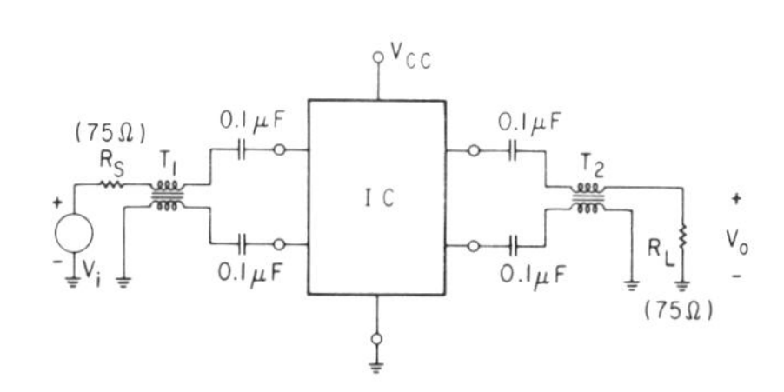

Fig. 8. Typical external connections to the IC.

p^+-diffusions, and the device structure is shown in Fig. 5. The breakdown voltage is 5.5 V and the mismatch between Z_1 and Z_2 is typically 50 mV. The measured noise-voltage density is shown in Fig. 6. This is essentially flat over the amplifier bandwidth, and is inversely proportional to operating current. At 5-mA bias current, the noise density is 10^{-16} V^2/Hz.

V. Circuit Performance

A die photograph of the final circuit is shown in Fig. 7. The output devices can be seen on the right, and the layout is symmetrical along the centerline to minimize thermal imbalance. This is important for the nulling of second-order distortion in the amplifier. The die size is 40 × 60 mil.

Typical external connections to the amplifier are shown in Fig. 8. Wide-band balun transformers T_1 and T_2 maintain balanced operation of the circuit and allow significant second-order intermodulation distortion cancellation. Amplifiers can be cascaded with negligible interaction, and if this is done an interstage transformer is not required. The IC has six pins and operates from a single power supply. Distortion performance can be traded against power dissipation by varying the supply voltage.

The measured gain of the amplifier is compared with computed values in Fig. 9, and both show a value of about 20 dB with a bandwidth of 250 MHz. Measurements at 100 and

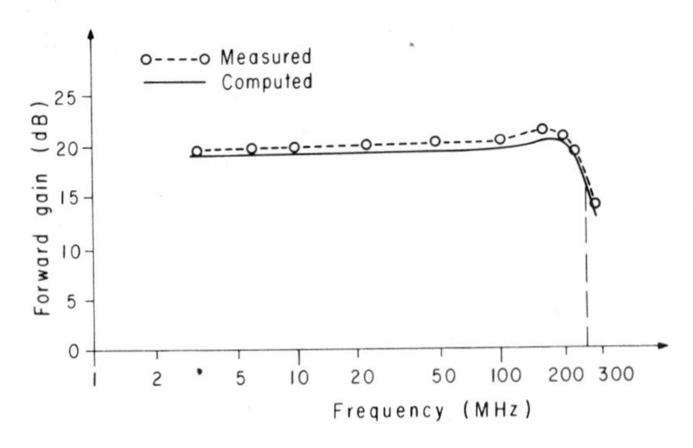

Fig. 9. Amplifier gain versus frequency.

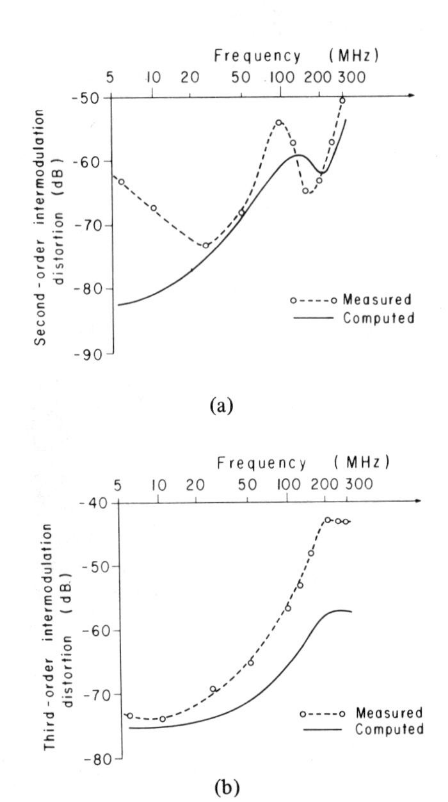

Fig. 10. Computed distortion in the monolithic circuit compared with measured values. (a) Second-order intermodulation. (b) Third-order intermodulation.

200 MHz showed gain sensitivity to a temperature of 0.004 and 0.02 dB/°C, respectively.

Fig. 10 shows measured circuit distortion compared with computed values at signal levels of 0 dBm. In Fig. 10(a) measured and computed values of IM_2 show good agreement and IM_2 is below -55 dB across the amplifier bandwidth. The distortion becomes worse as frequency increases because the feedback-loop gain of the amplifier diminishes. Below 20 MHz the measured IM_2 increases because the coupling transformers become unbalanced and the second-order cancellation degrades. This can be improved by using larger cores on the transformers.

Measured and computed values of IM_3 in Fig. 10(b) are fairly close except at high frequencies. The measured values are below -42 dB across the band and below -55 dB below 100 MHz. The high-frequency discrepancy could be due to inaccuracies in modeling capacitive parasites in the IC.

The noise figure of the amplifier has a worst case value of 9 dB at 250 MHz. Reverse transmission in the amplifier is -37 dB and measured input and output VSWR was less than 1.7 in a 75-Ω system. Total power dissipation was 1.65 W from a 14-V supply.

VI. Further Developments

The results just described were obtained using an 800-MHz IC process and a circuit optimized around those process parameters. However, computer simulation showed the circuit to be close to optimum for a range of f_T values. This was confirmed by fabricating circuits with the same mask set but using a more advanced IC process. This process uses shallower diffusions to achieve $f_T = 1.7$ GHz and gave amplifiers with gain characteristics similar to those of Fig. 9 but with 370-MHz bandwidth. Measured values of IM_2 and IM_3 were below -60 and -44 dB, respectively, below 300 MHz. For frequencies below 200 MHz all distortion is below -53 dB.

VII. Conclusions

A wide-band monolithic amplifier has been described. Computer optimization has yielded 20-dB gain and 250-MHz bandwidth using an 800-MHz integrated-circuit process. The process incorporates eight masks and also includes low-noise Zener diodes for bias purposes. Appropriate choice of device geometry and operating conditions, circuit topology, and epitaxial characteristics has yielded low distortion across the amplifier bandwidth. The circuit performance improves when more advanced IC processes are used for fabrication.

References

[1] R. G. Meyer, R. Eschenbach, and R. Chin, "A wideband ultra-linear amplifier from 3 to 300 MHz," *IEEE J. Solid-State Circuits*, vol. SC-9, pp. 167–175, Aug. 1974.

[2] K. A. Simons, "The decibel relationship between amplifier distortion products," *Proc. IEEE*, vol. 58, pp. 1071–1086, July 1970.

[3] J. B. Couglin, R. J. Gelsing, P. J. Jochems, and H. J. M. van der Laak, "A monolithic silicon wideband amplifier from DC to 1 GHz," *IEEE J. Solid-State Circuits*, vol. SC-8, pp. 414–419, Dec. 1973.

[4] S. H. Chisholm and L. W. Nagel, "Efficient computer simulation of distortion in electronic circuits," *IEEE Trans. Circuit Theory*, vol. CT-20, pp. 742–745, Nov. 1973.

[5] H. E. Abraham and R. G. Meyer, "Transistor design for low distortion at high frequencies," *IEEE Trans. Electron Devices*, vol. ED-23, pp. 1290–1297, Dec. 1976.

[6] W. F. Davis, "A five-terminal ±15 V monolithic voltage regulator," *IEEE J. Solid-State Circuits*, vol. SC-6, pp. 366–376, Dec. 1971.

Part V
Multipliers and Modulators

Modulators or controlled-gain circuits are one of the key building blocks of linear signal processing, data transmission and detection systems. The analog multipliers, in turn, correspond to a special class of modulators where the output signal is a *linear* product of the two input signals. Furthermore, if the polarity of the output product is preserved for all polarities of the two input signals, then the multiplier is referred to as a "four-quadrant" analog multiplier. The analog multipliers offer a wide range of applications in instrumentation, industrial controls, and telecommunications.

This part brings together a number of key technical papers which explain and demonstrate the basic design techniques for monolithic analog multipliers. Although a number of circuit techniques have been developed for analog multiplication, the method that has gained most acceptance in monolithic analog IC design is the "variable-transconductance multiplier". This circuit technique was originally introduced by Gilbert, as described in the first paper of this part. The second-paper, by Bilotti, covers the application of the basic transconductance multiplier as a balanced-modulator in a number of modulation and detection applications. In the third paper, Gilbert demonstrates methods of optimizing the performance of the basic variable-transconductance multiplier by use of active feedback. An alternate approach to the design of analog multipliers, using the pulsewidth-amplitude modulation technique, is described in the fourth paper by Holt. Although the response time of such a multiplier circuit is much slower than the variable-transconductance technique, it offers greatly improved accuracy and linearity characteristics. The final paper by Gilbert presents the fundamentals of analog multiplication based on the "carrier-domain" principle.

ADDITIONAL REFERENCES

1. H. Bruggemann, "Feedback Stabilized Four-Quadrant Analog Multiplier," *J. Sol. State Circuits,* vol. SC-5, pp. 150-159, August 1970
2. J. Smith, "A Second-Generation Carrier Domain Four-Quadrant Multiplier," *J. Sol. State Circuits,* vol. SC-10, pp. 448-457, December 1975.
3. E. Traa, "An Integrated Function Generator with Two-Dimensional Electronic Programming Capability," *J. Sol. State Circuits,* vol. SC-10, pp. 458-463, December 1975.
4. R.G. Erratico and R. Caprio, "An Integrated Expandor Circuit," *J. Sol. State Circuits,* vol. SC-11, pp. 762-772, December 1972.
5. L. Counts and D. Sheingold, "Analog Dividers: What Choices Do You Have?," *EDN Magazine,* pp. 55-61, May 5, 1974.
6. R. Allen, "Analog Multipliers: Great Design Tools Becoming Less Expensive," *EDN Magazine, pp. 34-41, February 20, 1974.*

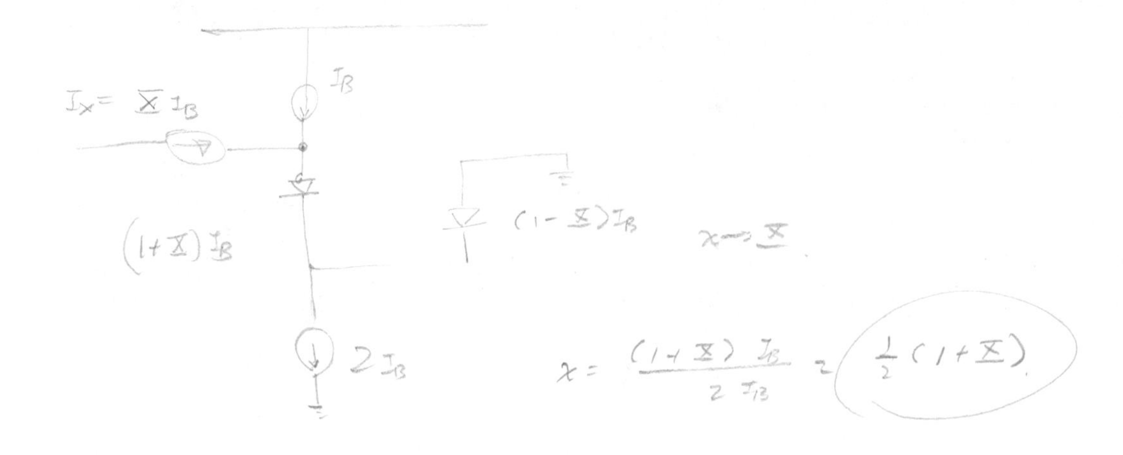

A Precise Four-Quadrant Multiplier with Subnanosecond Response

BARRIE GILBERT, MEMBER, IEEE

***Abstract*—This paper describes a technique for the design of two-signal four-quadrant multipliers, linear on both inputs and useful from dc to an upper frequency very close to the f_t of the transistors comprising the circuit. The precision of the product is shown to be limited primarily by the matching of the transistors, particularly with reference to emitter-junction areas. Expressions are derived for the nonlinearities due to various causes.**

I. Introduction

An IDEAL FOUR-quadrant multiplier would perfectly satisfy the expression

$$Z = \text{constant}, XY \tag{1}$$

for any values of X and Y, and produce an output having the correct algebraic sign. Ideally, there would be no limitation on the rate of variation of either input.

All practical multipliers suffer from one or more of the following shortcomings.

1) A nonlinear dependence on one or both of the inputs.
2) A limited rate of response.
3) A residual response to one input when the other is zero (imperfect "null-suppression").
4) A scaling constant that varies with temperature and/or supply voltages.
5) An equivalent dc offset on one or both of the inputs;
6) A dc offset on the output.

In the field of high-accuracy medium-speed multipliers, the "quarter-square" technique has gained favor [1]. This method makes use of the relationship

$$XY = \tfrac{1}{4}[(X + Y)^2 - (X - Y)^2] \tag{2}$$

and employs elements having bipolar square-law voltage–current characteristics, together with several operational amplifiers.

Much work has been put into harnessing the excellent exponential voltage–current characteristics of the junction diode for multiplier applications, either by using single diodes (or transistors) in conjunction with operational amplifiers [2], or, more recently, pairs of transistors connected as a differential amplifier [3]–[6]. In the majority of cases, the strong temperature dependence of the diode voltage proved a problem, and at least two commercially available multipliers are equipped with an oven to reduce this dependence.

Manuscript received June 28, 1968; revised September 16, 1968. This paper was presented at the 1968 ISSCC.
The author is with Tektronix, Inc., Beaverton, Ore.

Reprinted from *IEEE J. Solid-State Circuits*, vol. SC-3, pp. 365-373, Dec. 1968.

Another problem of the "differential-amplifier" multiplier, analyzed in [7], is the nonlinear response with respect to the base-voltage input. To achieve useful linearity, the dynamic range on this input must be restricted to a very small fraction of the full capability, leading to poor noise performance and worsened temperature dependence, including poor zero stability.

The problems associated with this type of multiplier can be largely overcome, however, by using diodes as current–voltage convertors for the base inputs, thus rendering the circuit entirely current controlled, theoretically linear, and substantially free from temperature effects. This paper is concerned mainly with the determination of the magnitude of the nonlinearities in a practical realization, and the analysis draws heavily on the groundwork laid in [7]; some mathematical expressions will be quoted directly from this paper without proof here.

II. The Basic Circuit

The basic scheme is shown in Fig. 1. It is comprised of two pairs of transistors, $Q2$-$Q3$ and $Q5$-$Q6$, having their collectors cross-connected, driven on the bases by a further pair of transistors, $Q1$-$Q4$, connected as diodes. It is the addition of this pair of diodes that linearizes the circuit. The X signal input is the pair of currents xI_B and $(1-x)I_B$. The Y signal is yI_E and $(1-y)I_E$, where x and y are dimensionless indexes in the range zero to unity.

It was previously shown [7] that the *ratio* of the emitter currents in the $Q2$-$Q3$ and $Q5$-$Q6$ pairs is the same as that in the $Q1$-$Q4$ pair and independent of the magnitudes of I_B and I_E (neglecting second order effects). We can thus write

$$\begin{aligned} I_{C2} &= xyI_E \\ I_{C3} &= (1-x)yI_E \\ I_{C5} &= x(1-y)I_E \\ I_{C6} &= (1-x)(1-y)I_E. \end{aligned} \tag{3}$$

The *differential*[1] output is

$$I_{\text{out}} = I_{C2} + I_{C6} - I_{C3} - I_{C5}. \tag{4}$$

Thus, the normalized output Z is

$$\begin{aligned} Z = \frac{I_{\text{out}}}{I_E} &= xy + (1-x)(1-y) \\ &\quad - (1-x)y - (1-y)x \\ &= 1 - 2y - 2x + 4xy. \end{aligned} \tag{5}$$

It is seen that the circuit is balanced when x and y are equal to 0.5. If we apply bias currents such that *bipolar* signals X and Y can be used as the inputs, and

[1] The output may also be taken as a single-sided signal from the collectors of $Q2$ and $Q6$, in which case it is $Z = \frac{1}{2}(1 + XY)$.

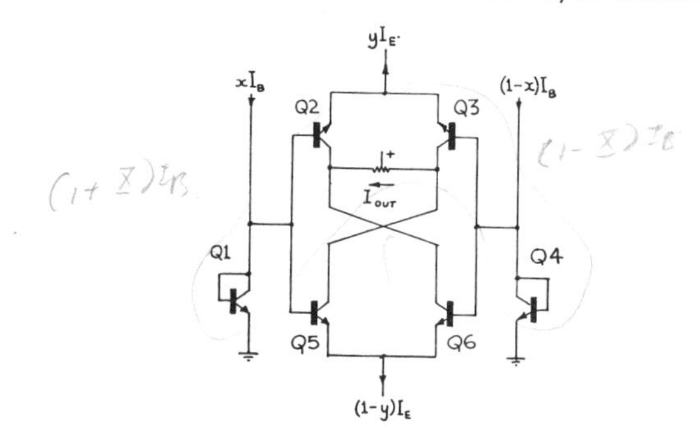

Fig. 1. The basic four-quadrant multiplier.

substitute

$$\begin{aligned} X &= 2x - 1 \\ Y &= 2y - 1 \end{aligned} \tag{6}$$

where X and Y are in the range -1 to $+1$, the output is

$$Z = XY. \tag{7}$$

This is an exact large-signal analysis, and makes no assumptions about temperature. It did, however, assume that the transistors had 1) perfectly matched emitter diodes, 2) perfect exponential characteristics (no ohmic resistance), and 3) infinite betas.

The extent to which departures from this ideal case impair the linearity will now be analyzed.

III. Distortion Due to Area Mismatches

In [7] it was found that "offset voltage"—the voltage required to balance the emitter currents of a pair of transistors in a differential amplifier—could be expressed more conveniently as a ratio of the saturation currents (or areas) of the two emitter junctions. For the four-transistor amplifier "cell" discussed in that paper, the mismatch ratio

$$\gamma = \frac{I_{S2}I_{S4}}{I_{S1}I_{S3}} \tag{8}$$

was defined. It was then shown that for $\gamma \neq 1$ (imperfect matching), the output currents were no longer simply in the same ratio x as the input currents, but had the form

$$a = \frac{\gamma}{1 + x(\gamma - 1)} \cdot x. \tag{9}$$

This can be expressed in a form that shows the nonlinearity due to area mismatches as a separate term D_A

$$a = x + D_A = x + \frac{x(1-x)(1-\gamma)}{1 + x(\gamma - 1)}. \tag{10}$$

For $\gamma \approx 1$, this simplifies to

$$D_A \approx x(1-x)(1-\gamma). \tag{11}$$

This is a parabolic function of x having a peak value $\hat{D}_A$ of $0.25(1-\gamma)$, which leads to the useful rule of thumb

$$\hat{D}_A(\text{percent}) \approx V_0(\text{mV}) \text{ at } 300°\text{K} \tag{12}$$

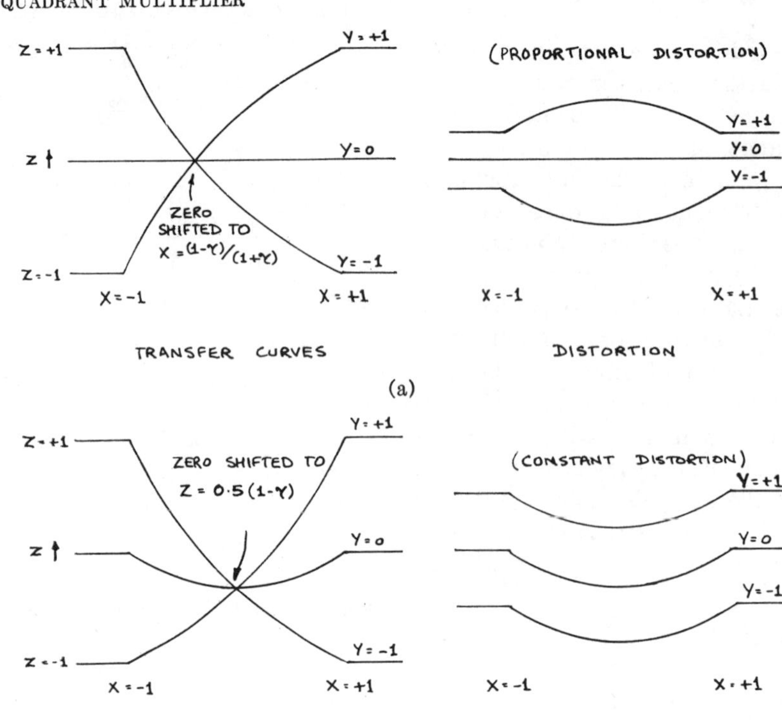

Fig. 2. Distortion introduced by area mismatches (exaggerated). (a) $\gamma_1 = \gamma_2$. (b) $\gamma_1 = 1/\gamma_2$.

where $V_0 = (kT/q) \log \gamma$, the total loop offset voltage. However, notice that $\hat{D}_A$ is not a function of temperature.

In the case of the four-quadrant multiplier, there are two such circuits working in conjunction, so we must define two area ratios

$$\gamma_1 = \frac{I_{S2} I_{S4}}{I_{S1} I_{S3}}$$

and (13)

$$\gamma_2 = \frac{I_{S5} I_{S4}}{I_{S1} I_{S6}}.$$

The total distortion (with respect to the x-input) will now be a function of y. For example if $Q1$-$Q2$-$Q3$-$Q4$ match perfectly ($\gamma_1 = 1$) but $Q1$-$Q5$-$Q6$-$Q4$ do not ($\gamma_2 \neq 1$), there will be no distortion when $y = 1$, becoming maximal when $y = 0$.

The output can be expressed as

$$Z = XY + 2yD_{A1} - 2(1 - y)D_{A2} \tag{14}$$

where

$$D_{A1} \approx x(1 - x)(1 - \gamma_1)$$

and

$$D_{A2} \approx x(1 - x)(1 - \gamma_2). \tag{15}$$

It will be seen that the linearity of Z with respect to the y input is not affected by area mismatches.

For the purposes of demonstration we can consider the cases where $Q1$ and $Q4$ match perfectly, but

1) $Q2$ and $Q3$ have the same mismatch as $Q5$ and $Q6$, that is $\gamma_1 = \gamma_2$;
2) $Q2$ and $Q3$ have the opposite polarity mismatch of $Q5$ and $Q6$, but the same magnitude, that is $\gamma_1 = 1/\gamma_2$, or $\gamma_1 \approx -\gamma_2$.

In the first case,

$$Z = XY + 2x(1 - x)(1 - \gamma)(2y - 1). \tag{16}$$

When the y input is balanced, $Y = 0$, $y = \frac{1}{2}$. Thus $Z = 0$ for all values of X. Stated differently, the null suppression with respect to the X input is unaffected by this mismatch situation.

The general form of the transfer curves and distortion products for this case is shown in Fig. 2(a), which also shows that the common point of intersection P (where $dZ/dy = 0$) is shifted to $X = (1 - \gamma)/(1 + \gamma)$, $Z = 0$. Notice also that the nonlinearity is always of the same sign as the output slope, and *varies in proportion* to it.

For the second case

$$Z = XY - 2x(1 - x)(1 - \gamma) \tag{17}$$

which corresponds to a constant parabolic distortion component *added* to the signal. In this case, when the Y input is balanced, there is a *residue* on the output of peak amplitude $0.5(1-\gamma)$. The point P is thus at $X = 0$, $Z = -0.5(1-\gamma)$, as shown in Fig. 2(b).

The general co-ordinates of P are

$$P(X, Z) = \left(\frac{1 - \sqrt{\gamma_1\gamma_2}}{1 + \sqrt{\gamma_1\gamma_2}}, \frac{\gamma_2 - \sqrt{\gamma_1\gamma_2}}{\gamma_2 + \sqrt{\gamma_1\gamma_2}}\right). \tag{18}$$

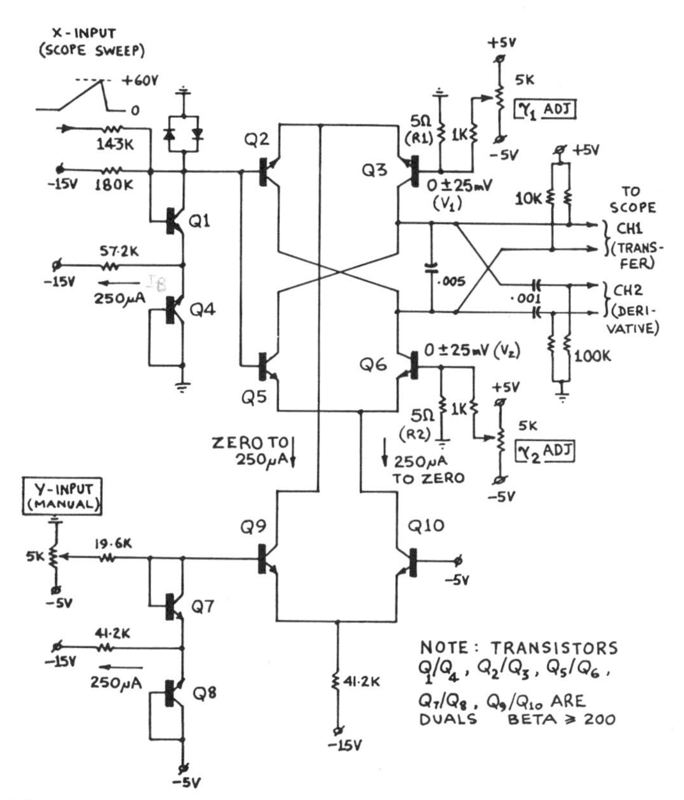

Fig. 3. Experimental circuit for investigation of nonlinear effects.

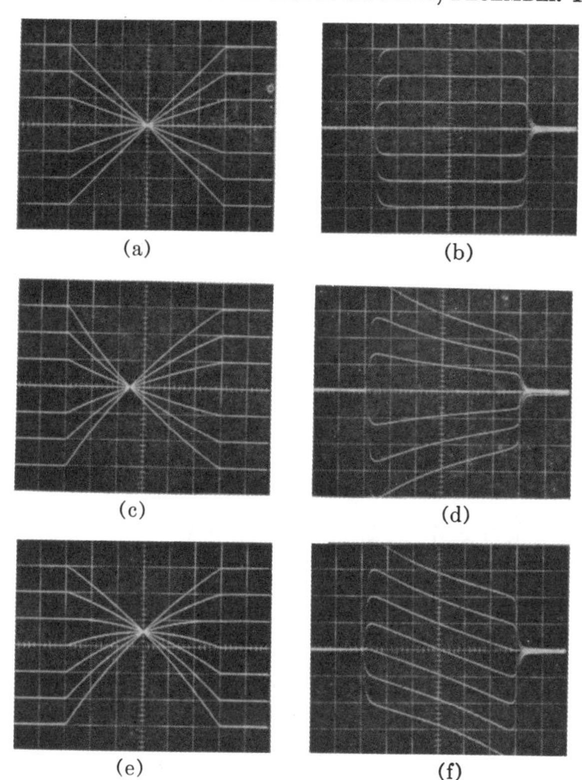

Fig. 4. Demonstration of distortion due to area mismatches. Scales are arbitrary.

Verification

To verify the above theory, and demonstrate the two cases discussed, a circuit was built as shown in Fig. 3. Use was made of the equivalence of area mismatches and offset voltage. The equivalent areas of $Q2$, $Q3$, $Q5$, and $Q6$ could be varied by the bias voltages V_1 and V_2, giving

$$\gamma_1 = e^{qV_1/kT}$$

and

$$\gamma_2 = e^{qV_2/kT}. \tag{19}$$

The devices were operated at low powers ($I_B = I_E = 250\ \mu A$, $V_C = 2.5$ volts) so that the junction temperatures were close to 300 °K. The use of low operating currents also eliminated the distortion due to ohmic resistances, discussed later.

To demonstrate the nonlinearity more clearly, a linear ramp was used as the X input, and a simple R–C differentiator produced a waveform corresponding to the incremental slope of the transfer function. This technique provides a very convenient sensitive measurement of distortion, and became a valuable tool during the investigation of improved multiplier designs, without which it would have been necessary to resort to tedious point-by-point DVM measurements to reveal the nonlinearities.

Fig. 4(a) shows the transfer curves with V_1 and V_2 adjusted for minimum distortion, and Fig. 4(b) are the derivatives. Seven static values of Y, from -1 to $+1$, are shown. These demonstrate the excellent linearity that can be achieved with well-matched transistors. The departure from constant slope is within $+0\ -1$ percent over 75 percent of the dynamic range. In terms of the nonlinearity term D_A (which is a measure of the deviation from the ideal line), this amounts to less than 0.3 percent at any point.

With $V_1 = V_2 = -10$ mV, ($\gamma_1 = \gamma_2 = 1.47$) the theoretical point of intersection is shifted to $X = -0.19$. The actual point is at -0.18, as shown in Fig. 4(c). Notice that the slope [Fig. 4(d)] falls as X varies from -1 to $+1$, starting 30-percent high and finishing 30-percent low. The deviation from the ideal line is now about 8 percent; of course, an offset voltage this large would be exceptional.

With $V_1 = +10$ mV, $V_2 = -10$ mV ($\gamma_1 = 1.47$, $\gamma_2 = 0.68$) the theoretical value of Z at $X = 0.0$ should be -0.197. This is close to the value of -0.185 measured from the waveforms of Fig. 4(e). An interesting feature of the derivatives shown in Fig. 4(f) is the one for $Y = 0$. Its linear form confirms the parabolic shape of the distortion term.

IV. Distortion Due to Ohmic Resistances

Fig. 5 shows the circuit with the addition of linear resistances in the emitters of all the transistors. These represent all the bulk resistances of the diffusions, particularly the base resistance, referred to the emitter circuit. In fact, these elements will be current dependent, due both to crowding effects and beta nonlinearities. However, if the device geometry is such that the cur-

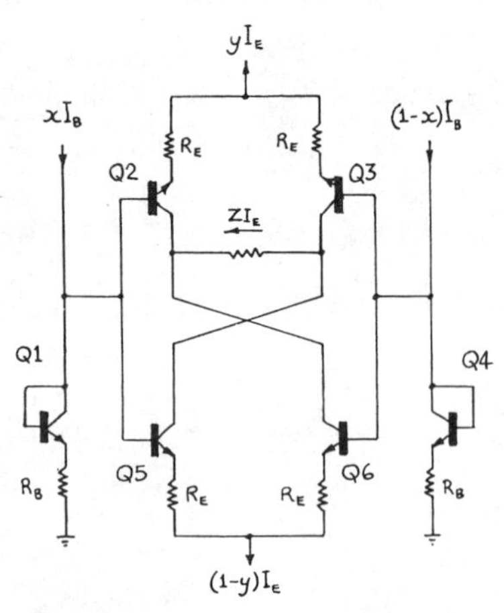

Fig. 5. Circuit having ohmic emitter resistances.

rent-density distribution is equalized in the appropriate sets of devices, the current dependence can be neglected.

Using the variables shown in Fig. 5, the loop equation for the quad $Q1$-$Q2$-$Q3$-$Q4$ under these conditions becomes

$$\frac{kT}{q}\log\left[\frac{x(1-a)}{a(1-x)}\right] = I_B R_B(1-2x) - yI_E R_E(1-2a) \qquad (20)$$

which has no explicit solution for a in terms of the other variables. However, guessing that the distortion will be small, we will make the substitutions

$$a = x + D_R \qquad (21)$$

and

$$\log(1 - D_R) \approx -D_R$$

where D_R is the fractional distortion due to resistances. Equation (20) simplifies to

$$\frac{kT}{q}\frac{D_R}{(x+D_R)(1-x)} = yI_E R_E(1-2x-2D_R) - I_B R_B(1-2x). \qquad (22)$$

Solving for the distortion term, and changing input variable from x to X,

$$D_R = \frac{q}{kT}(y\phi_E - \phi_B)\,\Delta(X) \qquad (23)$$

where

$$\phi_E = I_E R_E$$

and

$$\phi_B = I_B R_B, \qquad (24)$$

these representing the extra voltages in the emitter and base circuits due to resistances, and

$$\Delta(X) = \tfrac{1}{4}X(X^2 - 1) \qquad (25)$$

which describes the form of the distortion, and has peak values of ± 0.096 at $X = \pm 0.577$, and zeros at $X \pm 1$ and 0.

Equation (23) makes the reasonable assumption that ϕ_E and ϕ_B are small compared to kT/q. For example, assume $R_E = 1$ ohm and $I_E = 2.5$ mA, giving $\phi_E = 2.5$ mV, about 10 percent of kT/q at 300°K.

Using the above approximate analysis, we can state a rule of thumb for the peak magnitude of D_R, for the quad $Q1$-$Q2$-$Q3$-$Q4$:

$$\hat{D}_{R1} \approx \pm 0.37(y\phi_E - \phi_B) \qquad (26)$$

for ϕ_E, ϕ_B in millivolts, at 300°K, and $\hat{D}_{R1}$ in percent. Similarly, for the $Q1$-$Q5$-$Q6$-$Q4$ circuit, we have

$$\hat{D}_{R2} \approx \pm 0.37\,\{(1-y)\,\phi_E - \phi_B\}. \qquad (27)$$

The *net* nonlinearity will come from both circuits, and vary with the y input. The outputs of each quad (and hence the distortion terms) are also weighted by y and connected out of phase. Thus

$$\begin{aligned}\hat{D}_R &= \hat{D}_{R2} - \hat{D}_{R1}\\ &= \pm 0.37[\{y\phi_E - \phi_B\}y - \{(1-y)\phi_E - \phi_B\}(1-y)]\\ &= \pm 0.37(\phi_E - \phi_B)Y \quad \text{(percent)}, \qquad (28)\end{aligned}$$

with the substitution of $Y = 2y - 1$. The nonlinearities introduced by *balanced* emitter resistances can be summarized as follows.

1) The distortion with respect to the X input has a symmetrical form and is a fixed percentage of the output Z.
2) There is no distortion with respect to the Y input.
3) The common point of intersection of the transfer curves is always at $X = Y = Z = 0$.
4) No distortion arises when $\phi_E = \phi_B$.

Thus, quite large ohmic resistances can be tolerated (that is, it is possible to use devices with high base resistance and/or low beta), provided that the base and emitter voltage terms are balanced. By scaling the device geometrics in the ratio I_E/I_B, the closeness with which $\phi_E = \phi_B$ is then a matter of device matching.

In practice the resistors labeled R_B in Fig. 5 will not be equal. It can be shown that under these conditions there will be a residue in the multiplier output for $Y = 0$, having the S-shaped form described by (25), and having a peak amplitude (at 300°K) of

$$\hat{D}_R \approx \pm 0.05\, I_E(R_{E2} + R_{E3} - R_{E5} - R_{E6}) \qquad (29)$$

where $\hat{D}_R$ is in percent, I_E in milliamperes, R_E in ohms. Notice that this residue term is independent of ϕ_B, a fact that has been experimentally confirmed. It will be apparent that linear emitter resistances reduce the output swing capability because in the limit (when the diode voltages are small compared to the "ohmic" voltages) the circuit becomes completely cancelling for all values of X or Y. Also, the case where these resistances are unbalanced will give rise to an equivalent offset on one or both of the inputs.

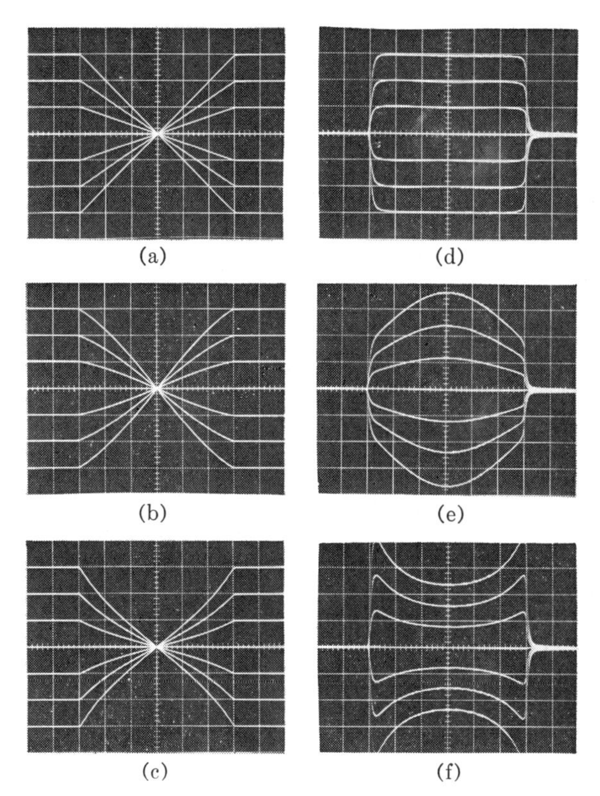

Fig. 6. Demonstration of distortion due to ohmic resistances.

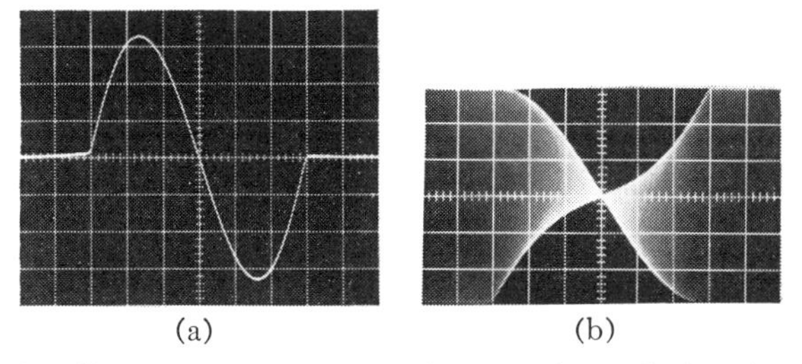

Fig. 7. Characteristic distortion due to mismatched resistances. (a) Vertical scale expanded to 0.33 percent/div. (b) 3.3 percent/div.

Verification

Using the test circuit of Fig. 3, to which emitter resistors were added, these nonlinearities were demonstrated. In Fig. 6(a), all resistors were 50 ohms and $I_B = I_E = 250$ μA; thus $\phi_E = \phi_B = 12.5$ mV. The derivative waveform, shown in Fig. 6(b), shows little degradation of linearity over the full dynamic range.

By omitting the resistors in the $Q2$-$Q3$ and $Q5$-$Q6$ emitters, a net error of $\phi_B = 12.5$ mV remains. Equation (28) predicts a nonlinear term of ± 4.5 percent at $Y = \pm 1$. The measured value is ± 3.3 percent. (Due to the approximations, (28) will err on the high side when ϕ_B or ϕ_E become comparable with kT/q). See Figs. 6(c) and (d). By omitting the resistors in the $Q1$-$Q4$ emitters, the distortion is of the opposite polarity, as Figs. 6(e) and (f) demonstrate.

The most typical distortion is due to the case where the ohmic voltages do not match, due probably to mismatches in r_b and beta. This can be demonstrated, too, by inserting the 50-ohm resistors in just the $Q2$-$Q3$ pair, when (29) predicts a peak distortion of ± 1.25 percent of full scale with the Y input balanced. The measured nonlinearity is shown in Fig. 7(a), in which the display was expanded vertically 50 times and the distortion has peak values of ± 1.1 percent. Fig. 7(b) gives the appearance of the distortion when the Y input was modulated to a depth of about 20 percent.

V. Distortion Due to Beta

The final imperfection to consider is that of finite beta. Three cases can be considered:

1) the transistors have identical, constant beta;
2) the transistors have differing, but still constant, beta;
3) the transistors have identical, current-dependent beta.

The first case was dealt with in [7] where it was shown that the only error is that the output current is reduced by the factor alpha (for I_E less than βI_B). In the version of the circuit driven by a single-sided input current (as, for example, the test configuration shown in Fig. 3), a small offset term also arises, and the dc output for $Y = 0$ is approximately

$$Z(X, 0) = (1 - \bar{\alpha}) \frac{I_E}{I_B} \tag{30}$$

where $\bar{\alpha}$ is the large-signal common-base current gain.

The second and third cases have not been completely analyzed, and it is doubtful whether explicit expressions involving all the betas and their nonlinearities would be of any value. Clearly, there is now the possibility for distortion terms to arise. However, the variations in beta from device to device, and over a small current range, are usually sufficiently small that no serious distortion should arise using typical transistors with betas in the neighborhood of 100.

VI. Thermal Distortion

The topic of thermal distortion in this category of circuits was dealt with in [7], where it was shown that theoretically no distortion arises due to the differential heating of devices if the power dissipation in the inner and outer pair are equal. This can usually be arranged, and, if necessary, the circuit can operate with I_E less than I_B.

In practice, using monolithic circuits the thermal distortion in response to a step input is very much less than 1 percent of the output amplitude, and persists for no more than a few microseconds.

VII. Transient Response

Because of the very small voltage swings at the inputs, and the cross connection of the transistors, the aberrations due to capacitances are very small, especially when properly balanced inputs are used. The main speed limitation is the f_t of the transistors.

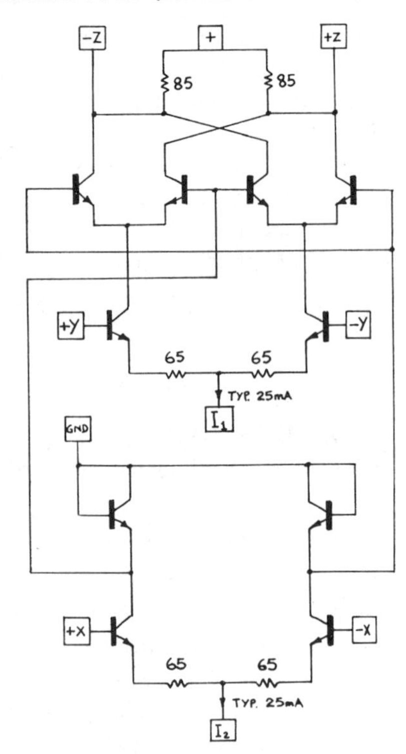

Fig. 8. Circuit used to examine high-frequency behavior.

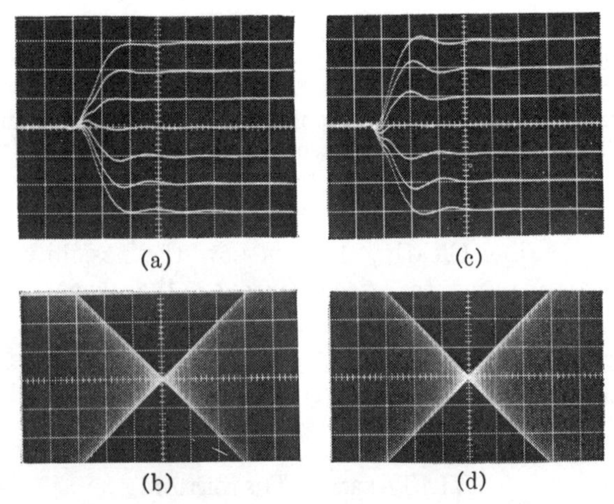

Fig. 9. Performance of integrated version of Figs. 8. (a) and (b). Transient response on X and Y inputs, respectively, at 1 ns/div. using dc control on other input. (c) and (d) 200-MHz carrier on X and Y inputs, respectively, staircase voltage on other input. Peak swing is 90 percent of full scale in all cases.

At $Y = \pm 1$, one of the transistor pairs $Q2$-$Q3$ or $Q5$-$Q6$ is producing all the output. The 3-dB bandwidth is, thus, about $f_t I_E / I_B$. At $Y = 0$, each pair receives $I_E/2$, and the bandwidth is doubled. We would, therefore, expect a risetime variation in response to a step on the X input of about two to one between these extremes. The step response to the Y input should be fairly independent of the X amplitude.

Measurements on an early integrated multiplier were made to examine the high-frequency behavior. The circuit, shown in Fig. 8, uses an "inverted" pair of input diodes [7], which are conveniently driven from pairs of emitter-degenerated stages for the X and Y inputs.

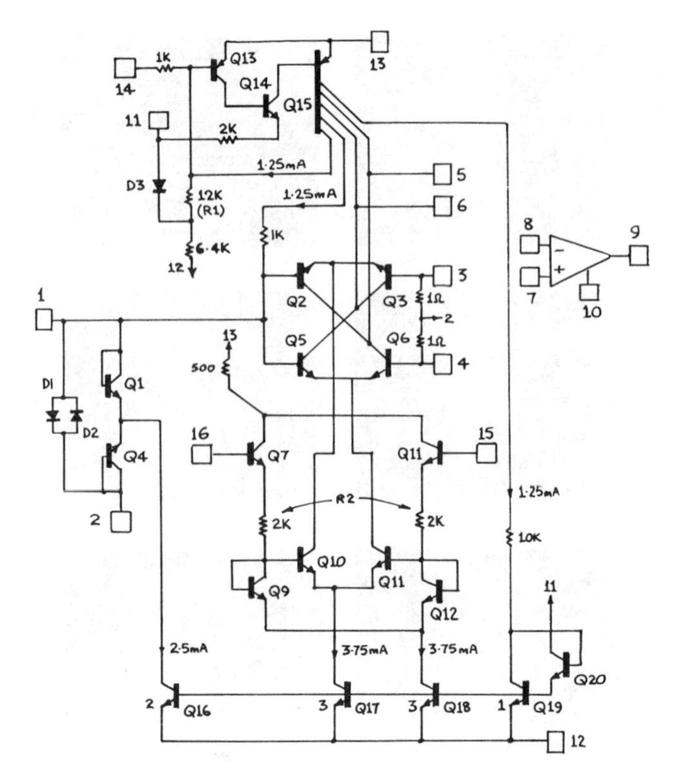

Fig. 10. Complete monolithic multiplier.

Transient response for each input is shown in Fig. 9(a) and (b). Figs. 9(c) and (d) show the CW response for a 200-MHz input, with the other input driven by the staircase output of the sampling time base in the oscilloscope used to examine the responses. The null suppression was better than 20 dB at 500 MHz.

VIII. A. Complete Monolithic Multiplier

Fig. 10 is the circuit of a complete multiplier suitable for integration. It is designed so as to be usable with a minimum of additional components to achieve medium-accuracy operation, or with extra components to perform at a higher accuracy. Wide-band operation (dc to >100 MHz) is available, or more versatility can be obtained by using the built-in operational amplifier to give division, squaring and square-rooting modes. These variations are possible by pin changes only. The X input is a single-sided current I_x into a summing point at ground potential, in the range 0 ± 1 mA. The y input is a differential voltage V_Y into a high impedance (approx. 400 kΩ) in the range 0 ± 5 volts. It can be shown that the output from pin 6 is

$$I_Z = \frac{I_x V_y}{5} \tag{31}$$

the scale factor being determined by the +15-volt supply and the ratio of R_1 to R_2. The diode D_3 ensures temperature stable scaling, and also makes the scaling factor proportional to the positive supply over a limited range.

Input and output current balance, and the rest of the circuit currents, are determined by the five-collector

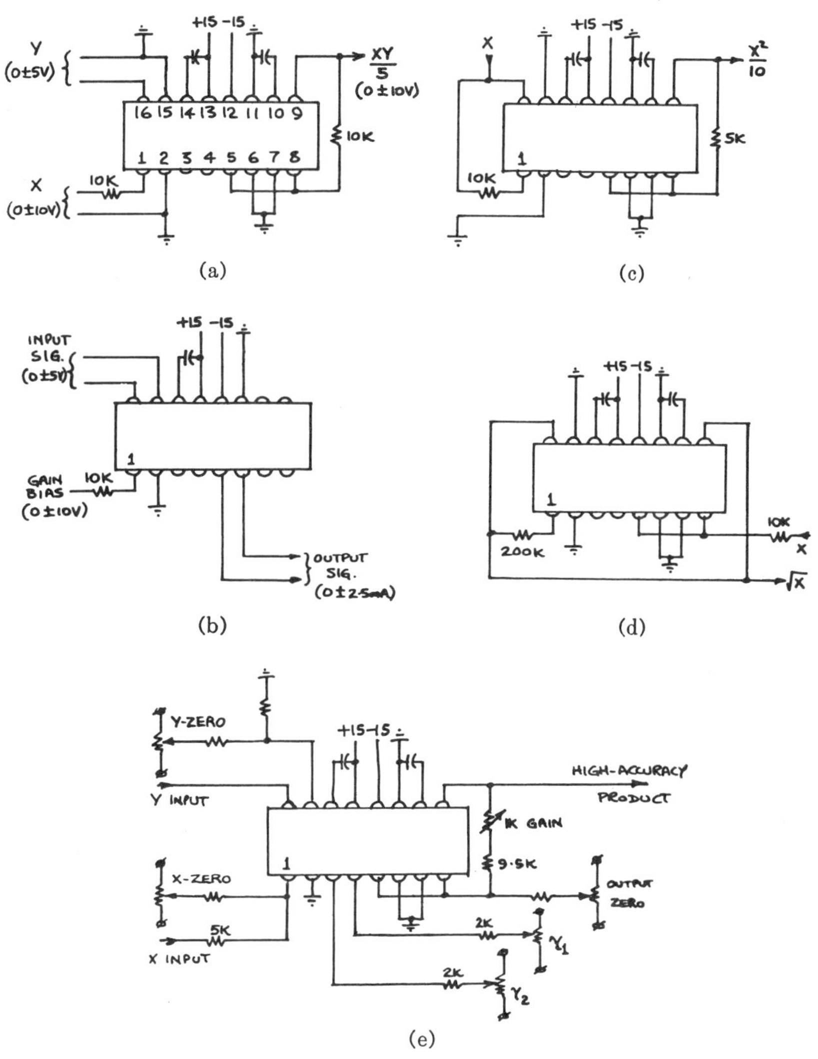

Fig. 11. Circuit of Fig. 9 connected as (a) medium-accuracy multiplier, (b) wide-band gain and polarity control, (c) squarer, (d) square-rooter, and (e) fully corrected multiplier.

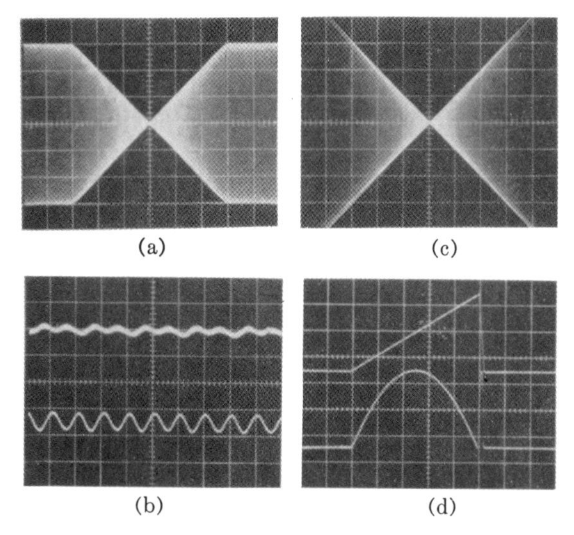

Fig. 12. Typical performance. (a) As balanced modulator, carrier frequency 5 MHz, peak output swing is 90 percent of full scale. (b) Output expanded ten times in vertical and horizontal axes—vertical now 1.67 percent of full scale/div. (c) Null suppression for full-scale 5-MHz carrier on X (upper trace) and Y (lower trace), expanded to 0.1 percent/div. (d) Offset ramp applied to both inputs produces the parabolic output, 1 μs/div.

lateral p-n-p, $Q15$. The matching of the currents to the five collectors (base diffusions) is vital to balanced operation. Measurements indicate that matching errors considerably less than ± 2 percent can be achieved. Notice that one of the collectors is connected in an operational configuration, through $Q13$ and $Q14$. This loop has to be stabilized by an external capacitor connected between pins 13 and 14. The second collector supplies a nominal 1.25 mA to balance the X input; collectors 3 and 4 supply the output balance currents; collector 5 sets up the current tails for the multiplier via $Q16$ through $Q20$.

Pins 3 and 4 give access to the bases of $Q3$ and $Q6$, allowing linearity-connection voltages to be applied. For perfect connection, these voltages should be proportional to absolute temperature. The aluminum 1-ohm resistors come close to this ideal, having a temperature coefficient of 0.38 percent per °K, slightly greater than the coefficient of kT/q at 300°K.

The operational amplifier increases the versatility of the device by permitting several modes to be implemented. Pin 7 is normally grounded, and the inverted output from the multiplier, pin 5, is connected to pin 8; pin 6 is also grounded. Hence, the multiplier block $Q2$, $Q3$, $Q5$, $Q6$ works with a collector-base voltage of 0 ± 100 mV (the base voltage swing). The overload diodes, $D1$ and $D2$, must, therefore, be Schottky-barrier diodes having negligible conduction for most of the working voltage range at the X-input summing point, but being able to conduct heavily before the collector diodes of $Q2$ and $Q5$ under overload conditions. These may now be fabricated along with the standard silicon circuitry.

Fig. 11(a) through (e) illustrate the versatility of the circuit. The waveforms in Fig. 12 show linearity and null suppression at 5 MHz, and the output in the squaring configuration.

IX. Summary

A technique has been described that overcomes the inherent temperature dependence and nonlinearity of a transistor four-quadrant multiplier, and the feasibility of producing a complete monolithic multiplier with a worst-case linearity error of the order of 1 percent on either input has been demonstrated. Better linearity is possible by adjustment of transistor offset voltages. Bandwidths of over 500 MHz have been measured.

Acknowledgment

Thanks are due to the Integrated Circuits Group at Tektronix for the fabrication of many experimental circuits, and to G. Wilson and E. Traa [8] for helpful discussions.

References

[1] G. A. Korn and T. M. Korn, *Electronic Analog Computers*. New York: McGraw-Hill, 1956, pp. 281–282.

[2] G. S. Deep and T. R. Viswanathan, "A silicon diode analogue multiplier," *Radio and Electron. Engrg.*, p. 241, October 1967.

[3] H. E. Jones, "Dual output synchronous detector utilizing transistorized differential amplifiers," U. S. Patent 3 241 078, June 18, 1963.

[4] A. R. Kaye, "A solid-state television fader-mixer amplifier," *J. SMPTE*, p. 605, July 1965.

[5] W. R. Davis and J. E. Solomon, "A high-performance monolithic IF amplifier incorporating electronic gain-control," *1968 ISSCC Digest of Tech. Papers*, pp. 118–119.

[6] A. Bilotti, "Applications of a monolithic analog multiplier," *1968 ISSCC Digest of Tech. Papers*, pp. 116–117.

[7] B. Gilbert, "A new wide-band amplifier technique," this issue, pp. 353–365.

[8] E. Traa, "An integrated analog multiplier circuit," M.Sc. thesis, Oregon State University, Corvalis, June 1968.

Applications of a Monolithic Analog Multiplier

ALBERTO BILOTTI, SENIOR MEMBER, IEEE

***Abstract*—A fully balanced analog multiplier using differential transistor pairs is briefly described. Several circuit functions usually required in communication systems can be derived from the basic circuit. In particular, the different modes of operation leading to FM detection, suppressed carrier modulation, synchronous AM detection, and TV chroma demodulation are discussed. Experimental data obtained with a monolithic analog multiplier are also presented.**

I. Introduction

MANY CIRCUIT functions required in communication systems can be derived by way of analog multiplication. Fig. 1 shows a functional block consisting of an analog multiplier, symmetrical input limiters, and an optional output low-pass filter. When properly combined with passive networks, this block can perform FM detection, phase comparison, synchronous AM detection, amplitude modulation, and other functions based on frequency translation. This paper considers an integrated circuit that can be represented by the block diagram in Fig. 1.

Manuscript received June 5, 1968; revised September 25, 1968. This paper was presented at the 1968 ISSCC.

The author was with Sprague Electric Company, North Adams, Mass. He is now with the Faculty of Engineering, University of Buenos Aires, Buenos Aires, Argentina.

II. Basic Circuit

Fig. 2 shows the fully balanced arrangement of the three differential transistor pairs Q_1, Q_2, and Q_3 forming an analog multiplier block. The essential features of this circuit have been discussed elsewhere [1], [2] and the objective here is to achieve an understanding of the modes of operation possible, and through these modes, to consider a number of possible system applications [3]. Assume for a moment low-level driving at the inputs of V_1 and V_2. The current I_b established in the current source transistor is split in proportion to the applied voltage in transistor pair Q_1. This current division determines the bias and, therefore, the gain of the pairs Q_2 and Q_3. The output collector current summed in the load resistor R is proportional to the product of the two applied signals V_1 and V_2. The circuit topology is such that if $V_1 = 0$, the output currents due to a signal V_2 are of equal magnitude and opposite instantaneous polarity, giving a zero net output. The same is true for an applied

Reprinted from *IEEE J. Solid-State Circuits*, vol. SC-3, pp. 373-380, Dec. 1968.

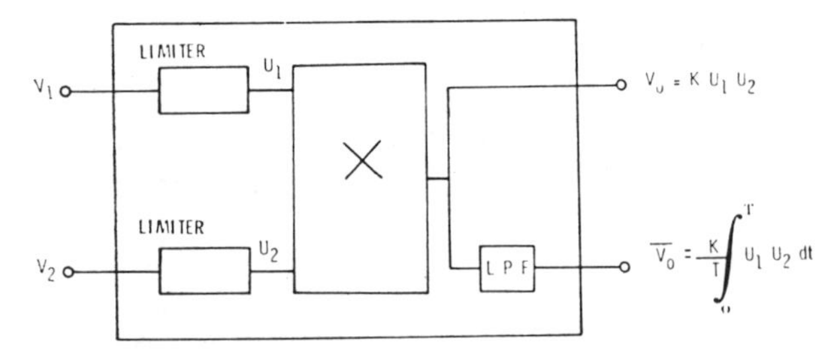

Fig. 1. Basic functional block.

signal V_1 when $V_2 = 0$. This fully balanced operation implies that each amplified input is cancelled at the output and only the product term is amplified. In this way, post-multiplier filtering is eased, and twice the output is available when compared to the output obtained from an unbalanced arrangement. Low-pass filtering is provided by the addition of the external capacitor C, which produces a 3-dB frequency in proportion to the inverse of the RC product.

At high input levels, symmetrical limiting occurs in the base-emitter junctions of the input devices. The limiting transfer characteristic is shown in Fig. 3. U_1 and U_2 are proportional to the collector currents resulting from inputs V_1 and V_2, respectively. For simplicity, U_1 and U_2 can be considered to represent the normalized inputs to an ideal analog multiplier after including the symmetrical limiting of the input devices.

For low input levels, operation is approximately linear, and U_1 and U_2 are proportional to V_1 and V_2, respectively. For high levels, hard limiting occurs, and U_1 and U_2 can be approximated by switching functions of frequency determined by V_1 and V_2, respectively.

Neglecting the base spreading resistance of the transistors, the hyperbolic relation of the Ebers and Moll equations holds between these extremes of operation. Straightforward application of these equations to the circuit of Fig. 2 yields a general expression for the output voltages $\bar{V}_0$ or V_0 at any input level V_1 and V_2 (bias term neglected):

$$\bar{V}_0 = \frac{V_R}{t}\int_0^t U_1 U_2\,d\tau \tag{1a}$$

or

$$V_0 = V_R U_1 U_2 \tag{1b}$$

where

$$U_1 = \tanh\frac{V_1}{2kT/q} \tag{2a}$$

$$U_2 = \tanh\frac{V_2}{2kT/q} \tag{2b}$$

and

$$V_R = \frac{I_b R}{2} \tag{3}$$

is the quiescent voltage drop across the collector load.

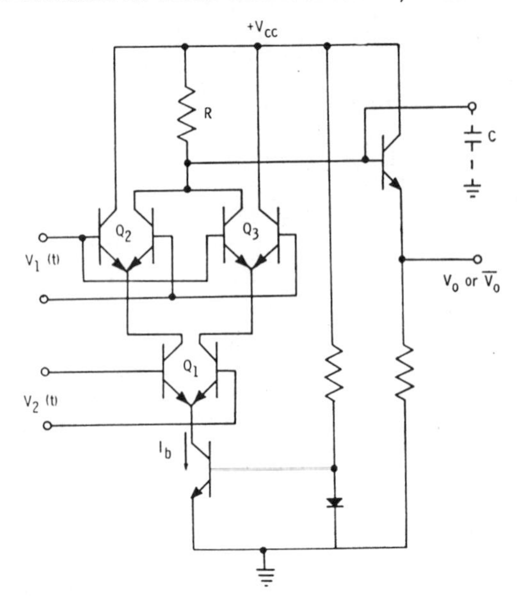

Fig. 2. Monolithic multiplier block.

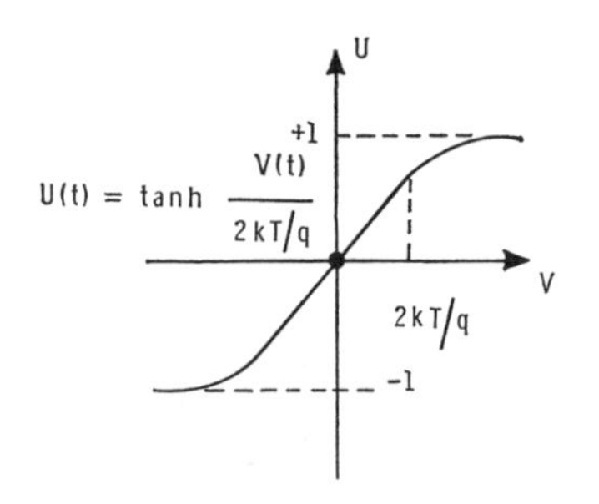

Fig. 3. Limiting transfer characteristic

III. Modes of Operation

Depending on the amount of limiting and the synchronous or asynchronous nature of the inputs, several modes of operation are possible. To clarify the performance of the multiplier in two of these modes, Fig. 4 shows the signal waveform associated with the application of two sinusoidal signals of the same frequency, differing in phase by an angle ϕ. One of the inputs is considered to be above the limiting threshold of the input devices and yields a switching function (in this case U_2). The two modes of operation result by considering the other applied signal (V_1) to be above or below the input limiting threshold.

When V_1 is above the limiting threshold (high level), U_1 is also a switching function. The phase difference between U_1 and U_2 results in an output of the multiplier whose filtered value is proportional to the phase angle difference ϕ. In fact, by expressing each switching function by its Fourier expansion,

$$U_1 = \frac{4}{\pi}\sum_1^\infty \frac{\cos n\omega t}{n} \qquad (n \text{ odd}) \tag{4a}$$

$$U_2 = \frac{4}{\pi}\sum_1^\infty \frac{\cos n(\omega t + \phi)}{n} \qquad (n \text{ odd}) \tag{4b}$$

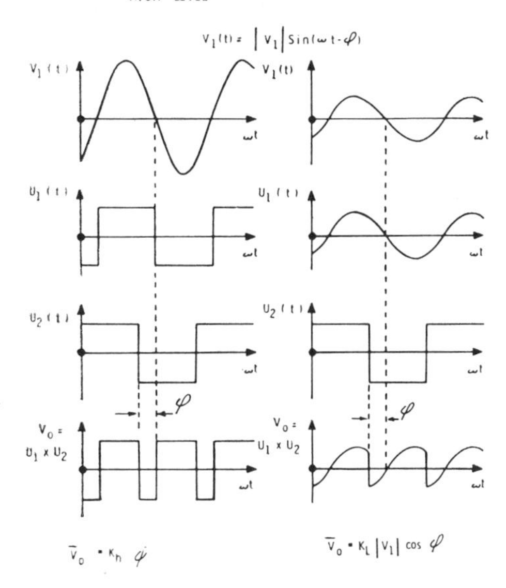

Fig. 4. Waveforms for two extreme operating levels for V_1, keeping V_2 under high-level conditions.

the average value of the product of both series becomes

$$\frac{\bar{V}_0}{V_R} = \frac{8}{\pi^2}\sum_1^{\infty}\frac{\cos n\phi}{n^2}$$

$$= \begin{cases} 1 - \dfrac{2}{\pi}\phi & \text{for} \quad 0 < \phi < \pi \\ -3 + \dfrac{2}{\pi}\phi & \text{for} \quad \pi < \phi < 2\pi \end{cases} \qquad (n \text{ odd}). \tag{5}$$

The transfer characteristic has therefore a triangular shape with maxima and minima equal to $\pm V_R$ and with a slope of $\pm 2/\pi V_R$ (volts/rad). For this mode of operation, the multiplier provides a zero-crossing coincidence detector for *linear phase detection* or *phase comparison*.

If, however, V_1 is below the limiting level (low level), the U_1 function is

$$U_1 = \frac{|V_1|}{2kT/q}\cos \omega t \tag{6}$$

and the average value of the product U_1U_2, where the switching function U_2 is given by (4b), becomes

$$\frac{\bar{V}_0}{V_R} = \frac{2}{\pi}\frac{|V_1|}{2kT/q}\cos\phi. \tag{7}$$

This last expression shows that the output of the multiplier has a filtered value proportional to the magnitude of V_1 times the cosine of the phase angle difference ϕ. This mode of operation, therefore, enables the analog multiplier to operate as a *synchronous detector*. Note that due to the fully balanced arrangement, each half cycle of the input contributes to the total output or, in other terms, full-wave detection occurs.

Expression (7) also includes the pure analog case where both inputs are below the limiting level, the $2/\pi$ factor being replaced by $\frac{1}{2}(|V_2|/2kT/q)$.

TABLE I
OPTIMUM OPERATING LEVELS FOR DIFFERENT APPLICATIONS OF THE ANALOG MULTIPLIER

	Operating Level V_1	Operating Level V_2
1) Balanced Product Mixer	Low	Low
2) FM Detector*	Low or High	High
3) Suppressed Carrier Modulator*	Mod/Low	Carrier/High
4) Synchronous AM Detector*	Low	High
5) Frequency Discriminator	Low or High	High
6) Phase Detector	High	High
7) TV Color Demodulator*	Chroma/Low	Carrier/High

The mode of operation chosen is, of course, dependent upon the circuit function desired. Table I gives a list of analog multiplier applications and the modes of operation appropriate for each. A pure analog multiplier requires low-level operation at both inputs. For FM detection, one input can be either high or low level; however, as shown later, the optimum choice depends on the system considered. Suppressed carrier modulation is obtained by high-level operation at one input and low at the other. We will consider, in the order listed, the applications indicated by asterisks in Table I. This will include the application of the analog multiplier as a synchronous detector for the demodulation of the color information in a TV receiver. This application is most interesting, since both amplitude and phase detection are simultaneously required.

A. FM Detector

The first application considered is the use of the analog multiplier circuit as an FM detector. Fig. 5 shows the circuit diagram for the integrated discriminator and limiter (Sprague ULN2111A) [1]. Added to the basic three-transistor-pair analog multiplier is a 60-dB voltage-gain limiter. This limiter provides amplitude modulation rejection over a wide dynamic range of input signal levels. For FM detection, the output of the limiter is split. The limited signal is applied to one of the multiplier inputs and a simple single-tuned external LC network provides the signal to the second multiplier input with an amplitude and phase angle that varies with frequency. Frequency variations are, therefore, first converted to amplitude and phase variations between signals at the inputs of the multiplier. The detection of these variations is then obtained in the analog multiplier as discussed earlier. Recall, however, that two modes of operation were possible, depending upon the level of the voltage applied at the second multiplier input (see Fig. 4).

Substitution of the amplitude and phase response of the network for $|V_1|$ and ϕ into (5) and (7) yields the frequency-transfer characteristics for each operating mode.

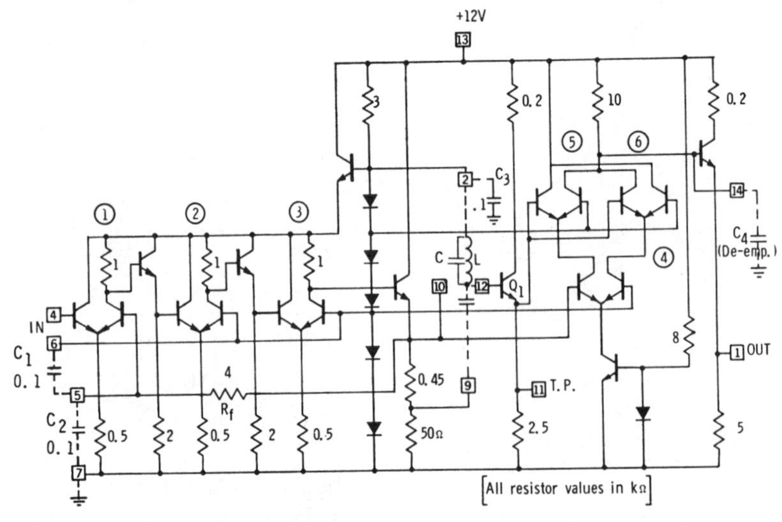

Fig. 5. Monolithic analog multiplier (ULN211A) with the appropriate external network required for operation as an FM detector. Included in the integrated circuit is a 60-dB limiter amplifier (stages 1, 2, and 3) directly coupled to one of the inputs of the multiplier (stages 4, 5, and 6).

$$\frac{\bar{V}_0}{V_R} = \frac{2}{\pi} \arctan a \qquad \text{(high level)} \tag{8}$$

$$\frac{\bar{V}_0}{V_R} = \frac{2}{\pi}\left(\frac{|V_1|_0}{2kT/q}\right)\frac{a}{1+a^2} \qquad \text{(low level)} \tag{9}$$

where

$$a = 2Q\frac{\Delta F}{F_0} = \text{normalized frequency deviation} \tag{10}$$

and

$$|V_1|_0 = \text{magnitude of } V_1 \text{ at the center frequency.}$$

Expressions (8) and (9) have been plotted in Fig. 6. For low-level operation, the amplitude of the bell-shaped response of the tuned circuit provides the response fall-off on either side of the center frequency, and the familiar S-shaped transfer function is obtained. The peak-to-peak separation of the S curve is directly related to the 3-dB bandwidth of the tuned circuit. For high-level operation and within a large range of frequency deviations, the amplitude response of the network transfer function is eliminated, and the response is directly that of the phaseshift properties of the LC network.

Fig. 6 shows the measured response at the same frequency scale and Q for the phase-shift network, but for the two levels of operation. In the low-level case, the S curve follows the simple analytic expression, and the bandwidth of the network determined by the tuned circuit is preserved. This latter property, bandwidth preservation, is important for reducing undesired Gaussian noise recovered by the multiplication of noise-frequency components lying outside the bandwidth of interest. Fig. 7 shows the output signal-to-noise ratio measured as a function of the operating level at one of the multiplier inputs. The noise level was adjusted to be above the limiting threshold of the high-gain limiter. For high-level operation at V_1, the effective bandwidth of the system is broadened, due to the post-filter limiting (the tuned response of the phaseshift network being the filter). In systems where "squelch" is not employed (this includes most TV receivers and FM receivers on the market), low-level operation means improved interchannel noise performance. When squelch is available, or where the noise level under "no received signal condition" is not important, operation under high-level conditions will provide increased conversion efficiency. In fact, the conversion efficiency, defined as the slope of the transfer characteristic at the center frequency, can be shown to be

$$\left(\frac{d\bar{V}_0}{da}\right)_{a=0} = \frac{2}{\pi} V_R \tanh \frac{|V_1|_0}{2kT/q}. \tag{11}$$

Fig. 8 shows the measured values of the conversion efficiency as a function of the magnitude of the driving voltage V_1 at the center frequency of 4.5 MHz.

The multiplier FM detector provides AM rejection at the center frequency, where quadrature conditions hold. In fact, (5) and (7) show that no output voltage is developed at $\phi = \pi/2$. Also, the fully balanced arrangement cancels the output due to any low-frequency component at the limiter output (i.e., due to parasitic pulse width or rise-time modulation).

The monolithic circuit affords a very wide-band operating capability due to the use of unsaturated current mode switching gates (rise times about 5–6 ns). In Fig. 9, the transfer characteristic of the detector is shown at two extreme center frequencies, 5 kHz and 50 MHz. The only circuit modification between the two was the use of the appropriate phaseshift network in each case. Note the high reproducibility and symmetry of the characteristics for different bandwidths and frequencies. This fea-

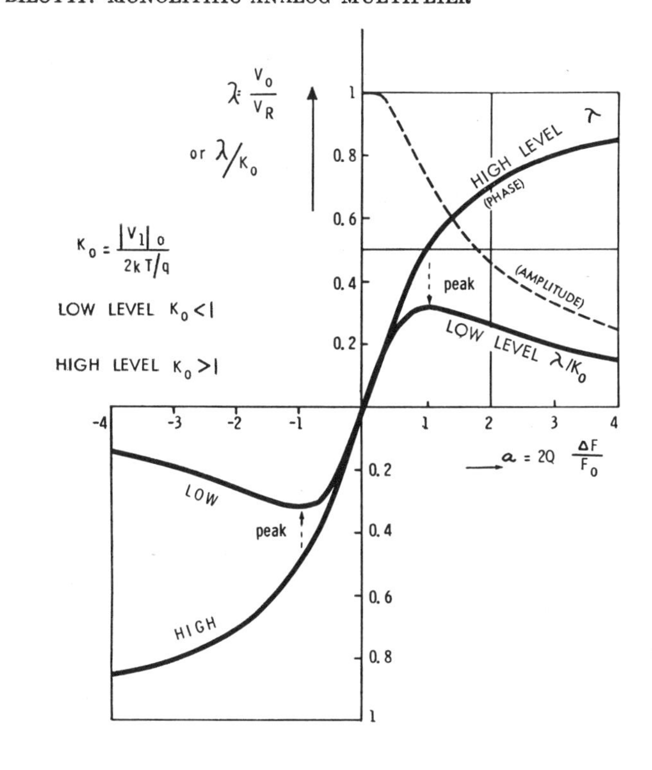
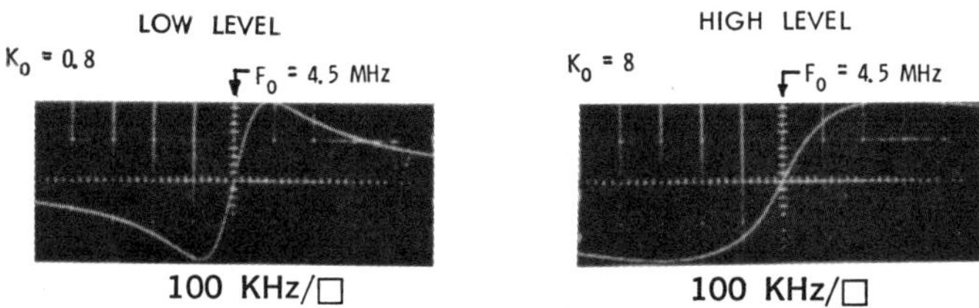

Fig. 6. Calculated and measured transfer characteristics for the low- and high-level cases.

ture makes the multiplier also attractive as a frequency discriminator in AFC systems.

B. Suppressed Carrier Modulator

The basic multiplication function (1b) allows straightforward operation of the circuit as a suppressed carrier or balanced modulator. As shown in Fig. 10, the modulating signal is applied to one of the inputs in the low-level mode and the carrier reference to the other under high- or low-level conditions. Carrier high-level operation provides a modulated output independent from the carrier level and should be preferred unless the harmonic spectral responses created by the switching carrier can not be tolerated by the system. Fig. 10(a) shows the modulated waveform obtained at a carrier frequency of 1 MHz with sinuosidal modulation. As another example, the monolithic analog multiplier was tested as a digital phase modulator. In this last case, an unmodulated carrier of 20 MHz and a square-waveform signal of 2.5 MHz were applied to each input under high-level operating conditions. The obtained output is shown in Fig. 10(b), where the typical carrier phase reversal for each modulation polarity is clearly seen.

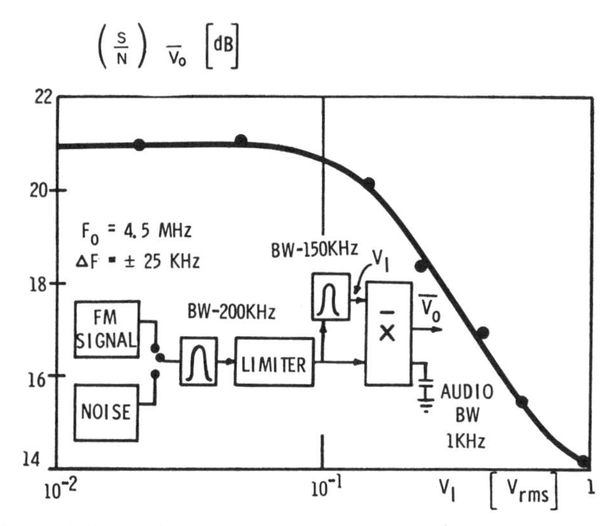

Fig. 7. Measured noise detection properties of the FM detector as a function of the operating level.

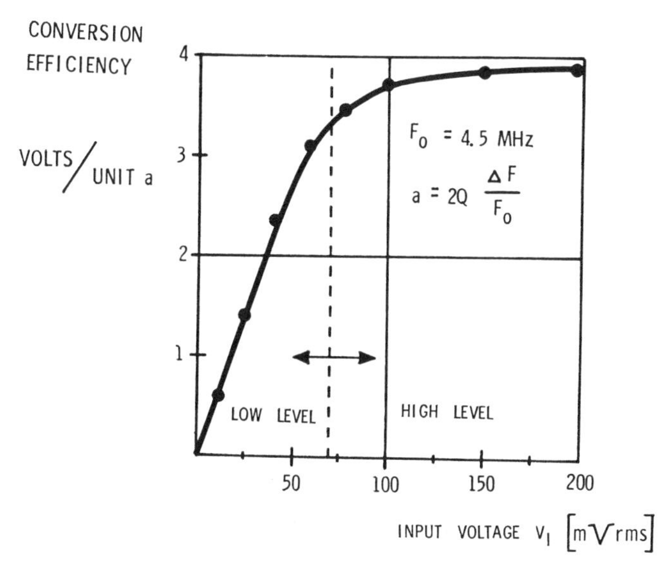

Fig. 8. Measured conversion efficiency of the monolithic FM detector.

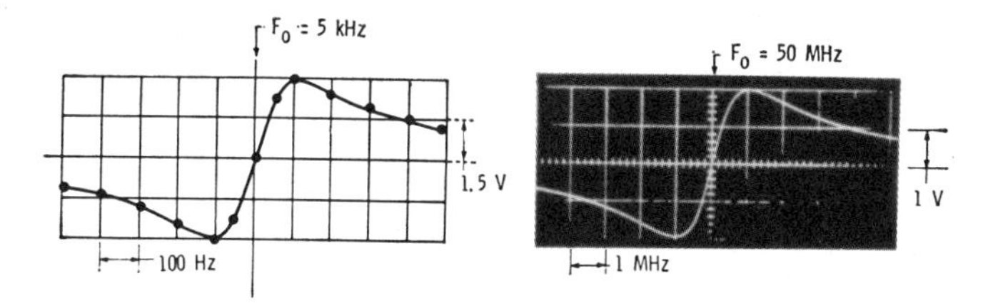

Fig. 9. Experimental transfer characteristic at 5-kHz and 50-MHz center frequencies.

The amount of carrier suppression is a function of the matching accuracy of the transistor differential pairs. By introducing the residual unbalance as an equivalent dc input voltage v and assuming $m = v/2kT/q \ll 1$, the unfiltered output voltage, previously given by (1b), becomes (neglecting quiescent terms):

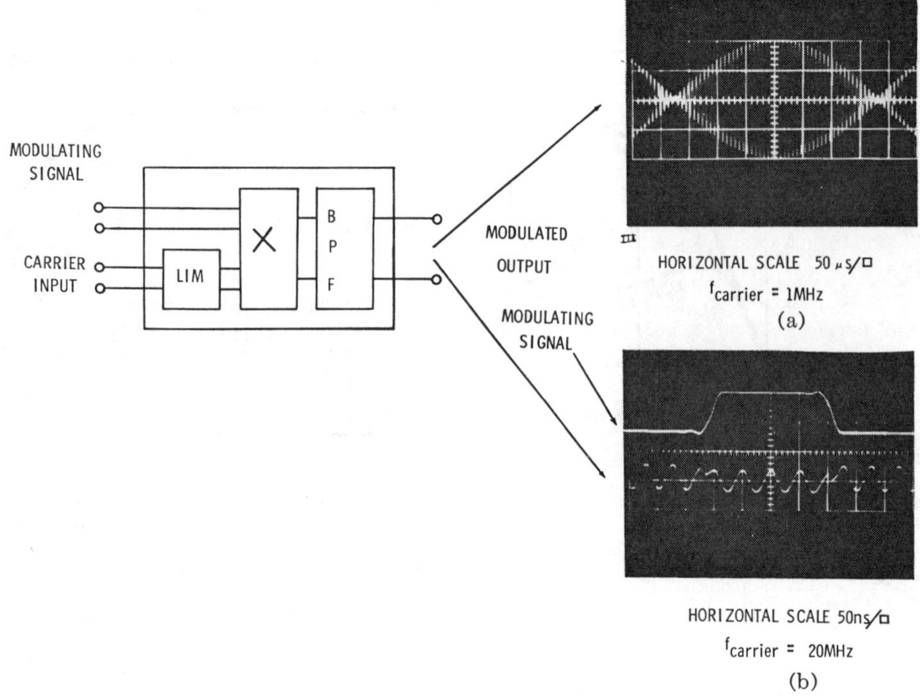

Fig. 10. Suppressed carrier modulation.

$$V_0 = V_R(U_1U_2 + U_1m_2 + U_2m_1) \tag{12}$$

where

$$m_1 = \frac{v_1}{2kT/q} \tag{13a}$$

$$m_2 = \frac{v_2}{2kT/q}. \tag{13b}$$

Expression (12) shows that the degree of cancellation of the signal applied at any input is a function of the unbalanced m at the other input. Therefore, the carrier can be fully suppressed.

C. Synchronous AM Detector

The envelope of an amplitude modulated signal can be recovered by means of a multiplication of the modulated signal with the reference carrier. This process is known as AM synchronous or coherent detection [4]. The analog multiplier can perform such a type of detection when the AM signal to be detected is substituted for the modulating signal shown in Fig. 10 and the bandpass filter is replaced by an appropriate low-pass filter. The multiplication action with the reference carrier will develop the modulating signal for any double or single sideband AM modulation with suppressed or nonsuppressed carrier.

The high-gain limiter included in one channel of the monolithic multiplier can be used to remove the amplitude modulation and provide an unmodulated carrier for the synchronous detection. Fig. 11 shows the waveforms for a 90-percent modulated carrier of 45-MHz and 100-mV signal level, proving the capability of the multiplier to provide distortionless AM detection at low signal levels. With this kind of carrier recovery technique, faithful detection of the envelope will only occur for the case of a double-sideband AM modulated signal and with the maximum amplitude variations of the available carrier kept within the range defined by the two channel limiting thresholds. Furthermore, (7) shows that the detection efficiency is proportional to cos ϕ; therefore, the phaseshift associated with the high-gain limiter at the carrier frequency should be equal or close to $k\pi$ with $k = 0, 1, 2 \cdots$.

D. TV Chroma Demodulator

The last application to be considered is the demodulation of the chroma signal in the color TV receiver. This application is particularly interesting since both AM and phase detection is involved. The basic circuit employed is shown in Fig. 12. Two analog multipliers operating as synchronous detectors (the chroma signal is applied under low-level conditions) recover the blue and red color information from the 3.58-MHz chroma subcarrier. To distinguish the blue information from the red, each detector operates with a switching voltage of a proper phase derived from the transmitted color burst signal.

The chroma subcarrier is simultaneously applied to the bottom transistor pairs where emitter degeneration resistors have been added. This degeneration improves the linearity of the transfer characteristics and allows the adjustment of the relative conversion efficiencies of both detectors. The gain and phase of the synchronous detectors have been chosen to match the chromaticity coordinates of the present sulphide phosphors minimiz-

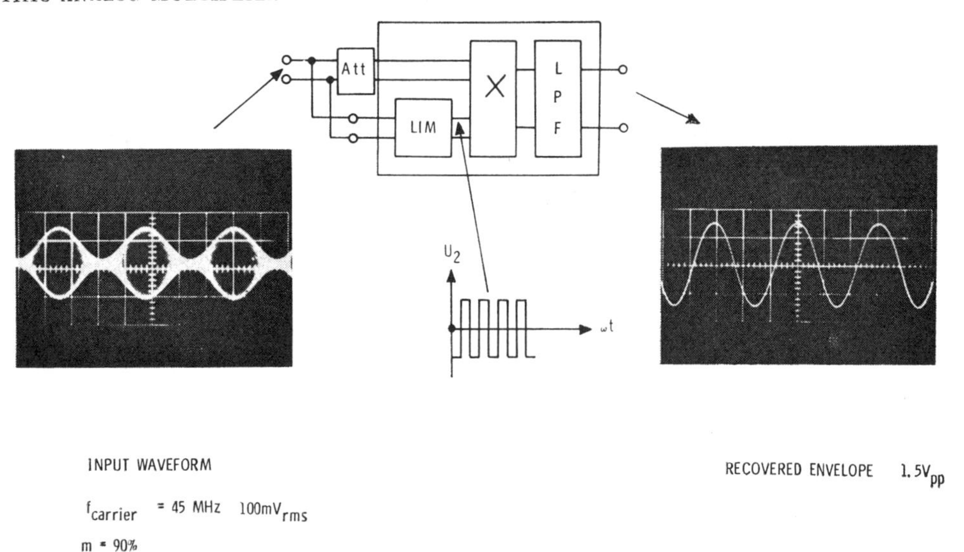

Fig. 11. AM synchronous detection with carrier recovery by symmetrical limiting.

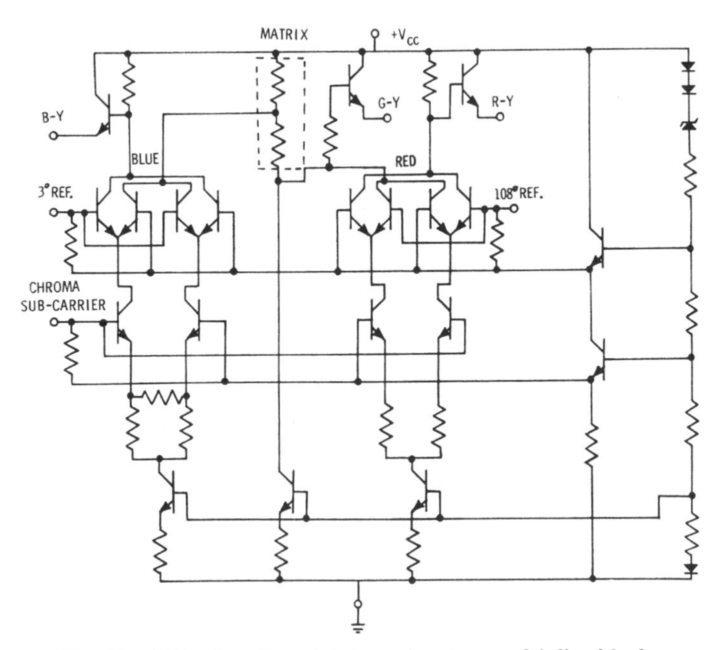

Fig. 12. TV color demodulator using two multiplier blocks.

ing, at the same time, gamma distortion effects as proposed by Parker [5]. The fully balanced arrangement increases the linear swing and cancels the fundamental chroma subcarrier frequency component at the output, simplifying post-detection filtering. Another advantage of the differential approach is the availability of the antiphase red and blue color-difference outputs. The green information is easily derived by matrixing the appropriate amounts of each complementary output as shown.

Fig. 13 shows the amplitude and phase characteristics for the synchronous detector at 3.58 MHz. The amplitude characteristic was obtained by keeping the phase constant at 0 or 180 degrees and varying the chroma amplitude. A 14-volt *p-p* swing is obtained using a 30-volt power supply. The phase characteristics were obtained by using a constant subcarrier amplitude of 200 mV rms and varying the phase from 0 to 180 degrees.

IV. Conclusions

A basic monolithic analog multiplier has been discussed. The component match of integrated circuit fabrication has been exploited to provide a reproducible fully balanced operation. Two basic modes of operation were defined, based on the input signal level. The application of the analog multiplier operating in these two modes was shown to provide FM detection, synchronous AM detection, suppressed carrier modulation, and TV color demodulation. Undoubtedly further work will lead to other areas of application in communication systems, as it appears that the analog multiplier is an example of a linear circuit function that has wide applicability.

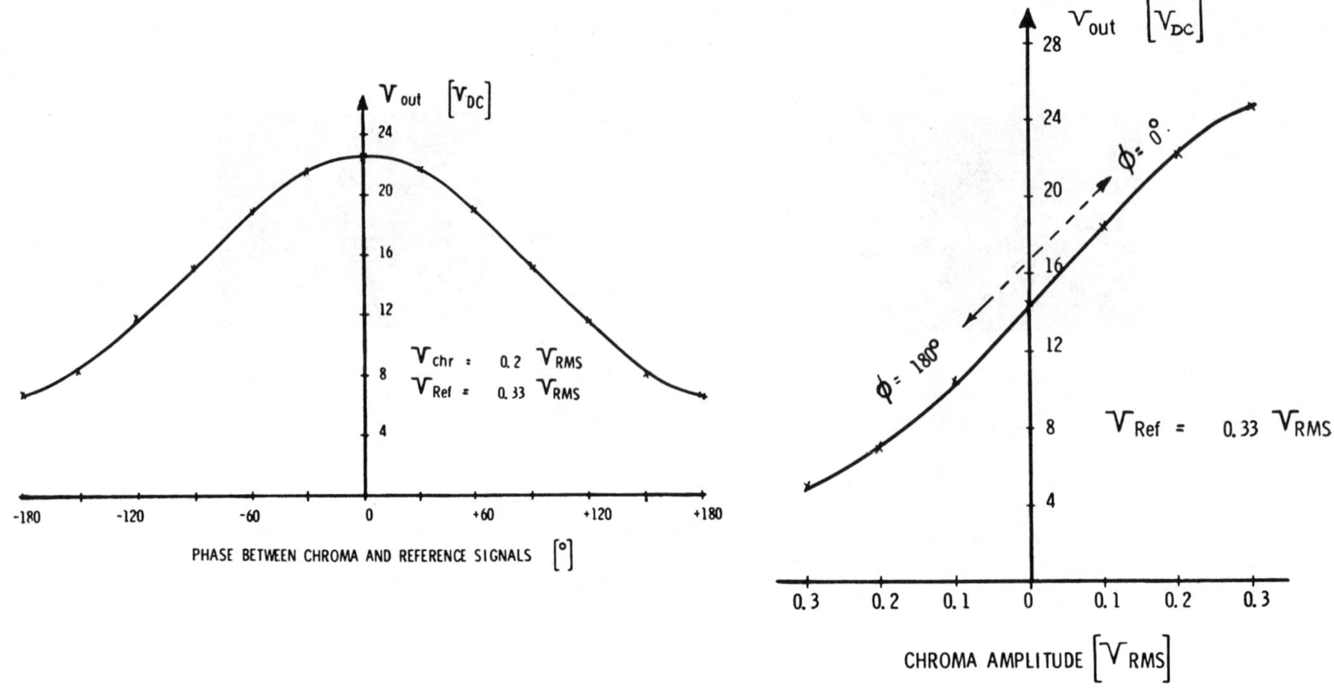

Fig. 13. Phase and amplitude characteristics of the chroma synchronous detectors.

Acknowledgment

The author gratefully acknowledges the valuable suggestions by and discussions with Dr. R. S. Pepper and G. W. Haines of the Sprague Electric Company.

References

[1] A. Bilotti and R. S. Pepper, "A monolithic limiter and balanced discriminator for FM and TV receivers," *1967 Proc. Nat'l Electron. Conf.*, vol. 13, pp. 489–494, October 1967, and *IEEE Trans. Broadcast and Television Receivers,* vol. BTR-13, pp. 54–65, November 1967.

[2] A. Bilotti, "FM detection using a product detector," *Proc. IEEE,* vol. 56, pp. 755–757, April 1968.

[3] Other circuit topologies also provide excellent analog multipliers, e.g., B. Gilbert, "A DC 500-MHz amplifier-multiplier principle," *1968 ISSCC Digest of Technical Papers,* p. 114. It is not the intent of this paper to compare circuits in terms of analog multiplier performance, but to point out the versatility of application provided by the circuit considered.

[4] J. P. Costas, "Synchronous communications," *Proc. IRE,* vol. 44, pp. 1713–1718, December 1956.

[5] N. W. Parker, "An analysis of the necessary decoder corrections for color receiver operation with nonstandard receiver primaries," *IEEE Trans. Broadcast and Television Receivers,* vol. BTR-12, pp. 23–32, April 1966.

A High-Performance Monolithic Multiplier Using Active Feedback

BARRIE GILBERT, SENIOR MEMBER, IEEE

Abstract—Since its conception in 1967, the linearized transconductance multiplier (LTM) has rapidly gained acceptance as the preferred approach to the realization of monolithic analog multipliers, and its simplicity has commended it for use in low-cost modular designs. Accuracies of these units have been limited to about 0.5 to to 2 percent, and drift and noise performance have generally been worse than that possible using the dominant alternative technique of pulse-width-height modulation. This paper shows that when careful attention is given to all the sources of error it is possible to attain a five-fold improvement in accuracy and corresponding reductions in the drift and noise levels. Odd-order nonlinearities can be reduced to negligible magnitudes by the use of active feedback, by substituting the usual resistive-bridge feedback path by an amplifier identical to that used as the input stages.

I. Introduction

THE LINEARIZED transconductance multiplier (LTM) technique[1] is now widely used in one or another of its several forms, the commonest of which are shown in Fig. 1. Using ideal transistors these circuits are fundamentally exact and insensitive to isothermal variations of temperature. Compared to alternative multiplication techniques, the LTM core is extremely simple and has intrinsically wide bandwidth. It is the basis of many other "functional" circuits, such

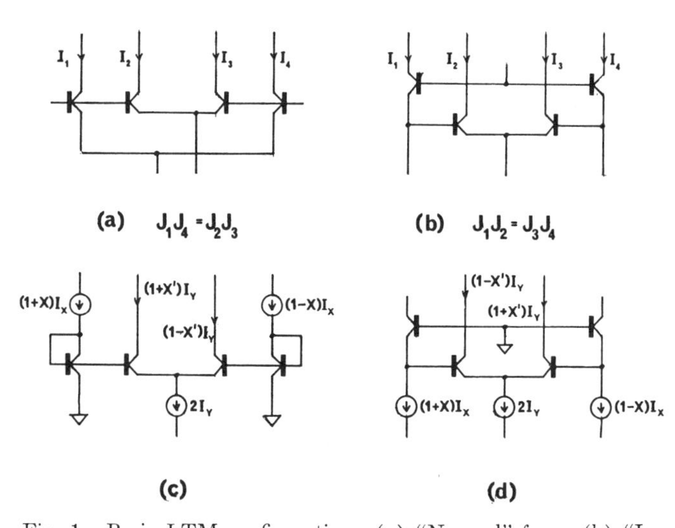

Fig. 1. Basic LTM configurations. (a) "Normal" form. (b) "Inverted" form. (c) Specific version of the normal form having low beta-sensitivity. (d) Common inverted form, which is convenient but has high beta-sensitivity.

as precise two-quadrant dividers, rms converters, vector sum modules, etc.

This paper describes a four-quadrant multiplier designed to make full use of the potential accuracy afforded by this technique using currently available monolithic processes. The basic error sources are reviewed, and methods presented to minimize their effects. A significant improvement has resulted from the use of an active feedback scheme which largely eliminates the nonlinearities introduced by the input amplifiers. Apart from this, the resulting design is similar to most transconductance

Manuscript received May 25, 1974; revised August 11, 1974. This paper was presented at the International Solid-State Circuits Conference, Philadelphia, Pa., February 1974.

The author is with Analog Devices Semiconductor, Wilmington, Mass. 01887.

[1] B. Gilbert, "A new wide-band amplifier technique," *IEEE J. Solid-State Circuits*, vol. SC-3, pp. 353–365, Dec. 1968.

Reprinted from *IEEE J. Solid-State Circuits*, vol. SC-9, pp. 364-373, Dec. 1974.

multipliers. The high performance is a result of the systematic elimination of individual sources of inaccuracy, through careful choice of operating conditions, the use of beta-compensation, and a precise layout. Accuracies of ±0.1 percent over all four quadrants have been measured on the best prototype devices. This has previously only been possible using the classical precision techniques such as pulse modulation.

II. Design Objectives

Completely self-contained monolithic circuits performing multiplication, division, squaring, square-rooting (MDSSR) functions have been available for over three years. They are fabricated using a high-quality bipolar process incorporating on-chip thin-film resistors. Accuracies[2] of better than ±1 percent can be consistently achieved from fully-trimmed units, and selected units have guaranteed accuracies of ±0.5 percent. Recently, laser-trimmed versions of these circuits have become available, relieving the user from the need to apply rather tedious trimming procedures to each unit.

However, this accuracy is usually not maintained over a wide temperature or supply-voltage range. A representative device has an accuracy specification of ±4 percent over the range −55 to +125°C. In fact, even premium-grade modular multipliers using pulse modulation have accuracy-versus-temperature specifications of typically 0.01 percent/°C.

A new design was therefore conceived whose primary objectives were as follows.

1) To retain the familiar MDSSR format (a complete, calibrated system with fully-stabilized reference voltage and a high-gain output op amp with uncommitted output connection) having the standard interface levels of ±10-V FS with ±15-V supplies.

2) Achieve a high production yield to ±0.25 percent accuracy at 25°C after nulling offset and scale errors (using an active-circuit laser-trimming technology).

3) Achieve a reliable yield-fraction to ±0.1 percent trimmed accuracy.

4) Maintain an accuracy of ±1 percent over the entire −55 to +125°C range.

5) Reduce supply-related errors to negligible proportions over the range ±10 to ±20 V.

Thus, the primary objectives were intended to fill the need for a high-accuracy "computing" element. Multipliers are often used, of course, in applications where absolute accuracy is of less importance, and consequently the secondary objectives took into account these other requirements.

6) High linearity. With proper design, errors can be held to very low levels, particularly on the Y input, for which the objective was <0.05 percent nonlinearity.

7) Low noise and drift. This necessitates using as

[2] A discussion of multiplier accuracy specifications is found in *Nonlinear Circuits Handbook,* Analog Devices, Inc., P.O. Box 280, Norwood, Mass., 1974, pp. 209–211.

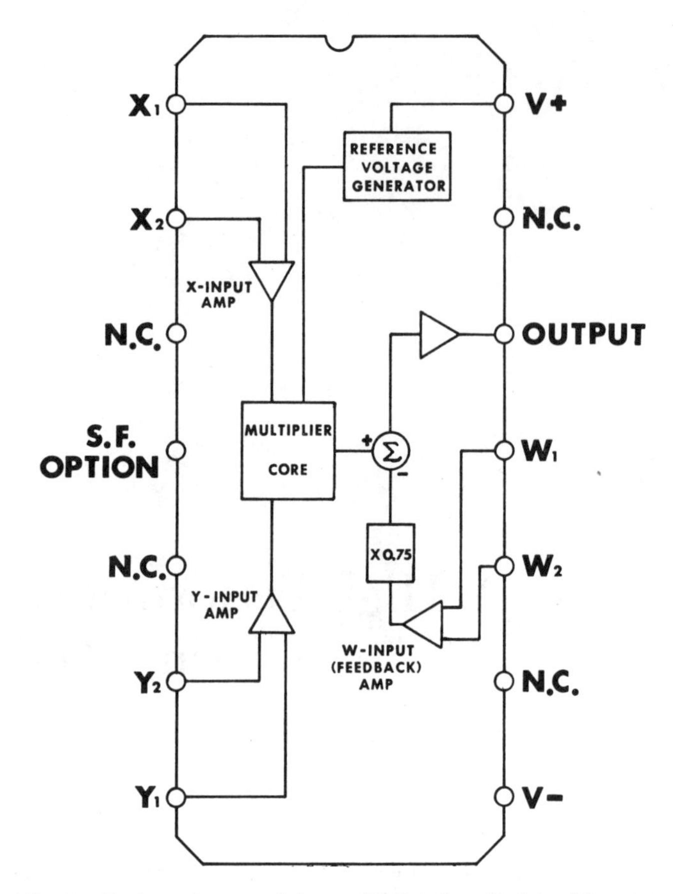

Fig. 2. Package layout of the multiplier described in this paper.

much of the inherent dynamic range of the core as possible, and is thus in conflict with the previous goal.

8) Good dynamics. As a multiplier, the −3 dBs bandwidth should be at least 1 MHz, and the slew rate greater than 25 V/μs. There should be little feedthrough at HF. The inputs should be free from foldover, and the circuit unharmed by large overloads.

The main elements of the complete device are shown in Fig. 2. Note that all inputs, including the feedback to the op-amp summing point, are differential and have high impedance and low bias currents. This allows the device to be used, for example, as a two-quadrant divider with differential numerator and denominator. The high open-loop gain of the op amp (~20 000) is valuable in this application.

III. Overall Design

Fig. 3 is a simplified schematic of the multiplier. The inputs V_x and V_y are converted to differential currents by the emitter-degenerated amplifiers. The products I_xR_x and I_yR_y are both equal to 13.3 V which results in the factors X and Y having peak excursions of ±0.75 at the nominal full-scale input of ±10 V. These peak excursions are larger than usual (a common range for X is ±0.4) and were used to achieve lower effective noise and offsets, and better drift performance. However, this calls for more careful design of the core (see Section IV) and results in greater nonlinearity in the input am-

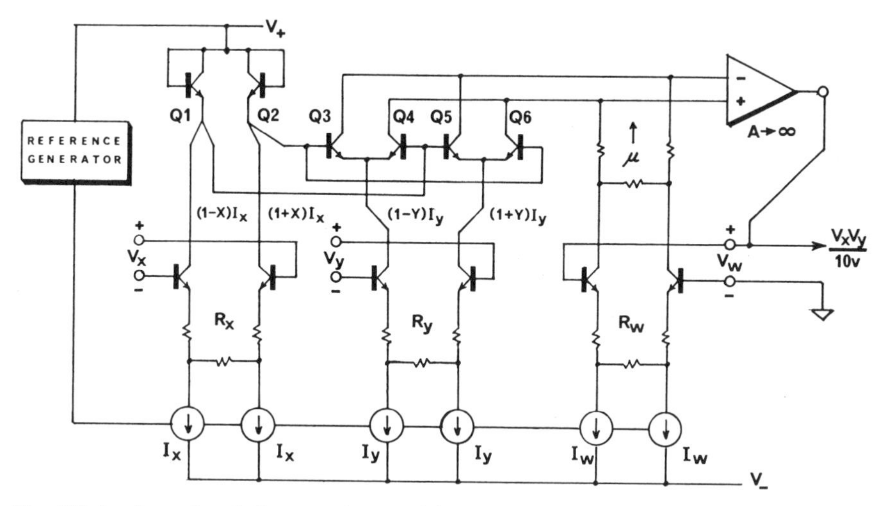

Fig. 3. Simplified schematic of the complete multiplier. Note the addition of an active feedback path.

plifiers. The latter can be substantially reduced by using a third amplifier to introduce the feedback round the op amp, having identical distortion characteristics to the X and Y stages. An analysis is given in Section VI.

The condition for loop balance is

$$\frac{V_x V_y}{I_x R_x R_y} = \mu \frac{V_w}{R_w} \tag{1}$$

where μ is the attenuation factor introduced by the π network in the collector circuit of the feedback stage, and is set equal to the peak-signal value of 0.75 for minimum distortion. $R_x = R_y = R_w = 88.5$ kΩ total (26 kΩ + 36.5 kΩ + 26 kΩ), and $I_x = I_y = I_w = 150$ μA nominal, controlled by the reference generator and trimmed to absorb resistor tolerances for a scaling voltage of exactly 10 V. Thus, in the multiplier connection the device conforms to the standard relationship $V_w = V_x V_y / 10$ V. The connections for dividing, squaring, and square-rooting do not change the scaling voltage.

In addition to reducing the nonlinearities and providing a differential input to the summing-point, the active feedback scheme has the following advantages.

1) The insertion of an alpha in the feedback path provides full compensation for the alpha-error introduced by the Y-input amplifier.

2) The circuit has a very small offset sensitivity to the supply voltages, in the presence of resistor mismatch. By contrast, the usual "bridge" feedback, using resistors direct to the op-amp inputs (which are near V+) has poor supply rejection unless the resistors are carefully matched.

3) The scaling-voltage can be easily changed (for example, to 1 V, giving a ×10 gain compared to the usual connection) by inserting an external attenuator in the feedback path.

4) The normally grounded $-V_w$ connection can optionally be used to sum in an additional signal (up to ±10 V) to the output.

Special care was taken to minimize errors due to variations of transistor current gain, which are particularly troublesome in the inverted form of the LTM core [Fig. 1(d)]. Details of the compensation method are contained in Section V.

The output op amp is conventional and will not be described here. It uses a lateral p-n-p input stage for level-shifting; no feedforward was used, resulting in a low unity-gain frequency point. However, the low transconductance of the first stage (which may be any one of the three emitter-degenerated amplifiers) allows a high slew rate to be achieved.

IV. Errors in the Multiplier Core

An analysis of the error sources has been previously given.[3] The most important of these is undoubtedly V_{BE} matching, and this is dealt with at further length here, because it is the key to a successful implementation. The errors caused by a departure from the ideal case of isothermal nonresistive infinite-beta transistors are reviewed briefly.

V_{BE} (Emitter-Area) Mismatch

The large-signal linearity[4] of multiples based on two-quadrant elements such as Fig. 1(c) and (d) is critically dependent on V_{BE} matching. It is possible to devise transconductance multipliers in which this is not the case, or alternatively, to introduce linearity-correction adjustments. These approaches are usually more complex; some have inferior dynamics, or introduce additional noise or nonlinear terms; in some cases, they require more complex trimming procedures. Therefore the simple six-transistor core was retained, and several precautions taken to minimize mismatch.

It is convenient to assume that all V_{BE} errors are due

[3] B. Gilbert, "A precise four-quadrant multiplier with subnanosecond response," *IEEE J. Solid-State Circuits*, vol. SC-3, pp. 365–373, Dec. 1968.

[4] Strictly speaking, a multiplier is a nonlinear circuit, but the generally-accepted meaning of linearity in connection with multipliers is that measured when one input is held at a fixed value.

to differences in emitter area. Other sources, such as local variations in doping, dislocations, etc., can be converted into an equivalent area factor. This permits the definition of area-mismatch factors, which are easier to incorporate into analyses and have the advantage of being temperature-invariant. Thus, referring to Fig. **3**, the emitter areas of Q_1 through Q_6 are designated A_1 through A_6.

It is immediately apparent that there are two quite independent mismatch factors to consider. When V_y is very negative transistors Q_5 and Q_6 are inoperative, and consequently cannot contribute to the output. The area mismatch factor of the quad Q_1-Q_2-Q_3-Q_4 is defined as

$$a_1 = \frac{1}{4}\left(1 - \frac{A_1 A_4}{A_2 A_3}\right). \qquad (2)$$

This is a small number, usually within the range ± 0.005. The equivalent V_{BE} error is given by

$$\Delta_1 V_{BE} = (V_{BE1} + V_{BE4} - V_{BE2} - V_{BE3}) \approx \frac{4kT}{q} a_1, \qquad \text{for } a_1 < 0.01. \qquad (3)$$

Likewise, the quad Q_1-Q_2-Q_6-Q_5 has an area mismatch factor

$$a_2 = \frac{1}{4}\left(1 - \frac{A_1 A_5}{A_2 A_6}\right) \qquad (4)$$

and

$$\Delta_2 V_{BE} \approx \frac{4kT}{q} a_2. \qquad (5)$$

Starting with the fundamental current-density equation shown in Fig. 1(b), the output of the core is found to be

$$Z = \frac{1}{X_{fs} Y_{fs}} [(X + a_1 + a_2)Y + (a_2 - a_1) - \{Y(a_1 + a_2) + (a_1 - a_2)\}X^2] \qquad (6)$$

where Z is the normalized output, full scale $= \pm 1$, X_{fs} is the chosen value for X at nominal FS input, and Y_{fs} is the chosen value for Y at nominal FS input. It is not essential for X_{fs} and Y_{fs} to be equal (they are often not) but in this design both are 0.75.

Equation (6) shows that in addition to the product term XY there is: 1) an X offset of $(a_1 + a_2)$; 2) an output offset of $(a_2 - a_1)$; 3) a parabolic error on the X input, which varies with the sign and magnitude of Y; 4) no explicit Y offset; 5) no Y nonlinearity.

Closer inspection reveals that the magnitude of the parabolic error is proportional to the design value X_{fs} and the offsets and noise (the latter not included specifically) are inversely proportional to X_{fs}.

Eliminating the offsets in (6) by putting $Z_0 = (a_2 - a_1)$ and $X_0 = -(a_1 + a_2)$, we can examine the residual nonlinearity. The squared term becomes $(X + X_0)^2$, and this has the effect of inducing small Y-offset and scale errors, but these are also trimmable. The remaining error is

$$\epsilon(X, Y) = \frac{X^2}{X_{fs} Y_{fs}} [(a_2 - a_1) - Y(a_1 + a_2)]. \qquad (7)$$

The "X feedthrough" of the fully trimmed device is simply the residual output when $Y = 0$ and X is varied from $-X_{fs}$ to $+X_{fs}$. Its peak value is therefore

$$\epsilon(X_{fs}, 0) = a_2 - a_1 \qquad \text{relative to a FS output of } \pm 1. \qquad (8)$$

Two points are of interest here. First, the X feedthrough can be zero, even though distortion is present for finite values of Y. Second, X feedthrough is not affected by the choice of X_{fs} and Y_{fs}.

When Y is at its FS value (± 0.75) the error is

$$\epsilon(X_{fs}, \pm Y_{fs}) = (a_2 - a_1) \mp 0.75\,(a_1 + a_2). \qquad (9)$$

The area mismatch factors can be of equal or opposite sign. In the first case, let $a_1 = a_2 = a$. Then the peak errors are $-1.5a$ for +FS and $+1.5a$ for −FS and the error is completely proportional to Y. In the second case, let $-a_1 = a_2 = a$. Then the error is a constant $2a$, independent of Y. This is clearly a worst case, and for ± 0.1 percent error, $a = 0.0005$, which in V_{BE} terms amounts to a mismatch of only 52 μV at 25°C.

It is worth stressing here that the existence of *two* independent mismatch factors prevents the use of a single adjustment to eliminate all nonlinearities. Some designs introduce a small voltage between the bases of Q_1 and Q_2 to eliminate X offset, but this trimming technique is dangerous, because it can actually introduce parabolic distortion when (as is often the case) all or some of this offset is caused by unbalances in the current-sources and X amplifier, requiring the core to be deliberately mismatched. Also, unless the correction voltage is proportional to temperature, the trim will introduce a drift term. Note that this technique also cannot change the parabolic feedthrough at $Y = 0$, and results in an X nonlinearity which is independent of the value of Y.

In the absence of adjustments, some means to ensure inherently low V_{BE} errors is sought. Quad layouts are sometimes used for this purpose.[5] At first sight, though, this solution would appear to require a total of twelve transistors, raising formidable layout difficulties. But (2) and (4) show a way to avoid this. Note that linearity is preserved so long as $A_1A_4 = A_2A_3$ and $A_1A_5 = A_2A_6$. So it is not necessary for all the pairs to match. Indeed, it is demonstrably possible to obtain good linearity even when one device of each pair (e.g., Q_3-Q_4) is twice that of the other. So, by arranging the transistors in two overlapping quads, sharing the inner pair Q_1-Q_2 and cross-connecting to the bases of the outer pairs, quite large processing gradients can be tolerated. This "six-pack" layout for the core is clearly visible in the

[5] M. A. Maidique, "A high-precision monolithic super-beta operational amplifier," *IEEE J. Solid-State Circuits,* vol. SC-7, pp. 480–487, Dec. 1972.

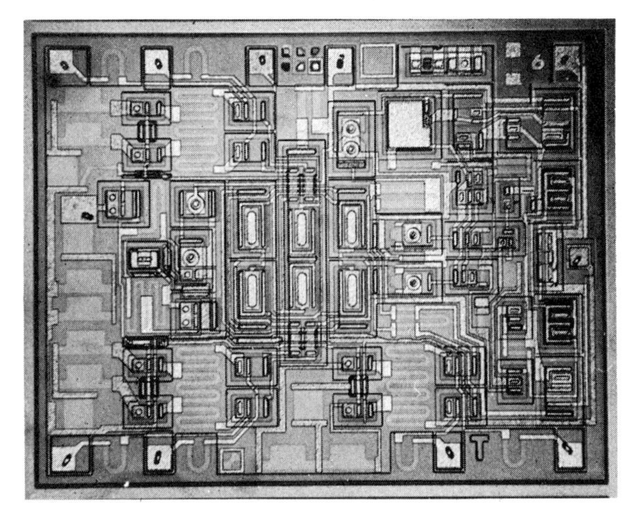

Fig. 4. Picture of the chip. The "six-pack" multiplier core is at the center, and the four output transistors of the op amp are on the far right.

picture of the chip, Fig. 4, which also shows the transistor geometry.

Optimum transistor design for this type of multiplier is almost totally a matter of preventing processing defects from disturbing the V_{BE} matching. As will be seen later, the devices do not need to have the very low ohmic resistances usually required of a logarithmic device. The large emitter areas used here are for the purpose of reducing the effects of fine-structure processing grain, such as poorly-etched edges; they have an area of 7100 μm^2 and operate at a peak current-density of 4.2 $A \cdot cm^{-2}$, at which level the intrinsic transistor is near ideal. In preparing the ×500 artwork, only one emitter outline was cut, then photorepeated six times and patched into the rubyliths.

A further possibility for improving V_{BE} matching is the use of a shallower emitter than usual, yielding betas with median values of about 200, rather than the 400–500 more typically achieved. This has the further advantage of improving the match of base resistance, and lowering the output conductance. The latter is a subtle source of parabolic nonlinearity in the input amplifiers, arising from variations of the current I_x as the input voltage varies through its full range of 20 V, when the multiplier is driven from a single-sided signal. Low-beta processing also increases the BV_{CEO} of those transistors operating at the full supply voltage, thus ensuring a good yield of devices to a high-voltage specification. The provision of full beta-compensation allows the use of low-beta processing without jeopardizing multiplier accuracy.

Thermal Gradients

The second most serious threat to accuracy is a difference of temperature between transistors in the core. Two types of errors can arise. The first is that due to a thermal gradient across the individual pairs Q_1-Q_2, Q_3-Q_4, Q_5-Q_6. This causes input offset errors of 400 mV/°C relative to the FS of ±10 V. Put another way, only 0.025°C can be tolerated between a single pair to maintain an error of ±0.1 percent of FS. The second kind of error is less serious. A gradient which causes the pairs Q_3-Q_4 and Q_5-Q_6 to be at a different net temperature to the pair Q_1-Q_2 will introduce a cubic nonlinearity on the X input whose magnitude becomes ±0.1 percent peak for a difference of less than 0.5°C. Thermal offsets do not cause Y nonlinearity.

The six pack minimizes the errors introduced by thermal gradients. If these are linear, offset errors for any vector direction exactly cancel (due to the cross connection). Linear gradients along the major axis introduce cubic errors into the two quads, but the low sensitivity prevents these being significant, and these cancel anyway near $Y = 0$. The low thermal sensitivity of the six pack is one of the improvements in chip design. Additionally, the core was located on what is effectively a thermal plateau at the center of the chip. The power dissipation of the surrounding circuits was arranged to be evenly distributed. The output transistors of the op amp were split into four separate devices and placed along the far edge of the chip, so that changes in load current have only a small effect on the thermal "wavefront."

Resistive Mismatch

It was shown in footnote 3 that it is not necessary for the transistors used in the core to have low resistances, as would be the case in many logarithmic circuits. It is interesting to note that this is an advantage of the "two-quadrant" forms, Fig. 1(c) and (d). By contrast, using the general forms of Fig. 1(a) and (b) in "one-quadrant" multiplier implementations (as in the "computing" rms converter, for example) small ohmic resistances introduce dramatic errors, whereas V_{BE} matching errors can be tolerated, as they affect only the scale factor.

While high ohmic resistances can be tolerated, these must match reasonably well, and since much of this resistance arises in the base circuit, this also implies that good beta-matching is required. Mismatch introduces cubic errors on the X input very similar in form to those caused by thermal gradients, in the case where the pairs Q_3-Q_4 and/or Q_5-Q_6 have a different ohmic resistance to the pair Q_1-Q_2. When the mismatch is between individual devices in the pairs, an input offset results. The latter is the more significant, for a 0.1 percent input error is caused by only 0.33 Ω of mismatch. The large transistors used in the six pack have inherently low resistance, ensured by the long emitter geometry and wrap-around base contact. Low-beta processing further improves the match.

Metalization resistances are often a cause of error. Typically, the sheet resistance of the aluminum is 50 mΩ per square, and the interconections must be carefully balanced. Note the way the emitters of Q_1 and Q_2

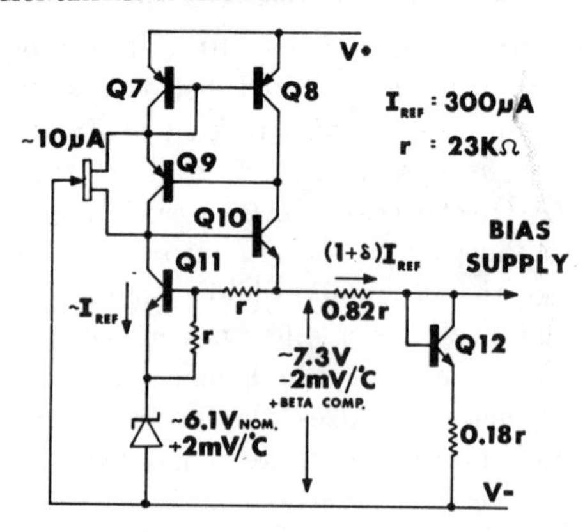

Fig. 5. Reference generator circuit. Q_{12} is part of the current-source circuits.

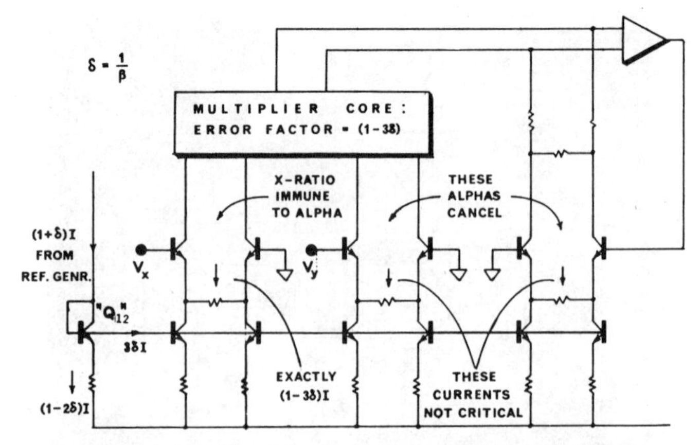

Fig. 6. Details of the beta-compensation technique.

are connected, and the dummy lengths of metal inserted in the emitter leads of Q_3-Q_4 and Q_5-Q_6.

Base Currents

Base currents in Q_3 through Q_6 are additive to the currents $(I + X)I_x$ and $(I - X)I_x$. This has the effect of increasing the scaling voltage, and the effect is not negligible. It is shown in the Appendix that the transfer function is multiplier by a factor $(1 - 3\delta)$, where $\delta = 1/\beta$. Assuming a worst-case beta of 100, this is a -3 percent scale error. This might be tolerable, but beta is not stable with temperature, and typically increases by 1 percent per °C, thus causing a scale-factor drift of $+0.03$ percent per °C max. This is clearly not compatible with a 1 percent accuracy over a 180°C range. Fortunately, use can be made of the dependence of scale factor on I_x to introduce an identical $(1 - 3\delta)$ factor in the denominator of the transfer function, and thus achieve close compensation. This will be discussed in the next section.

V. Scale-Factor Stability

The output of a two-signal multiplier having voltage inputs must be scaled by a denominator having the dimension of voltage. A self-contained unit must therefore include a voltage-reference element of adequate stability. Reverse-biased emitter-base junctions are often used, but these generate significant amounts of noise unless driven at high current densities, which appear as modulation noise at the output. They also are subject to long-term drift. This design uses the breakdown between the emitter diffusion and the base-enriched isolation diffusion, resulting in a planar bulk junction. The voltage is referenced to the substrate, resulting in a very simple structure. The breakdown voltage is nominally 6.1 V and the temperature coefficient (TC) approximately +2 mV/°C.

Fig. 5 shows the complete reference generator, which includes compensation for the TC's of the reference diode and the diode Q_{12} which forms part of the current sources for the multiplier. The resistors in the base circuit of Q_{11} multiply its V_{BE}. They are thin-film resistors, and may be adjusted for exact temperature compensation by changes to the metalization; as shown, they are correct for a breakdown TC of +2 mV/°C, but the range of adjustment can absorb any TC from zero to about +5 mV/°C. The bias current I_{REF} is supplied by Q_{10} and the collector current of this transistor is inverted by the Wilson current-mirror Q_7-Q_8-Q_9. Consequently, the bias current for the reference diode is self-stabilizing, resulting in a well-defined, supply-insensitive operating point. A small epi-FET is included to ensure start-up; its channel current is uncritical, and may range from a few nanoamps to about 100 μA. The finite channel conductance has very little effect on supply rejection, due to its location across Q_9, which has the effect of multiplying the slope resistance of the FET by the p-n-p beta. Typical devices show a variation of only 0.4 mV in the voltage at the emitter of Q_{12} for a supply voltage variation of 20 V. This corresponds to a scale factor variation of less than ±0.02 percent over the supply-voltage range ±10 to ±20 V.

Beta Compensation

Exact compensation for the errors caused by finite current-gain in all the signal-path transistors requires a careful assessment of their interactions. Fig. 6 shows the overall scheme in a way that highlights these errors. First note that bias currents in the Y and W amplifiers do not have any direct effect on scale factor; they are made equal to the X-amplifier bias current in order to minimize cubic distortion terms. Second, the alpha errors in the forward path through the Y amplifier are cancelled by those in the feedback path through the W amplifier. Third, it is important to realize that the alphas of the X amplifier transistors do not appear as scale-factor errors. This is because the X input of the multiplier core is *ratio dependent,* and the ratio is determined in the *emitter* circuit of the X amplifier and preserved exactly in the collector circuit even when alpha is low. A severe alpha mismatch can introduce parabolic distortion on this input, but the magnitude is small; a 10 percent mismatch in a beta of 100 causes less than 0.1 percent distortion.

These considerations show that only the basic $(1 - 3\delta)$ error of the core remains. If the bias current to the X amplifier is multiplied by the same factor, compensation is achieved. Now, the six current sources supply a total of 900 μA, and the bias current to Q_{12} is nominally 300 μA. Consideration of the current terms shows this amounts to an error of $(1 - 4\delta)$. To avoid raising the input bias to 450 μA and get the right factor this way, an additional compensation term is added to the input bias current. This is the reason for choosing the value of resistor in the base of Q_{11} to be equal to the sum of the bias resistors, because the base current of Q_{11} is $\delta\ I_{REF}$, and this raises the bias current to $(1 + \delta)I_{REF}$. For the compensation to be perfect, Q_{11} ought really to operate at exactly the same current-density as the transistors in the multiplier core; as a compromise, the emitter area of this transistor was raised to 2600 μm^2. Note that the reference diode actually operates at $I_{REF} + V_{BE}/r$, which is not temperature-stable. However, the TC of the breakdown voltage is not critically dependent on the current density, and a slight shift in the TC due to this effect is absorbed by the initial TC adjustment.

VI. Input-Amplifier Distortion

Simple emitter-degenerated voltage-to-current converters introduce odd-order distortion, due to the current-dependence of the two V_{BE}'s. Incidentally, this so-called "S-shaped" distortion is often incorrectly blamed on the transconductance core and described as "typical Y-input distortion." The fact is, it occurs on both X and Y inputs, but in the former case it is usually masked by other nonlinearities, or is less because of the smaller current swings used in the X amplifier.

While it is not possible to give a closed-form solution for the input-stage distortion, it is easy to relate the input voltage required to set up a particular current ratio, and analyze the distortion with this as the dependent variable. Referring to Fig. 7, which is a simplified form of the multiplier in which the core has been replaced by a variable attenuator, we can write the inputs as

$$V_y = YIR + \frac{kT}{q} \ln \frac{1 + Y}{1 - Y} \tag{10}$$

and

$$V_w = WIR + \frac{kT}{q} \ln \frac{1 + W}{1 + W} \tag{11}$$

In each case, the first term represents the linear relationship and the second term is the nonlinear voltage introduced by the difference of V_{BE}. Not all of this second term is nonlinear, however, for it includes the small signal r_e (the value when $Y = W = 0$), which is $2kT/qI$ in series with the large resistors R. Rewriting (10), we can separate the linear and nonlinear terms as

$$V_y = YI\left(R + \frac{2kT}{qI}\right) + \frac{kT}{q}\left(\ln \frac{1 + Y}{1 - Y} - 2Y\right) \tag{12}$$

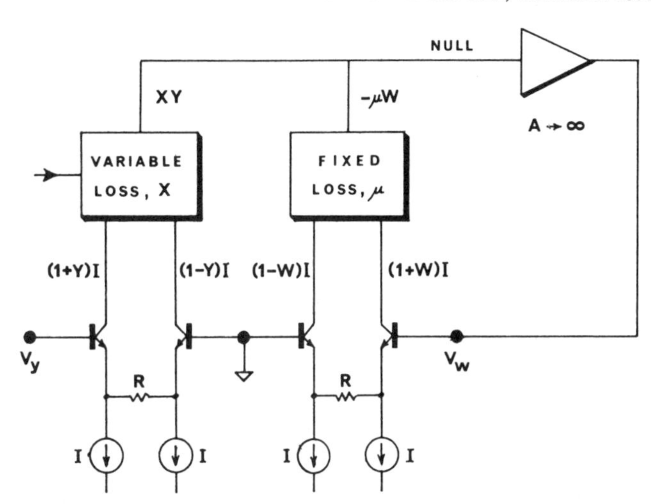

Fig. 7. Basic active feedback scheme. The "variable loss" element is the multiplier core, and the variable X corresponds to that input of the multiplier.

and a corresponding expression can be written for V_w. Inserting the peak values used for Y and W (± 0.75), the nonlinear component is found to be ± 11.6 mV at 25°C, or about 0.12 percent of FS.

It is intuitively obvious that if the fixed loss μ and the variable loss X are equal, the nonlinearities will exactly cancel. This has been experimentally confirmed, and nonlinearities of less than 0.001 percent of FS measured, using purely resistive elements to introduce the variable loss (or with both attenuators removed). This suggests the possibility of making a very precise unity-gain amplifier with differential input and output, having independent common-mode levels.

We must now determine how effective the cancellation is when X and μ are not equal, by deriving expressions for the distortion between the Y input and the W input with X as a parameter. The system is completely symmetrical in this respect, however, and the distortion from the X input with Y as a parameter merely requires a change of labels.

The input to the op amp is nulled when $W = XY/\mu$, so this value of W can be substituted in the equation for V_w to find the actual output voltage. The exact output voltage in the absence of distortion would be XV_y/μ. The error voltage is therefore

$$\epsilon(X, Y) = (\text{actual value of } V_w) - (\text{exact value of } V_w)$$

and making the substitution $X/\mu = K$ (where $K = 1$ at FS), and inserting these variables into (10) and (11), the linear terms cancel to leave

$$\epsilon(K, Y) = \frac{kT}{q} \ln \left(\frac{1 + KY}{1 - KY}\right)\left(\frac{1 - Y}{1 + Y}\right)^K. \tag{13}$$

This expression gives the absolute error as a function of the internal variable Y (which is very nearly linear with respect to V_y). By equating the derivative of $\epsilon(K, Y)$ to zero, the value of K at which the error is greatest is found to be

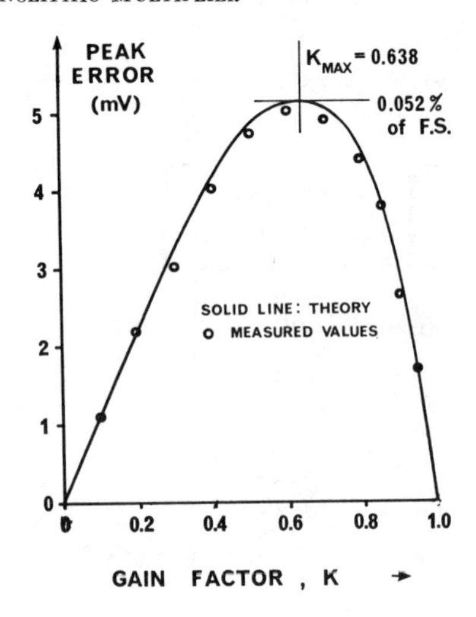

Fig. 8. Theoretical and measured peak error of the active feedback scheme.

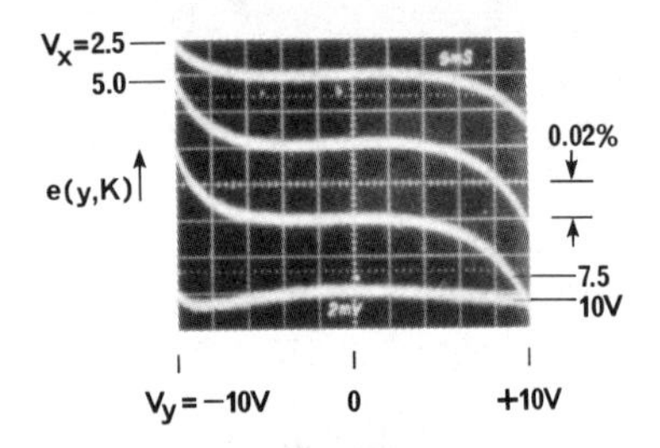

Fig. 9. Residual Y nonlinearity for four values of X_x. Multiple-exposure with vertical shift between exposures. Approximately 0.007 percent parabolic distortion is visible at $V_x = 10$ V, possibly due to a small beta mismatch.

$$K = \left(\frac{1}{Y^2} - \frac{1}{Y \tanh^{-1} Y}\right)^{1/2} \tag{14}$$

Using the FS value of 0.75 for Y yields $K = 0.638$, that is, $V_x = \pm 6.38$ V. With this value for K put back into (13) the maximum absolute error is found to be −5.18 mV; in other words, the output is never more than 0.052 percent of FS below the theoretical value. The nonlinearity is actually about half this because some of the error can be absorbed by a gain trim.

Fig. 8 shows the peak absolute error at $V_y = \pm 10$ V as a function of K. These results were obtained using a prototype monolithic device; the form of the error is shown in Fig. 9, for four values of V_x. Of course, when $V_x = 0$ there is no residual error. Note that the residual error at $V_x = 10$ V, amounting to about ±0.005 percent, is the *total* error for the complete device including the multiplier core and op amp. This validates the claim that there is inherently no distortion in the Y channel of the intrinsic transconductance multiplier.

VII. Performance

The performance of early prototype chips verifies the effectiveness of the error-reducing design. Most notable is the "theoretical" appearance of the residual errors when displayed on an oscilloscope. With the X input set to FS (±10 V) the Y nonlinearity is within ±0.01 percent for most samples; with the Y input set to ±10 V, the only visible errors are the residual parabolic terms. The available data indicate that the six pack layout is yielding better matching, some devices meeting the full test standards for a 0.1 percent multiplier. Further statistical data are needed to properly appraise this. Fig. 10 shows some representative cross plots; notice that the residual linear errors in Fig. 10(b) and (d) are the result of the parabolic X error, and conform to the predictions of (7).

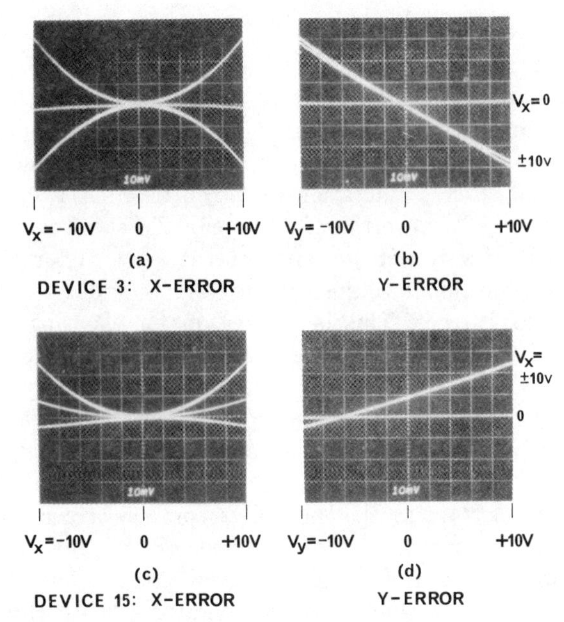

Fig. 10. Error cross-plots for two fully-trimmed units. Multiple-exposures with no shift between exposures. Device 3 has area-mismatch factors which are closely equal in sign and magnitude, hence cancel for $V_y = 0$. Device 15 shows one mismatch factor which is nearly zero (at $V_y = -10$ V).

The temperature dependence of scale factor has been measured; the results are shown in Fig. 11. The downward curvature at high temperature is believed to be due to a parasitic MOS effect, which has since been rectified. The untrimmed scaling-voltage spreads from approximately 9.5 to 11 V, and the measured TC over the range 0 to +70°C is fairly independent of the voltage, being about +75 ±25 ppm/°C with the initial compensation ratios.

The measured noise referred to the output is approximately 400 μV (rms) over 20 Hz to 250 kHz, and was found to be substantially independent of signal level. This suggests that the reference supply is noise free, and this was confirmed by measuring the open-circuit voltage at pin 4, where the reference voltage is about 1.2 V and the noise level is below 10 μV (rms).

Considerable improvement in the feedthrough of signals at HF is seen. Fig. 12 shows the swept-frequency response with one input being nulled and the other a 20 V (peak to peak) sine wave from about 10 to 500 kHz. The HF feedthrough is usually lower on the X input

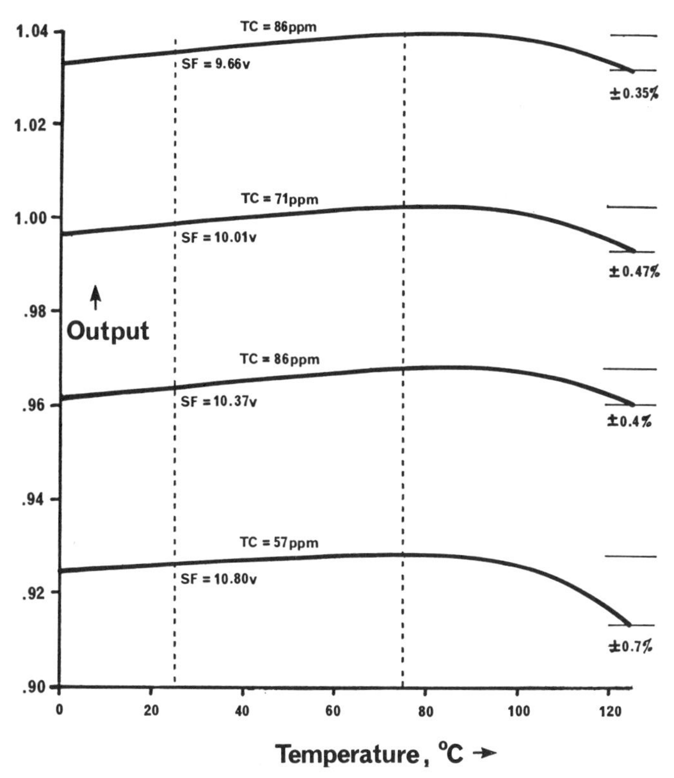

Fig. 11. Variation of scale-factor with temperature. Conditions were $V_x = +10$ V, $V_y = 1$ V (rms), $V_s = \pm15$ V. No trimming was applied, so some error may be due to drift of the X input.

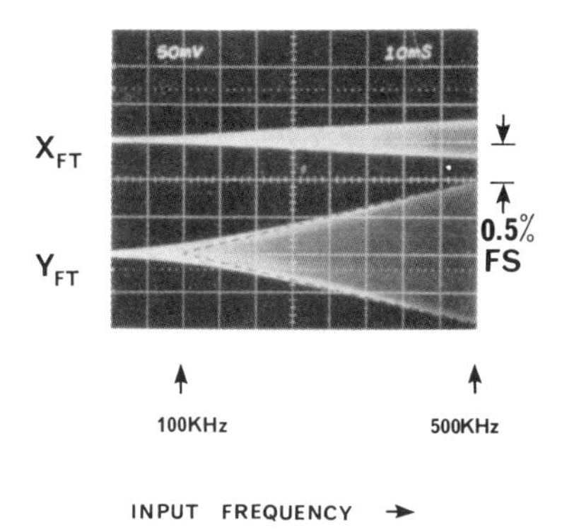

Fig. 12. Typical HF feedthrough; for one input nulled, the driven with 20 V (peak to peak).

TABLE I

PERFORMANCE (MULTIPLIER CONNECTION, $V_S = \pm15$ V, 25°C)

Overall accuracy	±0.25 percent of FS
Scale-factor temperature coefficient	+75 ppm/°C
Scale-factor supply sensitivity	<0.002 percent/V
Output noise (signal-insensitive)	
20 Hz–250 kHz bandwidth	400 μV(rms)
Small-signal bandwidth	1.2 MHz
Slew rate	30 V/μs
X nonlinearity	
($Y = +10$ V, $X = -10$ to $+10$ V)	0.15 percent of FS
Y nonlinearity	
($X = +10$ V, $Y = -10$ to $+10$ V)	<0.01 percent of FS
Input offset current	
(all input pairs)	40 nA
Input offset voltage drift	<100 μV/°C

of a transconductance multiplier, as is the case here. This is due to the cancellation of currents in the collector-base capacitances of the transistors Q_3 through Q_6 in the multiplier core.

Table I lists other key performance parameters.

VIII. CONCLUSION

It has been demonstrated that the linear transconductance multiplier is capable of accuracy specifications considerably higher than generally accepted, and the gap between this technique and the most important alternative high-accuracy technique of pulse-width-height modulation is narrowing. The wider bandwidth of the LTM, the absence of need for any external components, and the freedom from residual carrier components on the output will continue to be its main advantage. Future improvements in precision bipolar technology will almost certainly result in accuracies of 0.1 percent becoming commonplace.

When used as variable-gain elements, four-quadrant multipliers are typically limited to a control range of 20:1, because of the detrimental effect of drift. This design has reduced drift by a factor of approximately five, providing an extension of the control range to at least 100:1, and reduced feedthrough to about −60 dBs at 100 kHz. However, specially designed two-quadrant multiplier/divider units are better suited to variable-gain applications.

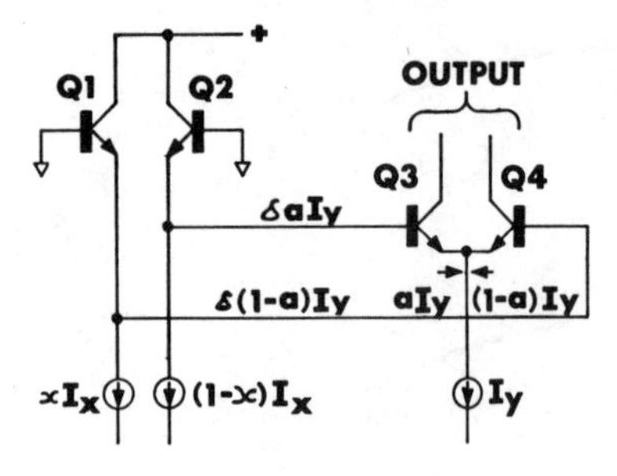

Fig. 13. Basic core for analysis of beta errors.

The one remaining fundamental drawback of the LTM technique is the intrinsically high noise level of the basic circuit. By using much more of the available dynamic range this design has achieved a reduction by a factor of about 3, to a level of −85 dBs referred to FS output. For applications where noise is critical, multipliers based on carrier-domain principles are promising.[6]

Appendix

Beta Errors in Core

Fig. 13 shows a "half-multiplier" for analysis; the results are easily extensible to the full circuit. The fundamental relationship is

$$J_1 J_2 = J_3 J_4. \tag{A1}$$

[6] B. Gilbert, "New planar distributed devices based on a domain principle," in *IEEE Int. Solid-State Circuits Conf. Tech. Dig.*, 1971, p. 166.

Assuming equal emitter areas, and noting that the emitter currents of Q_1 and Q_4 are increased by the base currents of Q_2 and Q_3, we can write

$$[xI_x + \delta(1-a)I_y][(1-a)I_y] = [aI_y][(1-x)I_x + \delta a I_y] \tag{A2}$$

where δ is the base-defect term, I_B/I_E.

This yields

$$a = \frac{x+\delta}{1+2\delta}. \tag{A3}$$

For $x = 0.5$ (input balanced), $a = 0.5$ (output is balanced) for any δ; that is, base currents do not cause offsets, unless the base-defect terms are substantially different. So, $da/dx = (1+2\delta)^{-1}$, or (since $\delta \ll 1$), $(1-2\delta)$.

Since the output is from the *collectors* of Q_2 and Q_3, an additional $(1-\delta)$ term is introduced, yielding an overall gain error of $(1-2\delta)(1-\delta) \approx (1-3\delta)$.

Acknowledgment

The author would like to thank A. P. Brokaw and M. A. Maidique for their valuable suggestions during the development of this circuit.

A Two-Quadrant Analog Multiplier Integrated Circuit

JAMES G. HOLT

Abstract—**A pulsewidth-amplitude-modulation (PWAM) approach is used in an integrated-circuit design of a two-quadrant analog multiplier. PWAM multiplier designs are available in hybrid form at a higher cost than single-chip transconductance multipliers. This higher cost is often justified in applications requiring higher accuracies and stabilities than those offered by transconductance multipliers. The circuit described is a single-chip PWAM multiplier of comparable cost to monolithic transconductance multipliers. It requires an external capacitor and a low-pass filter and achieves accuracies better than 0.5 percent without trimming. Temperature stability is typically 0.005 percent per degree Celsius.**

Introduction

THE INCREASED safety and environmental protection required for automobiles of the future will be achieved with the aid of sophisticated electronic systems. Considerable recent effort has been devoted to design of first-generation versions of monolithic linear integrated circuits for such automotive systems. Much of the work has been on design of custom special-purpose circuits. Design of more widely applicable building blocks that can be easily used for a variety of differing automotive systems has also been reported in [1], [2].

The two-quadrant analog multiplier discussed in this paper is intended for a specific automotive fuel injection system requiring good multiplication accuracy at moderate bandwidth. The inherent circuit flexibility extends its range of applications to many other automotive and industrial areas as well. The fuel injection system requirements for 1-percent multiplication accuracy without trimming, a common-mode range of 2 V while using a single-ended 4-V power supply, and a working bandwidth of a few kilohertz, are comfortably met by the monolithic design.

Existing integrated-circuit multipliers failed to meet the system objectives because of either cost or accuracy. Pulsewidth-amplitude-modulation (PWAM) multipliers, which are available as hybrid circuits [3], seemed ideally suited technically to the accuracies and bandwidths needed for automotive systems. The monolithic PWAM multiplier described here delivers a typical accuracy of 0.2 percent.

Principle of Operation

Pulsewidth Amplitude Modulation

The integrated PWAM multiplier (Fig. 1) produces a pulse output V_1 with an amplitude modulated by the Y input and with a pulsewidth modulated by the X and Z inputs. Ground is not included in the input common-mode range for single-supply operation; hence X and Y are referenced to X_1 and Y_1, respectively. The X_1 and Y_1 nodes are typically connected to a common reference voltage source and can be grounded when positive and negative supplies are used. The flexibility provided by bringing X_1 and Y_1 out to separate pins will be shown in a later section discussion a four-quadrant configuration.

The two-quadrant single-supply multiplier produces an output pulse having a duty cycle of $(X - X_1)/(Z - X_1)$. The pulse has a base voltage of Y_1 and a peak voltage of Y. It can be easily filtered as shown in Fig. 1 by a low-pass filter to produce a voltage V_{out} given by the equation

$$V_{out} = \frac{(X - X_1)(Y - Y_1)}{(Z - X_1)} + Y_1. \quad (1)$$

Manuscript received May 5, 1973; revised July 9, 1973.
The author is with Fairchild Semiconductor, Mountain View, Calif. 94040.

Reprinted from *IEEE J. Solid-State Circuits*, vol. SC-8, pp. 434-439, Dec. 1973.

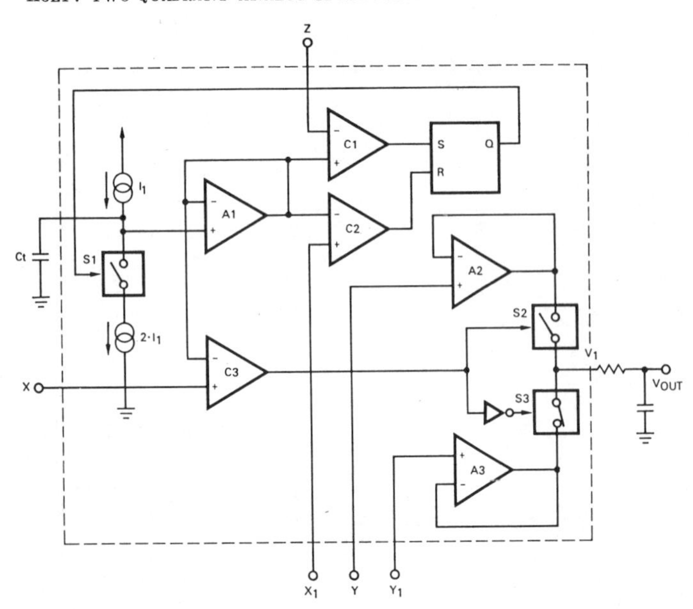

Fig. 1. Block diagram of the two-quadrant multiples.

The special case when positive and negative supplies are used to allow grounding of X_1 and Y_1 yields the simpler equation

$$V_{out} = \frac{XY}{Z}. \tag{2}$$

The bandwidth of the multiplier circuit is limited by either the cutoff frequency of the low-pass filter, which must be made low enough to filter the pulse frequency, or the constraints set by the sampling theorem. For the majority of cases, the low-pass cutoff frequency is made the dominant factor. With a pulse frequency of 20 kHz, the circuit delivers accuracies better than 0.5 percent full scale. Most automotive control applications need no more than simple single-pole RC filtering; more elaborate filtering techniques are necessary for higher bandwidth systems.

Block Diagram Description

The multiplier employs a triangle waveform generator and comparator to produce the pulsewidth modulation and switched voltage followers to produce the pulse-amplitude modulation. The two inputs to the pulsewidth-modulation comparator, shown in the upper waveform, Fig. 2, are a dc voltage from the X input and a triangle wave which slews between X_1 and Z. The output of the multiplier, shown in the lower waveform, Fig. 2, is a pulse train which switches between Y_1 and Y, with pulsewidth controlled by the X input and triangle wave. The triangle waveform generator is comprised of the two comparators $C1$ and $C2$, voltage follower $A1$, current sources I_1 and $2I_1$, RS flip-flop, and external capacitor C_t, as shown in Fig. 1. The $2I_1$ current source is controlled by the state of the flip-flop through symbolic switch $S1$ which is closed when the flip-flop is set by a signal from comparator $C1$.

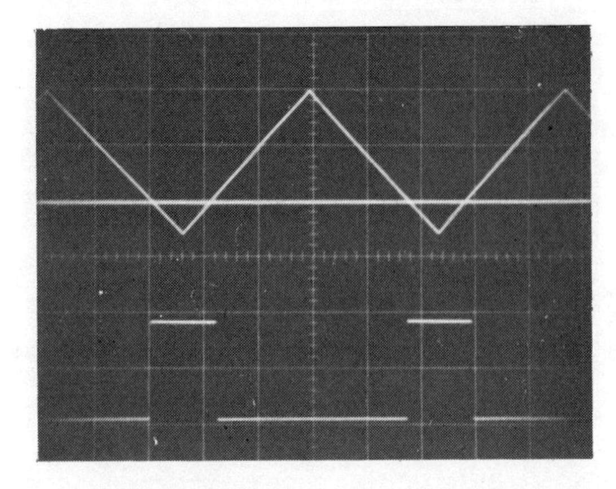

Fig. 2. Circuit waveforms: Top, triangle wave and X input; bottom, PWAM output at V_1 pin. Vertical, 2 V/div. Horizontal, 5 μs/div.

When $S1$ is open, capacitor C_t charges at a linear rate equal to I_1/C_t V/s. The capacitor voltage appears at the output of voltage follower $A1$ and the input of comparators $C1$, $C2$, and $C3$. The Z input voltage is compared to the positive voltage ramp on C_t by comparator $C1$; the flip-flop is set when the two voltages are equal. Setting of the flip-flop closes $S1$, causing C_t to be discharged at a linear rate equal to $-I_1/C_t$ until its voltage equals X_1. Comparator $C2$ then resets the flip-flop. This opens $S1$, and the cycle repeats producing a triangle wave on C_t with voltage extremes at X_1 and Z.

The triangle wave is compared to the X input by $C3$ after buffering by $A1$. The output of $C3$ is high when the triangle wave voltage is less than the X input voltage. The output of $C3$ is thus a pulse signal, modulated in width by X.

This pulsewidth-modulated signal controls switches $S2$ and $S3$ to selectively turn on voltage followers $A2$ and $A3$. When the output of $C3$ is high, which occurs during the time interval when the triangle wave voltage is less than X, $S2$ is closed and amplifier $A2$ drives the output to a voltage equal to Y. During the remainder of the triangle wave period, $C3$'s output is low and amplifier $A3$ drives the output to a voltage equal to Y_1. When integrated by a low-pass filter, output voltage V_1 becomes V_{out} as specified by (1) or (2).

Fig. 3 shows the 64 by 67 mil multiplier chip. The round bonding pads are 7 mil in diameter to accomodate solder bumps for flip-chip mounting, although the circuit shown in the figure was lead bounded for easy testing.

Design Considerations

The accuracy of this form of PWAM multiplier is strongly dependent upon the linearity of the positive and negative ramp voltages of the triangle waveform developed across capacitor C_t. Accurate analog multiplication also requires precise input signal control of the minimum and maximum voltage levels of the triangle wave. Good linearity in the triangle wave is achieved by using high output impedance current sources and by

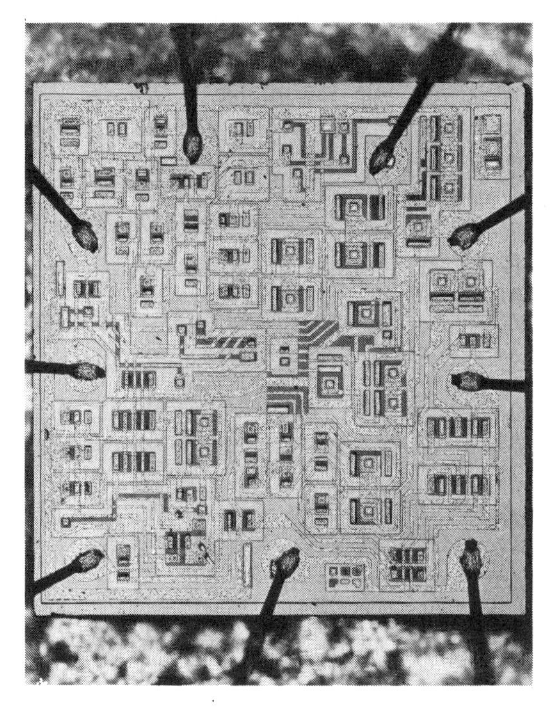

Fig. 3. Chip photograph of the two-quadrant multiplier.

loading the capacitor with a voltage follower buffer amplifier having high input impedance. The loading of the capacitor by the buffer amplifier input bias current has negligible effect on the linearity of the voltage ramp since the bias current remains almost constant with input voltage.

Sensing for comparators $C1$, $C2$, and $C3$ is taken at the output of the buffer amplifier. This eliminates any effect on accuracy caused by the input offset voltage of buffer amplifier $A1$.

Offsets in comparators $C1$, $C2$, and $C3$ affect accuracy and have been minimized by careful matching of device geometries. The delay between sensing of a minimum or maximum voltage limit for the triangle waveform and switching of the $2I_1$ current source also affects multiplication accuracy, since the delay can be equated to a comparator offset voltage by considering triangle slew rate. Positive and negative level shifts of 15 mV were added at the buffer amplifier output as compensation for the delay-induced errors, resulting in ideal compensation with a 0.0025-μF capacitor of the C_t terminal. Since the compensating level shift and triangle wave slopes track well with supply voltage variations, and since the circuit delays are reasonably repeatable from unit to unit, this compensation scheme effectively reduces multiplication errors in the usual circuit applications.

Circuit Description

Comparators $C1$ and $C2$ and the RS flip-flop appear in the left portion of Fig. 4. A design objective was high comparator and flip-flop speed with minimum circuit complexity. The comparator comprises a p-n-p differential stage with n-p-n collector load connected in a commonly used double-ended to single-ended conversion configuration. Emitter-follower buffers drive the inputs of the simple RS flip-flop, which uses base–collector diodes to isolate the input switching current from the saturated side of the flip-flop. The stored charge in the base collector diodes $D1$ and $D2$ removes the charge stored in the base of the appropriate saturated transistor in the flip-flop, Q_{35} or Q_{36}, resulting in storage delay times which were considerably shorter than the total switching circuit delays.

A very linear positive slope voltage ramp is developed across external timing capacitor C_t because of the high output impedance of the p-n-p I_1 current source (Fig. 5) connected in the commonly used cascode configuration [4]. Vertical transistors Q_{16} and Q_{19} buffer the base current of the typically low beta lateral p-n-p's Q_{14}, Q_{15}, Q_{17}, and Q_{18}. The 10-kΩ resistor sets current I_1. The current source accurately controls current for capacitor voltages up to $1V_{BE} + 1V_{sat}$ below the positive supply voltage, an important consideration since the design requirements include provision of a reasonable input common mode range at supply voltages as low as 4 V.

A similar configuration using n-p-n transistors is employed in the $2I_1$ current sink which discharges C_t, although beta buffering is not required. The $2I_1$ current is set by a 5-kΩ resistor connected to the same voltage as is the 10-kΩ resistor which sets the positive charging current I_1. The ratio of the two currents, and hence the symmetry of the triangle wave, depends predominantly on the matching of the 10-kΩ and 5-kΩ resistors. However, since the multiplication accuracy depends on the triangle wave linearity and amplitude, and is independent of symmetry, the typical 2 percent resistor mismatch is unimportant. Q_{38} switches the $2I_1$ current source off by shunting the current in the 5-kΩ resistor to ground when driven into saturation by the output of the flip-flop.

For a multiplier with an ideal triangle wave slewing between 0 and 4 V, the percentage pulsewidth modulation is given by

$$\text{percent pulsewidth modulation} = \frac{X}{4} \times 100.$$

For the practical case where the triangle wave is generated by current sources having finite output impedance, the expression for pulsewidth modulation becomes

$$\text{percent pulsewidth modulation} = \frac{R_1C_t \ln\left(\frac{I_1R_1}{I_1R_1 - X}\right) + R_2C_t \ln\left(\frac{I_1R_2 + 4}{I_1R_2}\right) - R_2C_t \ln\left(\frac{I_1R_2 + 4}{I_1R_2 + X}\right)}{R_1C_t \ln\left(\frac{I_1R_1}{I_1R_1 - 4}\right) + R_2C_t \ln\left(\frac{I_1R_2 + 4}{I_1R_2}\right)}$$

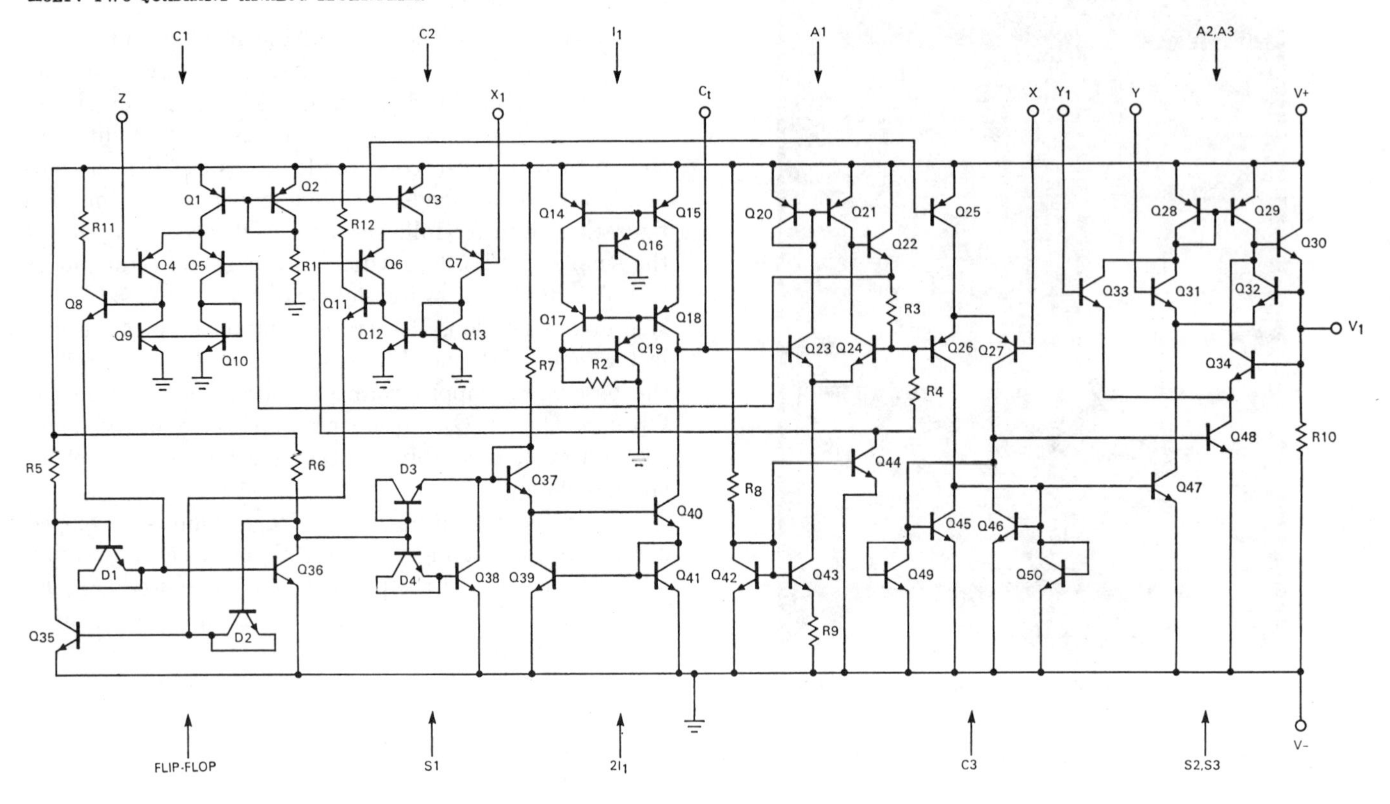

Fig. 4. Circuit schematic of the two-quadrant multiplier.

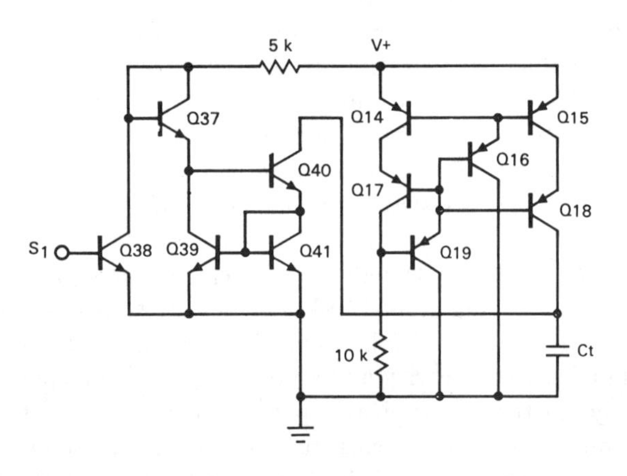

Fig. 5. I_1 and switched $2I_1$ triangle waveform current generators.

where R_1 is the charging current source output impedance, R_2 is the discharging current source output impedance, and $\pm I_1/C_t$ is the average positive and negative slew rate.

A comparison of the solutions of the two equations yields percentage error as a function of current source impedance. This error is shown in Table I for three values of R_1 and R_2 and at $I_1 = 400\ \mu$A. Error is independent of the value of C. The measured impedance, R_1 or R_2, was greater than 1 MΩ.

Comparator $C3$ and pulse-amplitude-modulation amplifiers $A2$ and $A3$ (Fig. 6) use a regenerative active load (Q_{45}, Q_{46}, Q_{49}, Q_{50}) for a p-n-p differential input stage (Q_{26} and Q_{27}). Regeneration is accomplished by taking the voltage bias for Q_{46} and Q_{50} from the collector of Q_{45}

TABLE I
PERCENT PULSEWIDTH-MODULATION ERROR

R_2 \ R_1	100 kΩ	1 MΩ	10 MΩ
100 kΩ	0.1 percent	0.6 percent	0.66 percent
1 MΩ	0.5 percent	0.001 percent	0.06 percent
10 MΩ	0.6 percent	0.06 percent	0.00001 percent

and the voltage bias for Q_{49} and Q_{45} from the collector of Q_{46}. This positive feedback results in zero current in one of the active load pairs, Q_{49}, Q_{45} or Q_{46}, Q_{50}, for either input condition. For equal Q_{26}-Q_{27} base voltages, there are two stable states for the collector load, one with Q_{45} cutoff and the other with Q_{46} cutoff. When the base voltage of Q_{26} rises from below to above the base voltage of Q_{27}, the stable state of the active load switches from Q_{45} cutoff to Q_{46} cutoff.

Output voltage V_1 is a pulse train whose instantaneous voltage equals Y when triangle voltage V_t is below X and equals Y_1 when V_t is above X. This occurs because current sources Q_{47} and Q_{48} for the voltage follower stage (Q_{28} through Q_{34}) are turned off and on alternately by the regenerative comparator load transistors.

Typical electrical characteristics are listed in Table II. The accuracy specification was measured using a 0.002-μF timing capacitor to produce the desired slew rate for cancellation of delay-induced offsets. Since the delay-induced offsets are in opposite directions at the top and bottom of the triangle wave, they have little

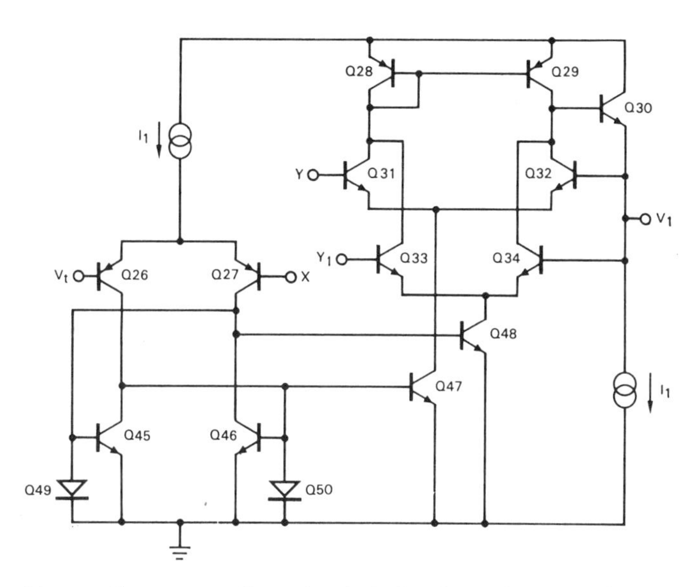

Fig. 6. Comparator $C3$ and switched voltage followers $A2$, $A3$.

TABLE II
ELECTRICAL CHARACTERISTICS
($V^+ = 5$ V, $V^- = -1$ V, $X_1 = Y_1 = 0$, $Z = Y = 4$ V)

Parameter	Conditions	Typical	Units
Error in percent of full scale	$X = 0$ to 4 V	0.2	percent
X, X_1, Z input current		−35	μA
Y, Y_1 input current		2.5	μA
Timing capacitor current		360	μA
Output sink current		400	μA
Output source current		1.6	mA
Usable bandwidth		10	kHz
Input voltage range ($V^+ = 6$ V, $V^- = 0$ V)		1–5	V

effect on accuracy near 50-percent modulation, but degrade accuracy near 0-percent and 100-percent modulation. The effect of the delay-induced offsets and compensation for several different capacitor values is shown in Table III at two modulation rates. As expected, the effect is reduced at 25-percent modulation as compared to 0-percent. The triangle wave frequencies in parentheses occur for the given capacitor value when Z-X_1 = 4 V and $B^+ = 6$ V.

The dominant error source is offset voltages in the input stages of the comparators and switched followers. The effect of offset voltages on accuracy can be determined by adding an offset term to each input term in (1). For example, when $X = Y = Z$ and $X_1 = Y_1$, 5-mV offsets added to each input in the appropriate direction for worst case effect results in an output error of 25 mV or 0.82 percent full-scale error. However, offsets do not always result in output error. Offsets of the same magnitude and sign in the X, X_1 and Z input cancel and do not degrade accuracy. Neither the fortunate nor unfortunate examples occur in very many units, and a small yield loss is incurred to realize the typical accuracy specification of 0.2 percent.

Since the dominant error source is offset voltage, temperature sensitivity can be related to the (temperature

TABLE III
TYPICAL ACCURACY

	0-Percent Pulsewidth Modulation (%)	25-Percent Pulsewidth Modulation (%)
0.0005 μF (80 kHz)	2.5	1
0.001 μF (40 kHz)	1.1	0.5
0.002 μF (20 kHz)	0.2	0.2
0.004 μF (10 kHz)	0.2	0.2
0.008 μF (5 kHz)	0.3	0.3

coefficient) of the offset voltages. A typical sensitivity of 0.005 percent per degree Celsius was measured over the 0°C to 100°C temperature range.

The usable bandwidth specification was determined by inputting identical ac signals to Y and Z and observing their expected cancellation at the output with a wave analyzer. With a triangle frequency of 20 kHz, 46-dB cancellation was achieved up to 10 kHz, the sampling theorem limit.

THE FOUR-QUADRANT MULTIPLIER

In (1), which describes the general two-quadrant multiplier transfer function, zero input conditions occur at $X = X_1$ or $Y = Y_1$. Values of Y less than Y_1 are permitted if the common-mode constraints at the input are not violated. The output voltage V_1 satisfies (1). Values of X less than X_1 result in 0-percent pulsewidth modulation and cannot produce outputs satisfying (1). These constraints on X and Y define the two quadrants of operation of the multiplier.

A four-quadrant multiplier can be realized if the zero input condition for X is made to occur at 50-percent pulsewidth modulation instead of 0-percent when $X = X_1$. Such a multiplier can be implemented by making $Y_1 = -Y$, which requires an inverter and dual power supplies. An output square wave, or 50-percent duty cycle output signal, integrates to 0 V after passing through the low-pass filter when $Y_1 = -Y$. At 0-percent modulation, $V_{out} = -Y$, and at 100-percent modulation, $V_{out} = +Y$.

The two-quadrant approach was chosen because it offered single-supply operation and good accuracy and was satisfactory for the fuel injection system for which the PWAM multiplier was designed. The accuracy of a four-quadrant version is reduced by the mismatching of the inverter feedback resistors.

The two-quadrant circuit can also be used as a four-quadrant multiplier by the addition of external inverters, as shown in Fig. 7. To preserve the divide function another inverter is used to set $X_1 = -Z$, which could be replaced by a resistive divider feeding a fixed voltage to Z and X_1. The Z input is limited to positive values since zero must be excluded from the set of acceptable Z input voltages.

The transfer function for the circuit in Fig. 7 can be

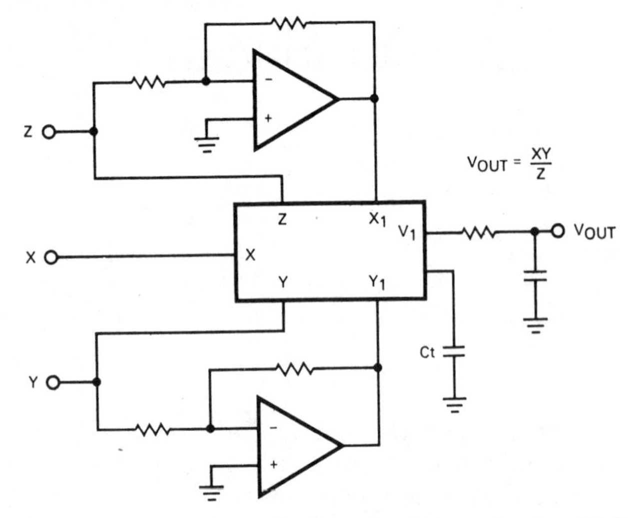

Fig. 7. Block diagram realization of a four-quadrant multiplier using the two-quadrant multiplier.

found by substituting $X_1 = -Z$ and $Y_1 = -Y$ into the general multiplier equation (1), yielding

$$V_{\text{out}} = \frac{(X + Z)(Y + Y)}{Z + Z} - Y \qquad (3)$$

which reduces to

$$V_{\text{out}} = \frac{XY}{Z}. \qquad (4)$$

Transconductance Multipliers

Numerous papers have been published on transconductance multipliers [5]–[7] that use the linear relationship between transconductance and collector current. Transconductance multipliers offer excellent linearity and wide bandwidth, but have marginal untrimmed accuracy. This is principally due to the large output offset voltage and to the dependence on resistor values and current in the transfer function (5) [8]

$$V_{\text{out}} = \frac{2R_L XY}{I(R_X + 2r_e)(R_Y + 2r_e)}. \qquad (5)$$

In contrast, (1) shows that the PWAM multiplier transfer function depends only on input voltages. Except for applications requiring small values on the divide input Z, output offset voltages are comparable to input offset voltages.

Conclusion

A monolithic integrated circuit which can provide accurate analog two-quadrant multiplication with reasonable bandwidth from a single-ended 4-V supply has been described. The circuit is ideally suited for many automotive and industrial control applications where accuracy and low cost is more important than bandwidth. In low-frequency systems, the multiplier can be used with only 3 noncritical external components. Since the system designer has the ability to choose the pulse frequency and the filtering means by selection of external components, the multiplier offers the flexibility of tailoring the circuit to the system accuracy and bandwidth requirements.

Acknowledgment

The author would like to thank G. Tammi and D. Long for their assistance in the development of the circuit.

References

[1] R. W. Russell and T. M. Frederiksen, "Automotive and industrial electronic building blocks," *IEEE J. Solid-State Circuits*, vol. SC-7, pp. 446–454, Dec. 1972.
[2] W. F. Davis and T. M. Frederiksen, "A precision monolithic time-delay generator for use in automotive electronic fuel injection systems," *IEEE J. Solid-State Circuits*, vol. SC-7, pp. 262–469, Dec. 1972.
[3] R. Panholyer, "Four quadrant analog multiplier," Fairchild Applications Brief 54, Aug. 1968.
[4] G. R. Wilson, "A monolithic junction FET-npn operational amplifier," *IEEE J. Solid-State Circuits*, vol. SC-3, pp. 343–344, Dec. 1968.
[5] B. Gilbert, "A precise four-quadrant multiplier with subnanosecond response," *IEEE J. Solid-State Circuits*, vol. SC-3, pp. 365–373, Dec. 1968.
[6] A. Bilotti, "Applications of a monolithic analog multiplier," *IEEE J. Solid-State Circuits*, vol. SC-3, pp. 373–380, Dec. 1968.
[7] C. Ryan, "Applications of a four-quadrant multiplier," *IEEE J. Solid-State Circuits*, vol. SC-5, pp. 45–48, Feb. 1970.
[8] M. Vander Kooi, "The μA795, A low-cost monolithic multiplier," Fairchild Applications Note 211, p. 7, May 1971.

A New Technique for Analog Multiplication

BARRIE GILBERT, SENIOR MEMBER, IEEE

Abstract—**A new method is described for performing accurate multiplication of analog signals using devices of special geometry but capable of fabrication with a standard bipolar process. A narrow region of current injection—a carrier domain—can be positioned on an emitter by one electrical input and controlled in magnitude by a second input. The resistive epi layer resolves this current into a differential output proportional to the product of the inputs. A key advantage of these multipliers is their low noise. The basic principle can be applied to many other nonlinear operations.**

I. Introduction

A WIDE VARIETY of methods have been used in the past to perform analog multiplication, many of which involved mixed technologies, for example, magnetic, optical, or electromechanical elements combined with electronics [1]. Currently, the translinear principle [2] is in widespread use, because it is ideally suited to monolithic fabrication, requires no external components, and combines good accuracy and bandwidth with simplicity and low cost. Recent improvements [3] have elevated the accuracy of translinear multipliers to that previously attained only by the best alternative techniques, notably those based on pulse modulation. However, for some applications the noise performance of the basic translinear cell is unsatisfactory [4].

The multiplier described here resulted when a well-known problem in transistor design—emitter current crowding [5]—was turned to advantage. It was the first of several integrated devices in which the function arises directly from the diffusion geometry, rather than by the interconnection of many individual components. The idea is simple: the minority-carrier current injected from a long narrow emitter can be deliberately crowded into a narrow zone, named a carrier domain, by introducing voltage gradients into the base region. By proper design of the base geometry it is possible to induce a voltage which exhibits a maximum potential point whose physical position can be controlled by signals applied to fixed contacts on the base, and thus provide direct control of the domain location in space. This basic concept [6] has obvious applications in A-D conversion, analog waveform generation and function fitting, infinite-hold samplers, and other nonlinear operations. Experimental devices are also being studied in which the domain, acting as a mobile filament of current, moves in response to a magnetic or optical input. Results in these areas will be the subject of future papers.

This paper is limited to a description of the original carrier-domain multiplier [7] which offers a possible alternative to the translinear cell. It can be fabricated with a standard bipolar IC process, occupies roughly the same chip area, has a response bandwidth extending from dc to about 30 MHz, and requires similar interfacing circuits for conversion from input voltages to differential drive currents and from a differential output current to an output voltage. Its main advantage is that the multiplication process is fundamentally quiet. A second advantage is that the accuracy is not dependent on close matching of transistor V_{BE} and is insensitive to thermal gradients, which by contrast require very careful fabrication and layout of monolithic translinear multipliers. A disadvantage of the design presented here is that relatively large currents are needed to bias the base (2–10 mA) in order to define a narrow domain, and thereby achieve satisfactory linearity. Even then, this design suffers from inherent nonlinearity as a result of departures from the idealized geometry, some of which are discussed in the companion paper by Smith [8], who also describes a layout modification which eliminates some of these sources of nonlinearity, although at the expense of doubling the device area.

Section II provides a description of a basic two-quadrant multiplier and a qualitative explanation of its operation. An analysis of the three main regions is given in Section III using material derived in the Appendixes. The extension to four-quadrant operation follows in Section IV, and a summary of the performance of the four-quadrant design is given in Section V.

II. Two-Quadrant Multiplier

Fig. 1(a) shows a plan view of the device, which consists of three main regions.

1) A rectangular n-type epitaxial collector, defined by a standard isolation diffusion, about 250 μm by 125 μm in area and of thickness 5 μm. A resistivity of 1 $\Omega \cdot$ cm was used, resulting in a nominal collector sheet resistance of 2 kΩ per square. Each end of the collector is contacted by n+ diffusions under which bars of buried layer diffusion extend as far as the ends of the emitter.

2) A semicircular p-type base diffusion, of junction depth about 2 μm and sheet resistance 200 Ω per square, contacted at the center and at the two corners.

3) An annular n+ emitter diffusion, whose junction depth was controlled to yield an h_{FE} of 40 to 200 and a peak f_T of about 600 MHz. The emitter is contacted by the metallization along its entire length.

The experimental device was fabricated alongside conventional bipolar transistors, and could equally well have used a thick epi process, allowing the concurrent fabrication of the interfacing circuits for operation with standard input and output levels of ±10-V full scale (FS) and ±15-V supplies.

In operation the collector-base and collector-substrate

Manuscript received June 23, 1975; revised August 18, 1975.
The author is with Analog Devices Semiconductor, Wilmington, Mass. 01887.

Reprinted from *IEEE J. Solid-State Circuits*, vol. SC-10, pp. 437-447, Dec. 1975.

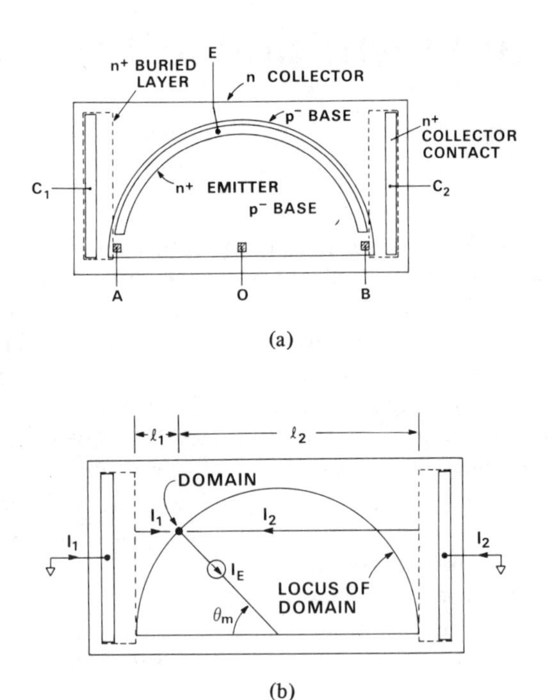

Fig. 1. (a) Layout of the basic two-quadrant multiplier. (b) Geometry of the collector region.

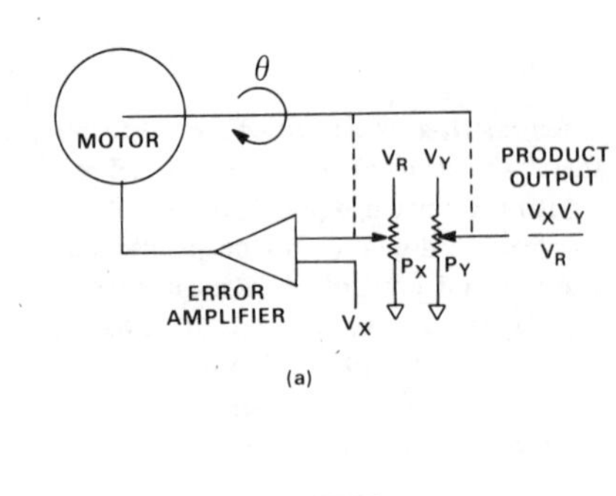

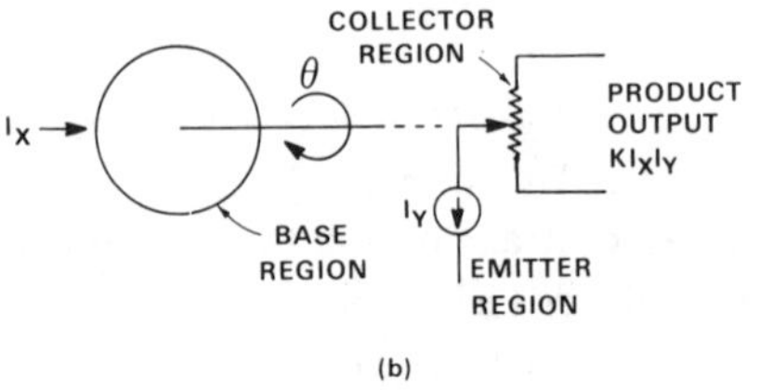

Fig. 2. (a) Classical servomultiplier compared with (b) the corresponding elements of a domain multiplier.

junctions are reverse biased. The epitaxial layer simultaneously serves as the normal sink for injected carriers and as a linear resistive current divider. The base diffusion simultaneously serves as the normal intrinsic base and as a means to generate the voltage gradients required to induce crowding in the emitter, which is driven by a current source I_E proportional to one of the signal inputs. In the absence of bias currents to contacts A and B, the base voltage would be nearly uniform, and injection would occur from all parts of the emitter. Application of negative currents to these contacts establishes a potential distribution in the base which has a pronounced maximum somewhere around the inner edge of the emitter-base junction, which confines emitter injection to that locality and causes most of the emitter to be reverse biased.

A key feature of the base design is its ability to predictably control the domain location. By applying differential currents $-(1 - X)I_B$ to contact A and $-(1 + X)I_B$ to contact B, where X is a dimensionless variable in the range -1 to $+1$, the angular position of the most positive point around the inner circumference of the base-emitter edge, and hence the angular position of the domain, occurs at

$$\theta_m = \cos^{-1} X \tag{1}$$

where θ_m is measured clockwise from contact A. Contact 0 supplies the constant current $2I_B$ plus the usual base current due to recombination in the intrinsic (and now mobile) transistor, and it is tied to a fixed voltage point.

Now consider the effect of a point of current entering the resistive collector sheet [Fig. 1(b)]. It splits into two components which subsequently emerge as I_1 and I_2. Intuitively, we would expect the division of current to be proportional to the perpendicular distances l_1 and l_2, and this is confirmed by analysis. Thus, the differential output current is

$$I_Z = (I_1 - I_2) = \frac{l_2 - l_1}{l_2 + l_1} I_E \tag{2}$$

where I_E is the current supplied to the emitter terminal. A consideration of the geometry shows that l_1 and l_2 may be replaced by expressions using θ_m which transforms (2) to

$$I_Z = I_E \cos \theta_m. \tag{3}$$

Combining this with the injection angle given by (1) yields

$$I_Z = XI_E \tag{4}$$

that is, the differential output is the linear product of the two inputs, and since X can have either sign two-quadrant operation is possible.

It is useful to compare this device to the classical servomultiplier, to which it bears some interesting similarities. Fig. 2(a) shows a typical electromechanical multiplier, the operation of which is well known [1]. One of the inputs is first converted to a shaft angle, and consequently has a long response time. The output potentiometer P_Y provides almost noise-free voltage division of the input V_Y. In the carrier-domain multiplier, the angular response is about eight orders of magnitude faster and the output potentiometer (formed by the collector sheet, over which the domain moves in analogy to the wiper of P_Y) provides low-noise current division of the input I_E [Fig. 2(b)].

III. Analysis

The first carrier-domain multipliers had decidedly heuristic beginnings. The unique properties of a semicircular resistive dee were discovered during experiments using large models cut from "Teledeltos" resistance paper to which voltage or current sources were attached and the resulting potential distributions

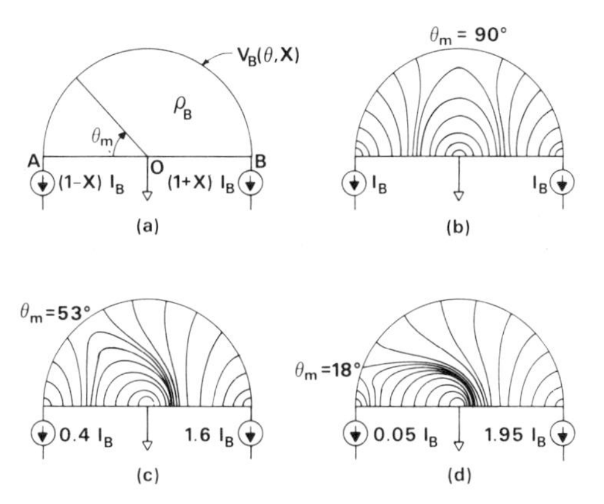

Fig. 3. (a) Idealized base region and approximate equipotential distributions for (b) balanced base drive, $X = 0$, (c) $X = 0.6$, and (d) $X = 0.95$.

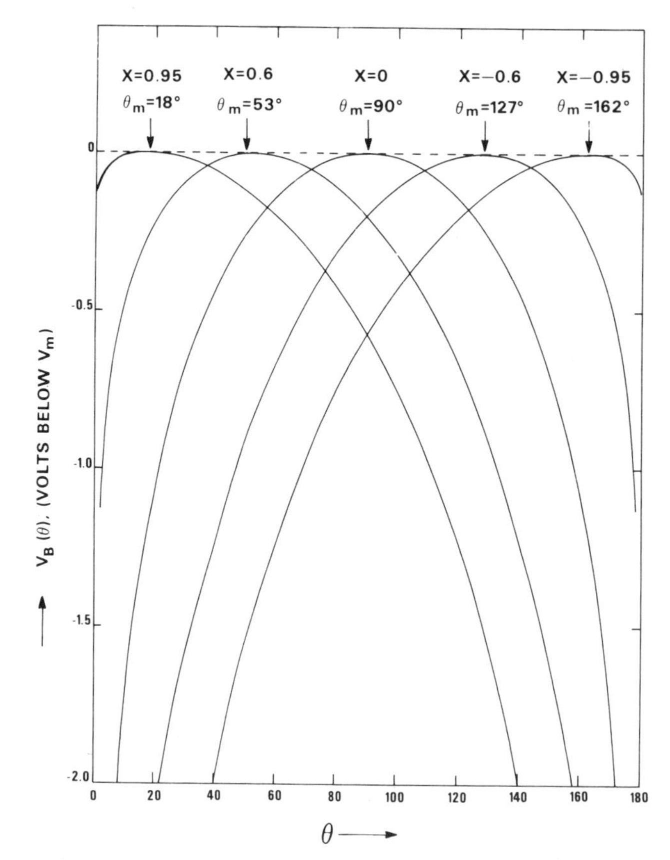

Fig. 4. Potential distribution at circumference of base region for various values of X.

measured with a digital voltmeter (DVM). The moving potential maximum at the circumference was first used as a means to steer a current into one of several transistors, as part of a monolithic character generator [9]. It then became apparent that the inverse-cosine relationship between θ_m and X could be combined with the cosine current-division relationship in the collector to form a linear multiplier. The detailed analysis followed much later. The base potential is the key to device behavior so it is discussed first. The emitter current density profile can then be calculated. Finally, the collector region is examined more closely.

Base Potential Distribution

Fig. 3(a) shows an idealized base dee, having point contacts. The conformal transformation given in Appendix I shows that the circumferential potential at a given value of the angular control input X is

$$V_B(\theta, X) = \frac{2\rho_B I_B}{\pi}\left[(1 - X)\ln \sin\frac{\theta}{2} + (1 + X)\ln\cos\frac{\theta}{2}\right] + C \tag{5}$$

where ρ_B is the base sheet resistance. The inverse-cosine relationship stated in (1) follows from equating the derivative of this expression to zero. True point contacts would result in infinite potentials at the contacts. In practice, the contact areas dictated by processing tolerances are sufficient to keep the contact voltages within sensible bounds, but it was later realized that larger contacts should have been used to avoid unnecessarily large drive voltages, and also to avoid mobility modulation due to the high current densities. The optimization of base contacts is discussed by Smith [8]. The constant C in (5) is determined by the contact radius at 0. Its value need not be known exactly, because the emitter current density profile is independent of the absolute base potential.

Fig. 3(b)-(d) show typical equipotential lines for three values of X. The potentials along the radius at 90° do not depend on X, due to the use of differential drive currents, so the set of curves for $V_B(\theta, X)$ intersect at $\theta_m = 90°$, and the potential at the maximum V_m must be a function of X. Its exact value is derived in Appendix I. In practice, it is convenient to refer the base potential distribution to this peak value and use the adjusted variable

$$V_B'(\theta) = V_B(\theta, X) - V_m. \tag{6}$$

The motion of θ_m from 0° to 180° as X is varied from −1 to +1 is clearly seen in the curves of Fig. 4, calculated for $\rho_B = 200\ \Omega$ per square and $I_B = 8$ mA.

Emitter Current Density

It is obvious from Fig. 4 that very severe crowding of emitter injection will occur near θ_m. Appendix II derives the expressions for current density in a straight emitter. For an annular emitter

$$J_E(\theta) = J_o \exp\frac{V_B'(\theta)}{V_o} \tag{7}$$

where

$$J_o = \frac{I_E}{R_E W_E \displaystyle\int_0^{\pi} \exp V_B'(\theta)/V_o\, d\theta} \tag{8}$$

and R_E is the emitter radius. Using the adjusted values for base potential J_o is also the peak current density. By substituting the expression for the actual base potential (5) a closed-form equation for $J_E(\theta)$ can be derived. However, it is complicated and not very helpful. It is more convenient to take

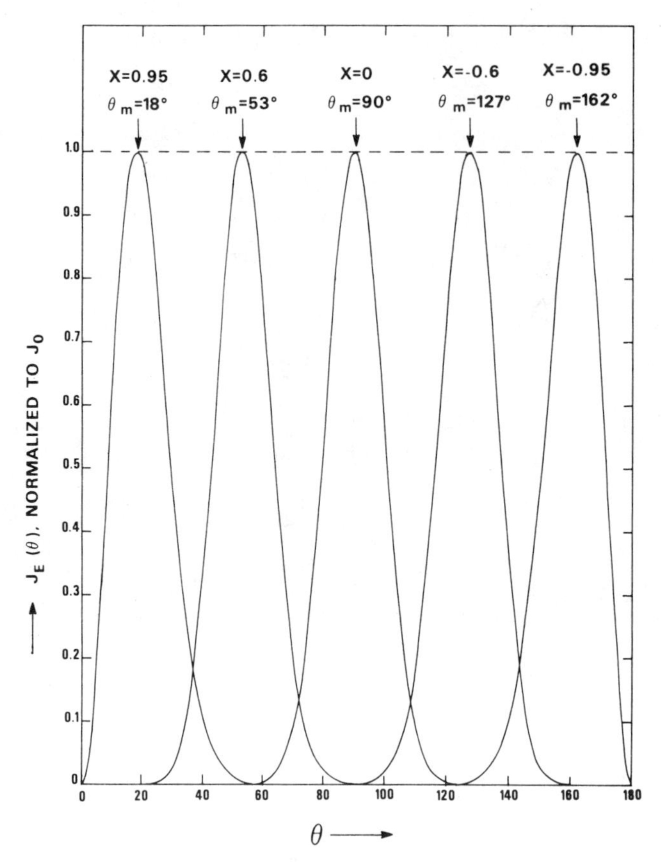

Fig. 5. Emitter current density distributions for bias conditions shown in Fig. 4 and T_A = 300 K.

the numerical values for $V_B'(\theta)$ and calculate $J_E(\theta)/J_o$ using (7). The absolute value of J_o can be found independently (if required) from the approximation given in Appendix III, which also demonstrates that in the region near θ_m the potential distribution is nearly parabolic, giving the domain a Gaussian form.

Fig. 5 shows curves for $J_E(\theta)/J_o$ using the values for $V_B'(\theta)$ plotted in Fig. 4, which actually correspond to a high base drive current, and the domain would typically be more diffuse. Nevertheless, the predicted Gaussian form is visible, and the domain is seen to retain its symmetry over most of the control range. The width of the domain at the half-magnitude points is about 35 μm for R_E = 100 μm, or 11 percent of the annular length.

The emitter junction will be reverse biased at points remote from θ_m, the greatest stress occurring when $X = \pm 1$ and all of I_B flows in either contact A or B. The drive current should not be allowed to cause emitter breakdown, since this will result in gross nonlinearities.

This simplified analysis assumed no transverse potential gradients in the base under the emitter. In fact, there is a strong transverse gradient. With reference to Fig. 1(a) note that the base can be partitioned into the dee below the emitter and the roughly annular part above it. The latter can be approximated by a simple resistive stripe separated from the dee by the sub-emitter region. If there were no conduction under the emitter, the potential around the annular part of the base would vary linearly from A to B, and a moment's consideration shows that this profile is always more negative than on the lower edge of the base-emitter junction. Consequently, most of the injection occurs along the lower edge, resulting in much higher current densities than predicted by simple theory. Furthermore, injection will tend to occur near the surface (where the base potential is highest). These mechanisms lower the effective alpha of the intrinsic transistor and make it sensitive to surface defects, which may not be uniform over the emitter length. The resulting modulation of alpha as the domain moves around the emitter may be a cause of observed variations in the nonlinearity of individual samples.

The intrinsic transistor operates in essentially a grounded-base configuration, because even though the base voltage is not fixed the emitter is driven from a high-impedance current source. This has several advantages. First, lot-to-lot variations in "V_{BE}" (which here refers to the base-emitter voltage measured with respect to V_m at a total current I_E) will have negligible effect on performance. Second, because the domain occupies very little space at a given time, this V_{BE} can vary considerably from one end of the emitter to the other without causing serious skewing, and linearity is therefore not a strong function of V_{BE} uniformity. Third, large thermal gradients along the emitter can be tolerated, by the same reasoning. Fourth, in the four-quadrant multiplier described later, the two emitters can likewise have a large V_{BE} mismatch and operate at different temperatures with little effect on accuracy. Contrast this with the extreme care required in the design of an accurate translinear cell [3].

The grounded-base operation also results in relatively low-noise operation since the thermal noise voltage of the base resistance does not contribute to the output noise for frequencies up to which the emitter current-source impedance remains high. Here also, the contrast with a translinear cell is noteworthy, for which circuit this noise voltage is multiplied by a factor $\sqrt{2}$ (due to use of the differential pair of transistors) and then translated to an output noise current by the high transconductance of the stage. In this connection, a diffuse domain behaves somewhere between an ideal point domain and a differential pair. Viewed another way, the centroid of the domain suffers a statistical fluctuation due to the thermal voltages in the base. This mechanism has not been analyzed, but it is likely that the noise contribution is reduced in proportion to the projected width of the domain to the perpendicular line between collector contacts.

Collector Current Division

In Section I some assumptions were made about the collector region. They were: 1) the outputs were connected to zero-impedance nodes; 2) the output currents divided in proportion to the distances l_1 and l_2; and 3) the domain behaved like a point source of current. These assumptions need closer scrutiny.

The output currents can be directly converted to voltages using load resistors, and the maximum load power is delivered under the usual matching conditions of equal source and load resistance. However, low-impedance current-summing nodes are preferable, because the current gain is maximized, and the supply-voltage requirements and collector circuit time constants are minimized. More important, the scaling accuracy is then independent of the absolute resistance of the collector layer, which may vary by as much as +1 percent/°C for a 5 $\Omega \cdot$ cm epi. The loading conditions can be met using an op amp as shown in Fig. 7, or by the use of a differential feedback circuit, such as the one used in [3].

The proportionality between the distances l_1 and l_2 and the output currents can be proved using another conformal transformation. Let the collector rectangle be placed in the z-plane with one long edge lying along the x-axis and one collector contact lying along the y-axis. Using $w = e^z$ this maps to the w-plane as an annulus, and the contacts become a pair of coaxial rings. The distances l_1 and l_2 become radial lengths and vertical position becomes angular position. Hence, the division of the input current is independent of vertical position in the z-plane, and by shrinking this dimension the collector layer reduces to a one-dimensional resistor, for which the proportionality is obvious.

This argument assumes that the collector sheet is uniform. In fact, the epi layer is subject to at least three distortions. First, the layer is thinner under the base dee than elsewhere. Second, the injected current will cause modulation of the majority-carrier mobility, and therefore of the local sheet resistance, particularly using a lightly-doped epi. Third, the voltages induced in the collector will modulate the depletion-layer width, again varying the local sheet resistance. These nonuniformities are not easily quantified, but of course their net effect is to introduce nonlinearity, both with respect to the angular signal and, to a lesser extent, to the emitter-current signal. The problem can be reduced by using a special light buried layer diffusion for the entire collector sheet; gridiron and peppered diffusions using a regular buried-layer concentration have been considered. More recent multiplier designs use a long straight emitter, and the domain slides over a narrow bar of regular buried layer forming a "resistance track" of about 1 kΩ end-to-end.

To deal with the fact that the domain is not a point of current, but really rather fat, we must show that there is an effective center of injection, whose angular position closely tracks θ_m. This is possible because the diffuse domain can be decomposed into elemental point currents, the total effect of which is found by their superposition in the resistive collector layer. This centroid can be defined in three ways: 1) the point in the emitter plane at which a point source of current, equal in magnitude to the total emitter current I_E, could be substituted without changing the outputs; 2) the intersection of the lines dividing the emitter plane on either side of which exactly half of I_E is injected, which provides the formal definition, 3) in terms of the solution to the equation

$$\int_0^{\theta_c} \int_0^{w_c} J_E(\theta, w)\, d\theta dw = \int_{\theta_c}^{\pi} \int_{w_c}^{W_E} J_E(\theta, w)\, d\theta dw \qquad (9)$$

where θ_c and w_c are the angular and radial coordinates of the centroid. If the transverse gradient is ignored, this equation can be used to derive an expression for the angular error $\theta_m - \theta_c$. However, a simpler approach can be used to demonstrate that this error is not a serious source of nonlinearity. First, note that the centroid occurs at exactly θ_m as long as the domain is symmetrical about this angle. Fig. 5 shows that this condition is reasonable for a line emitter, and it has already been stated that most of the injection does occur along the inner emitter edge. At $X = \pm 0.95$ there is some skewing, but even here it is mostly in the tails of the current-density distribution. Furthermore, as θ_c approaches 0° or 180° the sensitivity of output current to injection angle approaches zero, since ultimately the domain is moving parallel to the contacts. The central 80 percent of the X-input range translates into an angular range of only 60 percent (37° to 143°). This provides a good explanation of the observed fact that the X-linearity holds up quite well even for drive currents (I_B) down to about 2 mA, one fourth the value used to plot Figs. 4 and 5.

Finally, it should be pointed out that isothermal variations of temperature do not affect the position of the centroid as long as the symmetry condition holds. Probably the largest source of temperature dependence of scaling factor is the variation of alpha, amounting to about +0.01 percent/°C, and this can easily be compensated.

IV. Four-Quadrant Multiplier

Analog multipliers are usually expected to operate in all four quadrants, and this is achieved simply by putting two of the devices just described into a common collector region. Fig. 6 is a photomicrograph of both types. The four-quadrant multiplier has its base regions cross connected and driven from

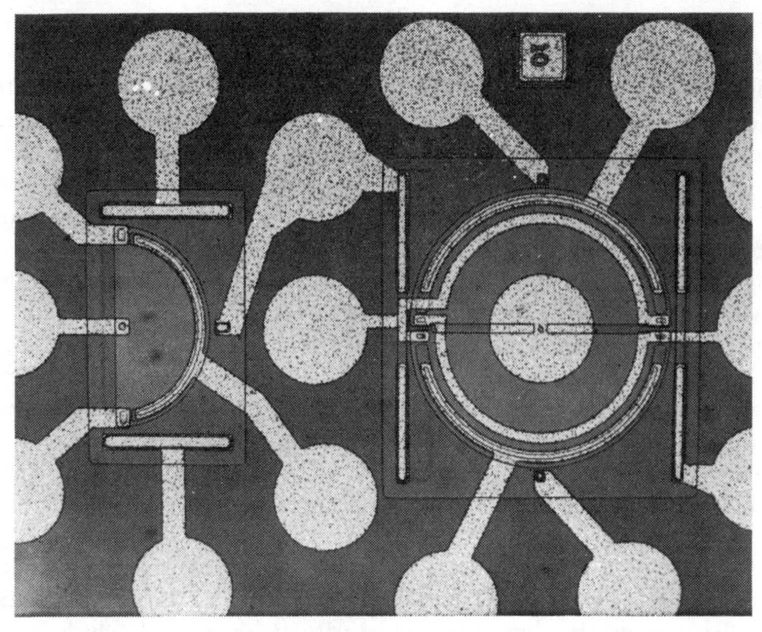

Fig. 6. Photomicrograph of the two-quadrant (left) and four-quadrant (right) carrier-domain multipliers.

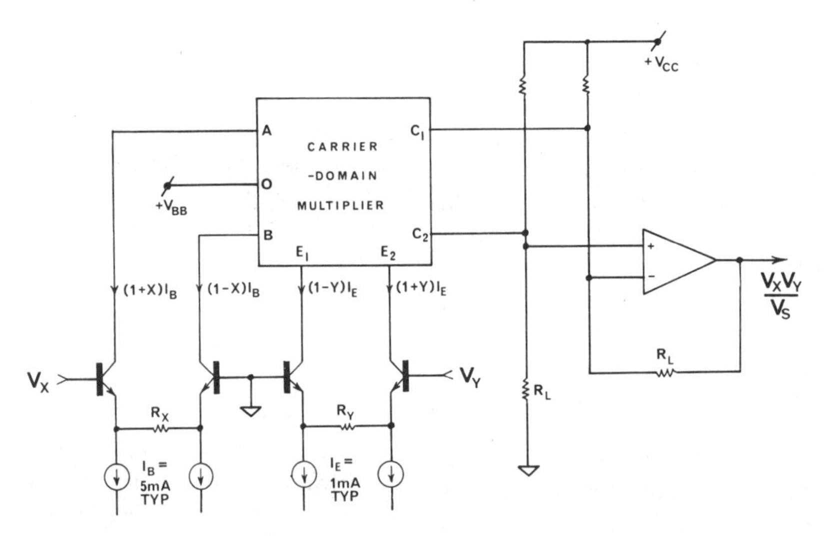

Fig. 7. Typical interface circuitry.

a pair of differential currents modulated by the X input. The common base connection is made to the circular bonding pad at the center; this was convenient, but some of the performance variations between samples may be traceable to the strain induced in the base diffusion by wire bonding to this pad. The two emitters are driven from a pair of differential currents modulated by the Y input. Fig. 7 shows a suitable interfacing circuit.

The operation is as follows. When the emitter currents are balanced ($Y = 0$) the two domains move on arcs at a constant 180° separation in response to the X input, but since they contribute equally to the outputs, the differential output is zero. When the base currents are balanced ($X = 0$) both domains inject into the center of the collector, and again the differential output is zero. The upper half-multiplier generates the output

$$I_{ZU} = X(1 + Y)I_E \tag{10}$$

and the lower half-multiplier generates

$$I_{ZL} = -X(1 - Y)I_E \tag{11}$$

the negative sign being due to the opposing motion of the domain. The total differential output is therefore

$$I_Z = 2XYI_E \qquad \text{for } -1 < X, Y < +1. \tag{12}$$

Using the bridge-feedback output converter shown in Fig. 7, the voltage output can be expressed simply as

$$V_Z = \frac{V_X V_Y}{V_S} \tag{13}$$

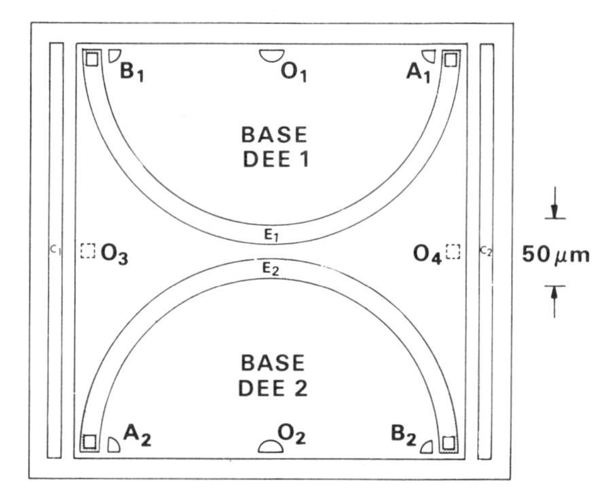

Fig. 8. A modified geometry for the four-quadrant multiplier.

where V_S is the scaling voltage for the complete multiplier, given by

$$V_S = \frac{R_X R_Y I_B}{2R_L}. \tag{14}$$

Feedthrough

In any practical four-quadrant multiplier, a small output is generated even when one input is nulled. This "feedthrough" signal is usually specified as the percentage of FS output which appears when one input is driven to ±FS and the other is adjusted to minimize this output. The carrier-domain multiplier exhibits very low feedthrough on the Y input, because of the good linearity of this channel. A small initial offset in the domain positions can be adjusted by a corresponding offset to the X input until near-perfect cancellation occurs in the linear collector resistance. X feedthrough is much worse, because now cancellation depends on exact matching of the nonlinearities of the two half-multipliers. Local defects can affect each half separately, especially when the domains are at 90°, and hence near the edge of the collector boundary, and this also is the position where the output is most sensitive to an angular error.

Fig. 8 shows a proposed rearrangement of the design, expected to have superior X-feedthrough characteristics, and generally lower distortion. The most conspicuous difference is that the emitter arcs have been reversed. Consequently, when the domains are at their point of maximum sensitivity ($\theta_m = 90°$) they move on very nearly parallel paths in close proximity, and nonuniformities in the epi layer, caused either by processing gradients or by some of the modulation effects described earlier, are much less damaging to linearity. The base region is now rectangular, so the collector is nearly uniform in thickness, and the semicircular base boundaries are now defined by the emitter arcs, using the high sub-emitter sheet resistance as a sort of isolation wall. Transverse crowding is not avoided, but preliminary studies show it may be possible to bias the contacts 0_3 and 0_4 so as to set up an image profile on the outer edges of the emitters. Note that there is now no need to put a bonding pad on the base diffusion. Finally, the two emitters are contacted just at their ends. This allows the use of a thinner emitter, thus reducing its transition capacitance. Provided that the emitter currents are no more than a milliamp or two, the voltage drop in the emitter series resistance can be tolerated. It will cause a broadening of the domain, but if both contacts are used, this will be symmetrical.

Transient Response

The step response of the domain multipliers described here is complicated, and can only be described in qualitative terms. The simplest situation occurs when the domains are parked in their limit positions, and so inject over the collector contacts. The elemental transistors then behave like conventional grounded-base stages with almost zero load resistance in the collector circuit. However, there is a large base resistance, formed by the dee, and a fairly large emitter transition capacitance due to the inactive portion of the emitter. These conspire to form a resonant circuit, and therefore a response overshoot or ringing.

When the X input is balanced the domains are stationary over the center of the collector. In addition to the effects just described, there is now a delay and rise time contribution due to the distributed RC network formed by the epi sheet resistance and the base- and substrate-junction capacitances. It may be concluded that the response on the Y input will be fast for $X = \pm 1$, and slowest for $X = 0$.

The response on the X input is more complex. Here, the input step must first change the potential distribution on the base dees, which are again slugged by distributed RC effects, although less seriously due to the order of magnitude lower base sheet resistance. Some of the resulting displacement currents appear quickly at the collector terminals, due to imperfect cancellation, and cause preshoot. Next, the domains have to follow the moving potential peak, and finally, their currents must traverse the collector network before appearing as output. No attempt has been made to analyze these mechanisms. However, in most practical cases the overall response time will be dominated by the interface circuitry, because as the mea-

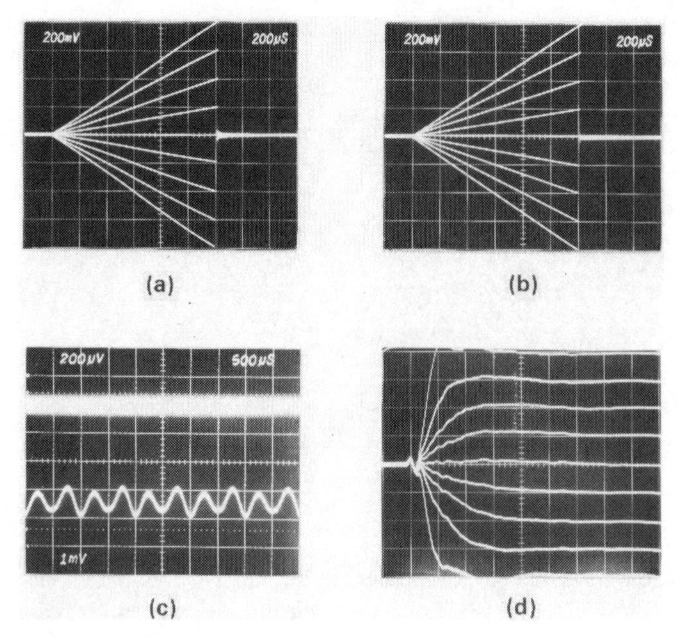

Fig. 9. Typical performance, showing output for (a) ramps applied to bases. (b) Ramps to emitters. (c) Feedthrough for Y (upper trace) and X (lower trace) inputs. (d) Transient response of Y input, 5 ns/div.

surements show, these responses are all in the nanosecond range.

V. PERFORMANCE

The results given here are intended only to demonstrate some of the main features of the four-quadrant device, rather than specify the performance of a complete multiplier circuit. In most cases, a test circuit similar to that shown in Fig. 7 was used, with the scaling factor adjusted to $V_S = 1$ V, and nominal FS inputs and outputs of ±1 V, clipping at about ±1.5 V. The response-time tests used the 50-Ω loads presented by the inputs of a sampling scope.

Fig. 9 shows the general behavior of the device when ramp inputs are applied to the bases [Fig. 9(a)] or the emitters [Fig. 9(b)]. In each case, the other input was a dc level between -0.8 V and +0.8 V, in 0.2-V steps. These waveforms confirm multiplier operation, but do not reveal nonlinearity, some indication of which is provided by Fig. 9(c), which shows the feedthrough for an input of ±1 V peak (sine wave at 1 kHz), with the other input nulled exactly. The upper trace shows the feedthrough on the Y input; as expected, it is very small, and hidden in the noise. The lower trace shows the X feedthrough, which has an amplitude of about 0.1 percent of FS, consisting mainly of second harmonic. This was the best sample of those available at the time, there being a large spread between samples, the worst having an X feedthrough of 0.6 percent of FS. Fig. 9(d) shows the step response of the Y input for a pulse of 1 V peak, with -0.8 V to +0.8 V on the X input. The expected variation in rise time is evident, and after factoring out the rise time of the measuring system was found to vary from 2 to 15 ns.

Fig. 10 illustrates the wide-dynamic range of this multiplier. In each case, the Y input was driven to FS with a 1 kHz sine wave, and the X input was adjusted to produce an output of

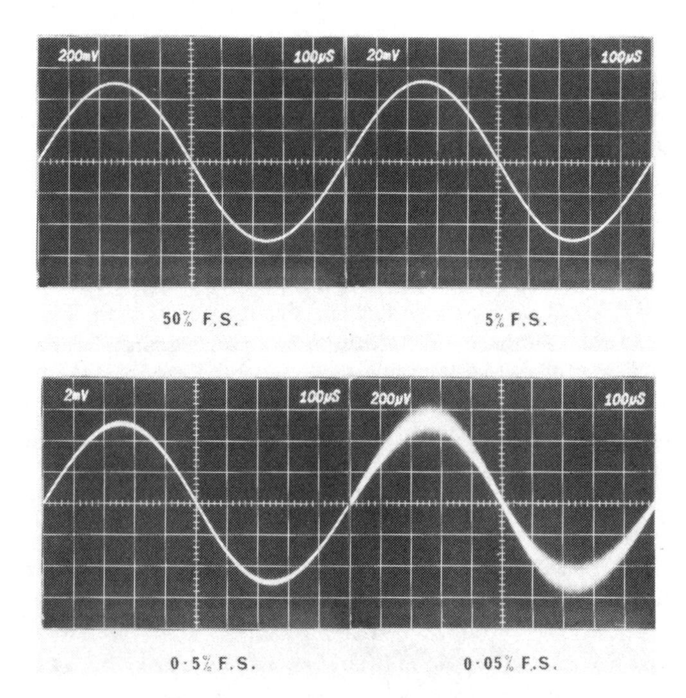

Fig. 10. Output for an FS 1-kHz sine wave input on Y and successively smaller dc inputs on X.

first 50 percent of FS, then reduced in decade steps to 0.05 percent of FS output. No serious distortion is visible over this control range of 60 dB, and the output noise level is still well below the signal in the last waveform. The signal bandwidth was 1 MHz. Using a low-noise op amp (AD504), and a Fluke 931B rms voltmeter, the noise voltage was measured for a 10 Hz to 300 kHz bandwidth, and found to be between 15 µV and 22 µV rms, corresponding to a noise spectral density of about 33 nV per root hertz, some of which is attributable to the op amp and the rest of the interface circuitry. Over the bandwidth 10 Hz to 30 kHz this amounts to a noise voltage of only 5 μV rms, which is 100 dB below the overload point at 1.06 V rms (±1.5-V peak).

Nonlinearity was originally revealed using a ramp drive to one input, derived from the oscilloscope, the vertical axis of which displayed the derivative of the output. This technique relies on a precisely linear ramp, and requires the use of a short differentiator time constant, with attendant noise problems. Also, the resulting displays are not easily translated into nonlinearity percentages. A more practical technique is to display the difference between input and output, using FS signals on X and Y, and an incremental adjustment to scaling to equalize the large-signal output to the input. This cross-plot technique showed that the Y nonlinearity was less than 0.1 percent, and limited by the interface circuitry. The X nonlinearity was much more gross, with an average value of about 1.5 percent, and a maximum value of 2.8 percent. The form of the distortion was roughly S-shaped, but varied from sample to sample.

No measurements were made of scaling accuracy and stability, since these are parameters determined to a large extent by the interface circuitry. However, it has been established that the domain location is not significantly affected by temperature variations from -20°C to +100°C, using fixed bias supplies and DVM measurements of output.

VI. Conclusions

The carrier-domain multiplier holds promise of nearly noise-free control of attenuation or gain. Improvements in the linearity of one of the inputs is required before the device equals the performance of prevalent translinear multipliers, which in spite of the problems referred to in this paper remains the preferred technique for wide-band monolithic applications. The relatively large supply currents and voltages needed to bias a domain multiplier, and the fact that its output current appears across a finite load resistor are further ways in which it differs from a translinear device. In most other respects, however, it is comparable, and may provide a complementary technique in the long run. Variations in design, notably devices using a slender geometry and linear motion of the domain, are being explored with a view to overcoming some of the disadvantages of the reported designs.

Appendix I
Voltage Distribution in a Semicircular Dee

We wish to find the voltage around the circumference of the dee shown in Fig. 3(a) for any value of the modulation index X and the angle θ_m at which this voltage is maximum. Assuming point contacts, the result will be independent of the size of the dee, which is therefore given unit radius. The magnitude of the voltages will be proportional to the sheet resistivity ρ_B and the drive current I_B but the form of the distribution will be strictly a function of X. The analysis is simplified by considering the effect of just the input at A with $X = 0$, that is, an input of I_B. The total distribution is then found by substituting the actual values for the inputs and summing.

To provide a transformable boundary the mirror image of the dee is added, and the resistivity doubled (to maintain the correct voltage scale). This unit circle is supposed to exist in the z-plane [Fig. 11(a)]. Using the transformation

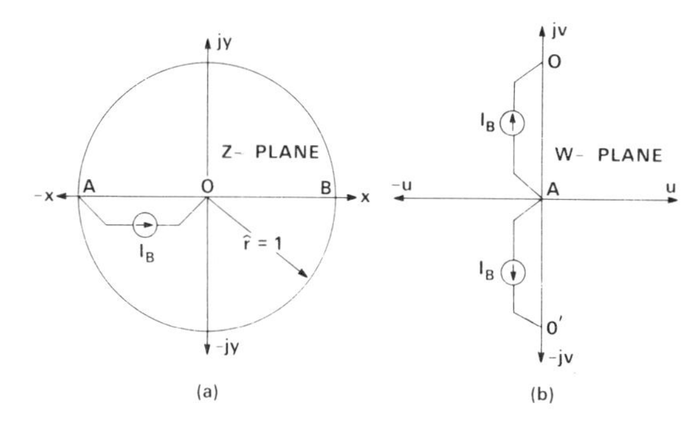

Fig. 11. Conformal transformation of circular base region for analysis.

$$w = j\,\frac{1 - z}{1 + z} \tag{15}$$

this maps to the infinite w-plane, as shown in Fig. 11(b). The origin in the z-plane moves to $w = j$, point A becomes the new origin and the point B splits outward to the circle at infinity. The boundary of the circle, across which no current flows, maps to the real axis of the w-plane, but the sources at 0 and A generate an asymmetric potential distribution, which would violate the original boundary condition. This problem is eliminated simply by adding an image source at 0', since the potentials are now balanced in the upper and lower half-planes. These three point sources generate similar distributions, differing only in location and magnitude and summing to

$$V(w) = \frac{2\rho_B}{2\pi}\,[2I_B \ln w - I_B \ln (w - j) - I_B \ln (w + j)]. \tag{16}$$

This simplifies to

$$V(w) = \frac{\rho_B I_B}{\pi} \ln \frac{w^2}{1 + w^2} \tag{17}$$

which transforms back to the z-plane as

$$V(z) = \frac{\rho_B I_B}{\pi} \ln \frac{4z}{(1 - z)^2} \tag{18}$$

which in cylindrical coordinates becomes

$$V(r, \theta) = \frac{\rho_B I_B}{\pi} \ln \frac{1 - 2r \cos \theta + r^2}{4r} \tag{19}$$

where r and θ are general points in this plane. Now applying the actual inputs to A and B and noting that $\cos (\pi/2 - \theta) = -\cos \theta$, the general solution is found to be

$$V(r, \theta, X) = \frac{\rho_B I_B}{\pi}\left[(1 - X) \ln \frac{1 - 2r \cos \theta + r^2}{4r} + (1 + X) \ln \frac{1 + 2r \cos \theta + r^2}{4r}\right]. \tag{20}$$

We are especially interested in the circumferential distribution, obtained by letting $r = 1$. This gives

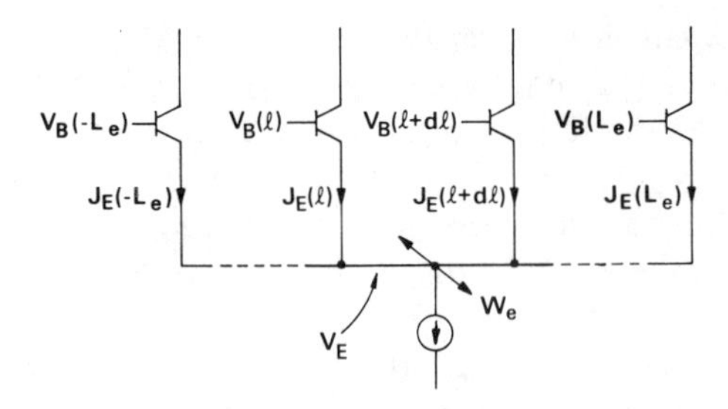

Fig. 12. Model of stripe emitter for analysis.

$$V_B(\theta, X) = \frac{2\rho_B I_B}{\pi}\left[(1 - X)\ln\sin\frac{\theta}{2} + (1 + X)\ln\cos\frac{\theta}{2}\right]. \tag{21}$$

The maximum of this function occurs where

$$\left.\frac{\partial V_B(\theta)}{\partial\theta}\right|_{x=\text{const}} = (1 - X)\cot\frac{\theta}{2} + (1 + X)\tan\frac{\theta}{2} = 0 \tag{22}$$

the solution of which is

$$\theta_m = \cos^{-1} X. \tag{1}$$

The peak voltage, $V_m(X) = V_B(\theta_m, X)$ is found by putting this value of θ_m in (21):

$$V_m(X) = \frac{\rho_B I_B}{\pi}\ln\left[x^x(1 - x)^{(1-x)}\right] \tag{23}$$

where

$$x = \frac{1 + X}{2}. \tag{24}$$

Appendix II
Current Density in a One-Dimensional Domain-Forming Structure

Fig. 12 shows the distributed structure quantized into elemental transistors for the purpose of analysis. Each elemental emitter is of width W_e and there is no transverse bias. Summation of current densities along the length axis gives the current at the emitter terminal:

$$I_E = W_e\int_{-Le}^{Le} J_E(l)\,dl. \tag{25}$$

Applying the well-known junction law

$$J_E(l) = J_s\exp\frac{V_{BE}(l)}{V_o} = J_s\exp\frac{V_B(l) - V_E}{V_o} \tag{26}$$

where

- J_s is the saturation current density for the particular diffusion profiles;
- $V_B(l)$ is the base voltage profile measured with respect to an arbitrary reference potential;
- V_E is the voltage at the emitter terminal with respect to the same reference potential;

and

- V_o is the thermal voltage kT/q.

Equation (26) can be written as

$$J_E(l) = \frac{J_s}{\exp\frac{V_E}{V_o}}\exp\frac{V_B(l)}{V_o} = J_o\exp\frac{V_B(l)}{V_o} \tag{27}$$

where J_o is a new scaling constant.

Inserting this value for $J_E(l)$ into (16) and rearranging terms:

$$J_o = \frac{I_E}{W_e\int_{-Le}^{Le}\exp V_B(l)/V_o\,dl}. \tag{28}$$

Appendix III
Approximate Shape and Magnitude of Domain

An explicit solution for the current density distribution $J_E(\theta)$ can be obtained by inserting the expression for $V_B'(\theta)$ into (7) and (8). As an aid to visualizing the domain, however, it is not very informative. By using a simple approximation to $V_B'(\theta)$ in the vicinity of V_m the domain can be shown to have a Gaussian distribution, and a workable approximation for the peak current density J_o can be obtained. The proof is given for $\theta_m = 90°$, but can easily be generalized to show that it holds over most of the range of θ_m.

$$\begin{aligned} V_B'(\theta) &= V_B(\theta, 0) - V_m(0) \\ &= \frac{2\rho_B I_B}{\pi}\left[\ln\sin\frac{\theta}{2} + \ln\cos\frac{\theta}{2} - \ln\sin\frac{\pi}{4} - \ln\cos\frac{\pi}{4}\right] \\ &= \frac{2\rho_B I_B}{\pi}\ln\sin\theta. \end{aligned} \tag{29}$$

Now let $V_B(\phi)$ be the voltage at angles $\pm\phi$ from 90°. Then

$$\begin{aligned} V(\phi) &= \frac{2\rho_B I_B}{\pi}\ln\sin\left(\frac{\pi}{2} \pm \phi\right) \\ &\approx -\rho_B I_B \phi^2 \quad \text{for } \phi \text{ in radians} \\ &\approx -3.05 \times 10^{-4}\rho_B I_B \phi^2 \quad \text{for } \phi \text{ in degrees.} \end{aligned} \tag{30}$$

The approximation is accurate to 1 percent up to $\phi = \pm 14°$. The curves of Fig. 5 were plotted from exact data, using $2\rho_B I_B = 1$ V, which requires $I_B = 7.85$ mA for $\rho_B = 200\ \Omega$ per square, and they show that the domain is contained within this angular range, so the approximation is realistic. Therefore,

$$J_E(\phi) \approx J_o \exp\left(-\frac{\rho_B I_B \phi^2}{V_o}\right) \tag{31}$$

which is simply a Gaussian distribution in ϕ. By the use of (8) the value of J_o is found to be

$$J_o \approx \frac{I_E}{\pi R_E W_E}\sqrt{\frac{2\rho_B I_B}{V_o}}. \tag{32}$$

Inserting typical values of $I_E = 1$ mA, $R_E = 100\ \mu$m, $W_E = 10\ \mu$m, $2\rho_B I_B = 1$ V, $V_o = 26$ mV, the approximate value of J_o works out as 197 A $\cdot$ cm^{-2}.

Acknowledgment

The original carrier-domain multipliers were designed while the author was with Tektronix Inc., Beaverton, Oreg., as part of a free-ranging program of research into superintegrated circuits. Thanks are due to W. Velsink for his encouragement, and to L. Larson for his penetrating observations and criticisms. G. Wilson and his staff in the IC Engineering Division made the devices come to life. The enthusiastic dedication of J. Smith, who wrote the next paper in this issue describing improvements to the multiplier, and who also generously assisted in the preparation of the figures for this paper, is worthy of special mention.

References

[1] G. A. Korn, *Electronic Analog and Hybrid Computers.* New York: McGraw-Hill, 1972.

[2] B. Gilbert, "Translinear circuits: A proposed classification," *Electron. Lett.*, vol. 11, pp. 14-16, 1975.

[3] ——, "A high-performance monolithic multiplier using active feedback," *IEEE J. Solid-State Circuits*, vol. SC-9, pp. 364-373, Dec. 1974.

[4] W. M. C. Sansen and R. G. Meyer, "An integrated wide-band variable-gain amplifier with maximum dynamic range," *IEEE J. Solid-State Circuits*, vol. SC-9, pp. 159-166, Aug. 1974.

[5] J. R. Hauser, "The effects of distributed base potential on emitter-current injection density and effective base resistance for stripe transistor geometries," *IEEE Trans. Electron Devices*, vol. ED-11, pp. 238-242, May 1964.

[6] B. Gilbert, "Resistive conversion device," U.S. Patent 3 524 998 assigned to Tektronix Inc., filed Jan. 26, 1968.

[7] ——, "New planar distributed devices based on a domain principle," in *ISSCC Dig. Tech. Papers*, 1971, p. 166.

[8] J. Smith, "A second-generation carrier domain four-quadrant multiplier," *IEEE J. Solid-State Circuits*, this issue, pp. 448-457.

[9] B. Gilbert, "Monolithic analog READ-ONLY memory for character-generation," *IEEE J. Solid-State Circuits*, vol. SC-6, pp. 45-55, Feb. 1971.

Part VI
Data Conversion Circuits

Data conversion circuits form an essential link between the analog world of direct measurements and the digital world of numerical computation, data transmission and storage. The advent of the microprocessor has made this class of circuits one of the most rapidly expanding areas of monolithic IC's. This part brings together a number of selected papers on the subject of data acquisition and conversion IC's which cover various areas of this rapidly evolving field.

The first group of papers cover the subject of digital-to-analog conversion. In the first paper by Kelson *et al.*, design of a 10-bit D/A converter is described, using ion-implanted resistors. In the next paper, Dooley presents an alternate approach to the same function, using only diffused resistors. This paper also demonstrates a high-speed current-switching technique which has become one of the basic building blocks of later designs. In the third paper, Schoeff describes the design and the performance of an 8-bit companding D/A converter which provides a dynamic range equivalent to 12-bits plus sign. Such nonlinear converters find a wide range of applications in PCM telecommunications, data recording and servo control systems.

In the next group of papers included in this part, the subject of analog-to-digital conversion is examined, and a variety of different design approaches are demonstrated. The fourth and the fifth papers form a two-part series describing the design of all-MOS A/D converters using charge-redistribution techniques. The fourth paper by McCreary and Gray discusses the charge-redistribution technique using a weighted-capacitor network; the fifth paper by Suarez, *et al.* demonstrate an alternative design, which sacrifices speed but conserves chip area by using only two switched MOS capacitors. The subsequent paper by Landsburg discusses the design of a 3-digit monolithic A/D converter which uses a charge-balancing algorithm. The entire converter, with the exception of the voltage reference, is implemented in CMOS technology. In the seventh paper, by van de Plassche and van der Grift describe the design of a bipolar 5-digit plus sign A/D converter, using the sigma-delta modulation principle.

The third group of papers in this part cover an alternate method of data conversion using voltage-to-frequency (V/F), conversion techniques. The method of conversion is particularly suited to direct interface with analog transducers. The paper by Gilbert describes a V/F converter design using a stable oscillator with wide and linear sweep range. In the paper by Schmoock, design of a V/F converter using the charge-dispensing technique is demonstrated.

In the final paper by Zuch, an overview of data converter types and their applications are presented to assist the user in choosing the right converter. The subject of data conversion circuits is a very broad one. The papers included in this part cover only a limited segment of this rapidly expanding field. For a more detailed and in-depth coverage of the monolithic data conversion circuits, the interested reader is referred to a companion book in this series, "Data Conversion Integrated Circuits," edited by D. Dooley and to be published by the IEEE Press.

ADDITIONAL REFERENCES

1. D. Dooley, *Data Acquisition and Conversion Integrated Circuits.* New York, IEEE Press, 1978.
2. H. Schmid, *Electronic Analog/Digital Conversions.* New York: Van Nostrand Reinhold, 1970.
3. D.F. Hoeschele, *Analog-to-Digital and Digital-to-Analog Conversion Techniques.* New York: Wiley, 1968.
4. D.T. Comer, "A Monolithic 12-Bit A/D Converter," *ISSCC Digest of Tech. Papers,* pp. 104-105, February 1977. (Reprinted in Part VIII of this book.)
5. C.E. Woodward, K.E. Konkle, and M.L. Naiman, "A Monolithic Voltage Comparator Array for A/D Converters," *J. Sol. State Circuits,* vol. SC-10, pp. 392-399, December 1975.
6. T. Hornak and J.J. Corcoran, "A Precision Component-Tolerant A/D Converter," *J. Sol. State Circuits,* vol. SC-10, pp. 386-391, December 1975.
7. R.B. Craven, "An Integrated Circuit 12-Bit D/A Converter," *ISSCC Digest of Tech. Papers,* pp. 40-41, February 1975.
8. D. Fullagar, P. Bradshaw, L. Evans, and B. O'Neill, "Interfacing Data Converters and Microprocessers," *Electronics,* pp. 81-89, December 1976.
9. R.J. van de Plassche, "Dynamic Element Matching for High-Accuracy Monolithic D/A Converters," *J. Sol. State Circuits,* vol. SC-11, pp. 795-800, December 1976.
10. L. Mattera, "Data Converters Latch onto Microprocessors," *Electronics,* pp. 81-90, September 1977.
11. P. Prazak, "What Designers Should Know About Data Converter Drift," *Electronics,* pp. 111-114, November 1977.
12. G. Grandbois and T. Pickerell, "Quantized Feedback Takes Its place in Analog-to-Digital Conversion," *Electronics,* pp. 103-107, October 1977.
13. J.A. Schoeff, "A Microprocessor Compatible High-Speed 8-Bit DAC," *ISCC Digest of Tech. Papers,* pp. 132-133, February 1978.
14. P.H. Saul and J.A. Jenkins, "A 10-Bit Monolithic Tracking A/D Converter," *ISSCC Digest of Tech. Papers,* pp. 138-139, February 1978.
15. A.P. Brokaw, "A Monolithic 10-Bit A/D Using I^2 L and LWT Thin-Film Resistors," *ISSCC Digest of Tech. Papers,* pp. 140-141, February 1978.
16. P. Pinter and D. Timm, "Voltage-to-Frequency Converter—IC Versions Perform Accurate Data Conversion at Low Cost," *EDN Magazine,* vol. 22, pp. 153-157, September, 1977.

A Monolithic 10-b Digital-to-Analog Converter Using Ion Implantation

GARY KELSON, MEMBER, IEEE, HANS H. STELLRECHT, MEMBER, IEEE, AND DAVID S. PERLOFF

Abstract—**The design and fabrication of a self-contained 10-b monolithic digital-to-analog converter is described. To overcome the limitations of standard bipolar processing and to achieve a circuit with reasonably low process sensitivities, a new process incorporating ion implantation is used. The circuit has been designed in a manner to fully utilize the characteristics of this process with the objective of high performance along with simple wafer processing. System considerations as well as the design of each component block are discussed.**

I. Introduction

A COMMON problem has plagued monolithic analog integrated circuits since their inception: poor component tolerance. However, early workers in the field quickly recognized the advantages of being able to fabricate large numbers of components, in close proximity to one another, on a single silicon substrate [1].

Designers began using the excellent matching characteristics of integrated components to their advantage as well as the relative economy of active components over passive components. During the ensuing years great strides forward in circuit design techniques and process sophistication have been made to the point where many monolithic circuits outperform their discrete counterparts, often at a lower cost [2]. In spite of these achievements, many circuit applications still exist to challenge both circuit and process engineer with virtually the same tolerance problems. The high-accuracy monolithic digital-to-analog converter (DAC) is such a circuit.

The monolithic circuit to be described in this paper is a completely self-contained 10-b DAC which requires no external precision components. Overall system accuracy (± 0.05 percent) is ensured through the use of well-matched internal components. The unique characteristics of ion implantation have been combined with standard bipolar process technology to achieve the high precision required by this circuit. Thus, the DAC can be manufactured on the same process line with ultrahigh volume circuits such as operational amplifiers with only a minimum of extra handling.

Manuscript received May 18, 1973.
The authors are with the Signetics Corporation, Sunnyvale, Calif. 94086.

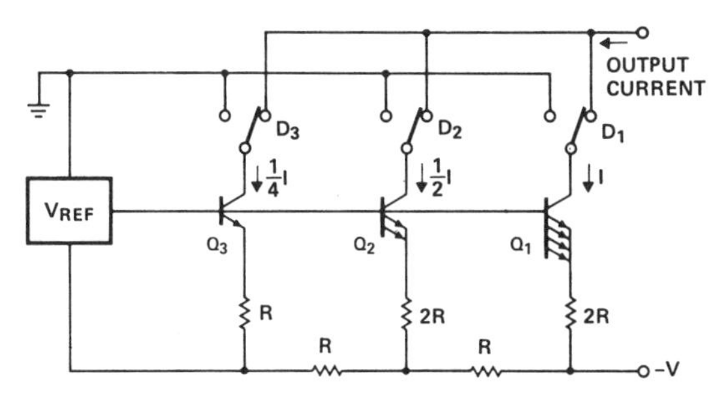

Fig. 1. Summing of binary weighted currents.

II. Circuit Approaches Suitable for Monolithic Realization

Many techniques are available for converting a parallel input digital signal into an analog signal. Most can be categorized into one of three groups:

1) pulsewidth modulation;
2) binary weighting;
3) binary attenuation.

The pulsewidth modulation method offers some advantages for high-resolution monolithic DAC's but is limited to low-speed applications [3].

The second and third categories are very similar but have some rather subtle differences especially when monolithic realization is considered. The binary weighting approach is shown in Fig. 1. Here a 3-b converter is formed by summing the collector currents of Q_1, Q_2, and Q_3 through three digitally controlled switches D_1, D_2, and D_3. These currents are weighted in a binary manner by the R-$2R$ resistor ladder network. Because each transistor operates with a different collector current, the geometry of each transistor must be scaled, as shown, to maintain equal emitter current density and therefore equal base–emitter voltage drops (V_{BE}). Differences in V_{BE} between current sources changes the termination voltage for the R-$2R$ ladder leading to errors in the binary weighting. The problem becomes especially acute for high-accuracy converters; a 10-b DAC requires a 512:1 current ratio. Due to the weighting of emitter currents, a binary weighted voltaged drop is generated across the series emitter resistors. This presents a subtle problem for monolithic integration since resistors are normally formed by reversed biased p and n-type semi-

Reprinted from *IEEE J. Solid-State Circuits*, vol. SC-8, pp. 396-403, Dec. 1973.

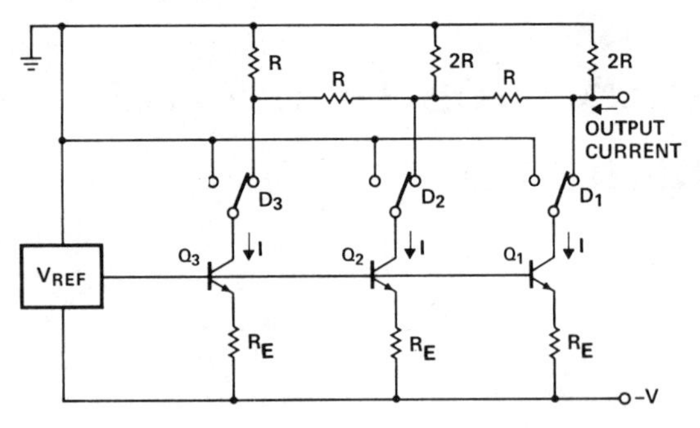

Fig. 2. Binary attenuation of equal valued currents.

conductor regions which are voltage sensitive. The sheet resistance ρ_s for these resistors increases with the voltage drop across the resistor which produces nonlinearities in the DAC output. This phenomenon, characterized by a voltage coefficient of resistance (VCR), is similar to channel pinchoff in depletion mode FET's. Since VCR is proportional to sheet resistance, one can minimize this problem by using low sheet resistance base diffusions for the ladder network. (VCR $\simeq$ 0.025 percent/V for $\rho_s = 135\ \Omega/\square$.) But the value of R in the ladder network cannot be made arbitrarily small since the voltage drop generated across the series emitter resistors $(I \cdot R)$ must be sufficiently large when compared to the V_{BE} mismatches of the current-source transistors. For a 10-b converter, this voltage drop should exceed approximately 1000 times the V_{BE} mismatch [4]. Thus, one must trade off tight transistor matching against increasing $I \cdot R$. Optimization of this tradeoff requires higher power dissipation and larger die areas or very tight masking and process control.

The binary attenuation method shown in Fig. 2 circumvents several of the problems of the binary weighted approach by separating the R-$2R$ ladder network and the series feedback current setting resistors. The current sources formed by Q_1, Q_2, and Q_3 along with the resistors R_E are now identical in value. These currents are switched into the binary attenuator to attain the weighted output current. With the voltage drop across R_E being the same for all bits, VCR will not produce any errors in current cell matching, thus allowing R_E to be made large using a high sheet resistance process. The resistors in the binary attenuator do have differing voltage drops but these can be made negligibly small since R can be made small without sacrificing accuracy. Thus, optimization of the $I \cdot R$ versus V_{BE} matching tradeoff can be made without requiring excess die area or high power dissipation. This scheme has two additional advantages stemming from the identical current cells. V_{BE} matching of identical transistors operating at the same current level is better than scaled transistors operating with scaled currents. Also, each cell has the same transient response when switched, giving faster settling for lower order bits and less output glitching [5].

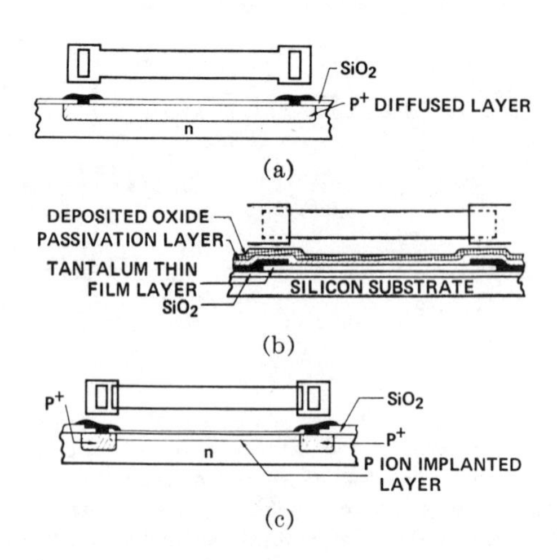

Fig. 3. Cross sections of monolithic resistor technologies. (a) Diffused resistor. (b) Thin-film resistor. (c) Ion-implanted resistor.

III. Monolithic Resistor Technologies

In order to realize the advantages which the equal valued current approach offers, large resistance values are needed. This requires a resistor process with high sheet resistance. Other requirements of the resistor process are high matching accuracy, good stability, and ease of fabrication. Since resistors are such an important factor in DAC's, a statistical comparison was made between the most widely used monolithic resistor technologies with respect to sheet resistance, matching, and temperature sensitivity. The technologies that were examined are

1) diffusion;
2) thin film;
3) ion implantation.

Cross sections of these are shown in Fig. 3.

Diffused resistors, as shown in Fig. 3(a), represent the least complicated technology from a processing standpoint since no additional process steps are required in a standard bipolar process. However, sheet resistance is usually limited to under 200 $\Omega/\square$.

Thin-film resistors, shown in Fig. 3(b), are made by depositing a thin layer of resistive material, tantalum in this case, on an oxidized substrate. Being surface devices, thin-film resistors require careful passivation to maintain long-term stability. Typically, two passivation layers are required to obtain stability sufficient for DAC applications. From a processing standpoint, this is the most complex resistor technology.

An ion-implanted resistor cross section is shown in Fig. 3(c). Usually, the contact beds are formed by the base diffusion. Then a p-type layer is formed by ion implantation. Finally a post-implant heat treatment is used to electrically activate the ions, thereby obtaining the desired sheet resistance. The sheet resistance of the implanted layer is typically much higher than the diffused contact beds so that the resistance value is almost completely determined by the implantation.

Since the ion dose can be controlled very accurately, excellent definition of the absolute value of resistors, as well as good matching between resistor pairs, can be obtained. Also, since the device, unlike a thin-film resistor, is formed within the semiconductor substrate, it is passivated by silicon dioxide in the same manner as a diffused resistor. In other words, the ion-implanted resistor combines the advantages of bulk diffused devices with the high sheet resistance of thin-film devices.

The details of the resistor comparison are given in the Appendix along with a summary of the results. These results indicate that ion-implanted resistors match as well or better than diffused resistors while having nearly an order of magnitude higher sheet resistance. This can be achieved with a process which contains only one additional masking step in conjunction with an easily controlled ion-implantation step. Additionally, since the ion-implanted resistors do not require any high-temperature processing, all previous diffusion steps are unaffected. This is a key feature of the process since it allows all active components in the circuit to be fabricated by a standard diffusion process and the high-accuracy resistors to be formed at a subsequent time without interaction [6].

IV. Digital-to-Analog Circuit Design

A block diagram of the monolithic DAC is shown in Fig. 4. The ten equal valued current cells are biased in parallel from a reference current cell which is part of a negative feedback loop. The feedback loop sets the reference current at a value given by

$$I_{\text{ref}} = \frac{V_{\text{ref}} - V_{\text{mult}}}{R_{\text{ref}}}.$$

In this configuration, the current of each cell will be the same as I_{ref} so long as each cell matches the reference cell and therefore absolute parameter variations are automatically compensated. The multiplying input V_{mult} may be used to obtain 100-percent modulation of the reference voltage and thus the current of each cell. The binary R-$2R$ attenuator generates the weighted output current, I_{OA}. This current may be converted into a voltage V_{OA} which is given by

$$V_{OA} = \frac{R_F}{R_{\text{ref}}}(V_{\text{ref}} - V_{\text{mult}}) \cdot \left(D_1 + \tfrac{1}{2} D_2 + \tfrac{1}{4} D_3 + \cdots + \frac{1}{512} D_{10}\right)$$

where D_n represents the binary inputs.

A simplified schematic drawing of one of the ten identical current cells and the emitter-coupled switch is shown in Fig. 5. The cell current is generated by transistor Q_5 and the series feedback resistor R_{E1}. The base of

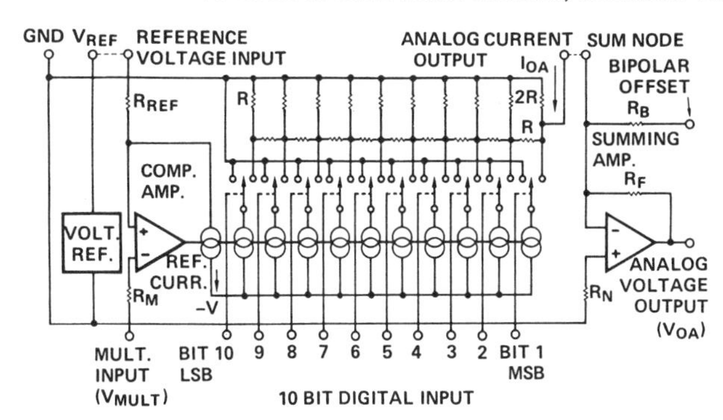

Fig. 4. Block diagram of monolithic 10-b DAC.

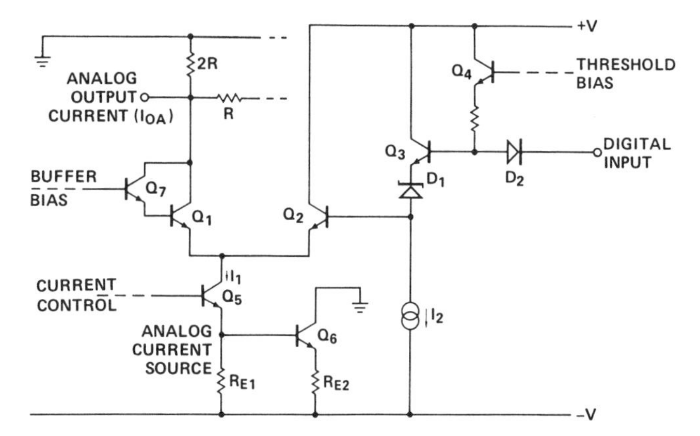

Fig. 5. Simplified schematic of current switching cell.

Q_5 is driven from a common current control bus by the compensation amplifier as shown in Fig. 4. Neglecting, for the moment, Q_6 and R_{E2},

$$I_1 = \frac{\beta}{\beta + 1}\left(\frac{V_{E5} + V}{R_{E1}}\right) = \alpha\left(\frac{V_{E5} + V}{R_{E1}}\right).$$

If the current cells in the DAC are to match with less than 0.1-percent error, the common-base current gain α must also match with less than 0.1-percent error. For a worst case β mismatch of 25 percent, a minimum absolute β of almost 400 would be required. While this is well within present technological limits, it does require some special processing and is not desirable on a circuit of this complexity. This minimum β constraint can be reduced by the addition of transistor Q_6 and resistor R_{E2} [7]. If I_{E6} is made equal to I_{E5} and $0.75 < \beta_{Q5}/\beta_{Q6} < 1.25$, then in the worst case, the error current flowing in R_{E1} due to the finite β of Q_5 is multiplied by at least 4 or the minimum absolute gain is reduced to $400/4 = 100$.

The emitter-coupled current switch, made up of transistors Q_1, Q_2, and Q_7, steers the cell current between the $+V$ bus and the binary attenuator. The Darlington connection of Q_1 and Q_7 minimizes current transmission errors to the attenuator and again reduces the minimum β for good cell matching. High switching speeds are achieved by limiting the voltage swing at the collector

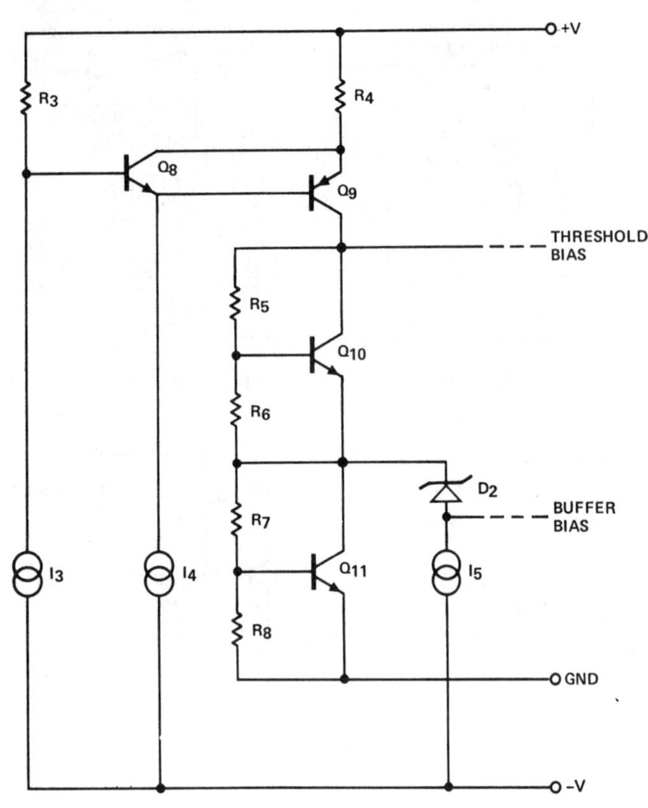

Fig. 6. Biasing circuit for current switching cells.

of Q_5. In order to ensure complete switching under these conditions, the buffer bias V_{BB} and threshold bias V_{TB} must track both temperature and absolute variations of the Zener level-shift diode D_1. These bias voltages are generated in the bias circuit shown in Fig. 6 where

$$V_{BB} = V_{BE}(1 + R_7/R_8) - V_Z$$

$$V_{TB} = V_{BE}(2 + R_5/R_6 + R_7/R_8).$$

The emitter-coupled pair Q_1 and Q_2 (Fig. 5) will switch when $V_{B1} \simeq V_{B2}$. Thus

$$V_{in}\ (\text{threshold}) - V_Z \simeq V_{BB} - V_{BE}$$

or

$$V_{in}\ (\text{threshold}) \simeq V_{BE}\ (R_7/R_8)$$

which is independent of V_Z. In addition, the voltage swing ΔV at the collector of Q_5 is given by

$$\Delta V = V_{TB} - 3V_{BE} - V_Z - (V_{BB} - 2V_{BE})$$

$$= V_{BE}(R_5/R_6).$$

For TTL compatibility V_{in} (threshold) $= 1.5$ V. Thus $R_7/R_8 = 2.15$. If $\Delta V = 450$ mV then $R_5/R_6 = 0.69$. Under these conditions the propagation delay and settling time (to 0.1 percent) for the output current I_{OA} is approximately 10 ns and 100 ns, respectively.

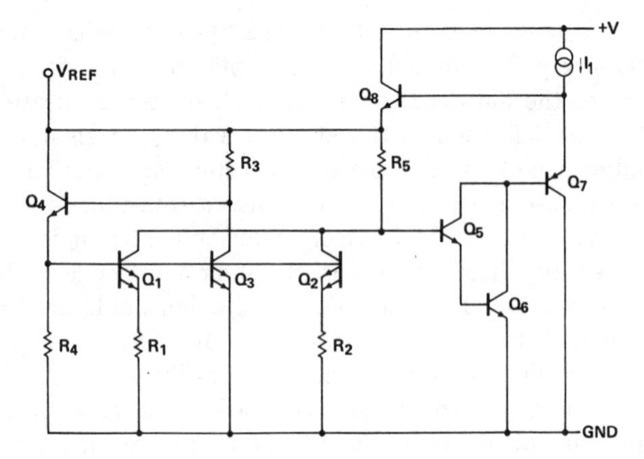

Fig. 7. Simplified schematic of voltage reference circuit.

The voltage reference for a 10-b DAC must have an extremely low temperature coefficient if the circuit is to have a wide operating temperature range. For example, if the DAC is to be ½ LSB accurate over the commercial temperature range (0 to 75°C) the complete converter should not drift more than 10 ppm/°C. Since such low drift is difficult to achieve with present technology, it is customary to simply specify the full-scale drift of the DAC output.

Two kinds of reference circuits which are suitable for monolithic integration are Zener diode circuits and circuits which make use of the band-gap principle [8]. A band-gap-type circuit was chosen for the 10-b DAC because the temperature compensation depends, to a first order, only on V_{BE} matching and resistor ratios. Zener diode references normally are more surface sensitive and depend on the absolute value of the base sheet resistance, which is more difficult to control.

A simplified circuit diagram of the voltage reference circuit is shown in Fig. 7. This circuit uses the positive temperature drift of the V_{BE} difference (ΔV_{BE}) between transistors operating at different current densities to cancel the negative drift of the absolute base–emitter voltage. The combination of transistors Q_1, Q_2, and Q_3 provide the positive drift component by forcing ΔV_{BE} to appear across the parallel combination of resistors R_1 and R_2. This voltage is multiplied by $-R_5\ (R_1 + R_2)/R_1R_2$ and compensates for the negative drift of the Darlington pair Q_5 and Q_6. The transistors Q_5, Q_6, Q_7 and Q_8 form a high gain feedback loop which provides regulation and low output impedance for the reference voltage. The approximate value of the reference voltage (V_{ref}) is given by

$$V_{ref} \simeq V_{BE_5} + V_{BE_6} + \frac{R_5(R_1 + R_2)}{R_1\,R_2}\frac{kT}{q}\ln\left(\frac{J_3}{J_{1,2}}\right)$$

where $J_{1,2}$ and J_3 are the current densities in transistors Q_1, Q_2, and Q_3, respectively. These current densities are determined by the resistors R_1, R_2, and R_3 as well as by the ratio of emitter areas between Q_1, Q_2, and Q_3. To ensure tight thermal tracking of devices within the reference circuit, and to minimize the effects of chip thermal gradients, the transistors Q_1, Q_2, and Q_3 are made from groups of parallel connected transistors which are topologically symmetric about Q_5 and Q_6. Similar layout considerations were used for the resistors R_1, R_2, R_3, and R_5. With this circuit, temperature drifts of less than 10 ppm/°C have been realized over the 0 to 75°C temperature range.

To convert the DAC output current I_{OA} into a voltage, a transimpedance amplifier is required, as shown in Fig. 4. The transimpedance

$$\frac{V_{OA}}{I_{OA}} = R_F$$

is realized with an operational amplifier connected in the current summing mode. In order not to degrade the accuracy and speed of the basic DAC, the summing amplifier requires several important features:

1) high open-loop gain;
2) low input offset voltage;
3) high slew rate and low settling time.

From Fig. 4 the transfer function for the summing amplifier is given by

$$V_{OA} = \frac{R_F(I_{OA} + I_{OS}) + V_{OS}(1 + R_F/R_N)}{1 + (1/A_{VOL})(1 + R_F/R_N)}$$

where I_{OS} and V_{OS} are the input offset current and voltage, respectively; A_{VOL} is the open-loop gain; and R_N is the R-$2R$ ladder source resistance.

With a full-scale output voltage of 10 V the total error introduced by the summing amplifier must be much less than 5 mV ($\frac{1}{2}$ LSB). For typical operational amplifiers, I_{OS} is negligible compared to I_{OA} for the least significant bit (1 μA). If the error introduced by the summing amplifier is split equally between the V_{OS} term and the finite open-loop gain, then for I_{OA} (full scale) $= 1$ mA, $R_F =$ 10 kΩ and $R_N = 2$ kΩ,

$$V_{OS} < 0.5 \text{ mV}$$

$$A_{VOL} > 90 \text{ dB}.$$

While this open-loop gain is not difficult to achieve in a two-stage amplifier, circuit yields would be low for a single operational amplifier requiring $V_{OS} < 0.5$ mV. Fortunately, one has the freedom to null this offset initially, then the total drift in V_{OS} with ambient changes must not exceed 0.5 mV. For the 0 to 75°C operating range

$$\frac{dV_{OS}}{dT} < \frac{0.5 \text{ mV}}{75°\text{C}} = 6.7 \ \mu\text{V}/°\text{C}.$$

However, for a differential amplifier with balanced loads, there is a direct correlation between dV_{OS}/dT and V_{OS} since tightly matched transistors will track each other more closely [9]. It has been shown that for each millivolt of offset voltage, the offset drift will be approximately 3.3 μV/°C [9]. Thus the greatest input offset voltage which can be tolerated for the summing amplifier is

$$V_{OS}' < 6.7 \ \mu\text{V}/°\text{C}/3.3 \ \text{V}/°\text{C}/\text{mV} \simeq 2 \text{ mV}.$$

This limit should not appreciably diminish yields.

A simplified schematic drawing of the summing amplifier is shown in Fig. 8. A modified second-generation design is used with feed-forward compensation to enhance the transient response without degrading dc stability. The lateral p-n-p transistors Q_3 and Q_4 have split collectors to reduce the first-stage transconductance and simplify biasing. This technique provides high slew rates by allowing the first-stage bias current to be increased without increasing the compensation capacitor C_2 [10]. The second stage is a resistively loaded common-emitter amplifier comprised of Q_9 and R_4. A low parasitic ion-implanted resistor is used for R_4 to maintain wide bandwidth. High-frequency input signals are fed forward around the narrow band, first stage by C_1 and R_9. A frequency-compensation capacitor C_2 of approximately 5 pF is used to control the damping factor or output settling time. While dissipating less than 25 mW the summing amplifier has a slew rate of 20V/μs and settles to 0.05 percent in 2 μs.

V. Integrated-Circuit Layout and Performance

The inherent accuracy of the DAC described here is heavily dependent on component matching. This has been stated repeatedly in the descriptions of each of the blocks for the converter system. Integrating all of these blocks on a single silicon substrate presents numerous problems. High-speed switching signals must be isolated from high-accuracy reference signals. Power dissipation on the chip must be distributed so that variations of the output level do not produce thermal feedback which can seriously affect output linearity. To overcome these potential problems, the current switches and R-$2R$ ladder are arranged in a horseshoe pattern as shown in Fig. 9. With this arrangement the center line of power dissipation is independent of input code. Power buses are symmetrically arranged and fed so that ohmic metal drops do not affect accuracy. The summing amplifier is located symmetrically with respect to the reference and feedback resistors as well as the voltage reference and compensation amplifier. This configuration makes the output linearity virtually independent of load current. The chip size is 113 mil by 124 mil and contains 140 transistors, 25 diodes, and over 530 kΩ of precision resistance.

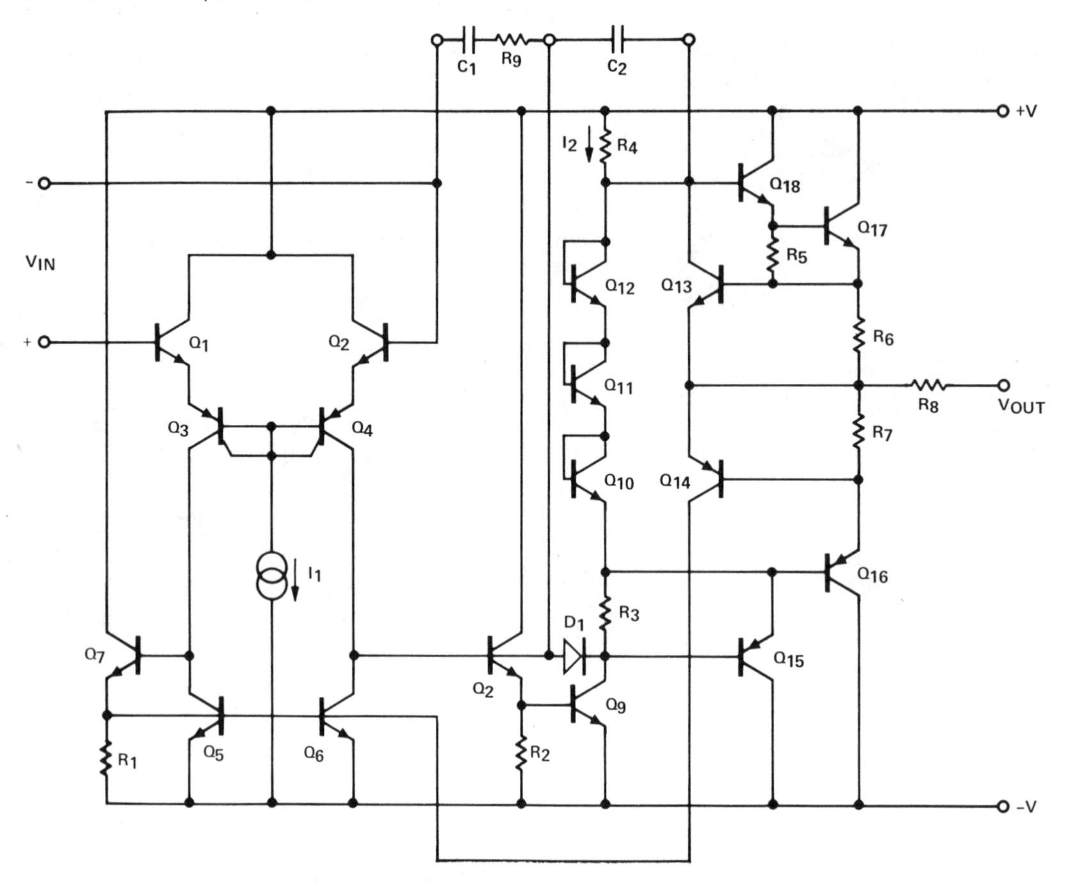

Fig. 8. Simplified schematic of summing amplifier.

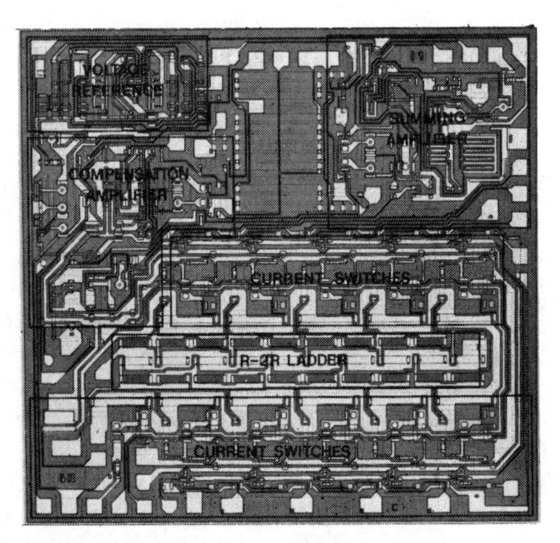

Fig. 9. Photomicrograph of 10-b DAC.

Fig. 10 shows the typical connection of external components for the DAC which requires only two external adjustments to provide 10-b accuracy. The zero adjustment nulls the system offsets, which are typically less than 5 mV, by varying the voltage on the noninverting input to the summing amplifier. Full scale is adjusted by varying the multipling input which changes the effective reference voltage. Due to the absence of thermal feedback, these two adjustments are sufficient to achieve $\frac{1}{2}$ LSB accuracy for the converter over a 0 to 50°C operating range. The performance of the system is summarized in Fig. 11. The temperature coefficient (TC) of the basic converter, with an external zero TC voltage reference, is less than 5 ppm/°C. The output slew rate when operating in the multiplying mode is limited to 5 V/μs by the compensation amplifier in the current cell bias system.

VI. Conclusion

The design and fabrication of a 10-b monolithic DAC has required the optimization of both circuit and process technologies. To meet the severe accuracy requirements ($\pm$0.05 percent) in a complex system, a circuit has been designed around a process which complements the design without overly complicating processing.

This has been accomplished through the use of ion implantation to form the precision resistor ladder networks. Electrical and thermal interactions between components of the DAC have been analyzed to ensure that the system performance is not compromised by integration. The realization of a high-accuracy (10-b) monolithic DAC indicates that integrated-circuit technology has not only moved toward large-scale systems but also toward systems which can provide the precision only previously available with discrete or hybrid techniques.

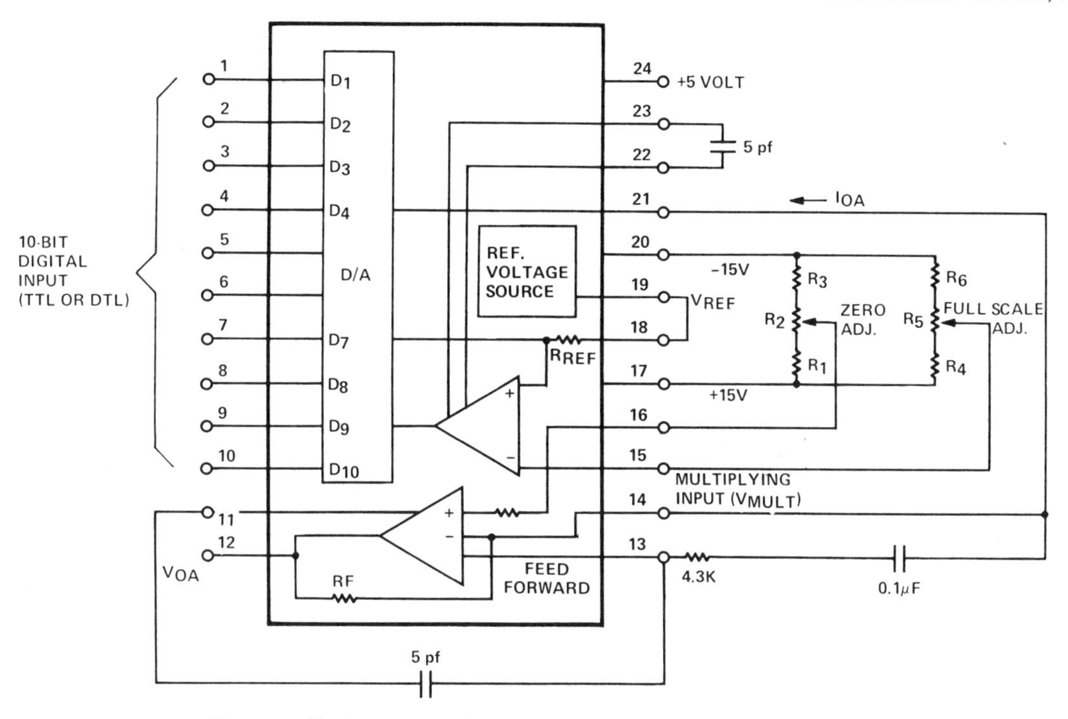

Fig. 10. Typical connection of external components for DAC.

OUTPUT VOLTAGE .... 0 TO ±5 V, 0 TO 10 V
LINEARITY 0.05%
FULL SCALE TEMPERATURE DRIFT
WITH INTERNAL REFERENCE .... ±15 ppm/°C
VOLTAGE REFERENCE DRIFT .... ±10 ppm/°C
POWER SUPPLY REJECTION .... .005%/%
SWITCHING PERFORMANCE
SETTLING TIME
CURRENT OUTPUT .... 200 ns TO .05%
VOLTAGE OUTPUT .... 2 μs TO .05%
OUTPUT SLEW RATE
DIGITAL INPUT .... 20 V/μs
ANALOG INPUT (V_{MULT}) .... 5 V/μs
POWER CONSUMPTION .... 350 mW

Fig. 11. 10-b DAC performance.

Appendix

In order to determine a suitable resistor process for high-precision linear integrated circuits such as DAC's, a comparative evaluation of resistors made by the standard-diffusion, by thin-film, and by ion-implantation processes, was carried out. The experiment was done with production processing, and a large number of devices was processed so that statistical analysis could be performed on the data. The statistical analysis included measurement of electrical parameters of resistor pairs with different geometries, calculation of distributions for absolute and matching tolerances, and calculation of percentile distributions. Fig. 12 shows a block diagram of the measurement procedure.

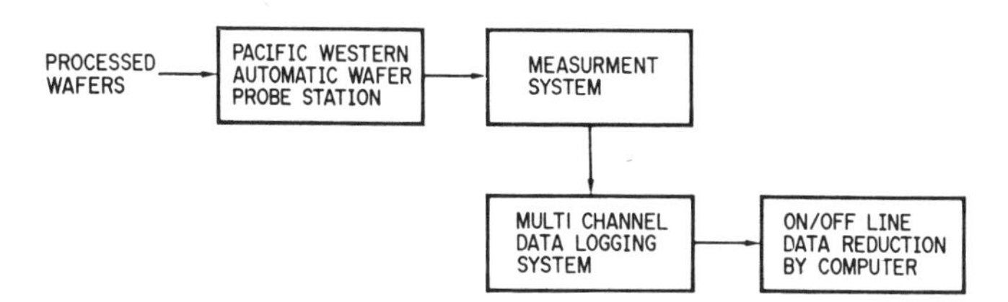

Fig. 12. Block diagram of measurement procedure.

The results of this experiment are summarized in Table I. The nominal sheet resistance values of 1000 and 1250 Ω/□ for the thin-film and ion-implant process, respectively, were chosen for both process and layout compatibility. A wide range of sheet resistance values are available for both processes. The mean and standard deviation for matching are defined as follows:

TABLE I
EXPERIMENTAL RESULTS OF COMPARATIVE RESISTOR EVALUATION

FABRICATION PROCESS	NOMINAL SHEET RESISTANCE OHMS/SQUARE	MATCHING TOLERANCE				TEMPERATURE COEFFICIENT ppm/°C
		σ (%)		MEAN (%)		
		10μ	40μ	10μ	40μ	
DIFFUSION	135	.44	.23	-0.1	0.07	1500
THIN FILM	1000	.24	.11	-0.1	-0.06	-200
ION IMPLANTATION	1250	.34	.12	-0.04	0.05	+400

$$\text{mean match} = \overline{\frac{\Delta R}{R}} = \frac{1}{N} \sum_{i=1}^{N} \frac{\Delta R_i}{R_i}$$

$$\sigma \text{ match} = \left[\frac{1}{N-1} \sum_{i=1}^{N} \left(\frac{\Delta R_i}{R_i} - \overline{\frac{\Delta R}{R}} \right)^2 \right]^{1/2}$$

where $\Delta R_i = R_{i1} - R_{i2}$ and $R_i = (R_{i1} + R_{i2})/2$, R_{i1} and R_{i2} being the ith matched pair. The four columns in Table I under matching tolerance give the sigma and the mean for resistors of 10 μ and 40 μ linewidth, respectively.

An independent calculation of mean and standard deviation allows a better comparison of the three processes. The mean reflects biased errors due to mask offsets and/or geometry and is relatively independent of the particular technology used. The standard deviation by contrast indicates statistical variations characteristic of the uniformity of the particular process. Also, the effect of resistor width on matching is more evident in this comparison in that wider resistors have significantly smaller standard deviations in all cases. The TC's of resistance which were obtained for the three different processes are given in the last column of Table I. It should be noted that the temperature sensitivity of ion-implanted resistors can be varied over a wide range, depending on the ion-implant process variables such as implant dose, energy, and anneal conditions [6].

Acknowledgment

The authors would like to thank Dr. J. Marley, Dr. J. T. Kerr, Dr. J. Conragen, J. Ennals, F. Coburn, and Dr. D. Kleitman for their technical contributions and support; H. Hart and R. Sahm for their characterization efforts; and M. Luft for her secretarial assistance.

References

[1] R. J. Widlar, "Some circuit design techniques for linear integrated circuits," *IEEE Trans. Circuit Theory,* vol. CT-12, pp. 586–590, Dec. 1965.

[2] H. Johnson, "The anatomy of integrated-circuit technology," *IEEE Spectrum,* vol. 7, pp. 56–66, Feb. 1970.

[3] W. S. Schopfer, "A variable pulsewidth D/A," in *ISSCC Dig. Tech. Papers,* Feb. 1971.

[4] D. F. Hoeschele, *Analog to Digital and Digital to Analog Conversion Techniques.* New York: Wiley, 1968, pp. 108–120.

[5] D. H. Sheingold, *Analog-Digital Conversion Handbook,* Analog Devices, Inc., Norwood, Mass., p. I-50, 1972.

[6] H. H. Stellrecht, D. S. Perloff, and J. T. Kerr, "Precision ladder networks using ion implantation," in *1971 WESCON Tech. Papers,* Session 28.

[7] M. B. Rudin, R. O'Day, and R. Jenkins, "System/circuit device considerations in the design and development of a D/A and A/D integrated circuits family," in *ISSCC Dig. Tech. Papers,* Feb. 1967.

[8] R. J. Widlar, "New developments in IC voltage regulators," *IEEE J. Solid-State Circuits,* vol. SC-6, pp. 2–7, Feb. 1971.

[9] J. G. Graeme, G. E. Tobey, and C. P. Huelsman, *Operational Amplifiers.* New York: McGraw-Hill, 1971, pp. 54–55.

[10] R. W. Russell and T. M. Frederiksen, "Automotive and industrial electronic building blocks," *IEEE J. Solid-State Circuits,* vol. SC-7, pp. 446–454, Dec. 1972.

A Complete Monolithic 10-b D/A Converter

DANIEL J. DOOLEY, MEMBER, IEEE

Abstract—A monolithic 10-b plus sign D/A converter has been developed that incorporates all necessary circuit functions including voltage reference and internally compensated high-speed output op amp in a single 82 × 148 mil chip. A unique logic switch and current source configuration achieves 0.05 percent nonlinearity with ±10 V compliance current output option as well as true or complementary binary coding. The design constraints and area requirements for scaling of current source emitter areas are reduced by using a new active current-splitting technique. The circuit features a 1.5 μs settling time voltage output and sign-magnitude coding.

I. Introduction

Monolithic D/A converters with 10-b resolution and accuracy have received considerable attention [1]–[3]. Although there are numerous approaches in both design and technology to obtain a high-performance D/A converter, the one which will be described uses an all-diffused approach to achieve a complete single-chip 10-b plus sign converter. The new converter uses positive logic sign-magnitude coding to provide resolution equal to a conventional 11-b offset binary converter while achieving 0.05 percent nonlinearity over the entire 0–70°C range.

To offer both excellent performance and functional simplicity to the user, the basic design goal was to provide maximum function on the die without sacrificing D/A converter performance. By incorporating a voltage reference, internally compensated high-speed output op amp, and R-$2R$ ladder network on the die, the need to specify, purchase, and mount external components to achieve a D/A function is eliminated. Other building blocks included are the reference buffer amp, temperature compensated current source driver amp and current sources, polarity (sign) inverting amp, and Schottky-clamp current switches.

II. Circuit Description

A. Basic Approach

A simplified diagram of the circuit organization is shown in Fig. 1. The R-$2R$ resistor ladder network determines the binary weighting of the precision current sources. The switched precision current sources have their collector currents steered either to digital ground or to the output current sum line as determined by the input logic condition. A buffered reference voltage input pin allows the user to select the internal reference or apply an external reference, and it also eliminates differential temperature coefficient problems between the external full scale adjust resistors and the internal diffused resistors. Full scale output is adjusted by dividing the reference voltage with a potentiometer and applying it to the reference input terminal. Internal voltage reference is generated by a temperature-compensated zener diode powered from a current source for good power supply rejection. Its output is brought to an external pin. The buffered reference voltage is used to set a voltage regulator which drives the bases of each binary current source. This regulator compensates for the V_{BE} and finite beta of the current sources by including an identical current source operating at the third bit's current in its feedback loop. The collector current of this current source is held equal to V_{REF}/R_{REF} by the regulator op amp. To insure V_{BE} and beta compensation of all bits, their operating current densities are made equal by scaling their emitter geometries. This compensation is effective over the full operating temperature and power supply voltage ranges.

The current sum line is steered under logic control either to the inverting or noninverting terminal of the sign bit op amp, while the output op amp and the sum resistor provide the current-to-voltage translation. Since the output summing resistor tracks R_{REF} over temperature, both being diffused resistors, the output has essentially the same low temperature coefficient as V_{REF}.

B. Current Switch/Current Source

The DAC uses current-steering logic in the collectors of the current source transistors to switch the bit currents; a high input will cause each bit's current to be drawn from the sum line, while a low input will cause it to be drawn from digital ground. A circuit diagram for the switched current source is shown in Fig. 2. The current switch consists of a Schottky-clamped ECL gate ($Q4$, $Q5$, $Q6$, $Q7$, $S1$, $S2$) and a current-steering network forward by $Q2$ and $Q3$. Since the gate output is clamped by Schottky diodes, its output is constrained to be a Schottky diode drop higher than bias line 2 (logic "0") or lower than bias line 2 (logic "1"). This, in turn, drives the base of the switch transistor $Q3$ above or below the base of sum line transistor $Q2$, therefore steering the collector current of $Q1$ either through the sum line transistor to the output line or through the switch transistor $Q3$ to digital ground. Examining the logic gate, a 1.4-V

Manuscript received May 21, 1973; revised August 16, 1973.
The author is with Precision Monolithics, Inc., Santa Clara, Calif. 95050.

Reprinted from *IEEE J. Solid-State Circuits*, vol. SC-8, pp. 404-408, Dec. 1973.

monoDAC-02 SIMPLIFIED SCHEMATIC

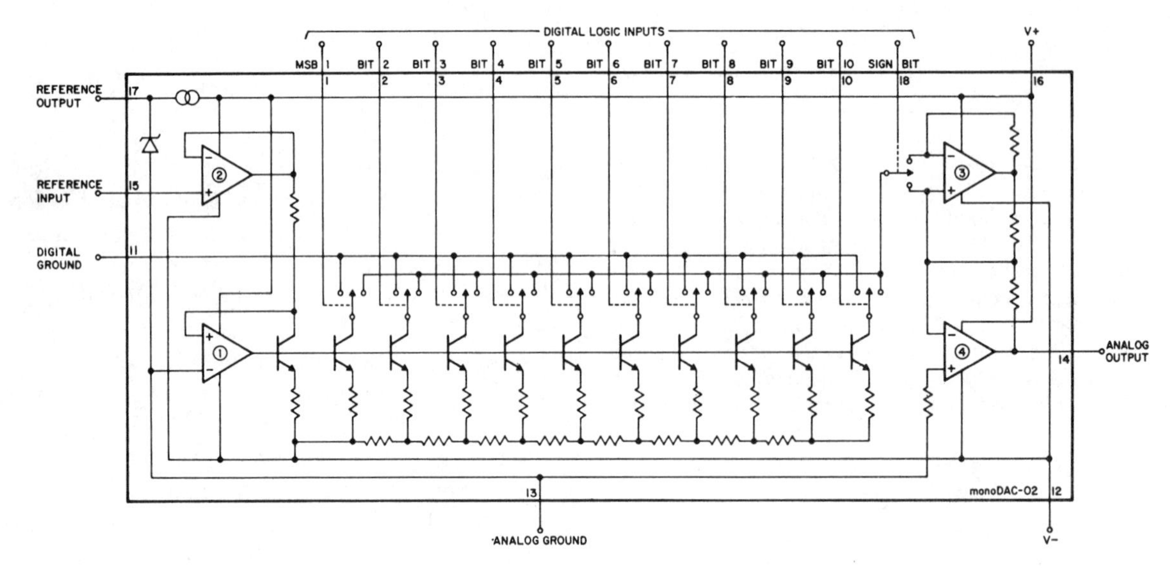

Fig. 1. monoDAC-02 simplified schematic.

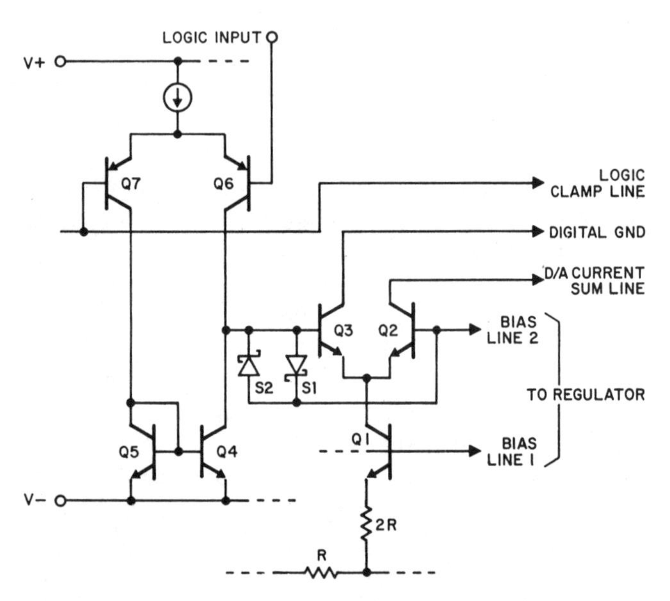

Fig. 2. Circuit diagram of the current source/switch.

logic clamp line is connected to the base of p-n-p transistor $Q7$. When a logic zero is applied at the input, the emitter of $Q6$ clamps the emitter of $Q7$, holding it off and steering the emitter current through $S1$. This will pull up the base of $Q3$, turning the sum line transistor $Q2$ off. When a logic "1" is applied, $Q6$ is clamped off and the emitter current is steered through $Q7$ and the diode-transistor current source $Q5$ and $Q4$. This steers the collector current of $Q4$ through $S2$, pulling the base of $Q3$ down turning the bit on.

This type of current-steering configuration has several advantages over other circuit approaches. It has excellent dynamic characteristics due to Schottky clamping the output swing to two diode voltage drops regardless of input logic levels. The gate has a typical turn-on time of 40 ns and turn-off time of 60 ns. The lateral p-n-p logic input devices have a $BV_{EBO} \geq 45$ V, permitting logic input voltage swings equal to the power supply voltages while maintaining compatibility with all standard forms of logic. This configuration has an additional advantage over conventional level shift current-steering approaches since it minimizes switch signal feedthrough to the sum line. This is due to isolating logic input swings from the output current line. Other advantages are 1) it avoids direct switching of precision current sources, thus preventing electrical transients from propagating on the ladder network; 2) it provides constant loading on bias line 1 which eases constraints on the regulator; 3) power dissipation is held constant in the R-$2R$ network and current sources, thus eliminating error-producing changes due to thermal gradients; and 4) it is compatible with the more easily integrated low-value R-$2R$ ladder network.

The logic switch current source design also allows flexibility for the chip. First a high-voltage compliance current output can be obtained by a simple metal mask option. A voltage swing of approximately ±10 V can be obtained using the current output to directly drive a resistive load. This permits realization of the dynamic performance (300 ns) which a current output allows. Since the diffused R-$2R$ ladder network does not need to track the sum resistor or R_{REF}, the temperature coefficient of the diffused resistors will not reflect into the current output. An additional feature which is easily achieved with the logic switch is positive or negative input logic options. The input logic gate is a simple comparator circuit which compares the logic signal with the 1.4-V logic clamp line. By merely interchanging the two inputs, the output coding will either be "true" binary or the complement.

C. R-2R Resistor Ladder Network

An R-$2R$ resistor ladder network was chosen to determine the binary weighted bit currents because of the greater ease in matching similar values in monolithic form and the lower total resistance required. For a 10-b D/A converter, a straight binary resistor network would require 500–1000 MΩ of total resistance, while partitioning into 4-b sections would require approximately 250 kΩ of total resistance. It can be seen that these approaches compare unfavorably to the 40 kΩ required for an R-$2R$ ladder network. However, there are several disadvantages to an R-$2R$ configuration. The first disadvantage is that voltage drops across the resistor elements in the LSB's become small. This results in V_{BE} mismatch contributing a greater error to the total nonlinearity of the converter. In addition, the R-$2R$ resistor network must be terminated in its characteristic impedance, which could lead to the need for additional components. However, these disadvantages can be circumvented with a simple circuit technique discussed in Section II-D.

Because the precision resistor ladder network required to binarily weight the current sources represents the greatest single error source in a monolithic D/A converter, considerable attention [1]–[3] has been focused on the technology used to obtain the resistive layer. Thin-film and diffused resistors are traditional approaches, with ion implant being relatively new. However, the photolithography process used to define the resistor length and width is common to all these technologies and represents the primary source of error in resistor matching. If the total error for a 10-b D/A converter (0.05 percent) were assigned to matching the width of adjacent 1-mil-wide resistors, the maximum difference in width allowable would be 0.013 μ. This minute value illustrates the degree of dimensional control necessary independent of the resistor technology used.

Although there are advantages to be realized with thin-film and ion implant resistors through a wider range of available sheet resistances [2], improved absolute temperature coefficient of resistance, and voltage coefficient of resistance, none of the three resistor technologies yields an inherently superior resistor match due to photomasking limitations.

D. Active Current-Splitting Network

A simple active current-splitting technique shown in Fig. 3 solves the difficulties discussed above and simultaneously increases the speed of the least significant bits. The circuit adds a device ($Q12$) in parallel with the sum line transistor ($Q11$) with its collector connected to digital ground. The added device has its emitter area precision scaled such that current from the constant current source is split between the sum line transistor and the current-splitting transistor in the ratio of their emitter areas. Therefore, the constant current source can be operated at higher current levels to terminate the ladder, maintain the same current density as other precision binarily weighted current sources, yet obtain the correct weighted current in the output sum line. Since the speed of the current output of the D/A is limited by the bit current in the LSB available to drive the nodal capacitance, the speed will be increased by using current splitting and operating the LSB current source at a higher current. Another problem that current splitting solves is the range of scaled emitter areas necessary to achieve V_{BE} matching and compensation of transistor beta and V_{BE} over temperature. To scale all emitter areas from the most significant bit to the least significant bit would require the MSB emitter area to be 512 times larger than the LSB emitter area. Since this would be prohibitive with state-of-the-art photomasking technology, another solution must be found. The current-splitting technique provides the soltuion. Since a scaling ratio of 32:1 can be achieved in a reasonable area with today's technology, the first six current sources can be scaled normally. By keeping the remaining current sources the same current density as bit 7 and using current splitting, the scaling ratio can be effectively increased to 1024:1, but the actual implementation is with 32:1 scaling of emitter areas.

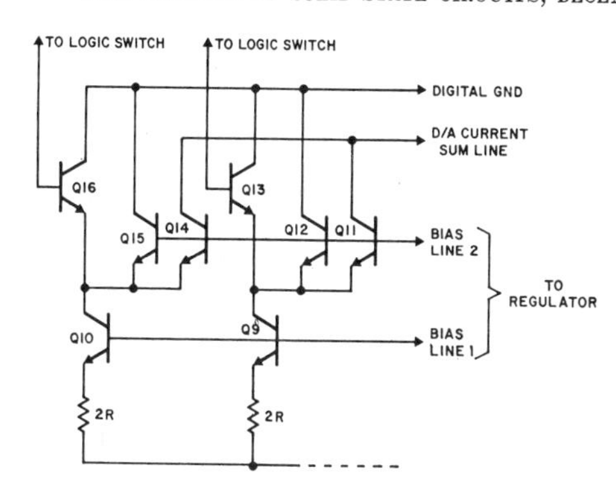

Fig. 3. Active current-splitting network/current source.

E. Sign-Bit Op Amp/Switch

The DAC uses positive logic for both the magnitude bits and the sign bit, i.e., an input of all "ones" produces a full scale output of positive polarity, while an input code of "0" on the sign bit and "ones" on the magnitude bits produces a negative full scale output. A block diagram of the sign bit circuit implementation is shown in Fig. 4. The current sum line is switched under logic control either to the inverting or noninverting input of the sign bit op amp (op amp 3). When the switch is connected to the noninverting terminal, the sum line current is drawn directly from the output op amp through the sum resistor R_s. The D/A will now be in the positive sign condition with all logic codes producing a positive output voltage. When the sign bit is switched to a logic "0," the sum line is moved to the inverting terminal of the sign bit op amp. Sum line current will now be drawn through R_1 of the sign bit amplifier, producing a positive voltage at its output equal to $I_S R_1$. This voltage will then

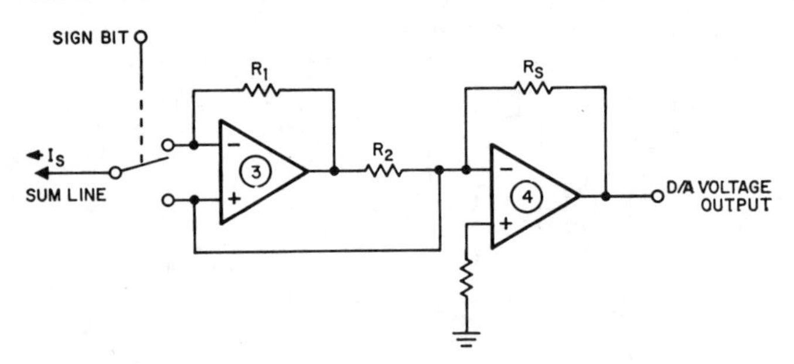

Fig. 4. Sign bit current-inverting amplifier block diagram.

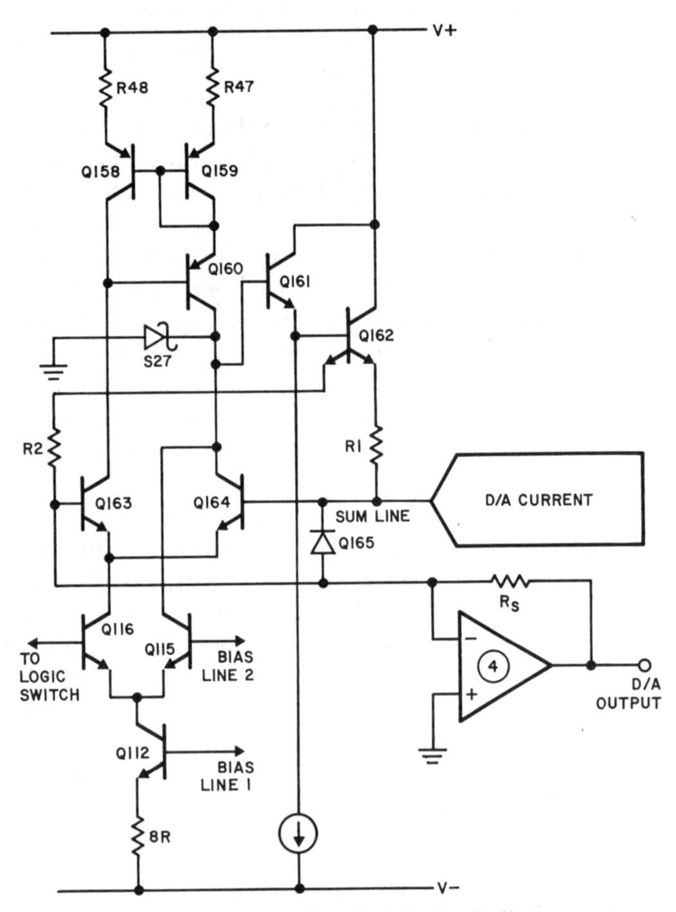

Fig. 5. Sign bit amplifier/switch circuit diagram.

be applied to R_2. With R_1 equal to R_2, a current equal to the sum line current will then be generated of opposite polarity. This current applied to the inverting terminal of the output amplifier will result in negative output voltages for all input codes.

The circuit configuration for implementing the block diagram discussed above is shown in Fig. 5. Since accurate voltage switching is difficult to implement, current steering is used. The current source for the gain stage and logic switch configuration is exactly the same as used in the current source/switch section of the D/A converter. However, instead of steering the current source output ($Q112$) between the sum line and digital ground, it is now steered between the emitters of the differential stage and the collector of $Q164$. This provides the control not only to turn the op amp gain stage on and off, but also pull the op amp off with the collector current of $Q115$. This permits the output of the sign bit op amp to be turned off with only 50 ns additional delay in the DAC output. In the off condition, the collector of $Q164$ is clamped a Schottky diode drop below ground, holding $Q164$ and $Q162$ off, steering the D/A current through the sum line diode $Q165$ to the output amplifier. This produces a positive sign for all input codes. When the sign bit op amp is turned on, the sign bit op amp drives the bias across $Q165$ to 0 V, turning it off. The D/A current is then supplied by the sign bit op amp, which in turn delivers an equal and opposite polarity current to the output op amp. Thus a negative output is produced for all logic input codes.

TABLE I
PERFORMANCE CHARACTERISTICS FOR THE MONOLITHIC D/A CONVERTER

Resolution	10 b + sign (11 b)
Nonlinearity	0.05 percent
Full scale tempco—internal ref	±60 ppm/°C
Output voltage options	0 to ±5 V, 0 to ±10 V
Settling time (±5 V out)	1.5 μs to 0.05 percent
Reference input bias current	100 nA
Reference input impedance	200 MΩ
Zero offset	±2 mV
Zero scale symmetry	±0.5 mV
Power supplies	±12 V to ±18 V
Power supply rejection	0.015 percent FS/V
Power consumption at ±15 V	300 mW max
Logic input current	1 μA
Logic levels	V_{IH} = 2.0 V min, V_{IL} = 0.8 V max

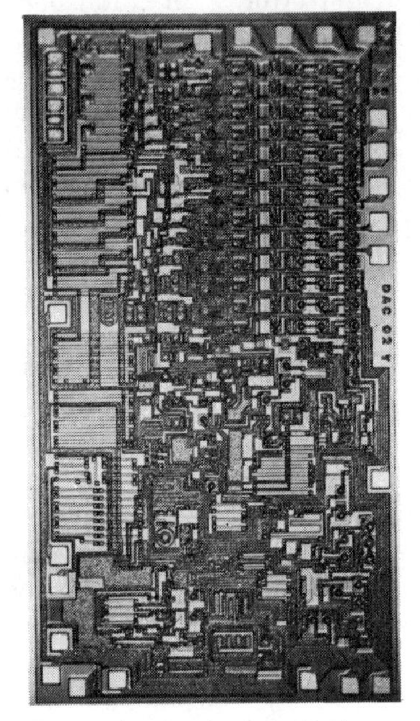

Fig. 6. Photomicrograph of the D/A converter chip.

III. COMPLETE 10-b PLUS SIGN D/A CONVERTER

Typical performance characteristics for the monolithic D/A converter are shown in Table I. The circuit functions described are obtained on a 82 × 148 mil chip which contains 120 n-p-n transistors, 62 p-n-p transistors, 6 capacitors, 33 Schottky diodes, and 280 kΩ of resistance. A photomicrograph of the DAC chip is shown in Fig. 6.

IV. Conclusions

The techniques described show that a high-performance monolithic D/A converter can be achieved using all-diffused technology. A unique current-steering network provides the required level shifting of the input logic levels, yet provides a high-voltage compliance current output. By using a simple active current-splitting technique, device geometry constraints for V_{BE} and beta compensation can be eased while simultaneously increasing the dynamic performance. The device provides maximum D/A function without sacrificing D/A converter performance.

Acknowledgment

The author wishes to express his appreciation to H. J. Bresee for the process development; W. E. Rinde for the breadboarding, test, and evaluation; Dr. G. H. Wilson for valuable discussions on the circuit design; and G. Erdi for originating active current-splitting.

References

[1] W. R. Spofford, Jr., "A 10-bit linear monolithic D/A converter," in *ISSCC Dig. Tech. Papers,* Feb. 1973.
[2] H. H. Stellrecht and G. Kelson, "A self-contained 10-bit monolithic D/A converter," in *ISSCC Dig. Tech. Papers,* Feb. 1973.
[3] D. J. Dooley, "A complete monolithic 10-bit D/A converter," in *ISSCC Dig. Tech. Papers,* Feb. 1973.

A Monolithic Companding D/A Converter

John A. Schoeff

Precision Monolithics, Inc. *

Santa Clara, CA

COMPANDED PULSE CODE MODULATED (PCM) transmission of voice signals has become standardized through widespread use of the Bell system μ-law and the CCITT** A-law transfer characteristics. Until now, all codecs (coder/decoders) for these communication systems have been fabricated in either discrete or hybrid form and have been relatively expensive. This paper will present a newly developed monolithic digital-to-analog converter specifically designed for compression and expansion of signals according to the existing PCM standard. This converter, however, is not limited to PCM communication, but may be used in other areas such as data acquisition, servo controls, data recording, telemetry, voice synthesis, log attenuation, secure communications, sonar, and many other applications which require a 12-bit plus sign dynamic range and the convenience of an 8-bit digital code.

When used in a telecommunications application, the companding DAC is a complete PCM decoder, with metal options for μ-law and A-law. A one-half step decision level for encoding is provided within the circuit and controlled with the encode/decode logic input. This current offsets the entire transfer characteristic one half step, regardless of the value of the output current. The outputs are multiplexed for time sharing of one DAC for both encode and decode operation. The DAC settling time is 500 ns, and it will decode more than 32 PCM channels in 125 μs, which is the sampling period at 8 kHz. In a shared encoder it will convert eight channels, assuming a 10 μs sample and hold acquisition time. The outputs are high impedance, high compliance current sources and will interface with most balanced loads. The reference inputs will accept a fixed reference or a positive or negative multiplying input.

The transfer characteristic of the companding DAC is shown in Figure 1. The output consists of eight positive chords and eight negative chords, each containing sixteen steps. The slopes of these chords are binarily related with the chord at the origin having steps equivalent in size to those in a 12-bit converter. The step size is a nearly constant 3.2% of reading throughout most of the dynamic range, which corresponds to approximately 0.3 dB per step. Each successive chord endpoint is 6 dB below the next higher endpoint for every chord in the A-law specification, and follows this for most chords in the μ-law. The dynamic range, or ratio, of the full scale to the smallest step size is 72 dB for the μ-law version and 66 dB for the A-law unit. The electrical specifications for the circuit are summarized in Table I.

Before the circuit is discussed in detail, it is useful to review the basic principle involved in synthesizing the transfer characteristic. Figure 2 illustrates how each chord current is transferred to the output at the chord carry, and used as a *pedestal* upon which to build the next chord. Thus, the output will be monotonic regardless of the value of the chord current. This assumes that the group of sixteen steps are monotonic, which is easy to achieve in monolithic technology. Each of the 128 output current steps can be identified by the notation $I_{c,s}$ where c = chord number 0 through 7 and s = step number 0 through 15 within the chord.

The block diagram of the circuit is shown in Figure 3. A one of eight decoder, consisting of multicollector differential PNP switches, selects one of the currents in the chord generator, which is fed to the step generator and split into 33 parts. The step generator is shown in greater detail in Figure 4. Thirty of these 33 parts of current make up the 15 steps within the chord, while two of them are the remainder and the last is used as the half step offset for encoding. When a chord carry occurs, a change of 1-1/2 steps of current is seen at the output since the 30 parts of current are removed from the step generator and the entire chord, 33 parts, is transferred to the output.

The pedestal generator selects all the lower order chords and sums them at the output switch to form the pedestal current. In the output switch, the encode/decode input steers the current to one pair of output terminals, adding in the half step offset if the encode pair is selected. The sign bit input then selects either the positive or negative output, both of which are sink current outputs. When a pair of encode or decode terminals is connected to a balanced load, a signed-magnitude output format is generated.

The chord generator is shown in greater detail in Figure 5. Eight chord currents are generated with an overall dynamic range of 128 to 1. Since it is not practical to scale transistor emitters over a range this wide, a new approach called a master-slave ladder is used. The remainder current of the master ladder is fed to the slave ladder, which divides the current into the four lower order chord currents. The base current of this ladder network is compensated by additional circuitry. In a conventional binary DAC, the least significant bits can be less accurate than the most significant bits, so very few resistors would be necessary in the slave ladder and scaled emitters would generate the bit currents. Thus, when applied to a binary DAC, this technique requires the fewest total resistors. The companding DAC, however, must have 12-bit accuracy in the lowest chord, which corresponds to a match of ± 0.01% with respect to full scale. Thus a large non-R-2R resistor array was necessary in the slave network.

The step generator uses a network of scaled devices to generate the steps. Since the step generator performance requirements are similar to that of a four-bit DAC, no resistors are necessary to guarantee monotonicity within the chord. A much greater challenge, however, is presented in the need for high speed switching of current levels as low as 0.5 μA. A fully differential current switch, which eliminates the need for capacitive charge and discharge of low current nodes, provides a propagation delay of approximately 40 ns over a current range of 1 mA to 0.5 μA. This is a dynamic range of more than three decades.

The use of the companding DAC in a codec, or encoder/decoder, is shown in Figure 6, where the addition of a successive approximation register (SAR) and a precision comparator enable the DAC to be used on a time-shared basis for both encoding and

Reprinted from *1977 IEEE Internat. Solid-State Circuits Conf. Digest of Tech. Papers,* Feb. 16-18, 1977, pp. 58-59.

decoding. In instrumentation applications, this circuit would be used as a transceiving converter which is multiplexed between A/D and D/A conversion. The current mode output switch allows the successive approximation loop to be broken, and the SAR becomes a serial to parallel register which feeds the DAC for decoding an incoming data word. The decoding operation does not require an additional time frame, because it is done while the encode sample and hold amplifier is sampling for the next conversion.

The companding DAC, when used with both a precision comparator and sample and hold amplifier, provides exceptional end-to-end system performance. It yields values of signal to quantizing distortion over the full dynamic range which are so close to the theoretical ideal that they cannot be distinguished by standard measurement techniques. The theoretical limits are shown in Figure 7. The signal-to-noise ratio is 20 dB at −60 dBm0, which corresponds to encoding a sine wave on a converter with 3 positive and 3 negative steps and achieving less than 10% distortion. Compandor tracking performance is also excellent with typical deviations of ± 0.1 dB end-to-end in a system using two companding DACs. Idle channel noise contributed by the DAC is negligible. This parameter is governed by other components and by system layout. The full scale matching of the converter is very tight, so that calibration is unnecessary when used with a known voltage reference and matched reference and output resistors.

Another application for the converter is as a logarithmic digitally programmed attenuator. If an audio signal is fed into the multiplying inputs, the resulting distortion is in the range of 0.01 to 0.02% from dc to over 50 kHz. The output level is programmed over a 72 dB range with negligible effect on the distortion.

The circuit has been fabricated on an 84 by 119-mil silicon chip using a standard linear process and a diffused, untrimmed resistor ladder. True 12-bit accuracy is achieved in the first chord, and the circuit meets all listed specifications with high yield over the military temperature range.

---

*Current address: Advanced Micro Devices, Inc., Sunnyvale, CA.

**Consultative Committee on International Telephone and Telegraph.

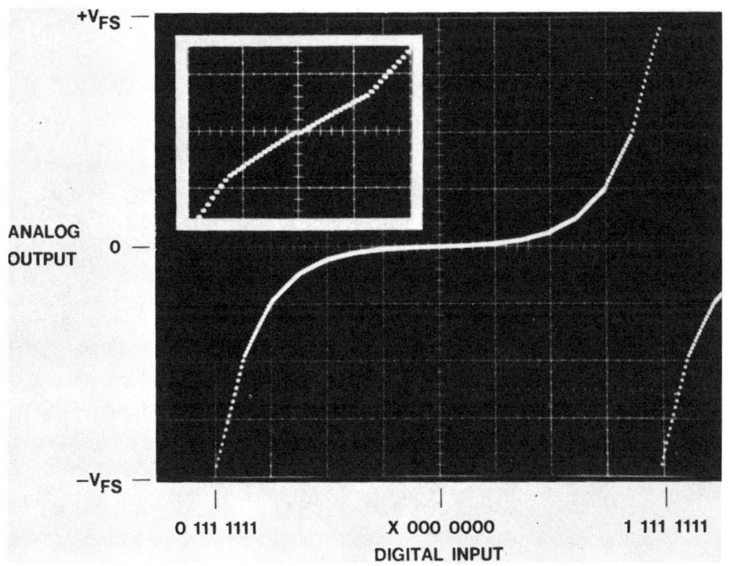

FIGURE 1 – Companding DAC transfer characteristic.

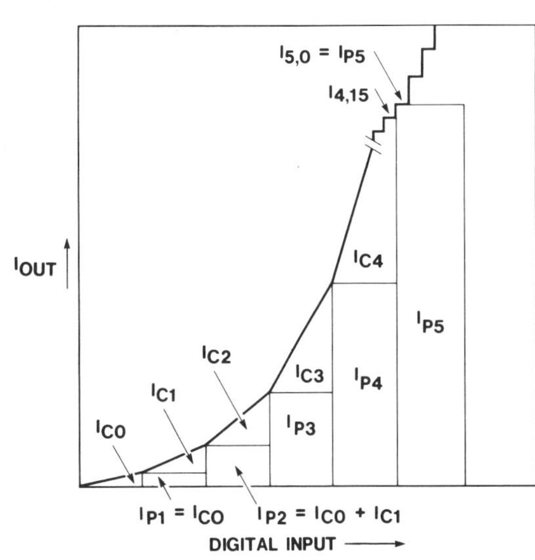

FIGURE 2 – Synthesis of the companding curve.

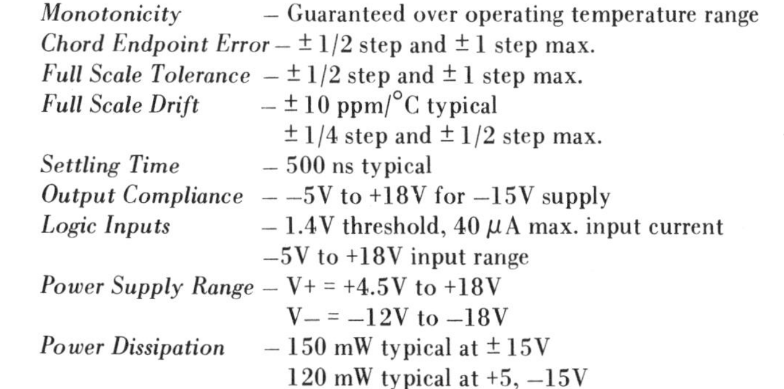

Resolution	– Sign +7 bits (± 128 steps)
Monotonicity	– Guaranteed over operating temperature range
Chord Endpoint Error	– ± 1/2 step and ± 1 step max.
Full Scale Tolerance	– ± 1/2 step and ± 1 step max.
Full Scale Drift	– ± 10 ppm/°C typical ± 1/4 step and ± 1/2 step max.
Settling Time	– 500 ns typical
Output Compliance	– −5V to +18V for −15V supply
Logic Inputs	– 1.4V threshold, 40 μA max. input current −5V to +18V input range
Power Supply Range	– V+ = +4.5V to +18V V− = −12V to −18V
Power Dissipation	– 150 mW typical at ± 15V 120 mW typical at +5, −15V

TABLE 1 – Electrical specifications

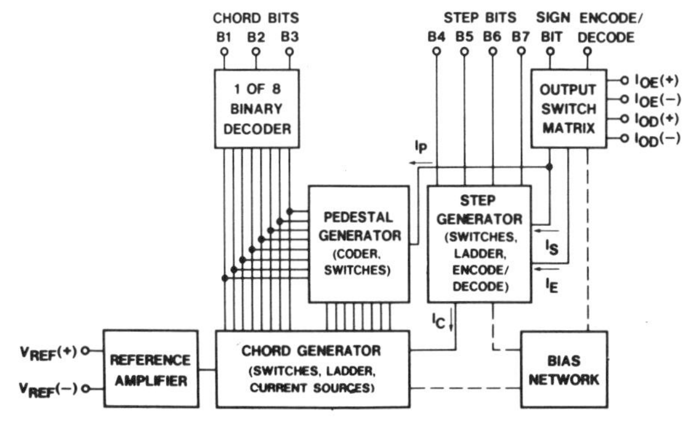

FIGURE 3 – Companding DAC block diagram.

[see page 366 for Figures 4, 5, 6, 7]

All-MOS Charge Redistribution Analog-to-Digital Conversion Techniques—Part I

JAMES L. McCREARY, STUDENT MEMBER, IEEE, AND PAUL R. GRAY, MEMBER, IEEE

Abstract—This two-part paper describes two different techniques for performing analog-to-digital (A/D) conversion compatibly with standard single-channel MOS technology. In the first paper, the use of a binary weighted capacitor array to perform a high-speed, successive approximation conversion is discussed. The technique provides an inherent sample/hold function and can accept both polarities of inputs with a single positive reference. The factors limiting the accuracy and conversion rate of the technique are considered analytically. Experimental results from a monolithic prototype are presented; a resolution of 10 bits was achieved with a conversion time of 23 μs. The estimated die size for a completely monolithic version is 8000 mil^2.

The second paper [3] describes a two-capacitor successive approximation technique, which, in contrast to the first, requires considerably less die area, is inherently monotonic in the presence of capacitor ratio errors, and which operates at a somewhat lower conversion rate. Factors affecting accuracy and conversion rate are considered analytically. Experimental results from a monolithic prototype are presented; a resolution of 8 bits was achieved with an A/D conversion time of 100 μs. Used as a D/A converter, a settling time of 13.5 μs was achieved. The estimated total die size for a completely monolithic version including logic is 5000 mil^2.

I. Introduction

MOST conventional techniques for analog-to-digital (A/D) conversion require both high-performance analog circuitry, such as operational amplifiers, and digital circuitry for counting, sequencing, and data storage. This has tended to result in hybrid circuits consisting of one or more bipolar analog chips and an MOS chip to economically perform the digital functions [1]. This paper describes a new, all-MOS technique which is realizable in a single chip and performs a 10-bit conversion in 23 μs [6]. It includes an intrinsic sample-and-hold function, accepts both bipolar and unipolar inputs, and is realizable with standard N-channel metal gate technology.

For the realization of a fast, successive-approximation A/D converter in MOS technology, conventional voltage driven R-$2R$ techniques are cumbersome since diffused resistors of proper sheet resistance are not available in the standard single-channel technology. A complex thin-film process must be used. Furthermore, these approaches require careful control of the "ON" resistance ratios in the MOS switches over a wide range of values.

In contrast to its utilization as a current switch, the MOS device, used as a charge switch, has inherently zero offset voltage and as an amplifier has very high input resistance. In addition, capacitors are easily fabricated in metal gate technology. Therefore, one is led to use capacitors rather than resistors as the precision components, and to use charge rather than current as the working medium. This technique, referred to as charge-redistribution, has been used in some discrete component A/D converters for many years [2], [7]. However, these converters have required high-performance operational amplifiers which are difficult to realize in single-channel MOS technology.

This paper describes a new A/D conversion technique using charge redistribution on weighted capacitors, while a companion paper [3] describes another application of the charge redistribution concept using two equal capacitors. In Section II of this paper, the binary-weighted capacitor capacitor array and its operation is described. In Sections III and IV, the limitations on the speed and resolution are analyzed. Ex-

Manuscript received March 28, 1975; revised July 30, 1975. This research was sponsored by the National Science Foundation under Grant GK-40912. This paper was presented at the International Solid-State Circuits Conference, Philadelphia, Pa., February 1975.

The authors are with the Department of Electrical Engineering and Computer Sciences and the Electronics Research Laboratory, University of California, Berkeley, Calif. 94720.

Reprinted from *IEEE J. Solid-State Circuits, SC-10, pp. 371-379,* Dec. 1975.

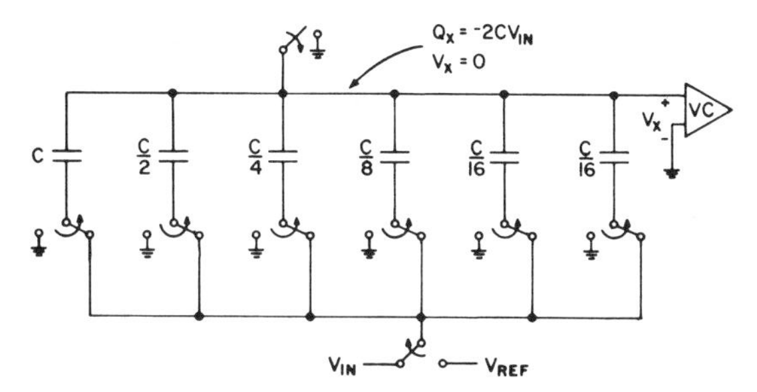

Fig. 1. Conceptual 5-bit A/D converter illustrating the sample mode operation.

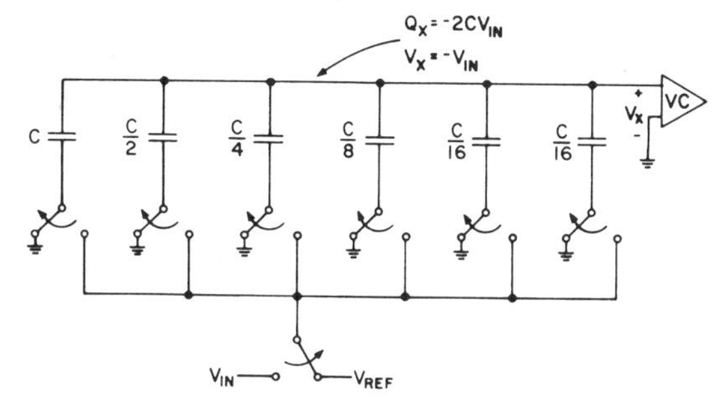

Fig. 2. Pre-redistribution hold mode operation.

perimental results from a prototype system are presented in Section V.

II. Charge Redistribution A/D Conversion Technique Using Binary Weighted Capacitors

The new A/D conversion technique is illustrated with a conceptual 5-bit version of the converter shown in Fig. 1. It consists of a comparator, an array of binary weighted capacitors plus one additional capacitor of weight corresponding to the least significant bit (LSB), and switches which connect the plates to certain voltages. A conversion is accomplished by a sequence of three operations. In the first, the "sample mode" (Fig. 1), the top plate is connected to ground and the bottom plates to the input voltage. This results in a stored charge on the top plate which is proportional to the input voltage V_{in}. In the "hold mode" of Fig. 2, the top grounding switch is then opened, and the bottom plates are connected to ground. Since the charge on the top plate is conserved, its potential goes to $-V_{in}$. The "redistribution mode," shown in Fig. 3, begins by testing the value of the most significant bit (MSB). This is done by raising the bottom plate of the largest capacitor to the reference voltage V_{ref}. The equivalent circuit is now actually a voltage divider between two equal capacitances. The voltage V_x, which was equal to $-V_{in}$ previously, is now increased by $\frac{1}{2}$ the reference as a result of this operation.

$$V_x = -V_{in} + \frac{V_{ref}}{2}.$$

Sensing the sign of V_x, the comparator output is a logic '1' if $V_x < 0$ and is a '0' if $V_x > 0$. This is analogous to the interpretation that

$$\text{if } V_x < 0 \quad \text{then } V_{in} > \frac{V_{ref}}{2};$$

hence the MSB = 1; but

$$\text{if } V_x > 0 \quad \text{then } V_{in} < \frac{V_{ref}}{2};$$

therefore the MSB = 0. The output of the comparator is, therefore, the value of the binary bit being tested. Switch $S1$ is returned to ground only if the MSB b_4 is a zero. In a similar manner, the next MSB is determined by raising the bottom plate of the next largest capacitor to V_{ref} and checking the polarity of the resulting value of V_x. In this case, however,

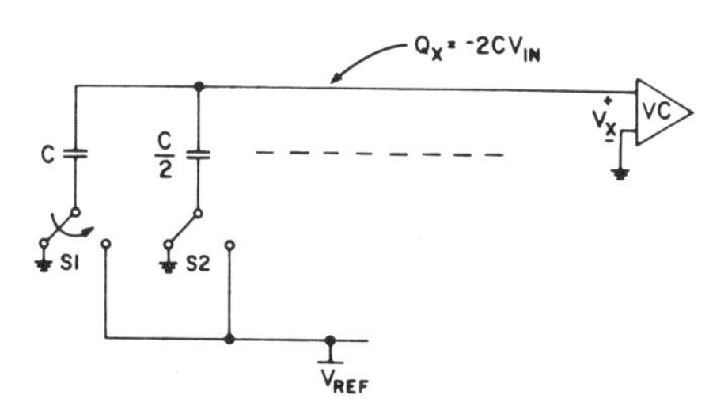

Fig. 3. Redistribution mode operation.

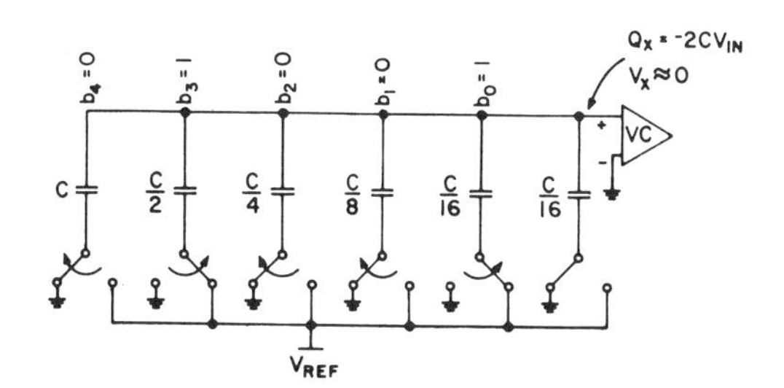

Fig. 4. Example of a final configuration.

the voltage division property of the array causes $V_{ref}/4$ to be added to V_x:

$$V_x = -V_{in} + b_4 \frac{V_{ref}}{2} + \frac{V_{ref}}{4}.$$

Conversion proceeds in this manner until all the bits have been determined. A final configuration is illustrated in Fig. 4 for the digital output 01001. Notice that all capacitors corresponding to a '0' bit are completely discharged. The total original charge on the top plates has been redistributed in a binary fashion and now resides only on the capacitors corresponding to a '1' bit. N redistributions are required for a conversion resolution of N bits. In contrast to earlier charge redistribution techniques, the capacitance of the lower plate switch does not affect the accuracy of the conversion [4]. This fact is evident since the switch capacitance is either discharged to ground or is charged by V_{ref} but never absorbs charge from the top plate. Therefore, the switch devices can be quite large permitting rapid redistributions. On the other hand, the top plate of the array is connected to all the capacitors and to a switch and to the comparator resulting in a large parasitic capacitance from the top plate to ground. The nature of the conversion process, however, is such that V_x is converged back towards zero–its initial value. Hence, the charge on this parasitic is the same in the final configuration as it was in the sample mode. Therefore, the error charge contributed by this parasitic is very near zero. For the 10-bit converter the parasitic capacitance at the top plate can be 100 times the value of the smallest capacitor and still cause only 0.1 bit offset error. This is equivalent to saying that the smallest capacitor may be much smaller than the parasitic and consequently the largest capacitor may be reduced in value proportionally. Furthermore, the initial value of V_x need not necessarily be zero but can be the threshold voltage of the comparator. This fact allows cancellation of comparator offset by storing the offset in the array during the sample mode. The linearity then is primarily a function of the ratio accuracy of the capacitors in the array.

By only a slight modification of the array switching scheme bipolar voltage inputs can be encoded while still using only the single positive reference. This is achieved by connecting the bottom plate of the largest capacitor to V_{ref} during the sample mode resulting in a stored charge:

$$Q_x = -C_{TOT}\left(\frac{V_{in}}{2} + \frac{V_{ref}}{2}\right).$$

Each bit is then tested in sequence just as before except that the largest capacitor is switched from V_{ref} to ground during its test, while all the other capacitors are switched from ground to V_{ref}. Also, as before, a bit value is true if V_x is negative after the test. The expression for V_x again converges back towards zero:

$$V_x = -\frac{V_{in}}{2} + V_{ref}\left(-\frac{b_{10}}{2^1} + \frac{b_9}{2^2} + \frac{b_8}{2^3} + \cdots + \frac{b_1}{2^{10}}\right) \approx 0.$$

For a 10-V reference, b_{10} is '0' for $0 \leqslant V_{in} \leqslant 10$ V, but is '1' for -10 V $\leqslant V_{in} < 0$. Therefore, b_{10} represents the sign bit and its function is to level shift V_x in order to accommodate negative inputs. Hence, a 10-bit conversion is achieved over the input range ±10 V with negative numbers expressed in 1's complement.

III. Factors Limiting Accuracy

While monolithic circuit technology has had great impact on the cost of many analog circuit functions, such as operational amplifiers, the impact on the cost of A/D and D/A converters has not been as great. This is due to the complexity of a complete converter, and more importantly, to the problem of component matching. Because of the difference between the aspect ratios of diffused resistors of practical value versus those of capacitors, the attainable matching accuracy is higher for capacitors given the same overall die area. The flexibility of capacitor geometry allows them to be made square or even circular so as to optimize matching accuracy, as discussed further in the companion paper [3].

In this technique a mismatch in the binary ratios of capacitors in the array causes nonlinearity. It cannot cause a gain error because the end points of the transfer function curve, V_{out} versus V_{in}, are not dependent on capacitor matching. This results from the fact that no net charge redistribution between capacitors occurs for either zero or full-scale input since all capacitors are either fully discharged or fully charged, respectively. For the same reason no offset error can arise from capacitor mismatch since the mismatch cannot be manifested unless a charge redistribution exists in the final configuration.

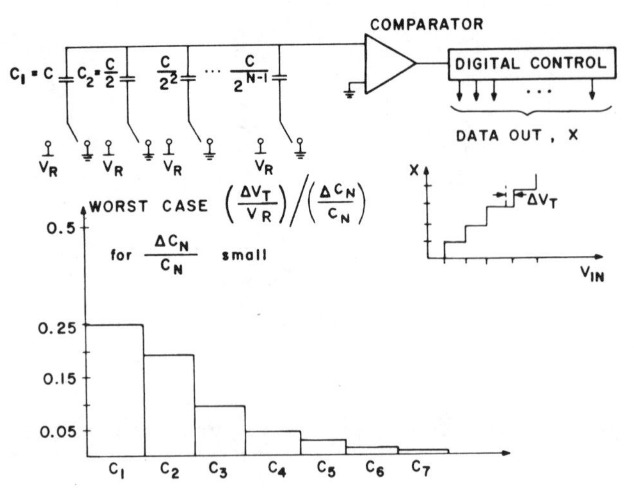

Fig. 5. Sensitivity of A/D conversion linearity to individual capacitor value.

First consider the ideal case in which all the capacitors of the general N-bit converter shown in Fig. 5 have the precise binary weight values. For this case the digital output x is a regular staircase when plotted against V_{in}, and every transition occurs at a precise value of V_{in} designated V_T. On the other hand, changing one capacitor from its ideal value by a small amount ΔC_N causes all transition points to shift somewhat, but there will be one worst case transition. The ratio $\Delta V_T/V_R$ is the normalized worst case fraction deviation in transition point from the ideal. This is also a measure of the nonlinearity. The ratio of this deviation to the fractional change in capacitor value $\Delta C_N/C_N$ represents the sensitivity of linearity to individual capacitor value. The plot of sensitivity, also shown in Fig. 5, shows that linearity is very sensitive to a fractional change in the large capacitors, but not very dependent upon similar fractional changes in the smaller capacitors. Therefore, the smaller capacitors have great allowable tolerances. It should be pointed out that actually all capacitors have simultaneous deviations which cause ratio errors and the worst case combination of these must always be considered. Thus, the optimization of the ratio accuracy in the array is a prime consideration.

Capacitor ratio error results from several causes, one of which is the undercutting of the mask which defines the capacitor. Consider two capacitors C_4 and C_2, shown in Fig. 6, which are nominally related by a factor of 2: $C_4 = 2C_2$. During the etching phase of the photomask process a poorly controlled lateral etch occurs called undercut. Let Δx be the undercut length and L_4 be the side length of C_4, also shown in Fig. 6. Then a ratio error is produced between C_4 and C_2 which is proportional to the undercut length:

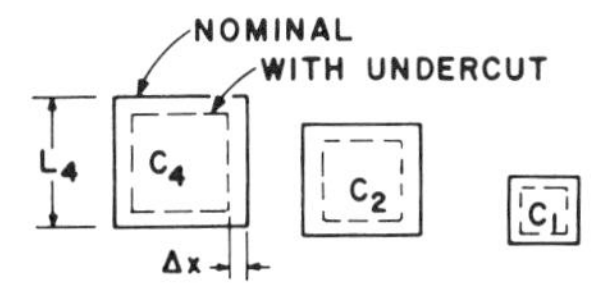

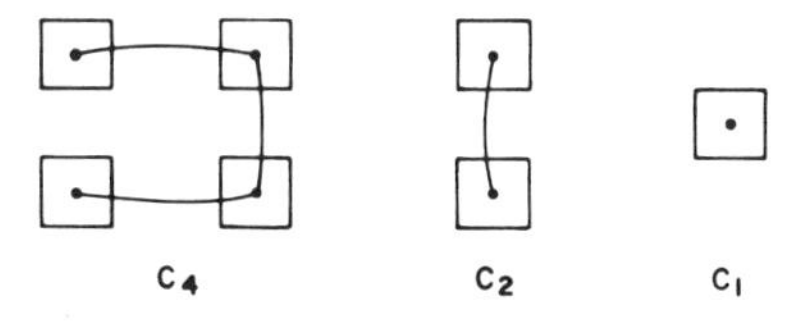

Fig. 6. Capacitor ratio error due to photomask undercut.

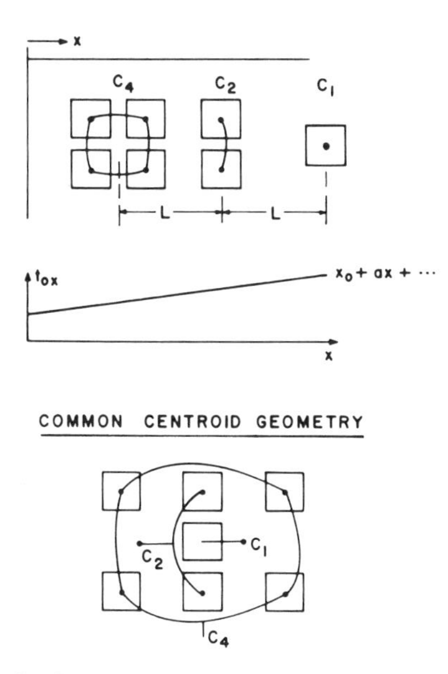

Fig. 7. Capacitor ratio error due to oxide gradients.

$$C_4 = 2C_2(1 + \epsilon); \qquad \epsilon \simeq 4\frac{\Delta x}{L_4}.$$

This problem can be circumvented by a geometry such that the perimeter lengths as well as the areas are ratioed. This can be done, as seen in Fig. 6, by paralleling identical size plates to form the larger capacitors. Thus the capacitor ratios are not affected by undercut.

Long range gradients in the thin capacitor oxide can also cause ratio errors. These gradients arise from nonuniform oxide growth conditions. If this variation in oxide thickness is approximated as first-order gradient, as shown in Fig. 7, then the resulting ratio error is proportional to the fractional variation in oxide thickness:

$$C_2 = 2C_1(1 + \epsilon L)$$

$$C_4 = 4C_1(1 + 2\epsilon L); \qquad \epsilon = \frac{a}{x_0}.$$

Experimentally, values of 10–100 ppm/mil have been observed for the factor ϵ. Error from this source can be minimized by improved oxide growth techniques and by a common centroid geometry. This is done in Fig. 7 by locating the elements of the capacitors in such a way that they are symmetrically spaced about a common center point. In this way the capacitor ratios may be maintained in spite of first-order gradients. Although undercut and oxide gradient errors can be minimized, a random edge variation will still exist and cause small ratio errors.

Any significant variation of small signal capacitance with dc terminal voltage would limit the accuracy because a nonlinearity would result. For MOS capacitors on heavily doped N+ back plates voltage coefficients of less than 24 ppm/V have been observed. This is insignificant at the 10-bit converter level.

Dielectric absorption is a phenomenon in which a residual voltage appears on a capacitor after it has been rapidly discharged. However, MOS capacitors display a relaxation which is unimportant for the 10-bit converter.

The voltage comparison process is fundamental to A/D conversion. The offset voltage of the comparator is usually manifested as an offset error in the digital conversion. Because of the relatively large gate-source voltage mismatch in MOS differential amplifiers, the offset voltage of the all-MOS comparator must be eliminated as a source of error. This can be accomplished either by digital means or by offset cancellation techniques. In this circuit the offset is stored and then subtracted at a later time by the sequence of events illustrated in Fig. 8. During the sample mode, V_{in} remains connected to the bottom plates but switch $S1$ is also ON and precharges the top plate of the array to the threshold voltage of stage $A1$. $S1$ then turns off, but since $A2$ is identical to $A1$ its input is also at the threshold. Since $S2$ is ON in the "up" position the output of $A2$ is saved at a storage node at one input of $A3$. During the subsequent redistribution mode $S1$ always remains OFF but $S2$ turns ON in the "down" position after each redistribution loading the output of $A2$ at the other storage node input of $A3$. This provides a first-order cancellation of switch feedthrough at both storage nodes. Since $A3$ is a difference amplifier, the offset of $A1$ and $A2$ together with the feedthrough of $S1$ have been cancelled. The offset of $A3$ is

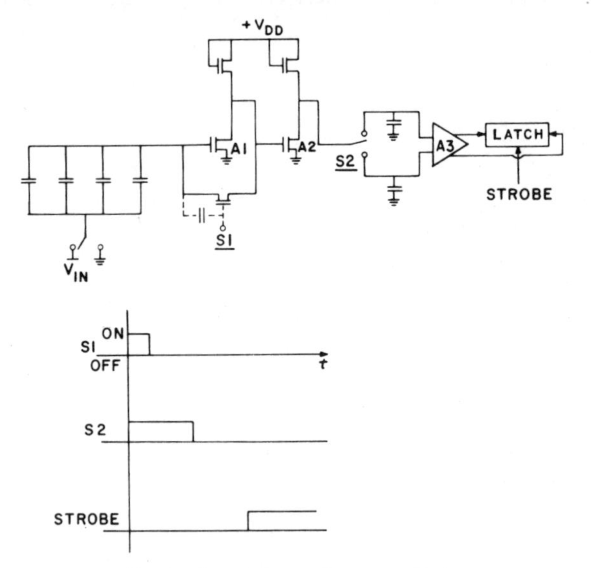

Fig. 8. Comparator offset cancellation.

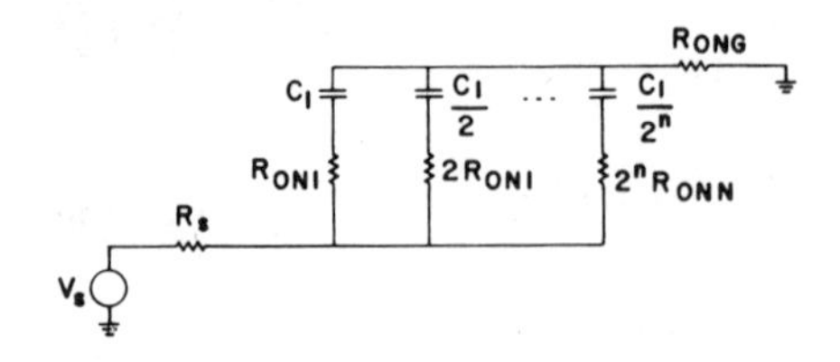

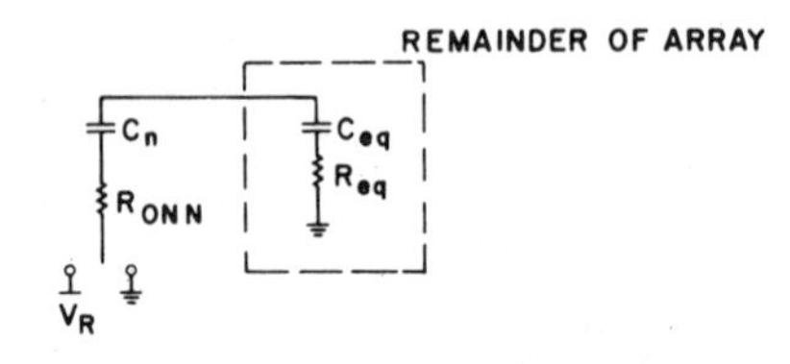

Fig. 9. Factors limiting conversion rate.

therefore the only significant offset in the comparator. This offset is reflected back through the gain stages $A1$ and $A2$ giving an effective offset [5]:

$$V_{OS_{\text{effective}}} = \frac{V_{OS_{A3}}}{G_{A1} \cdot G_{A2}}.$$

The gain product $G_{A1} \cdot G_{A2} \approx 50$ for this circuit.

The conversion technique as described thus far produces a transfer characteristic like that shown in Fig. 5 in which the first transition occurs at one LSB voltage away from the origin. Actually, for minimum quantization error this first transition should occur only $\frac{1}{2}$ LSB voltage away from the origin. The cancellation of this offset may be accomplished by returning the lower plate of the smallest capacitor to $V_{\text{ref}}/2$ rather than ground after the sample mode.

A great advantage of the S_iO_2 dielectric capacitor is its very low temperature coefficient of approximately 20 ppm/°C in contrast with approximately 2000 ppm/°C for diffused resistors and several hundred ppm/°C for thin film and implanted resistors. Thus, capacitor ratios are less sensitive to temperature gradients than resistor ratios. Since component mismatch usually has a greater effect upon linearity than upon gain or offset errors, the temperature coefficient of nonlinearity is lower for this conversion technique in comparison with R-$2R$ approaches.

IV. Factors Limiting Conversion Rate

Compared with many conversion techniques, the successive approximation method used in this circuit is capable of rapid conversion. The conversion requires two distinct operations. The first is the charging of the capacitor array to the input voltage during the sample mode, and the second is the redistribution of the charge on the capacitor array during the successive approximation conversion process. The equivalent circuit during the sample mode is shown in Fig. 9. The total array capacitance C_T is charged through the source resistance R_s, and equivalent ON resistance of the top plate grounding switch R_{ON_G}, and the equivalent ON resistance of the binarily scaled switches. The time constant is thus:

$$\tau_1 = \left[R_S + \frac{R_{\text{ON}_1}}{2} + R_{\text{ON}_G}\right] C_T.$$

During the redistribution cycle, also illustrated in Fig. 9, the time constant is the series combination of the capacitor being tested and its switch resistance plus the equivalent capacitance and resistance of the remaining elements of the array:

$$\tau_2 = \frac{R_{\text{ON}_1}}{2} C_T.$$

The maximum conversion rate is determined by the time required for one sample mode precharge and subsequently for ten redistribution cycles to go to completion, with time allowed for one comparison after each redistribution. In the experimental circuit to be described later, each of these delays contribute to the total conversion time, but it is instructive to consider the fundamental limitations on the conversion rate of this technique.

Assuming that the source resistance R_s can be made small, the acquisition time τ_1 is dependent upon R_{ON_1} and R_{ON_G}. The former can be made small compared to R_{ON_G} since the gate-source and gate-drain capacitance of the bottom plate switches do not result in conversion errors. The gate-drain capacitance of the grounding switch, however, can affect the accuracy of the conversion [8]. In the realization described later this error is cancelled, but assume for the time being that this was not done. The time constant during precharge for this limiting case is

$$\tau_1 = R_{\text{ON}_G} C_T = \frac{C_T}{\mu \frac{W}{L_C} C_0(V_{GS}(\text{ON}) - V_T)}.$$

Assuming an optimum gate drive signal on the grounding

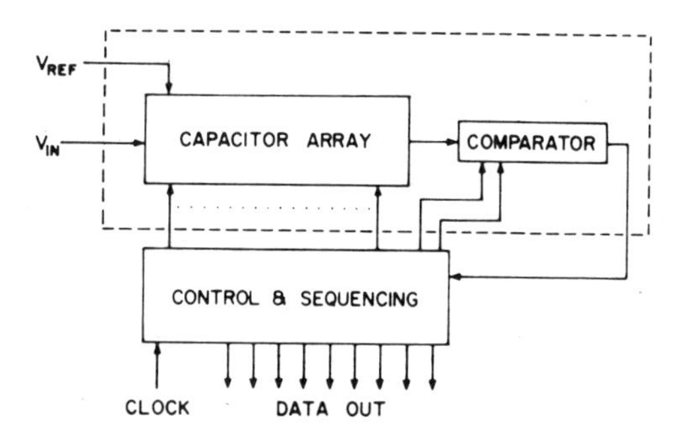

Fig. 10. Complete A/D converter.

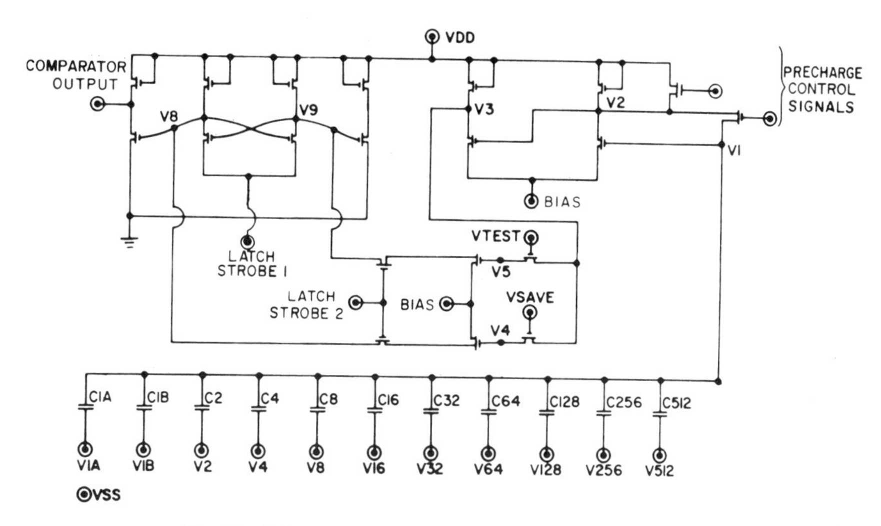

Fig. 11. Schematic of the experimental integrated circuit.

switch, the error voltage is given by

$$V_E = \frac{C_{GS}}{C_T}(V_{GS}(\text{ON}) - V_T)$$

and therefore

$$V_E \tau_1 = \frac{C_{GS}}{\frac{W}{L_C} C_0} \simeq \frac{L_C^2}{\mu}.$$

Here the overlap capacitance has been neglected, and it has been assumed for simplicity that the error charge transferred to C_T is simply the channel inversion layer charge, represented by the quantity $C_{GS}[V_{GS}(\text{ON}) - V_T]$. This result implies for example, that for a 0.1 percent error, 10-μ channel length, and N-channel devices, a minimum acquisition time of about 3 μs is required.

Actually, this source of error is not a fundamental limitation on accuracy since it is simply an offset error and can be cancelled by using either analog or digital techniques. These techniques, however, require considerable time for the sampling and processing of the offset voltage so that this problem does place a practical limit on the conversion rate.

The redistribution time τ_2 is dependent on R_{ON} of the bottom plate switches which can be made arbitrarily small at the cost of the die area. For very large devices, the redistribution time will approach that required for the switches to change their own drain-bulk capacitance.

Following each redistribution, a time interval is required for the voltage comparator to settle to the correct state. This time interval becomes longer as the resolution becomes higher and the minimum overdrive signal becomes smaller. Comparator delay is a fundamental limiting factor in the rate at which conversions can be accomplished.

V. Circuit Description and Experimental Results

Fig. 10 shows a complete A/D converter using the technique described in this paper. In block diagram form the system is composed of a capacitor array, a comparator, and control and sequencing logic. The feasibility of this technique was investigated by fabricating only the critical portions of the circuit, the array and the comparator, in monolithic form using standard N-channel MOS technology. Although the control and sequencing logic was composed of TTL gates, the monolithic realization of this circuit in MOS technology is straightforward.

The experimental integrated circuit schematic is shown in Fig. 11. During the sample mode, offset cancellation is accomplished by the precharge control signals. After precharge is completed, the initial value of V_1 is saved at a storage node V_4 by the action of a transmission gate. After each redistribution, the amplified value of V_1 is transmitted to node V_5 in

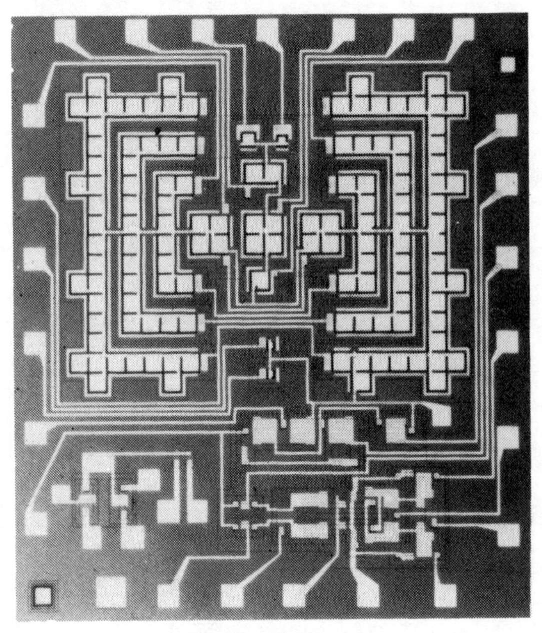

Fig. 12. Die photo.

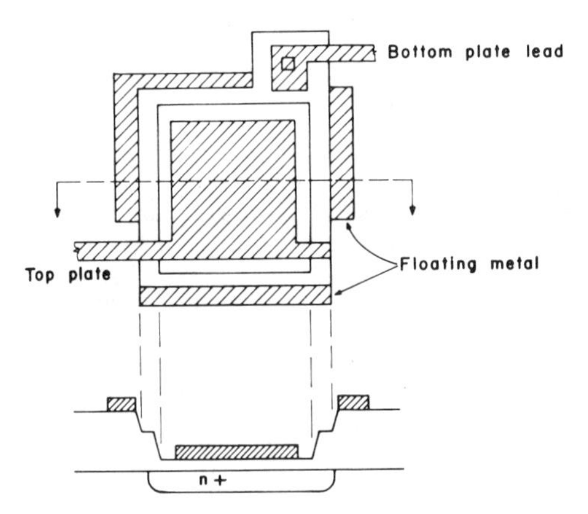

Fig. 13. MOS capacitor structure.

preparation for the subsequent test of the bit value. Since both latch strobe signals are initially high the difference signal at the latch is given by

$$(V_8 - V_9) = (V_4 - V_5) \cdot G_{A3}$$

where G_{A3} is the gain of amplifier $A3$ previously discussed. The total signal gain from the comparator input to the latch input is about 200, hence a 10-mV change in V_1 produces a 2-V differential signal at the latch. When the latch strobe signals go low the cross-coupled pair becomes regenerative and the full gain of the comparator is realized. The final state of the latch is then buffered off chip.

The die photo of Fig. 12 shows the capacitor array totaling 240 pF positioned above the comparator. The extreme right and left segments of the array are connected in parallel to form the largest capacitor. The next two segments on each side compose the next largest capacitor and so on, so that the common centroid is the center of the array. Note that the larger capacitors are composed of small squares 3 X 3 mils to reduce the sensitivity to undercut. The capacitor array has dimensions 75 X 58 mils, and an estimated die size including the digital control is 90 X 90 mils.

The MOS capacitor structure is shown in Fig. 13. Notice that only the metal edge defines the capacitor area, thus taking advantage of the better resolution properties of some positive photoresists. The floating redundant metal strips maintain a relatively constant etchant concentration at every capacitor edge, thereby assuring nearly uniform undercutting effects for all capacitors. The capacitor interconnect spans the thin oxide of each capacitor in a single direction so that capacitor area is insensitive to mask alignment errors. The area of the interconnect over the thin oxide is included in capacitor area calculations, while the area over the thicker oxide is neglected. Interconnect that passes over field oxide causes a parasitic capacitance from the top plate to ground which does not affect circuit performance as previously shown. A first-order cancellation of corner-rounding effects is made by designing each capacitor with an equal number of 90° and 270° corners.

The central question of the feasibility of this approach is the

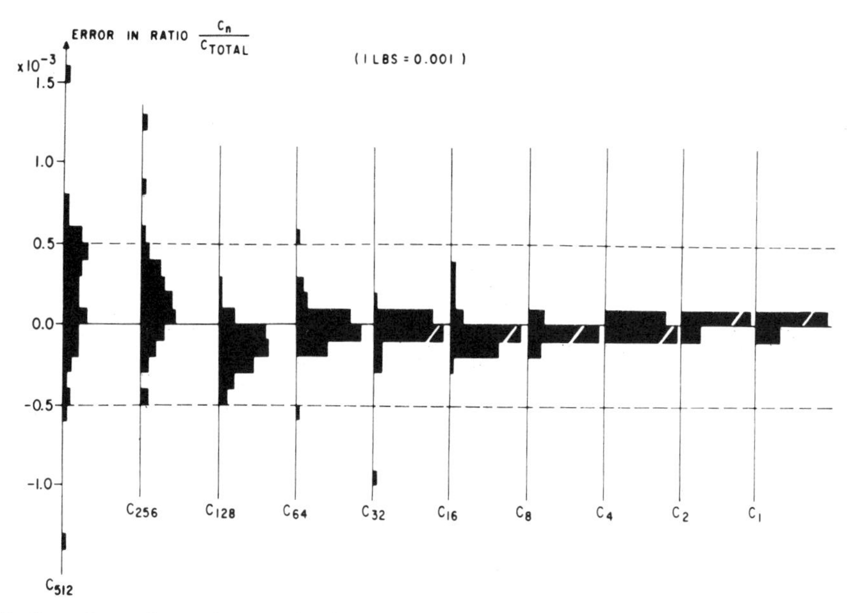

Fig. 14. Distribution of capacitor ratio errors for 47 10-bit converter arrays. Vertical axis is the fractional ratio error while the horizontal axis is the frequency of occurrence for each capacitor.

accuracy with which the capacitor ratios can be maintained using conventional photomasking. Extensive data have been gathered on the ratio accuracy of the array and typical results from three different wafers are shown in Fig. 14. This is the distribution of measured errors in the ratio of each individual capacitor to the total array capacitance. The horizontal axis represents the frequency of occurrence. Assume for a moment that capacitor ratio error were the only factor affecting yield. Then from these data the yield for $\pm\frac{1}{2}$ LSB linearity for 8, 9, and 10 bits resolution would be 98 percent, 94 percent, and 45 percent, respectively, for this sample of 47 functional arrays.

The feasibility of the technique was further studied by operating the experimental chip with external logic as a complete A/D converter. The linearity and offset of a sample of devices were evaluated using the experimental procedure shown in Fig. 15. The output of the A/D was connected to a 12-bit D/A converter. Since V_{in} was a long period ramp, the output of the D/A was a staircase. The difference between these signals is a representation of the total conversion error, and is connected to the y-axis of the plotter. The zero-to-full-scale recording of the output is shown also in Fig. 15. Since the 10-mV marks correspond to the $\frac{1}{2}$ LSB error levels the error is seen to be less than $\frac{1}{2}$ LSB. An expanded recording which permitted the visual resolution of all 1024 states was used for actual verification of the results.

A summary of the measured results is shown in Table I. The sample mode acquisition time is the minimum precharge time required for an accurate conversion of a 5-V step change at the input. The total conversion time corresponds to a 44-kHz sampling frequency.

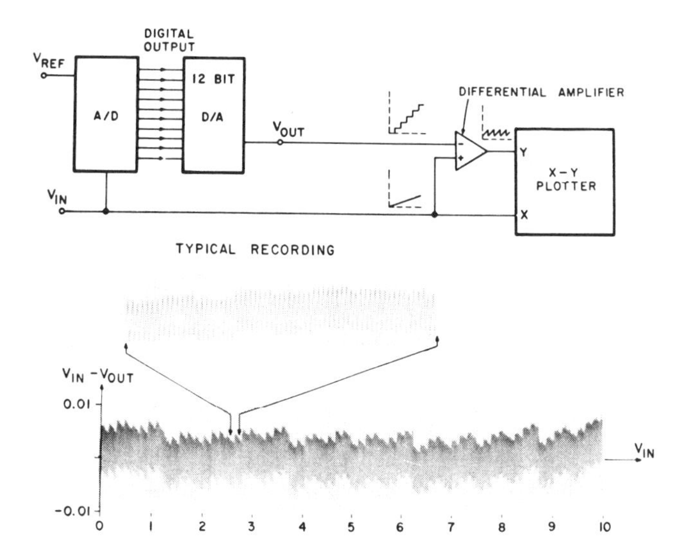

Fig. 15. Experimental measurement of A/D converter nonlinearity.

TABLE I
MEASURED PERFORMANCE DATA

Resolution	10 bits
Linearity	±1/2 LSB
Input voltage range	0–10 V
Input offset voltage	2 mV
Gain error	<0.05 percent (external reference)
Sample mode acquisition time	2.3 μs
Total conversion time	22.8 μs

VI. SUMMARY

A new, all MOS A/D conversion technique has been demonstrated which, with the addition of an external reference, can be used to realize a standard-process A/D converter on a single chip. Experimental data were presented which indicated that 8-bit resolutions can be attained at very high yield and low cost using standard N-channel MOS technology, and that 10-

bit resolution can be achieved at somewhat lower yield. It is believed by the authors that more careful control of photolithographic processing would result in very high yield at the 10-bit level and significant yield at even higher resolutions.

Acknowledgment

The authors wish to acknowledge the contributions of D. A. Hodges and R. E. Suárez to the work described in Part I.

References

[1] J. A. Schoeff, "A monolithic analog subsystem for high-accuracy A/D conversion," in *ISSCC Dig. Tech. Papers*, Feb. 1973, pp. 18–19.

[2] H. Schmidt, *Analog-Digital Conversion*. New York: Van Nostrand-Reinholt, 1970.

[3] R. E. Suarez, P. R. Gray, and D. A. Hodges, "All-MOS charge redistribution analog-to-digital conversion techniques–part II," this issue, pp. 379–385.

[4] ——, "An all-MOS charge-redistribution A/D conversion technique," in *ISSCC Dig. Tech. Papers*, Feb. 1974, pp. 194–195.

[5] R. Poujours et al., "Low level MOS transistor amplifier using storage techniques," in *ISSCC Dig. Tech. Papers*, Feb. 1973, pp. 152–153.

[6] J. McCreary and P. R. Gray, "A high-speed, all-MOS successive approximation weighted capacitor A/D conversion technique," in *ISSCC Dig. Tech. Papers*, Feb. 1975, pp. 38–39.

[7] C. W. Barbour, "Simplified PCM analog to digital converter using capacity charge transfer," in *Proc. 1971 Telemetering Conf.*, pp. 4.1–4.11.

[8] J. McCreary, "Successive approximation analog-to-digital conversion in MOS integrated circuits," Ph.D. dissertation, Univ. California, Berkeley, 1975.

All-MOS Charge Redistribution Analog-to-Digital Conversion Techniques—Part II

RICARDO E. SUÁREZ, MEMBER, IEEE, PAUL R. GRAY, MEMBER, IEEE, AND DAVID A. HODGES, SENIOR MEMBER, IEEE

Abstract—This two-part paper describes two different techniques for performing analog-to-digital (A/D) conversion compatibly with standard single-channel MOS technology. In the first paper, the use of a binary weighted capacitor array to perform a high-speed successive approximation conversion was discussed.

This second paper describes a two-capacitor successive approximation technique, which, in contrast to the first, requires considerably less die area, is inherently monotonic in the presence of capacitor ratio errors, and which operates at somewhat lower conversion rate. Factors affecting accuracy and conversion rate are considered analytically. Experimental results from a monolithic prototype are presented; a resolution of eight bits was achieved with an A/D conversion time of 100 μs. Used as a digital-to-analog (D/A) converter, a settling time of 13.5 μs was achieved. The estimated total die size for a completely monolithic version including logic is 5000 mil^2.

Manuscript received May 19, 1975; revised July 30, 1975. This research was sponsored in part by the National Science Foundation under Grant GK-40912.

R. E. Suárez was with the Department of Electrical Engineering and Computer Sciences and the Electronics Research Laboratory, University of California, Berkeley, Calif. He is now with the Instituto Venezolano de Investigaciones Cientificas (I.V.I.C.), Caracas, Venezuela.

P. R. Gray and D. A. Hodges are with the Department of Electrical Engineering and Computer Sciences and the Electronics Research Laboratory, University of California, Berkeley, Calif. 97420.

I. INTRODUCTION

AS DISCUSSED in Part I [9] of this paper, widespread application of techniques for digital processing of analog signals has been hindered by the unavailability of inexpensive functional blocks for analog-to-digital (A/D) conversion. Traditional approaches to A/D conversion have required the simultaneous implementation of high-performance analog circuits, such as operational amplifiers, and of digital circuitry

Reprinted from *IEEE J. Solid-State Circuits*, vol. SC-10, pp. 379-385, Dec. 1975.

for counting, control, and data storage. Consequently, current A/D converter realizations have tended to be multiple-chip approaches [1] wherein the advantages offered by bipolar and MOS fabrication technologies are separately exploited.

In contrast, the charge-redistribution A/D conversion technique to be described in this paper requires a minimum number of precision components and is realizable on a single low-cost MOS chip.[1] Compared with the weighted capacitor technique described in Part I [9], the technique requires only two equal grounded capacitors of moderate value [11]. As a result, it can be realized in a relatively small die area compared with the weighted capacitor approach. On the other hand, ultimate resolution of this technique is sensitive to parasitic capacitances associated with the switch transistors, and the conversion rate is lower for a given clock rate because of the conversion algorithm required. The technique is thus most suitable when moderate conversion rate and moderate resolution are required in a small die area.

II. Serial D/A Converter

A simplified schematic diagram of a serial digital-to-analog (D/A) converter (DAC) circuit is shown in Fig. 1. Capacitors C_1 and C_2 are nominally of equal value. Conversion is accomplished serially by considering the least significant bit (LSB) d_1 first. Capacitor C_1 is precharged either to the reference voltage V_R by a momentary closure of S_2 if $d_1 = 1$ or to ground through S_3 if $d_1 = 0$. Simultaneously, C_2 is discharged to ground through S_4. With S_2, S_3, and S_4 open switch S_1 is then closed momentarily to redistribute the charge, and the resulting capacitor voltages are

$$V_1(1) = V_2(1) = \frac{d_1}{2} V_R. \tag{1a}$$

Holding the charge on C_2, the precharging of C_1 is repeated, this time considering the next least significant bit d_2. After redistribution the capacitor voltages are

$$V_1(2) = V_2(2) = \frac{1}{2}\left(d_2 + \frac{d_1}{2}\right) V_R. \tag{1b}$$

The process continues in this fashion, and at the end of K redistributions the existing charge on C_2 is divided by two and a charge increment corresponding to

$$\frac{d_K}{2} V_R \tag{1c}$$

is added. Therefore, for a K-bit D/A conversion the voltage on the capacitor is

$$V_1(K) = V_2(K) = \sum_{i=1}^{K} \frac{2^i d_i}{2^{K+1}} V_R \tag{1d}$$

which is the desired output. A total of K precharge steps and K redistributions are required to complete a K-bit D/A conversion. The conversion sequence for the input word 1101 is illustrated in Fig. 2.

[1] Patent pending.

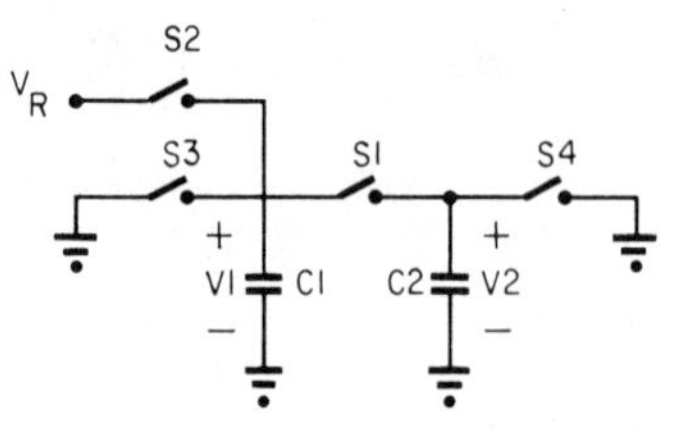

Fig. 1. Serial charge–redistribution digital-to-analog (D/A) converter.

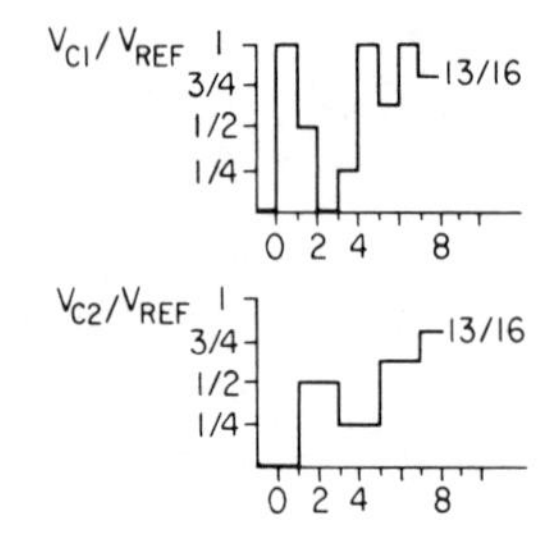

Fig. 2. Illustration of D/A conversion sequence for the input word 1101.

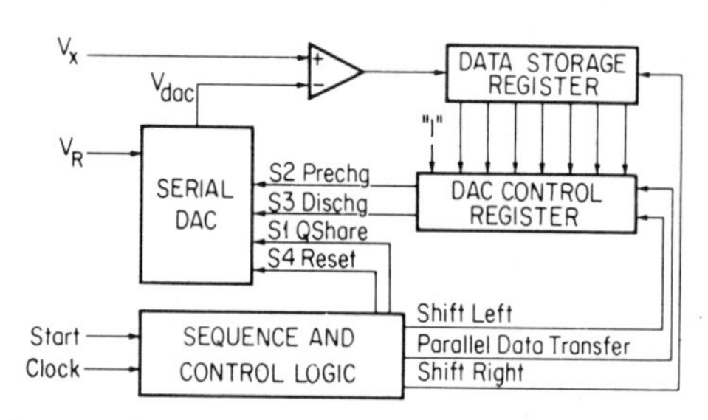

Fig. 3. Complete analog-to-digital (A/D) converter.

III. Successive Approximation A/D Converter

With the addition of a voltage comparator, storage registers, and sequencing logic the serial DAC can be used to construct a successive approximation A/D converter (ADC), as shown in Fig. 3. For an A/D conversion, the most significant bit (MSB) a_N must be determined first.[2] The control logic then takes on a particularly simple form since the DAC input string at any given point in the conversion is just the previously encoded word taken LSB first. For example, consider a point during the A/D conversion in which the first K MSB's have been decided. To decide the $(K + 1)$th MSB, a $(K + 1)$-bit word is formed in the DAC Control Register by adding a "1" as the LSB to the K-bit word already encoded in the Data Storage Regsiter. A $(K + 1)$-bit D/A conversion then establishes the value of a_{N-K} by comparison with the unknown voltage V_X. The bit is stored in the Data Storage Register and the next serial D/A conversion is started. The conversion sequence is detailed in Table I, and Fig. 4 illustrates a 4-bit A/D conversion

[2] The jth bit is denoted d_j in a D/A conversion and a_j in an A/D conversion.

TABLE I
SUCCESSIVE APPROXIMATION A/D CONVERSION SEQUENCE

D/A Conversion Number	DAC Input Word						Comparator Output	No. of Charging Steps
	d_1	d_2	d_3	...	d_{N-1}	d_N		
1	1	--	--	...	--	--	a_N	2
2	1	a_N	--	...	--	--	a_{N-1}	4
3	1	a_{N-1}	a_N	...	--	--	a_{N-2}	6
...	...	...	...	...	...	...	...	...
N	1	a_2	a_3	...	a_{N-1}	a_N	a_1	2N
							TOTAL..	N(N+1)

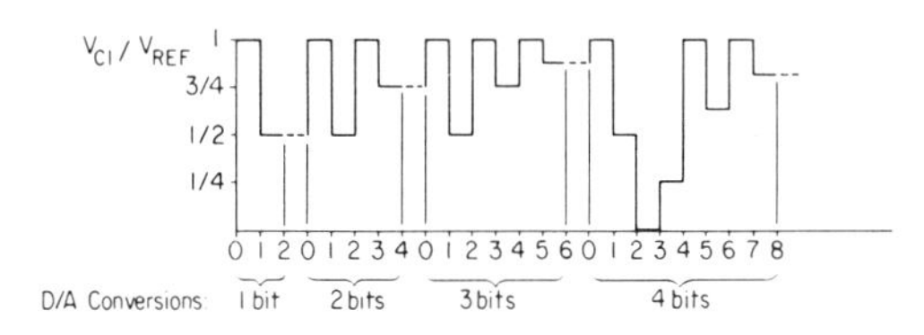

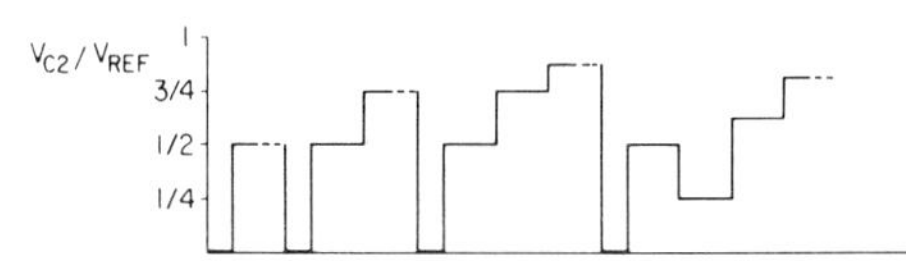

Fig. 4. Illustration of A/D conversion sequence for V_X/V_R = 13/16.

for $V_X = 13/16\ V_R$. Altogether $N(N+1)$ charging steps are required for an N-bit A/D conversion. The total A/D conversion time also includes N comparator settling times.

IV. ACCURACY CONSIDERATIONS IN MONOLITHIC PASSIVE COMPONENTS

In a successive approximation ADC it is necessary to generate accurate fractions of the reference voltage to establish the value of the bits $a_1, \cdots, a_N$. These fractions are obtained in an integrated circuit with resistor or capacitor dividers or by scaling of active device geometries. The relatively poor matching of active device characteristics, however, makes it difficult to achieve accurate voltage or current ratios over a wide range of values and operating conditions. For this reason voltage or current ratios are normally defined through fractions of passive components.

A major source of component mismatches in integrated circuits are uncertainties in photolithographic edge definition. For a pair of nominally identical resistors, as shown in Fig. 5(a), an edge uncertainty in the resistor length (ΔL) and width (ΔW) results in a mismatch

$$\frac{\Delta R}{R} = \frac{\Delta L}{L_R} - \frac{\Delta W}{W_R}. \tag{2}$$

The limited range of sheet resistances available in integrated resistors usually demands fairly large length-to-width (L/W)

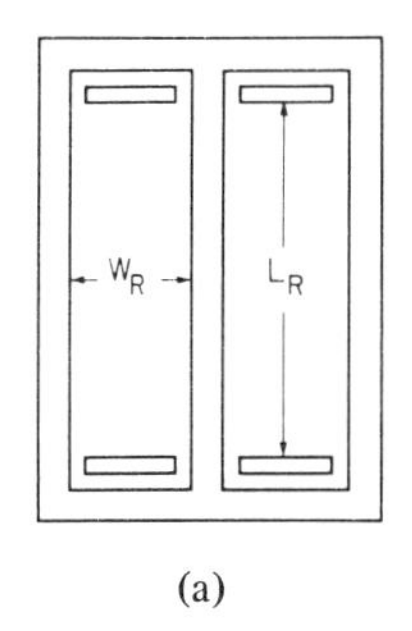

(a)

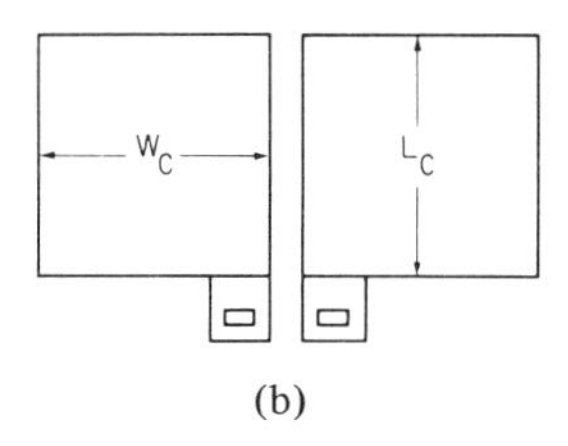

(b)

Fig. 5. Surface geometries for integrated passive components. (a) Resistor pair. (b) Capacitor pair.

TABLE II
CHARACTERISTICS OF INTEGRATED RESISTORS AND CAPACITORS

Component	Fabrication technique	Standard Deviation Matching	Derived σ_X (µm)	Temperature Coefficient (ppm/°C)	Voltage Coefficient (ppm/V)
Resistors	Diffused (W=40 µm)	± 0.23% [2]	0.1	+1500 [2]	-200*
	Ion-implanted (W=40 µm)	± 0.12% [2]	0.05	+400 [2]	-800**
Capacitors	MOS $(t_{OX}=0.1$ µm L= 10 mils N^+ substrate)	± 0.06%***	0.1	26	-10

ratios. Hence

$$\frac{\Delta R}{R} \cong -\frac{\Delta W}{W_R}. \tag{3}$$

For a pair of nominally identical capacitors, as shown in Fig. 5(b), the mismatch due to edge uncertainties is

$$\frac{\Delta C}{C} = \frac{\Delta L}{L_C} + \frac{\Delta W}{W_C}. \tag{4}$$

Since the capacitance value is determined by the capacitor area (as opposed to L/W ratio for resistors) the freedom exists to optimize capacitor geometries to reduce the mismatch sensitivity to uncertainties in edge definition. Thus for a given area it is possible to have $L_C \simeq W_C > W_R$. Hence

$$\frac{\Delta C}{C} < \frac{\Delta R}{R}. \tag{5}$$

One further consideration with regard to mismatch error is the effect of oxide thickness gradients for large capacitor areas. As discussed in Part I [9], layout of the capacitors about a common centroid minimizes the effect of long-range gradients in oxide thickness.

Table II lists published matching data on integrated resistors [2] and measured data on MOS capacitors. The data suggest a standard deviation edge uncertainty of approximately 0.1 µm. Also listed in the table are voltage and temperature coefficients on these components. The data indicate that in all of these aspects MOS capacitors over highly doped silicon offer an attractive alternative for the generation of accurate voltage ratios for monolithic A/D conversion.

V. Factors Limiting Linearity and Offset

In a practical implementation of the D/A converter presented in Section II the output voltage will have the form:

$$V'(K) = V(K) + V'_\epsilon \tag{6}$$

where $V(K)$ is the ideal voltage level given in (1d), and V'_ϵ is an error voltage. The primary sources of error are: feedthrough voltages ($V_{f\epsilon}$), capacitor mismatches ($V_{m\epsilon}$), voltage and temperature coefficients of capacitance ($V_{V\epsilon}$ and $V_{T\epsilon}$), and leakage currents ($V_{L\epsilon}$).

For an A/D converter, the input offset of the voltage comparator (V_{OS}) contributes an additional error component. The total error voltage for an ADC can be expressed as a sum:

$$V_\epsilon = V_{f\epsilon} + V_{m\epsilon} + V_{V\epsilon} + V_{T\epsilon} + V_{L\epsilon} + V_{OS}. \tag{7}$$

In order to keep the worst case conversion error below the converter resolution the following condition must be maintained:

$$\max |V_\epsilon| < \frac{V_R}{2^{N+1}} \tag{8}$$

for N bits.

A. Mismatches

For a mismatch ΔC between capacitors C_1 and C_2

$$\Delta C \equiv C_2 - C_1 \tag{9a}$$

and letting

$$C \equiv (C_1 + C_2)/2 \tag{9b}$$

an analysis of the precharge and charge redistribution sequence outlined in (1) indicates an error voltage

$$V_{m\epsilon} = \frac{\Delta C}{4C} \sum_{i=1}^{N} \frac{N - i - 1}{2^N} 2^i d_i V_R. \tag{10}$$

The worst cases occur when $d_N = 1$ and $d_i = 0$ for $i \neq N$ or vice versa. At these points, the mismatch error voltage is

$$V_{m\epsilon} \simeq \pm \frac{\Delta C}{4C} V_R \tag{11}$$

and at the $(01 \cdots 1)$ to $(10 \cdots 0)$ transition a discontinuity occurs with magnitude

$$\frac{\Delta V_{m\epsilon}}{V_R} \simeq \frac{\Delta C}{2C}. \tag{12}$$

An extrapolation of the data in Table II indicates that a capacitance value of 25 pF for C_1 and C_2 is adequate for an 8-bit converter.

B. Feedthrough

The asymmetry introduced into the circuit by the nonlinear capacitances associated with the charge-sharing MOS transistor switch results in a net feedthrough error voltage during re-distribution. With respect to the circuit in Fig. 6, after application of a pulse to the gate of the transistor the resulting capacitor voltages are

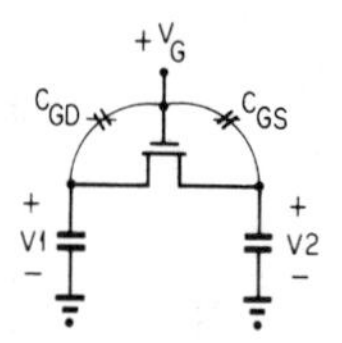

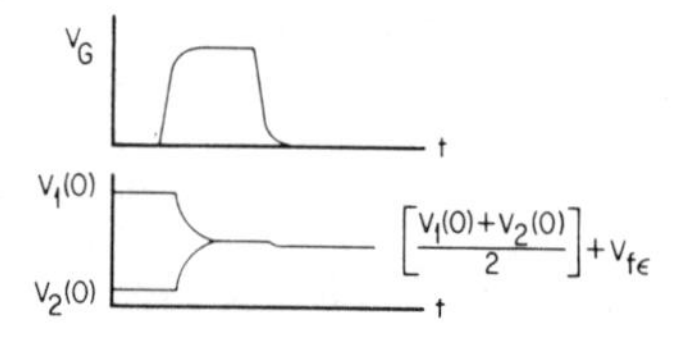

Fig. 6. Feedthrough error voltage during charge redistribution.

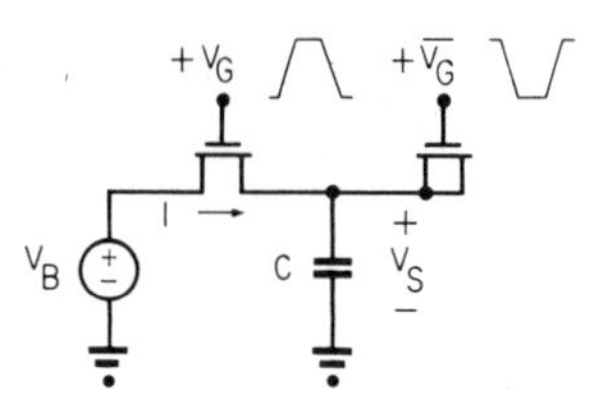

Fig. 7. Charge-canceling device for feedthrough compensation.

$$V_1 = V_2 = \frac{V_1(0) + V_2(0)}{2} + V_{f\epsilon 1}. \tag{13}$$

The feedthrough error voltage can be reduced by reading the capacitor voltages only after the gate voltage has been returned to zero. In this manner, the negative falltime feedthrough partially cancels the positive risetime feedthrough, but an exact cancellation is not achieved. The magnitude of this error component is proportional to the channel capacitance of the MOS transistor C_0 and inversely proportional to the sum of the charge-sharing capacitors

$$V_{f\epsilon 1} \propto \frac{C_0}{(C_1 + C_2)}. \tag{14}$$

An additional feedthrough error component is caused by the precharge transistors. After a capacitor has been charged to V_R or to ground, a negative feedthrough error voltage is introduced during the falltime of the gate voltage. In this case, the addition of a "charge-canceling" device [3], as shown in Fig. 7, reduces the net precharge feedthrough error voltage to approximately the same level as the feedthrough error voltage during charge redistribution. The charge-canceling transistor is a dummy device with the drain and source short-circuited to prevent dc current flow and with one-half the channel area of the precharge devices. The voltage applied at the gate of the charge-canceling device is the complement of the precharge transistor voltage. Since charge-redistribution divides the voltage equally between the two capacitors, the cumulative feedthrough error voltage is halved after each redistribution. Thus the maximum total feedthrough error voltage is approximately twice the worst case value of $V_{f\epsilon}$. In view of (14), therefore, the allowable error constrains the size of the MOS transistor switches in relation to the charge-sharing capacitors. This restriction, in turn, limits the magnitude of the charging currents and thus determines conversion speed. For $C_1 = C_2 =$ 25 pF and a channel length $L = 7\ \mu$m, a device W/L ratio of 10 produces a worst case cumulative feedthrough error voltage under 10 mV, for $V_R = 5$ V and gate voltages of 10 V. The charging time under these conditions is approximately 800 ns. These parameters are in agreement with simulations using the program ISPICE [4].

C. Capacitor Voltage and Temperature Variations

To characterize the error voltage due to capacitor nonlinearities, let

$$\alpha = \gamma_V^C = \frac{1}{C}\frac{\partial C}{\partial V}$$

where γ_V^C is the effective voltage coefficient of the MOS capacitors. The charge-sharing capacitors then have the form:

$$C_1 = C(1 + \alpha V_1) \tag{15a}$$

and

$$C_2 = C(1 + \alpha V_2). \tag{15b}$$

Here it is assumed that the nonlinearity is small enough that the higher order terms in $C(V)$ can be neglected. After precharging the capacitors to $V_1(0)$ and $V_2(0)$ and charge redistribution to a final voltage $V(f)$, we have:

$$V(f)\,2C\,\{1 + \alpha V(f)\} = C\,\{V_1(0) + V_2(0) + \alpha V_1^{\,2}(0) + \alpha V_2^{\,2}(0)\}. \tag{16}$$

Solving for $V(f)$ in the worst case, when $V_1(0) = V_R$ and $V_2(0) = 0$, then

$$V(f) \cong \frac{V_1(0) + V_2(0)}{2} + \frac{\alpha}{4} V_R. \tag{17}$$

Hence

$$V_{V\epsilon} \cong \frac{1}{4}\gamma_V^C V_R. \tag{18}$$

Consideration of the data listed in Table II indicates that for a heavily doped ($5 \times 10^{20}\,\text{cm}^{-3}$) lower capacitor plate the resulting error voltage due to capacitor nonlinearities is negligible. Moreover, since γ_T^C is also very small, temperature gradients of several degrees have a minimal effect on the capacitor voltages after redistribution.

D. Leakage Currents

With respect to conversion errors caused by leakage currents, a calculation for $C_1 = C_2 = 25$ pF indicates that 10 nA of leakage current will produce a voltage loss of 10 mV over 24 μs. Since room-temperature leakage currents in the order of 10 pA/mil^2 are typical in MOS circuits this source of error is also negligible and will allow operation at temperatures up to at least 100°C.

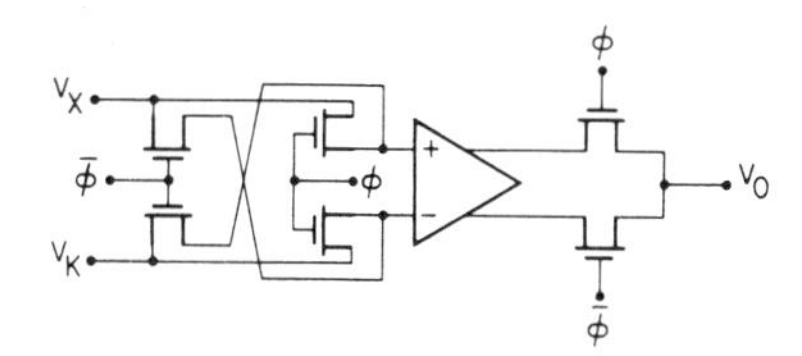

Fig. 8. Circuit for offset cancellation by subtraction.

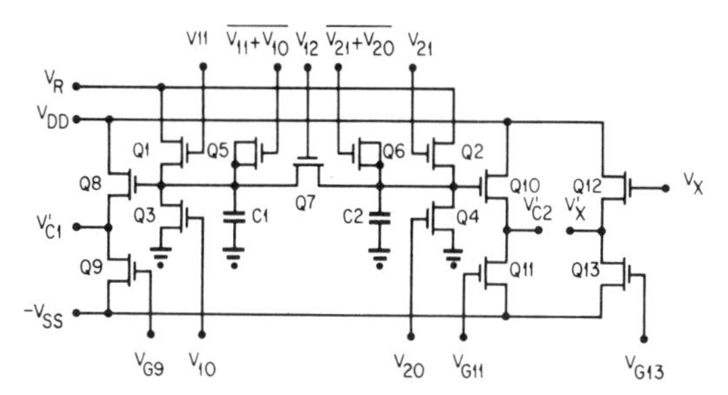

Fig. 9. Circuit schematic of monolithic DAC.

E. Offset Voltage

MOS transistor differential pairs exhibit substantial input offset voltages, typically in the order of 100 mV [5]. Several offset compensation schemes have been proposed [6], [7]. The problem of cancellation of offset voltage in MOS comparators was discussed in Part I of this paper [9], and by other authors. These techniques involve storage of the offset voltage of the comparator as a whole or the individual stages thereof on capacitors during a sample period prior to the conversion itself. As discussed in Part I, effective input offset voltages of less than 5 mV are readily achieved experimentally with this technique.

An alternative approach which is readily implementable in an MOS integrated circuit is illustrated in Fig. 8. Two conversion sequences are required to cancel the offset voltage. During the first A/D conversion pulse Φ is high and an offset voltage V_{OS} is encoded with the voltage difference $V_X - V'(K)$. On the second A/D conversion Φ is low, the input terminals of the comparator are inverted, and the comparator output is taken from the complementary output terminal. The encoded bits now include a contribution of $-V_{OS}$ to the difference $V_X - V'(X)$. The two binary words are added and divided by 2 to null out the offset voltage. Although this approach requires twice the conversion time, it can be used to cancel simultaneously the effect of capacitor mismatches. The mismatch error can be compensated by precharging all bits onto capacitor C_1 on the first conversion, and onto C_2 on the second conversion. The resulting mismatch errors are equal in magnitude and opposite in sign. The subsequent addition and division by 2 therefore also cancels the mismatch error component.

VI. Experimental Results

To verify experimentally the feasibility of the conversion technique, the two-capacitor DAC was fabricated by a five-mask n-channel aluminum-gate process. Fig. 9 is a schematic diagram of the D/A converter circuit, and Fig. 10 is a photo-

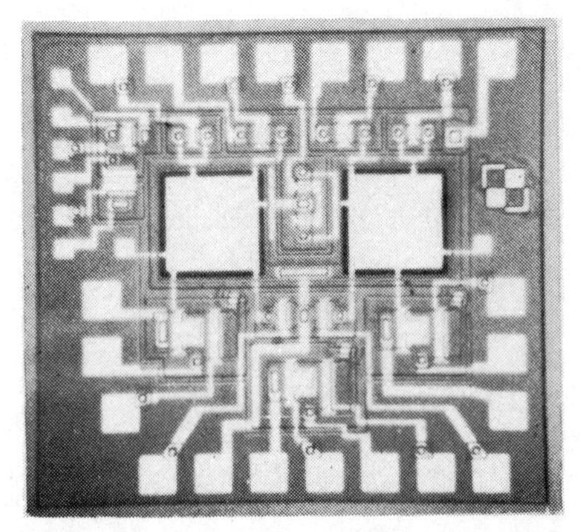

Fig. 10. Photomicrograph of experimental DAC.

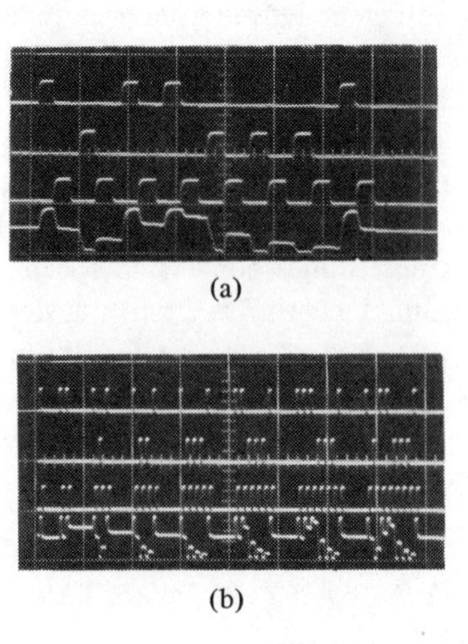

(a)

(b)

Fig. 11. Oscilloscope photographs of conversion sequence. (a) D/A conversion input word 10001101. Top to bottom: V_{11}, V_{10}, V_{12}, V'_{C1}. Vertical: 20 V/div (except V'_{C1}, 5 V/div). Horizontal: 2 μs/div. (b) A/D conversion output word 10001101. Top to bottom: V_{11}, V_{10}, V_{12}, V'_{C1}. Vertical: 20 V/div (except V'_{C1}, 5 V/div). Horizontal: uncalibrated (8 div $\simeq$ 100 μs).

micrograph of the 8-bit experimental die. The total chip size is 48 × 52 mil^2.

All transistors were designed with a nominal channel length $L = 7$ μm and an oxide thickness of 1000 Å. The switch devices, $Q1$–$Q4$ and $Q7$, were assigned a $W/L = 10$ based on feedthrough error considerations. The charge-canceling devices $Q5$ and $Q6$ have $W/L = 5$. The source followers $Q8$–$Q11$ are output buffers for the capacitor voltages. Source follower $Q12$–$Q13$ are included to level shift the voltage V_X to the same level as the DAC output voltage. The pull-down devices $Q9$, $Q11$, and $Q13$ have $W/L = 10$ and are biased in saturation. A quiescent current of 200 μA flows through these devices to drive a 10-pF load capacitance. The pull-up transistors $Q8$, $Q10$, and $Q12$ have $W/L = 15$. No offset-cancellation circuitry was included on the chip. The gate terminals of the pull-down transistors are taken off-chip to permit offset nulling. The feasibility of offset cancellation has been verified separately as discussed in Part I [9]. The integrated DAC was tested in a successive approximation ADC test circuit following the configuration shown in Fig. 3. Approximately 40 bits of shift register plus decoding gates were implemented with TTL logic. A bipolar integrated circuit was used for the voltage comparator.

The oscilloscope photographs in Fig. 11(a) and (b) correspond to an 8-bit D/A and A/D conversion, respectively. The control waveforms include a 5-μs time interval after each D/A conversion to take into account the estimated settling time of an MOS differential voltage comparator. The settling time estimate is based on computer simulation of a cascade of three differential pairs, each with a voltage gain of 10. The D/A conversion time for 8 bits is 13.5 μs and the total A/D conversion time is 100 μs. The A/D conversion time is the sum of a 1-bit D/A time, 2-bit, etc., up to 8 bits, plus assorted comparator and logic delays.

Fig. 12 includes a plot of the error voltage for an 8-bit ADC versus input voltage. The error has been defined as the difference in the measured transition points from the ideal staircase transfer function to eliminate the inherent quantization distortion from the error description. The maximum error is 9 mV which is less than $\frac{1}{2}$ LSB for a reference voltage of 5.120 V.[3]

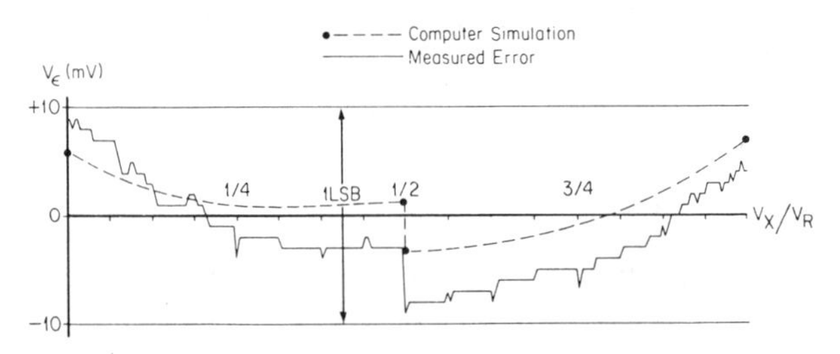

Fig. 12. Error voltage versus input voltage for the A/D converter.

The general features of this error curve can be explained in terms of the error components also shown in Fig. 12. A masking misalignment on the chip was observed which reduced the gate–source overlap capacitance of transistors $Q1$ and $Q2$ by approximately 15 percent. Since the capacitance components of the charge-canceling devices $Q5$ and $Q6$ are unchanged, this misalignment results in a net positive error voltage on $C1$ and $C2$ when the most significant bits are "1." The masking misalignment also causes a reduction in the gate–drain overlap capacitance of $Q3$ and an increase in the gate–drain overlap capacitance of $Q4$. The change in the capacitances of $Q3$ and $Q4$ causes a net positive error voltage on $C1$ and $C2$ when the most significant bits are "0." A third error component arises from the mismatch between capacitors $C1$ and $C2$, which results in a discontinuity of one-half the value of the capacitor mismatch at the $(011 \cdots 1)$ to $(100 \cdots 0)$ transition. The composite diagram of the three error components shown in Fig. 12 is based on computer simulations and is in general agreement with the measured error curve.

Of the three error components isolated above the first two can be minimized by a self-aligned process or, alternatively, with a mask layout which is more tolerant to masking misalignments. The mismatch error can be compensated along with the input offset voltage of the comparator with the two-conversion approach previously described. The most important source of error in the circuit, therefore, is the feedthrough error voltage.

[3]This voltage level was chosen for convenience to make 1 LSB = 20 mV.

TABLE III
PERFORMANCE CHARACTERISTICS OF EXPERIMENTAL ADC

Resolution	8 Bits
Conversion Time	100 μs
Power Supplies	+10v,-5v
Input Range	0-5v

VII. CONCLUSIONS

An MOS compatible A/D conversion technique utilizing charge redistribution on equal capacitors has been developed. For a given level of resolution, the technique can be realized in considerably less area than in the case of a weighted capacitor technique. The principal reason for this is that fabrication of precisely ratioed capacitors requires that the large valued capacitors be made up of many small elements, and the use of such techniques as dummy metal strips. These geometrical requirements enlarge the capacitor array by approximately a factor of 2 compared to the area required for two equal capacitors adding up to the same total capacitance. While the total amount of capacitance required to achieve a given level of resolution using the 2-capacitor approach is roughly the same as in the weighted approach for the same resolution, the capacitance can be realized in a much smaller area.

An experimental 8-bit D/A converter was fabricated in *N*-channel metal gate MOS technology. This was used in an experimental A/D converter, and the experimental evidence gathered indicates that the 8-bit level of resolution can readily be achieved with the circuit and capacitors used. The observed performance is summarized in Table III. With an improved layout to minimize sensitivity of feedthrough to mask alignment, and with a common centroid geometry for the capacitors, the area required to achieve 8-bit resolution in a complete single chip A/D converter would be approximately 5000 mil^2.

REFERENCES

[1] J. A. Schoeff, "A monolithic analog subsystem for high-accuracy A/D conversion," *ISSCC Dig. Tech. Papers*, pp. 18-19, Feb. 1973.

[2] G. Kelson, H. H. Stellrecht, and D. S. Perloff, "A monolithic 10-b digital-to-analog converter using ion implantation," *IEEE J. Solid-State Circuits* (*Special Issue on Analog Circuits*), vol. SC-8, pp. 396–403, Dec. 1973.

[3] K. R. Stafford, R. A. Blanchard, and P. R. Gray, "A completely monolithic sample/hold amplifier using compatible bipolar and silicon-gate FET devices," *IEEE J. Solid-State Circuits*, vol. SC-9, pp. 381-387, Dec. 1974.

[4] National CSS, Inc., Norwalk, Conn., *ISPICE Reference Guide*, 1974. (This program is based on the SPICE Circuit Analysis Program developed at the University of California at Berkeley.)

[5] H. C. Lin, "Comparison of input offset voltage of differential amplifiers using bipolar transistors and field-effect transistors," *IEEE J. Solid-State Circuits* (Corresp.), vol. SC-5, pp. 126-129, June 1970.

[6] H. Schmidt, *Electronic Analog/Digital Conversions.* New York: Van Nostrand-Reinhold, 1970.

[7] R. Poujors, B. Baylac, D. Barbier, and J. M. Ittel, "Low-level MOS transistor amplifier using storage techniques," *ISSCC Dig. Tech. Papers*, pp. 152-153, Feb. 1973.

[8] J. McCreary and P. Gray, "A high-speed all-MOS successive approximation weighted-capacitor A/D conversion technique," *ISSCC Dig. Tech. Papers*, Feb. 1975.

[9] J. McCreary and P. R. Gray, "All-MOS charge redistribution analog-to-digital conversion techniques–Part I," this issue, pp. 371-379.

[10] R. E. Suárez, P. R. Gray, and D. A. Hodges, "An all-MOS charge redistribution A/D conversion technique," *ISSCC Dig. Tech. Papers*, Philadelphia, Pa., Feb. 1974.

[11] C. W. Barbour, "Simplified PCM analog-to-digital converter using capacitor charge transfer," in *Proc. 1961 Nat. Telemetering Conf.*, pp. 4.1-4.11.

A Charge-Balancing Monolithic A/D Converter

GEORGE F. LANDSBURG, MEMBER, IEEE

Abstract—The quest for a minimum-parts-count DPM led to the development of this monolithic, low power analog-to-digital converter. It incorporates the analog and digital functions historically implemented separately with specialized process technologies into a chip with full ±3 digit accuracy. The integration of resistors, compensation capacitors, and an oscillator reduces the external component complement to three capacitors and one adjustable reference; with the addition of a BCD decoder driver and a display, the device becomes a ±1 V DVM, operating up to 50 samples per second. TTL compatible outputs include sign, overrange, and underrange information in addition to the three digit strobes and the BCD data outputs. The logic operates between +5 V and ground, the linear section between +5 V and -5 V.

The conversion algorithm makes the monolithic converter realizable, due to its achievement of effective signal amplification without an extreme linear range requirement, and its autozero feature. As the five CMOS amplifiers on the chip are characterized by moderate gain and significant offsets, and, as the integrated resistors and analog switches are nonideal, algorithm tolerance of analog performance is vital. Modified CMOS logic which times and controls the conversion comprises about two thirds of the 131 × 133 mil integrated circuit.

The paper describes the conversion algorithm and its CMOS implementation, emphasizing the analog design of this innovative device.

I. Introduction

THIS paper describes an analog-digital converter which makes low cost digital multimeters possible and is a practical replacement for analog meters, extending digital metering into areas in which they previously were too expensive. It provides full three digit accuracy in a single 18-pin package and requires minimal external parts. Section II of the paper describes the design objectives, and the circuit is described in Section III. In the last section, the circuit layout and experimental results are discussed.

Manuscript received April 15, 1977; revised September 7, 1977.
The author is with Siliconix, Inc., Santa Clara, CA 95054.

II. Design Objectives

In order to be useful in a wide variety of the envisioned applications, the circuit required the following properties:

1) multiplexed BCD output;
2) internal oscillator and clock generator;
3) automatic zeroing so that only full scale adjustment is required;
4) single reference for bipolar operation;
5) high impedance signal and reference outputs;
6) ±1 V input range;
7) ±5 V supplies;
8) overrange and underrange indications;
9) low power;
10) low cost.

The inclusion of both analog and digital circuitry inevitably leads to less than ideal operating conditions for one section or the other (or both), so the algorithm chosen to meet the above requirements had to be somewhat lenient in its circuit requirements. Extremely high speed logic forcing high power and/or large silicon area was out of the question as were linear circuits having to achieve performance levels of isolated devices. A

Reprinted from *IEEE J. Solid-State Circuits,* vol. SC-12, pp. 662-673, Dec. 1977.

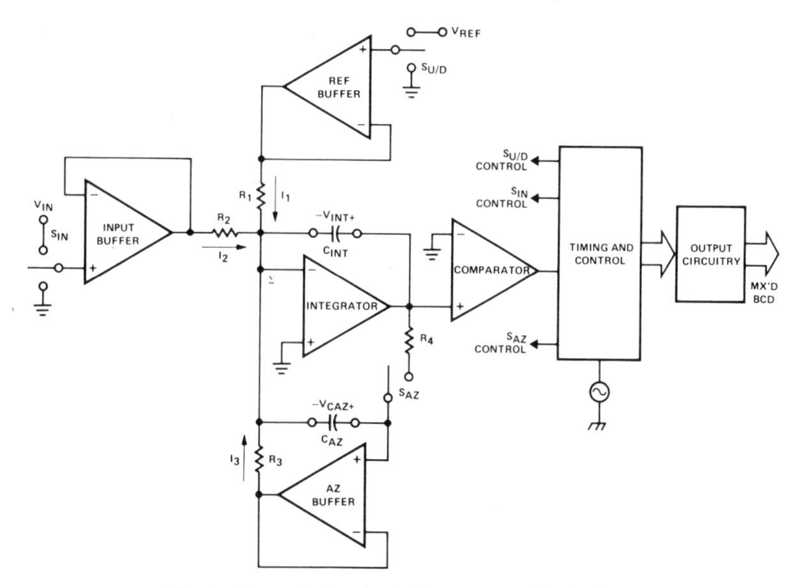

Fig. 1. Charge balancing A/D converter block diagram.

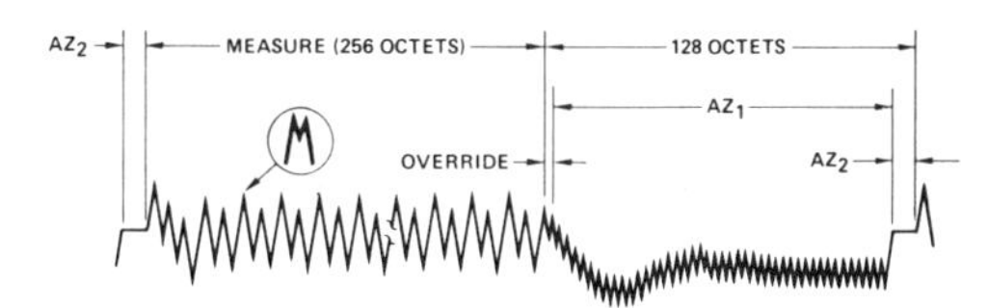

Fig. 2. LD130 coverter cycle.

number of techniques were considered, with the charge-balancing (or quantized feedback) [1] method emerging as the most likely to succeed. It had demonstrated high accuracy capability in two-chip form, and needed no exotic digital or linear circuitry. Its built-in autozero feature was deemed indispensable for a monolithic analog-to-digital converter.

The low cost target made chip size and processing complexity prime concerns. The semiconductor technology had to provide both a basic differential amplifier structure and low power 5-V logic. A standard low voltage CMOS process was chosen. The task became one of best implementing the charge-balancing algorithm as a monolithic CMOS device.

III. The System

General Description

Fig. 1 presents the system in functional block form. The analog section is composed of three unity-gain buffers, an integrator, a comparator, resistors, and analog switches. Standard logic blocks time and control the conversion and finally output the coded analog signal. The system timing diagram of Fig. 2 shows the three intervals, Measure, Override, and Autozero which comprise the conversion cycle. A two-phase system clock (nominally 20 kHz) times operations. Groups of eight clocks are referred to as octets. The Measure interval lasts for 256 octets; the Override interval and both segments of the Autozero interval together consume 128 octets. The length of the Override interval is two octets or less. Since the second segment of the Autozero interval is four octets, the first is at least 122 octets long. Each conversion takes a total of 384 octets, so the conversion rate is $f_{\text{clock}}/3072$.

The Autozero interval has two purposes: to establish a negative reference (the external reference is positive) and to account for the errors in the analog circuitry while so doing. This negative reference manifests itself as the voltage on the capacitor C_{AZ}, and is generated with the input at ground by closing the switch S_{AZ}, forming a closed autozero loop. The three buffers and the integrator provide inputs to or are included in this loop, so their offsets are accounted for. However, the system ignores the comparator output at this time, so its offset is ignored until the fixed length second segment of the Autozero interval. Then, the system locates the true comparator threshold, thereby nulling any such offset. This also defines the initial integrator voltage for the Measure interval. The unknown input signal is connected through the input buffer and resistor R_2 to the integrator during the Measure interval. Charge accumulated on the integrating capacitor C_{INT}, due to this input, is continually nulled by adding or subtracting fixed amounts of charge. The quantity of balancing charge, provided via the reference buffer (negative since this integrator inverts) or the autozero buffer (positive), and resistors R_1 and R_3, respectively, is recorded by the BCD counter. The balance maintained during the Measure interval is approximate, being achieved by the deposition of quantized reference charges; it is made exact during the Override

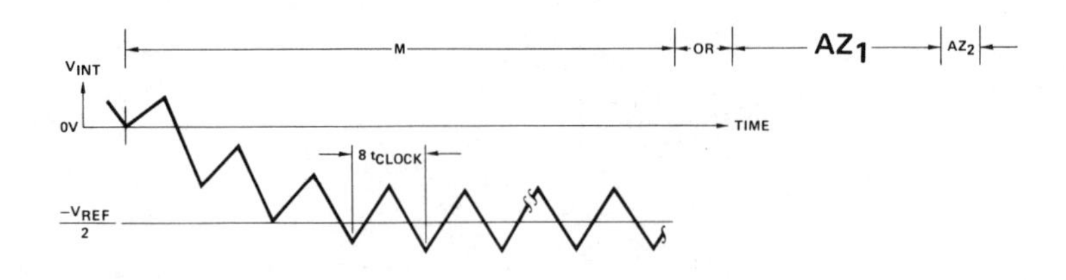

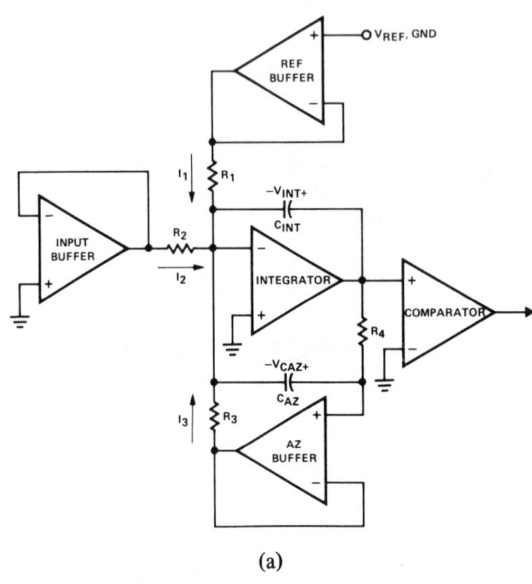

(a)

Fig. 3. (a) Autozero AZ_1.

interval when the input signal is disconnected and any net charge remaining on the integrating capacitor is canceled. As this residual charge is a function of the input signal, the Override interval is of variable duration; the BCD count is adjusted according to its length.

Detailed Description

The conversion process will be explained in detail with the aid of Fig. 3 which presents the integrator waveform–unique for each interval. The Autozero interval, which prepares the system for a measurement, is examined first.

During the first segment of the Autozero interval, AZ_1 in Fig. 3(a), the input buffer is connected to analog ground and the switch S_{AZ} is closed, so, as can be seen by referring to Fig. 1,

$$I_2 = \frac{V_{OS_{ib}} - V_{OS_{int}}}{R_2},$$

where $V_{OS_{ib}}$ and $V_{OS_{int}}$ are the offset voltages of the input buffer and the integrator, respectively. The switch $S_{U/D}$ alternately connects V_{REF}, then analog ground to the reference buffer for four clocks each (50 percent duty cycle) so the integrator current due to this source is alternately

$$I_1 = \frac{V_{REF} + V_{OS_{rb}} - V_{OS_{int}}}{R_1} \text{ and } I_1 = \frac{V_{OS_{rb}} - V_{OS_{int}}}{R_1},$$

where $V_{OS_{rb}}$ is the reference buffer offset. The third integrator current is sourced from the autozero buffer whose input is always the voltage on the capacitor C_{AZ} and whose offset is $V_{OS_{az}}$:

$$I_3 = \frac{V_{C_{AZ}} + V_{OS_{az}}}{R_3}.$$

In this closed-loop situation, a dc balance is attained; by summing the average integrator currents to zero, an expression for $V_{C_{AZ}}$ results.

$$\bar{I}_1 + \bar{I}_2 + \bar{I}_3 = 0$$

$$V_{C_{AZ}} = \frac{R_3}{R_2}(V_{OS_{int}} - V_{OS_{ib}}) - \frac{R_3}{2R_1} \cdot V_{REF} + \frac{R_3}{R_1}(V_{OS_{int}} - V_{OS_{rb}}) - V_{OS_{az}}.$$

This demonstrates the inclusion of buffer and integrator offsets in $V_{C_{AZ}}$, and a lucid expression emerges when these offsets are ignored:

$$V_{C_{AZ}} = -\frac{R_3}{2R_1} \cdot V_{REF}.$$

In this circuit, $R_2/2.048 = R_1 = R_3 = R$, so $V_{C_{AZ}} = -V_{REF}/2$. In Fig. 3(a) it can be seen that the integrator ramps equal distances above and below the nominal value of $V_{C_{AZ}}$, following the regular operation of the switch $S_{U/D}$ during the Autozero interval. A similarly shaped signal, attenuated and

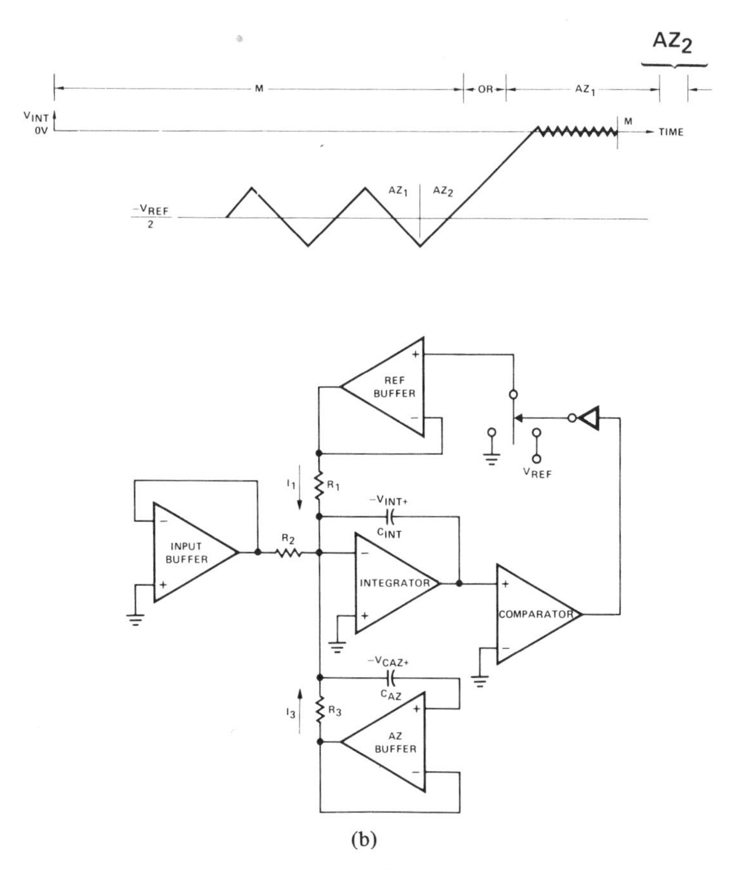

(b)

Fig. 3 (continued). (b) Autozero AZ_2.

lagging the integrator by 90° appears across C_{AZ}. Hence, to accurately retain the average value of $V_{C_{AZ}}$ when the autozero loop is opened, the switch S_{AZ} must be operated precisely when the ac component of $V_{C_{AZ}}$ is at a zero crossing, or when the integrator voltage is at a peak. The timing of the opening of the switch S_{AZ} is critical, but its closing is not. During the signal conversion the integrator operates around ground, and slews rapidly toward $V_{C_{AZ}}$ when the autozero loop is closed. Approximately ten time constants ($\tau = R_4 \cdot C_{AZ}$) are allowed for loop settling.

The final four octets of the Autozero interval, AZ_2 in Fig. 3(b), take place with the autozero loop reopened and the complement of the comparator driving the switch $S_{U/D}$ to form an oscillating loop: $S_{U/D}$-reference buffer-R_1-integrator-comparator-logic inverter. If the comparator output is low, ground is connected to the reference buffer, and, as the input buffer remains grounded, I_3 drives the integrator positive. When the comparator switches, so does the switch $S_{U/D}$, and V_{REF} drives the integrator low via the reference buffer and R_1 as $|I_1| = 2|I_3|$ and $I_2 = 0$. This oscillation, which is shown in Fig. 3(b), tends to center the integrator output between the rails, and places it within a few millivolts (the oscillation amplitude) of the true comparator threshold. The Autozero interval ends with the offsets of all five amplifiers taken into account, a negative reference established, and the integrator voltage set at the comparator threshold.

The Measure interval, Fig. 3(c), is characterized by a notched triangular waveform. The signal charge added to C_{INT} each octet is

$$Q_S/\text{octet} = -\frac{V_{in}}{R_2} \cdot t_{octet} = -\frac{V_{in}}{2.048R} \cdot t_{octet}.$$

As in the second segment of the Autozero interval, the logic operates in response to the comparator to continually nullify this charge; however, the null is sought not by letting a loop free-run, but rather by supplying fixed amounts of charge, $\pm Q_R$. Q_R is generated by operating the switch $S_{U/D}$ in one of two modes for each octet, depending on the comparator state just prior to each octet. If it is low, an "up" octet ensues, with the switch $S_{U/D}$ connecting the reference buffer to ground for the first seven clocks of the octet and to V_{REF} for the final clock. A high comparator output indicates a "down" octet, with ground selected as the reference buffer input for one clock, followed by seven clock cycles with V_{REF} connected. The reference charge supplied to C_{INT} during each octet is (remembering that $V_{C_{AZ}} = -V_{REF}\, R_1/2R_3$):

$$\text{Up Octet: } (-V_{REF}/8R_1)t_{octet} - (V_{C_{AZ}}/R_3)t_{octet} = 3V_{REF}/8R)t_{octet} = Q_R/\text{octet}$$

$$\text{Down Octet: } (-7V_{REF}/8R_1)t_{octet} - (V_{C_{AZ}}/R_3)t_{octet} = -(3V_{REF}/8R)t_{octet} = -Q_R/\text{octet}.$$

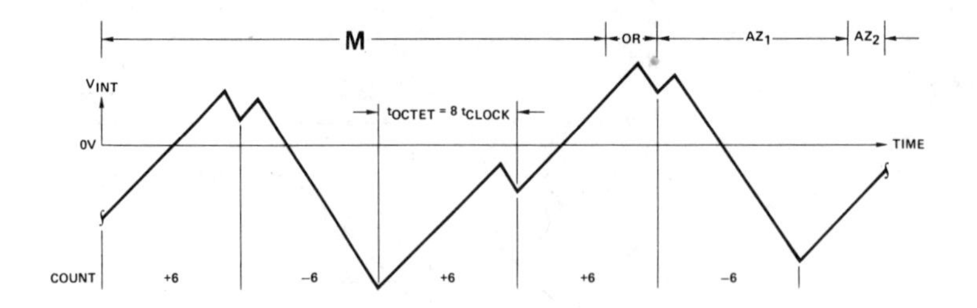

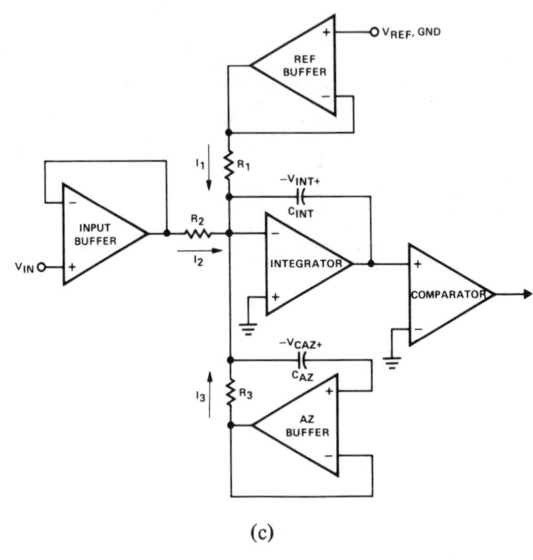

(c)

Fig. 3 (continued). (c) Measure interval.

Thus, $Q_R/\text{octet} = (3V_{REF}/8R)t_{octet}$ and $Q_S/\text{octet} = (-V_{IN}/2.048R)t_{octet}$. Note that the integrator is continually driven by the reference source resulting in similar integrator voltage excursions for all input signals. Corresponding to the operation of the switch $S_{U/D}$, the BCD counter is indexed—by +6 counts (7 up, −1 down) for an up octet and by −6 counts (1 up, −7 down) for a down octet; each count therefore represents $Q_R/6$, or

$$1 \text{ count} = Q_R/6 = 3V_{REF}/8R_1 \cdot \frac{1}{6} \cdot t_{octet} \cdot \frac{8t_{clock}}{t_{octet}}$$

$$= V_{REF} \cdot t_{clock}/2R_1.$$

If, for a given input signal, the Measure interval consists of Nu up octets and Nd down octets, the count at the end of the Measure interval is

$$C_M = 6(Nu - Nd) \quad \text{where} \quad Nu + Nd = 256.$$

The residual charge at the end of the Measure interval is

$$Q_{residual} = \left(\frac{-256V_{IN}}{2.048R} + Nu \cdot \frac{3}{8} \cdot \frac{V_{REF}}{R} - Nd \cdot \frac{3}{8} \cdot \frac{V_{REF}}{R}\right) t_{octet}$$

$$= \left(\frac{-128V_{IN}}{1.024R} + C_M \cdot \frac{V_{REF}}{R}\right) t_{clock}$$

$$= \left(\frac{-1000V_{IN}}{R} + C_M \cdot \frac{V_{REF}}{2R}\right) t_{clock}.$$

If V_{IN} is chosen such that $Q_{residual} = 0$, that is $V_{IN} = \pm 3n/500 \cdot V_{REF}$, $n = 0, 1, 2, \cdots$, it can be seen that C_M directly represents V_{IN}:

$$C_M = 2000 \cdot V_{IN}/V_{REF}.$$

In such cases, the Override interval, Fig. 3(d), has no net effect and consists of equal length positive and negative ramps. For the general case where $Q_{residual} \neq 0$, the Override interval has the role of canceling the net charge accumulated during the Measure interval and adjusting the BCD count by C_{OR}. The input buffer is grounded and the switch $S_{U/D}$ is driven to achieve equilibrium. The total integrator charge change per count is now just the reference charge per count,

$$\frac{\pm V_{REF}}{2R} \cdot t_{clock} \text{ (as before)},$$

so

$$Q_{residual} = C_{OR} \frac{(\pm V_{REF})t_{clock}}{2R} =$$

$$= \left(\frac{-1000V_{IN}}{R} + C_M \frac{V_{REF}}{2R}\right) t_{clock}.$$

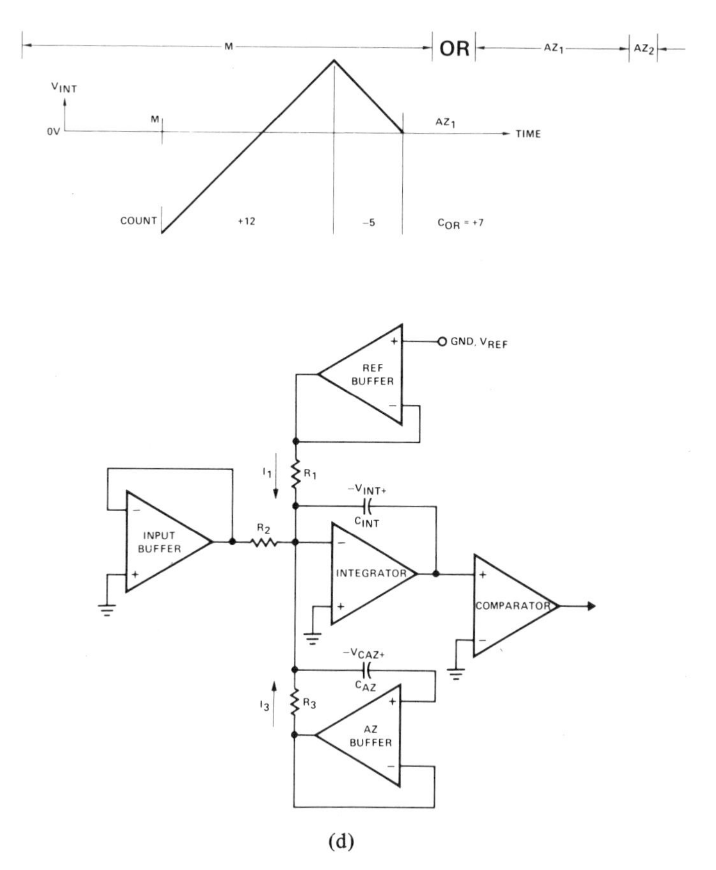

(d)

Fig. 3 (continued). (d) Overide interval.

Expressing the total count C_T as $C_M + C_{OR}$,

$$C_T = C_M + C_{OR} = 2000\, V_{IN}/V_{REF}.$$

If V_{REF} is set to 2 V, $V_{C_{AZ}}$ becomes -1 V, and C_T represents V_{IN} in millivolts. The analog-to-digital conversion is complete.

Once the Override interval ends, the system proceeds to autozero and then begins another conversion. Some fine points of the scheme need elaboration. Fig. 3 shows how the charge balancing algorithm produces a large integrator slope, $\Delta V_{C_{INT}}/\Delta t$, without extreme absolute voltages by interspersing up and down octets. The continual infusion of balancing charge during the Measure and Override intervals holds the average integrator voltage near ground while producing steep threshold crossings for the comparator to sense. During the Override interval, when the most critical comparator transition occurs,

$$\Delta V_{C_{INT}}/\Delta t \pm \frac{V_{REF}}{2RC_{INT}} = 2000 \text{ V/s} = 0.1 \text{ V}/t_{clock},$$

so a 100-mV comparator switching inaccuracy causes only one count of error. This illustrates how the algorithm effectively amplifies the signal to reduce the tolerances on the analog circuitry and to limit the effect of comparator noise. Also, the two-phase clock is split to accurately determine the time at which the charge on C_{INT} reaches zero (when the integrator voltage trips the comparator) signifying the end of the Override interval. Quantizing error is thereby reduced to ±1/2 count. The Override interval always consists of a positive integrator voltage ramp followed by a negative one, so that every conversion culminates in a one to zero comparator transition.

Note that C_{AZ} is returned to the integrator virtual ground rather than to circuit ground in Fig. 1. This has the dual benefit of making the autozero loop unconditionally stable and reducing errors arising from the imperfection in that virtual ground. C_{AZ} effectively bootstraps the integrator summing node via the autozero buffer, so that the integrator error current due to a nonzero voltage on that node is reduced. The choice of 7-up/1-down and 1-up/7-down octets during the Measure interval (rather than 8/0 and 0/8) ensures a constant number of switch $S_{U/D}$ operations for every Measure and Override interval, and further balances the number of up-to-down and down-to-up transitions. Charge coupled into the integrator due to $S_{U/D}$ operations over the course of a conversion is canceled to a first approximation and minimized absolutely. Further, any net charge injected due to switch asymmetry is accounted for during the Autozero interval where the numbers of each sense transition are also equal.

The linearity of this technique is exceptional. When the various voltage offsets are included in the conversion relationship derived above, they can be shown to cancel perfectly for any input voltage. The Autozero interval ensures stability of full-scale calibration as well as zero stability.

The algorithm is forgiving in that switch transients and

TABLE I
CMOS DEVICE CHARACTERISTICS

	N-CHANNEL	P-CHANNEL	UNITS
THRESHOLD VOLTAGE, V_{th}	1.2	1.4	Volts
MOBILITY, μ*	650	300	cm^2/Volt-Sec
OUT DIFFUSION, xj	0.1	0.05	Mils
DYNAMIC IMPEDANCE, r_{DS}*	750	620	KΩ

*3.0 MIL WIDE x 0.4 MIL LONG DEVICES OPERATED @ V_{DS} = 5 V, I_{DS} = 100 μA, V_{GSN} = 1.85 V, V_{GSP} = 3.3 V

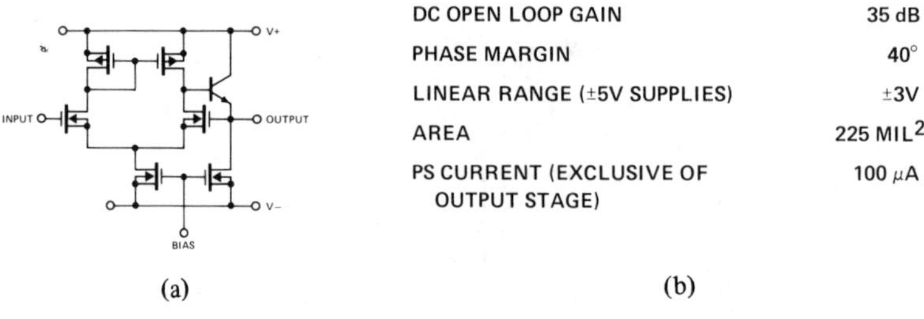

Fig. 4. (a) Single-stage buffer amplifier. (b) Characteristics (typical).

delays are nulled because of the similar functioning of the system during the Measure and the Autozero intervals. The centering of the integrator operation between the supplies guarantees maximum linear swing. Still, there are sources of error to be considered in the circuit design and chip layout.

1) The opening of the autozero loop must be precisely timed to minimize error in $V_{C_{AZ}}$.

2) The oscillation amplitude during the second segment of the Autozero interval (AZ_2) is direct integrator offset and must be small.

3) Digital noise (electrical and thermal) must be considered in the chip layout.

4) A compromise must be reached between maximum integrator linear swing and comparator sensitivity.

IV. CIRCUIT DESIGN

Analog Circuitry

Table I presents some typical characteristics of the n- and p-channel devices of the low-voltage CMOS process. The voltage gain expression for the common source structure of a differential amplifier is $Av = g_m Z_{out}$, where gm is a parameter of the n- or p-channel driver and Z_{out} if the parallel combination of r_{ds} of the driver and the p- or n-channel load, respectively.

$$g_m \propto \mu \cdot \left(\frac{W_{drawn}}{L_{drawn} - 2xj}\right) \cdot (V_{gate} - V_{source} - V_{threshold})\Big|_{driver}$$

$$z_{out} = r_{ds_{driver}} \| r_{ds_{load}}.$$

The n-channel devices have better amplifier characteristics than do the p-channel devices (excepting noise), but since ±5-V supplies preclude cascode structures, the p-channel deficiencies cannot be avoided. A single-stage differential amplifier is shown in (Fig. 4.) The open-loop voltage gain may be expressed as:

$$Av_{ol} = \frac{g_{mN}}{2}(r_{ds_N} \| r_{dsp}) \left(\frac{g_{mp}(r_{ds_N} \| r_{ds_p})}{g_{mp}(r_{ds_N} \| r_{ds_p}) + 1} + 1\right)$$

which approaches

$$g_{mN}(r_{ds_N} \| r_{ds_p})$$

when

$$g_{mp}(r_{ds_N} \| r_{ds_p})$$

greatly exceeds unity. The current-mirror load biasing constrains the drain to source voltage to equal the gate to source for device $P1$ in Fig. 4, which does not maximize its output impedance r_{dsp}. The best operating conditions include a low p-channel gate to source voltage to achieve high load impedance, strong enhancement of the drivers $N1$ and $N2$ to boost g_{mN}, and adequate compliance on the current source, which operates at about 100 μA. Under such conditions, the parameters of Table I suggest a gain of roughly 35 dB; this limits this simple amplifier to applications where high input impedance and specific dc signal levels are the prime concerns. The structure is stable in closed-loop configurations without compensation. Fig. 4 also lists typical characteristics.

The two-stage amplifier of Fig. 5 is more akin to an operational amplifier. The keys to its improvement over the simple differential stage are the second gain stage and the bias loop of $P3$, $P4$, $N3$, $N4$, and $P5$. This loop provides common-mode feedback to the bias of the load transistors of the differential stage, and thereby multiplies the common-mode gain of that stage by the factor

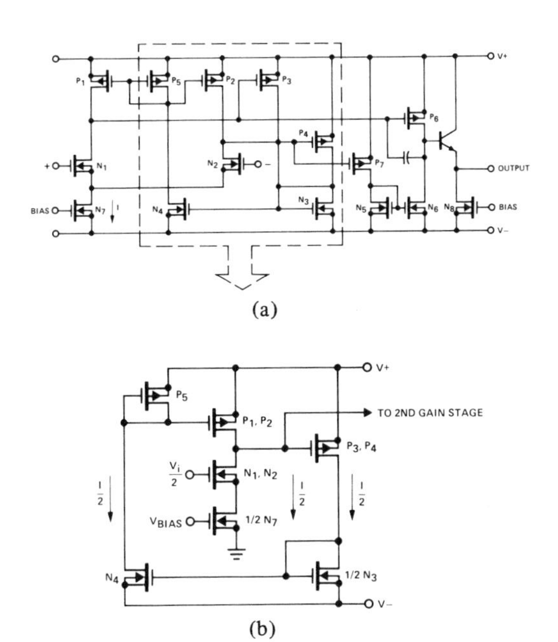

Fig. 5. (a) Two-stage differential amplifier. (b) Common-mode bias loop.

$$\frac{1}{1 - A_{CML}},$$

where

$$A_{CML} = \frac{g_{mp1} \cdot g_{mp4} \cdot g_{mN4} \cdot r_{ds_{p1}}}{(g_{mN3}/2 + 1/2r_{ds_{N3}} + 1/r_{ds_{P4}})(g_{mp5} + 1/r_{ds_{N4}} + 1/r_{ds_{p5}})}$$

$$= g_{mp} r_{ds_p}$$

under the assumption that:

$$g_{mN4} = g_{mN}/2 = g_{mN}, g_{mp5} = g_{mp1} = g_{mp4} = g_{mp}, r_{ds_{N4}} = 2r_{ds_{N3}} = r_{ds_N},$$

and

$$r_{ds_{p5}} = r_{ds_{p1}} = r_{ds_{p4}} = r_{ds_p}.$$

The first stage common-mode output voltage, which arises from the finite output impedance of the biasing current source under common-mode input, is thus greatly reduced. The output of the second stage can be expressed as

$$V_{2_{out}} = A_{2_d} \cdot V_{1_{out_d}} + A_{2_{cn}} \cdot V_{1_{out_{cm}}}$$

where cm and d denote common mode and differential, respectively, where A_2 is the second-stage gain, and V_1 is the output of the first stage. $A_{2_{cm}}$ can be shown to be approximately $-g_{mp}/g_{mN}$, where the g_m's are those of the second-stage transistors. Since $V_{1_{out_{cm}}}$ has been minimized, the only significant effect of a common-mode input to the composite amplifier is a resultant differential first stage output, caused by mismatches in the driver and load transistors and amplified by the differential second-stage gain A_{2_d}. Any noise pickup on the double rail input to the second stage is of course primarily common mode.

In addition to the common-mode improvement, the bias loop of Fig. 5 also allows some enhancement of the first stage-differential gain. The loop allows freedom of bias (gate) voltage for the load transistors $P1$ and $P2$, not available with conventional current mirror biasing where gate-source voltage is constrained to equal drain-source voltage at balance. This can be seen from the isolated bias loop of Fig. 5. Since their bias and geometries are identical, the matched transistors $P5$ and $P1$ carry the same current $I/2$ to a first approximation. The unity-current mirror, $N3$ and $(1/2)N4$, forces $P3$ also to carry $I/2$. Adjusting the size of $P3$ thus allows control of its gate voltage. In particular, if its aspect ratio is less than that of $P1$ and $P5$, its gate voltage will be below that of $P1$ and $P5$. This means that the drain to source voltage of $P1$ exceeds its gate to source voltage, biasing it farther into saturation with a corresponding increase in output impedance and concomitant differential gain.

The second stage of the amplifier adds perhaps 35 dB of gain to the 40 dB of the first stage and converts the signal

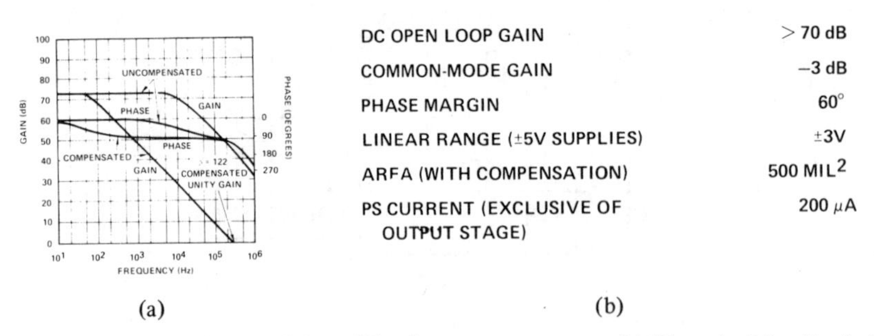

DC OPEN LOOP GAIN	>70 dB
COMMON-MODE GAIN	−3 dB
PHASE MARGIN	60°
LINEAR RANGE (±5V SUPPLIES)	±3V
AREA (WITH COMPENSATION)	500 MIL^2
PS CURRENT (EXCLUSIVE OF OUTPUT STAGE)	200 μA

(a) (b)

Fig. 6. (a) Two-stage differential amplifier frequency response. (b) Characteristics (typical).

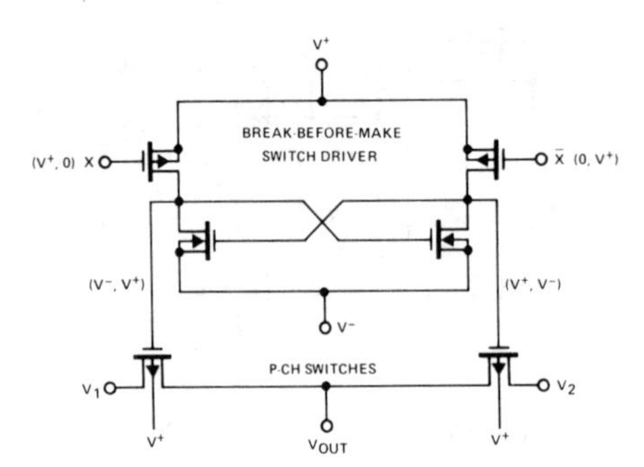

Fig. 7. Analog switch driver.

to a single-ended output. The two-stage structure adds phase shift requiring compensation for stable closed-loop operation. An integrated 15 pF capacitor produces 60° phase margin as shown by the frequency response of the two-stage amplifier plotted in Fig. 6. Other characteristics of this amplifier are also listed.

Both amplifiers make use of the substrate n-p-n of the CMOS process, with caution to avoid saturation, in either single or Darlington configurations as output drivers. A small series-base resistance effectively avoids emitter follower oscillation. A constant sink current n-channel MOSFET is the output pulldown. The choice of this output structure minimizes area for the drive currents of 0.1 to 0.5 mA, and since the emitter followers have h_{FE}'s in excess of 100, offset caused by their loading the gain stage is inconsequential, especially in Darlington form.

Amplifier Functions

The purpose of the Autozero buffer is to provide a negative reference current while presenting no appreciable load to the stored voltage $V_{C_{AZ}}$. Its operating point remains virtually constant at -1 V, with slight corrections being made during each Autozero interval. Its gain and offset are not critical as the system acts to produce the proper current through R_3, regardless of how the voltage at the autozero buffer output and $V_{C_{AZ}}$ differ. The single-stage differential structure is used; judicious sizing of the transistors centers its linear range around -1 V, its quiescent operating point.

If positive reference current were provided directly, the reference source would have to be capable of several hundred microamps. Also, the p-channel switch $S_{U/D}$ would carry this current and, as its on-resistance is a function of the substrate voltage (V+), effective modulation of R_1 would occur with supply variations. The reference buffer, a single-stage differential amplifier, sized to be most linear around +1 V, solves these problems. It must consistently achieve two levels spaced by V_{REF} (nominally 0 V and +2 V) but its gain and offset and linearity are not critical. The only drawback to using this simple structure, where offset depends somewhat on common-mode voltage, is when V_{REF} and V_{IN} are both variables, such as in ratio-type A/D applications.

The input buffer must recreate the input signal over its ±1 V range. In addition to providing a high impedance input, it has a strict linearity requirement—high gain and common-mode independent offset are vital. The two-stage amplifier of Fig. 5 provides input tracking over a 6-V range, as indicated by Fig. 6.

A program written to identify the error sources of the converter showed integrator open-loop gain to be a critical factor; again, the two-stage amplifier was employed.

The comparator must have fast response for small overdrive. The most critical comparator transition is at the end of the Override interval, when the BCD counter must be halted as soon as the integrating capacitor charge is reduced to zero. The uncompensated version of the two-stage amplifier has a maximum slew of ~10 V/μs and swings about 5 V/μs with 3-mV overdrive. This means that the integrator output can be sensed within 5 μs of its zero crossing, since it swings approximately 2 mV/μs during the Override interval. As t_{clock} = 50 μs, typically, and the BCD counter indexes once per clock time, less than one-tenth of a count (0.01 percent) error is caused by comparator response time.

Analog–Digital Interface

There are four interface routes. The digital-to-analog interface is composed of the logic signals controlling the analog switches S_{IN}, $S_{U/D}$, and S_{AZ} via the break-before-make switch drivers illustrated in Fig. 7. These drivers translate the 0-V to 5-V logic signal to ±5 V levels and ensure that both p-channel

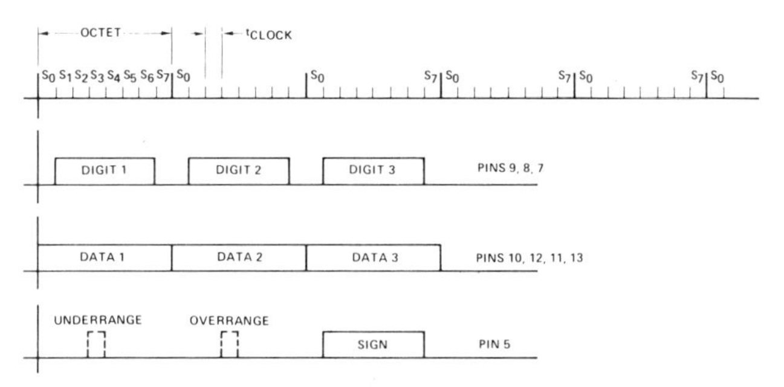

Fig. 8. LD130 output.

switches are off before either turns on, eliminating transient loading on the reference and input signals and companion disturbance of analog ground. The analog-to-digital interface is the comparator output, translated from ±5 V to 0 V and +5 V by a simple logic inverter.

The aforementioned oscillation loop set up in the second part of the Autozero interval AZ_2 involves both the analog and digital sections. Most of the propagation delay is in the analog section, specifically in the integrator where the frequency compensation required for stability also causes a response delay. Under normal operating conditions, the loop oscillates at ~50 kHz, resulting in a peak integrator displacement from the comparator threshold of ±10 mV upon entry into the Measure interval. This amounts to less than a tenth of a count of error (0.01 percent).

Digital Circuitry

A symmetrical, two-phase clock, emanating from a toggle flip-flop following the oscillator, drives the digital circuitry. As high speed was not a requisite, a modified CMOS logic approach was chosen. Gates with more than two inputs are composed of single p-channel active loads with n-channel pulldowns for each input. For the bulk of the logic, the slight speed and power penalties of this implementation were outweighed by its area advantage. Where speed was critical, full CMOS structures were employed. Such situations include the logic in the oscillation loop and that controlling the opening of the switch S_{AZ} at the end of the first segment of the Autozero interval AZ_1; the delay here is less than 20 ns.

Most of the digital functions–timing, analog switch control, BCD up/down counting, output formatting, and driving–are commonplace, but the determination of the sign of the input signal bears some explanation. The state of the sign flip-flop is updated whenever the BCD count is zero, which may occur once or many times during a conversion. A zero count indicates that equal amounts of positive and negative reference charge have been expended to cancel the charge on C_{INT}. Any residual charge is a manifestation of the input signal and forces the comparator to a low state for positive inputs and vice versa. Hence, a signal derived from the complement of the comparator is the sign flip-flop input. Large signals are balanced only by large expenditures of one sense reference charge; in such cases, the BCD counter quickly accumulates counts and departs from zero, meaning the sign is fixed early in the Measure interval. Small inputs result in slow net count accumulation and frequent counter values of zero, so the sign is not determined until the integrated result of the input signal is large enough to force the counter to depart from zero. Often, for low values of V_{IN}, the sign is not set until the Override interval is complete.

The BCD count is latched upon the completion of the Override interval and is output in multiplexed form. The digit strobe rate is 1/24th that of the system clock, and interdigit blanking amounting to 25 percent of the digit time facilitates the use of various display types. If the count is less than 080, an underrange signal is generated; if it exceeds 999, an overrange signal occurs and the digit strobes are disabled during the next Measure interval, causing the display to blink. Fig. 8 is concerned with the output format.

V. IC Layout

Fig. 9 is a photo of the converter. The regular cell group structures distinguish the digital section; the large devices around the chip periphery are output drivers. The three two-stage amplifiers–input buffer, comparator, and integrator–are arranged left to right in the center of the bottom third of the chip. They are identifiable because of their similar structure and their compensation capacitors. The amplifiers are laid out isothermally with respect to the main heat sources, and are completely guard-ringed to allow extension of the supplies to at least ±8 V. While offsets per se present no system problems, they must not force $V_{C_{AZ}}$ to exceed the linear range of the autozero buffer; paired transistors are identical, rather than mirror images, to reduce mismatching and to enhance tracking. In this manner, individual amplifier offsets are kept below 50 mV. The resistors R_1, R_2, and R_3 cannot exhibit measurable voltage sensitivity, and so are identical p^+ diffusions. The gross filter resistor R_4, being of a much larger value and requiring less precision, is a p^--type diffusion. The n-type substrate is tied to the positive supply, so decoupling of the analog and digital supply lines can be accomplished only by ensuring that all current carrying paths join at the positive supply pad. Perfect decoupling is not possible, but careful bussing reduces interaction to an acceptable level. Analog and digital ground are separate pins; in most applications, they are tied together externally, but need not be, and so long as they do not differ by more than a volt, operation of the circuit is not degraded.

VI. Conclusion

Table II summarizes the performance of the converter, which meets the requirements presented in the introduction

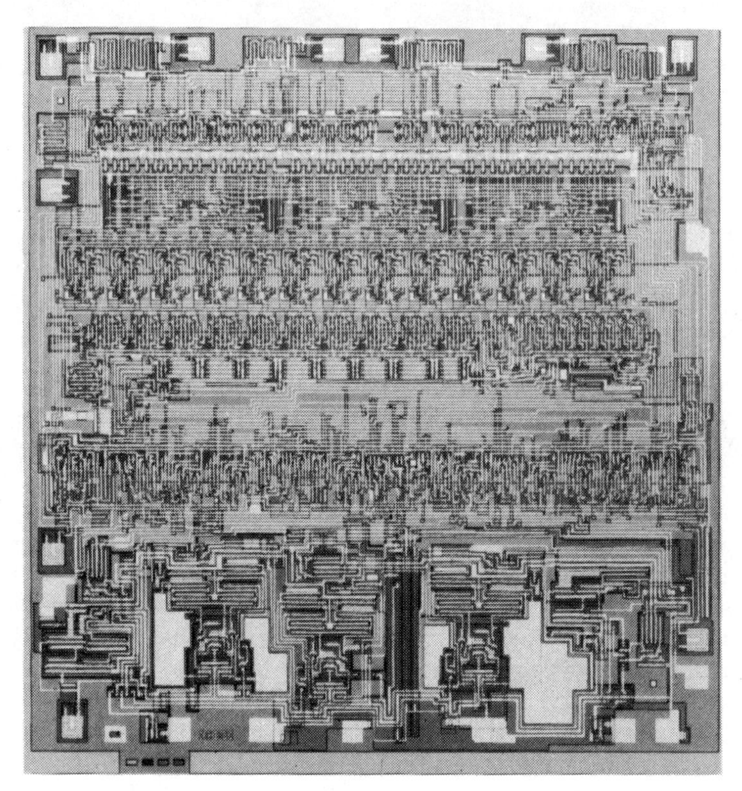

Fig. 9.

TABLE II
ELECTRICAL CHARACTERISTICS

#	Group	Symbol	Characteristic	MIN	TYP	MAX	UNIT	TEST CONDITIONS $V_1 = +5$ V, $V_2 = -5$ V, $V_{REF} = 2.000$ V, $T_A = 25°C$, $C_{INT} = 0.068$ μF, $C_{strg} = 0.1$ μF
1	GENERAL		Linearity			0.1	% rdg	
2	GENERAL		Noise		0.3		LSB	Peak-to-Peak noise apparent when going from one steady reading to another.
3	GENERAL		Zero T. C.		10		μV/°C	$0 \leq T_A \leq 70°C$
4	GENERAL		Gain T. C.		15		ppM/°C	$0 \leq T_A \leq 70°C$
5	GENERAL		NMR Normal Mode Rejection	36			dB	$f_{series} = 60$ Hz, $f_{osc} = 24$ kHz
6	INPUT	I_{IN}	V_{IN} Input Bias Current		7		pA	$T_A = 25°C$
7	INPUT	I_{IN}	V_{IN} Input Bias Current		90		pA	$T_A = 70°C$
8	INPUT	I_{REF}	V_{REF} Input Bias Current		7		pA	$T_A = 25°C$
9	INPUT	I_{REF}	V_{REF} Input Bias Current		90		pA	$T_A = 70°C$
10	INPUT	f_{CLK}	Clock Frequency		30		kHz	
11	INPUT	$D.C._{CLK}$	Clock Duty Cycle	30/70		70/30	%	
12	INPUT	I_{INL}	Clock Input Current Low			-1	mA	$V_{INL} = 1.0$ V
13	INPUT	I_{INH}	Clock Input Current High			1	mA	$V_{INH} = 4.0$ V
14	OUTPUT	V_{OL}	All Outputs			0.4	V	$I_{OL} = 1.6$ mA
15	OUTPUT	V_{OH}	Sign, Digits	2.4			V	$I_{OH} = -400$ μA
16	OUTPUT	V_{OH}	Data Bits	2.4			V	$I_{OH} = -100$ μA
17	SUPPLY	I_1	Supply Current			6	mA	
18	SUPPLY	I_2	Supply Current			-4	mA	
19	SUPPLY	$PSRR_1$	V_1 Supply Rejection		0.6		mV/V	
20	SUPPLY	$PSRR_2$	V_2 Supply Rejection		1.4		mV/V	

Typical values are for Design Aid Only, not guaranteed and not subject to production testing. ICBG

to this paper. Production experience to date has demonstrated good yields, making it a true alternative to analog metering. Its accuracy is better than 0.1 percent, its linearity well within ±1/2 LSB, and both extend perfectly through zero. A later version achieves extended operation to ±1500 counts and a ±1.5-V input range. Accuracy of ±0.1 percent is maintained, the range being limited solely by the capacity of the BCD counter.

Acknowledgment

The circuit described in this paper resulted from the efforts of a number of people at Siliconix, especially L. Hill, the Analog Design Manager. Their help is gratefully acknowledged.

References

[1] T. W. Pickerrell and G. Grandbois, "A/D conversion techniques," *Electronics*, Oct. 13, 1977.

[2] W. N. Carr and J. P. Mize, *MOS/LSI Design and Application.* New York: McGraw-Hill, Texas Instruments Electronics Series, 1972.

[3] A. B. Grebene, *Analog Integrated Circuit Design.* New York: Van Nostrand Reinhold, 1972.

A Five-Digit Analog-Digital Converter

RUDY J. VAN DE PLASSCHE AND ROB E. J. VAN DER GRIFT

Abstract—**The sigma-delta modulation principle is used in a five-digit plus-sign A/D converter. A simple auto-zero circuit is added to eliminate input-offset problems. Two input voltage ranges can be chosen in a cold-switched range-setting mode. The analog subsystem is implemented into two integrated circuits and contains nearly all necessary elements. Only a timer and up/down counter circuit are required to obtain the digital output signal.**

I. Introduction

THE APPLICATION of large-scale integrated circuits such as microprocessors, programmable logic arrays, etc., for instrumentation and control purposes opens up a new market for low-speed high-accuracy A/D converters. Most designs in this area make use of the dual-slope conversion principle in a MOS technology. In these designs the analog part plays a minor role with respect to the digital part. Furthermore, in the input and reference source circuits the MOS technology shows some drawbacks with respect to a bipolar technology. Therefore, an optimum in A/D converter performance can be obtained by using an analog processor containing all accurate functions in a bipolar technology and a digital controller performing control and counting functions in a MOS technology. The capability to be implemented into a bipolar integrated circuit determines the choice of a conversion principle. The sigma-delta modulator seems to be a good alternative to be implemented in a bipolar technology compared to the dual-slope converter in a MOS technology. The analog integrated circuits which will be described contain nearly all necessary elements. The digital controller, which basically contains a timing and counting function, is implemented in a LOC-MOS technology and will not be discussed.

Manuscript received May 20, 1977.

The authors are with Philips Research Laboratories, Eindhoven, The Netherlands.

II. Basic Sigma-Delta Modulator Circuit

The basic circuit diagram of a sigma-delta modulator is shown in Fig. 1(a). The most important circuit parts are the capacitor C which stores the charge of the input-signal current source I_{in} and the switch S which turns on the charge or discharge current sources I. A control loop consisting of the comparator, the flip-flop, and the clock generator f, determines the position of the switch S. This control loop acts in such a way that the average charge in the capacitor C is kept zero. When no input signal is applied, then the number of charge and discharge pulses must be equal. Therefore, the charge- and discharge-current sources I must be exactly equal. As a result, a triangular waveform synchronous with the clock will appear across the capacitor C. When an input signal I_{in} is applied, then a difference in the symmetrical up/down pulse pattern is found. This difference between

Reprinted from *IEEE J. Solid-State Circuits,* vol. SC-12, pp. 656-662, Dec. 1977.

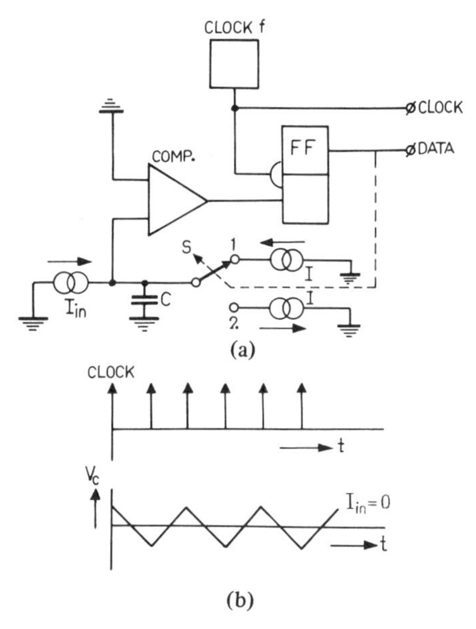

Fig. 1. (a) Basic sigma-delta modulator. (b) Pulse patterns as a function of time.

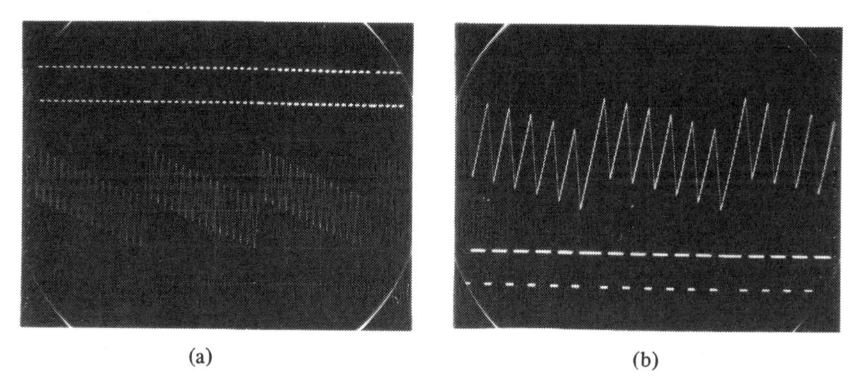

Fig. 2. (a) Photograph of data output and capacitor voltage with an input signal of 6 percent of full scale. (b) Same as (a), but with an input signal of $\frac{1}{3}$ of full scale.

the number of up and down pulses is proportional to the analog input signal. In fact, this circuit without digital storage acts as a clock-synchronous current to frequency or pulse number converter.

In Fig. 2(a), a photographic picture taken from a cathode-ray tube shows a pulse pattern in the case when a small input signal is applied to the converter. In the upper part of the photograph the digital output signal of the converter is shown, while the lower part shows the voltage across the capacitor C as a function of time. The zero level of the comparator is in between two graticule lines of the picture. At about every sixteen clock pulses, giving a symmetrical up/down pattern, an additional up pulse is added to keep the capacitor charge zero. In this way, an input signal of about 6 percent of the full-scale value is measured.

In Fig. 2(b), a photograph of the same signals as a function of time are shown. Here a larger input signal, about one-third of full scale, is applied. The picture shows that now the up/down pattern itself changes into a two-up-one-down pattern which again is interrupted after every six clock pulses and an extra up pulse is added to get the required fine control. With increasing input signals, the number of succeeding up pulses increases until at full scale only up pulses are generated by the converter.

III. Digital Output-Signal Generation

How the digital output value is obtained is shown in Fig. 3. For this purpose, a timer and an up/down counter with a gating function are required. In the timer circuit the total number of pulses $N = n_{up} + n_{down}$ are counted and the conversion time is fixed. During this conversion time the difference between the number of up and down pulses is counted in the up/down counter. After the conversion has been finished, the contents of the up/down counter equals $n =$

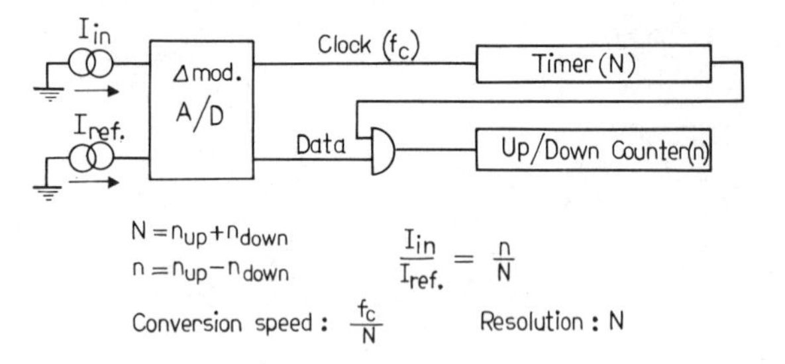

Fig. 3. Digital controller circuit.

$n_{up} - n_{down}$. It can be shown that the ratio between the input current I_{in} and the reference current I_{ref} only depends on the ratio between the stored pulses n in the up/down counter and the length N of the timer circuit

$$\frac{I_{in}}{I_{ref}} = \frac{n_{up} - n_{down}}{n_{up} + n_{down}} = \frac{n}{N}.$$

The resolution of the converter is determined by the length N of the timer circuit, while the conversion time T_c at a certain resolution depends on the clock frequency f_c, so

$$T_c = \frac{N}{f_c}.$$

A high normal-mode rejection ratio can be obtained by adjusting the clock frequency to a multiple of the mains frequency. With a phase-locked loop, this clock frequency can be coupled with the mains frequency to obtain the maximum possible normal-mode rejection. It is clear that during one conversion the integration must take place over exactly one cycle of the mains frequency to eliminate this disturbing effect. Therefore, the minimum conversion time depends on the mains frequency ($T_c = 1/f_{mains}$). With respect to the normal-mode rejection ratio no great differences between the sigma-delta modulator and the dual-slope converter are found.

IV. Basic Analog Subsystem

A simplified diagram of the analog circuit part is shown in Fig. 4. It consists of a voltage-to-current converter, the reference-current source I_{ref}, and the switch S, controlled by the comparator and the flip-flop at the events of the clock generator f. An additional operational amplifier is inserted for bias-current control. The voltage-to-current converter converts the input voltage signal into a current flowing through the resistor R_1 and charging the capacitor C. For simplicity this converter is shown as an emitter degenerated amplifier pair T_1, T_2 with degeneration resistor R_1. The equal charge and discharge currents are generated by the reference-current source I_{ref} and the differential p-n-p pair T_3, T_4 acting as the switch S. This circuit configuration allows the high matching accuracy between the charge and discharge currents (I in Fig. 1). The flip-flop controls at clock events the switching of the pair T_3, T_4, depending on the output signal of the comparator, which maintains the net charge in the capacitor at zero. The bias-current sources I_1 and I_2 are controlled by the operational amplifier to obtain a linear operation of the circuit with the voltage level of the capacitor around zero. With a high resolution at low input signal levels, the offset voltage of the voltage-to-current converter becomes a problem. To overcome this problem, an auto-zero circuit must be added.

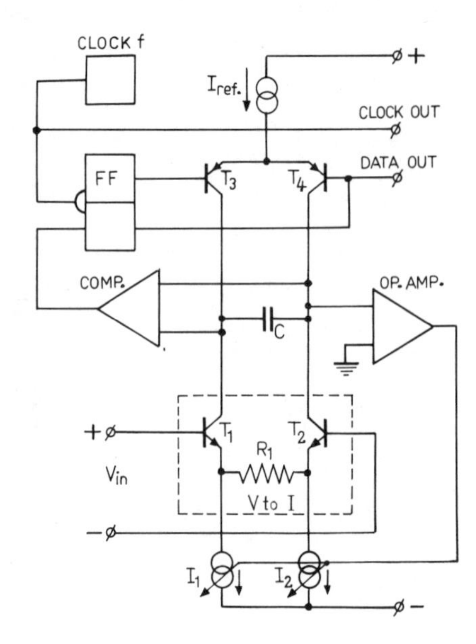

Fig. 4. Simplified analog subsystem.

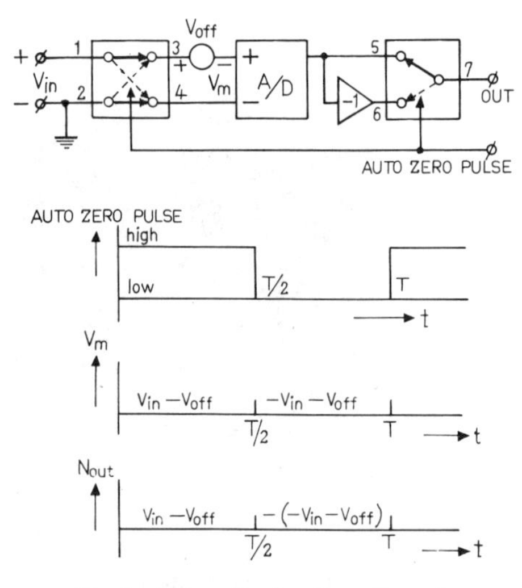

Fig. 5. Auto-zero circuit configuration.

V. Auto-Zero Circuit

The auto-zero system is shown in Fig. 5. Here the differential input configuration of the voltage-to-current converter with a set of switches is used. V_{off} is the total offset voltage of the converter referred to the input of the voltage-to-current converter. During the auto-zero pulse, which divides the total measuring time into two equal parts, terminals 1 and 2 are interchanged with respect to terminals 3 and 4. In the first half of the measuring time, a voltage V_m equal to $V_{in} - V_{off}$ is converted. During the second half of the measuring time, termi-

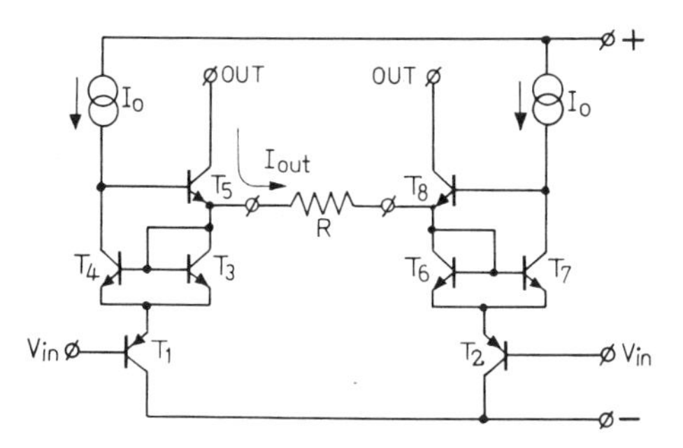

Fig. 6. Basic voltage-to-current converter.

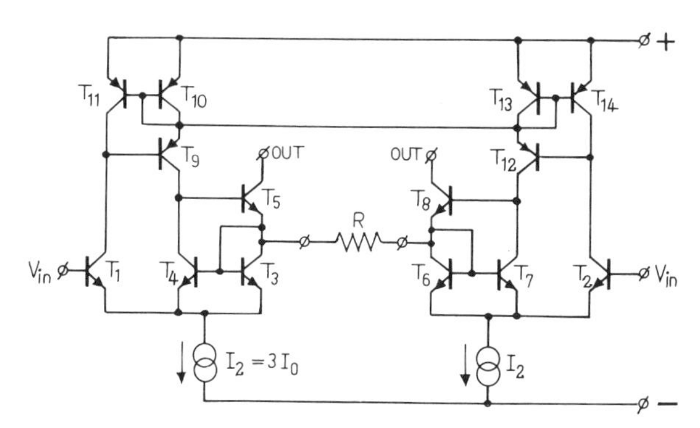

Fig. 7. Modified voltage-to-current converter.

nals 3 and 4 are interchanged, resulting in a voltage $V_m = -V_{in} - V_{off}$ to be converted. The offset voltage will cancel if both results are subtracted. However, a subtraction is more difficult to implement in a circuit. With an extra inverter stage between the output data line and the up/down counter, controlled by the auto-zero pulse, this subtraction changes into an addition which means continue with counting up. Theoretically, the auto-zero circuit cancels the offset voltage *without* increasing the conversion time. In practice, problems due to the finite time series of pulse patterns may cause an additional instability. This problem can be overcome by introducing some waiting time and an increase in measuring time.

VI. Voltage-to-Current Converters

The basic circuit of the voltage-to-current converter is shown in Fig. 6 [1]. It consists of a p-n-p input pair T_1, T_2 and two current mirrors (T_3, T_4, T_5 and T_6, T_7, T_8) with the bias-current sources I_0. With $V_{in} = 0$, then due to the current-mirror action a constant current I_0 must flow through T_3 and T_4 as well as through T_6 and T_7. Therefore, the currents through T_1 and T_2 are constant and have a value $2I_0$. When an input voltage is applied, this voltage is exactly reproduced across the conversion resistor R, because the voltage drops across T_1 and T_3 and T_2 and T_6 remain constant. The converted output current I_{out} can only flow through the transistors T_5 and T_8.

To reduce the input bias currents the p-n-p transistors can be replaced by n-p-n types. The modified circuit is shown in Fig. 7. The n-p-n transistors T_1 and T_2 form the input pair and the p-n-p current mirrors T_9, T_{10}, T_{11} and T_{12}, T_{13}, T_{14}

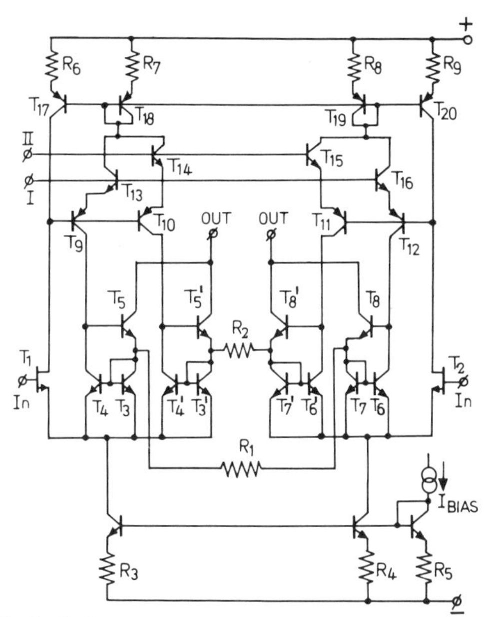

Fig. 8. Voltage-to-current converter with two gain settings.

replace the bias-current sources. These bias-current sources are connected between the emitters of the input transistors and the negative supply voltage and have a value $3I_0$. The coupling between the p-n-p current mirrors (the emitter of T_9 is connected with the emitter of T_{12}) increases the current gain for differential input signals and results in a higher conversion accuracy.

A switching of the conversion gain can be obtained by applying different conversion resistors. In Fig. 8 a circuit with two gain settings is shown. The gain setting is obtained by making either line I high with respect to line II to select resistor R_1 or line II high with respect to line I to select the conversion resistor R_2. The input transistors are common for both ranges together with the coupled p-n-p current mirrors (T_{17}, T_{18}, T_{19}, T_{20}). Every range resistor operates with its own n-p-n current mirrors (T_3, T_4, T_5 and T_6, T_7, T_8 for resistor R_1 and T_3', T_4', T_5' and T_6', T_7', T_8' for resistor R_2). The gain setting is now accomplished by switching the bias currents through T_9, T_{13} and T_{12}, T_{16} to the n-p-n current mirrors to select resistor R_1 or through T_{10}, T_{14} and T_{11}, T_{15} to select resistor R_2. This gain setting has no influence on the dynamic performance of the circuit and, due to the current switching, no special requirements have to be set for the switches.

VII. Comparator and Flip-Flop

In Fig. 9 the simplified circuit diagram of the comparator and the flip-flop is shown. It consists of an input differential pair T_9, T_{10} which senses the voltage across the capacitor C and a clocked master–slave flip-flop. When no current flows through the master flip-flop T_1, T_2, the differential amplifier T_9, T_{10} with the resistors R_1, R_1' acts as a linear amplifier, increasing the sensitivity of the comparator. At the moment a decision must be made, the bias current through the master flip-flop is switched on. The bias current I_0 flows through T_8 to the flip-flop T_1, T_2. Then this flip-flop acts as a very

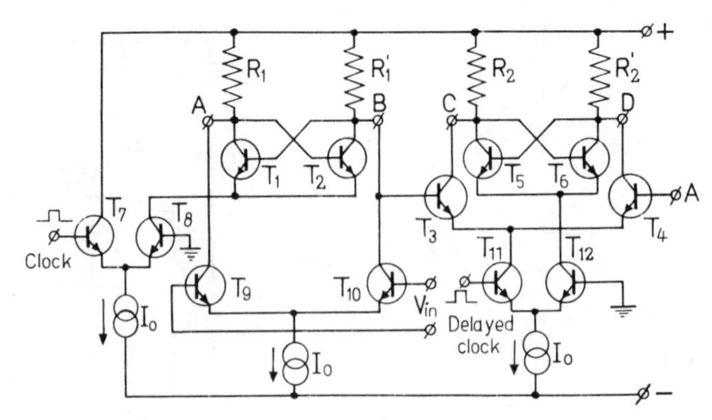

Fig. 9. Comparator and flip-flop circuit diagram.

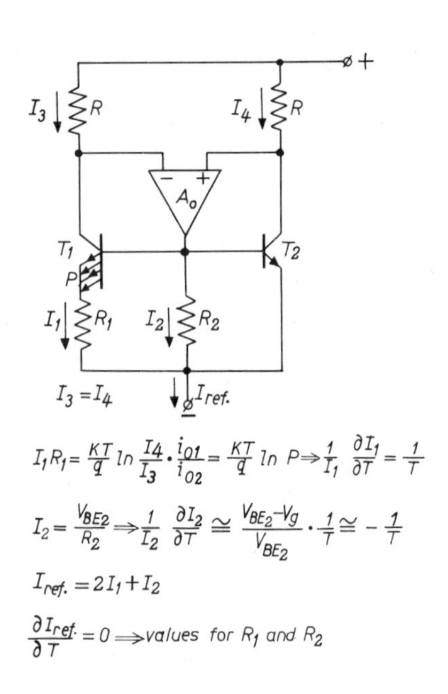

Fig. 10. Basic reference-current source circuit diagram.

TABLE I
SUMMARY OF CONVERTER PERFORMANCE

Input voltage :	100mV and 1V (bipolar)
Resolution :	5 digits plus sign. (f_{clock} = 200 kHz)
Linearity :	$\pm 10^{-5}$ of F.S.
Zero stability :	$< 10^{-6}$/°C (with auto zero)
T.C. of ref. source :	$5 \cdot 10^{-6}$/°C over $\Delta T = 80$ °C
Voltage stab. of ref. source :	$5 \cdot 10^{-6}$/V
Power supply :	$\pm$ 7.5V, 27mW.
Chip sizes :	1.8 x 2.9mm^2 and 2x3.2mm^2

sensitive comparator and takes over the condition of which a "high" represents the smallest current through resistors R_1 or R_1'. After a decision has been made, this condition is transferred to the slave flip-flop by a delayed clock signal. The output signals C and D of the slave flip-flop control the switching of the charge and discharge currents in the total circuit.

VIII. Basic Reference-Current Source

The basic diagram of the reference-current source based on the bandgap of silicon is shown in Fig. 10. If the resistor R_2 is deleted, the circuit behaves like a simple current stabilizer and operates as follows. The resistors R with the operational amplifier force equal currents to flow through transistors T_1 and T_2. Transistor T_2 has a p-times larger emitter area then T_1, so a difference in base-emitter voltage between these transistors, equal to $\Delta V_{BE} = (kT/q) \ln p \; (I_{c1} = I_{c2})$, is generated across resistor R_1. The current I_1 through this transistor becomes $I_1 = (1/R_1)\,(kT/q) \ln p$. The temperature coefficient of this current is positive and equals

$$\frac{I}{I_1}\frac{\partial I_1}{\partial T} = \frac{1}{T}.$$

Now resistor R_2 is inserted. Then an extra current equal to $I_2 = (V_{BE_2}/R_2)$ is added. The temperature coefficient of this current can be estimated, resulting in

$$\frac{1}{I_2}\frac{\partial I_2}{\partial T} \cong \frac{V_{BE_2} - V_g}{V_{BE_2}} \cdot \frac{1}{T} \cong -\frac{1}{T}, \quad \text{if} \quad V_{BE_2} \cong \frac{1}{2}V_g$$

and V_g is the bandgap voltage of silicon. The output reference current is equal to the sum of $2I_1 + I_2 = I_{ref}$. By choosing proper values for R_1 and R_2, the temperature coefficient can be adjusted to zero at room temperature [2]. The resistors R_1 and R_2 are off-chip resistors.

IX. Complete Analog Subsystem

The complete analog system is implemented on two chips. The first chip contains the voltage-to-current converter with level-conditioning logic for the gain setting and a stable reference-current source. The second chip contains the current switches, the comparator with the flip-flop, the clock generator, and the control operational amplifier. Furthermore, a bias-current stabilizer and additional level-conditioning circuitry for the data and clock output lines are incorporated.

In Table I a summary of the converter performance is given. A maximum resolution of 5 digits plus sign with an auto-zero compensation circuit at a low power dissipation has been reached. The high linearity (10^{-5} of FS) at two gain settings

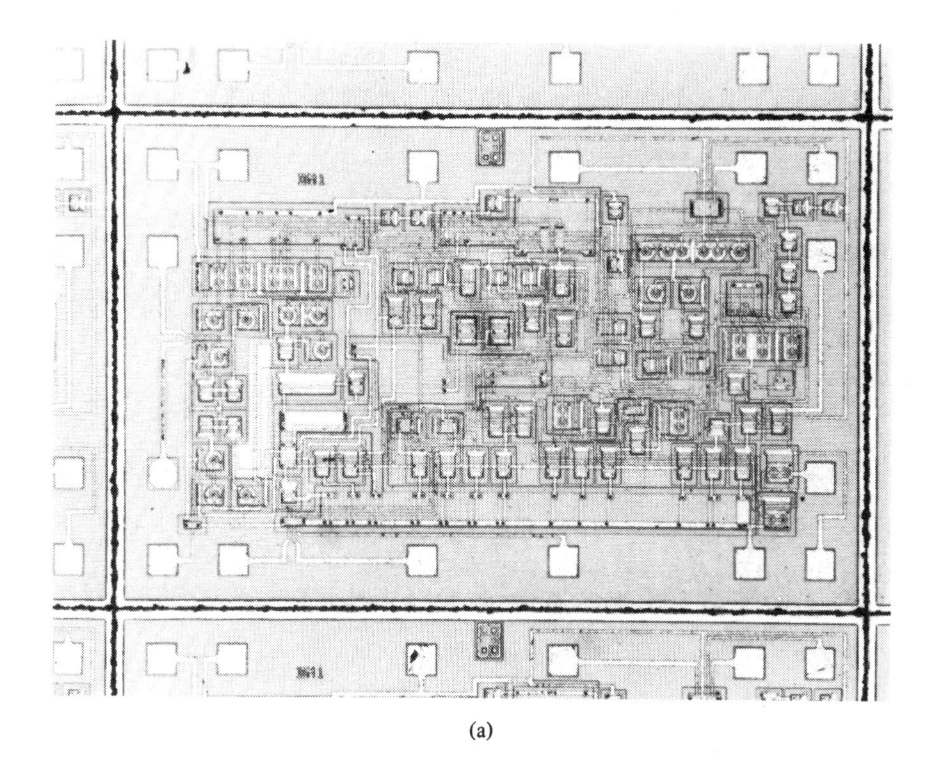

(a)

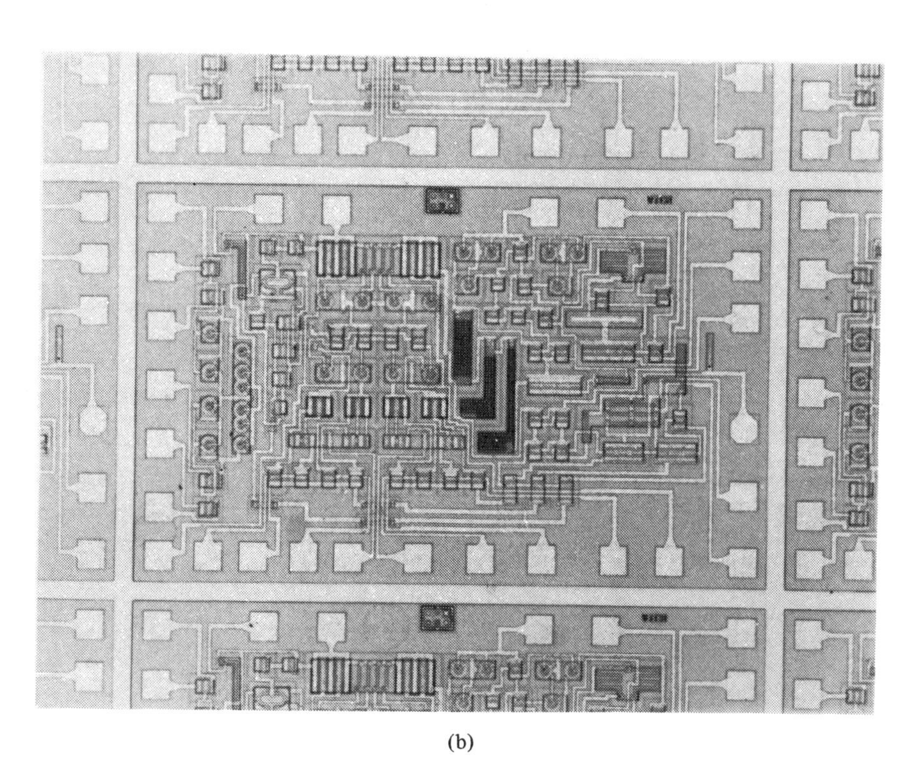

(b)

Fig. 11. (a) Photomicrograph of the voltage-to-current converter chip. (b) Photomicrograph of the switching and control chip.

and the high stability of the reference source allows the application of the two chips in battery-operated five-digit A/D converters. The two chips are designed to operate with a minimum of required external components. Output clock and data lines are CMOS compatible. The two-chip analog function block is chosen to obtain a maximum in input flexibility; e.g., multi-input circuit with scanning capability in data-acquisition systems. In Fig. 11(a), a photomicrograph of the voltage-to-current converter chip is shown, and in Fig. 11(b), a photomicrograph of the switching and control chip is shown.

X. Conclusion

The sigma-delta modulator principle implemented in a bipolar technology shows a good alternative to the dual-slope converter. A high accuracy with a simple auto-zero compensation can be reached. Furthermore, the chosen solution is very attractive as a multiinput data-acquisition system, while the low power consumption allows the battery operation of the complete system.

Acknowledgment

The authors wish to thank A. Schmitz for processing the circuits and H. Leenen for layout and programming.

References

[1] R. J. van de Plassche, "A wideband monolithic instrumentation amplifier," *IEEE J. Solid-State Circuits*, vol. SC-10, pp. 424-431, Dec. 1975.

[2] —, "Dynamic element matching for high accuracy D/A converters," *IEEE J. Solid-State Circuits*, vol. SC-11, pp. 795-800, Dec. 1976.

A Versatile Monolithic Voltage-to-Frequency Converter

BARRIE GILBERT, SENIOR MEMBER, IEEE

Abstract–**Voltage-to-frequency (*V–F*) conversion is achieved with high linearity (0.02 percent) using a precise multivibrator with an 80-dB dynamic range. The IC operates from a single, low-current supply and can accept millivolt signals. A unique thermometer output permits direct conversion of temperature to frequency.**

I. Introduction

A MONOLITHIC circuit has been developed which provides accurate voltage-to-frequency (*V–F*) conversion at low cost and with the minimum number of external parts: a timing capacitor and a scaling resistor. Any full-scale (FS) frequency up to 200 kHz and FS input from 100 mV to over 30 V can be chosen, with a simple scaling relationship. Good linearity has been achieved using a new multivibrator design which, in conjunction with a bandgap reference circuit, is temperature-compensated to ±30 ppm/°C. The multivibrator uses adaptive biasing to operate over a 20 000:1 frequency range with a given timing capacitor, and generates a square-wave output. This is often more useful than the narrow pulse generated by charge-dispensing *V–F* converters; for example, it allows the circuit to be used as the voltage-controlled oscillator (VCO) in precision phase-locked loops. It also ensures constant chip dissipation with variation in frequency, important when driving the maximum load current of 30 mA at the maximum supply of 36 V. Single-supply operation down to 4.5 V at a quiescent current of only 1.2 mA are valuable assets in battery-powered equipment. The high sensitivity and low drift of the input amplifier permit direct interfacing with low-level transducers such as thermocouples and strain gauges, while the high input impedance (250 MΩ typical) accommodates such signals as generated by potentiometric transducers. Either positive or negative inputs can be accepted; dual supplies allow operation from differential signals above or below ground level.

A unique feature is the provision of an output voltage scaled +1 mV/K. The IC can thus operate as a thermometer delivering a frequency directly proportional to absolute temperature. The low dissipation is an essential prerequisite in this mode. A fixed reference-voltage output of 1.00 V is also provided, having a variety of uses. For example, in the thermometer mode it can be used to offset the scale to read directly in degrees Fahrenheit or centigrade with a typical scaling factor of 10 Hz/degree. It may also be used to power resistive transducers, set up other scale offsets (for example, in converting a 4-20 mA input signal to a 0-10 kHz output), or define a stable fixed frequency. A controlled temperature coefficient (TC) of up to ±250 ppm/°C or more can be introduced to compensate for some external temperature sensitivity, using these outputs in combination.

Very few external components are required to realize all of the major operating modes. However, by adding a quad op amp, very exact ratiometric operation can be achieved (±0.1 percent accuracy in the division of two analog inputs over a 10:1 denominator range and 10 000:1 numerator range). A precision phase-locked loop requires the addition of just a quad NAND gate, providing ultralinear frequency-to-voltage conversion (Section VIII).

Fig. 1 shows the main sections of the *V–F* converter in block-diagram form.

II. Background

A *V–F* converter must exhibit high linearity (<0.1 percent of FS deviation from zero-based straight line) over a dynamic range of at least 10 000:1, with good temperature stability (<100 ppm frequency drift per degree centigrade) and supply-rejection (<100 ppm per volt). Since the introduction of the precision charge-dispensing technique [1], which can meet these requirements with ease, alternative approaches based on classical relaxation oscillator techniques have been somewhat in eclipse. Nevertheless, one of these, the symmetrical emitter-timed astable multivibrator, has attractive features. It is simple, needing only a single capacitor for timing, and ideally suited to monolithic fabrication; its push–pull configuration results in large charging voltages being available across the capacitor even at low supply voltages for improved stability and jitter; its square-wave output is generally more useful than the pulse output of a charge-dispensing converter; the tight coupling within the regenerative loop minimizes nonlinearity caused by transit-time and switching delays.

Prevalent designs have tended to cast doubt on the applicability of multivibrators to *V–F* conversion, and their use has been most successful as the current-controlled oscillator in monolithic phase-locked loops. The earliest IC's had poor temperature stability and totally inadequate dynamic range and linearity for *V–F* applications [2]. With improved attention to the mechanisms which define these characteristics a dynamic range of about 1000:1 and a stability of ±20 ppm/°C was achieved by the author in a second-generation PLL circuit [3]. Concurrently (1971) the design of a 15 MHz emitter-timed circuit using precision collector clamping was in progress, and the design presented in this paper is an extension of that work. Recently, a design bearing some similarity has been published by Cordell and Garrett [4], and a brief comparison is merited. In their circuit, the junction voltages developed across the collector-catching diodes are used to define the capacitor charging voltage directly (that is, no precise reference voltage is supplied in the collector circuit) which consequently

Manuscript received May 26, 1976; revised July 26, 1976.
The author is with Analog Devices, Inc., Wilmington, MA 01887.

Reprinted from *IEEE J. Solid-State Circuits*, vol. SC-11, pp. 852-864, Dec. 1976.

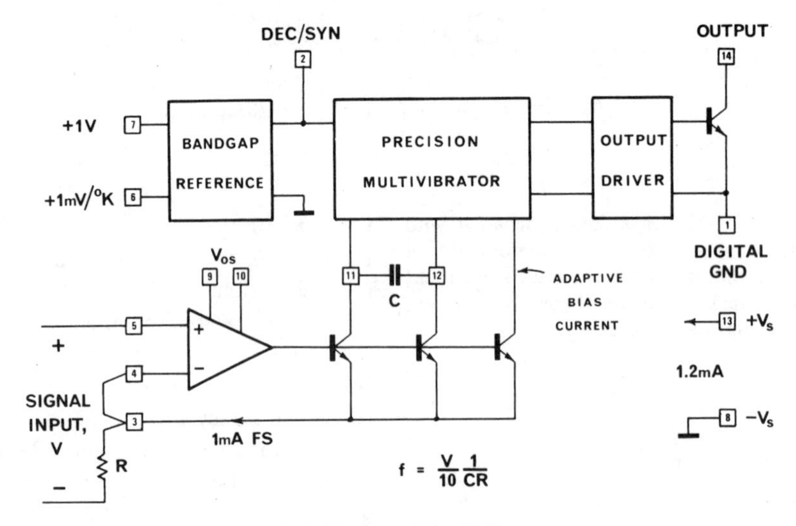

Fig. 1. Block diagram of the V–F converter.

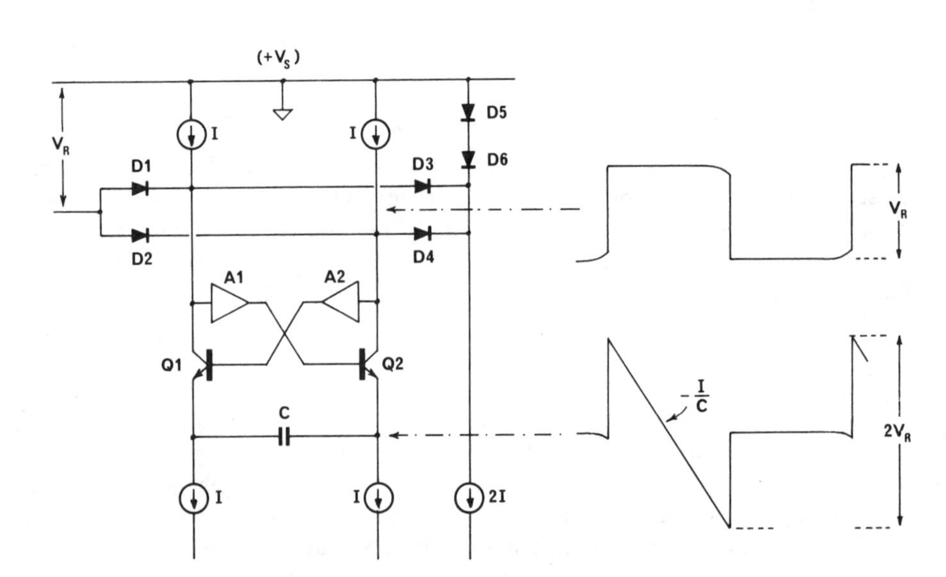

Fig. 2. The basic multivibrator showing precision collector clamping.

has a very large temperature dependence (about −3000 ppm/°C). A special current generator, having the same nominal TC, is then used to charge the capacitor. The compensation is exact only at the center frequency, determined by an external resistor, since the control current which modulates the frequency is not compensated, and even quite small deviations degrade the TC (typically ±300 ppm/°C at a ±10 percent deviation). The technique is suitable for some PLL applications, where the resulting temperature dependence of the control characteristic can be tolerated, but in order to use the circuit in V–F conversion it would be necessary to interpose a one-quadrant current-mode multiplier in the control-current path, to scale the timing current by a junction voltage. The approach adopted in the circuit to be described ensures that the collector clamping potentials are exact, and controlled by a known reference voltage. The capacitor charging current is provided by a very linear voltage-to-current interface. The only temperature compensation required is to eliminate a small (230 ppm/°C) drift of fundamental origin. This is a point of detail which seems to have been overlooked in previously published work on the emitter-timed species, although the exact magnitude of the drift depends on the particular configuration.

III. Precision Collector-Clamped Multivibrator

The basic wide-range oscillator is shown in Fig. 2. $A1$ and $A2$ are two-stage emitter-followers which provide high current-gain (hence, loadable output nodes) and level-shifting. The input transistors are biased with currents which track the timing current, resulting in a small, ratiometric loading of the collector nodes. The collector-clamping arrangement transfers a precise reference voltage V_R to the timing capacitor via $Q1$

and $Q2$. This voltage is normally provided by the bandgap circuit, but can alternatively be supplied from an external source, allowing the linear control of period.

The operation can be followed with the aid of the waveforms in Fig. 2. (The $+V_s$ supply is shown grounded for simplicity.) Assume $Q1$ is conducting. Its emitter current is $2I$, and the charging current I flows from left to right in the capacitor. $D1$ is conducting, clamping the collector of $Q1$ at $-(V_R + \phi)$, where ϕ is the base-emitter voltage of this diode-connected transistor at the current I. $Q2$ is nonconducting, so its collector is clamped by $D4$, and since the identical diodes $D5$ and $D6$ all conduct the current I, the collector of $Q2$ is at $-\phi$. Thus, a differential voltage of exactly V_R appears across the collectors, independent of the current, the temperature, the absolute value of ϕ, or the supply voltage. This ideal is impaired by junction offset-voltages, inaccuracies in the secondary current-sources, and loading by $A1$ and $A2$. However, the effects are minimal; for example, a composite error of 3 percent in the current-ratio factor that includes all of the forward-biased junctions causes an absolute error of about 0.1 percent in V_R and a TC-offset of 3 ppm/°C, a figure readily attained by the full design. Errors in the collector clamping potentials are practically independent of current, and do not jeopardize linearity.

In the emitter circuit the capacitor charges until $Q2$ begins to conduct. At a critical point regenerative switching occurs and the polarity of the differential collector voltage and the timing current reverse, beginning the next half-cycle. Neglecting second-order effects the capacitor charges through $2V_R$ twice each cycle, so the frequency of oscillation is

$$f = \frac{I}{4CV_R}. \tag{1}$$

The current I is actually one third of the control current, that is, $I = V/3R$, where V is the control voltage and R is the external scaling resistor. With $V_R = 833$ mV, the overall control relationship becomes

$$f = \frac{V/3R}{4C \times 0.833} = \frac{V}{10CR}. \tag{2}$$

For the common case of a 10 V FS input, the FS frequency is simply $1/CR$. The design is optimized for a control current of 1 mA FS, that is, a timing current range of 33 nA to 333 μA for an 80 dB dynamic range.

A. Dynamic Range

The "current-mode" nature of the collector-clamping system, and the high current-gain of the buffer stages $A1$ and $A2$, result in a dynamic range which is limited more by the quality of processing (affecting such parameters as low-current beta and surface leakage paths) and the allowable nonlinearity, than by the configuration. At very low values of timing current combined with high temperatures, the junction saturation currents cause frequency errors and, ultimately, cessation of oscillation. This is not expected to be a problem of much practical concern, and very similar effects limit the dynamic range of charge-dispensing converters (for example, the input bias-current and offset voltage at the current-summing input node). Measurements on the monolithic parts have shown that the period can be as long as 20 s on the 10-kHz range, corresponding to a timing current of only 1.5 nA (control-current input of 4.5 nA).

The upper limit of the dynamic range is determined mainly by power dissipation constraints, if linearity errors are of no consequence. In the complete design, the control current is deliberately limited to 5 mA. This suggests that a dynamic range of over one million is possible. However, in V-F applications a maximum input of 1.5 mA is recommended, and the circuit is not specified for an input of less than 100 nA, a guaranteed dynamic range of 15 000 : 1.

B. Temperature Coefficient

The simplified analysis took no account of the logarithmic junction characteristics and assumed abrupt switching after a capacitor rundown of $2V_R$, which would be true only at absolute zero temperature. The regenerative process under less frigid conditions is quite complex, and Appendix I shows that switching occurs after a transitional phase during which the charging current falls to $(\sqrt{3} - 1)I$. This causes the junction currents to be unequal at the switching instant, and the voltage swing across the capacitor is less than $2V_R$ by an amount proportional to absolute temperature (PTAT), due to the kT/q term in the junction offset voltages. While this shortens the period, the falloff in charging current during the transitional phase lengthens it. Overall, there is a net reduction in period with increasing temperature by an amount which is theoretically 230 ppm/K and in practice has been measured as 270 to 330 ppm/K. The compensation technique consists of adding a small PTAT component to the temperature-stable V_R, both of which are generated by the bandgap circuit which tracks the oscillator temperature. Initially, the compensation voltage was set at 84 mV at 300 K. Although there is some disparity in the measured and theoretical drifts, it is significant that a single PTAT correction can result in very low residual drift with only slight curvature.

C. Linearity

The linearity of a V-F converter is specified as a percentage of the FS frequency, so that the effect of absolute errors becomes progressively less down-scale. It has been found that the most troublesome operating region is from about midscale upward, where the logarithmic conformance of the junctions is less exact. The peak capacitor swing is only 1.666 V, and a 500-μV error therefore amounts to a nonlinearity of 0.03 percent. Thus, very low excess ohmic resistance can be tolerated, particularly in $Q1$ and $Q2$, which are six-emitter transistors. Excess voltage in the diode-connected transistors $D1$ through $D6$ does not affect the transfer of V_R into the collector circuit because their forward voltages cancel, but their resistance will have a small effect on the switching point by raising the loop gain and thus causing regeneration to occur slightly earlier.

In practice, resistance in series with the capacitor is of more serious concern. It can be shown that

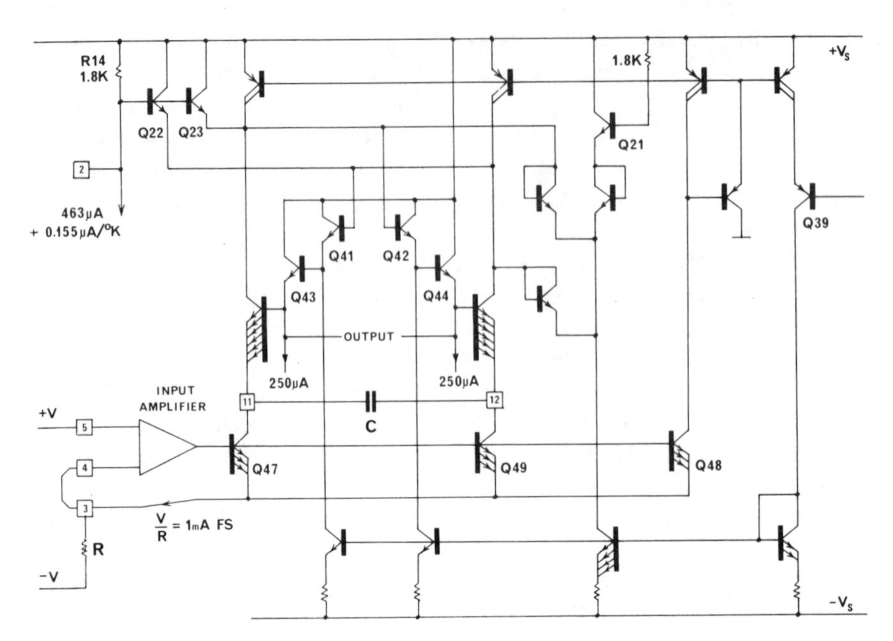

Fig. 3. Main components of complete precision multivibrator.

$$f' = \frac{1}{CR}\left(\alpha + \frac{2r}{R}\alpha^2\right) \tag{3}$$

where $\alpha = V/10$ and r is the total resistance in the capacitor leads. Thus, for $r = 1\ \Omega$ and $R = 10\ \text{k}\Omega$, the frequency would be 0.02 percent high at FS. Normally, the converter would be trimmed for exact frequency at FS, and there would be a residual parabolic error, having a peak value of -50 ppm/Ω at midscale.

D. Complete Circuit

Fig. 3 shows the complete multivibrator and the interface with the voltage-to-current converter. The peak negative voltage at the capacitor nodes is typically 3.5 V below the $+V_S$ supply, so the timing currents can be coupled directly from the output transistors of the input interface ($Q47$ and $Q49$) avoiding the extensive use of current-mirrors with their attendant scaling errors, nonlinearity, and supply sensitivity. The less crucial secondary currents are derived with current mirrors, using the collector current of $Q48$ as the primary supply. The input interface is discussed in the next section.

The clamping diodes $D1$ and $D2$ (Fig. 2) become $Q22$ and $Q23$, whose current gain is utilized; the temperature-corrected reference voltage is generated by a composite current from the bandgap generator and converted to a voltage across $R14$ (1.8 kΩ), providing a node (pin 2) which can be driven by an external current or voltage to alter the scaling. Base current errors are minimized by an equal resistor in the base of $Q21$ (=$D5$ of Fig. 2). Beta mismatch in these transistors causes a parabolic nonlinearity: a 10 percent mismatch in a nominal beta of 400 introduces a peak error of ±0.005 percent at midscale. Note that the reference-voltage input is differential (between the bases of $Q22/23$ and $Q21$), and a preferable compensation method would be to apply the (opposite-polarity) compensation voltage to $Q21$. This would have the advantage that the control of period on pin 2 would be linear through zero, and the control voltage would require no compensation.

IV. Input Interface

A current-controlled oscillator, however linear, is of no use as a voltage-to-frequency converter. An input interface is needed for voltage-to-current conversion, the performance requirements of which include the following.

1) Operation in either a high-impedance ("voltage-follower") or low-impedance ("current-summing") mode.

2) An input range which extends down to the $-V_S$ supply.

3) Good common-mode rejection to ensure linearity in the high-impedance mode.

4) Sufficient open-loop gain to ensure linearity at high scale sensitivities.

5) Low drift, commensurate with millivolt sensitivities.

6) Short-circuit protection of the current-summing node.

In addition, the interface circuit should have small bias currents, a current transfer ratio very close to unity, and be internally stabilized for ease of use. Response speed can be relatively low.

Fig. 4 shows the main details of the interface amplifier. In most applications the current input node (pin 3) and the voltage sense node (pin 4) are strapped and the scaling resistor is returned to ground. When the device is used at high sensitivities, separate force and sense connections can be used to eliminate lead-resistance effects. Ideally, the output transistors, $Q47$–49, would be Darlington connected (with the intermediate bases strapped to maintain good accuracy of current division) to minimize the effect of beta modulation as V_{ce} varies when the device is used in the high-impedance mode. However, there is only 400 mV of V_{ce} available when using a 5-V supply and the input is at +1.1 V, which not only prevents the use of Darlingtons but also leaves no room for current-

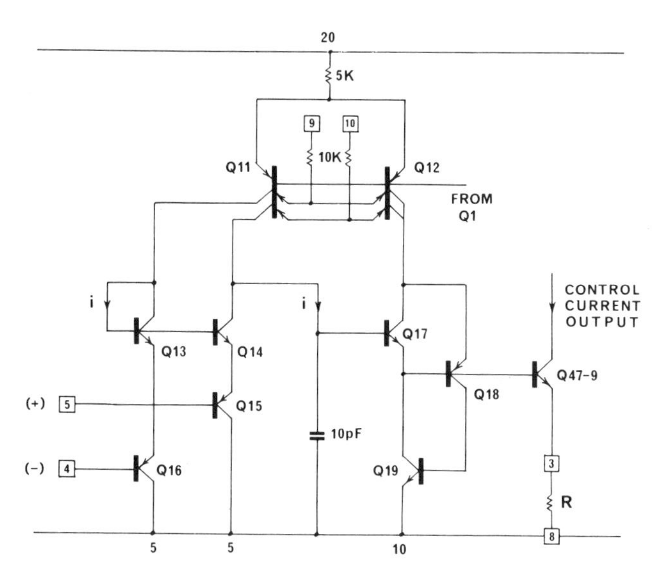

Fig. 4. The input interface (voltage to current converter).

limiting using an emitter sense resistor. Single transistors are therefore used, with controlled base-current drive to limit the short-circuit current to a value between 2.5 and 10 mA, this being a tight enough control for the application which must guarantee an input-current capacity to 1.5 mA and sensibly limit dissipation in the event of the current-summing input being grounded while pin 5 is elevated. The alpha error of the trio is compensated by the equal error of $Q9$ in the reference generator (Fig. 7); all of these transistors experience the same variation of collector voltage with supply variations. Emitter-area mismatch in $Q47$–49 can have several effects, depending on how the mismatches are distributed, including a scaling error, waveform asymmetry, and a small temperature drift (due to imperfect balance in the clamping currents). None of these effects has been troublesome, but to reduce the possibility of errors the transistors are multiemitter devices and operate at the same mean V_{ce}.

A. Gain Requirement

Nonlinear variation in the V_{be} of the trio is reduced by the open-loop gain of the amplifier, but it is not intuitively obvious how large this gain must be to achieve a given peak nonlinearity after scale and zero adjustments. This is analyzed in Appendix II, the main result of which is this expression for the peak error

$$\hat{e}_n = \frac{kT}{qG_o}\left[\ln \frac{\sigma}{\ln \sigma} - 1\right] \tag{4}$$

where σ is the ratio of FS frequency to the frequency at which the "zero" adjustment is made, and G_o is the open-loop gain. A typical example is given in Fig. 5 for the case where the circuit is trimmed at FS and one-thousandth FS (that is, $\sigma = 1000$). The peak open-loop error is 103 mV at 300 K, and to reduce this to 0.01 percent of a FS input of 1 V requires $G_o = 1000$. To take another example, assume an FS input of 100 mV and a zero adjustment made at 1 mV. The peak open-loop error is then 54 mV, requiring a gain of 5400 to be reduced to 10 μV, again 0.01 percent of FS.

B. Input Stage

Vertical p-n-p transistors ($Q15$ and $Q16$) provide an input range down to $-V_s$ and also contribute to the low-drift performance when operating with appreciable source resistance. The beta of these transistors is only slightly temperature dependent (typically -0.15 percent/°C) and if biased with PTAT emitter currents the base current, and thus any resistance-induced offset voltage, has the same TC as the junction offsets (which vary with kT/q). An adjustment made to null the composite offset at one temperature is therefore correct at all temperatures.

$Q13$ through $Q16$ are arranged in a thermally-symmetric quad to minimize the effect of temperature gradients generated by the high-power output stage of the V-F. All of the voltage gain is provided by $Q14$, which has the split-collector lateral p-n-p $Q11$ as the dominant load. This does not provide the highest gain configuration, but the scheme was used partly for simplicity and partly to achieve a good initial balance in the 5-μA PTAT bias currents. A current balance of ± 1 percent is easily achieved in this way, whereas if separate p-n-p transistors with independent emitter resistors of 20 kΩ were used considerable chip area would be required to achieve the same matching.

To maintain a low temperature drift after nulling, the ratio of the bias currents for $Q13$ and $Q14$ must be temperature independent and the bias for $Q17$ must be equal to the sum of these, to preserve base-current cancellation. Fig. 6 shows a simple way to achieve this in a small area. A pair of auxiliary emitters are located in the base (epi pocket) of $Q11/12$ and connected through the external trim resistors to $+V_s$, to set up small correction currents. Since the same proportion of the injected current goes to $Q12$ as to $Q11$, the second-stage balance is maintained. Since both the main and auxiliary

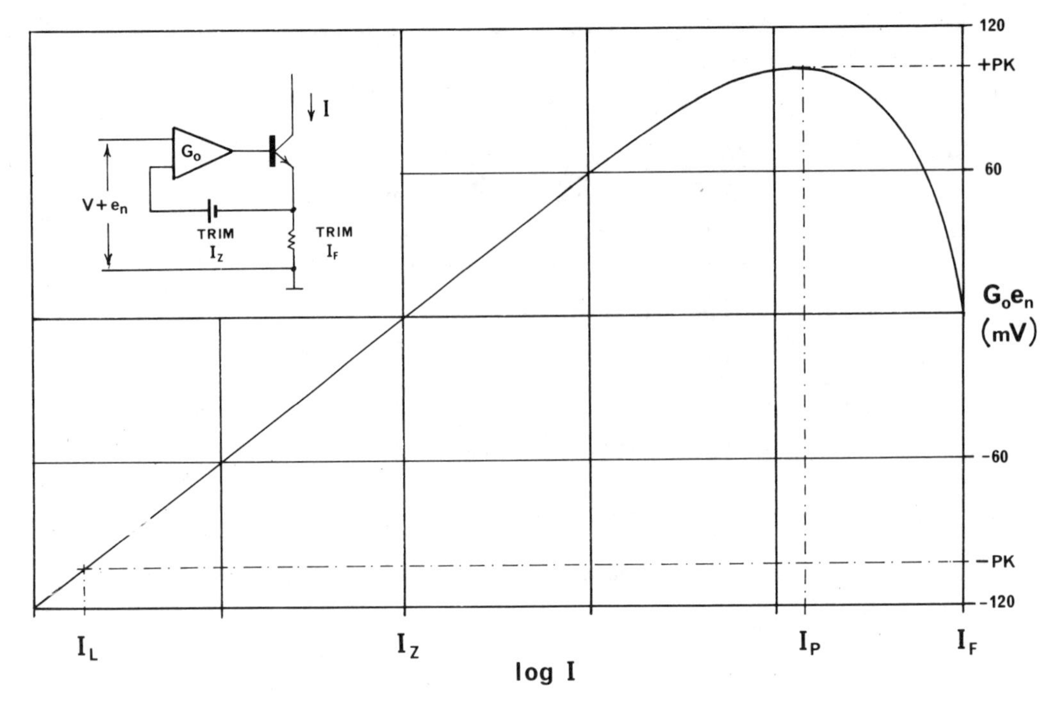

Fig. 5. Error voltage due to V_{be} of $Q47/48/49$. I_F is the FS current and I_Z the current at which the input offset is nulled.

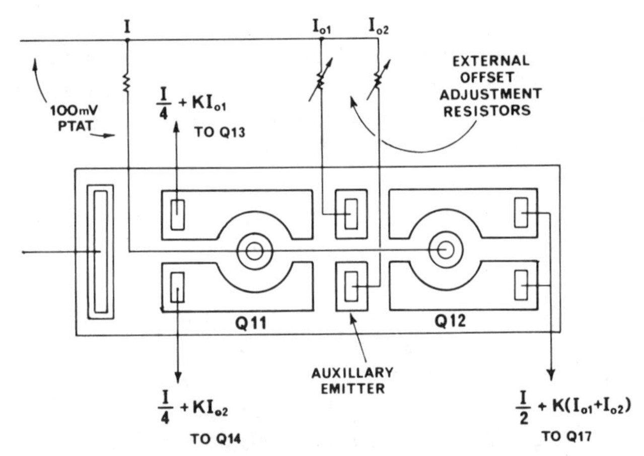

Fig. 6. Layout of $Q11/12$ showing auxiliary emitters.

currents are (very nearly) PTAT, ratios are maintained over temperature.

C. Intermediate Stage

$Q17$, $Q18$, and $Q19$ form a unity-gain buffer with a current gain of β^2 (because $Q17$ has its collector current forced by the local loop), and a further current gain of beta is provided by $Q47/48/49$. Assuming a minimum beta of 250 and a minimum value of 10 Ω for the scaling resistor (10 mV FS), the input resistance at the base of $Q17$ is still more than 150 MΩ and the gain is unimpaired. $Q18$ serves to bootstrap the collector of $Q17$ and also keep its V_{ce} near zero (so improving the beta matching with $Q13$ and $Q14$), while providing level shifting down to the base of $Q19$. HF feedforward around $Q18$ is provided by a 3-pF capacitor from the collector of $Q17$ to the base of $Q19$, with a 100-kΩ pinch resistor in the emitter of $Q18$; a small current source at the base of $Q19$ provides a charging current for this capacitor during slewing (not shown in Fig. 4).

The base-current drive to the output trio cannot exceed the nominal 10 μA supplied by $Q12$. This provides the short-circuit protection at pin 3, since the maximum current is limited to beta times 10 μA, which for a beta range of 250 to 1000 corresponds to 2.5 to 10 mA. If $Q17$ were a simple emitter-follower additional protection circuitry would be needed.

V. Reference Generator

The reference voltage generator, Fig. 7, is based on Brokaw's cell [5]. The eight emitters of $Q7$ are of 1.3 mils diameter, while the single emitter of $Q8$ is of 1.4 mils diameter. The current-density ratio for equal collector currents is thus 6.90 which produces a delta-V_{be} of 50 mV at 300 K and allows the use of integrally-ratioed (hence, close-tolerance) resistors to generate exactly 600 mV at the emitter of $Q8$, half of which provides the +1 mV/K output. An even number of emitters in $Q7$ allows them to be laid out symmetrically with respect to $Q8$, to improve the ratio accuracy in the presence of processing and thermal gradients. The offset of 0.1 mil in the diameters can also be shown to mitigate the variation of output voltage which is due to the dependence on absolute V_{be}.

The voltage at the bases of $Q7$ and $Q8$ is nominally 1.212 V, which results in a zero TC at 300 K for $m = 1.27$ [5]. This is attenuated by $R7$ and $R8$ to provide a short-circuit proof 1.00-V output. These resistors also supply the primary reference current to the multivibrator, and are made up of integral

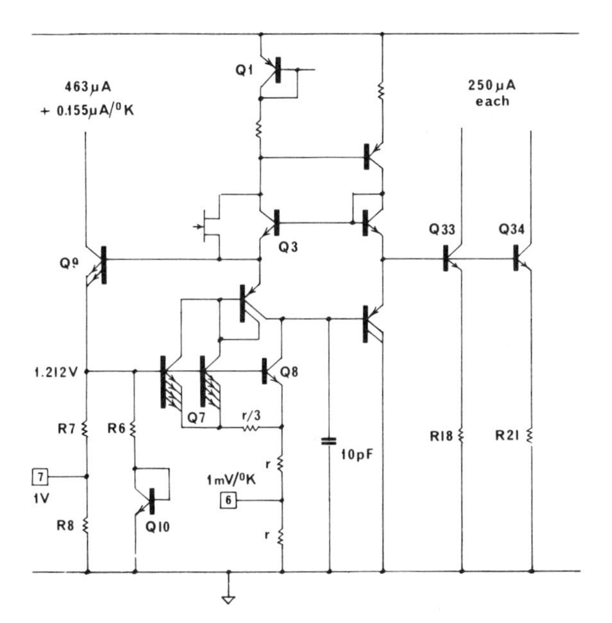

Fig. 7. Bandgap reference generator supplies fixed and PTAT bias currents.

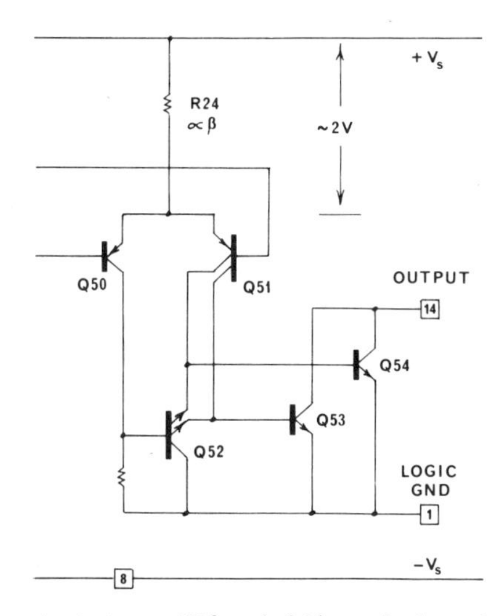

Fig. 8. The output stage. $Q53$ and $Q54$ are the large devices seen at the right of the photomicrograph (Fig. 9).

units of 1.8 kΩ (to ensure good ratio accuracy with $R14$). The additional PTAT component of the composite reference current is provided by $R6$ and $Q10$.

Loading of the 1.00-V output will lower the oscillator frequency by constant proportion, without impairing the TC compensation, and external resistances down to 5 kΩ can be used with a correction to the scaling resistor. This allows the use of a variety of variable-resistance devices to be driven by the reference output. In many applications requiring a stable fixed frequency the 1.00-V output will be applied to the nonloading input, pin 5.

The two 250-μA currents used to bias the inner emitter followers in the multivibrator are provided by $Q33$ and $Q34$. Their base currents cancel that of $Q9$, so preserving balance in the bandgap cell without unnecessary gain. An epi-FET provides startup; connected across $Q3$ it has little effect on the balance of the current mirror. HF compensation is provided by $C1$, using the design rules given in [5].

VI. Output Stage

The utility of the V–F converter is enhanced by provision of a generous load-driving capability. High-current devices such as LED's (for optical coupling of the signal) and long cables can be driven. The circuit is shown in Fig. 8. Two output transistors, $Q53$ and $Q54$, are used to improve the thermal symmetry of the layout and lower the thermal resistance (under worst-case conditions the peak dissipation is 1 W). A low saturation voltage is provided by the use of devices similar in geometry to those described by Frederiksen and Howard [6]. Short-circuit protection is provided by a controlled base-current supplied independently to the two output transistors by the split-collector lateral p-n-p, $Q51$, which forms a differential pair with $Q50$. This driver stage operates at a current controlled by the pinch resistor $R24$, which has similar processing variations as beta. Consequently, the base drive is automatically adjusted to maintain a short-circuit current which is reasonably independent of beta; a range from 30 to 40 mA is considered acceptable. The inverted transistor $Q52$ forms a cheap dual current mirror giving approximately equal turn-on and turn-off drive.

The output saturation voltage is typically 35 mV at 1 mA and less than 250 mV at 20 mA. Using a 500-Ω load the rise and fall times are 200 ns for a control current of 1 mA (FS frequency) rising to 1 μs at 1 μA. Note that several V–F converters can be multiplexed on to a common output bus using the emitters (pin 1) for chip selection.

VII. Performance

The monolithic circuit is fabricated on a standard linear process with 1 kΩ/□ thin-film resistors. Laser trimming was considered, but since the scaling accuracy is determined by external timing components and a total system offset adjustment will be available, the advantage is marginal. Instead, several steps were taken to achieve a basically accurate product, including the use of integral-ratio photomatched resistors and emitters and a well-balanced layout. Fig. 9 is a chip photograph.

The performance is appraised in terms of the circuit's primary application of V–F conversion, for which the most important specification is linearity. Because there are many subtle sources of nonlinearity, the error does not have a consistent form over the dynamic range, except at high frequencies when the dominant error is due to switching-time delays. The best parts show less than ±0.01 percent nonlinearity when used in the current-summing mode with an FS input of 10 V and a FS frequency of 10 kHz, and can be guaranteed to be less than ±0.05 percent. Fig. 10 shows typical errors for 10 and 100-kHz FS ranges, the positive peak on the 100-kHz range being due to the fact that the circuit was trimmed at FS, where switching delays are already causing a reduction in frequency. The behavior for small inputs is dominated by offset, but it is noteworthy that a carefully-trimmed device con-

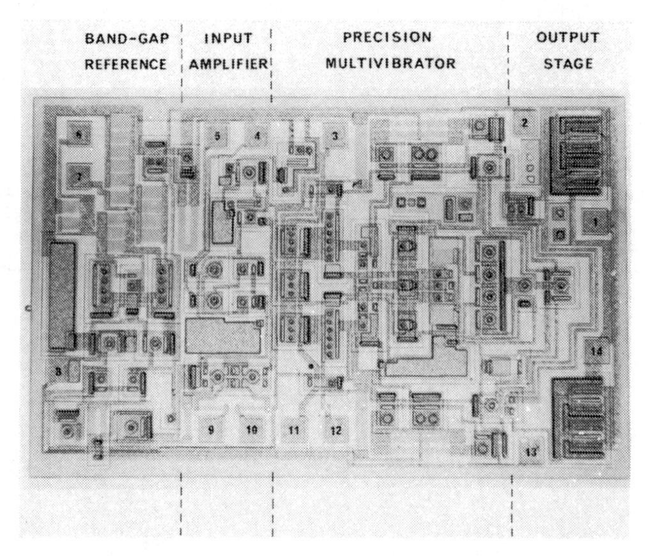

Fig. 9. Photomicrograph of the V-F converter.

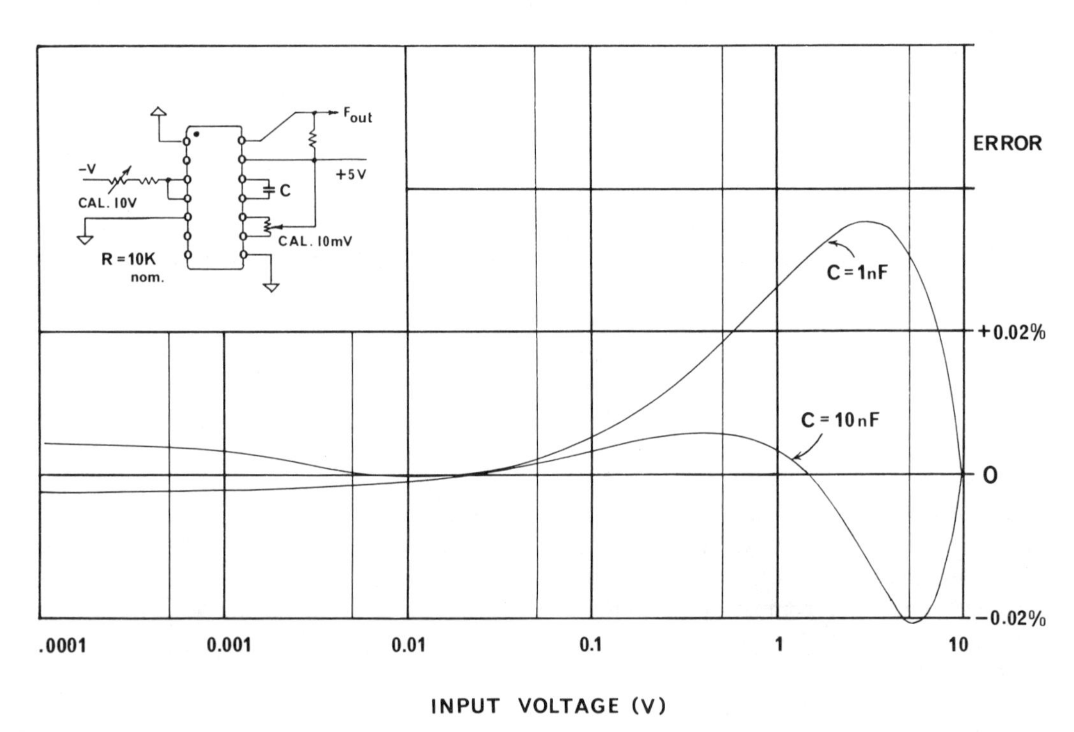

Fig. 10. Typical nonlinearity showing operation over a 100 000:1 dynamic range.

tinues to oscillate below 100 μV of input. The linearity in the voltage-follower mode is degraded by the beta variation with V_{ce} discussed in Section VI. Some samples show ±0.03 percent nonlinearity but the average is about ±0.06 percent, and can be guaranteed to be less than ±0.1 percent for all ranges up to 100 kHz. The oscillator can be operated beyond 1 MHz, but the response of the output stage limits use to 200 kHz, where linearity is already less than desirable. The basic multivibrator will have the same high-frequency capabilities as any other emitter-timed circuit, the main contribution of this design being the accurate clamping arrangement and the temperature compensation.

Frequency drift over temperature was measured for 12 samples from two early production lots, and averaged +70 ppm/°C at FS (1 mA control current, $f = 10$ kHz), +130 ppm/°C at one-tenth FS (equal to +13 ppm/°C relative to FS), and +150 ppm/°C at one-hundredth FS (equal to an insignificant +1.5 ppm/°C relative to FS). The scatter of the TC's was about ±30 ppm/°C, and a small design trim has been made to center the distribution, which will provide an FS temperature-coefficient of less than 100 ppm/°C over the -55 to +125°C operating range.

Frequency stability with supply variations is typically +100 ppm per volt at FS, rising to +1500 ppm per volt at one-hundredth FS, which although undesirably high, is only +15 ppm per volt relative to FS. It is believed that this sensitivity

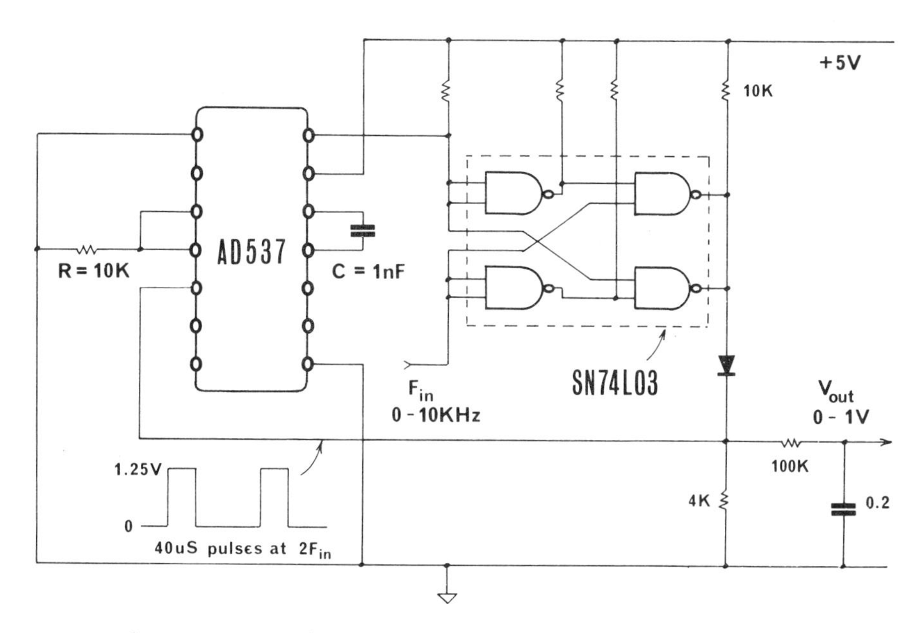

Fig. 11. Frequency-to-voltage converter. The circuit is a first-order precision phase-locked loop, using the quad-NAND gate as phase comparator.

is due to the design of the secondary current sources which use emitter-degeneration resistors (Fig. 3). These serve to improve rejection at FS, where the voltage across them is about 1 V, but are of no help down scale. This is also a contributing factor to nonlinearity.

The application of a 10-Ω load between the output and $+V_s$ causes a peak dissipation of about 1 W for a supply of 30 V, and shifts the frequency by about -0.15 percent immediately, followed by a slow drift due to general chip heating. No significant effect on waveform symmetry is observable, evidence of the good thermal balance of the layout.

The drift performance of the interface amplifier is comparable with that of precision op amps. Initial offsets are less than ±2 mV and after offset-nulling drifts of less than 1 μV/°C are typical, with long-term drifts in the range 0.2 to 0.5 μV per week. Supply rejection is typically 20 μV per volt. Wide-band noise averages 15 nV per root-hertz. The input bias current is below 0.1 μA, and the short-circuit limit on pin 3 is typically 7 mA.

Cycle-to-cycle jitter is less than ±100 ppm, and can be reduced to ±20 ppm using a decoupling capacitor from pin 2 to $+V_S$. General supply decoupling is also necessary in applications where timing jitter is critical, and the leads to the timing capacitor must be kept short to avoid induced noise voltages. The 1.00-V reference output shows a small spread of about ±1 percent and has a typical TC of -20 ppm/°C. The thermometer output is very linear (available measurement techniques were unable to reveal any curvature) over the range -30 to +130°C, and the output is within ±4 mV of nominal.

VIII. Applications

The versatility of the circuit is impressive and many unusual applications have been found. For example, with the help of a quad op amp and using the period control facility (pin 2) analog division of two variables can be performed, giving an output frequency proportional to the quotient with an error of less than ±0.1 percent. The two applications described here demonstrate the usefulness of the square-wave output and the outputs on pins 6 and 7.

A. *F–V Converter*

Phase-locked loops are not generally used for precision *F–V* conversion because VCO's with adequate dynamic range and linearity have been hitherto unavailable, and the usual second-order loop has limited lock and capture range. Fig. 11 shows a first-order phase-locked loop that can lock to any frequency from zero to full scale, shown here as 10 kHz, with an over-range of about 10 percent. Signal acquisition occurs within a few cycles, so the response is determined mainly by the averaging filter; as in most *F–V* converters, this can either be a simple single-pole filter, as shown here, or it may be a more efficient multipole design. The circuit requires only a single +5-V 4-mA supply and provides an output of +1.00 V at 10 kHz.

A low-power TTL quad-NAND gate with open-collector outputs is used as the phase comparator, producing a positive pulse at the voltage input to the *V–F* converter having a lower level of exactly zero and an upper level of +1.25 V (not critical). The *V–F* runs in a start–stop mode: when the input pulse is high a half-cycle of 40 μs is generated; when the pulse goes low the timing current drops to zero and the charge on the timing capacitor is held in readiness for the next half-cycle. The average value of the pulse is forced to equal that required to run the *V–F* at the input frequency. It is interesting to note that very high linearity can be achieved (typically ±0.007 percent) even with a *V–F* converter having much poorer linearity! The explanation of this apparent paradox lies in the fact that the V–F circuit operates at only one value of timing current (when the drive pulse is high) and behaves more like a monostable multivibrator dispensing fixed charge packets.

B. Two-Wire Centigrade Thermometer

The circuit can be used to convert ambient temperature to a frequency, simply by connecting the 1 mV/K output to the V–F input. This provides an absolute scale, and hence can convert temperatures below 0°C or 0°F. However, it may be useful to read temperature directly in a familiar scale, using a simple frequency-meter for the display. Of course, the scales then terminate at 0°, and the frequency becomes low near this temperature, so in many cases the absolute scale is preferable.

Fig. 12 shows how the offset is introduced, using the 1.00-V output. The component values shown are for the centigrade scale, requiring an offset of 273 mV, and include corrections for the loading effect (Section V). The frequency range is 0 to 1000 Hz for 0° to 100°C. Also shown is a method for conveying the output and the supply power over a single pair of conductors. The load resistance modulates the supply current at the output frequency, and this modulation is easily recovered at the supply end of the cable. Using a +5-V supply the self-heating effect due to 6.5 mW dissipation causes an offset of about +1°C in free-air conditions, which can be compensated if necessary by a slight adjustment to the 3-kΩ and 8-kΩ resistors.

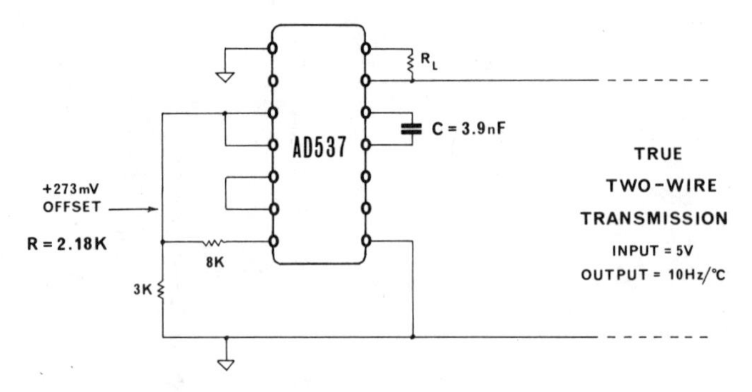

Fig. 12. Centigrade-reading temperature-to-frequency converter capable of operation at the remote end of a simple two-wire connection.

IX. Conclusions

A V–F converter based on an accurate multivibrator has been described, having a dynamic range at least as high as a charge-dispensing converter, and a nonlinearity which is satisfactory for a large number of applications. A versatile input interface provides high sensitivity for either voltage or current inputs. The converter is complete with a stable voltage reference and can operate from a single supply. Several design improvements, notably in the secondary current sources, are planned, and there is a good prospect for achieving linearities of ±0.02 percent or better on a production basis.

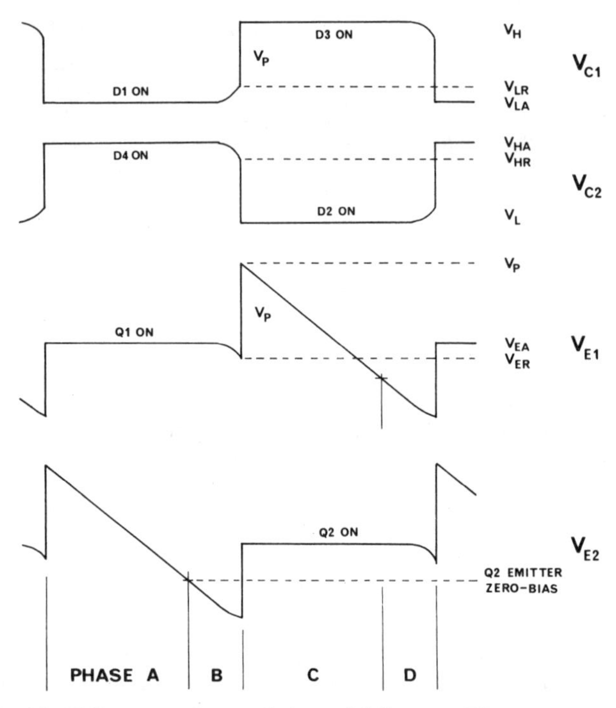

Fig. 13. Voltage waveforms of the multivibrator. The curvatures are exaggerated for clarity.

Appendix I
Analysis of Multivibrator Temperature Coefficient

Fig. 2 is used for analysis. The variable λ describes the modulation of the capacitor charging current λI. During most of the time λ is either 1 (current from left to right in capacitor) or -1, but for a brief period preceding each switching phase, λ falls in value until a critical point $\lambda = \lambda_c$ is reached. There are four timing phases in a complete cycle.

Phase A: $Q1$ has just switched fully on, $Q2$ fully off, and $\lambda = 1$. This phase ends after a period t_A when the base-emitter junction of $Q2$ is exactly zero biased.

Phase B: $Q2$ begins to conduct, and $\lambda_c < \lambda < 1$. This has two effects: the capacitor charges progressively more slowly, and the collector clamping potentials shift. After a further period t_B the critical point $\lambda = \lambda_c$ is reached and the circuit switches regeneratively into Phase C.

Phase C: As Phase A, but with $Q1$ off, $Q2$ on, $\lambda = -1$.

Phase D: As Phase B, but $-\lambda_c > \lambda > -1$, ending with $\lambda = -\lambda_c$, returning to Phase A.

Only the first two phases need be analyzed, the total period being $2(t_A + t_B)$. Fig. 13 shows the voltage waveforms, with some of the details exaggerated. The analysis is in four parts: determination of λ_c; determination of voltages; duration of Phase A; duration of Phase B. The transistors are supposed to have no dynamic limitations, infinite beta, zero-ohmic emitter resistance, and equal junction areas. Although the last assumption is unnecessary (and in fact, untrue), it simplifies the analysis without invalidating it. The buffers $A1$ and $A2$ are assumed to have infinite current gain and introduce a level-shifting of E volts.

Determination of λ_c

The circuit enters regenerative switching when the loop gain reaches unity. This occurs when the sum of the collector load resistances equals the sum of the emitter load resistances, using a small-signal analysis at this stage. This criterion is valid for reasonable time scales; at high frequencies the analysis would have to include dynamic factors which are beyond the scope of this discussion. The equation is

$$r_{e1} + r_{e2} = r_{c1} + r_{c2}. \tag{5}$$

During Phase B, $Q2$ is slowly turning on, and its emitter current is $(1 - \lambda)I$. Consequently, $r_{e2} = kT/qI(1 - \lambda)$. $Q1$ conducts $(1 + \lambda)I$, so $r_{e1} = kT/qI(1 + \lambda)$. The load resistance r_{c1} is generated by $D1$ (Fig. 2) conducting λI, and r_{c2} consists of $D4$ conducting λI, and $D5$ and $D6$ conducting $(2 - \lambda)I$, all in series. Therefore, the critical point occurs when

$$\frac{1}{1-\lambda_c} + \frac{1}{1+\lambda_c} = \frac{1}{\lambda_c} + \left[\frac{1}{\lambda_c} + \frac{2}{(2-\lambda_c)}\right]. \tag{6}$$

This has the solution $\lambda_c = \pm\sqrt{3} - 1 = 0.732$ (the solution $\lambda_c = 2.73$ being inapplicable).

There is evidently a large current imbalance in $Q1$ and $Q2$ at the point of regeneration, so a temperature-proportional drift term will appear in the timing calculations.

Determination of Voltages

In the following, ϕ is used to denote the forward-biased junction voltage at a given temperature and at a current I to avoid confusion with V_{be}, which is not constant. At a constant temperature we can write $V_{be} = \phi + (kT/q)\ln(i/I)$, where i is a general current, and since all currents are related to I through the variable λ, the recurrent form $V_{be} = \phi + (kT/q)\ln f(\lambda)$ will appear.

With reference to Fig. 13 we can develop expressions for the important voltage levels. The "high" collector-clamping voltage is

$$V_H = -2\left[\phi + \frac{kT}{q}\ln(2-\lambda)\right] + \left[\phi + \frac{kT}{q}\ln\lambda\right]. \tag{7}$$

During Phase A, $\lambda = 1$ ($D4$, $D5$, and $D6$ all conduct I) so

$$V_{HA} = -\phi. \tag{8}$$

At the end of Phase B (brink of regeneration) $\lambda = \lambda_c$ so

$$V_{HR} = -\phi + \frac{kT}{q}\ln\frac{\lambda_c}{(2-\lambda_c)^2}. \tag{9}$$

The general expression for the "low" collector-clamping voltage is

$$V_L = -V_R - \left(\phi + \frac{kT}{q}\ln\lambda\right). \tag{10}$$

During Phase A, $\lambda = 1$ ($D1$ conducts I), so

$$V_{LA} = -V_R - \phi. \tag{11}$$

At the end of Phase B, $\lambda = \lambda_c$, so

$$V_{LR} = -V_R - \phi - \frac{kT}{q}\ln\lambda_c. \tag{12}$$

To ascertain the voltage through which the capacitor must charge we first find the emitter voltages at the beginning of Phase A and the end of Phase B. During these times the emitter of $Q1$ is at

$$V_E = V_H - E - \left[\phi + \frac{kT}{q}\ln(1+\lambda)\right]. \tag{13}$$

For Phase A [using (8) and noting $\lambda = 1$]

$$V_{EA} = -E - 2\phi - \frac{kT}{q}\ln 2. \tag{14}$$

For Phase B, V_E can be calculated using (7). At the regenerative point

$$V_{ER} = -E - 2\phi - \frac{kT}{q}\ln\frac{(2-\lambda_c)^2(1+\lambda_c)}{\lambda_c}. \tag{15}$$

Now, the step labeled ΔV_p in Fig. 13 is simply $V_{HA} - V_{LR}$ and this step is transmitted to the emitter of $Q2$ during switching into Phase C. Thus the peak positive emitter voltage is

$$V_p = V_{ER} + \Delta V_p$$

$$= -E - 2\phi + V_R + \frac{kT}{q}\ln\frac{\lambda_c^2}{(2-\lambda_c)^2(1+\lambda_c)}. \tag{16}$$

Duration of Phase A

The capacitor charges down from this potential until $Q2$ is zero biased. The base of $Q2$ is at $V_L - E$ at this instant, so the charging voltage during this phase is

$$V_A = V_p - (V_L - E) \tag{17}$$

$$= 2V_R - \phi + \frac{kT}{q}\ln\frac{\lambda_c^2}{(2-\lambda_c)^2(1+\lambda_c)}. \tag{18}$$

Consequently, the duration of Phase A is simply

$$t_A = \frac{C}{I}V_A. \tag{19}$$

By continuing in this way, the peak negative emitter voltage can be shown to be

$$V_N = V_{LR} - E - \left[\phi + \frac{kT}{q}\ln(1-\lambda_c)\right] \tag{20}$$

and the total capacitor charging interval is

$$V_{\text{TOT}} = V_p + V_N = 2V_R + \frac{kT}{q}\ln\frac{\lambda_c^3}{(2-\lambda_c)^2(1+\lambda_c)}. \tag{21}$$

Thus, at absolute zero temperature the total time for all four phases is

$$t_{\text{TOT}} = 4\frac{CV_R}{I} \tag{22}$$

which is consistent with the fact that at this temperature switching would occur abruptly, as assumed in the earlier analysis in this paper. In fact, proper allowance must be made for the gradual reduction in charging current during Phases B and D.

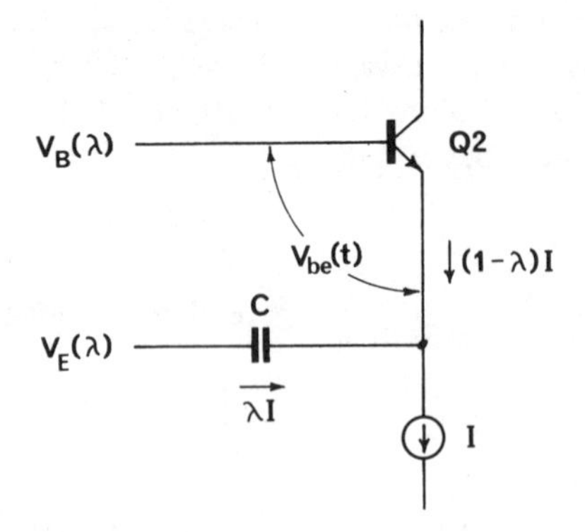

Fig. 14. Model for analysis of the transitional charging phase.

Duration of Phase B

Since the charging current is not constant during this phase (actually, during just a very small interval before regeneration) we must perform an integration. Fig. 14 shows the model for analysis of this phase. Note that there is a "moving target" in the sense that neither the base of $Q2$ nor the voltage at the "fixed" end of the capacitor is constant, and this must be included in the analysis. Using (10)

$$V_B(\lambda) = V_L - E = -V_R - \phi - \frac{kT}{q} \ln \lambda - E \tag{23}$$

and using (15)

$$V_E(\lambda) = -E - 2\phi - \frac{kT}{q} \ln \frac{(2 - \lambda_c)^2 (1 + \lambda_c)}{\lambda_c}. \tag{24}$$

The constant terms E, V_R, and ϕ can be eliminated, since at the onset of Phase B the V_{be} of $Q2$ is zero and only the logarithmic variations due to λ are important. The V_{be} is therefore

$$V_{BE}(t) = V_B(\lambda) - V_E(\lambda) + \frac{I}{C} \int_0^{t_B} \lambda \, dt. \tag{25}$$

V_{be} can also be equated to the emitter current of $Q2$

$$V_{BE}(\lambda) = \frac{kT}{q} \ln \frac{(1 - \lambda) I}{I_s} \tag{26}$$

where

$$I_s = I \exp\left(-\frac{\phi q}{kT}\right). \tag{27}$$

We can thus make λ a function of V_{be}

$$\lambda = 1 - \frac{I_s}{I} \exp\left(\frac{V_{BE}(t)}{kT/q}\right). \tag{28}$$

Equation (25) is not amenable to a closed-form solution for t_B and the integration was mechanized for solution on a programmable calculator in the form

$$x = a(\lambda) + b \sum \lambda(x) \tag{29}$$

iterated until $\lambda = 0.732$, where

$$x = q V_{BE}(t)/kT$$

$$a(\lambda) = \ln \frac{(2 - \lambda)^2 (1 + \lambda)}{\lambda}$$

and

$$b = \frac{Iq}{CkT} \cdot \Delta t$$

with

$$x_0 = 0, \; \lambda_0 = 1.$$

Numerical values of $I = 100$ μA and $C = 0.01$ μF were used, with an initial time step, Δt, of 1 μs reducing to 10 ns for $\lambda < 0.932$. Confirmation of the small-signal analysis of the switching point was provided by the fact that λ changes rapidly beyond the critical value of 0.732; after several hundred iterations to reach this value a further eight were sufficient to change the sign of λ, corresponding to a reversal of current in the timing capacitor. This also suggests that the regenerative phase lasts for about 80 ns for the stated parameters.

Calculations

The periods t_A and t_B were calculated for $T = 150$ K to 450 K in 50 K steps. An exact expression [7] for ϕ was used which includes the major temperature effects

$$\phi = E_{g0} \left(1 - \frac{T}{T_0}\right) + \frac{T}{T_0} \phi_o - \frac{mkT}{q} \ln \frac{T}{T_0}. \tag{30}$$

The coefficient m was determined from previous work [5] related to achieving a nominally zero TC in a bandgap generator, and a median value of 1.25 was used. This value of ϕ is used directly in the calculation of t_A and provides I_s for the solution of t_B. A reference value of $\phi_o = 650$ mV at $T = 300$ K and 100 μA was used.

The two sets of seven points were found by linear regression to accurately fit

$$t_A = 43.23 \text{ μs} + 181.2 \text{ ns/K}$$

and

$$t_B = 123.4 \text{ μs} - 219.4 \text{ ns/K}.$$

The total period is therefore

$$2(t_B + t_A) = 333.3 \text{ μs} - 76.4 \text{ ns/K}.$$

This result agrees exactly with the simplified theory at $T = 0$ K, and predicts a frequency drift of 230 ppm/K. This would be corrected by adding to the basic V_R of 833 mV a PTAT component of +192 μV/K. Measurements on the first prototypes show that this figure is insufficient, and a small correction was required to center the TC. The discrepancy can be attributed to an oversimplification of the model, which neglected transistor dynamics, ohmic resistance, and the effect of internal noise voltages. The latter could conceivably be responsible for an increased TC by causing the regeneration point to be earlier than predicted.

Appendix II

Input Amplifier Nonlinearity

The V_{be} of the trio $Q47$–49 varies logarithmically with the control current I, and introduces a nonlinearity which is reduced by the amplifier open-loop gain G_o, which appears at the input in addition to the linear term IR and a zero-adjust voltage E_Z (injected using an external V_{os} potentiometer connected to pins 9 and 10). Thus

$$V = IR + E_Z + \frac{kT}{qG_o} \ln \frac{I}{I_m} \tag{31}$$

where I_m is the FS value of I. Two adjustments are made to calibrate a V-F converter. First, the FS input voltage V_m is applied and the scaling resistor R adjusted for FS frequency f_m. Then a small voltage, typically one-thousandth of FS, is applied and E_z adjusted for correct frequency f_z. This process may need some iteration, but finally

$$R = \frac{1}{I_m}\left[V_m - \frac{kT}{qG_o} \ln \sigma\right] \tag{32}$$

and

$$E_z = \frac{kT}{qG_o} \ln \sigma \tag{33}$$

where $\sigma = f_m/f_z = I_m/I_z$. Using these expressions the input error voltage is found to be very nearly

$$e_n = \frac{kT}{qG_o}\left[\ln \frac{f}{f_z} - \frac{f}{f_m} \ln \sigma\right] \tag{34}$$

the peak value of this being

$$\hat{e}_n = \frac{kT}{qG_o}\left[\ln \frac{\sigma}{\ln \sigma} - 1\right] \tag{35}$$

which occurs for $f_p = f_m/\ln \sigma$. Fig. 5 shows typical results.

At some current below I_z the input error will be equal in magnitude but opposite in sign to the value at f_p. This current is of interest, since it defines the total dynamic range for a given plus-minor error, and is given by

$$I_0 = I_z \exp\left[\ln \frac{\sigma}{\ln \sigma} - 1\right]. \tag{36}$$

Acknowledgment

The author thanks L. Counts for his valuable assistance during the development of the monolithic circuit, P. Brokaw for the idea of using unequal emitter areas in the bandgap cell, and the staff of Analog Devices Semiconductor for their overall support of the development.

References

[1] Pease, Teledyne Inc., "Amplitude to frequency converter," U.S. Patent 3 746 968, filed Sept. 1972.

[2] A. B. Grebene and G. A. Rigby, "Phase-locked integrated circuits," in *1969 NEREM Rec.*, vol. 11, pp. 86–87.

[3] B. Gilbert, "A stable second-generation phase-locked loop," in *1972 ISSCC Dig. Tech. Papers*, 1972, pp. 78–79.

[4] R. R. Cordell and W. G. Garrett, "A highly-stable VCO for applications in monolithic phase-locked loops," *IEEE J. Solid-State Circuits*, vol. SC-10, pp. 480–485, Dec. 1975.

[5] P. Brokaw, "A simple three-terminal IC bandgap reference," *IEEE J. Solid-State Circuits*, vol. SC-9, pp. 388–393, Dec. 1974.

[6] T. M. Frederiksen and W. M. Howard, "A single-chip monolithic sonar system," *IEEE J. Solid-State Circuits*, vol. SC-9, pp. 394–403, Dec. 1974.

[7] J. S. Brugler, "Silicon transistor biasing for linear collector current temperature dependence," *IEEE J. Solid-State Circuits* (Corresp.), vol. SC-2, pp. 57–58, June 1967.

A Precision Voltage-to-Frequency Converter

James C. Schmoock

Raytheon Co.

Mountain View, CA

A SECOND-GENERATION monolithic Voltage-to-Frequency Converter (VFC), which incorporates an ion-implanted zener reference, a high-speed, stable one-shot multivibrator circuit design, and a precision voltage-to-current converter and current switch will be described.

Figure 1 shows the top view and cross-section of the ion-implanted zener reference. The zener junction is formed by the emitter N^+ diffusion and by a buried P^+ layer which is ion-implanted below the surface of the wafer to avoid surface-induced long term drift effects. Ion-implantation has been found to offer significant advantages over previously-described diffusion techniques for producing low-drift buried zener references. Diffused zeners fabricated using base, isolation, or other deep P^+ diffusion require careful control to achieve variations of less than ± 250mV from run-to-run in a wafer fabrication production line[2]. The ion-implantation technique results in variations of less than ± 100mV run-to-run with likewise improvement in the ability to control and specify temperature coefficient of the resulting reference voltage.

Figure 2 shows a block diagram of the complete IC chip configured with all the passive components necessary to make a complete VFC. In operation, a positive voltage applied to the input results in an input current I_I to the integrator causing the output to ramp downward until it reaches the trigger threshold voltage. This fires the one-shot, which switches the precision current source into the integrator summing node for a fixed period of time, T. The integrator output ramps upward by an amount of voltage proportional to the difference between the input current, I_I, and the precision current source output, I_O, thus completing one cycle of operation. In the steady-state mode, the feedback loop thus formed forces the frequency of these pulses to be proportional to the input voltage. It will be noted that the voltage reference output is tied through an external connection to the reference voltage input, allowing ratiometric operation or even better temperature stability through the use of an external reference. Additionally, the use of positive and negative supplies allows implementation of the one-shot with all-NPN current mode logic in critical signal paths.

Figure 3 shows a simplified schematic of the high speed one-shot. The dotted lines separate the one-shot further into an R-S latch, ramp generator, and voltage comparator. The R-S latch also uses the implanted zener diodes to level shift input and output levels for convenient drive of following stages. In this application the control of dc levels provided by the buried zener diodes offers significant performance advantages over the reverse biased emitter-base junction. The one-shot operates in the following sequence. A positive trigger voltage temporarily applied to the base of Q5 sets the R-S latch to the on-state causing the base of Q11 to be pulled low, turning on Q12, which causes a sharp negative-going step at the base of Q1. This turns off Q1 and the timing current, I_T charges capacitor C_O linearly in the negative direction. When the voltage on capacitor C_O reaches the threshold voltage, V_T, the comparator resets the latch, completing the one-shot cycle. This all-NPN bipolar circuit design serves to minimize delays by having a minimum number of components in the critical signal path. Minimum delay means minimum drift of resultant VFC scale factor at high frequency.

The precision voltage-to-current converter and current switch are shown in Figure 4. The transistors Q21 – Q38 form a differential input transconductance amplifier, with the output current being the collector current of Q35 and Q36. The reference voltage, V_R is applied to R60, thus causing a current, I_{IN} in R60. The feedback loop formed by tying the collector of Q35 through Q38 back to the summing node at the base of Q22 forces an identical current to flow in the current switch, Q7 and Q8, neglecting input bias current in Q22. The purpose of Q38 is to compensate for base current lost in the current switch. The transconductance amplifier itself uses a complementary paraphase input composed of Q21 – Q26, with a current mirror formed by Q27 – Q30, which converts from differential to single-ended output. Level shift diodes Q32 and Q34 and emitter follower Q31 bootstrap the emitters of mirror devices Q29 and Q30 to increase gain and lower input offsets which would otherwise be caused by unbalanced collector voltages on Q23 and Q26. Matching emitter currents in Q35 and Q36 are assured by degeneration resistors R3 and R4, which each drop 4V under standard operating conditions. These and other resistors on the chip, which must be matched are 25μ wide or wider, and are also fabricated with ion implant[3].

Table 1 summarizes overall performance, reflecting the improved temperature stability at 100kHz operation. Die size of the monolithic VFC is 90 x 108 mils. Figure 5 is a photomicrograph of the high-speed one-shot portion of the die.

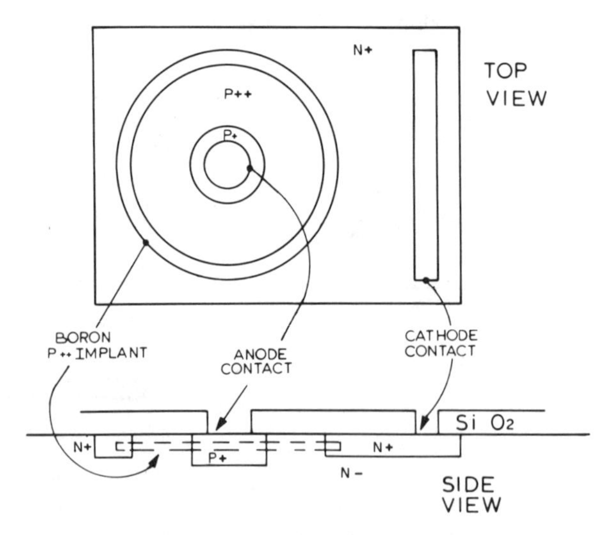

FIGURE 1—Ion implanted reference zener.

[1] Pinter, P. and Timm, D., "Voltage-to-Frequency Converter — IC Versions Perform Accurate Data Conversion (and much more) at Low Cost", *EDN*, Vol. 22, p. 153-157; Sept. 5, 1977.

[2] Dobkin, R., "Monolithic Temperature Stabilized Voltage Reference with 0.5ppm/°C Drift", *ISSCC DIGEST OF TECHNICAL PAPERS*, p. 108-109; Feb., 1976.

[3] Todd, C. and Stellrecht, H., "A Monolithic Analog Compandor", *ISSCC DIGEST OF TECHNICAL PAPERS*, p. 26-27; Feb., 1976.

Reprinted from *1978 IEEE Internat. Solid-State Circuits Conf. Dig. Tech. Papers*, Feb. 15-17, 1978, pp. 136-137.

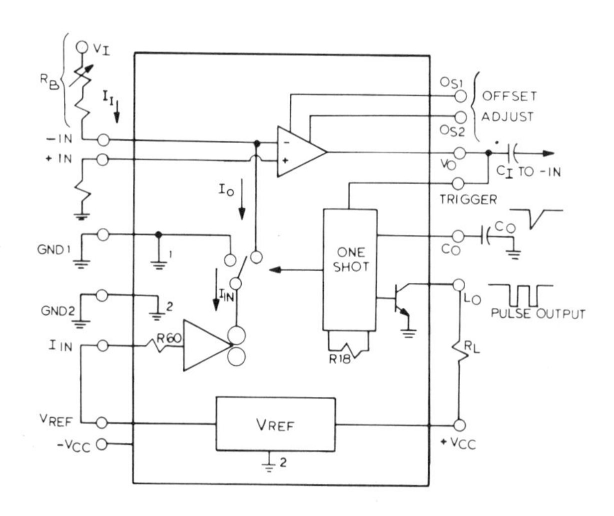

FIGURE 2—Precision VFC block diagram.

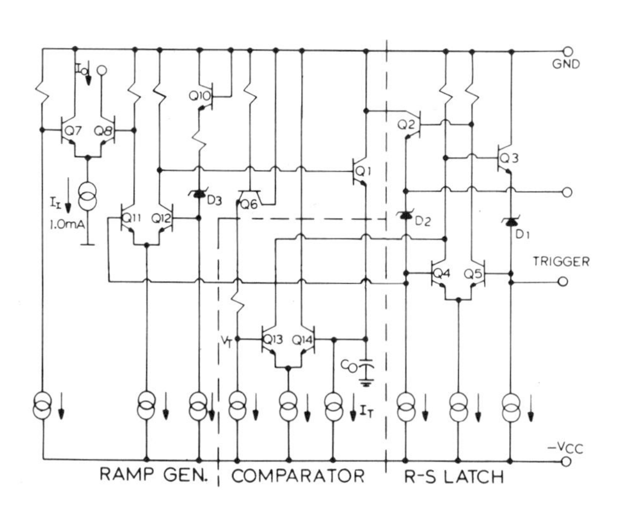

FIGURE 3—Simplified high-speed one-shot.

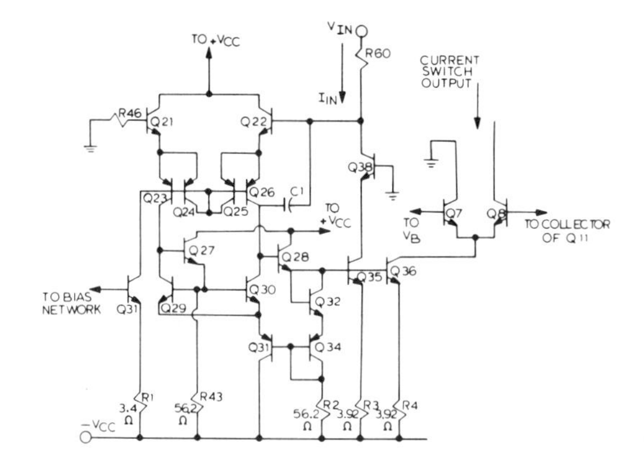

FIGURE 4—Precision voltage-to-current converter.

T = +25°C, V_{CC} = ± 15V			
PARAMETER	FULL SCALE FREQ 1.0 kHz	FULL SCALE FREQ 100 kHz	UNITS
Input:			
Voltage Range	0 to +10	0 to +10	V
Overrange	50 %	50 %	—
Impedance	20k	20k	Ω
Nonlinearity	.015	.025	% F.S.
Scale Factor TEMPCO	25	35	ppm/°C
TEMPCO VREF	20	20	ppm/°C
Offset Voltage	1.0	1.0	mV
V_{OS} Drift Vs.Temp	1.0	1.0	μV/°C
Settling Time +10V Step	1000	10	μs
Output:			
Pulse Width	500	5	μs
Pulse Polarity	NEG	NEG	—
VSAT	0.1	0.1	V@I=5mA
Supply:			
Voltage Range	±12 to ±18	±12 to ±18	V
Current	+5, -7	+5, -7	mA

TABLE 1—VFC typical performance.

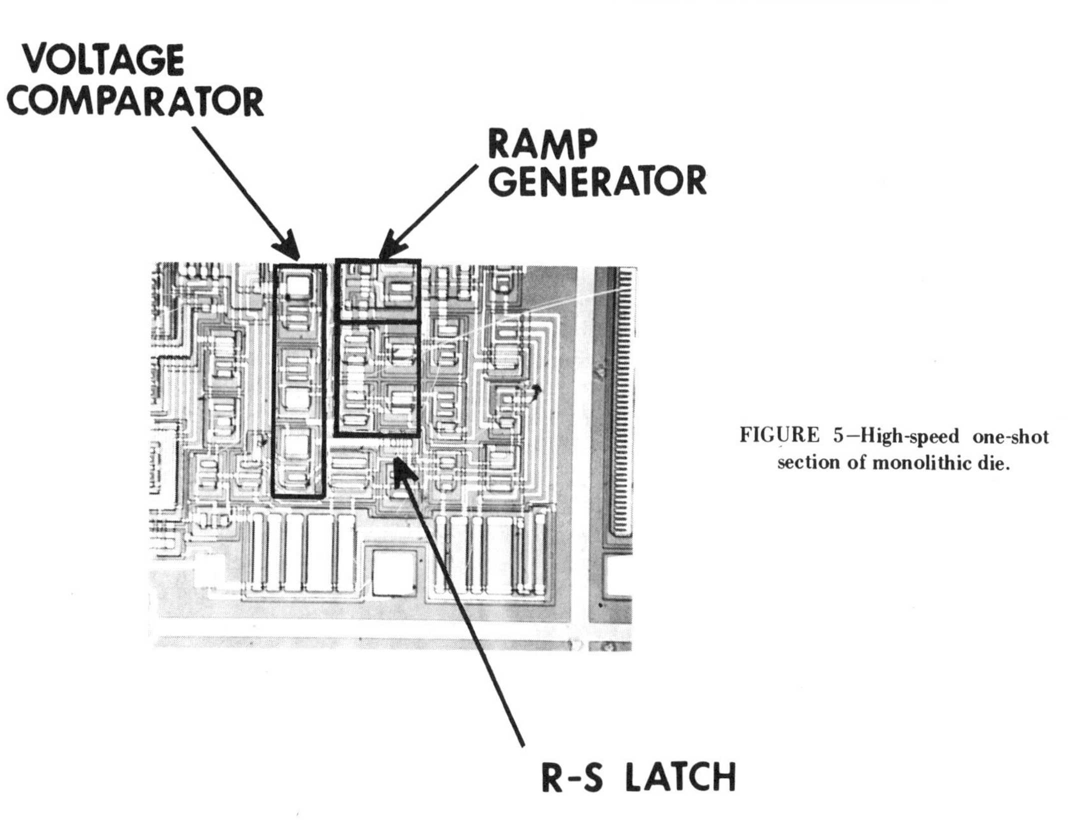

FIGURE 5—High-speed one-shot section of monolithic die.

Where and when to use which data converter

A broad shopping list of monolithic, hybrid, and discrete-component devices is available; the author helps select the most appropriate

With the commercial availability of data-converter products—the result of both hybrid (multichip) and monolithic (single-chip) technologies—users of analog-to-digital (A/D) and digital-to-analog (D/A) converters now have an impressive array of designs from which to choose. In addition, the older discrete-component designs still remain a viable choice for many high-performance applications, particularly those where broad operating characteristics and specialized features are important. Unfortunately, rather than helping the users to match the proper products to their needs, data-converter manufacturers confuse the issue by arguments over the relative merits of the different technologies. A closer look at the various types of data converters may help to clear up some of this confusion.

How data converters are used

Data converters are the basic interfaces between the physical world of analog parameters and the computational world of digital data processing. They are used in many industries in a wide variety of applications, including data telemetry, automatic process control, test and measurement, computer display, digital panel meters and multimeters, and voice communications, as well as in remote data recording and video signal processing.

As a typical example of the use of A/D and D/A converters, Fig. 1 illustrates how an entire industrial process can be controlled by a single digital computer, which may be located at a considerable distance from the process site. To communicate with the process, data inputs to the computer must be converted into digital form and the outputs reconverted into analog form.

Physical parameters of temperature, pressure, and flow are sensed by appropriate transducers and amplified to higher voltage levels by operational or instrumentation amplifiers. The various amplifier outputs are then fed into an analog multiplexer for sequential switching to the next stage—a sample-hold circuit that "freezes" the input voltage of a sequentially switched input for a fixed period of time, long enough for the following A/D converter to make a complete conversion cycle. In this manner, a single A/D converter is time-shared over a large number of analog input channels. Each channel is sampled periodically at a rate that is relatively fast when compared with any change in the process.

After receiving data from the process, the computer calculates the existing "state" of the process and compares it with the "desired state" stored in its memory. From this comparison, corrections are determined for the process variables. This information is fed to D/A converters that convert the digital data into analog form, and are then used to supply inputs to the process to bring it to the desired state.

Eugene L. Zuch Datel Systems, Inc.

The most common D/A and A/D conversion techniques are the parallel-input scheme for D/A conversion (A), the successive-approximation A/D conversion technique (B), the popular dual-slope A/D conversion circuit (C), and the parallel, or flash, A/D conversion circuit for ultrahigh speeds (D).

Types of data converters

Of the many different techniques that have been employed to perform data conversion, only a few are in wide use. Most D/A converter designs utilize a parallel-input circuit. In this scheme, the converter accepts a parallel binary input code and delivers an analog output voltage by means of binary weighted switches that act simultaneously upon application of the digital input. In the opening illustration, a representative parallel-input D/A-converter circuit is presented (A) in which binary weighted pnp-transistor current sources are controlled by emitter-connected diodes. For simplicity, a 4-bit converter is shown. The inputs operate from transistor–transistor logic (TTL) levels. The output current changes rapidly with a change in the digital input code. Since a voltage output is desired in most cases, the current from the pnp transistors is fed to an operational amplifier current-to-voltage converter. An internal voltage reference, which may be a Zener-diode or band-gap reference circuit, completes the circuit.

The most common A/D-conversion technique is the successive-approximation method, used in 70–80 percent of all present-day applications. As shown in the opening illustration, this circuit (B) incorporates the parallel-input D/A converter circuit previously described along with a successive-approximation register, a comparator, and a clock. The D/A converter's output, controlled by the successive-approximation register, is compared, one bit at a time, against the input signal, starting with the largest or most significant bit. A complete conversion is always accomplished in n steps for an n-bit converter, regardless of the input signal value. Successive-approximation A/D converters have the desirable characteristics of high conversion speed as well as excellent accuracy and stability, provided the circuit is well designed.

The next most popular A/D-conversion method is the dual-slope technique found in most digital panel meters and digital multimeters and commonly used in measurement and numeric display systems, also shown in the opening illustration (C). This converter circuit operates on an indirect principle, whereby the input voltage is

Reprinted from *IEEE SPECTRUM,* June 1977, pp. 39-42.

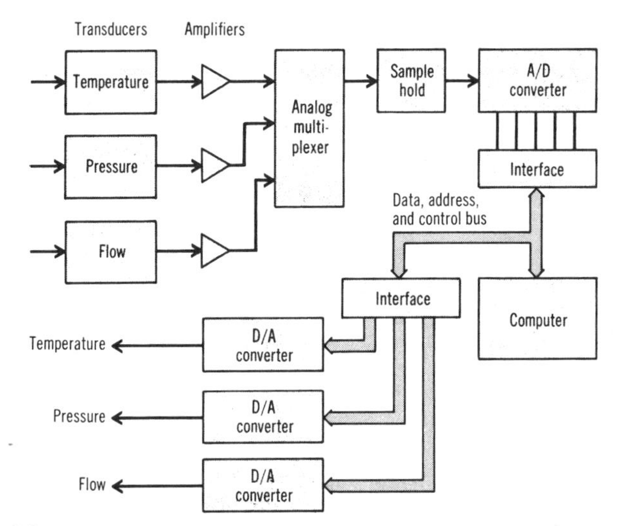

[1] A/D and D/A conversion products are a process-control system's basic interface elements between its physical variables and digital controlling computers.

converted to a time period measured by means of a reference and a counter. First, the input voltage is integrated for a fixed period of time, determined by the circuit's clock and counter. The integrator is then switched to the reference, causing integration in the opposite direction, until the output is back to zero, as determined by the comparator. The resultant digital-word output of the counter is proportional to the input voltage. The dual-slope method is very accurate, as its accuracy and stability depend only on the accuracy and stability of the circuit's reference. Its disadvantage is a much slower conversion time than with successive-approximation converters.

A third A/D-conversion method is the less frequently used ultrahigh-speed parallel, or flash, technique. As shown in the opening illustration, this circuit (D) employs $2^n - 1$ comparators to make an n-bit conversion. The comparators are biased by a tapped resistor connected to the reference voltage. The input signal is fed to the other comparator inputs all tied together. The result is a circuit that acts as a quantizer with 2^n levels, where n is the number of bits. For a given input-voltage level, all comparators biased below that level trip ON, and those biased above it remain OFF. The $2^n - 1$ digital outputs from the comparator must then be decoded into binary outputs. Since the complete conversion cycle occurs in only two steps, very-high-speed conversions are possible. The limitation of this technique is that it is difficult to realize high resolution, because of the large number of comparators required.

A comparison of technologies

Traditionally, data converters have been of the discrete-component type, first becoming available in instrument cases and later in compact, encapsulated modules. The advantage of this approach is that optimum components of all types can be combined. For example, for very high levels of speed and precision, a high-speed comparator may be combined with precision, high-speed current switches utilizing very-low-temperature-coefficient resistors and a very-low-temperature-coefficient reference. The result of such flexible component selection has been the development of some very impressive high-performance discrete-component converters over the past several years. For example, Table I shows that 12-bit current-output D/A converters are available with settling times as low as 50 ns, with voltage-output units achieving settling times down to 600 ns. Resolution can be as high as 16 bits. Excellent stabilities are also possible; ultralow drifts of 1 ppm/°C are achievable.

Equally impressive performance is also obtainable from discrete-component A/D converters. The fastest are the parallel types with resolutions of 8 bits at a 17-MHz conversion rate. Successive-approximation A/D converters offer rapid conversion times at various resolutions. As can be seen in Table II, conversion times of 0.80, 1.0, 2.0, and 10 μs at respective resolutions of 8, 10, 12, and 14 bits are possible. Slower but higher-resolution 14-bit dual-slope and 16-bit successive-approximation A/D converters with excellent stabilities are also available.

Hybrid converters

Although hybrid-converter design is almost as flexible as that of discrete-component converters, it does have two limitations: Not all of the semiconductor components used in a hybrid converter are readily available in chip form. Moreover, the number of chips used in a hybrid converter must be kept to a minimum for the converter to be economically producible. Minimizing the chip count minimizes the number of bonds required, which in turn minimizes labor content and maximizes both end-product yield and reliability.

Three factors have played a role in the emergence of new low-cost high-performance hybrid converters. The first is the availability of low-cost quad-current switches in chip form, which has simplified the circuitry required for the binary weighted current sources used. Second, new monolithic successive-approximation registers have minimized the logic circuitry required. And third, stable thin-film resistors with tight temperature tracking characteristics, and trimmable with fast laser trimming techniques, have made it possible to achieve economical 12-bit, and higher, resolutions. In fact, the excellent tracking characteristics of thin-film resistors (tracking within 1–2 ppm/°C is considered routine) provides hybrid converters with an advantage over modular discrete-component units.

As can be seen in Table I, hybrid devices include 8-, 10-, 12-, and 16-bit D/A converters with excellent performance characteristics. The 12-bit D/A converters have temperature coefficients as low as 10 ppm/°C, and some certain input registers. Most hybrid D/A converters do not require external output amplifiers.

Hybrid A/D converters with resolutions of 8, 10, and 12 bits are available, with respective conversion times of 0.9, 6.0, and 8.0 μs. An attractive feature of such A/D converters is their low price; for example, 12-bit models are now selling for under $100. Such units are complete converters and, except for calibration adjustments, require no external circuitry.

Monolithic converters

Monolithic data converters are generally a step below discrete and hybrid units in performance. In addition, various external components are usually required for proper operation, although this may not be viewed as a serious limitation since the attractive low prices of monolithic converters may more than compensate for the cost of the added components.

One of the obvious fabrication difficulties is making stable monolithic resistors for 10- and 12-bit monolithic

I. Representative D/A converters

Resolution (bits)	Settling Time (μs)	Output Type	Gain TC (ppm/°C)	Comments
Discrete-component converters				
10	0.025	Current	15	Ultrafast
10	0.25	Voltage	60	Ultrafast
12	0.05	Current	20	Ultrafast
12	2	Voltage	20	Fast; contains input register
12	5	Voltage	7	Low drift
12	0.60	Voltage	35	Ultrafast; deglitched
14	2	Voltage	10	Fast; low drift
16	25	Voltage	1	Ultralow drift
Hybrid converters				
8	2	Voltage	—	Contains input register
10	3	Voltage	—	Fast
12	3	Voltage	10	Fast
12	3	Voltage	20	Contains input register
16	50	Current	7	High resolution; requires external output amplifier
Monolithic converters				
8	0.085	Current	—	Fast; requires external reference
8	0.50	Current	—	Companding type for communication applications
10	1.5	Voltage	60	Complete unit; includes reference and output amplifier
10	0.50	Current	—	Multiplying type made from CMOS technology; requires external reference and output amplifier
10	0.25	Current	60	Uses thin-film resistors; has internal reference, but requires external output amplifier
12	0.50	Current	—	Multiplying type made from CMOS technology; does not have 12-bit linearity; requires external reference and output amplifier
12	0.30	Current	—	Bipolar type with true 12-bit linearity; requires external reference and output amplifier

II. Representative A/D converters

Resolution (bits)	Conversion Time (μs)	Conversion Type	Gain TC (ppm/°C)	Comments
Discrete-component converters				
5	0.01	Parallel	—	Ultrafast; 100-MHz rate
8	0.80	Successive approximation	20	Very fast; 1.2-MHz rate
8	0.06	Parallel	100	Ultrafast; 17-MHz rate
10	1	Successive approximation	20	Very fast; 1-MHz rate
12	2	Successive approximation	30	Very fast; 500-kHz rate
14	10	Successive approximation	6	Fast
16	400	Successive approximation	8	High resolution; very low drift
14	230 000	Dual slope	8	High resolution; low drift; Ratiometric with front-end isolation
Hybrid converters				
8	0.9	Successive approximation	—	Fast
10	6	Successive approximation	30	Fast
12	8	Successive approximation	20	Fast
Monolithic converters				
8	40	Successive approximation	—	Requires external reference and clock
8	1800	Charge balancing	—	CMOS; requires external reference and other components
10	40	Successive approximation	—	CMOS; requires external comparator, reference, and other components
10	6000	Charge balancing	—	CMOS; requires external reference and other components
12	24 000	Charge balancing	—	CMOS; requires external reference and other components
13	40 000	Dual slope	—	CMOS; has auto-zero circuit; requires external reference and other components
3½-digit BCD	40 000	Dual slope	—	CMOS; has auto-zero circuit; requires external reference and other components

converters. It is possible to use diffused resistors for 8-bit, and sometimes 10-bit, converters, but tracking requirements at the 12-bit level are severe, necessitating the use of thin-film resistors and the additional step of depositing the thin-film resistors onto the monolithic chip.

Monolithic-converter designers have been quite successful in employing ingenious circuit techniques to achieve what would have been difficult to do in a straightforward manner. This is one of the challenging aspects of monolithic circuitry. For example, although monolithic 12-bit successive-approximation A/D converters are quite difficult to make, equivalent accuracy can be readily achieved by use of the slower dual-slope conversion technique. A number of low-cost 12-bit monolithic units are on the market that offer good performance characteristics. They utilize either the dual-slope or charge-balancing technique.

Charge balancing involves switching, in discrete time intervals, the output of a current source fed into the summing junction of an operational integrator. The switching is controlled by a comparator, which is controlled, in turn, by the output of the operational integrator. The input signal, which is also fed into the operational integrator's summing junction, determines the switching current's pulse rate. As the input signal increases in magnitude, the switched current's pulse rate (controlled by the comparator) increases in proportion to the input signal, until a current balance is achieved at the operational integrator's summing junction.

Whereas earlier monolithic devices used a two-chip approach to separate the analog and digital portions of the circuit, newer converters are one-chip units. Never-

A brief look backwards

High-performance data converters first became available in 1955, when the Epsco Corp. unveiled its Datrac B-611 A/D converter (one of the historical exhibits of last year's ELECTRO in Boston). This vacuum-tube-based instrument, together with its companion power supply, weighed 150 pounds (68 kg) and cost $8500. Yet it offered impressive performance, even by today's standards: 11-bit resolution at a 44-kHz conversion rate.

The development of early converters was spurred in part by the then infant U.S. space program, which used them for high-speed pulse-code-modulation (PCM) data-telemetry and computer data-reduction applications, and also for digitizing radar signals.

By 1958–1959, packaged transistors replaced vacuum tubes to produce 12-bit A/D converters that were substantially smaller in size than their predecessors. At least three such converters were introduced at that time—by Adage, Epsco, and Packard-Bell; they ranged in conversion times from 13 to 48 μs. Selling price was still quite high (about $5000) and by 1960, only about 2000 of these converters were in use.

A breakthrough occurred in 1966 when Epsco introduced its Datrac 3, a small hand-held 12-bit discrete-component A/D converter constructed on just two circuit boards in a metal-case module. The unit had 24-μs conversion and sold for $1200. Similar devices soon followed and, by 1968, the Redcor Corp. had introduced the first encapsulated discrete-component 12-bit A/D converter with 50-μs conversion at a price of $600.

The next year saw rapid improvement in discrete-component-converter performance, with 12-bit A/D-converter conversion times dropping down to 12 μs. During that same period, the Beckman Instrument Co. unveiled a new-generation data converter—the first hybrid converter, an 8-bit D/A unit made from multiple monolithic IC chips and a thick-film resistor network. An 8-bit D/A with a thin-film resistor network was produced in 1970 by Micro Networks.

Monolithic data converters were also being developed in the late '60s. In 1968, Fairchild Semiconductor was able to build a monolithic 10-bit D/A converter based on its model μA722, although this unit was a basic building block requiring an external reference, resistor network, and output amplifier. By 1970, Analog Devices had manufactured the industry's first monolithic quad-current switches, to be used as building blocks for hybrid A/D and D/A converters up to 16 bits in resolution. It was also in 1970 that Precision Monolithics produced the first complete D/A converter in monolithic form—a 6-bit unit that included a reference and an output amplifier, and required no additional components for operation (model DAC-01).

Meanwhile, discrete-component converters continued to be improved in performance characteristics. By 1971, conversion times for 12-bit A/D units had dropped to just 4 μs, though prices still hovered around the $600–$700 mark. Hybrid-converter prices continued to drop, with an 11-bit D/A converter (Beckman Instruments' model 848) selling for $155 in 1971 and a 12-bit unit (Micro Networks' model MN312) dropping to $100 by the next year.

Monolithic units also continued to be improved. In 1972, Motorola announced an 8-bit monolithic D/A converter (model MC1408), and Precision Monolithics produced a 10-bit unit (model DAC-02) the next year.

By 1975, the price of hybrid data converters such as Datel Systems' 12-bit model ADC-HY12B A/D converter had dropped below $100. At the same time, 10-bit monolithic D/A converters were selling for as low as $20. And performance of discrete-component A/D converters such as Datel's ADC-EH12B3—a 12-bit 2-μs unit—has reached a new high at a record low price of $300, half the former price.

Monolithic-converter prices have continued to drop (now down to about $10 for a 10-bit D/A unit requiring external components) while performance is up (300-ns for a 12-bit D/A converter from Precision Monolithics requiring only an external reference and an op amp).

theless, external components, such as an integrating capacitor, a reference, and some compensation parts, are needed for proper operation.

CMOS circuitry has been used to fabricate monolithic converters with BCD coding, for digital panel meters and small instruments. Among the other popular monolithic A/D converter types is an 8-bit design that employs a variation of the successive-approximation technique. This device is made with an ion-implanted p-channel MOS technology. Instead of the conventional eight switches, it uses 255 switches connected to a 256-series-resistor chain. In hybrid or discrete form, this would be a gross waste of components, but not so in monolithic form. This approach results in a monotonic 8-bit A/D converter.

Another successful monolithic approach has been to use bipolar technology to make an 8-bit D/A converter with a companding characteristic for use in voice PCM systems. A typical device has eight inputs, which select eight chords (straight-line approximations to a curve), each with 16 equal steps. As part of an A/D converter, this device compresses a signal (provides high gain at low creases in magnitude, the switched current's pulse rate (controlled by the comparator) increases in proportion to the input signal, until a current balance is achieved at the operational integrator's summing junction.

Whereas earlier monolithic devices used a two-chip approach to separate the analog and digital portions of the circuit, newer converters are one-chip units. Never-converters are also available, one of which can be obtained with a reference and an output amplifier. And one multiplying-type 10-bit unit can have a variable reference applied to it. The CMOS 12-bit monlithic D/A converter model available at this time is a multiplying type, and does not provide full 12-bit linearity. Another bipolar 12-bit D/A converter has been introduced that has true 12-bit linearity, and requires an external reference and output amplifier. ◆

The author wishes to acknowledge the help of the following in tracing some early-model data converters: Nicholas Tagaris, John Gallagher, and James Zaros of Datel Systems; Richard Tatro of Micro Networks; Richard Snyder of Beckman Instruments; Donn Soderquist of Precision Monolithics.

[continued from page 311]

A Monolithic Companding D/A Converter

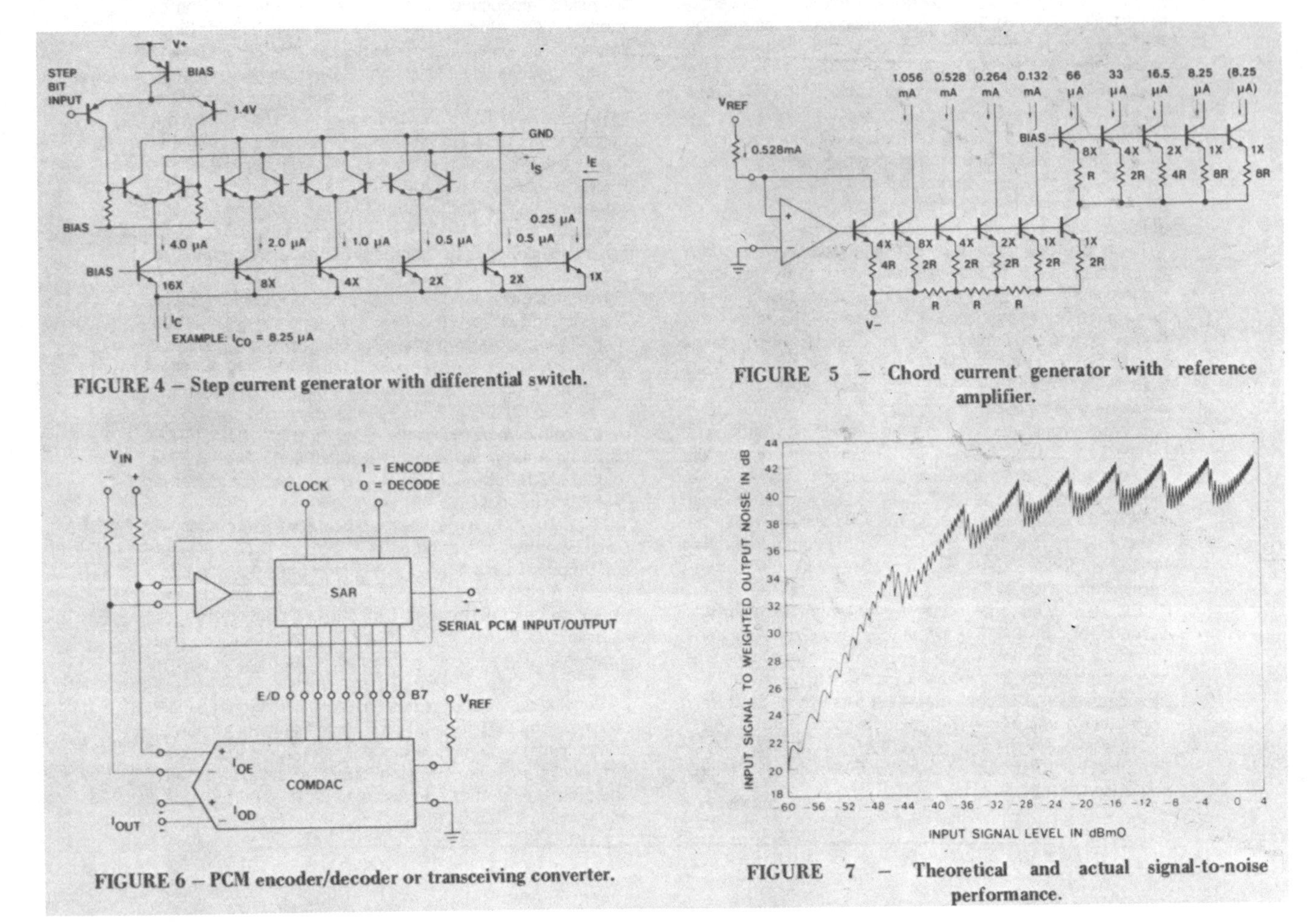

FIGURE 4 – Step current generator with differential switch.

FIGURE 5 – Chord current generator with reference amplifier.

FIGURE 6 – PCM encoder/decoder or transceiving converter.

FIGURE 7 – Theoretical and actual signal-to-noise performance.

Part VII
Communication Circuits

Over the recent years, we have experienced a very rapid influx of monolithic IC's into the field of data and voice communications, particularly in the field of telephony. This part brings together a number of selected technical and tutorial papers on the subject of communication IC's, with a particular emphasis on telecommunications.

The first paper, by Grebene, presents a tutorial discussion of the monolithic phase-locked loop (PLL) circuits which have become one of the key building blocks of communications systems. In the second paper, Gilbert describes the design and performance of monolithic PLL system which exhibits excellent temperature stability characteristics. In the third paper, the fundamentals of monolithic waveform generation and waveshaping is discussed, and a novel triangle-to-sinewave conversion technique described. Nonlinear amplification, i.e., compression or expansion of voice signal is often used in telecommunication systems, to accomodate signals exhibiting a very wide dynamic range. The next paper, by Todd, addresses this problem and describes the design of a monolithic compressor/expandor (compandor) circuit. In the subsequent paper, Janssen *et al.,* describe a dual-frequency tone generator IC for telephone dialing. In the sixth paper of this part, by Price and Kelley, design of a precision polarity-switch circuit is described for delta-modulation.

Pulse-code modulation, or PCM, has been the most widely accepted transmission technique for modern telephone systems and exchanges. The remaining three papers of this part describe a number of complex coder, decoder, and repeater IC's specially designed for PCM systems. The paper by Tsividis *et al.* describes a commpanding PCM encoder circuit using N-MOS technology. In the subsequent paper by Huggins *et al.,* a complete self-contained PCM coder/decoder (CODEC) system is described, again utilizing N-MOS technology. In the final paper of this section, the design and performance of a monolithic regenerative repeater IC is described, which is especially designed for T-1 type PCM lines.

ADDITIONAL REFERENCES

1. P.G. Erratico and R. Caprio, "An Integrated Expandor Circuit," *J. Sol. State Circuit,* vol. SC-11, pp. 762-772, December 1976.
2. M.J. Callahan and C.B. Johnson, "Integrated Dual-Tone Multi-Frequency Telephone Dialer," *ISSCC Digest of Tech. Papers,* pp. 64-65, February 1977.
3. J.B. Cecil, E.M.W. Chow, J.A. Flink, J.E. Solomon, C.G. Svala, and T. Svensen, "A Two -Chip PCM Codec for Per-Channel Application," *ISSCC Digest of Tech. Papers,* pp. 116-117, February 1978.
4. G.F. Landsburg and G. Smarandoiu, "A Two-Chip CMOS CODEC," *ISSCC Digest of Tech. Papers,* pp. 180-181, February 1978.
5. T.M. Frederiksen and W.M. Howard, "A Single-Chip Monolithic Sonar System," *J. Sol. State Circuits,* vol. SC-9, pp. 394-403, December 1974.
6. R.R. Cordell and W.G. Garrett, "A Highly-Stable VCO for Applications in Monolithic Phase -Locked Loops," *J. Sol. State Circuits,* vol. SC -10, pp. 480-485, December 1975.
7. M.P. Timko, "A Two-Terminal IC Temperature Transducer," *J. Sol. State,* vol. SC-11, pp. 784-788, December 1976.
8. J.M.R. Danneels and W.M. Sansen, "A New 4 x 4 Array of Active Bipolar Transistor Cross-points," *J. Sol. State Circuits,* vol. SC-11, pp. 779-783, December 1976.
9. J. Terry, T. Caves, C.H. Chan, S.D. Rosenbaum, L. Sellars, and J.B. Terry, "A PCM Voice CODEC with On-Chip Filters," *ISSCC Digest of Tech. Papers,* pp. 182-183, February 1978.
10. J.L. Henry and B.A. Wooley, "An Integrated PCM Encoder Using Interpolation," *ISSCC Digest of Tech. Papers,* pp. 184-185, February 1978.

The monolithic phase locked loop: a versatile building block

Here's a rundown on the many ways the PLL can be used as well as a look at the kind of devices now available.

Alan B. Grebene, Exar Corp.

The basic concept of the phase locked loop (PLL) has been around since the early 1930's and has been used for a variety of applications in instrumentation and space telemetry. However, before the advent of monolithic integration, cost and complexity considerations limited its use to precision measurements requiring very narrow bandwidths. In the past few years, the advantages of monolithic integration have changed the phase locked loop from a specialized design technique to a general-purpose building block. Therefore, what is "new" at this point is not the concept of the PLL, but its availability in a low-cost self contained monolithic IC package.

In many ways, this is similar to the case of the monolithic operational amplifier, which, until less than a decade ago, was an expensive building block. Today, with the advent of monolithic technology, it has become a basic building block in nearly every system design. The monolithic phase locked loop also offers a similar potential. In fact, many of the applications of the PLL outlined in this article become economically feasible only because the PLL is now available as a low-cost IC building block.

Today, over a dozen different integrated PLL products are available from a number of IC manufacturers. Some of these are designed as "general-purpose" circuits, suitable for a multitude of uses; others are intended or optimized for special applications such as tone detection, stereo decoding and frequency synthesis. This article is intended as a brief survey of the expanding field of monolithic phase locked loops. Its purpose is to familiarize the reader with their individual characteristics, capabilities and applications.

Applications for PLLs abound

As a versatile building block, the PLL covers a wide range of applications. Some of the more important are the following:

FM demodulation: In this application, the PLL is locked on the input FM signal, and the loop-error voltage, $V_d(t)$ in **Fig. A** (see Box), which keeps the VCO in lock with the input signal, represents the demodulated output. Since the system responds only to input signals within the capture range of the PLL, it also provides a high degree of frequency selectivity. In most applications, the quality of the demodulated output (i.e., its linearity and signal/noise ratio) obtained from a PLL is superior to that of a conventional discriminator.

FSK demodulation: Frequency-shift keyed (FSK) signals are commonly used to transmit digital information over telephone lines. In this type of modulation, the carrier signal is shifted between two discrete frequencies to encode the binary data. When the PLL is locked on the input signal, tracking the shifts in the input frequency, the error voltage in the loop, $V_d(t)$, converts the frequency shifts back to binary logic pulses.

Signal conditioning: When the PLL is locked on a noisy input signal, the VCO output duplicates the frequency of the desired input but greatly attenuates the noise, unde-

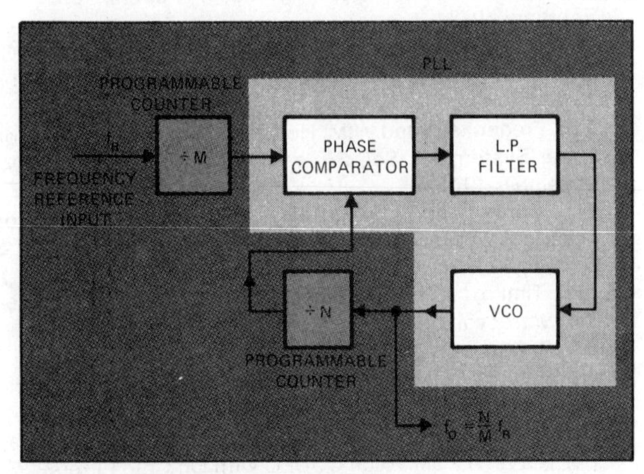

Fig. 1—A frequency multiplier/divider can be constructed using a phase locked loop.

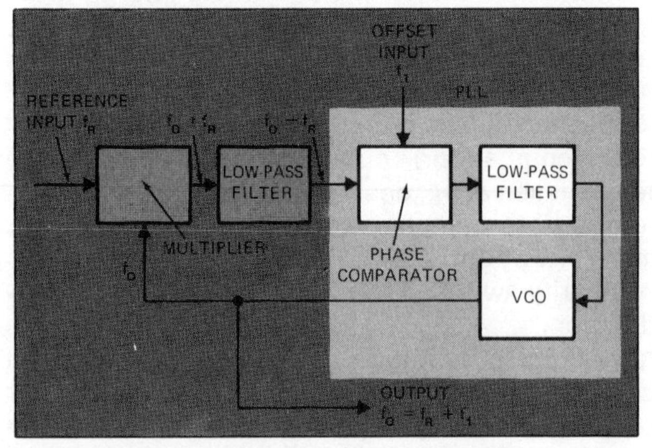

Fig. 2—Frequency translation can be accomplished with a phase locked loop by adding a multiplier and an additional low-pass filter to the basic PLL.

Reprinted with permission from *EDN Mag.*, pp. 26-33, Oct. 1, 1972.

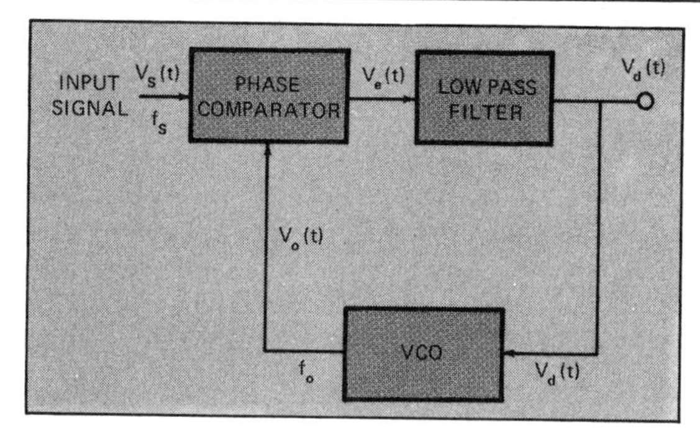

Fig. A – The basic phase locked loop consists of three functional blocks.

Basics of phase locked loops

The phase locked loop provides frequency selective tuning and filtering without the need for coils or inductors. As shown in **Fig. A**, the PLL in its most basic form is a feedback system comprised of three basic functional blocks: a phase comparator, low-pass filter and voltage controlled oscillator (VCO).

The basic principle of operation of a PLL can briefly be explained as follows: With no input signal applied to the system, the error voltage V_d is equal to zero. The VCO operates at a set frequency, f_o, which is known as the free-running frequency. If an input signal is applied to the system, the phase comparator compares the phase and frequency of the input signal with the VCO frequency and generates an error voltage, $V_e(t)$, that is related to the phase and frequency difference between the two signals. This error voltage is then filtered and applied to the control terminal of the VCO. If the input frequency, f_s, is sufficiently close to f_o, the feedback nature of the PLL causes the VCO to synchronize, or lock, with the incoming signal. Once in lock, the VCO frequency is identical to the input signal, except for a finite phase difference.

Two key parameters of a PLL system are its lock and capture ranges. They can be defined as follows:

Lock range: The range of frequencies in the vicinity of f_o, over which the PLL can maintain lock with an input signal. It is also known as the tracking or holding range. Lock range increases as the over-all gain of the PLL is increased.

Capture range: The band of frequencies in the vicinity of f_o where the PLL can establish or acquire lock with an input signal. It is also known as the acquisition range. It is always smaller than the lock range and is related to the low-pass filter bandwidth.

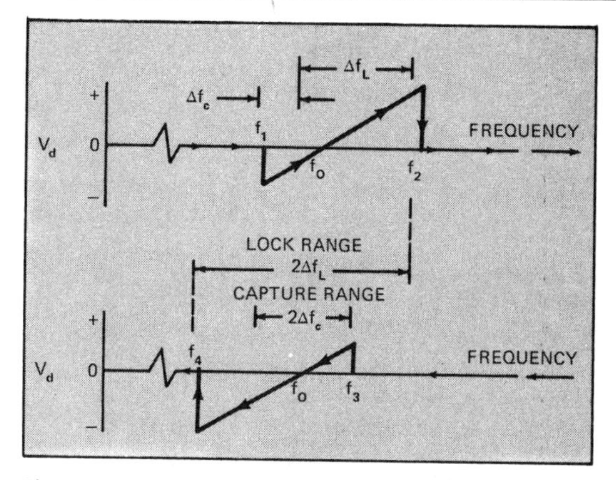

Fig. B – Typical PLL frequency-to-voltage transfer characteristics are shown for increasing (upper diagram) and decreasing (lower diagram) input frequency.

It decreases as the filter bandwidth is reduced.

The lock and the capture ranges of a PLL can be illustrated with reference to **Fig. B**, which shows the typical frequency-to-voltage characteristics of a PLL. In the figure, the input is assumed to be swept slowly over a broad frequency range. The vertical scale corresponds to the loop error voltage.

In the upper part of **Fig. B**, the loop frequency is being gradually increased. The loop does not respond to the signal until it reaches a frequency f_1, corresponding to the lower edge of the capture range. Then, the loop suddenly locks on the input, causing a negative jump of the loop error voltage. Next, V_d varies with frequency with a slope equal to the reciprocal of the VCO voltage-to-frequency conversion gain, and goes through zero as $f_s = f_o$. The loop tracks the input until the input frequency reaches f_2, corresponding to the upper edge of the lock range. The PLL then loses lock, and the error voltage drops to zero.

If the input frequency is now swept slowly back, the cycle repeats itself as shown in the lower part of **Fig. B**. The loop recaptures the signal at f_3 and traces it down to f_4. The frequency spread between (f_1, f_3) and (f_2, f_4) corresponds to the total capture and lock ranges of the system; that is, $f_3 - f_1$ = capture range and $f_2 - f_4$ = lock range. The PLL responds only to those input signals sufficiently close to the VCO frequency, f_o, to fall within the "lock" or "capture" range of the system. Its performance characteristics, therefore, offer a high degree of frequency selectivity, with the selectivity characteristics centered about f_o.

sired sidebands and interference present at the input. It is also a tracking filter since it can track a slowly varying input frequency.

Frequency synthesis: The PLL can be used to generate new frequencies from a stable reference source by either frequency multiplication and division, or by frequency translation. **Fig. 1** shows a typical frequency multiplication and division circuit, using a PLL and two programmable counters. In this application, one of the counters is inserted between the VCO and phase comparator and effectively divides the VCO frequency by the counter's modulus N. When the system is in lock, the VCO output is related to the reference frequency, f_R, by the counter moduli M and N as:

$$f_o = \left(\frac{N}{M}\right) f_R$$

By adding a multiplier and an additional low-pass filter to a PLL (**Fig. 2**), one can form a frequency translation loop. In this application, the VCO output is shifted from the reference frequency, f_R, by an amount equal to the offset frequency, f_1, i.e., $f_o = (f_R + f_1)$.

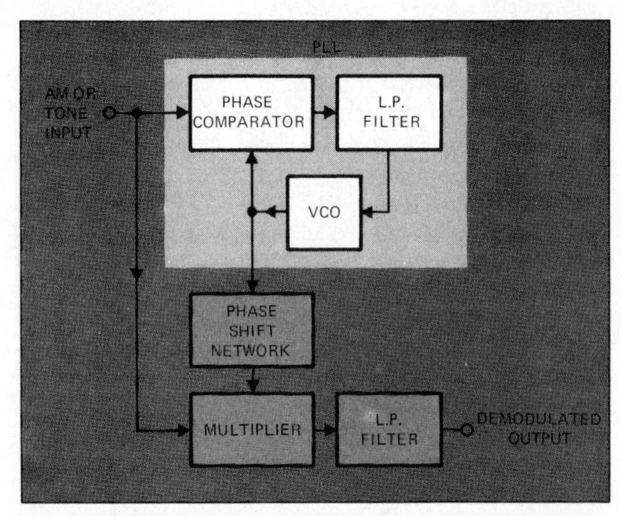

Fig. 3—AM and tone detection are possible by adding three functional blocks to the basic phase locked loop.

Data synchronization: The PLL can be used to extract synchronization from a composite signal, or can be used to synchronize two data streams or system clocks to the same frequency reference. Such applications are useful in PCM data transmission, regenerative repeaters, CRT scanning and or drum memory read-write synchronization.

AM detection: The PLL can be converted to a synchronous AM detector with the addition of a non-critical phase-shift network, an analog multiplier and a low-pass filter. The system block diagram for this application is shown in **Fig. 3**.

In this application, as the PLL tracks the carrier of the input signal, the VCO regenerates the unmodulated carrier and feeds it to the reference input of the multiplier section. In this manner, the system functions as a synchronous demodulator with the filtered output of the multiplier representing the demodulated audio information.

Tone detection: In this application, the PLL is again connected as shown in **Fig. 3**. When a signal tone is present at the input, within a frequency band corresponding to the capture range of the PLL, the output dc voltage is shifted from its tone-absent level. This shift is easily converted to a logic signal by adding a threshold detector with logic-compatible output levels.

Motor speed control: Many electromechanical systems, such as magnetic tape drives and disk or drum head drivers, require precise speed control. This can be achieved using a PLL system, as shown in **Fig. 4**. The VCO section of the monolithic PLL is separated from the phase-comparator and used to generate a voltage controlled reference frequency, f_R. The motor shaft and the tachometer output provide the second signal, frequency f_M, which is compared to the reference frequency. The controller is a power amplifier which drives the speed-control windings of the motor. Thus, the motor and tachometer combination essentially functions as a VCO which is phase locked to the voltage controlled reference frequency, f_R.

Stereo decoding: In commercial FM broadcasting, suppressed carrier AM modulation is used to superimpose the stereo information on the FM signal. To demodulate the complex stereo signal, a low-level pilot tone is transmitted at 19 kHz (1/2 of actual carrier frequency). The PLL can be used to lock onto this pilot tone, and regenerate a coherent 38 kHz carrier which is then used to demodulate the complete stereo signal. A number of highly specialized monolithic circuits have been developed for this application. Typical examples of monolithic stereo decoder circuits using the PLL principle are the MC 1310 and CA 3090E, manufactured by Motorola and RCA, respectively.

A survey of monolithic PLLs

Presently, a relatively wide choice of monolithic PLL circuits are commercially available, and this list is bound to grow rapidly in the coming months. The user of a monolithic PLL is usually faced with two questions at the onset of his design: (1) What is available? (2) What is the most suitable PLL circuit for the job? These questions are answered in the following paragraphs, based on available circuits at the time of this writing.

What is available?

Table 1 is a comparative listing of some of the presently available monolithic PLL circuits. In the table, the circuits are listed in ascending order of their product numbers. The table also lists some of the key performance characteristics and major applications of these circuits. All data listed are typical values unless indicated otherwise. It should be noted that such a brief table does not do justice to the full potential of a device for a specific application, since the measurement techniques for some of the specifications vary from manufacturer to manufacturer.

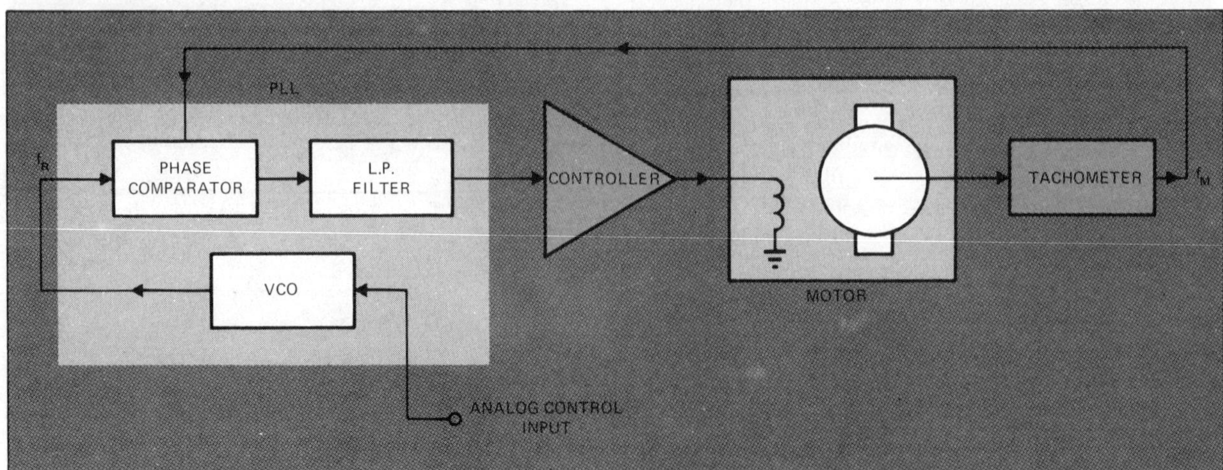

Fig. 4—Very precise motor speed control is possible with a phase locked loop system of this type.

The monolithic PLL circuits listed in **Table 1** also illustrate the wide range of design philosophies used by IC manufacturers. In one extreme are the "uncommitted," or multi-function designs, such as the XR-S200 and the SL-650, which strive for maximum versatility by keeping each of the functional blocks on the same substrate independent of each other. In the other extreme are the "committed" special-function devices, such as the SE-567 and the XR-2567 tone decoders. Each of the PLLs of **Table 1** are briefly reviewed on the basis of their functional block diagrams and most common applications

The XR-S200 (Exar Integrated Systems) is a multi-function PLL containing a four-quadrant analog multiplier, a high-frequency VCO and an operational amplifier (**Fig. 5A**), and is housed in a 24-pin package. Each of the functional blocks and their control inputs are independent of each other. In other words, they can be externally connected in any order. Thus, it is essentially a custom, or do-it-yourself, PLL. The user determines the function and performance characteristics by his choice of external connections and components.

The XR-210 (Exar Integrated Systems) was designed for FSK modulation/demodulation applications (**Fig. 5B**). In addition to the basic PLL, it contains a voltage comparator and and a RS232-C compatible output logic driver. The VCO section can be used for FSK modulation and has independent mark/space adjustments.

The XR-215 (Exar Integrated Systems) is a general-purpose PLL circuit particularly suited for FM demodulation, frequency synthesis, and tracking filter (**Fig. 5C**). The high-gain amplifier section can be used as an active filter, or can function as an audio preamplifier for FM detection. The VCO section has sweep and gain control options. A single PLL circuit can be time-multiplexed between two input channels by applying a binary input to its range-select control.

The SE-560 (Signetics) was the first monolithic PLL circuit to be introduced (**Fig. 5D**). It is a high-frequency circuit suitable for FM demodulation or signal conditioning. It contains a limiter block in series with the VCO section to control tracking range. The VCO output is internally connected to the phase comparator.

The SE-561(Signetics) is similar in basic design and performance to the SE-560 (**Fig. 5E**). However, it contains an additional multiplier/modulator section which is internally connected to the VCO. Thus it can be used for AM demodulation as well as for FM detection. AM demodulation is achieved by placing an external RC phase-shift network between the AM and FM input channels.

The SE-562 (Signetics) has basic features similar to their SE-560, except that the VCO output is not internally connected to the phase comparator input (**Fig. 5F**). This allows the circuit to be used for frequency-synthesis applications by inserting an external binary-counter into the loop. The VCO section has buffered differential outputs which have higher voltage swings than the 560 and 561 circuits (4.5V p-p., vs 0.6V p-p).

The SE-565 (Signetics) is a general-purpose PLL designed for low-frequency FM and FSK detection and frequency synthesis (**Fig. 5G**). Its frequency range is limited to 500 kHz (typical); however, its temperature stability is better than the other 560-Series circuits. The VCO section provides both square and triangular wave outputs, not internally connected to the phase-comparator inputs. However, the outputs can be dc coupled by shorting two adjacent pins in the package. (Signetics' SE-565, National Semiconductor's LM-565, and Raytheon Semiconductor's RM-565 are electrically-equivalent and interchangeable.

The SE-567 (Signetics) monolithic tone-decoder contains an internal current controlled oscillator (CCO) and two separate phase-detectors driven from the same oscillator (**Fig. 5H**). The quadrature phase detector, along with

Table 1 - A Summary of Commercially Available PLL Circuits

Product designation	Package	Operating supply range	High Frequency Limit	VCO stability		Primary Applications
				Power supply (%/V)	Temp. (ppm/°C)	
XR-S200	24 Pin DIP	6V to 30V ±3V to ±15V	30 MHz	0.08 (typ) 0.5 (max)	300 (typ) 650 (max)	Multi-function building block for FM/FSK detection, frequency synthesis
XR-210	16 Pin DIP	5V to 26V	20 MHz	0.05 (typ) 0.5 (max)	200 (typ) 550 (max)	FSK modem, frequency synthesis, data synchronization
XR-215	16 Pin DIP	5V to 26V	35 MHz	0.1 (typ) 0.5 (max)	250 (typ) 600 (max)	General purpose PLL. FM demodulation, tracking filter, frequency synthesis
SE-560	16 Pin DIP	Not specified	30 MHz	0.3 (typ) 2 (max)	600 (typ) 1200(max)	FM demodulation, signal conditioning tracking filter
SE-561	16 Pin DIP	Not specified	30 MHz	0.3 (typ) 2 (max)	600 (typ) 1200 (max)	AM and FM detection
SE-562	16 Pin DIP	Not specified	30 MHz	0.3 (typ) 2 (max)	600 (typ) 1500 (max)	Frequency synthesis, FM detection, signal conditioning
SE-565, LM565, RM565	14 Pin DIP 10 Pin TO	±5 to ±12V	500 kHz	0.1 (typ) 1 (max)	100 (typ) 525 (max)	FM, SCA detection, FSK demodulation
SE 567, XR567 LM567, RM567	8 Pin DIP 8 Pin TO	4.75V to 9V	500 kHz	0.5 (typ) 1 (max)	30 ± 140 (typ)	Tone detection
SL-650	24 Pin DIP	±3V to ±12V	1 MHz	0.05 (typ)	20 (typ)	Multi-function building block for signal generation/detection; FSK modem, FM detector
XR-2567	16 Pin DIP	4.75V to 15V	600 kHz	0.05(typ) 0.2 (max)	80 (typ)	Dual tone decoder, (Dual 567 equivalent)
HI-2800	16 Pin DIP	±5V to ±15V	30 MHz	0.1 (typ)	250 (typ)	FM demodulation, frequency synthesis, signal conditioning
HI-2820	14 Pin DIP	±6V to ±12V	3 MHz	0.1 (typ)	100 (typ) 250 (max)	FSK modems, data synchronization, motor speed control

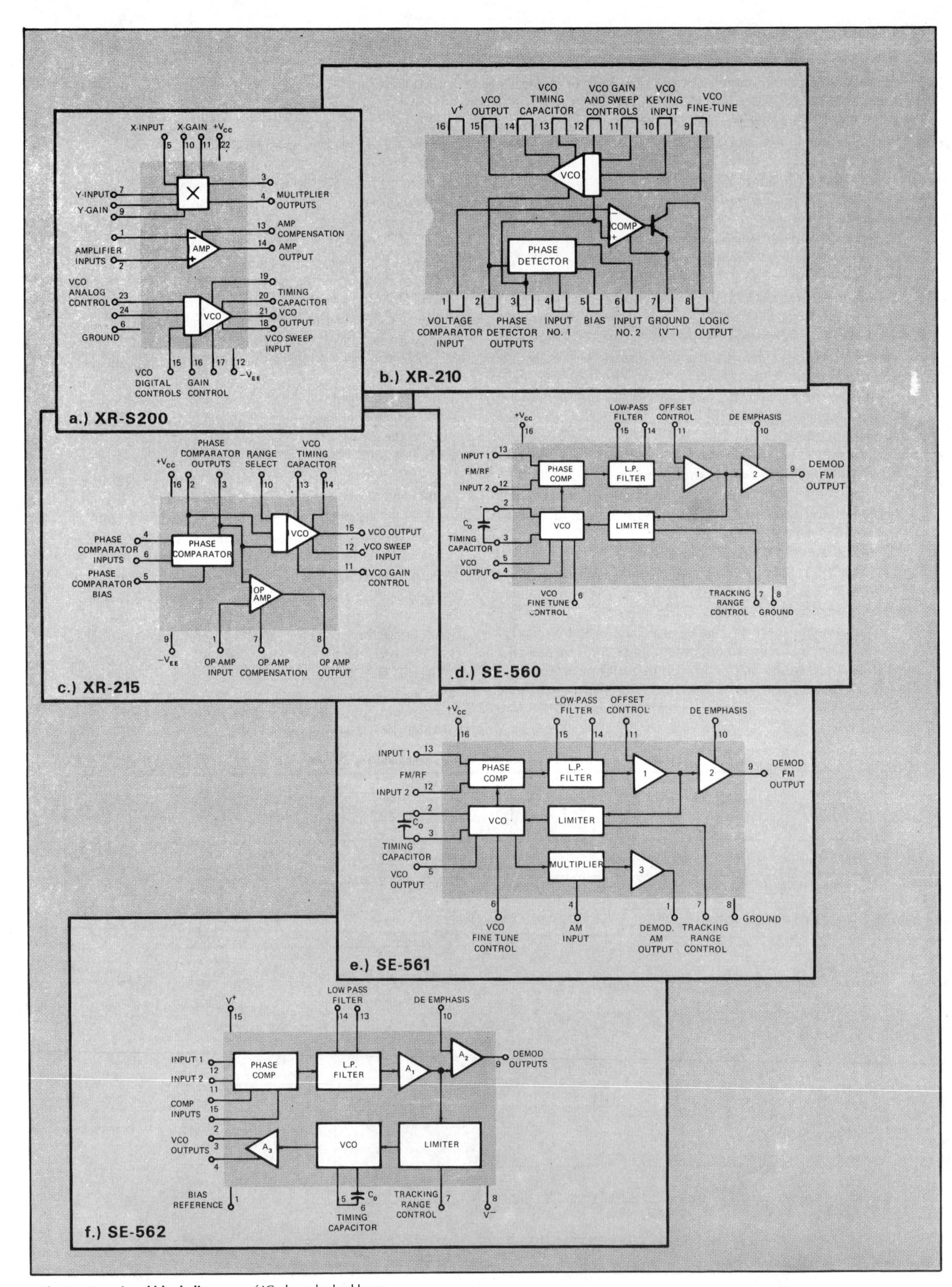

Fig. 5 – Functional block diagrams of IC phase locked loops.

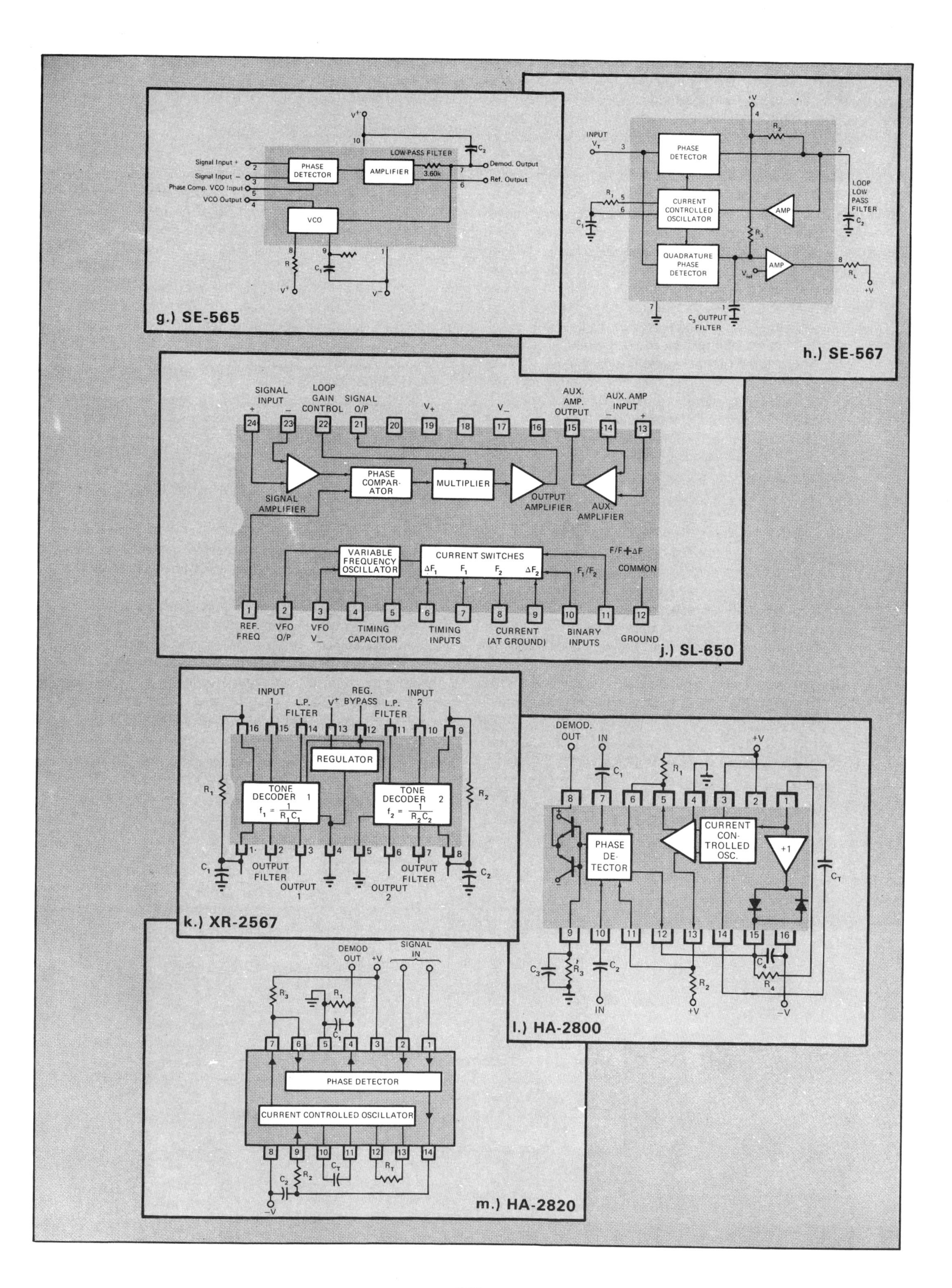

g.) SE-565
Signal Input +
Signal Input −
Phase Comp. VCO Input
VCO Output
PHASE DETECTOR
AMPLIFIER
LOW-PASS FILTER
3.60k
Demod. Output
Ref. Output
VCO
h.) SE-567
INPUT
PHASE DETECTOR
CURRENT CONTROLLED OSCILLATOR
QUADRATURE PHASE DETECTOR
AMP
LOOP LOW PASS FILTER
OUTPUT FILTER
j.) SL-650
SIGNAL INPUT
LOOP GAIN CONTROL
SIGNAL O/P
AUX. AMP. OUTPUT
AUX. AMP INPUT
SIGNAL AMPLIFIER
PHASE COMPAR-ATOR
MULTIPLIER
OUTPUT AMPLIFIER
AUX. AMPLIFIER
VARIABLE FREQUENCY OSCILLATOR
CURRENT SWITCHES
COMMON
REF. FREQ
VFO O/P
TIMING CAPACITOR
TIMING INPUTS
CURRENT (AT GROUND)
BINARY INPUTS
GROUND
k.) XR-2567
INPUT 1
L.P. FILTER
REG. BYPASS
INPUT 2
REGULATOR
TONE DECODER 1
TONE DECODER 2
OUTPUT FILTER
OUTPUT 1
OUTPUT 2
l.) HA-2800
DEMOD. OUT
IN
PHASE DE-TECTOR
CURRENT CON-TROLLED OSC.
m.) HA-2820
DEMOD OUT
SIGNAL IN
PHASE DETECTOR
CURRENT CONTROLLED OSCILLATOR

the buffer amplifier, is used to generate a binary output pulse if a signal tone at the input is within the pass-band of the system. Its detection bandwidth (capture range) and response time are controlled by external filter capacitors. It has high-current (100 mA) logic driver output. (National, Exar and Raytheon also manufacture this device.)

The SL-650 (Plessey Microelectronics) multi-function IC contains a current controlled oscillator (CCO), a phase comparator, a two-bit binary interface circuit and an auxillary amplifier in a 24-pin package (**Fig. 5J**). The circuit can be used in a variety of signal generation or modulation applications; or it can be connected as a phase-locked loop for demodulation or signal conditioning. The oscillator frequency can be swept over a 1000:1 range in frequency by an external control current, and exhibits excellent temperature stability (typically ±20 ppm/°C).

The XR-2567 (Exar Integrated Systems) dual tone-decoder system contains an equivalent of two independent 567 decoders on the same chip (**Fig. 5K**). It is particularly well suited for decoding multiple-tone inputs, such as those used in telephone dialing systems. Its operating voltage range is wider than that of the 567, and can switch two simultaneous 100 mA loads at the outputs. If only one of the two decoders is used, the remaining one can be deactivated to minimize power dissipation.

The HI-2800 (Harris Semiconductor) high-frequency PLL uses a current controlled oscillator (CCO). The oscillator and the phase-detector sections are not internally connected (**Fig. 5L**). The phase comparator section provides two independent high-impedance outputs. The loop bandwidth and the demodulation output bandwidth can be independently controlled. The CCO section has differential outputs with 1.2V p-p minimum swing.

The HI-2820 (Harris Semiconductor) low-frequency PLL has TTL compatible VCO outputs (**Fig. 5M**). It contains independent phase detector and oscillator sections and offers higher frequency stability (100 ppm/°C, typical) than its high-frequency counterpart, the HI-2800.

What is the best PLL for the job?

The PLL circuits listed in **Table 1** cover a wide range of applications and also a relatively broad range of parameter distribution. It is often difficult, if not impossible, to determine at a glance the best circuit for a given application, particularly in the case of the general-purpose or multifunction PLLs.

Table 2 gives a brief listing of some major classes of PLL applications and lists a number of "recommended" circuits for each. It should be noted that **Table 2** reflects the professional opinion of the author and is not an endorsement of one product over any other. The products are listed in numerical order by part numbers.

FM demodulation: Essentially all the PLL circuits listed in **Table 1** can be used for FM demodulation. However, it is often possible to narrow the choice down to 2 or 3 circuits, based on the particular performance criteria. In general, there are three key performance parameters which should be examined:

- ☐Quality of demodulated output: This is normally measured in terms of the output level, distortion, and signal/noise ratio for a given FM deviation.
- ☐VCO frequency range and frequency stability: For reliable operation, VCO upper frequency limit (see **Table 1**) should be at least 20% above the FM carrier frequency. VCO frequency stability is important, especially if a narrow-band filter is used in front of the PLL, or multiple input channels are present. If the VCO exhibits excessive drift, the PLL can drift out of the input signal band as the ambient temperature varies.
- ☐Detection threshold: This parameter determines mini-

TABLE 2.
Applications for commercially available monolithic PLL circuits

MAJOR APPLICATION / PRODUCT AND MANUFACTURER	EXAR XR-S200	EXAR XR-210	EXAR XR-215	SIGNETICS SE-560	SIGNETICS SE-561	SIGNETICS SE-562	SIGNETICS SE-565	NATIONAL LM-565	RAYTHEON RM-565	SIGNETICS SE-567	EXAR XR-567	NATIONAL LM-567	RAYTHEON RM-567	PLESSEY SL-650	EXAR XR-2567	HARRIS HI-2800	HARRIS HI-2820
FM DEMODULATION HIGH FREQUENCY (>5 MHz)	●	●	●	●	●	●										●	
LOW FREQUENCY (<5 MHz)	●	●	●	●	●	●	●	●	●					●		●	●
FREQUENCY SYNTHESIS HIGH FREQUENCY (>5 MHz)	●	●	●			●										●	
LOW FREQUENCY (<5 MHz)	●	●	●			●	●	●	●					●		●	●
FSK DEMODULATION	●	●	●				●	●	●					●		●	●
SIGNAL CONDITIONING	●	●	●	●	●	●	●	●	●					●		●	●
TONE DETECTION					●					●	●	●	●		●		
AM DETECTION					●												
MOTOR SPEED CONTROL	●	●	●											●		●	●

mum signal level necessary for the PLL to lock and demodulate an FM signal of given deviation.

In most FM demodulation applications, it is also desirable to control the amplitude of the demodulated output. This feature is provided in some of the PLL circuits (such as the XR-215, SL-650 or HI-2800) by means of a variable-gain amplifier contained on the chip.

For the low-frequency FM demodulation applications (in the kHz range) the following circuits are recommended: XR-215, SE-565 (or its equivalents), SL-650, HI-2820.

For high-frequency FM demodulation (1 MHz and above), the preferred circuits are the XR-215, the SE-560 and the HI-2800.

FSK decoding: Frequency-shift keying used in digital communications is very similar to analog FM modulation. Therefore, any PLL IC can be used for FSK decoding, provided that its input sensitivity and the tracking range are sufficient for a given FSK signal deviation. Some of the basic requirements and desirable features for a PLL used in FSK decoding are:

- □Center frequency stability.
- □Logic compatible output.
- □Control of VCO conversion gain.

Center frequency stability is essential to insure that the VCO frequency range stays within the signal band over the operating temperature range. A logic compatible output is desirable to avoid the need for an external voltage comparator (slicer) to square the output pulses. It is particularly convenient if the output conforms to RS-232C standard, thereby eliminating the need for a separate line/driver circuit. Control of the VCO's conversion gain allows the circuit to be used for both large deviation FSK signals (such as 1200 baud operation) as well as for small deviation (75 baud) FSK signals.

Concerning frequency stability, the recommended PLL circuits for FSK decoding are the XR-210, the SE-565 (and equivalents), the SL-650 and the HI-2820. Of these, the SL-650 has the highest frequency stability. Both the XR-210 and the SL-650 contain internal slicer and logic driver sections, whereas the SE-565 and the HI-2820 need an external voltage comparator to provide logic compatible outputs. In the XR-210, VCO conversion gain is controlled by means of an external resistor. With the SE-565 this control can be achieved by using an external PNP transistor.

Frequency synthesis: This application requires a PLL circuit with the loop opened between the VCO output and the phase comparator input. This includes most of the circuits listed in **Table 1**; namely: XR-S200, XR-210, XR-215, SE-562, SE-565, (or its replacements, LM-565, RM-565) SL-650, HI-2800 and HI-2820. Most frequency synthesis applications require high-frequency operation, which narrows the choice to the five devices shown in **Table 2**. One of the basic requirements for frequency synthesis applications is that the TTL logic-compatible output swings from the VCO. This requirement is met by all of the PLL circuits listed for frequency synthesis on **Table 2**. However, depending on the choice of circuit and supply voltages, dc level shifting may be necessary to interface with TTL logic.

Signal conditioning: Most signal conditioning applications require very narrow-band operation of the PLL. This in turn may require the use of active filters within the loop (between the phase detector and the VCO). The PLL circuits which allow active filters to be inserted into the loop are the XR-S200, the XR-215, the SL-650, the HI-2800 and the HI-2820. XR-S200, XR-215 and SL-650 already contain an op amp on the chip for active filtering.

Time decoding: The two different PLL circuits especially designed for this application are the SE-567's and the XR-2567. Both circuits have comparable performance characteristics. The XR-2567 contains two independent 567-type circuits on the same chip; therefore, it may be more economical to use in multiple tone-detection systems. Both circuits have relatively high input threshold (≈20 mV, RMS), and may require input preamplification.

AM detection: This application requires an additional multiplier section to be added to the basic PLL. Only Signetics SE-561 has this particular design feature. However, all the other PLL circuits can be modified for this application by adding an external balanced modulator (such as Motorola's MC-1596) to the basic PLL.

Motor speed control: In most speed-control applications, a tachometer connected to the motor shaft is used as the VCO in the loop and the actual VCO on the monolithic chip is either not used or used to generate a reference frequency (**see Fig. 6**). Thus, a PLL system which can be broken between the low-pass filter and the VCO is needed. The circuits which fit the need are the XR-S200, the XR-215, the SL-650, the HI-2800.

There is no simple definative answer to the question, "What is the best PLL for the job". The answer depends on the many specifics of the application and the overall performance requirements. Therefore, the recommendations given in this section are no more than a rough guide line. □

Manufacturers of monolithic PLLs

Signetics
811 East Arques Ave.
Sunnyvale, CA 94086

National Semiconductor
2900 Semiconductor Dr.
Santa Clara, CA 95051

Raytheon Semiconductor
350 Ellis St.
Mountain View, CA 94040

Harris Semiconductor
Post Office Box 883
Melbourne, FL 32901

Plessey Semiconductor Div.
170 Finn Court
Farmingdale, NY 11735

Exar Integrated Systems
733 North Pastoria Ave.
Sunnyvale, CA 94086

A Stable Second-Generation Phase-Locked Loop

B. Gilbert

Plessey Co., Ltd.

Dorset, England

WITH THE ADVENT of commercial monolithic phase-locked-loop (PLL) circuits[1] interest in a relatively dormant technique has been reawakened, and the PLL has rapidly become an important linear building block, not only for the recovery of signals embedded in noise, but as a simple and accurate means of wideband demodulation and frequency translation. Data modems using frequency-shift-keying (FSK) modulation can take advantage of the excellent demodulation linearity of a PLL to maintain low error rates under adverse line conditions.

However, prevalent PLL circuits are of inadequate stability to meet fully the environmental requirements of modems. The circuit to be described was originally intended to provide the major elements of a complete FSK modem having the necessary stability, which among other things demands a temperature coefficient in the variable frequency oscillator (VFO) of an order of magnitude below that previously obtainable. The same package serves both modulator and demodulator functions and contains much of the remaining circuitry found in typical modems. By careful choice of interfaces a wide range of other applications can be readily met with the one device, only some of which actually use the PLL configuration.

In addition to the usual VFO and phase comparator the chip (Figure 1) contains:

1)—A two-bit *universal* binary interface selecting four frequencies pre-determined (typically) by resistors connected between pins 6, 7, 8 and 9 (within millivolts of ground) and the −V supply. This feature permits generation of two channels of FSK or the selection of four demodulation channels without any additional circuitry. Because the control inputs are at ground the output of a PLL will be zero for signal frequencies equal to the VFO center-frequency.

2)—A signal amplifier which reduces the input level required to load the phase comparator (sufficient to retain 90% of full lock-range) to about 1.5 mW rms (−54 dBm) yet can tolerate severe overload (typically, + 15 dBm) without distortion. The differential input has a wide common-mode range.

3)—A linear multiplier operating on the gain of the phase comparator can be used as an AM modulator, a gating input, to level the signal at a remote point or to alter the characteristics (for example, lock-range or pole location) of a PLL.

4)—An auxiliary amplifier/comparator having low-input bias currents and a CCITT-compatible output, having the specific application of post-demodulation slicing in a modem, but also useful in active filters, as a means to invert the frequency-voltage slope, etc.

The device can operate on supply voltages down to 3 V at basic power-levels below 20-mW, but can be programmed to drive over 200 mW into the loads.

Figure 2 shows a circuit which has these useful properties: first, the differential gain a/x is about beta, so that only a small unbalance in the emitter-currents of a preceding stage fully switch the output; second, the common-mode current-gain I_2/I_1 is almost unity, even for low betas; third, the circuit is well-behaved when overdriven (no transistors saturate and the common-mode gain is unaffected). The bandwidth of the stage is of course limited to the common-emitter cut-off of the lateral PNP transistors (typically 1 MHz).

The circuit, Figure 3, has been designed to have nominally zero temperature- and supply-voltage-sensitivities. Measured values are typically 0 ± 20 ppm/°C and +30 ppm/per cent, respectively. In the state shown, timing-current I flows in Q5, C, Q8, inverted by a current-balance and establishes an identical emitter-current-density in Q2. The voltage at the control input (point *B*) is thus held at zero over a wide range of I. At an optimum value of current point *A* is also zero, and the TC introduced in this area of the circuit is eliminated. At other values of I the TC is well-characterized, and is +30 ppm/°C for $I = 10 \cdot I_{opt}$ and −30 ppm/°C for $I = I_{opt}/10$.

The instability introduced by shift in the regeneration point arising from dV_{be}/dT in Q7 and Q8 is minimized by the current-balance which forces the currents in these transistors to be equal at the switching-point. Q3 and Q6 compensate for the already small TC and absolute error introduced by the base current i in Q8. Exact current-density in Q2 at microamp levels of I is insured by Q1, which also provides a loadable output.

The design relationships are simple, and hold for $10\,\mu A < I < 3\,mA$. Unadjusted frequency is typically within ±1% of nominal, mark-space ratio accurate to within ±0.5%. Timing-current selection is achieved by making Q5 and Q10 with four emitters, and diverting unselected currents by decoder transistors in the logic section (not shown). In this way no additional offsets are introduced in the selection process.

Figure 4 shows a simplified schematic. The input stage hard-limits for inputs of $\pm 26 \ln(\beta + 2)/\beta$ millivolts, where β is that of the lateral PNP transistors. For $\beta = 10$, current-limiting occurs at ±5 mV. Q5 through Q8 form the linear multiplier[2], the gain being proportional to the current into pin 22. The full-scale output current from Q18 or Q20 is well-balanced and is approximately $10 \cdot I_{22}$. Thus, for a 10-kΩ load and $I_{22} = 250\,\mu A$, the full-scale output voltage could reach ±25V. However, assuming typical supply-voltages of ±5V, the output limits for input voltages of ±1 mV.

Acknowledgment

The author is pleased to acknowledge the valuable help given by J. Sandell and J. Stockdale during the development of this circuit.

---

1 Grebene, A.B., "The Monolithic Phase-Locked-Loop - A Versatile Building Block", *IEEE Spectrum*, p. 38; March, 1971.

2 Gilbert, B., "A New Wideband Amplifier Principle", *Journal of Solid State Circuits*, p. 353; Dec., 1968.

Reprinted from *1972 IEEE Internat. Solid-State Circuits Conf. Dig. Tech. Papers,* Feb. 16-18, 1972, pp. 78-79.

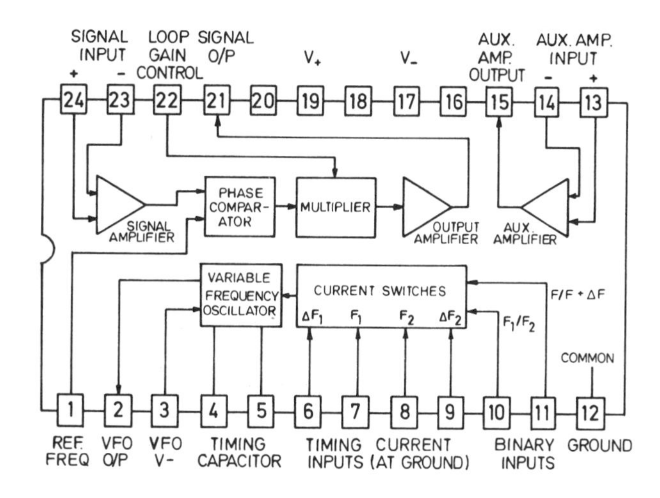

FIGURE 1—Basic elements of the circuit. Pad locations are arranged to facilitate packaging of simplified 10- and 16-pin versions.

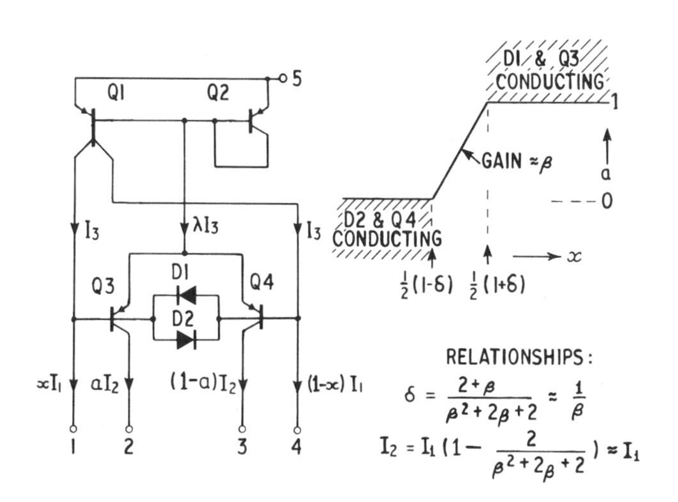

FIGURE 2—A useful *current-balance* circuit. Q1 is a dual-collector lateral PNP transistor. D1 and D2 are integral with Q3 and Q4.

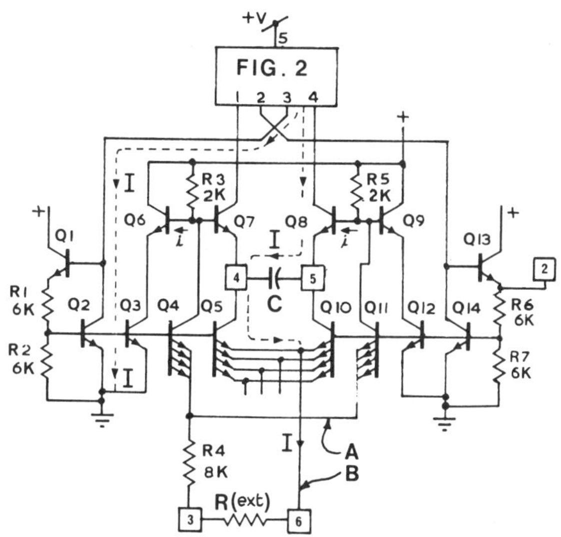

FIGURE 3—Stable VFO. Design equation is $f = 1/CR$. If R is returned to a separate supply V_1, $f = V_1/V_3CR$ or $f = I/CV_3$. Points A and B are at ground potential.

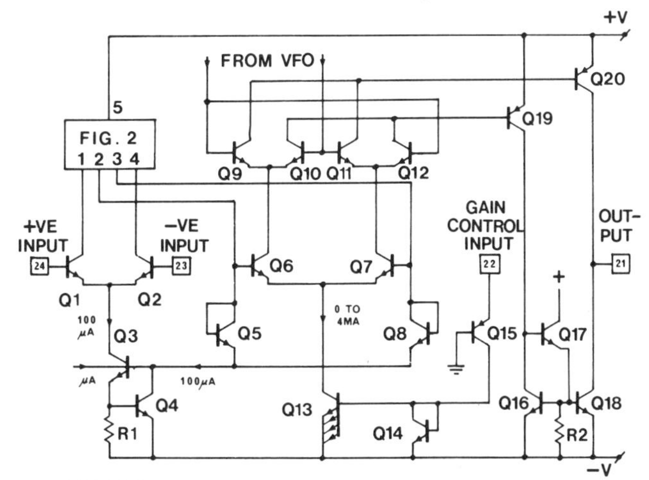

FIGURE 4—Simplified circuit of signal-path. Q3 and Q4 form a self-sustaining supply-insensitive current-source. Q9 through Q12 comprise the phase-comparator.

Monolithic waveform generation

It is now possible to duplicate the capabilities of a complex waveform or function generator in a single monolithic chip

Alan B. Grebene Exar Integrated Systems, Inc.

Waveform or function generators find a wide variety of applications in communications and telemetry equipment, as well as for testing and calibration in the laboratory. Almost all of the waveform generators available at present use discrete (nonintegrated) circuit design techniques for waveform generation and shaping. The basic generating, modulating, and shaping methods used in most of these instruments are either directly suitable for, or adaptable to, monolithic integration.

Monolithic integrated circuits offer some inherent advantages to the circuit designer, such as the availability of a large number of active devices and close matching and thermal tracking of component values. By making efficient use of the capabilities of integrated components and the batch-processing advantages of monolithic circuits, it is now possible to design integrated waveform generator circuits that can provide a performance comparable to that of complex discrete generators, at a very small fraction of the cost. Some of these design techniques and capabilities will be described, including design and performance characteristics of a monolithic waveform generator with sinusoidal, triangular, ramp, sawtooth, and pulse output waveforms capable of amplitude, phase, and frequency modulation.

Basic monolithic waveforms

A basic waveform generator consists of three fundamental sections: (1) a controlled-oscillator section, which generates the periodic waveform; (2) wave-shaping circuitry, which determines the output waveform; and (3) a modulator section, which is used to modulate the amplitude, phase, or frequency of the output. Monolithic design techniques for each of these sections will be reviewed briefly.

The oscillator section. The oscillator portion of a waveform generator can be of either the harmonic or relaxation type. Harmonic oscillators generate nearly sinusoidal waveforms, but often require the use of inductors or large numbers of precision components and so are not readily suitable for monolithic integration. Typical

FIGURE 1. Simple oscillator circuit for generating square and linear ramp waveforms.

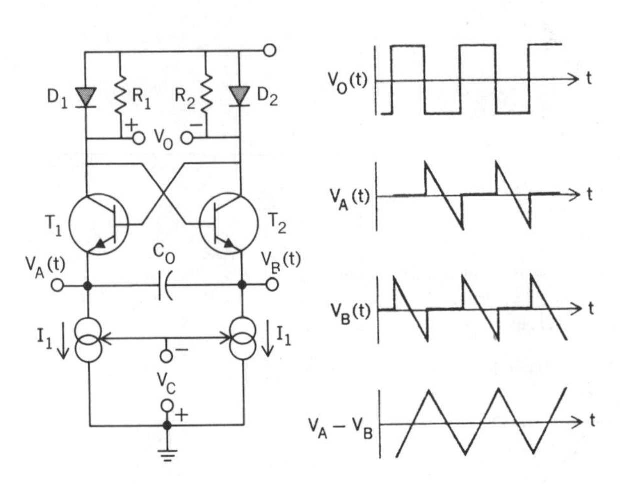

FIGURE 2. Ramp-to-triangular-wave conversion by means of a differential-gain stage.

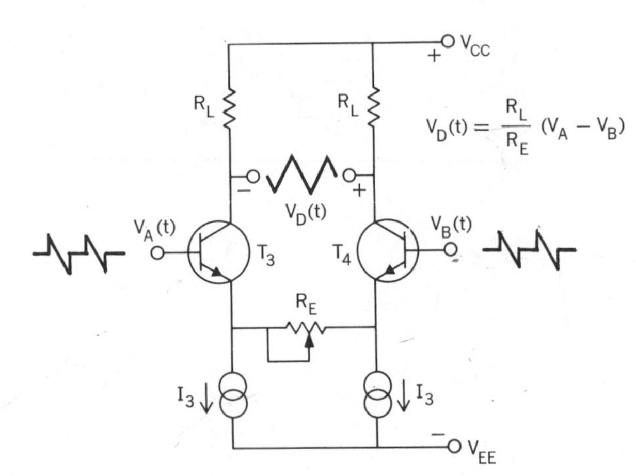

Reprinted from *IEEE SPECTRUM*, pp. 34-40, April 1972.

well-known examples of these types of circuits are the Hartley, the Colpitts, the Wien-bridge, and the twin-T oscillators.

Relaxation or switching oscillators are positive-feedback circuits that operate as self-triggered flip-flops. Typical examples are multivibrators and blocking oscillators. Among these, the basic *RC*-coupled multivibrator circuit is best suited for integration since it requires a minimum number of energy-storage elements (often only one capacitor) and its operation is not critically dependent upon device parameters. A multivibrator-type oscillator can provide three general classes of waveforms: (1) exponential ramp; (2) linear ramp; (3) square wave or pulse. The first two types of waveform can be obtained by charging or discharging a timing capacitor through either a resistor (for an exponential ramp) or a current source (for a linear ramp). The square-wave or pulse output waveform can be obtained from the first two by a trigger or level-detector circuit. For wave-shaping purposes, the linear ramp and square wave are the most convenient waveforms to work with since their harmonic content is predictable and can be easily controlled.

Linear ramp generators require the use of accurate constant-current stages. Since accurate and well-matched current sources can easily be fabricated in monolithic form, linear ramp generators are well suited to integration. These oscillator circuits have a generalized frequency expression of the form:

$$f_0 = \frac{KI_1}{C_0}$$

where K is a constant of proportionality in volts^{-1}, C_0 is the timing capacitance in farads, and I_1 is the charging current in amperes. In an oscillator of this type the frequency range is selected by the choice of C_0, and the frequency control can be achieved by varying the current I_1. Figure 1 is a simplified diagram of a relaxation oscillator circuit demonstrating this principle.[1] The circuit is derived from the emitter-coupled multivibrator configuration and can generate a square-wave, as well as a linear ramp, output. Its operation can be briefly explained as follows: At any given time either T_1 and D_1 or T_2 and D_2 are conducting, such that the capacitor C_0 is alternately charged and discharged by constant current I_1. The output across D_1 and D_2 corresponds to a symmetrical square wave, with a peak-to-peak amplitude of $2V_{BE}$, where V_{BE} is the transistor base–emitter voltage drop. The output V_A is constant when T_1 is on, and becomes a linear ramp with a slope equal to $(-I_1/C_0)$ when T_1 is off. The output $V_B(t)$ is the same as $V_A(t)$, except for a half-cycle delay. Both of these linear ramp waveforms have peak-to-peak amplitudes of $2V_{BE}$. The frequency of oscillation, f_0, can be expressed as

$$f_0 = \frac{I_1}{4V_{BE}C_0}$$

and can be controlled by varying the charging current I_1 by means of a control voltage V_C.

The ramp output voltages $V_A(t)$ and $V_B(t)$ can be subtracted from each other, to give a linear triangular waveform. This can be done by using a simple differential amplifier stage as shown in Fig. 2, where the output voltage $V_D(t)$ is a triangular waveform. The symmetry of the triangular and square-wave outputs can be offset by replacing one of the current sources in Fig. 1 by a current source I_2, where $I_2 \neq I_1$. Thus, the duty cycle of the output waveforms can be expressed as

$$\text{Duty cycle} = \left(\frac{50I_1}{I_2}\right) \text{ percent}$$

In this manner, the triangular and square-wave outputs can be converted to sawtooth or pulse waveforms.

Wave-shaping networks. After the generation of basic waveforms such as the triangle and the square wave, the next step is to convert one of these into a low-distortion sine wave. For this application, a triangular waveform is preferred over a square wave since its harmonic content is significantly lower. This initial harmonic content is further minimized by using a symmetrical triangular waveform that has a negligible even-harmonic content. This waveform can be converted into a sine wave by rounding off its peaks with the aid of a diode–resistor network as shown in Fig. 3, which also shows two practical diode–resistor circuits that can be used for this application. The distributed diode–resistor circuit of Fig. 3C can be integrated by making use of the distributed nature of the p-n junctions along a diffused resistor structure.

The wave-shaping networks of Fig. 3 can convert a symmetrical triangular-wave input into a sinusoidal output with less than 0.5 percent harmonic distortion if eight or more diodes are used. However, these wave-shaping circuits also have the disadvantage that the input-signal level, diode characteristics, and resistor ratios need to be very accurately controlled. If a higher harmonic content (of the order of 2 to 3 percent) is acceptable, the wave-shaping circuitry can be greatly simplified, and the triangle-to-sine conversion can be achieved with only two diode–resistor combinations.

As an alternate approach, the gradual cutoff of an

FIGURE 3. Conversion of a symmetrical triangular waveform into a sinusoid by the use of a diode–resistor clipping network. A—Input and output waveforms. B—Lumped diode–resistor chain. C—Distributed diode–resistor chain. (Sections shown for positive half-cycle clipping.)

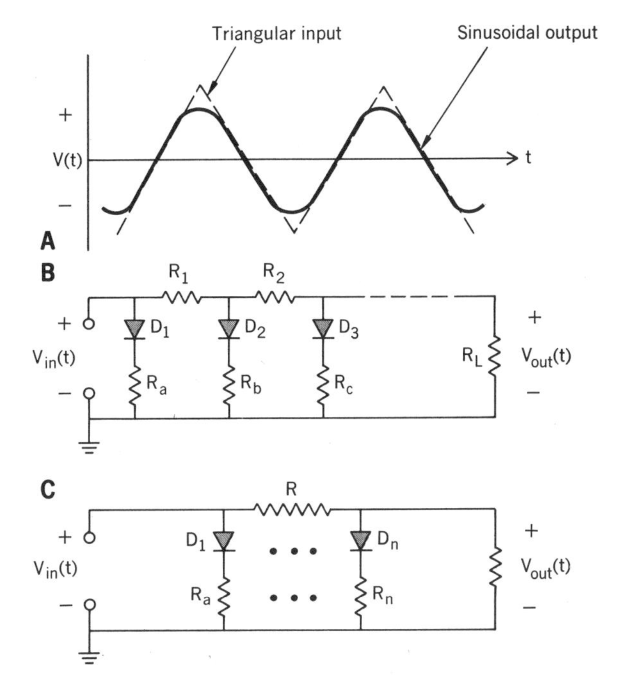

overdriven transistor stage can also be used for wave-shaping purposes. For example, in the differential gain stage of Fig. 2, if the differential input signal drive level $(V_A - V_B)$ is increased or if R_E is decreased, an overdrive condition can exist wherein the input transistors T_3 and T_4 are driven into cutoff. Under this condition, the voltage transfer characteristic of the circuit will appear as in Fig. 4. The gradual transition between the active region and cutoff is exponential in nature and can be used to round off the sharp peaks of the input signal. In the case of the circuit shown in Fig. 2, where the input drive level is constant, this gradual cutoff can be brought about by reducing the value of the emitter feedback resistor R_E. In this manner, the triangular output of the differential stage in Fig. 2 can be converted to a sine wave with a total harmonic distortion (THD) of 2.5 percent or less. This particular wave-shaping technique is used in the circuit design example described later in this article.

Modulator circuits. For waveform generator applications, balanced modulators are preferred over conventional mixer-type modulators.[2] The reason is that the balanced modulator circuits can offer a high degree of carrier suppression and thus can be used for suppressed carrier modulation as well as for conventional double-sideband AM generation. A number of balanced modulator configurations for monolithic circuit applications have been described in the literature.[3,4] The simplified circuit diagram is shown in Fig. 5. In addition to amplitude modulation, this circuit also functions as a linear four-quadrant multiplier. The output voltage $V_{out}(t)$ is a linear product of the signal input $V_S(t)$ and the modulation input $V_M(t)$:

$$V_{out}(t) = \frac{R_L}{R_E R_X I_E} V_S(t) V_M(t)$$

Typical normalized gain and phase characteristics of the balanced modulator circuit are shown in Fig. 6, as a function of the bias voltage V_M applied across the modu-

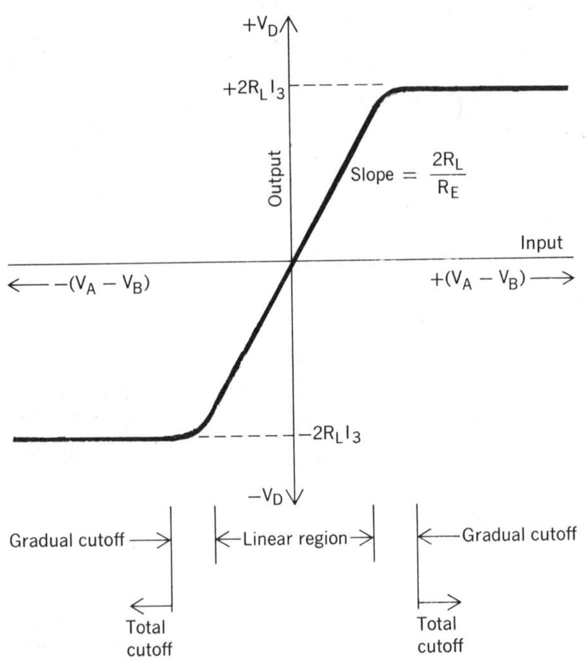

FIGURE 4. Voltage-transfer characteristic for the differential amplifier of Fig. 2 for large-signal operation.

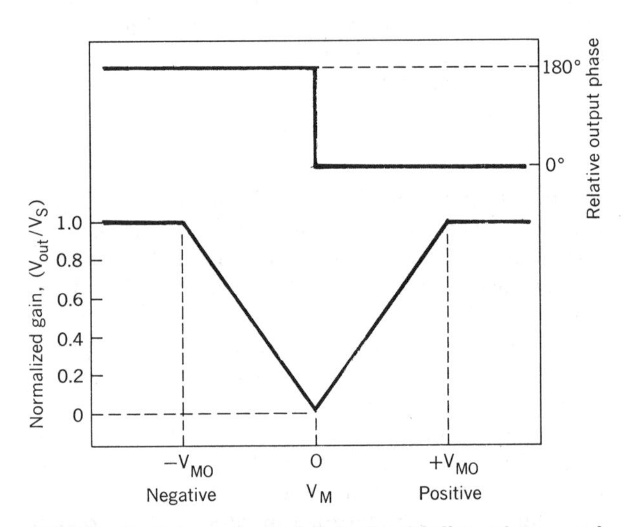

FIGURE 6. Balanced modulator/multiplier phase and amplitude transfer characteristics.

FIGURE 5. Balanced modulator/multiplier circuit suitable for monolithic integration.

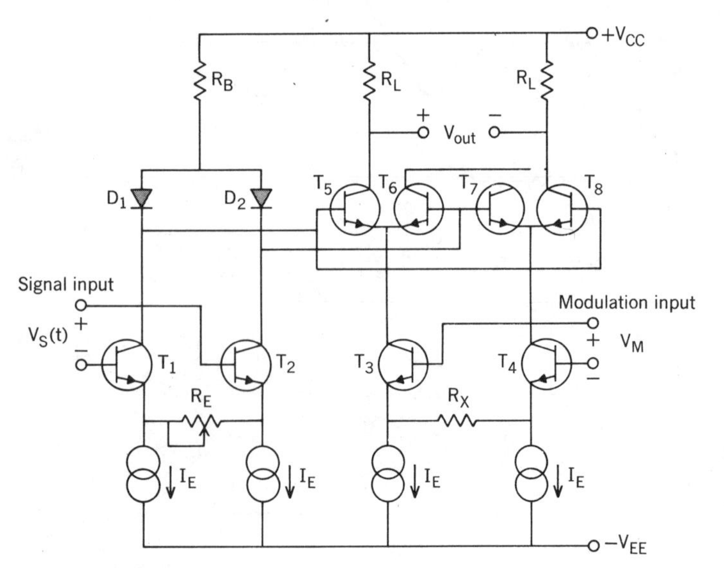

FIGURE 7. Functional block diagram of XR-205 monolithic waveform generator.

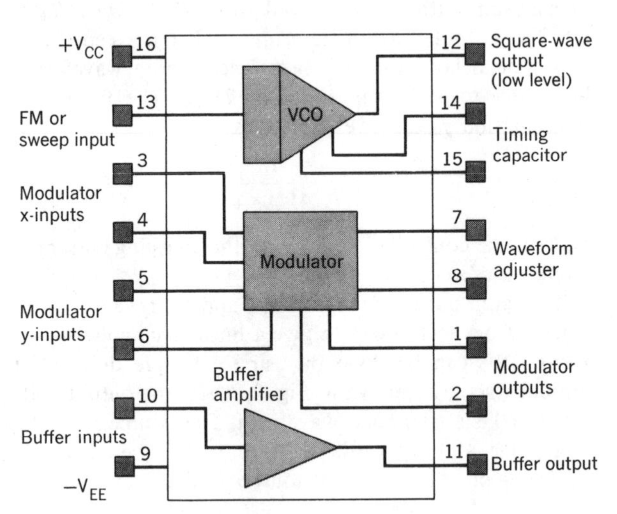

Typical applications for a monolithic waveform generator

Waveform generation
- Sinusoidal — Sawtooth
- Triangular — Ramp
- Square — Pulse

AM generation
- Double sideband
- Suppressed carrier

Sweep generation
Tone-burst generation
Simultaneous AM/FM
FSK and PSK signal generation
On/off-keyed oscillation

lation input terminals. The output amplitude is zero for $V_M = 0$ and increases linearly with V_M for $0 < V_M < V_{MO}$ or $-V_{MO} < V_M < 0$. For modulation inputs greater than V_{MO}, either T_3 or T_4 of Fig. 5 is cut off, and the relative output amplitude remains unaffected by V_M. For $-V_{MO} < V_M < V_{MO}$, the slope of the transfer characteristic is inversely proportional to R_X. Note that the output phase is inverted when the polarity of V_M is reversed. This property of the circuit can be used for phase-shift keyed (PSK) modulation.

The balanced multiplier/modulator circuit of Fig. 5 also has a unique advantage for wave-shaping applications. The signal input stage formed by T_1 and T_2 can function as the differential gain stage of Fig. 2, to convert two out-of-phase ramp input signals into a symmetrical triangular wave. This wave can then be converted into a low-distortion sine wave using the "gradual cutoff" of T_1 and T_2, as described in the section on wave-shaping networks, by adjusting the emitter degeneration resistor R_E. Thus, by means of a balanced modulator circuit of the type shown in Fig. 5, wave-shaping and modulation functions can be combined in a single circuit block.

Circuit design

Each of the three basic blocks can be designed in integrated form. An entire waveform generator system, therefore, is itself well suited to monolithic integration. Figure 7 is a functional block diagram of such a monolithic generator, in terms of the circuit terminals available in a 16-pin standard IC package. For added versatility, this waveform generator system, designated XR-205, also contains a buffer amplifier that, though uncommitted to the rest of the circuit, can be connected to any one of the outputs to boost the output current drive capability. The monolithic waveform generator is designed to be suitable for one, or a combination of several, of the applications listed in the box on this page.

A simplified schematic circuit diagram of the entire monolithic generator system is shown in Fig. 8. The functional blocks and the external terminals of the circuit are also identified. The oscillator section, shown on the left side of the diagram, was designed as an emitter-coupled multivibrator, using the basic circuit configuration of Fig. 1. The frequency is set by an external timing capacitor connected across the emitters of T_1 and T_2 (terminals 14 and 15). Transistors T_1 and T_2 form the

FIGURE 8. Simplified circuit diagram of XR-205 monolithic waveform generator.

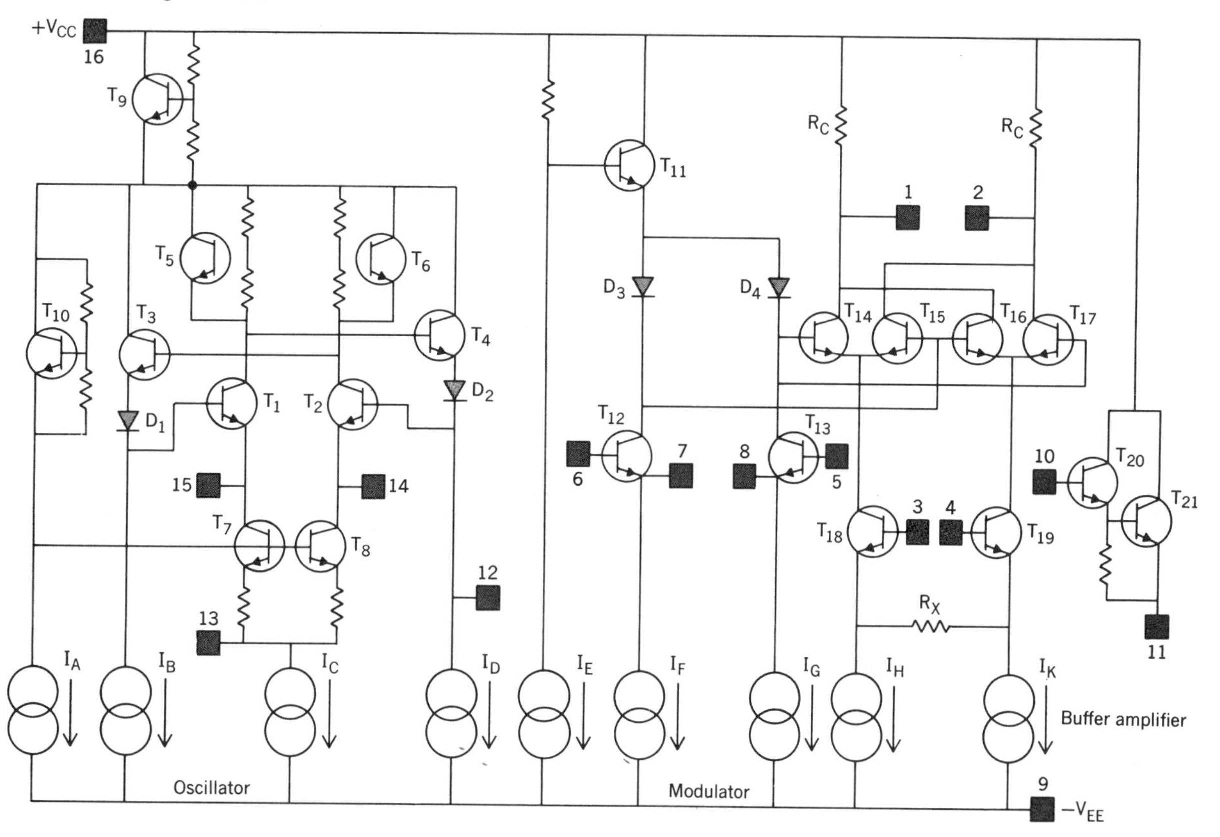

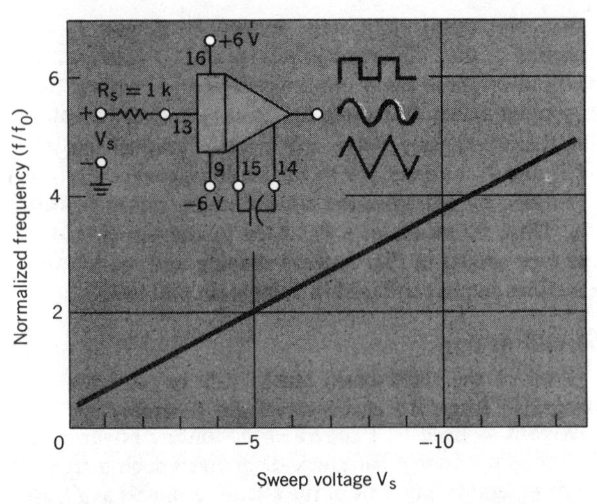

FIGURE 9. Waveform generator sweep characteristics.

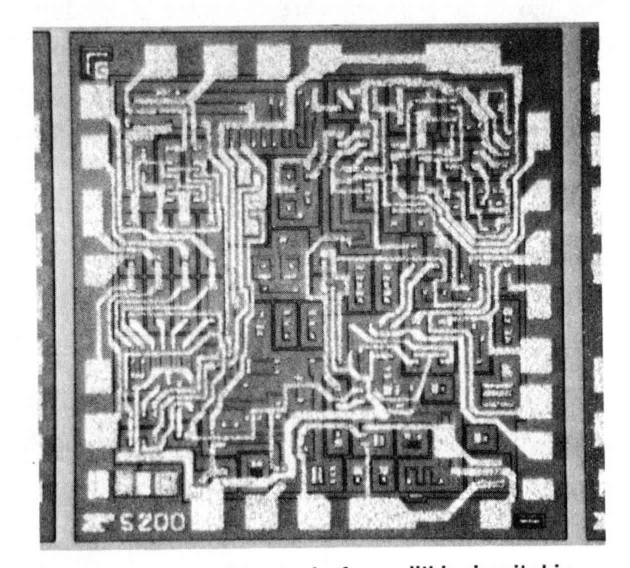

FIGURE 10. Photomicrograph of monolithic circuit chip.

FIGURE 11. Generalized circuit connection diagram for monolithic waveform generator.

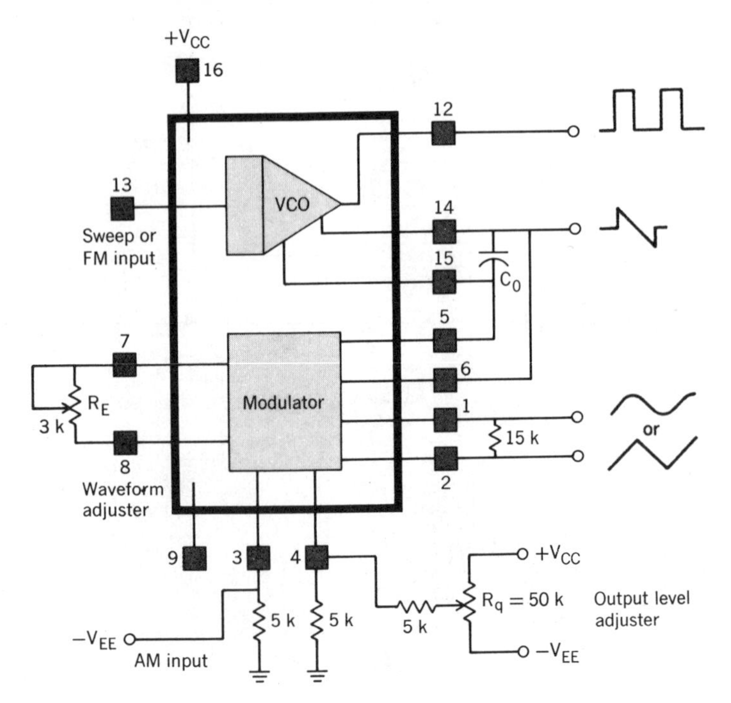

gain stage for the oscillator and are interconnected to emitter followers T_3 and T_4. Transistors T_7 and T_8 serve as the current sources shown in Fig. 1, and divide the total current I_C equally between them. The oscillator frequency can be voltage-controlled by applying a dc bias to terminal 13 (see Fig. 9).

The modulator section was designed using the basic multiplier/modulator configuration of Fig. 5. With reference to Fig. 8, the internally generated ramp waveforms from the emitters of T_1 and T_2 can be directly applied to the bases of T_{12} and T_{13} and can be converted to a symmetrical triangular or sinusoidal waveform by the choice of the external resistor (R_E in Fig. 5) connected across the emitters of these transistors. The modulation input (V_M in Fig. 5) is applied to the bases of T_{18} and T_{19}, which are designated as x-inputs in the block diagram of Fig. 7. The output of the modulator section is obtained from the collectors of the cross-coupled transistors T_{14} through T_{16} (terminals 1 and 2).

The buffer amplifier section is comprised of the Darlington-connected transistors T_{20} and T_{21}. In normal operation of the circuit, the input (pin 10) of this buffer amplifier can be connected to terminals 1, 2, 12, or 14,

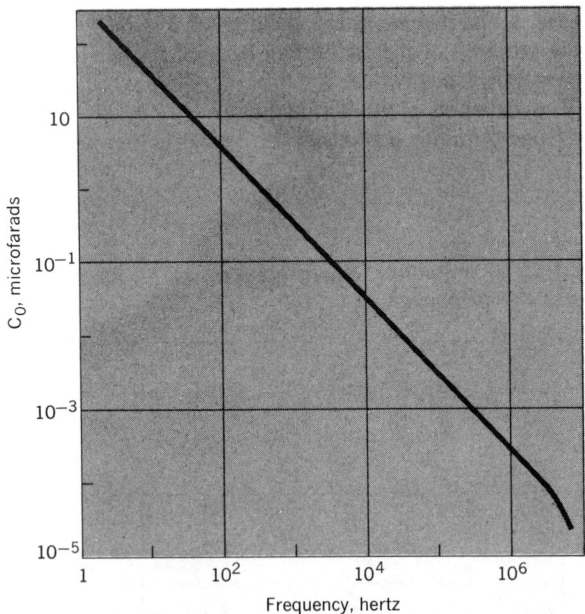

FIGURE 12. Timing capacitance vs. oscillator frequency.

FIGURE 13. Distortion vs. frequency for sinusoidal output.

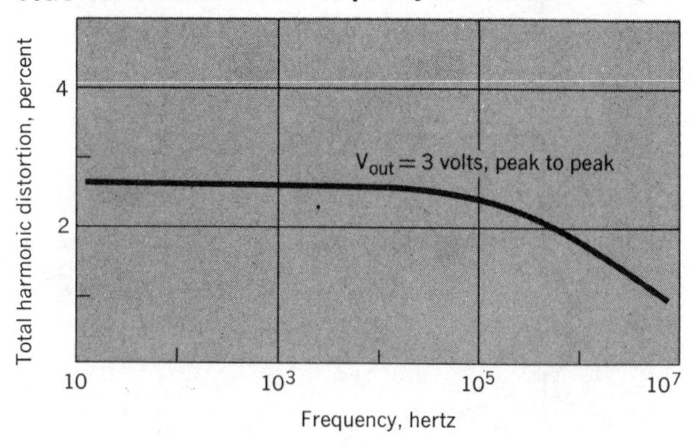

XR-205 performance characteristics

Supply voltage:
- Single supply, 8 to 26 volts
- Double supplies, ±5 to ±13 volts

Supply current: 10 mA
Operating temperature range: −55°C to +125°C
Frequency stability:
- Temperature, 300 ppm/°C
- Power supply, 0.2 percent per volt

Frequency range:
- Sine wave and square wave, 0.1 Hz to 5 MHz
- Triangle and ramp, 0.1 Hz to 500 kHz

Frequency sweep range: 10 to 1
Sweep (FM) nonlinearity: < 0.5 percent for 10 percent deviation
Output swing:
- Single-ended, 3 volts peak to peak
- Differential, 6 volts peak to peak

Output impedance:
- With buffer stage, 50 ohms
- Without buffer stage, 4000 ohms

Output amplitude control range: 60 dB
Sinusoidal output distortion: < 2.5 percent (THD)
Triangular, ramp, and sawtooth outputs:
- Nonlinearity, 1 percent for $f_0 < 200$ kHz

Square-wave output:
- Amplitude, 3 volts
- Duty cycle, 50 percent (±3 percent), adjustable from 20 to 80 percent

Modulation capability (all waveforms):
- AM, FM, FSK, PSK, tone burst

FIGURE 14. Basic periodic waveforms from monolithic waveform generator. A—Sinusoidal. B—Triangular. C—Linear ramp. D—Square-wave. E—Sawtooth. F—Pulse.

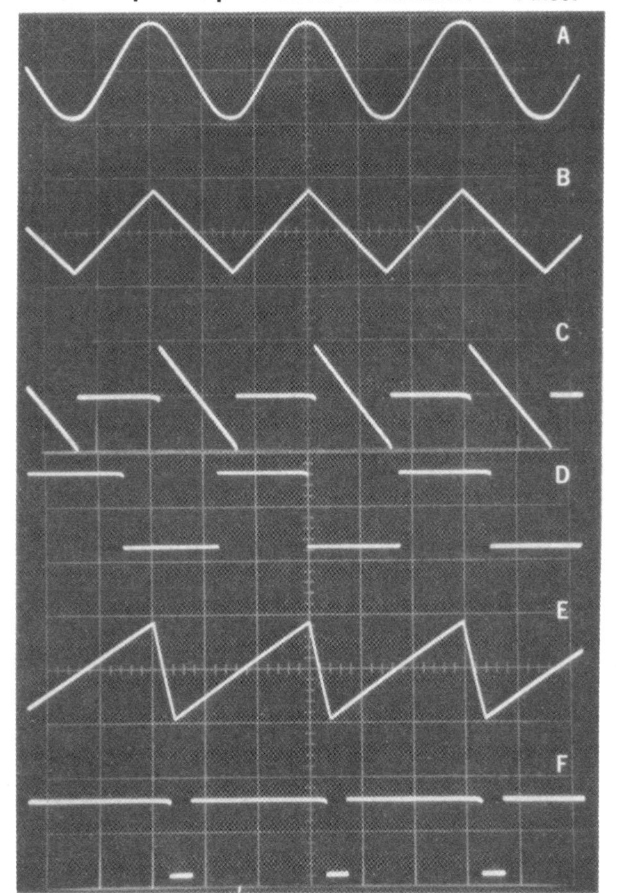

depending on the desired output waveform; and the output of the buffer amplifier is connected to ground or to $-V_{EE}$ through load resistor R_L.

The circuit of Fig. 8 was designed to take full advantage of the close matching and tracking properties of integrated components. For this reason, a differential rather than single-ended circuit configuration is used for each of the circuit blocks. This approach also provides a high common-mode range for the circuit and facilitates its operation over a range of supply voltages from ±5 volts to ±13 volts.

The waveform generator circuit was integrated on a 78-mil-square (2-mm-square) monolithic chip of silicon, shown in the photomicrograph of Fig. 10. A generalized circuit connection diagram for the XR-205 is shown in Fig. 11. Various output waveforms directly available from the circuit are also identified in the figure, but the buffer amplifier section is not explicitly shown. The capacitor C_0 across terminals 14 and 15 sets the free running frequency of the oscillator, as shown in Fig. 12. The output across pins 1 and 2 is either triangular or sinusoidal, depending on the setting of the resistor R_E connected across pins 7 and 8. The potentiometer R_q controls the dc bias applied to the modulation input. It

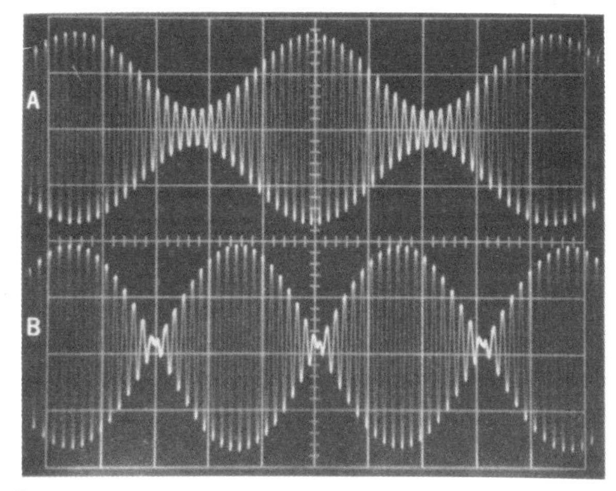

FIGURE 15. Amplitude-modulated waveforms. A—Double-sideband. B—Suppressed-carrier.

FIGURE 16. A—Linear frequency modulation. B—Square-wave simultaneous amplitude/frequency modulation.

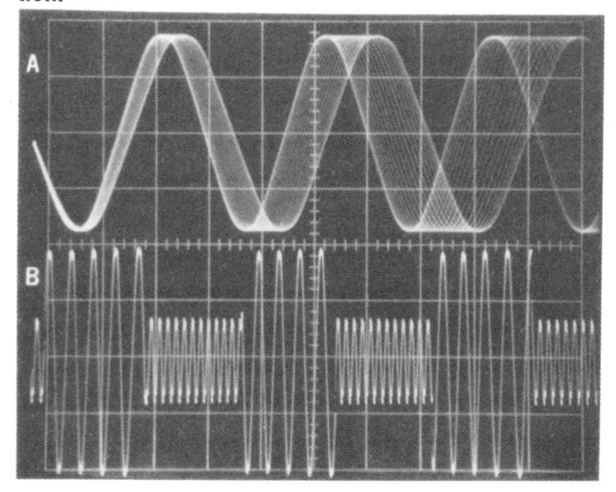

can be used to set the output signal amplitude in accordance with the modulator section transfer characteristics shown in Fig. 6. The output frequency can be voltage-controlled or modulated by applying a control signal to the sweep input terminal (pin 13), as shown in Fig. 9.

The harmonic content of the sinusoidal output has been found to be 2.5 percent or less and relatively independent of the modulation input and frequency of operation. Figure 13 shows the harmonic content of the output waveform as a function of frequency. Typical performance characteristics of the monolithic waveform generator are listed in the box on page 39.

Figures 14 through 18 show some of the output waveforms available from the monolithic generator circuit. The waveforms in Fig. 14 correspond to the six basic periodic waveshapes generated by the XR-205 circuit. The first four waveforms (sinusoidal, triangular, ramp, and square) are directly available from the circuit connection of Fig. 11. The asymmetrical waveforms, such as the sawtooth and pulse outputs, can be derived from the triangular and square-wave outputs by offsetting the oscillator duty cycle. In the circuit connection diagram of Fig. 11 this can be achieved by connecting a 1-kilohm resistor between terminals 13 and 14.

Figure 15 shows the amplitude-modulated output waveforms for double-sideband and suppressed-carrier modes of operation. In suppressed-carrier AM generation, the circuit offers a carrier suppression of approximately 50 dB for frequencies to 5 MHz. The linear frequency-modulated waveform of Fig. 16A can be obtained by applying a modulation input to the sweep terminal. The nonlinearity of the FM characteristic is less than ±0.5 percent for ±10 percent FM deviation. Simultaneous AM/FM capabilities of the circuit also permit generation of the complex waveforms of Figs. 16B and 17A by application of the same modulating signal to both the AM and FM terminals (pins 3 and 13) shown in Fig. 11.

The oscillator can be on/off-keyed for tone-burst generation by applying a positive pulse of greater than 3 volts to the sweep terminal. A typical tone-burst output waveform obtained in this manner is displayed in Fig. 17B. The phase-shift characteristics of the doubly balanced modulator section (see Fig. 6) of the generator can be used for PSK modulation of the output. Applying a pulse input between terminals 3 and 4 of Fig. 11 causes a 180-degree phase reversal of the output waveform, as shown in Fig. 18.

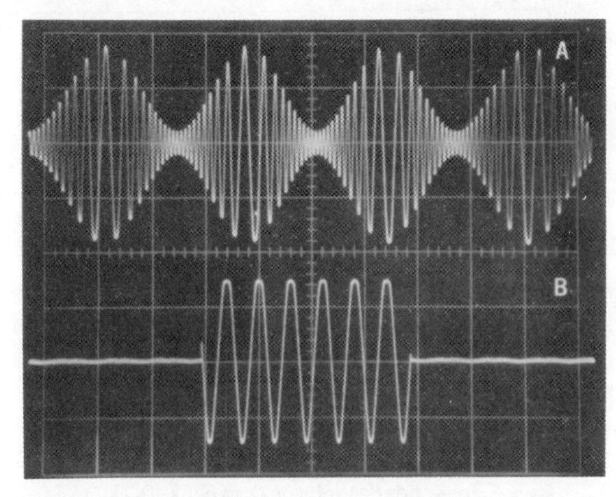

FIGURE 17. A—Sinusoidal amplitude/frequency modulation. B—Sinusoidal tone-burst generation.

FIGURE 18. Sinusoidal phase-shift keyed output. A—PSK output signal. B—Keying pulse.

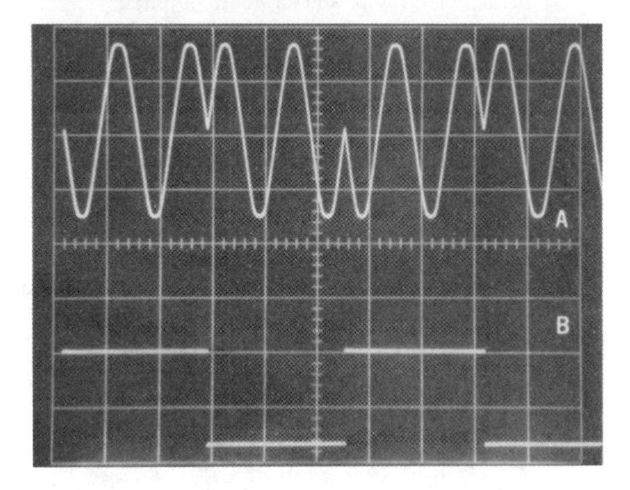

REFERENCES

1. Grebene, A. B., "The monolithic phase-locked loop—a versatile building block," *IEEE Spectrum*, vol. 8, pp. 38–49, Mar. 1971.
2. Meyer, R. G., "Integrated circuit mixers," *1970 NEREM Record*, vol. 12, pp. 62–63.
3. Grebene, A. B., *Analog Integrated Circuit Design*. New York: Van Nostrand Reinhold, 1972, chap. 7.
4. Gilbert, B., "A precise four-quadrant multiplier with subnanosecond response," *IEEE J. Solid-State Circuits*, vol. SC-3, pp. 365–373, Dec. 1968.

A Monolithic Analog Compandor

CRAIG C. TODD

Abstract—**An IC analog dynamic range compandor is described. The approach taken to implement the compressor and expandor functions is discussed, and the active full wave rectifier and variable gain cell are described in detail. Performance easily exceeds telecommunications requirements and is suitable for use in high quality audio systems. The full wave rectifier has excellent accuracy over a 60-dB dynamic range. The variable gain element offers low distortion (<0.1 percent) attenuation with high signal to noise ratio (90 dB). Ion-implanted resistors have been used for lower temperature sensitivity.**

I. Introduction

COMPANDING is a widely used technique for gaining a signal-to-noise ratio improvement in systems where an analog signal is passed through a noisy transmission medium. This paper will describe a monolithic analog compandor which was designed to satisfy the requirements of the telephone network, but which is also suitable for use in communications systems and many high quality audio systems.

The technique of companding is illustrated [1], [2] in Fig. 1, where a compandor is shown improving the signal-to-noise ratio of a signal sent through a limited dynamic range channel. The compressor is a variable gain amplifier which provides increasing gain as the input signal drops in amplitude. In the example shown in Fig. 1, an input signal range of +20 dB to -80 dB is shown undergoing a 2:1 compression, where every 2-dB change in input amplitude is compressed to a 1-dB change in output amplitude. The original 100 dB of input dynamic range is thus compressed down to a 50-dB (+10 dB to -40 dB) range for transmission through the channel which has a limited dynamic range of 55 dB (+10 dB to -45 dB). This channel could consist of a radio-link, a telephone carrier system, a tape recorder, a bucket brigade delay line, or even a uniformly coded PCM system.

At the receiving end of the system, a 1:2 expandor, which is an exact complement of the compressor, restores the original signal level range of +20 to -80 dB. For signal levels of less than 0 dB, the expandor will be operating with a gain less than unity. For an input signal level of -80 dB, the expandor will operate with a loss of 40 dB, which will lower the channel noise by that amount yielding a 40-dB improvement in signal-to-noise ratio. For larger signals the S/N ratio improvement will be less, and for signals above the crossover level of 0 dB there will be an actual impairment of S/N ratio. The actual improvement in the S/N ratio of a companded system is equal to the instantaneous value of the expandor loss.

II. Design Objectives

The compandors presently used in telephone trunk carrier systems [3] are costly discrete circuits. Rectification is performed by stepping up the signal voltage with a transformer and then rectifying with simple diodes. Gain control is performed by varying the dynamic impedance of a matched pair of "variolosser" diodes, which requires another transformer and is temperature sensitive due to the dependence of diode impedance on temperature. A complete compandor thus requires four transformers (in addition to the input and output transformers) two matched diode variolossers, more than a dozen diodes and transistors, and numerous resistors and capacitors.

The goals in producing a monolithic compandor were to make a single chip compandor superior in performance to present telephone trunk compandors, use few external components, and be inexpensive enough to be used in subscriber carrier systems (where there are two compandors per telephone and low cost is essential). Another design goal was to make the circuit as general purpose and flexible as possible to allow for widespread use in other applications.

III. Circuit Architecture

Fig. 2 shows the block diagram of one half of the IC: there are two identical channels on the chip to allow both a compressor and an expandor to be made using just one device. The full wave averaging rectifier measures the ac amplitude of a signal and develops a control current which sets the gain of the variable gain (ΔG) cell. For good performance, the rectifier must be accurate over a wide dynamic range. The ΔG cell accepts a signal input and produces an output current with a gain controlled by the rectifier. Ideally, this circuit should have a perfectly linear gain versus control current characteristic, be insensitive to temperature, and produce no distortion. A simple variable transconductance multiplier fails the temperature and distortion criteria, so a more elaborate linearized transconductance multiplier was designed. This circuit has the added advantage in that its residual distortion may be trimmed out, allowing the circuit to be used for high quality audio signal processing.

To drive the output, a simple 741 equivalent operational amplifier is provided, along with most of the resistors required to set up gain and biasing. An internal band gap reference supplies a stable 1.8 V reference for biasing the internal summing nodes, and for biasing current sources inside the ΔG cell.

Fig. 3 shows the circuit hookup required to realize an expandor. The input signal is applied to both the rectifier and ΔG cell inputs. Gain is thus proportional to the input signal amplitude. When the input signal drops 6 dB in amplitude, the rectifier gain control current will drop a factor or two, so the gain of the ΔG cell will drop 6 dB. The output signal will thus drop 12 dB in amplitude which gives us the desired 1:2 expansion. The gain equation shows the gain to be set entirely by resistor ratios, and the internal band gap reference voltage,

Manuscript received June 7, 1976; revised August 2, 1976.
The author is with Signetics Corporation, Sunnyvale, CA.

Reprinted from *IEEE J. Solid-State Circuits*, vol. SC-11, pp. 754-762, Dec. 1976.

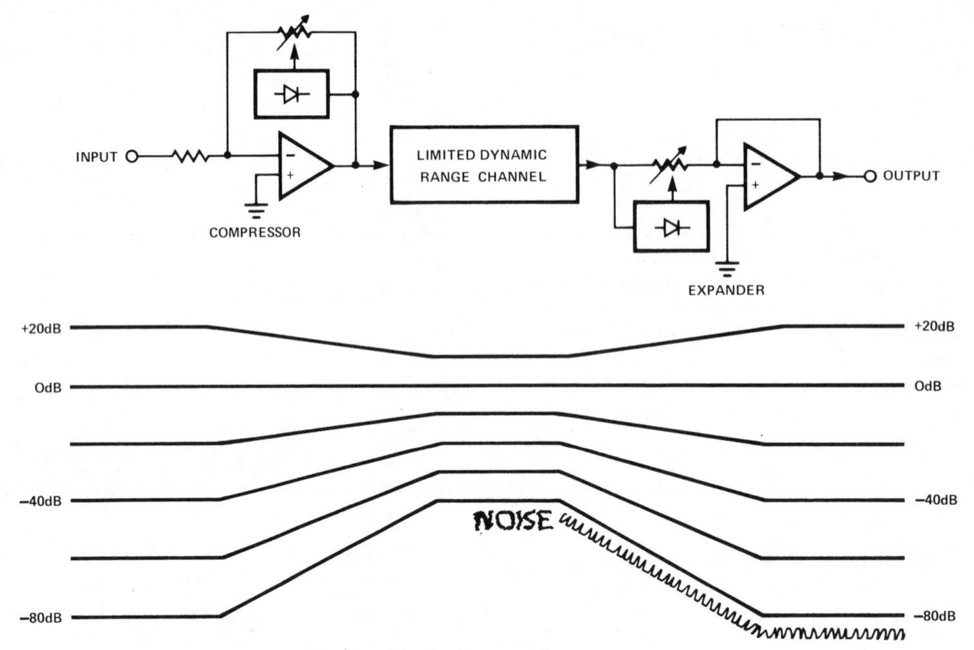

Fig. 1. Companding.

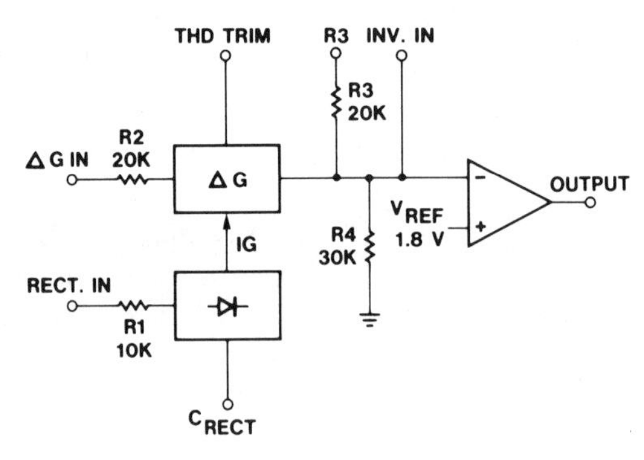

Fig. 2. Chip block diagram (1 of 2 channels).

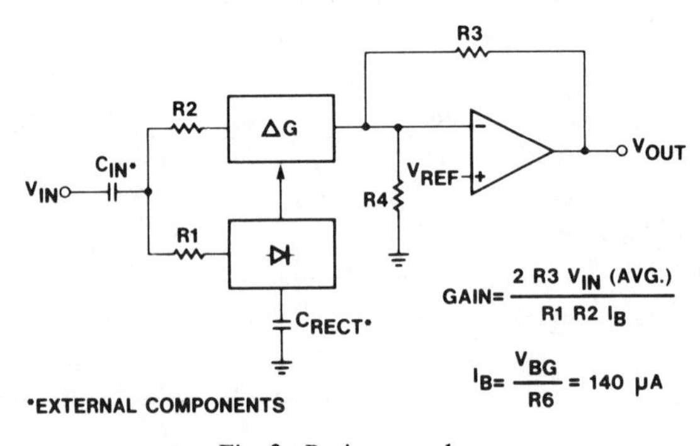

Fig. 3. Basic expandor.

V_{BG}, which along with resistor $R6$, generates the reference current I_B.

The compressor, shown in Fig. 4, is essentially the expandor placed in the feedback loop of the op amp. Gain is inversely proportional to the output level, which yields the desired 2 : 1 compression. The ΔG cell supplies only ac feedback around the op amp in this configuration, so a separate dc feedback loop is required consisting of the *R-C-R* network. The true complementary nature of the compressor and expandor yields a nearly ideal complementary symmetry of the compression-expansion characteristics, both statically and dynamically.

IV. Rectifier

The accuracy of the rectifier determines the input-output transfer curve accuracy of the compressor and expandor. A full wave rectifier is desirable so that less ripple will appear on

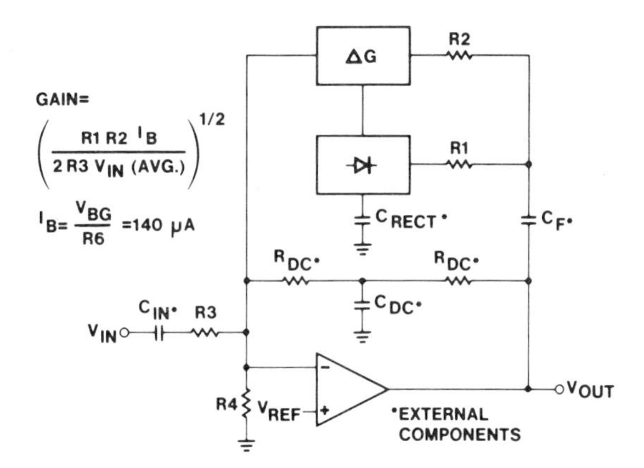

Fig. 4. Basic compressor.

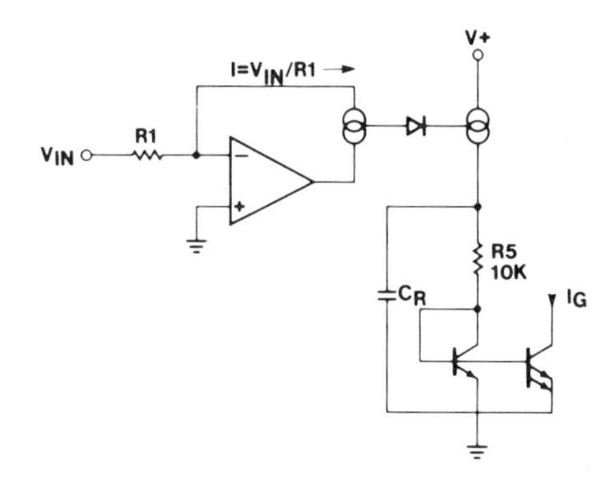

Fig. 5. Rectifier concept.

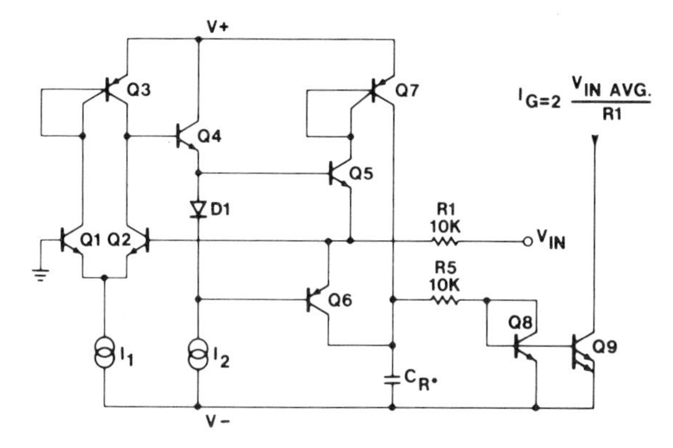

Fig. 6. Simplified rectifier schematic.

the filter capacitor, and in turn in the gain control current. The basic concept behind the rectifier is illustrated in Fig. 5. An input voltage of V_{in} will cause a current of V_{in}/R_1 to flow in the feedback loop of the op amp. Mirroring this current into a unipolar current produces an ideal rectifier. The unipolar current is averaged by R_5, C_R and then mirrored again to form the gain control current I_G.

The circuit implementation of this concept is shown in Fig. 6. The op amp is a simple one-stage circuit biased so that only one output device ($Q5$ or $Q6$) is on at a time. The op amp output current is nearly equal to the collector currents of the output transistors, so these currents are mirrored to form the rectified output current. Since the emitter current is the true op amp output current, the rectifier output current will be too small by a factor of α. The error will be small for negative signal swings since the n-p-n $Q5$ will be conducting and the n-p-n α is greater than 0.99 which gives less than 1-percent error. The major source of error is due to the lateral p-n-p $Q6$ which conducts during positive signal swings with a typical α of 0.97. The drop in p-n-p α at high and low currents will cause this error to increase for both large and small signals. The average error, considering both positive and negative signal swings, is about −2 percent or −0.2 dB, with a slight variation over signal level.

The dynamic range of the circuit is limited for large signals by the saturation of $Q7$. This circuit is internally powered by 4.5 V and $Q7$ will saturate when 300 μA flows through $R5$. This limits the maximum input signal to 3.3 V rms or +12.6

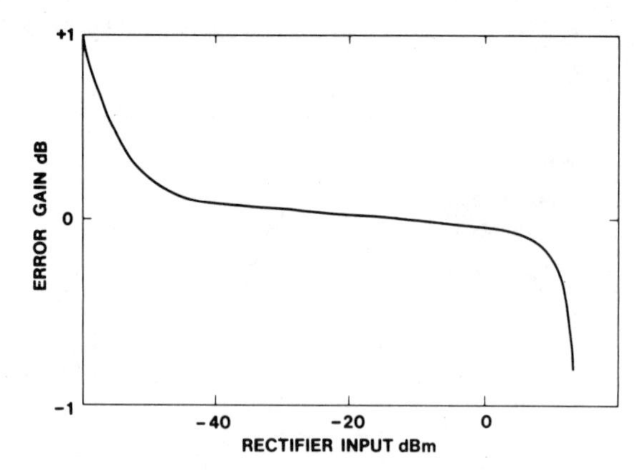

Fig. 7. Rectifier accuracy versus input level (1 kHz).

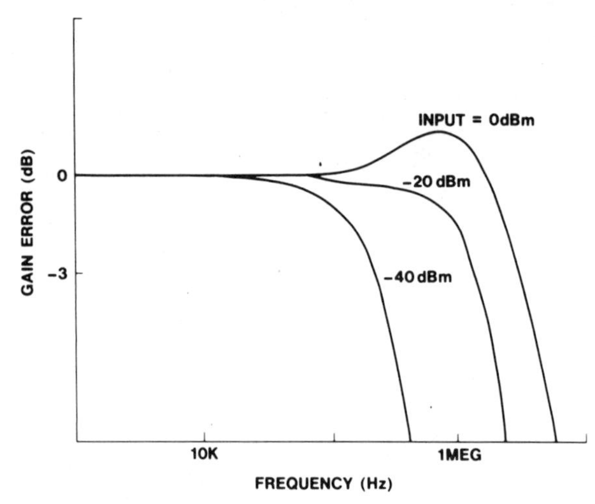

Fig. 8. Rectifier frequency response versus input level.

dBm. At the low end, dynamic range is limited by the bias current (typically 50 nA) of $Q2$ which must be supplied by $Q5$. Fig. 7 shows the rectifier accuracy versus input level relative to 0 dBm (0.775 V) at 1 kHz.

An important feature of this circuit is that for a capacitively coupled input, the performance at low signal levels is not affected by the op amp offset voltage. If the input is dc coupled and an offset exists between the op amp inverting input and the V_{in} pin, a small current will flow which will degrade the low-level accuracy of the circuit.

The finite gain of the rectifier op amp causes a decrease in frequency response at low signal levels. As the input signal amplitude drops, more gain is required to produce the 0.7 V p-p swing necessary to switch between $Q5$ and $Q6$ conducting. Since the op amp gain drops with increasing frequency, and more gain is required for smaller signals, the overall rectifier bandwidth falls with decreasing input amplitude. This is shown in Fig. 8 where rectifier frequency response is plotted for three different input amplitudes.

V. Variable Gain Cell

The ΔG cell is a linearized two-quadrant transconductance multiplier [4], [5]. A simplified schematic is shown in Fig. 9, and the operation of the circuit is as follows. An input voltage

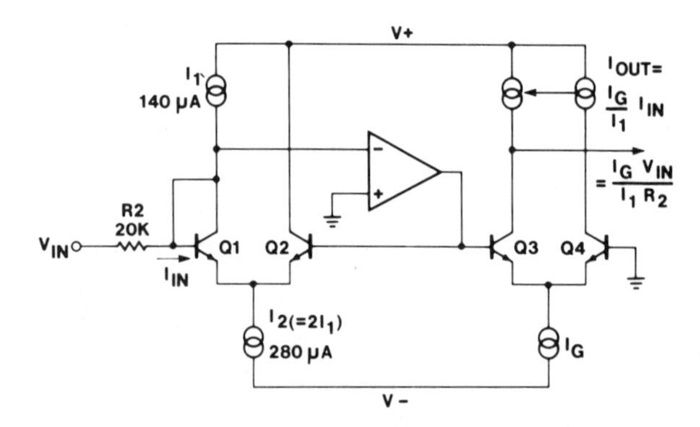

Fig. 9. Simplified ΔG cell schematic.

V_{in}, causes an input current $I_{in} = V_{in}/R_2$ to flow into the summing node which the op amp creates at the collector of $Q1$. The input current I_{in} is thus forced through $Q1$, along with the bias current I_1. We have the relations

$$I_{c1} = I_1 + I_{in} \tag{1}$$

$$I_{c2} = I_2 - I_1 - I_{in}. \tag{2}$$

Since the base of $Q1$ is held at virtual ground by the op amp, and the bases of $Q2$ and $Q3$ are tied together, the differential pairs $Q1$, $Q2$ and $Q4$, $Q3$ experience the same ΔVbe, and thus

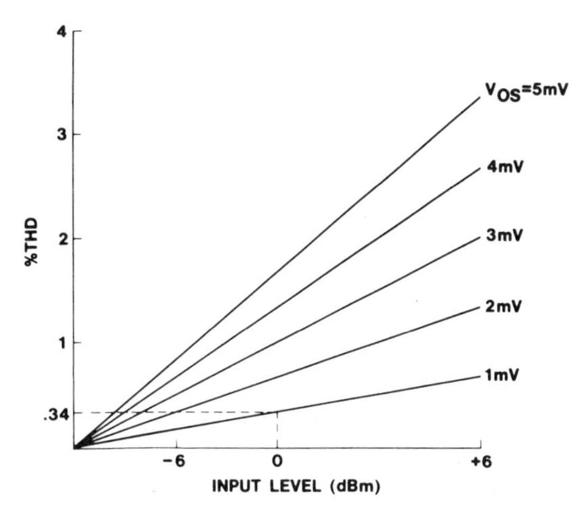

Fig. 10. ΔG cell distortion versus offset voltage.

(ideally) have identical collector current ratios. This gives us

$$\frac{I_{c1}}{I_{c2}} = \frac{I_{c4}}{I_{c3}}. \tag{3}$$

The output current mirror takes the difference of I_{c3} and I_{c4} so that

$$I_{out} = I_{c4} - I_{c3}. \tag{4}$$

The gain control current I_G is split between $Q3$ and $Q4$ so that

$$I_G = I_{c3} + I_{c4}. \tag{5}$$

Manipulation of these equations yields the basic multiplier transfer function

$$I_{out} = \frac{2}{I_2} I_{in} I_G + \frac{(2I_1 - I_2)}{I_2} I_G. \tag{6}$$

The multiplier output contains the product of I_{in} and I_G with $2/I_2$ as a scaling factor, and an output current proportional to the product of IG and $(2I_1 - I_2)/I2$. The first term is the desired product and the second term is undesired control signal feedthrough. The second term vanishes if $I_1 = \frac{1}{2}I_2$, so it is desirable to make these current sources match with a 1 : 2 ratio as accurately as possible. This also suggests a method of externally trimming the control signal feedthrough. An external current source tied to the V_{in} terminal can effectively trim I_1 and eliminate control signal feedthrough.

Ideally, the circuit is perfectly linear and insensitive to temperature but the use of nonideal components can degrade performance in several areas. More extensive analysis reveals gain to be proportional to transistor α, so that the gain can be in error by 1 percent for devices with $\beta = 100$, and the error will change as β changes over temperature.

Imperfect device matching leads to both gain control current feedthrough, and to a parabolic nonlinearity which produces virtually pure second harmonic distortion in audio signals. The magnitude of the distortion for a given input level and overall offset voltage is shown in Fig. 10. The distortion is proportional to input signal amplitude and offset voltage. An input level of 0 dBm (8 dB below clipping) with a 1 mV offset will produce 0.34 percent of second harmonic distortion. This distortion can be eliminated by trimming out the offset voltage, and a pin has been provided to allow the voltage at the op amp noninverting input to be trimmed. Use of this technique allows the second harmonic to be nulled to less than 0.01 percent, and one is left with a residual of 0.03 percent third harmonic component caused by finite op amp gain and transistor base resistances. The *THD* trim network will actually trim both distortion and control signal feedthrough, but in general, the nulls in distortion and feedthrough will be different (they will be identical only if the cell has perfect current source matching). If both parameters are being trimmed, the distortion should be trimmed first, followed by the feedthrough trim.

A two-quadrant multiplier such as this has the desirable feature that the output noise drops as the gain is reduced. The noise performance of the IC is shown in Fig. 11 where the maximum output signal and the noise are plotted as a function of gain. As expected, the noise drops as the gain is reduced, until it hits a floor which is caused not by the ΔG cell, but by the internal voltage reference and the output op amp biasing scheme. The signal-to-noise ratio at large gains is 90 dB, and total dynamic range is 110 dB. This good performance, along with the low distortion achievable with trimming, makes the circuit very attractive for use as an electronic gain control element in high quality audio processing systems. For even better performance, an external op amp may be used for higher slew rate and an even wider dynamic range (>120 dB).

VI. Output Operational Amplifier

The op amp used at the output is equivalent to a 741 type with 1 MHz bandwidth and 0.5 V/μs slew rate. The circuit is shown in Fig. 12. Split collectors are used for the input p-n-p's for lower transconductance so that a smaller capacitor can be used for compensation [6]. The quiescent output current has been set to a low value for lower power dissipation. This results in some crossover distortion when driving heavy loads at high frequencies. Short circuit protection is provided by limited base current drive, and a β dropoff at high current.

VII. Power Supply and Bias System

Since overall circuit gain is proportional to an internal voltage, a stable reference is required. A requirement for a minimum Vcc of 6 V rules out a Zener diode reference, so a band gap reference is used [7]. A simplified schematic is shown in Fig. 13.

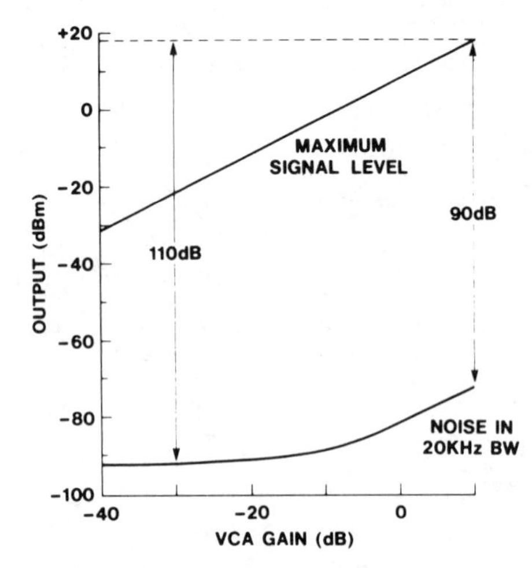

Fig. 11. Dynamic range of NE570.

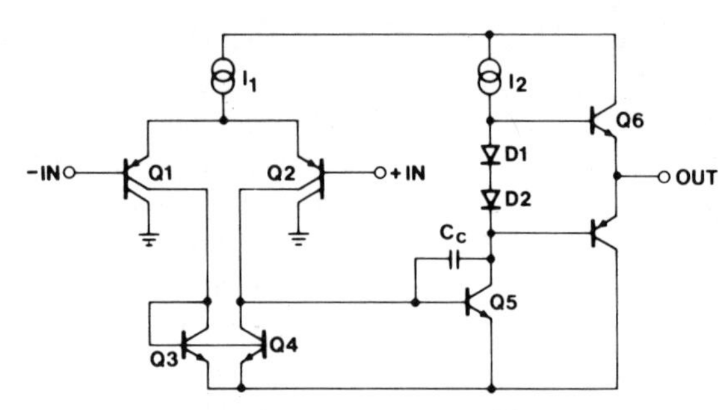

Fig. 12. Operational amplifier.

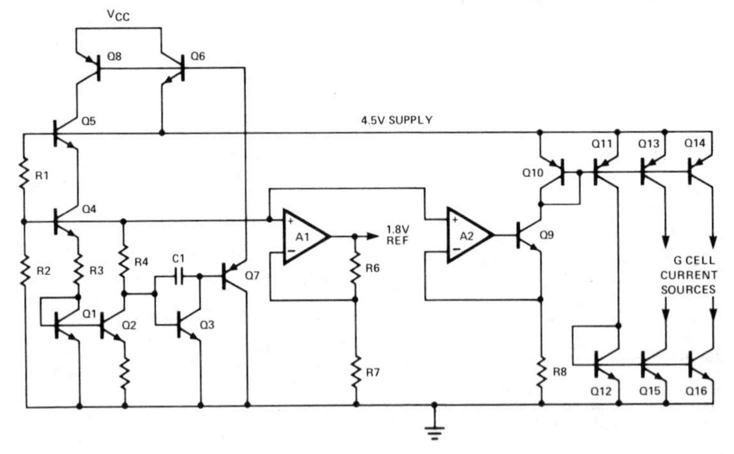

Fig. 13. Band gap reference and bias system.

Q_1 and Q_2 generate the ΔVbe, and $Q3$ provides the Vbe for the band gap reference. The stable 1.2 V appears at the emitter of $Q4$. The circuit also generates a supply decoupled 4.5-V supply which is used to power the rectifiers, the ΔG cells, and the ΔG-cell current sources. The 1.2-V band gap voltage is too low for biasing all of the internal summing nodes, so it is stepped up to 1.8 V by $A1$ and $R6, R7$. $A2, Q9$, and $R8$ form a band gap voltage to current converter to supply the 140 μA reference current to the ΔG cell current sources.

VIII. Resistors

In addition to the reference voltage, gain is proportional to resistor ratios. For the simplest hookups, the significant resistors are all internal to the IC so that ratio matching is very good and absolute accuracy and temperature drift are compensated. However, in many system applications, the gain equation must be modified with the use of external padding resistors. This means the internal resistors must have accurate absolute value and low temperature coefficient in order to maintain good overall system accuracy.

Resistors superior to standard diffused types are clearly desirable so an ion implantation process has been used which allows better control of absolute accuracy, and a smaller tempco over the important temperature range. Fig. 14 shows a comparison of the typical diffused resistor temperature drift and that of the ion implanted resistors which have been used. The ion-implanted resistors vary less than 1 percent over the critical 0–70°C temperature range.

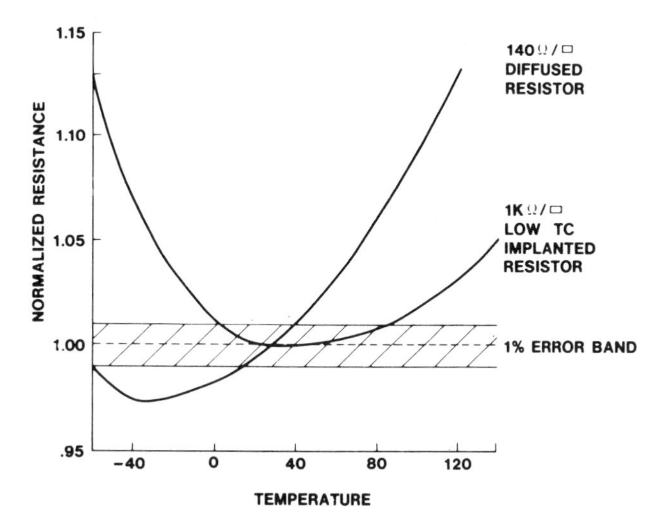

Fig. 14. Resistance versus temperature for implanted and diffused resistors.

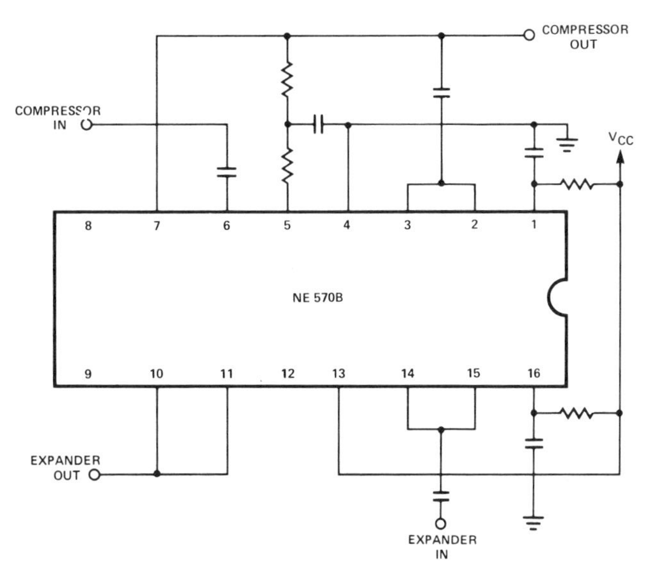

Fig. 15. Simple compandor system.

IX. System Performance

The application of the circuit as a simple compandor is illustrated in Fig. 15. The compressor and expandor are exactly as shown in Figs. 3 and 4 with the exception of two resistors which feed a small bias current from the power supply into the rectifier capacitor pin. This produces a tracking error for very weak signals because the rectifier output current becomes smaller than the external bias current. At low signal levels the circuit thus quits expanding or compressing and becomes linear. This low-level mistracking is necessary in order to be compatible with older discrete compandors which are still in service [3].

The dynamics of the compandor system, the attack and decay times, are determined by the value of the external rectifier capacitor. For a step level change in input, the gain change is a simple exponential function. If a compressor is operating at a gain $G1$, and a step change forces it to change gain to $G2$, the gain as a function of time is given by

$$G(t) = (G1 - G2)\, e^{-t/\tau} + G2 \tag{7}$$

where τ is the rectifier time constant set by the rectifier capacitor Cr, and the internal resistor $R5$ which is 10K. The CCITT [8] recommends a time constant of 20 ms which is met with a value of 2 μF for Cr. This yields a CCITT-defined attack time of 3 ms and release time of 13.5 ms. The compressor attack and release envelopes for a 12-dB step change are shown in Fig. 16. The combined compressor, expandor response to a 12-dB step, and an infinite step (no signal to full signal) are shown in Fig. 17.

X. Conclusion

A high performance analog compandor has been described with performance which easily exceeds the requirements of the telecommunications industry. The simple full wave rectifier has excellent accuracy over a 60-dB range of input signals. The variable gain cell offers low distortion (<0.1 percent with trimming), low noise (90 dB SNR) and wide dynamic range (110 dB). The IC has been configured to allow a wide variety of applications.

A photograph of the 88 by 92 mil die is shown in Fig. 18.

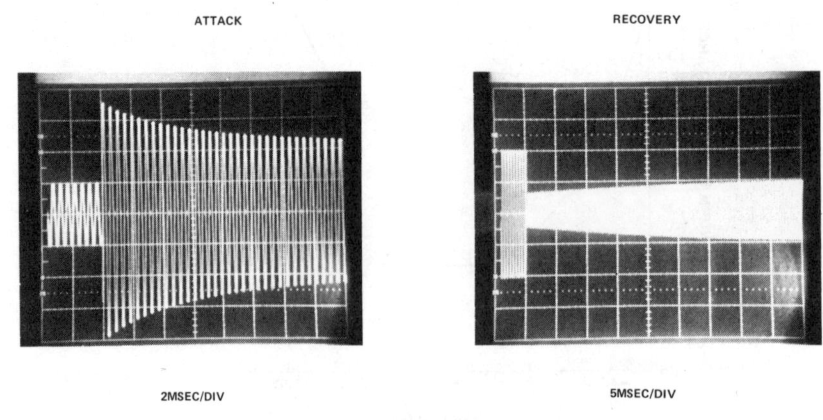

Fig. 16. Compressor transient response.

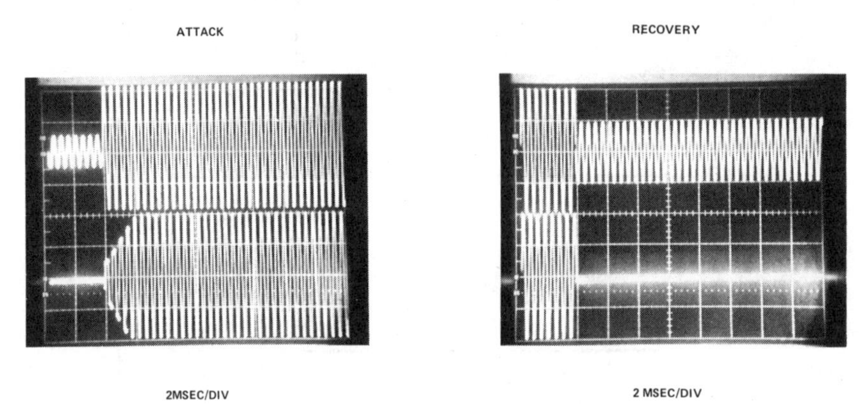

Fig. 17. Compandor transient response.

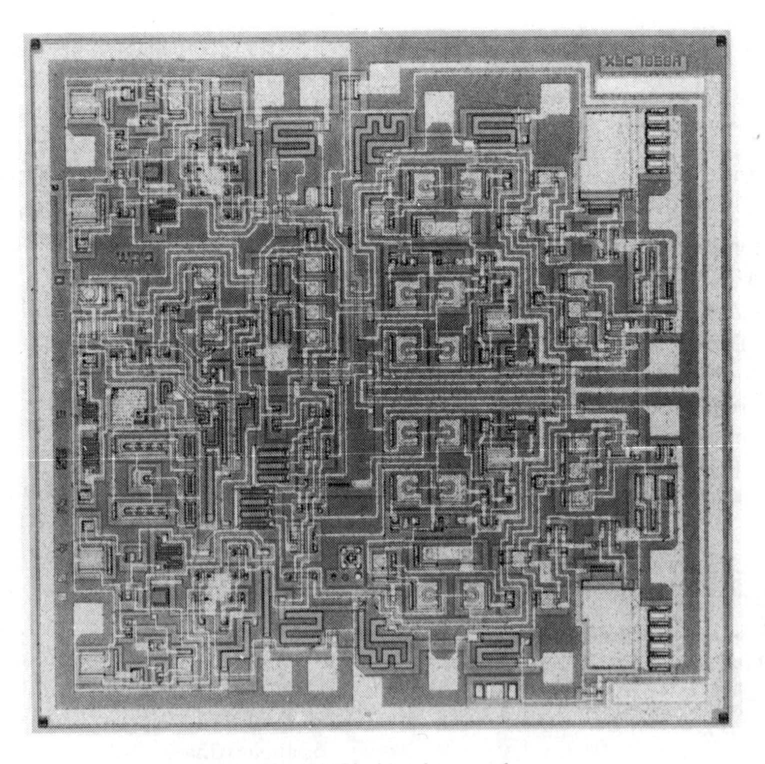

Fig. 18. NE570 chip photograph.

Acknowledgment

I would like to thank H. Stellrecht for his help and guidance. Thanks are also due to D. Perloff, F. Wahl, and F. Adamic for their help in developing the low T.C. ion implant resistor process.

References

[1] *Transmission Systems for Communications*, Bell Telephone Laboratories. Winston-Salem, NC: Western Electric Tech. Publ., 1970, pp. 677–682.

[2] E. M. Rizzoni, "Compandor loading and noise improvement in frequency division multiplex radio relay systems," *Proc. IRE*, pp. 208–220, Feb. 1960.

[3] R. C. Boyd *et al.*, "N2 carrier system," *Bell Syst. Tech. J.*, vol. 44, pp. 731–822, May–June 1965.

[4] B. Gilbert, "A precise four-quadrant multiplier with subnanosecond response," *IEEE J. Solid-State Circuits*, vol. SC-3, pp. 365–373, Dec. 1968.

[5] W. Jung, "Get gain control of 80 to 100 dB," *Electron. Des.*, pp. 94–99, June 21, 1974.

[6] R. Russel and T. Frederiksen, "Automotive and industrial building blocks," *IEEE J. Solid-State Circuits*, vol. SC-7, pp. 446–454, Dec. 1972.

[7] R. Widlar, "New developments in IC voltage regulators," *IEEE J. Solid-State Circuits*, vol. SC-6, pp. 2–7, Feb. 1971.

[8] *The International Telegraph and Telephone Consultative Committee (CCITT) Blue Book*, Vol. 3, Third Plenary Assembly, Geneva, May 25–June 26, 1964, Recommendation G-162, p. 62.

The TDA1077—An I²L Circuit for Two-Tone Telephone Dialing

DAAN J. G. JANSSEN, JEAN-CLAUDE KAIRE, AND PHILIPPE GUÉTIN

Abstract—**The circuit-design aspects of an integrated circuit to perform two-tone telephone dialing are described. The circuit is believed to be unique in that it combines both the crystal-controlled frequency synthesizer and the output amplifier on the same chip. Moreover, no external power supplies are required; the circuit is powered by the telephone-line current. Designed to require a minimum number of external components, this LSI chip provides an economical and accurate two-tone dialing unit. A typical application circuit for existing telephone apparatus is shown and aspects of future development are discussed.**

Introduction

CONVENTIONAL telephone apparatus uses rotary dials to produce a series of interrupting pulses in the line current. The number of these pulses corresponds to the number dialed. Push-button dialing units have been made to produce the pulses electronically. The series-pulse system is, however, both slow and liable to error. Interference on the lines may be seen as data by the exchange decoding circuits, resulting in an incorrect connection.

Frequency encoding of the dialed numbers allows faster dialing and gives a greater immunity to noise. The CCITT[1] has therefore recommended the use of a system in which each number is represented by a combination of two tones from a possible eight, see Fig. 1. Both tones representing a number are transmitted simultaneously, thus shortening the data transmission time. The tones have been carefully chosen to avoid problems caused by harmonics, while the use of two tones eliminates the possibility of line whistles causing dialing.

New decoding circuits are required in the exchange for the two-tone dialing system, but this can be done while retaining compatibility with existing series-pulse apparatus.

Design of the Integrated Circuit

Two-tone dialing systems are already in existence, using *LC* and *RC* tone generators to obtain the required frequencies. These conventional oscillators suffer from long-term instability and the need for a number of accurate components. The TDA1077 overcomes these problems by the use of the principle of frequency synthesis within the LSI chip. All the components except the quartz crystal, a polarity guard, one resistor and two capacitors are contained in the integrated circuit. Fig. 2 shows the functional diagram of the TDA1077.

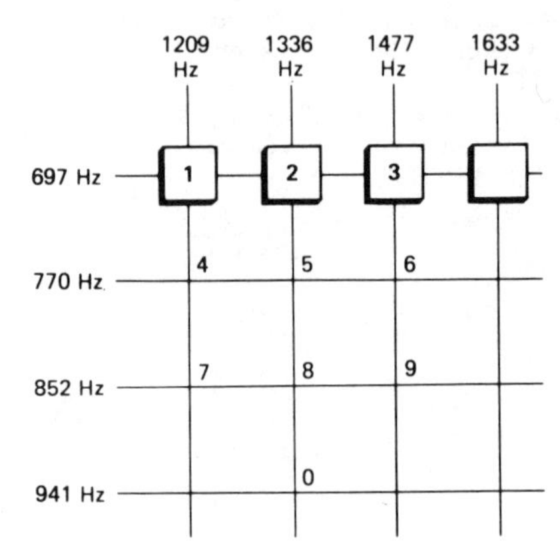

Fig. 1. Identification of the dialed numbers using two out of eight tones.

The choice of technology to produce the integrated circuit was influenced by the following factors: 1) low operating voltage; 2) low minimum current; 3) high maximum current; 4) implementation of both analog and digital elements; 5) compatibility with conventional fast logic elements.

Consideration of these points had led to an I²L design[2] with a small part in current mode logic for the high-frequency clock circuits, thus incorporating both fast logic and analog circuitry on one chip. The standard process for the ECL 10 000 family is used to achieve this.

Line Adaptor

The dc voltage across the integrated circuit is regulated by the shunt amplifier *A*1 in Fig. 2. The line voltage is compared with a reference voltage via the low-pass filter formed by R_3C_2 to provide the regulator with a low output impedance for dc without attenuating signals in the audio band of 300–3400 Hz. All the other circuits in the IC connected to the line present a high impedance to ac signals so that the ac input impedance of the IC is determined by resistors R_2 and R_4.

Fig. 3 shows the voltage across the IC. The dc component is about 3 V due to the action of the line adaptor. As the ac signal can be up to 3 V_{p-p}, all the circuitry must be designed for a minimum supply voltage of 1.3 V and must be unaffected by the presence of audio signals on the line. I²L technology is ideally suited to this application, providing low-voltage circuits driven by constant-current sources.

Keyboard Operation

The integrated circuit has four outputs corresponding to the lower tones and four inputs to select the upper tones. Depression of a key links an output to an input and selects two

Manuscript received November 17, 1976.

D. J. G. Janssen is with N. V. Philips Gloeilampenfabrieken, Eindhoven, The Netherlands.

J.-C. Kaire and P. Guétin are with R. T. C. La Radiotechnique-Compelec, Caen, France.

[1]CCITT—the International Telephone and Telegraph Consultative Committee, Green Book VI-1, Recommendation Q23.

[2]C. M. Hart and A. Slob, "Integrated injection logic: A new approach to LSI," *IEEE J. Solid-State Circuits*, vol. SC-7, p. 346, Oct. 1972.

Reprinted from *IEEE J. Solid-State Circuits*, vol. SC-12, 238-242, June 1977.

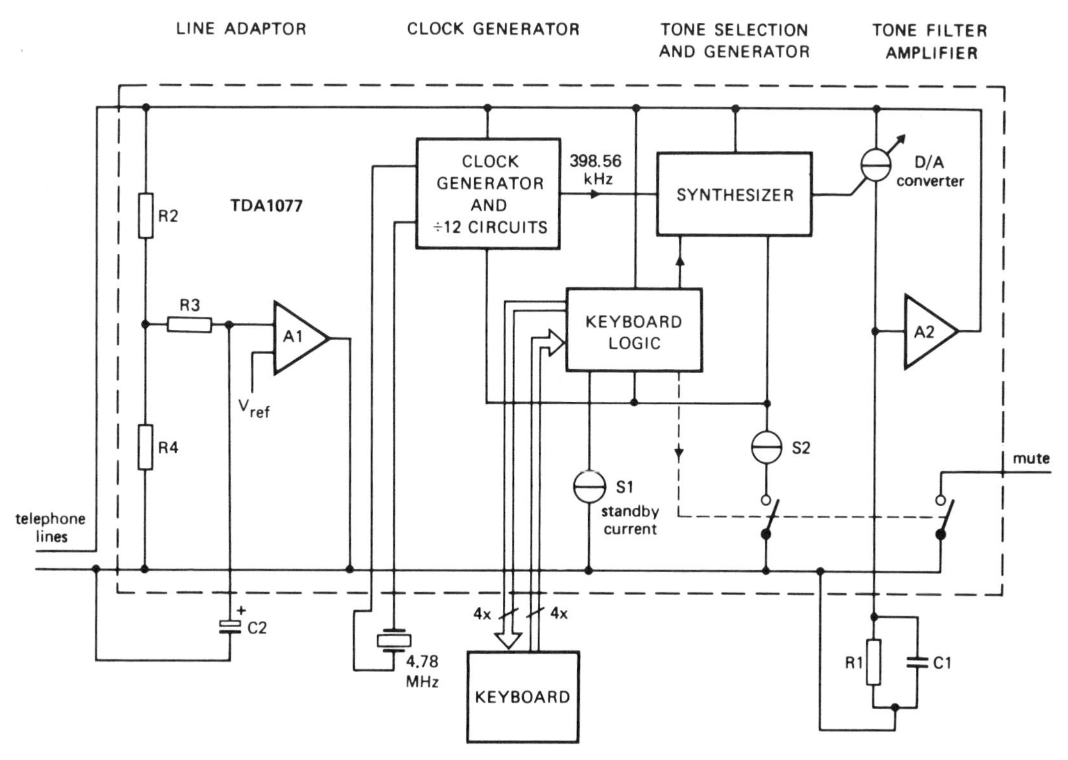

Fig. 2. Functional diagram of the TDA1077 and its external components.

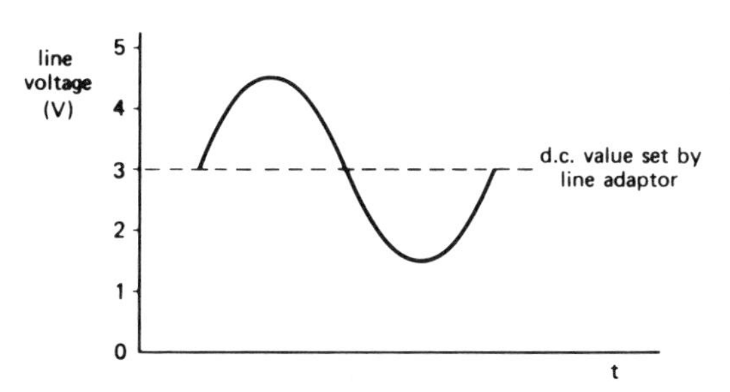

Fig. 3. The line voltage across the IC, after regulation by the line adaptor, showing the effect of dialing or audio signals.

tones, see Figs. 1 and 9. Unused tone combinations allow for future additions to the system.

Frequency Synthesizer

The synthesizer is implemented in I^2L as a variable divider followed by a D/A converter. As each dialed number requires the simultaneous generation of two tones, the circuitry is duplicated with only small differences between the upper and lower tone circuits.

Ideally, the clock frequency should be chosen as the lowest common denominator of the eight required tones. This frequency (14×10^{13} Hz) is impractical. However, calculations have shown that close approximations (within 0.11 percent) to the required frequency can be achieved using a clock frequency of 199.28 kHz and integer divisors. As some of these divisors are odd numbers, the generated tones would possess unequal half-cycles and even harmonics. The half-cycles can be made symmetrical, and the even harmonics removed, by doubling the clock frequency and divisors. Table I shows the divisors, the resulting tones, and the errors.

TABLE I
Generation of Tones from a 398.56-kHz Clock

Tone required (Hz)	Dividing factor	Tone generated (Hz)	Absolute error (Hz)	Relative error %
697	572	696.78	- 0.22	- 0.03
770	518	769.42	- 0.58	- 0.08
852	478	851.62	- 0.38	- 0.04
941	424	940.00	- 1.00	- 0.11
1209	330	1207.76	- 1.24	- 0.10
1336	298	1337.45	+ 1.45	+ 0.11
1477	270	1476.15	- 0.85	- 0.06
1633	244	1633.44	+ 0.44	+ 0.03

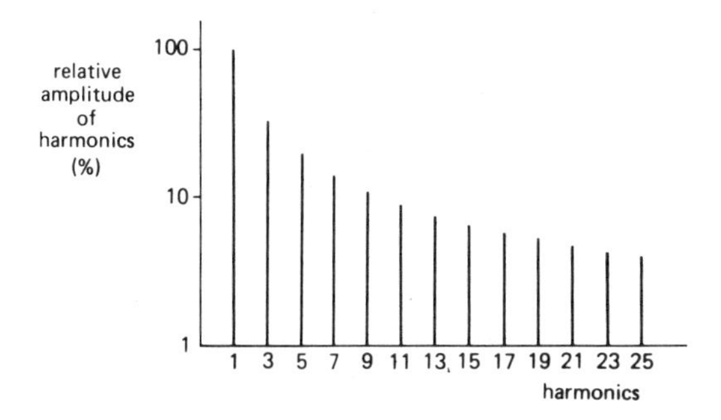

Fig. 4. Harmonic content of a symmetrical squarewave.

If a simple frequency divider is used, producing a squarewave, the output frequency spectrum contains a high percentage of odd harmonics, see Fig. 4. The higher harmonics (13th and above for the higher tones, 17th and above for the

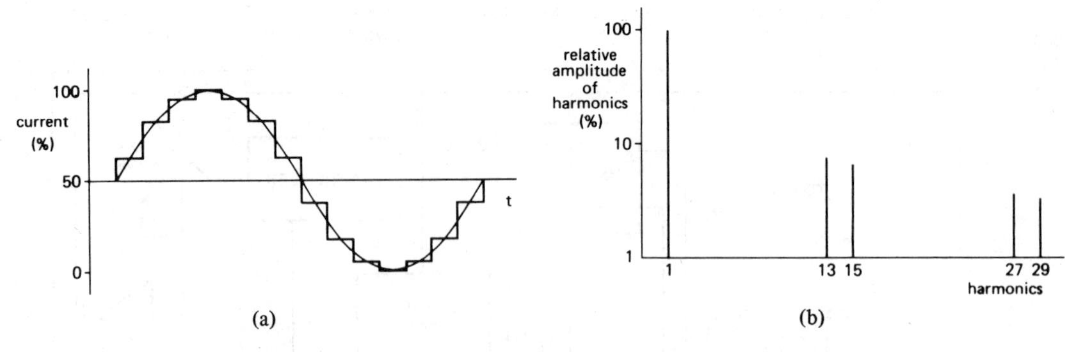

Fig. 5. Form (a) and spectrum (b) of the output of the D/A converter using seven current sources and 14 steps.

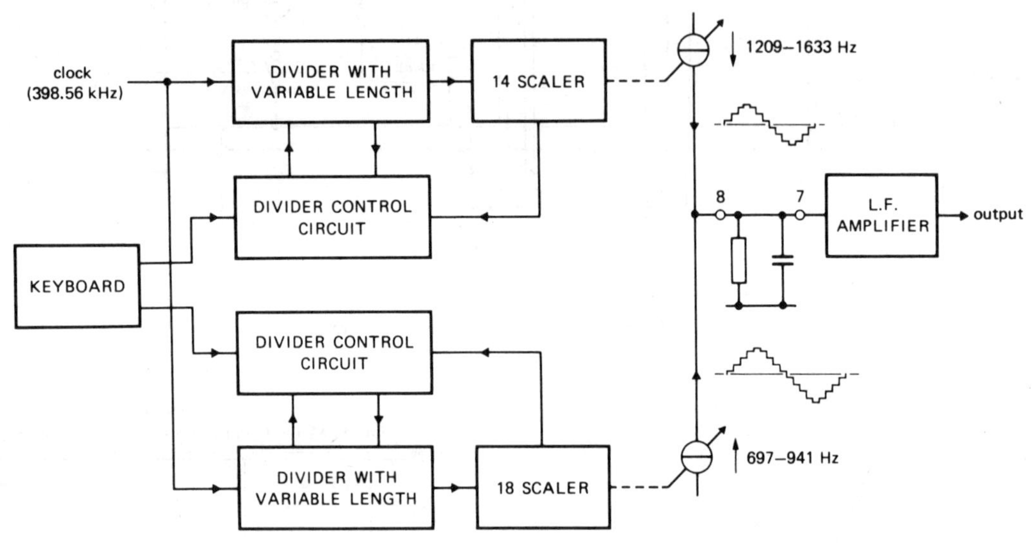

Fig. 6. Schematic diagram of the tone synthesis.

lower tones) are sufficiently attenuated (-36 dB) by the action of filter R_1C_1 in Fig. 2.

To prevent generation of the harmonics up to and including the 11th for the higher tones, the output waveform is generated as a crude sine wave in a series of 14 steps, see Fig. 5. Seven current sources are used, providing a symmetrical approximation to the required tone. The frequency of the output is selected by the duration of the steps.

To provide this stepped output, the divisors for the higher tones in Table I must themselves be divided by 14; this results in noninteger divisors, which are not realizable in the physical circuits. Thus some of the steps must be one clock period longer so that the generated tone is of the correct frequency. For example, the 1209-Hz tone requires a division of the clock frequency by 330. The divisor now becomes 330/14 = 23 4/7. This means that each of the 14 steps should have a length of 23 4/7 clock periods. The problem is solved by giving 8 of the steps a length of 24 and the remaining 6 steps a length of 23, making a total of 330.

The lower tones are similarly produced, using nine current sources to produce a sinewave with 18 steps for each cycle of the generated tone.

Fig. 6 shows the total scheme of the synthesizer and the D/A converter. The currents representing the lower and higher tones are generated in parallel and added at the input of the low-frequency amplifier. The current sources are switched on/off by the synthesizer logic.

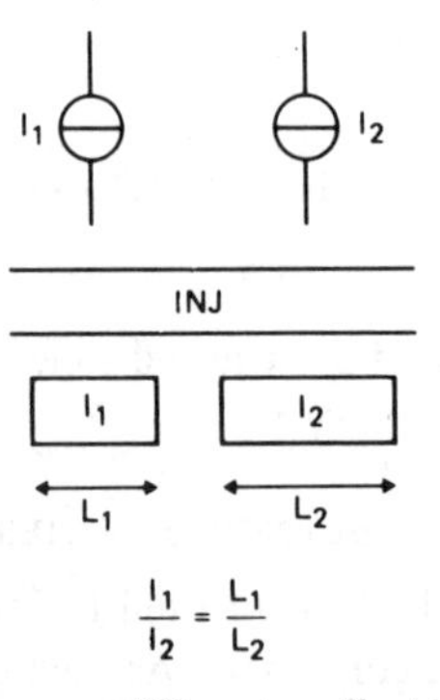

Fig. 7. Current sources use different collector widths to produce different currents in I^2L technology.

The current sources are lateral p-n-p transistors of the same type as those used for the I^2L injectors, see Fig. 7. The value of the current is proportional to the width of the collector, as the current sources are mirrors of the special constant current source S_2 in Fig. 2. This has been designed to provide a constant current independent of the dc line voltage and the ac signals present on the line. Because of this construction, the

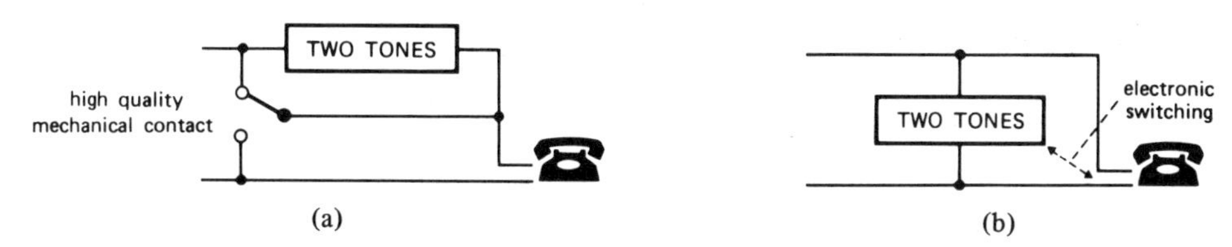

Fig. 8. Modes of connection for the two-tone dialing unit, serial connection (a), parallel connection (b).

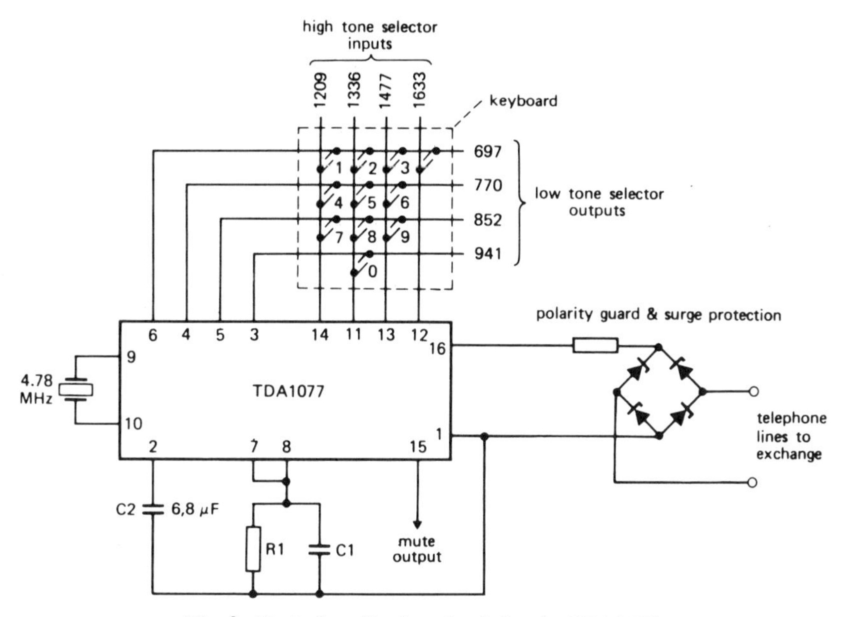

Fig. 9. Typical application circuit for the TDA1077.

circuits fulfill the requirement for operation at the minimum supply voltage of 1.3 V.

Timing Generator

The design of the frequency synthesizer has determined that the clock frequency must be 398.56 kHz, or a multiple thereof. To achieve the required long-term stability the timing generator is controlled by a quartz crystal, which only becomes an economic proposition at frequencies above about 4 MHz. Accordingly, a crystal with a nominal frequency of 4.78272 MHz has been chosen. A divide-by-12 circuit then produces the required 398.56 kHz.

Mode of Operation

It is expected that many tone-dialing units will be produced as replacements for existing rotary dials. These units must therefore be suitable for series operation, as shown in Fig. 8. A great disadvantage of this mode is the requirement for a special contact short circuiting the dialing unit during standby. This contact must be a high-quality long-life low-resistance unit, which is relatively expensive. Moreover, this contact must be operated whenever any of the push-buttons are depressed, requiring a complicated and expensive mechanical link to the keyboard.

In the future, telephones are expected to operate in the parallel mode shown in Fig. 8. With the dialing unit in parallel to the microphone and speaker electronics, the expensive short-circuiting contact can be replaced by an electronic inhibit control. The TDA1077 has been designed with parallel operation as a possibility, incorporating the following facilities.

1) A mute signal to inhibit the action of the microphone and receiver electronics during dialing and reduce current consumption. This has the added advantage of eliminating receiver noise due to dialing.

TABLE II
Characteristics of the TDA1077

Parameter	Minimum	Maximum	Unit
Supply current "ON" (including output amplifier)	10	100	mA
Supply current "OFF"	3	100	mA
D.C. line voltage (I_{16} = 30 mA)		3	V
Generated frequency accuracy	- 0.5	+ 0.5	%
Nominal output, lower tones (adjustable)	- 11 to	- 7	dBm
Nominal output, higher tones (adjustable)	- 9 to	- 5	dBm
Pre-emphasis (high tones)	1	3	dB
Total distortion		- 23	dB

Note that the values for the nominal outputs are adjustable between the limits given by selection of the impedance of the output low-pass filter.

2) The dialing circuitry is inhibited while the keyboard is not in use, further reducing current consumption. Only a small standby current is then required to allow scanning of the keyboard.

Application Circuit

Fig. 9 shows the completed circuit for a tone-dialing unit. The polarity guard and surge protection circuit are functions always required by all electronic circuitry in telephone apparatus. R_1 and C_1 form the low-pass filter at the output of the D/A converter. C_2 and an internal resistor form the low-pass filter for the line adaptor. Table II gives a summary of the characteristics of the TDA1077 integrated circuit.

Conclusions

The combination of both complex analog and digital circuits on a single LSI chip has led to the production of an economical two-tone generator that fully complies with the CCITT recommendations and satisfies most of the national European requirements. The TDA1077 provides a unique solution for the design of two-tone dialing apparatus: it is a complete system in one chip, working on a low voltage which permits the use of the maximum line length currently in use. The only critical external component required is the crystal, allowing simple production of two-tone dialing units with accurate tones.

Acknowledgment

The authors wish to thank G. C. Groenendaal and C. M. Hart of the Philips Research Laboratories, Eindhoven, for their cooperation in the development of the TDA1077.

A Precision Slope Polarity Switch for a Monolithic Telephone-Quality Deltamodulator

John J. Price
Motorola, Inc.
Mesa, AZ

Stephen H. Kelley
Motorola, Inc.
Phoenix, AZ

THIS PAPER will discuss a monolithic continuously variable slope deltamodulator capable of telephone quality digital voice transmission; Figure 1. It functions as either encoder or decoder, selectable by a logic input. The circuit is fabricated with a standard linear process which includes I^2L, ion-implanted resistors, high gain PNPs, and laser trimming. A unique laser trimmed precision slope polarity switch is used to achieve telephone quality performance.

A block diagram of the CVSD voice encoder is shown in Figure 2. A simple deltamodulator is formed by the inner control loop[1]. The encoder compares the audio input with the integrator output and loads a one-bit quantization of the difference into a clocked flip-flop. The flip-flop output is transmitted to a similar system at the remote location. The open loop integrator of the decoder reproduces the audio signal from the received digital bit stream.

Simple deltamodulators have a limited dynamic range due to the constant gain of the integrator[1]. The outer control loop of a CVSD encoder consists of a level detector and low-pass filter which generate a dc voltage proportional to the input level[2]. That voltage is used to control the gain of the integrator through the slope magnitude controller, increasing the encoder's dynamic range. The slope control voltage is converted to two matched currents of opposite sign by the slope polarity switch. One of the two currents is fed to the integrator depending upon the digital output state. At the receiver, the gain control information is recovered from the digital bit stream by an identical level detector and slope magnitude controller.

In the absence of an input signal to the encoder, the slope control voltage diminishes and a very small integrator current is produced. The two polarities of this current must be matched and the offsets of all the analog circuitry must be nulled so that a perfect idle channel output pattern (1010...) is produced, to give minimum idle channel noise[1,3].

The offsets of the comparator and integrator are nulled by a laser trim which adjusts the load resistors in the comparator. However, the slope polarity switch can still contribute an effective dc offset due to mismatch of opposite currents. A second laser-trim step nulls this mismatch.

Given a completely trimmed CVSD chip, the circuit in Figure 3 will produce the signal-to-noise performance and slope overload characteristics in Figure 4. The circuit in Figure 3 utilizes a complex gain control algorithm which takes maximum advantage of the available dynamic range.

The resulting system performance is dependent on the matching and range of the current slope polarity switch shown in Figure 5. Except for Q2, Q3, R1, R2 and R3, it is a conventional sink/source switch comprised of a differential pair and a PNP mirror. Resistors R2 and R3, the trim resistors, are each composed of several binarily scaled resistors connected in series. Initially, they are all shorted together by metal links. Trimming is performed by cutting the appropriate links with a laser to add resistance to either the left or the right side.

When V1 is high with respect to V2, Q1 and Q2 are both on; Q3 and Q4 are off. Since Q2 is cross-coupled to the output, it subtracts current from the output making I_{OUT} less than I_{IN} by twice the current flowing through Q2. In the other state of the switch, the same is true for Q3 and Q4. If a portion or all of R2 is added into the circuit by laser trimming, the current through Q2 decreases, and the output current increases by twice the change in Q2's current. By adding in some of R3 instead, the current of opposite polarity flowing into the output node increases. Thus, with this simple trim procedure, mismatches in the PNP mirror can be compensated so that the slope polarity switch contributes little to the total loop offset at idle channel.

As I_{IN} increases beyond the idle channel level, the relative error currents contributed by Q2 and Q3 of Figure 5 decrease. At higher currents, enough voltage appears across R4 and R5 for these resistors to dominate in the matching of the mirror, and adequate performance for high currents is achieved.

The laser trimming is done at wafer probe, and allows other convenient uses of the metal cutting capability. This die is manufactured with a three-bit shift register in the level detector for low bit rate systems. A fourth flip-flop in the register can be enabled by two additional trimmable links. Thus, depending upon the performance of the die being tested and upon the market demands, the unit may be left with the three bit register, or may be changed to four bits[1].

The resourceful use of innovative circuits utilizing metal link trimming and process compatible I^2L allows the design of complex linear monolithic circuits that can solve stringent real world requirements in a way that digital LSI cannot.

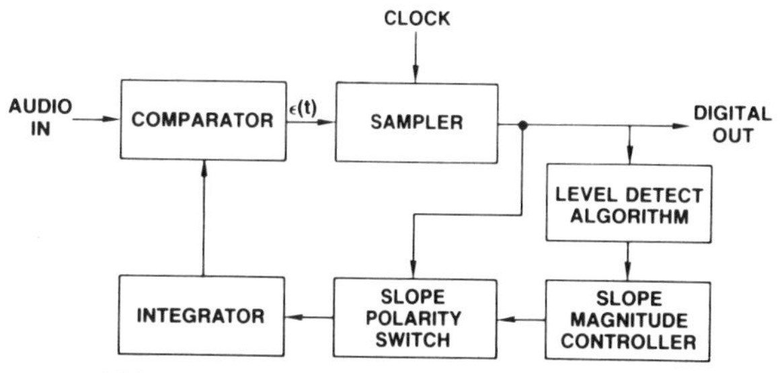

FIGURE 2 – Block diagram of CVSD encoder.

[1] Schindler, H.R., "Delta Modulation", *IEEE Spectrum*, p. 69-78; Oct., 1970.

[2] Chakravarthy, C.V. and Faruqui, M.N., "Two Loop Adaptive Delta Modulation Systems", *IEEE Trans. Communications*, p. 1710-1713; Oct., 1974.

[3] Brolin, S.J. and Brown, J.M., "Companded Delta Modulation For Telephony", *IEEE Trans. Communication Technology*, p. 157-162; Feb., 1968.

Reprinted from *1977 IEEE Internat. Solid-State Circuits Conf. Dig. Tech. Papers*, Feb. 16-18, 1977, pp. 60-61, 234.

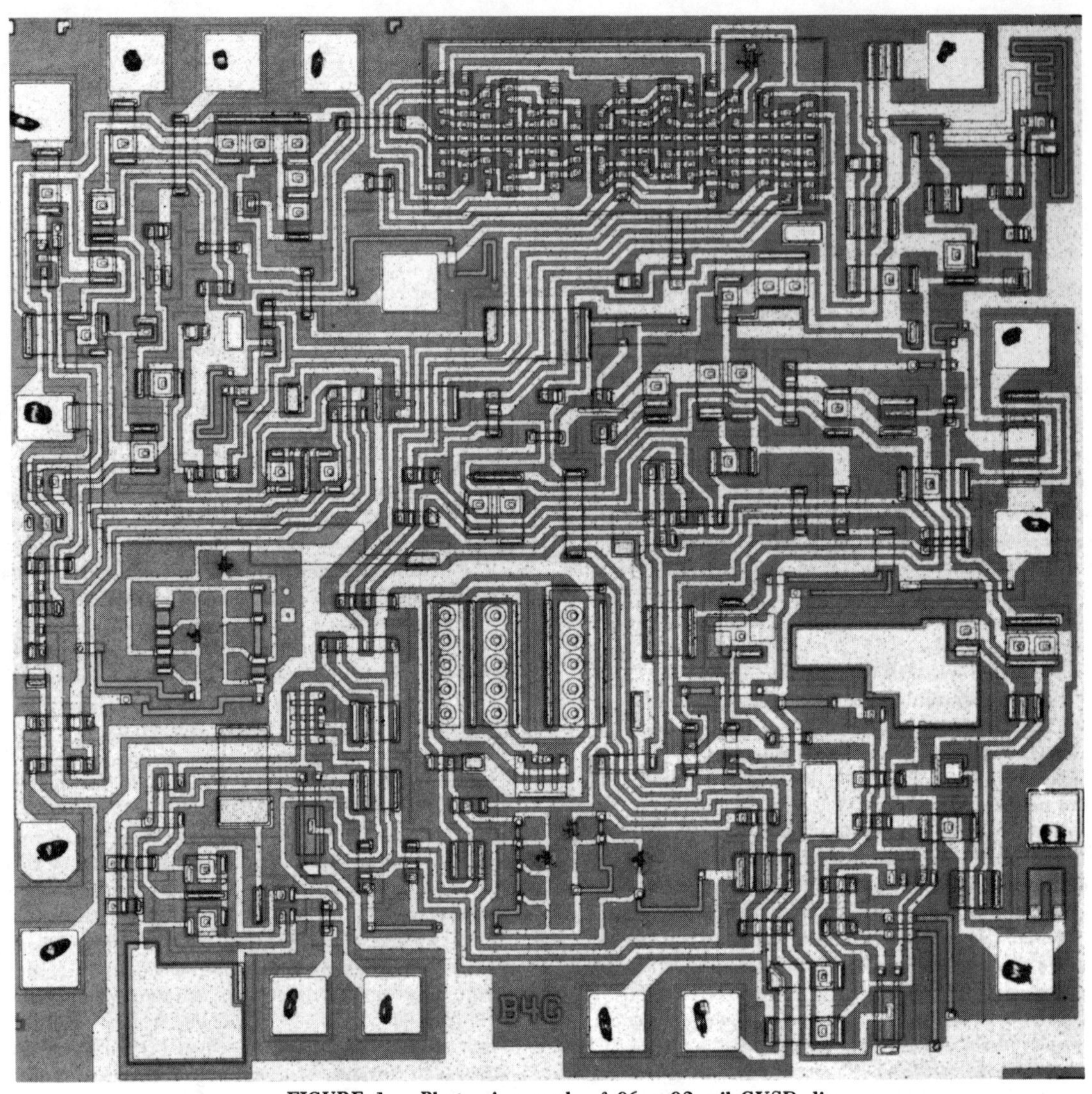

FIGURE 1 – Photomicrograph of 96 x 92 mil CVSD die. Trimmed metal links appear in three areas. I^2L circuitry is located at the top center.

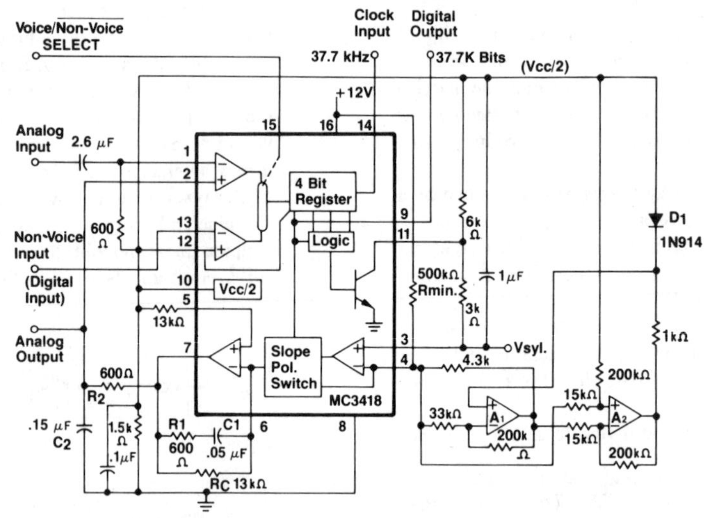

FIGURE 3 – Schematic of telephone-quality deltamod coder. Performance is shown in Figure 4; page 234.

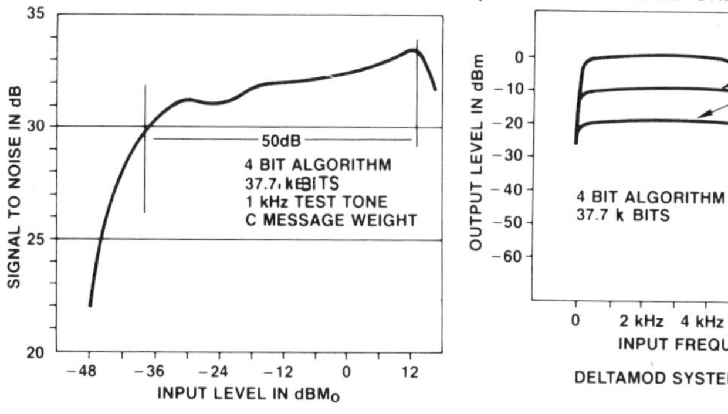

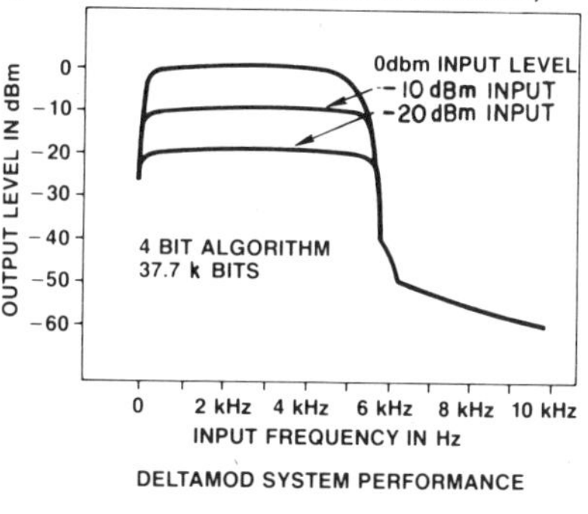

FIGURE 4 – Performance of deltamod system in Figure 3, which has 50-dB dynamic range and 5kHz usable bandwidth.

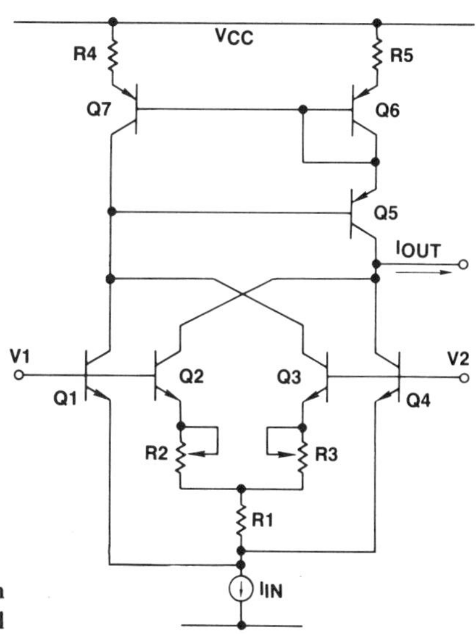

FIGURE 5 – Laser-trimmed slope polarity switch which provides the required dynamic range and idle channel performance.

A Segmented μ-255 Law PCM Voice Encoder Utilizing NMOS Technology

YANNIS P. TSIVIDIS, MEMBER, IEEE, PAUL R. GRAY, SENIOR MEMBER, IEEE,
DAVID A. HODGES, SENIOR MEMBER, IEEE, AND JACOB CHACKO, JR., MEMBER, IEEE

***Abstract*—A technique for pulse code modulation (PCM) encoding according to the 15-segment approximation to the μ-255 law is described. The technique makes possible the realization of a one-channel encoder together with the associated sample/hold function as a single NMOS integrated circuit. The requirements imposed on the individual components for satisfactory system performance are investigated, and it is shown that a performance can be achieved which is comparable with that required in toll quality PCM telephone transmission. Experimental results from a partially integrated prototype are presented.**

INTRODUCTION

UNTIL recently, digital transmission of voice signals in telephony has been implemented almost exclusively by means of a high-speed A/D converter and D/A converter which was multiplexed over many analog channels. However, recent technological developments have made it economically feasible to encode each channel separately and then multiplex the resulting digital signals. This amounts to replacing a single costly, high-speed encoder with several low-cost, low-speed ones. In addition, the analog multiplexer is replaced by a digital one, which is simpler and less vulnerable to crosstalk. The low encoder cost required for this approach to be cost-competitive can best be achieved by using some form of large-scale integration (LSI), and a single-chip encoder implementation would be desirable.

Per-channel PCM encoders will find application in private branch exchange equipment, channel bank equipment, and subscriber loop carrier systems.

Several recent papers discuss techniques for possible single-channel encoder realizations [1]-[3]. In all techniques, both analog functions and digital logic are required. In realizations using two or more chips, more than one integrated circuit technology can be used. For example, the high-performance analog functions can be realized using a bipolar chip, and the digital functions using an MOS chip. For the single-chip realization discussed below, an all-bipolar approach is possible using I^2L technology. We have, however, chosen the all-MOS approach for several reasons.

First, the process is simple and requires only five masking steps. Second, it has been demonstrated [4] that voltage dividers of accuracy sufficient for this application can be constructed in MOS technology by using MOS capacitors. The same capacitors can be used to implement the decoder on the same chip with the encoder, on a time-share basis. Third, a sample-and-hold function is easily realizable in MOS. The fourth, and perhaps the most important reason is that the process is compatible with charge-coupled device (CCD) and bucket brigade sampled-data filters. Thus the possibility exists of incorporating the encoder anti-aliasing filter and the decoder output filter on the same chip with the encoder/decoder using transversal or recursive realizations. The process is also compatible with MOS microprocessor technology.

The efforts of this research have, therefore, been directed towards a single-channel, low-cost chip which performs PCM encoding according to the 15-segment approximation to the 255-μ law utilizing NMOS technology.

Manuscript received June 14, 1976; revised July 28, 1976. This research was supported in part by NSF Grant ENG73-04184-A01 and in part by a grant from Bell Northern Research Ltd. This paper was presented at the International Solid-State Circuits Conference, Philadelphia, PA, February 1976.

Y. P. Tsividis was with the Department of Electrical Engineering and Computer Sciences and the Electronics Research Laboratory, University of California, Berkeley, CA 94720. He is now with the Department of Electrical Engineering and Computer Science, Columbia University, New York, NY 10027.

P. R. Gray and D. A. Hodges are with the Department of Electrical Engineering and Computer Sciences and the Electronics Research Laboratory, University of California, Berkeley, CA 94720.

J. Chacko, Jr. was with the Department of Electrical Engineering and Computer Sciences and the Electronics Research Laboratory, University of California, Berkeley, CA 94720. He is now with Bell Laboratories, Holmdel, NJ 07733.

THE SEGMENTED μ-255 ENCODING LAW

The 15-segment approximation to the μ-255 encoding law is shown in Fig. 1. There are 8 segments for each input polarity. Each segment consists of 16 equal steps, as, for example, is shown for segment *AB*. The step size within one segment is constant, but doubles as one goes from one segment to the next, starting from the segment adjacent to the origin and going towards higher amplitudes. Finally, the two steps adjacent to and symmetrical about the origin are merged into a single step, with the origin as its middle point. There are a total of 255 steps in the characteristic, and the ratio of the size of the largest step to that of the smallest is $2^7 = 128$. The decoding characteristic is complementary to that of the encoding law just described.

We will now consider the format of the digital word at the output of the encoder. Since there is a total of 255 intervals, an 8-bit word is needed since $2^8 = 256$. The format used for this word is shown in Fig. 1(b). The first bit indicates the sign of the input, being 1 for positive and 0 for negative inputs. The next three bits indicate the number of the segment to which the input corresponds, with 000 representing the segment closest to the origin and 111 the longest segment. The last four bits represent the number of the step within the segment, to which the input corresponds, with 0000 being the first step and 1111 the last. For example, if the input corresponds to segment *CD* in Fig. 1, the word at the output of the encoder will be 1 100 0111. This coding format is sometimes

Reprinted from *IEEE J. Solid-State Circuits*, SC-11, pp. 740-747, Dec. 1976.

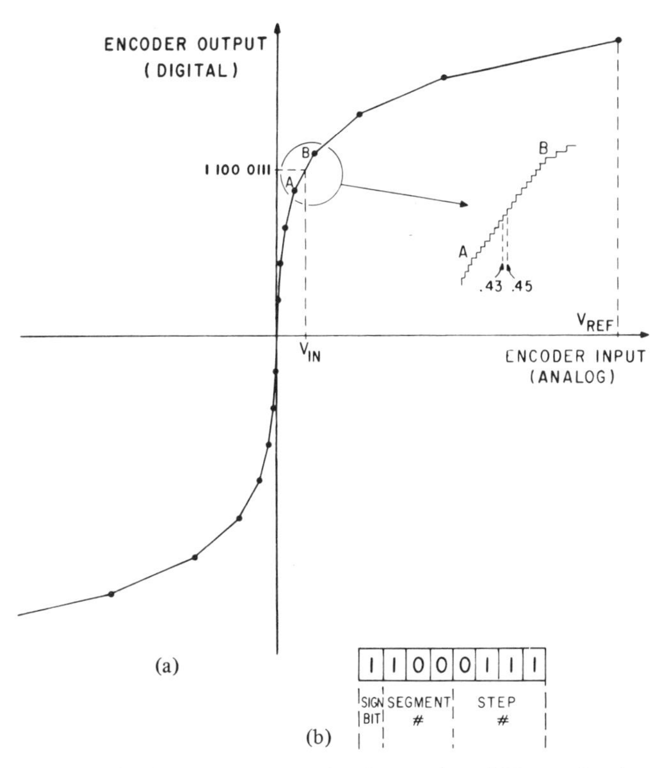

Fig. 1. (a) The 15-segment approximation to the μ-255 encoding law. (b) Encoding format.

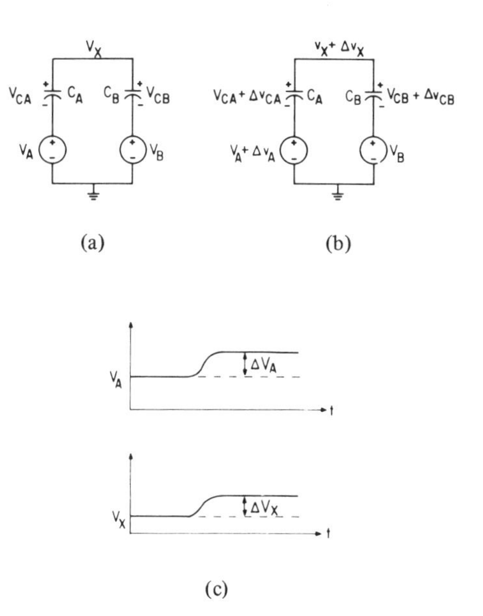

Fig. 3. Principle of charge redistribution. (a) Circuit before V_A transition. (b) Circuit after V_A transition. (c) Waveforms which result from the transition.

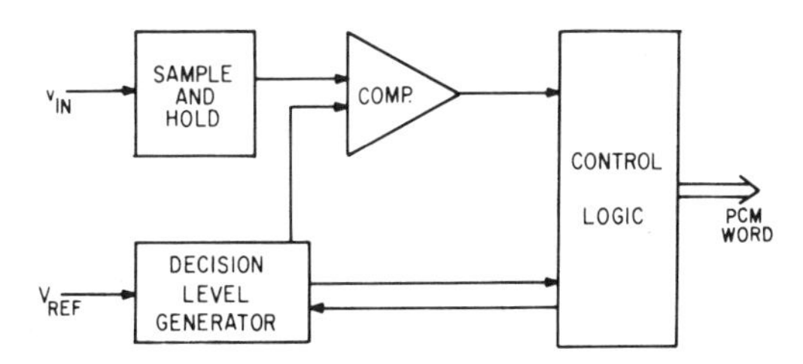

Fig. 2. Block diagram of encoder.

modified for actual transmission. We will, however, use it throughout our discussion, since it is completely straightforward. Other coding formats can always be implemented with additional logic at the output of the encoder.

Encoder Description

The principle of the encoder is illustrated in Fig. 2. Samples of the input signal are compared to fractions of the reference voltage V_{REF} which are developed by a "decision level generator" and correspond to the decision levels of the encoding law in Fig. 1. The logic then develops the PCM word corresponding to the interval in which the input lies. We now discuss the individual elements of the encoder.

Charge-Redistribution Decision Level Generator

To generate fractions of the reference voltage, some form of voltage divider must be employed. Accurate capacitive voltage dividers can be constructed using the principle of charge redistribution [4]. This principle is illustrated in Fig. 3. In the circuit shown in Fig. 3(a), equilibrium has been achieved. Next, V_A is assumed to undergo a change of magnitude ΔV_A. At the end of the transition, the capacitor voltages are as shown in Fig. 3(b). The change of the voltage of the top plates will be given by

$$\Delta V_x = \left(\frac{C_A}{C_A + C_B}\right) \Delta V_A. \tag{1}$$

The following aspects of this relation are important.

1) The capacitive divider action displayed by (1) relates only *changes* of voltages. The total values of the voltages of the sources and the initial capacitor voltages do not enter in the equation.

2) The shape of $V_A(t)$ during the transition is completely irrelevant. It is only the change $V_{A,\text{FINAL}} - V_{A,\text{INITIAL}}$ that determines ΔV_X.

3) If resistors are present in series with the capacitors to represent practical limitations, such as switch resistances and source resistances, the voltages will still take on the correct values once the transients have died out.

This principle of charge redistribution can be employed in the generation of the decision levels corresponding to the end points of the segments of the encoding law shown in Fig. 1. For the moment, we will consider only the 8 segments corresponding to positive voltages. All segments have 16 steps each except for the first, which has $15\frac{1}{2}$ steps. This last fact is of minor importance, and for the moment we will assume that segment also has 16 steps. The modification needed to get $15\frac{1}{2}$ steps will be considered separately later.

The voltages corresponding to the endpoints of the segments can be developed using the capacitor array of Fig. 4 which utilizes the principle of charge redistribution. The switches shown in this figure are electronic switches thrown by control

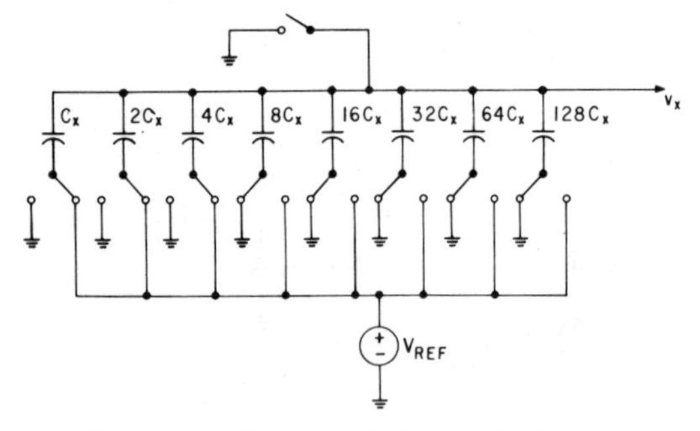

Fig. 4. Capacitor array used to generate the decision levels corresponding to the end-points of the segments.

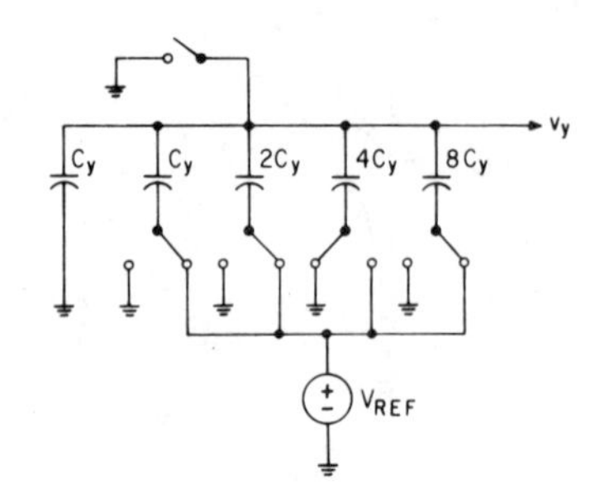

Fig. 5. Capacitor array used to develop multiples of $V_{\mathrm{REF}}/16$.

logic, and are for the moment assumed ideal. The system works as follows. Initially, all switches are thrown to ground, including that connected to the capacitors' top plates, so that all capacitors are discharged. Next, the top switch is opened, without affecting the capacitor voltages. Assume now that the first n-switches, starting from the left, are thrown to the reference voltage. Applying (1) gives

$$V_x = \frac{\sum_{k=1}^{n} 2^{k-1} C_x}{\sum_{k=1}^{8} 2^{k-1} C_x} V_{\mathrm{REF}} \tag{2}$$

Simplifying this expression,

$$V_x = \left(\frac{2^n - 1}{255}\right) V_{\mathrm{REF}}. \tag{3}$$

It is easily verified that V_x corresponds to the upper end-point of the nth segment. Thus, for example, to develop a voltage corresponding to the upper endpoint of the third segment, the bottom plates of the first three capacitors must be connected to V_{REF}.

The problem of developing voltages corresponding to the endpoints of each one of the sixteen equal steps of any segment will now be considered. This is easily done if the bottom plate of each capacitor can be thrown to not only V_{REF}, but to any integer multiple of $V_{\mathrm{REF}}/16$ from $1 \times (V_{\mathrm{REF}}/16)$ to $15 \times (V_{\mathrm{REF}}/16)$. For example, to make V_x correspond to the end of the 7th step of the 5th segment, the first 4 switches should be thrown to V_{REF}, and the 5th one to $7 \times (V_{\mathrm{REF}}/16)$.

These multiples of $V_{\mathrm{REF}}/16$ must be developed separately. Another capacitor array can be used for this purpose, as shown in Fig. 5. Notice that two equal capacitors of the smallest size are used to bring the total capacitance to $16\,C_y$. Any integer multiplier of $V_{\mathrm{REF}}/16$ can now be developed at the top plate, if after discharging all capacitors an appropriate combination of them is thrown to V_{REF}. More specifically, assume that $m \times (V_{\mathrm{REF}}/16)$ is desired, where m is an integer between 1 and 15. The number m can be expressed as a sum of powers of 2:

$$m = \sum_{k=1}^{4} b_k 2^{k-1} \tag{4}$$

where b_k is 1 or 0. The bottom plates of the capacitors corre-

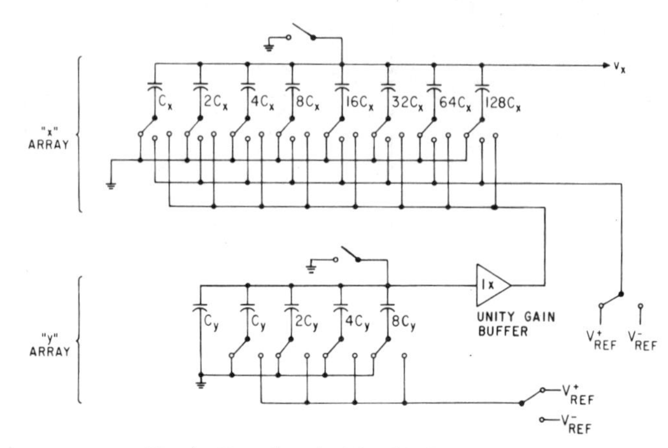

Fig. 6. Complete decision level generator.

sponding to $b_k = 1$ are then thrown to V_{REF}. The voltage V_y then can be found from (1):

$$V_y = \left(\frac{\sum_{k=1}^{4} b_k 2^{k-1} C_y}{16 C_y}\right) \times V_{\mathrm{REF}} = \sum_{k=1}^{4} b_k 2^{k-1} \times \frac{V_{\mathrm{REF}}}{16}. \tag{5}$$

For example, if $V_y = 7 \times (V_{\mathrm{REF}}/16)$ is desired, then, since $7 = 1 + 2 + 4$, the bottom plates of the capacitors of values C_y, $2C_y$, and $4C_y$ must be thrown to V_{REF}.

The fractions of V_{REF} developed by this array must now be fed to the bottom plates of the x array of Fig. 4, as mentioned above. The proper operation of these arrays is based on the fact that they are unloaded, and therefore V_y cannot be directly fed to the bottom plate of the x array. Therefore, a unity gain buffer of very high input resistance must be placed in the path. The complete decision level generator is then as shown in Fig. 6. Using this system, voltages corresponding to any decision level of the encoding law of Fig. 1 can be generated. Negative decision levels can be generated using the principle explained above, but a negative reference voltage is required. To generate a decision level, all capacitors of both arrays are first discharged, by throwing all switches to ground. Next, the switches connected to the top plates of both arrays are opened. If the desired decision level corresponds to the end of the mth step of the nth segment, the first $n - 1$ capacitors of the x array are thrown to V^{+}_{REF}(or V^{-}_{REF}), the nth capacitor of the x array is thrown to the output of the buffer,

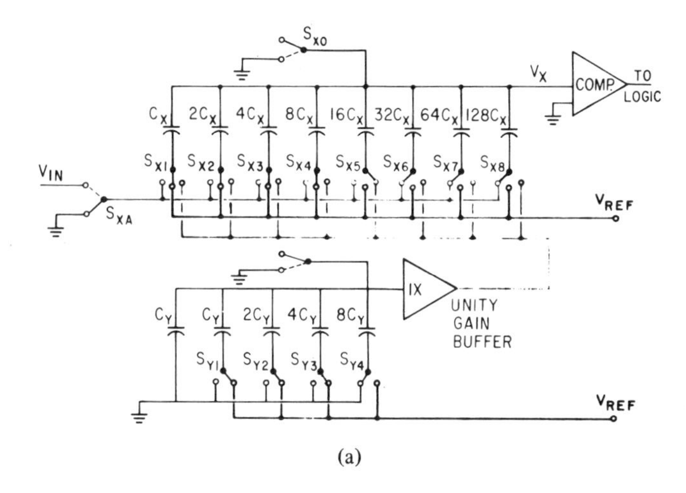

(a)

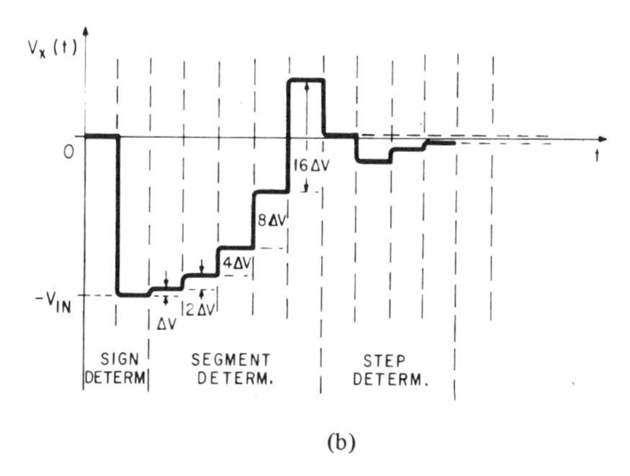

(b)

Fig. 7. (a) Diagram of the complete encoder. (b) Comparator input waveform.

and in the y array these capacitors which result to $V_y = m(V_{REF}/16)$ are thrown to V_{REF}^+ (or V_{REF}^-).

Sample-and-Hold Circuit and Comparator

The sample-and-hold function is realized by using the x-array of Fig. 6 as the store capacitor. The array is then used to subtract the input voltage sample from the decision levels, and the difference thus formed is compared to zero by the comparator. This process is explained in detail in the following section.

Operation of Complete Encoder

The complete PCM encoding principle is shown in Fig. 7(a), and the voltage at the input of the comparator in Fig. 7(b). The encoding sequence will be described by means of example. Assume that at the time of the sampling, V_{IN} is positive and lies in the 8th step of the 5th segment. This is determined by the system as follows.

A) Sampling and Sign Determination: Switch S_{x0} is thrown to ground, and S_{x1} through S_{x8} to the input voltage (through S_{xA}). Let V_{IN} be the value of the input at the end of this step. This value is thus sampled and stored on the x array. Next, S_{x0} is opened and S_{xA} is thrown to ground, making V_x equal to $-V_{IN}$, as shown in Fig. 7(b). The sign of V_x is sensed by the comparator, whose state, therefore, determines the sign of V_{IN}. This information is fed to the logic, which connects a positive or a negative reference voltage to the buses labeled V_{REF}, according to whether the sign of V_{IN} was sensed as positive or negative. For the particular example, a positive reference voltage is connected. The first bit of the 8-bit PCM word, corresponding to the sign of V_{IN}, has been determined at the end of this step.

B) Segment Determination: Switch S_{x1} is thrown to V_{REF}, making V_x increase by an amount ΔV corresponding to the first segment, as shown in Fig. 7(b). Since V_x does not change sign after this, the next switch (S_{x2}) is thrown, then the third, and so on, until the comparator senses a change of sign in V_x, which happens just after S_{x5} is thrown. This information is interpreted by the logic as meaning that V_{IN} lies in the 5th segment. The logic then develops the next three bits of the PCM word, corresponding to segment 5.

C) Step Within Segment Determination: The system must now determine within which one of the sixteen equal steps of the 5th segment the input lies. To this end, S_{x5} is thrown to the output of the buffer, and fractions of V_{REF}, developed by the y array, are fed to the bottom plate of capacitor $16C_x$. This is done in a successive approximation sequence as follows: Initially, all capacitors of the y array are discharged by throwing all switches to ground. Next, S_{y0} is opened, and S_{y4} is thrown to V_{REF}, thus making $V_y = V_{REF}/2$. As a consequence V_x drops, as seen in Fig. 7(b), but not enough to make its sign change. This is interpreted by the logic as meaning that V_{IN} must be in the first half of the segment. To determine to which quarter of the segment the input corresponds, S_{y4} is thrown back to ground and S_{y3} is thrown to V_{REF}, thus making V_y equal to $V_{REF}/4$. This makes V_x change sign, meaning that V_{IN} is in the second quarter of the segment. S_{y3} is left at V_{REF}, and S_{y2} is also thrown to V_{REF}. No sign change occurs in V_x, meaning that V_{IN} is in the fourth eighth of the segment. Finally, to check in which sixteenth of the segment V_{IN} lies, S_{y3} and S_{y2} are left undisturbed and S_{y1} is thrown to V_{REF}, which does not result in a sign change in V_x. At the end of this, therefore, V_{IN} has been bounded by the values $V_{REF}/2$ and $7(V_{REF}/16)$, and therefore must lie in the 8th step of the segment. The last 4 bits of the PCM word are thus determined by the logic, and the conversion is complete. For this particular example, and for the code format shown in Fig. 1(b), the logic develops the PCM word 1 100 0111 at the output of the encoder, either in parallel form or sequentially. The final position of the switches at the end of the encoding process is as shown in Fig. 7(a).

In the encoding scheme described, one step is needed for sampling, one for sign determination, up to eight steps for segment determination, and four steps for step determination. A maximum of fourteen steps is therefore needed for the complete encoding of each input value. The number of steps can be reduced to 9 by using successive approximation in segment determination.

The values developed by the decision level generator described above are slightly different from the analog values at the output of the corresponding decoder. However, by splitting one of the two smallest capacitors of the y-array in two, the decision level generator can be used to also generate the decoder output levels, thus making possible the realization of both the encoder and the decoder on the same chip, on a time-share basis.

The negative reference voltage can be generated on the chip from the positive reference by charging a capacitor to the positive reference, inverting its polarity and connecting it to the input of a unity of a gain buffer.

Effect of Nonidealities on Coder Performance

In an implementation of the encoder scheme nonidealities in the active and passive components deteriorate the system performance from the idealized case discussed above. These non-idealities include parasitic capacitance, capacitor value errors, offset voltage in the comparator, offset voltage and gain error in the buffer, and mismatch between positive and negative reference voltages. In order to discuss these effects, we will consider a practical encoder coupled to an ideal decoder, and we will investigate the effect of the nonidealities on the complete codec. Errors in the transfer characteristic of this codec will be manifested as degradation in the signal-to-distortion ratio, gain tracking, and single-frequency distortion performance.

A computer program has been written [6] for simulating the performance of coder-decoder combinations with user-specified characteristics. The effect of nonidealities on the performance of the codec has been studied using the program. The most important results will now be summarized.

Parasitic Capacitances

There are several parasitic capacitances associated with the encoder of Fig. 7. These are due to junction capacitance, interconnect and gate overlap capacitance of the switch devices and the input devices of the comparator and the buffer. Of these, the parasitic capacitances appearing between the bottom plates of the x and y array capacitors and ground have no effect in the performance. As the bottom plates of the array capacitors are being switched between the various dc voltages required, these parasitics will only affect the shape of the voltage waveforms during the transition. It is, however, clear that the final dc voltages at the bottom plates *after* the transients cannot be affected by the parasitics. Since the final values of V_x and V_y only depend on the initial and final values of these voltages, these parasitics cannot affect the performance of the system.

The parasitic capacitance between the top plates of the x array and ground does not have any significant effect on the performance of the circuit. The encoding sequence discussed in the previous section starts with $V_x = 0$ and finishes with $V_x \simeq 0$. This means that the parasitic starts and finishes with the same charge and, therefore, it does not contribute a net amount of charge to the array's top plates over the encoding cycle. This will be the case even if this parasitic capacitance is nonlinear. For this reason, the value of the parasitic can be large compared with C_x without affecting the encoder's performance.

The parasitic capacitance between the top plates of the y array and ground will have the effect of reducing the values of the voltage V_y produced by that array. Notice that this parasitic has no effect on the boundaries between segments, the latter being determined by the x array. It has been found that parasitics larger than C_y can be tolerated.

Buffer Gain Error

In the absence of buffer offset, the effect of buffer gain error is the same as that of parasitic capacitance at the top plates of the y array; the boundaries of the segments are unaffected, but the step size within the segments is reduced (if the gain is below unity). The result is a single long step at the top end of the segment. The parameter most strongly affected by this error is the signal-to-distortion (S/D) ratio. The effect becomes large only when the gain error approaches the resolution of the y array, which is only one part in sixteen.

Buffer Offset

The effect of an input offset voltage in the buffer is to add to all values of V_y a constant offset equal to that value. The segment boundaries again do not move but each decision level within the segments is changed. The principal effect is to degrade signal-to-noise ratio, and the effect does not become significant until the offset becomes comparable to the step size for the voltage V_y, which is 310 mV for a reference voltage of 5 V.

Comparator Offset

The effect of an input offset voltage in the comparator is to shift the transfer characteristic of the encoder so that it no longer passes through the origin. If the size of the shift is comparable to that of the first segment, then small-amplitude inputs will fall partially into segments of coarser resolution than would otherwise occur, and the signal-to-distortion ratio will be degraded. Since the first segment has a span of approximately 20 mV for a reference of 5 V, the comparator input offset voltage should be small compared to 20 mV.

Reference Source Mismatch

Mismatch between positive and negative reference voltages destroys the symmetry of the encoder characteristic. The different slopes of the positive and negative transfer characteristics have the principal effect of increasing the second-harmonic distortion. Since the second-harmonic component must be kept more than 40 dB down in toll-quality transmission systems, the coder should contribute a second-harmonic component which is small compared to 1 percent of the fundamental. This implies a reference voltage mismatch which is small compared to 2 percent.

Capacitor Value Errors

From the description of the encoding principle given, it is apparent that what is really important for accurate encoding is not the absolute values of the capacitors in the arrays, but rather the ratios between them and the total capacitance. Capacitor errors in the x array will result in average slope errors for the segments in the encoding characteristic which will degrade the gain tracking and single-frequency distortion. Errors in the y array will result in unequal step sizes for each segment, leading to degradation in S/D ratio.

Effect of the Half Step at the Origin

Simulation shows that using sixteen equal steps in the smallest segment rather than the $15\frac{1}{2}$ required for smooth transition at the origin, does not result in any significant degradation in the encoder performance when compared to the degradation due to other nonidealities. The S/D ratios for the two cases differ by a fraction of a decibel even for low amplitudes.

Performance of the Encoder in the Presence of Simultaneous Nonidealities

In an actual coder realization, all the errors mentioned above are present. In addition, the finite settling time of the array, noise, pickup, and so forth will degrade the performance of the coder. While it is not feasible to simulate all possible combinations of nonidealities, it is possible to choose one set of worst case values, and demonstrate by simulation that if these performance levels are met in the comparator, buffer, and capacitor arrays, the performance of the coder will be acceptable. What constitutes acceptable performance for an encoder for use in toll quality telephone transmission is not easily determined. The most widely used specification, the Bell System D3 channel bank specification [5], is one which describes the performance of a complete transmission system and depends on many degrading factors other than the performance of the coder. For the purpose of this study, we have assumed that the encoder and decoder contribute equally to the various errors, so that the coder design objective when evaluated with an *ideal* decoder is an S/D ratio 3 dB above the D3 specification, a single-frequency distortion 3 dB below the D3 specification, and a gain tracking error of one-half that allowed in the D3 system. Computer simulation of the coder has shown that these requirements will be met if the following performance levels are achieved in the individual components, assuming a reference voltage of 5 V:

a) comparator offset: <5 mV;
b) buffer offset: <100 mV;
c) buffer gain error: <2 percent;
d) reference sources mismatch: <0.5 percent;
e) x array capacitor value errors: 10 percent for smallest capacitor, 0.2 percent for largest;
f) y array capacitor value errors: 10 percent for smallest capacitor, 1 percent for largest.

The result of a computer simulation of S/D ratio as a function of signal level for a typical worst case combination of these nonidealities is shown in Fig. 8.

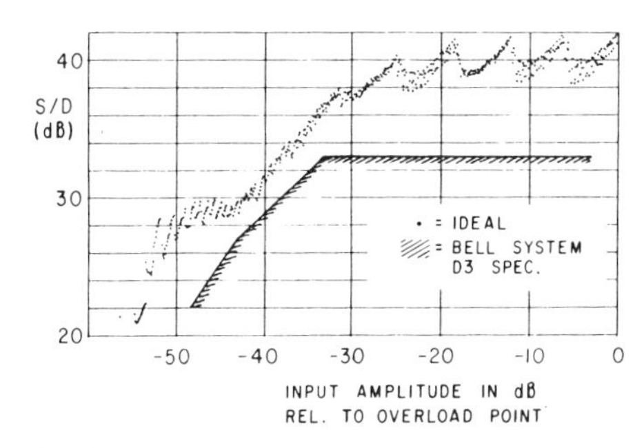

Fig. 8. Simulated S/D ratio versus input amplitude for typical combination of worst-case comparator offset, buffer offset, buffer gain error, reference source mismatch, and capacitor ratio errors.

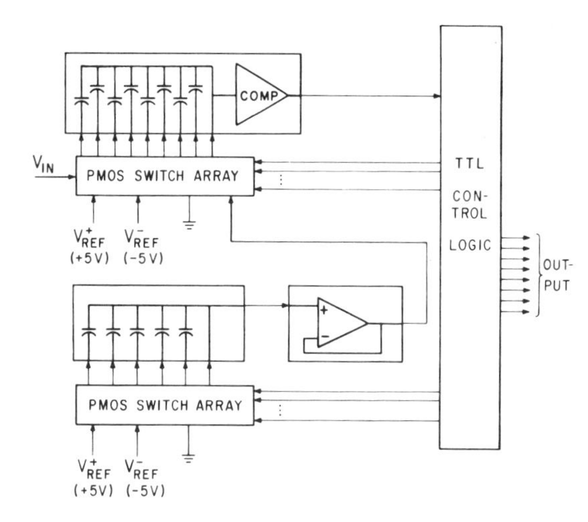

Fig. 9. Block diagram of the experimental encoder.

Experimental Prototype

The PCM encoder prototype is shown in block diagram form in Fig. 9. Here, the logic controls the state of the electronic analog switches and develops the output PCM word corresponding to the input analog voltage value to be encoded. The logic is in turn controlled by the output of the comparator.

To implement the x- and y-capacitor arrays and the comparator, two of the A/D converter chips developed by McCreary have been modified and used, with the total capacitance in each array being 64 pF. A detailed description of these chips is given in [7]. The op amp used to implement the unity gain buffer was fabricated using the same technology with the one used for the comparator and capacitor array chips, and is described in detail in [6] and [8].

The control logic consists of standard logic gates and has been implemented using discrete TTL integrated circuits. It consists of 150 gates and 50 flip-flops. Integrating the logic in NMOS is straightforward; for simplicity this was not attempted.

Although there are no problems in fabricating the analog switches required in NMOS, the number of package pins required for their connection to the rest of the system would be prohibitive, since most of them are equivalent to single-pole triple-throw types. Commercially available PMOS switches have been used, their performance being similar to that attainable in NMOS.

In the encoder prototype, the top plates of the x-array are not grounded at the beginning of the conversion cycle. Instead, all capacitors of the x-array are charged to the equilibrium voltage of the comparator [7]. Since the charge redistribution principle described above is based on changes of voltages rather than their absolute values, the operation of the encoder remains basically as described.

The performance of the system is representative of what can be achieved if the complete encoder is fabricated on a single chip, since all critical parts, mainly the two capacitor arrays, the comparator, and the operational amplifier have been fabri-

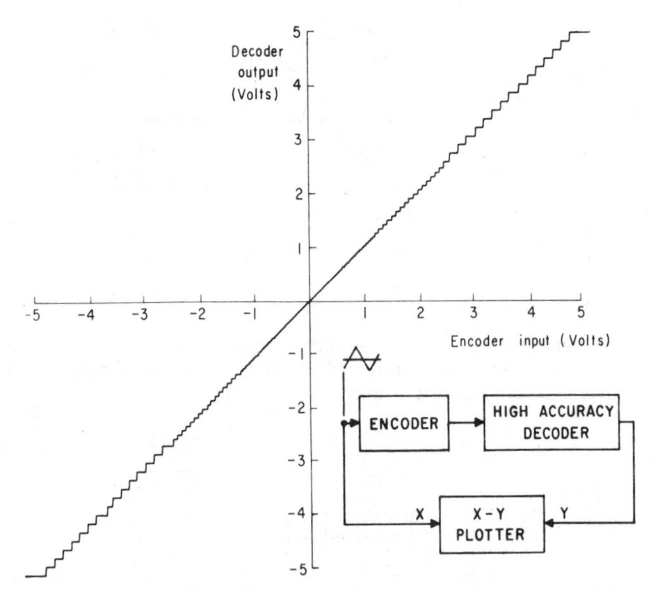

Fig. 10. DC transfer characteristic of the coder together with an ideal decoder.

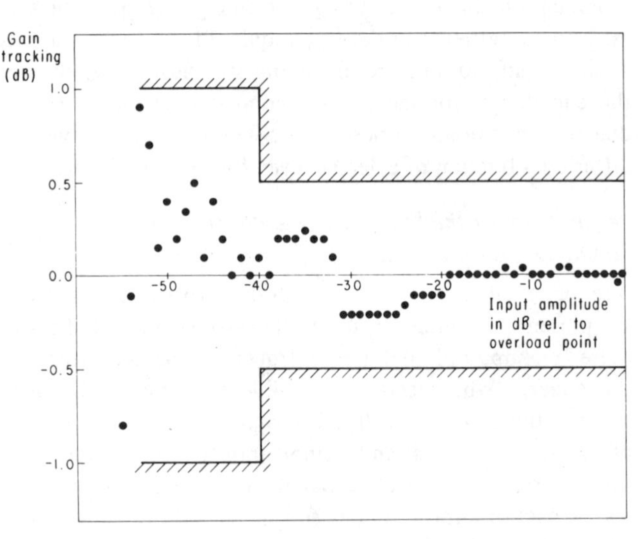

Fig. 12. Measured gain tracking versus input amplitude.

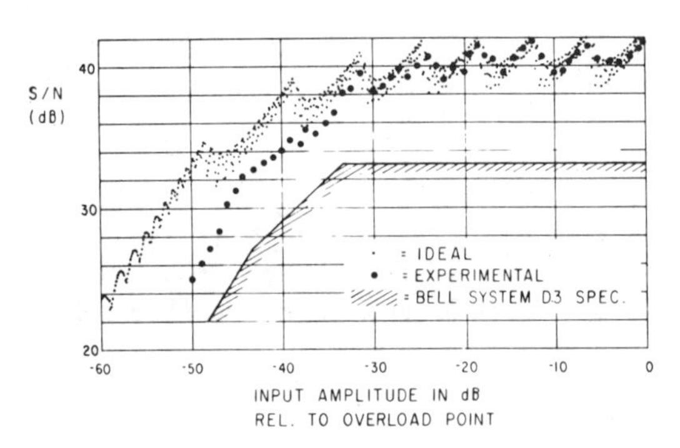

Fig. 11. Measured S/D ratio versus input amplitude.

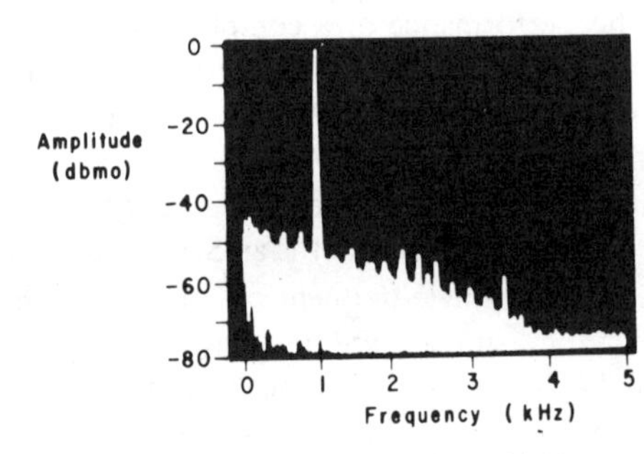

Fig. 13. Output spectrum for a 1.02-kHz sinusoidal input of amplitude −3 dB relative to overload.

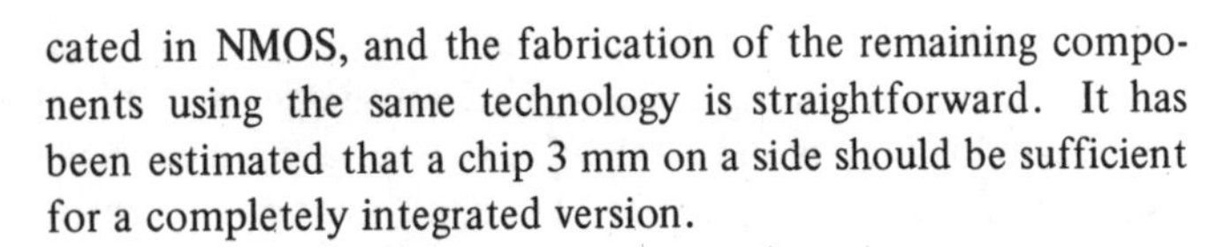

cated in NMOS, and the fabrication of the remaining components using the same technology is straightforward. It has been estimated that a chip 3 mm on a side should be sufficient for a completely integrated version.

Experimental Results

The input-output characteristic of the encoder-decoder combination has been obtained using an *x-y* plotter and is shown in Fig. 10. In this and all subsequent experiments a near-ideal decoder was used. The continuity at segment boundaries and the monotonicity are apparent. The reader is cautioned that in the following results the overload point of the system has been used as a 0-dB reference. In the D3 channel bank specifications, the value of the voltage corresponding to overload is defined as +3 dBm.

The *C*-message weighted S/D ratio was measured for input amplitudes over a 50-dB dynamic range at a frequency of 1.02 kHz. The results are shown in Fig. 11, along with the computer simulated performance of an ideal codec and the D3 channel bank specification bound. It is seen that the behavior of the encoder is practically ideal for high amplitudes. At low amplitudes, S/D still exceeds the specification bound by a margin sufficient to allow for about 5 dB additional degradation by the other components of a practical PCM system. The idle channel noise of the system was 18 dB rnc. The gain tracking versus amplitude is shown in Fig. 12 along with the D3 channel bank system specification bounds.

Fig. 13 shows the observed output spectrum for a sinusoidal input of frequency 1.02 kHz, an amplitude of −3 dB relative to overload, and with an output filter but no *c*-message filter. The vertical scale is in decibels relative to the amplitude of the fundamental. The gradual reduction of the amplitude for increasing frequency is partly due to the effect of a sample-and-hold circuit which was used at the output of the decoder for these measurements. When corrected for this droop, the second and third harmonic components are more than 43 dB below the fundamental.

Conclusion

A technique for PCM encoding has been described which makes use of the charge redistribution principle in order to realize the 15-segment approximation of the μ-255 encoding law. A prototype has been built, in which all critical functions are performed by circuits which have been integrated using

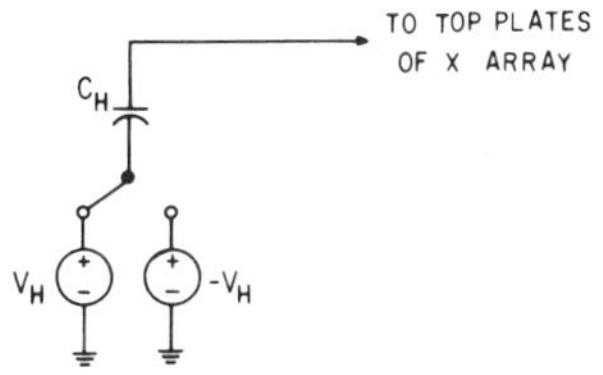

(a)

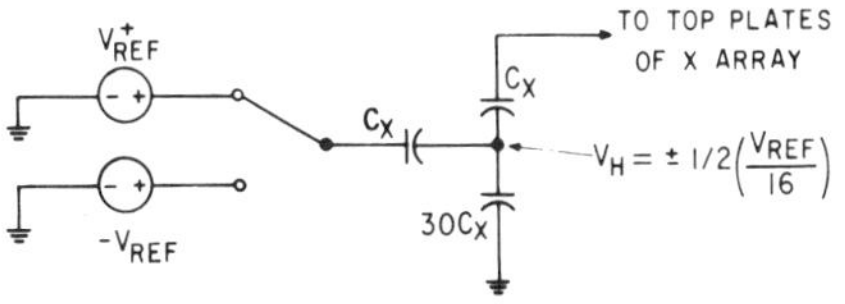

(b)

Fig. 14. Methods for introducing a one-half step offset in the first segment.

n-channel Al-gate MOS technology. Both the performance of the prototype and computer simulation show that a single chip, one-channel PCM encoder is realizable.

APPENDIX

The scheme described implements an encoding law in which there are sixteen equal steps in the first segment. In the actual encoding law of Fig. 1, however, there are $15\frac{1}{2}$ steps rather than sixteen in the first segment, and half-step being adjacent to the origin. This half-step, along with the symmetrical half-step for negative inputs, result in a whole step which has the origin as its middle point. As has been explained in the main text, whether there are 16 or $15\frac{1}{2}$ steps in the first segment is of minor importance. If it is desired that the half-step detail be implemented, one can introduce an offset to the *x*-array at the beginning of the conversion, which is equivalent to the half step. To do this, a ninth capacitor can be connected to the top plate, as shown in Fig. 14(a). A positive or a negative voltage V_H is connected to this system according to whether the sign of the input voltage has been determined to be negative or positive, respectively, at the beginning of the conversion sequence. One particular realization, which derives V_H from V_{REF}, is shown in Fig. 14(b).

ACKNOWLEDGMENT

The authors wish to thank Dr. J. L. McCreary for fabricating and making available to them a special version of the capacitor arrays, Dr. D. L. Duttweiler and Dr. R. M. Scarlett for useful discussions and suggestions. The technical advice and help of J. Albarrán and R. McCharles is appreciated.

REFERENCES

[1] J. C. Candy, W. H. Ninke, and B. A. Wooley, "A per-channel A/C converter having 15 segment μ-255 companding," *IEEE Trans. Communications*, vol. COM-24, pp. 33-42, Jan. 1976.
[2] K. Euler, "PCM codecs for communication," in *Proc. Int. Symp. Circuits and Systems*, pp. 583-586, 1976.
[3] A. Rijbroik, "Design approaches for a PCM-codec per channel," in *Proc. Int. Symp. Circuits and Systems*, pp. 587-590, 1976.
[4] J. McCreary and P. R. Gray, "All-MOS charge-redistribution analog-to-digital conversion techniques, Part I," *IEEE J. Solid-State Circuits*, vol. SC-10, pp. 371-379, Dec. 1975.
[5] *The D3 Channel Bank Compatability Specification*, Issue 2, Engineering Department, American Telephone and Telegraph Company, Oct. 14, 1974.
[6] Y. P. Tsividis, "Nonuniform pulse code modulation encoding using integrated circuit techniques," Ph.D. dissertation, University of California, Berkeley, 1976.
[7] J. L. McCreary, "Successive approximation analog-to-digital conversion in MOS integrated circuits," Ph.D. dissertation, University of California, Berkeley, 1975.
[8] Y. P. Tsividis and P. R. Gray, "An integrated NMOS operational amplifier with internal compensation," this issue, pp. 748-753.

A Single-Chip NMOS PCM CODEC for Voice

John M. Huggins, Marcian E. Hoff and Ben M. Warren

Intel Corp.

Santa Clara, CA

THE LOGIC DENSITY inherent in N-channel MOS technology has been combined with high density analog techniques in the development of a per-channel, single-chip PCM coder-decoder. The CODEC performs all sample/hold, A/D, D/A, and voltage reference functions with no external precision components. The chip also incorporates a user-programmable timeslot decoder which gives the device local switching capability. This paper will describe the system interface and several significant NMOS-compatible analog circuit techniques incorporated in the design.

Figure 1 shows a block diagram of the CODEC. Independent channels are provided for send and receive paths. Sharing of the DAC necessitates a transparent internal interrupt for asynchronous operation. The PCM data format, both send and receive, consists of 8-bit serial data blocks multiplexed to a high-speed serial bus which is shared by as many as 32 CODECs. Each CODEC controls the multiplexing and demultiplexing of its own data via user assigned time-slots. A pair of user accessible time-slot registers (one for each direction) are loaded via an 8-bit serial control word. The CODEC counts clock pulses from the framing pulses and accesses the send and receive busses only in the designated time-slots. The send and receive time-slots may be independently and dynamically reassigned in switching, concentrator, and control environments[1]. The same control word can be used to place the CODEC in a reduced power standby mode. A CODEC generated time-slot strobe identifies the window in which PCM data is being transmitted. This signal can be used for data bus isolation, or it can be used to distinguish PCM channels from data channels, permitting external common zero code suppression in channel banks. The three-state output to the PCM bus is capable of driving up to 500pF at bit rates in excess of 2MHz.

Digital signaling (control) information can be multiplexed to and from the PCM bit stream during user specified signal frames via device ports. On a signaling frame the CODEC send side will encode the incoming analog signal and substitute the signaling data for the 8th-bit in the PCM word. Signaling operation on the receive side is independently controlled; the CODEC decodes the resulting 7-bit PCM code word, while latching the 8th bit to the signaling output pin.

The resistor string D/A converter provides one resistor and one transmission gate for each discrete DAC level. A weighted resistor voltage divider employs one weighting resistor per segment to produce the desired companding law. Metal mask selection of weighting resistors allows implementing either μ or A-law companding. Each weighting resistor is shunted by a loop of 32 diffused unit resistors which interpolate between the segment end points. The 16 encode levels in each segment are interlaced with the 16 decode levels, producing mid-tread or *decision level assignment* decoding. Access to any of the 32 resistive nodes in each segment and hence to any DAC value is obtained by selection of its associated transmission gate.

An NMOS compatible voltage reference has been developed. The reference circuit is based on the difference in thresholds between two suitably implanted MOS devices. Control of the geometry and biasing of the two devices produces a reference voltage with very low temperature and bias voltage coefficients and insignificant long term drift. The reference voltage is buffered to provide current drive for the resistive DAC. Because process-related variations in the initial reference value can be substantial, the reference voltage is trimmed at wafer sort to provide the desired full scale DAC voltage of 3.155V.

The resolution and offset demands of the CODEC voltage comparator are beyond the capabilities of simple NMOS differential sense amplifiers; hence the techniques of chopper stabilization were employed to obtain a comparator with input offset below 100μV. The high gain requirements of the comparator (50,000V/V) are easily obtained with cascaded amplifiers.

A photograph of the packaged 22,000 mil^2 die is shown in Figure 2. The analog functions of D/A conversion, compare, voltage reference, and linear buffering comprise less than 50% of the die area, with the remaining area used for the logic interface. This partitioning of complex analog and digital functions on a single MOS substrate demonstrates the versatility of the technology. Table 1 contains typical data taken on fabricated μ-law devices.

The specification derived from worst case data meets accepted transmission and switching standards[2].

Acknowledgments

The authors wish to acknowledge the contributions of D. Senderowicz to the buffer amplifier and general amplifier techniques. The expert assistance of D. Melvin in product definition and D. Royce in test design were major factors in the project. The dedicated support of M. Bisgood and G. Simko was invaluable.

[1] Melvin, D.K., "Microcomputer Applications in Telephony", *Proceedings of the IEEE*, Vol. 66, No. 2; Feb., 1978.

[2] "The D3Channel Bank Compatibility Specification", *American Telephone and Telegraph Company*, Issue 2; Oct., 1974.

Reprinted from *1978 IEEE Internat. Solid-State Circuit Conf. Dig. Tech. Papers*, Feb. 15-17, 1978, 178-179.

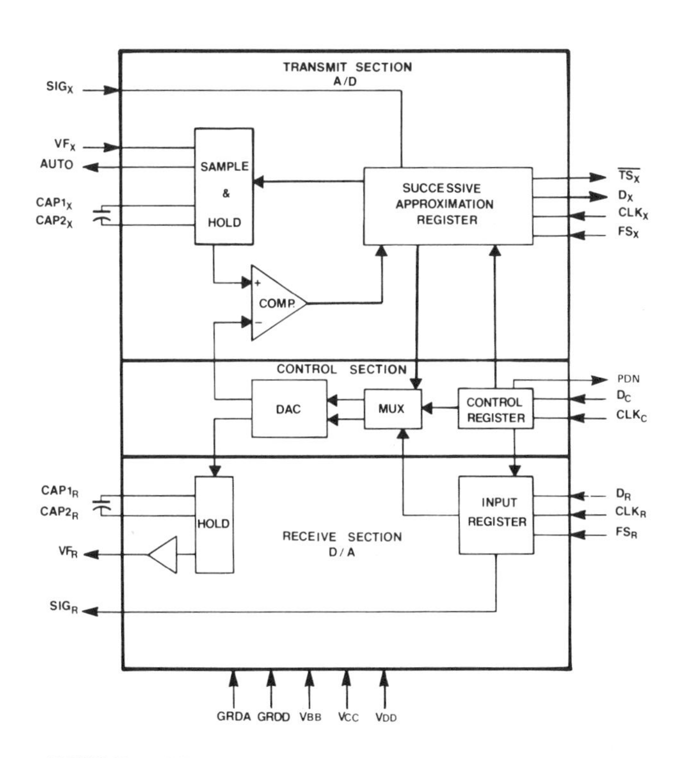

FIGURE 1—CODEC block diagram. The three sections can be operated with asynchronous clocks; the control word on D_C is clocked in via CLK_C.

FIGURE 2—The μ-255 version assembled in a 24-lead package.

IDLE CHANNEL NOISE	12 dBrnc0
GAIN TRACKING	0.2 dB @ −37 dBm0
	0.5 dB @ −50 dBm0
	1 dB @ −55 dBm0
SIGNAL/DISTORTION	33 dB @ −39 dBm0
	39 dB @ −29 dBm0
	42 dB @ 0 dBm0
SINGLE FREQUENCY DISTORTION	−48 dB

TABLE 1—Typical CODEC performance.

XR-C277 Low-Voltage PCM Repeater IC

INTRODUCTION

The XR-C277 is a monolithic repeater circuit for Pulse-Code Modulated (PCM) telephone systems. It is designed to operate as a regenerative repeater at 1.544 Mega bits per second (Mbps) data rates on T-1 type PCM lines. It is packaged in a hermetic 16-pin CERDIP package and is designed to operate over a temperature range of -40°C to +85°C. It contains all the basic functional blocks of a regenerative repeater system including Automatic Line Build-Out (ALBO) and equalization, and is insensitive to reflections caused by cable discontinuities.

The key feature of the XR-C277 is its ability to operate with low supply voltages (6.3 volts and 4.3 volts) with a supply current of less than 13 mA. Compared to conventional repeater designs using discrete components, the XR-C277 monolithic repeater IC offers greatly improved reliability and performance and provides significant savings in power consumption and system cost.

FUNDAMENTALS OF PCM REPEATERS

Figure 1 shows the block diagram of a bi-directional PCM repeater system consisting of two identical digital regenerator or repeater sections, one for each direction of transmission. These repeaters share a common power supply. The DC power is simplexed over the paired cable and is extracted at each repeater by means of a series zener diode regulator. The XR-C277 monolithic IC replaces about 90% of the electronic components and circuitry within the "digital repeater" sections of Figure 1. Thus, a bi-directional repeater system would require two XR-C277 IC's, one for each direction of information flow.

Figure 2 shows the functional block diagram of one of the digital repeater sections, along with the external zener regulator. The basic system architecture shown in the figure is the same as that utilized in the design of the XR-C277 monolithic IC.

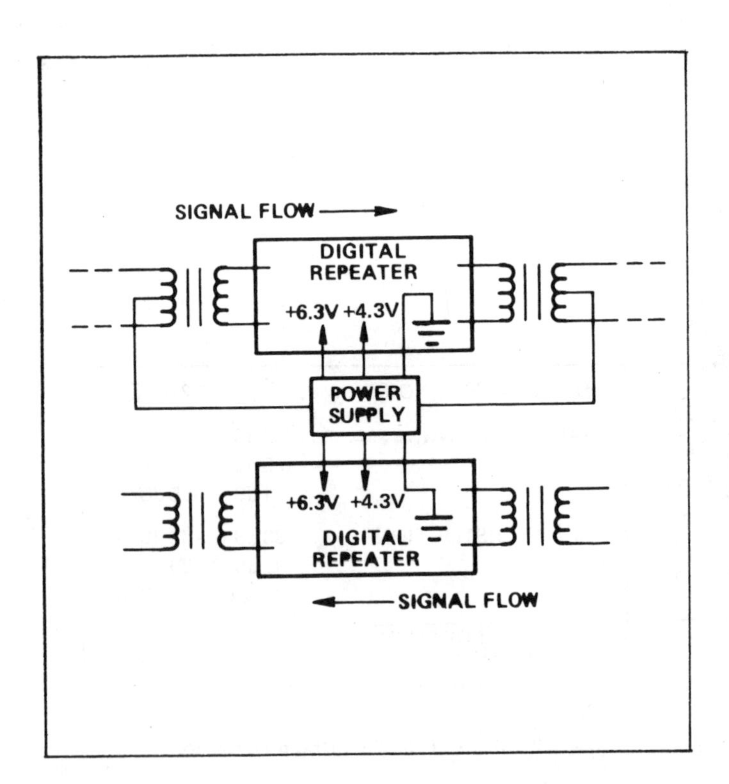

Figure 1. Block Diagram of a Bi-Directional Digital Repeater System

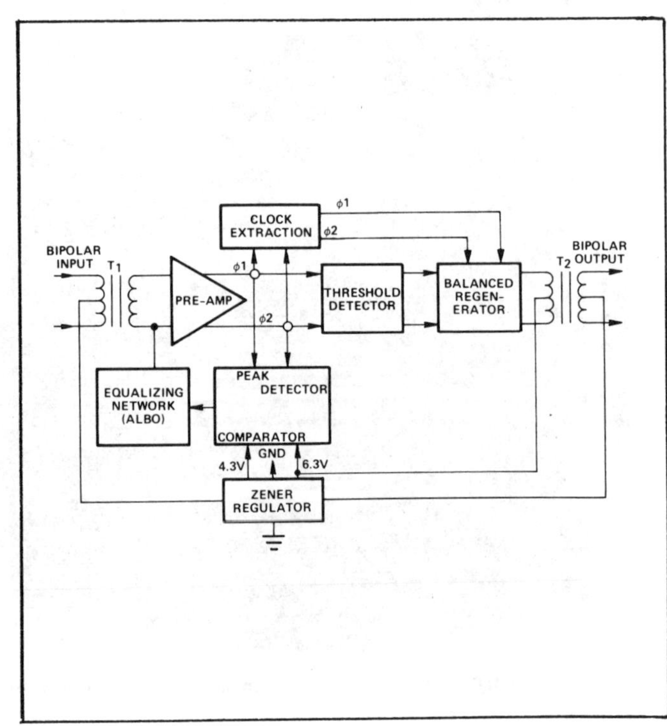

Figure 2. Functional Block Diagram of a Digital PCM Repeater Section

EXAR INTEGRATED SYSTEMS, INC.
750 Palomar Ave., P.O. Box 62229, Sunnyvale, CA 94088
(408) 732-7970 TWX 910-339-9233

Reprinted with permission from *Exar Integrated Systems Application Note AN-04,* Feb. 1978.

In terms of the functional blocks shown in Figure 2, the basic operation of the repeater can be briefly explained as follows:

> The bipolar signal, after traversing through a dispersive, noisy medium, is applied to a linear amplifier and automatic equalizer. It is the function of this circuit to provide the necessary amount of gain and phase equalization and, in addition, to band limit the signal in order to optimize the performance of the repeater for near-end crosstalk produced by other systems operating within the same cable sheath.
>
> The output signals of the preamplifier which are balanced and of opposite phases are applied to the clock extraction circuit and also to the pulse regenerator. The signals applied to the clock extraction circuit are rectified and then applied to a high-Q resonant circuit. This resonant circuit extracts a 1.544 MHz frequency component from the applied signal. The extracted signal is first amplified and then used to control the time at which the output signals of the preamplifier are sampled and also to control the width of the regenerated pulse.
>
> It is the function of the pulse regenerator to perform the sampling and threshold operations and to regenerate the appropriate pulse. The regenerated pulse is in turn applied to a discrete output transformer which is used to drive the next section of the paired cable.

REFERENCES OF PCM REPEATERS:

1. Mayo, J. S., "A Bipolar Repeater for Pulse Code Signals," B.S.T.J., Vol. 41, January, 1962, pp. 25-97.
2. Aaron, M. R., "PCM Transmission in the Exchange Plant," B.S.T.J., Vol. 41, January, 1962, pp. 99-143.
3. Davis, C. G., "An Experimental Pulse Code Modulation System for Short-Haul Trunks," B.S.T.J., Vol. 41, January, 1962, pp. 1-25.
4. Fultz, K. E. and Penick, D. B., "The T-1 Carrier System," B.S.T.J., Vol. 44, September, 1965, pp. 1405-1452.
5. Tarbox, R.A., "A Regenerative Repeater Utilizing Hybrid IC Technology," Proceedings of International Communications Conference, 1969, pp. 46-5 – 46-10.

OPERATION OF THE XR-C277

The XR-C277 combines all the functional blocks of a PCM repeater system in a single monolithic IC chip. The pin connections for each of the functional circuits within the repeater chip are shown in Figure 3, for a 16-pin dual-in-line package.

The circuit is designed to operate with two positive supply voltages, V^{++} and V^{+} which are nominally set to be 6.3V and 4.3V, respectively. Figure 4 gives the recommended power supply connection for the circuit.

The supply currents I_A and I_B drawn from the two supply voltages applied to the chip are specified to be within the following limits:

a. Current from 6.3V supply voltage, I_A:

$$2.5 \text{ mA} \leqslant I_A \leqslant 4.0 \text{ mA}$$

b. Current from 4.3V supply voltage, I_B:

$$7 \text{ mA} \leqslant I_B \leqslant 9 \text{ mA}$$

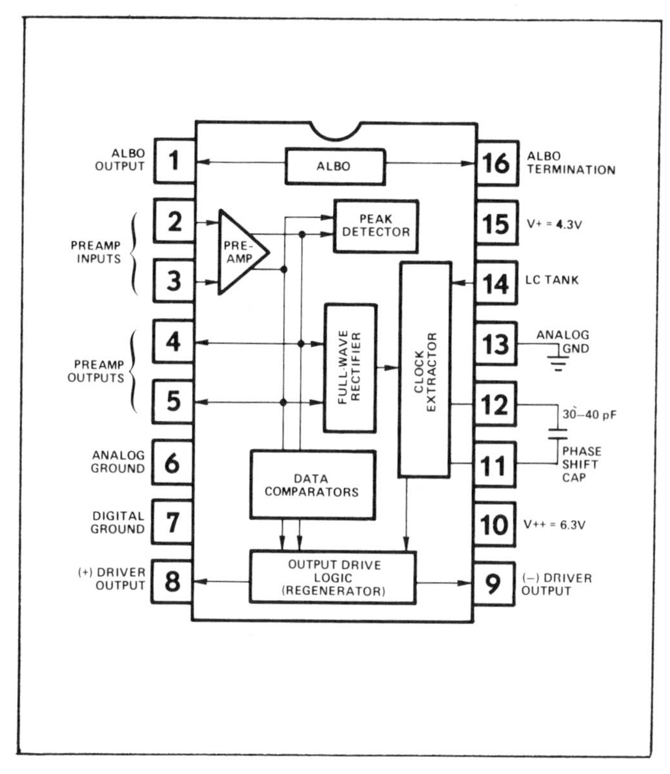

Figure 3. Package Diagram of XR-C277 Monolithic PCM Repeater

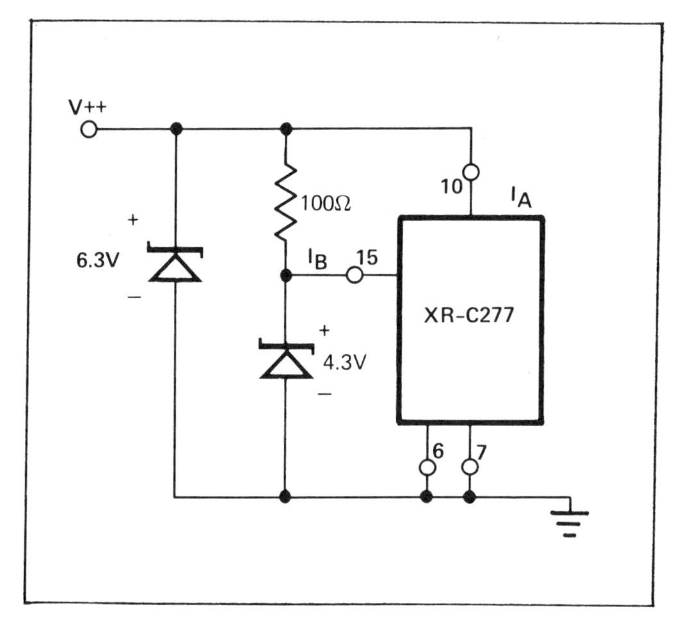

Figure 4. Recommended Supply Voltage Connection for XR-C277 (Note: See Figure 6 for Recommended Bypass Capacitors)

The external components necessary for proper operation of the circuit are shown in Figure 5, in terms of the system block diagram. Note that all the blocks shown in Figure 5 are a part of the monolithic IC; and the numbered circuit terminals correspond to the IC package pins (see Figure 3).

Figure 6 shows a practical circuit connection for the XR-C277 in an actual PCM repeater application for 1.544 Mbps T-1 Repeater application. For simplification purposes, the lightning protection circuitry and the second repeater section are not shown in the figure.

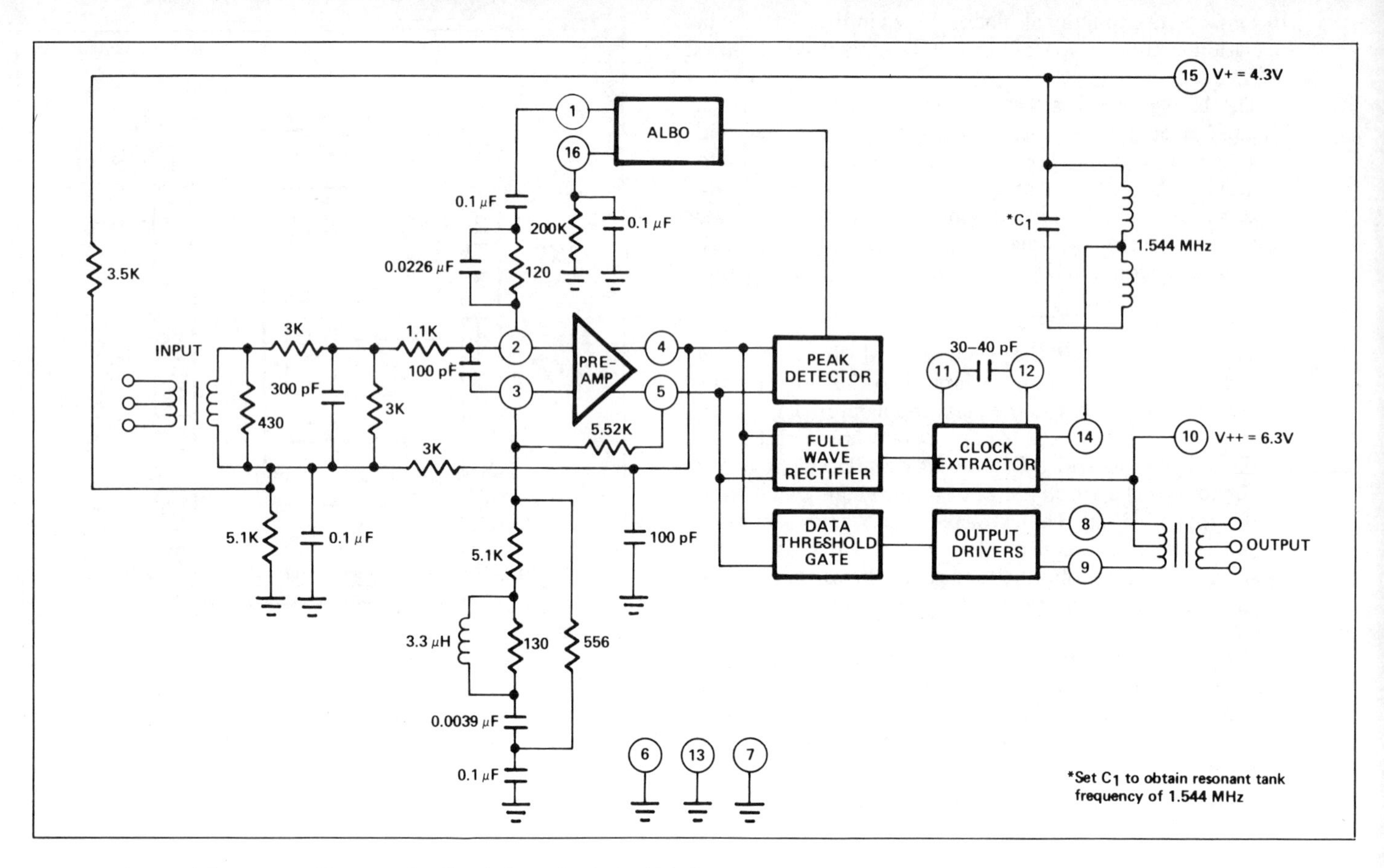

Figure 5. External Components Necessary for Circuit Operation in 1.544 MHz T-1 Repeater

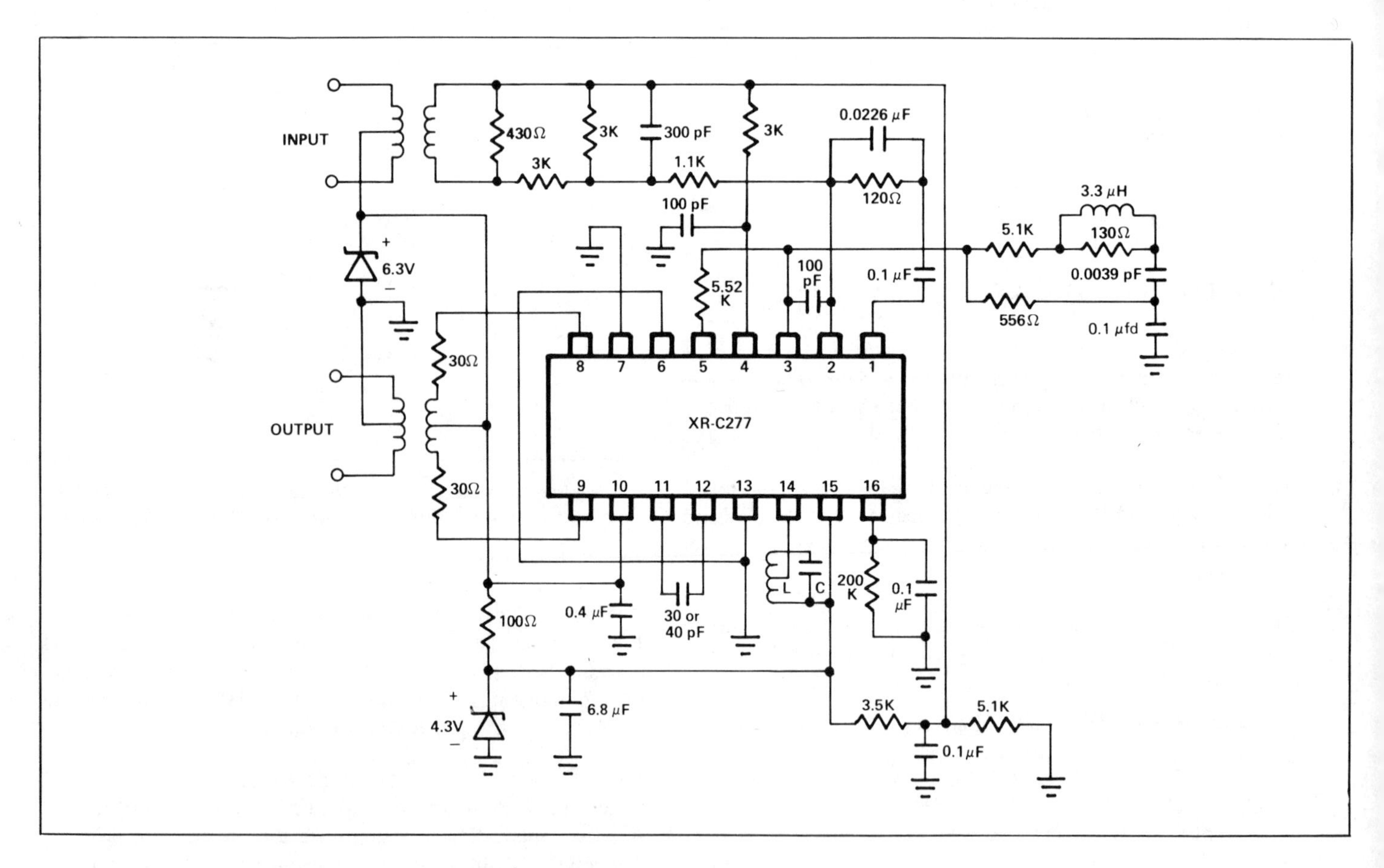

Figure 6. Typical Circuit Connection of XR-C277 in 1.544 MHz T-1 Repeater System. (Note: Set L and C to form a high Q tank resonant at 1.544 MHz. It is recommended that Q > 100, and C ≈ 200 pF for most applications).

DESCRIPTION OF CIRCUIT OPERATION

Preamplifier Section (Fig. 7):

The circuit diagram of the preamplifier section is shown in Figure 7. This section is designed as a two-stage differential amplifier with a broadband differential voltage gain of 52 db. The differential outputs of the preamplifier (pins 4 and 5) are internally connected to the peak-detector, full-wave rectifier and the data threshold detector sections of the XR-C277.

Automatic Line Build-Out (ALBO) Section (Fig. 8):

The ALBO function is achieved by controlling the dynamic impedance of ALBO diodes (Q_{21} and Q_{22}). The current whic sets this dynamic impedance is supplied through Q_{21} and is controlled by the peak-detector output level applied to base of Q_{23}.

Data-Threshold Detector; Full-Wave Rectifier and Peak Detector Sections (Figure 9):

The level detector and peak rectifier sections of the XR-C277 are made up of two sets of gain stages which are driven differentially with the (A^+) and (A^-) outputs of the preamplifier section. The outputs of the data threshold comparators, D^+ and D^- activate the data latches shown in Figure 11.

The peak-detector output (terminal B of Figure 9) is internally connected to the Automatic Line Build-Out (ALBO) section of the circuit and controls the DC bias current through the ALBO diodes Q_{21} through Q_{22}, as shown in Figure 8.

The full-wave rectifier output is internally connected to the clock-extractor section of the repeater and provides the excitation signal for the L-C tuned tank circuit (pin 14) of the injection locked oscillator. The detection thresholds of the comparators are set by the resistor chains (R_{45}, R_{47}, R_{51}, R_{55}) and (R_{46}, R_{48}, R_{52}, R_{56}). The resistor ratios are chosen such that the data threshold is 50% of the ALBO threshold; and the clock extractor threshold is 86% of the ALBO threshold.

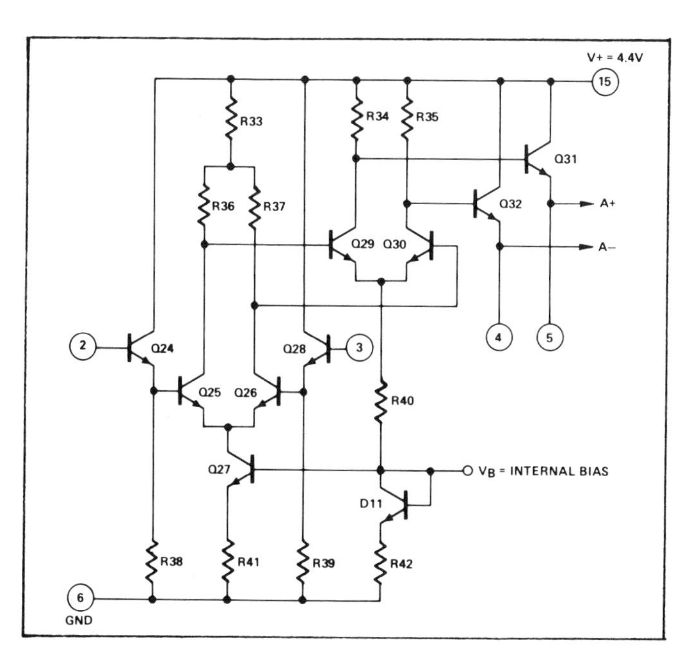

Figure 7. Circuit Diagram of Preamplifier Section

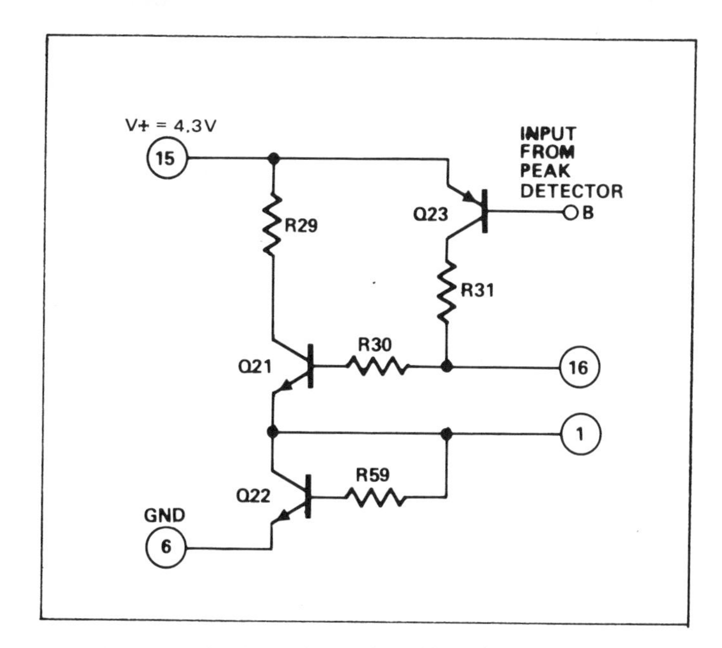

Figure 8. Automatic Line Build-Out (ALBO) Section

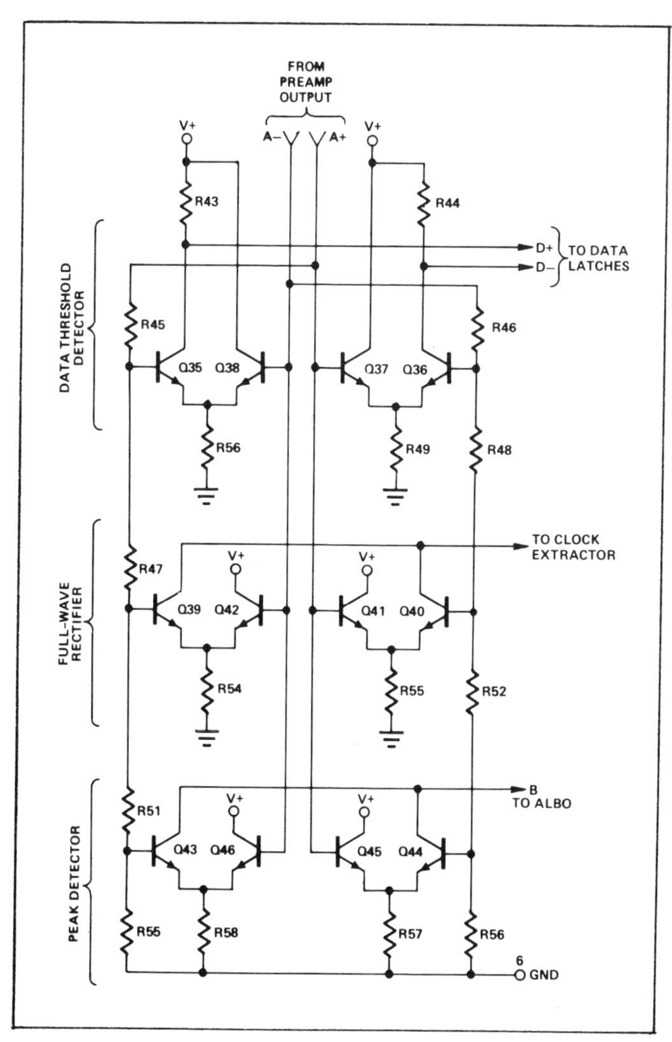

Figure 9. Data-Threshold Detector, Full-Wave Rectifier and the Peak Detector Sections of XR-C277

Clock Extractor Section (Figure 10):

The clock-extractor section of XR-C277 is designed as an injection locked oscillator as shown in the circuit schematic of Figure 10. The excitation is applied to the emitter of Q_{1B}, from the output of the full-wave rectifier. This signal in turn controls the current in the resonant L-C tank circuit connected to pin 14. The sinusoidal waveform across the tank is then amplified and squared through two cascaded differential gain stages made up transistors Q_3 through Q_9. The output swing of the second gain stage is "integrated" by the phase-shift capacitor, C_1, externally connected to pins 11 and 12. (See timing diagrams of Figure 13.) The nominal value of this capacitor is in the 30 to 40 pf range. The triangular waveform across pins 11 and 12 is at quadrature phase with the sinusoidal voltage swing across the L-C tank circuit. This waveform is then used to generate the "strobe" signal, C_p, and the clock pulse C_ϕ, which are applied to the data latches of the logic section.

Data-Latch and Output Driver Sections (Figures 11 and 12):

The data-latch section consists of two parallel flip-flops, driven by the (D^+) and (D^-) inputs from the data-threshold detector. When the D+ input is at a "low" state, the sampling or strobe pulse, C_p, is steered through Q_{47A} and sets flip-flop 1, on the leading edge of C_p. Conversely, when D^- input is at a "low" state, the sampling pulse is steered through Q_{47B} to set flip-flop 2. Each flip-flop section is then reset at the trailing edge of the clock pulse input, C_ϕ. The flip-flop outputs, (F_1, $\bar{F}_1$) and (F_2, $\bar{F}_2$) are then used to drive the output drivers. This logic arrangement results in an output pulse width which is the same as the extracted clock pulse width (see timing diagram of Figure 13).

The outputs of the two data latches drive the two output driver stages shown in Figure 12. The high-current outputs of the driver stage (pins 8 and 9) are connected to the center-tapped output transformer as shown in Figure 5. The voltage swing across the output is one diode drop (V_{BE}) less than the supply voltage at pin 10. The output stages are designed to work into a nominal load impedance of 100 ohms, and can handle peak load currents of 30 mA.

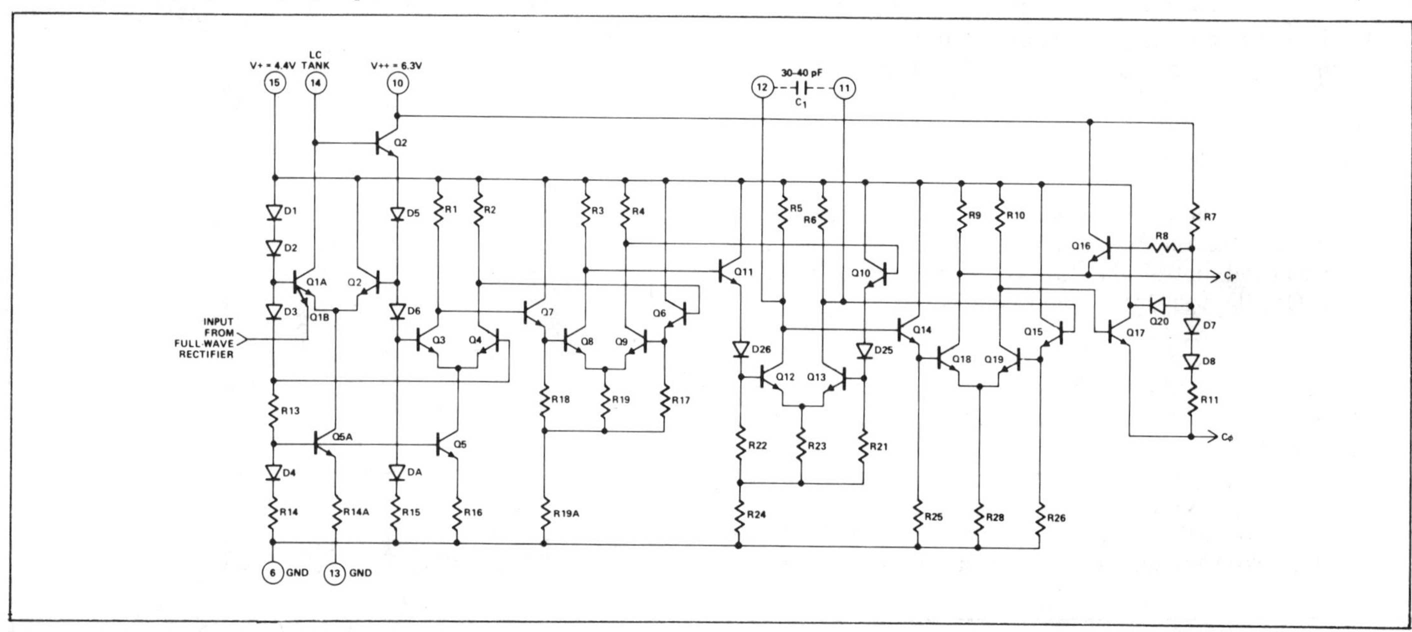

Figure 10. Circuit Diagram of Clock Extractor Section

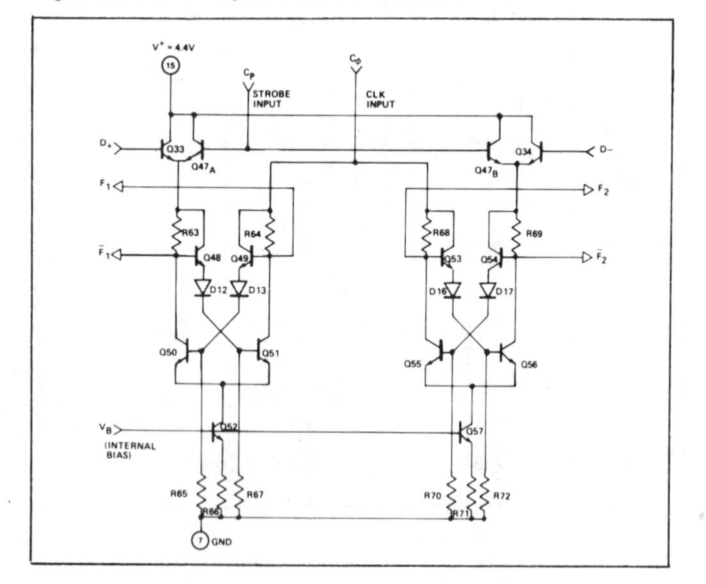

Figure 11. Data-Latch Section of XR-C277

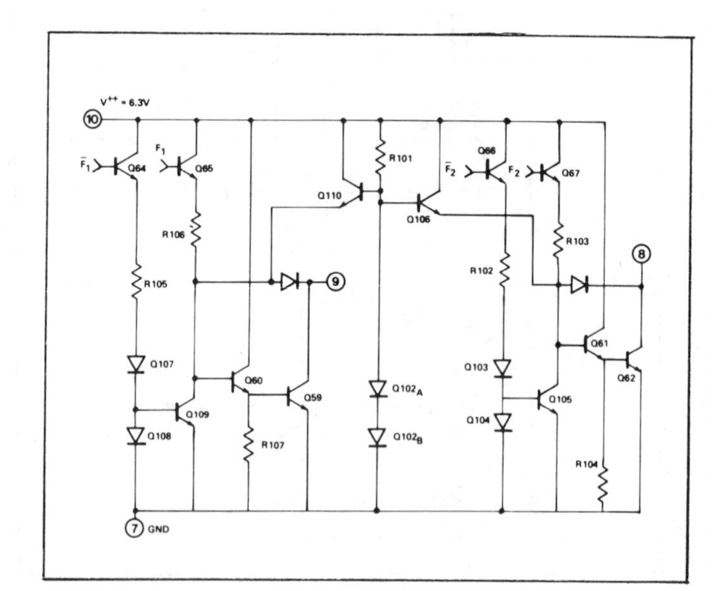

Figure 12. Output-Driver Section

Figure 13 shows the typical timing sequence of the circuit waveforms. For illustration purposes, a "one-zero-one" input data pattern is assumed.

ABSOLUTE MAXIMUM RATINGS

Supply Voltage	+10V
Power Dissipation	750 mW
Derate above +25°C	6 mW/°C
Storage Temperature Range	-65°C to +150°C

AVAILABLE TYPES

Part Number	Package	Operating Temperature
XR-C277	CERDIP	-40°C to +85°C

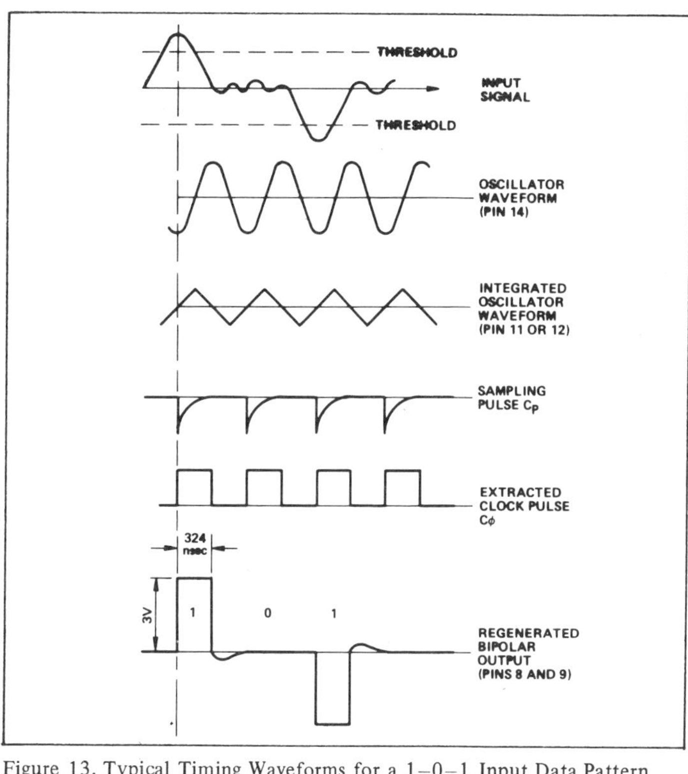

Figure 13. Typical Timing Waveforms for a 1–0–1 Input Data Pattern

ELECTRICAL CHARACTERISTICS

(-40°C to +85°C, V++ = 6.3V ±0.5V, V+ = 4.3V ±0.5V, unless specified otherwise.)

PARAMETER	LIMITS				CONDITIONS
	MIN.	TYP.	MAX.	UNITS	
Supply Current					See Figure 4
I_A		3.5		mA	Measured at Pin 10
I_B		7.5		mA	Measured at Pin 15
Total Current	8	11	13	mA	$(I_C + I_B)$
Preamplifier					See Figure 7
Input Offset Voltage		1.5	15	mV	Measured at Pins 2 and 3
Input Bias Current		0.3	4	μA	Measured at Pins 2 and 3
Voltage Gain	44	48	51	dB	Single-ended gain
Preamp Output Swing					Measured at Pins 4 and 5
High Swing	3.45	3.6	3.75	V	Maximum voltage swing
Low Swing	1.25	1.4	1.55	V	Minimum voltage swing
Output DC Level	2.47	2.55	2.72	V	
ALBO Section					See Figure 8
ALBO "Off" Voltage		10	75	mV	Measured from Pin 1 and Pin 16 to ground
ALBO "On" Voltage	0.6	0.87	1.1	V	Measured at Pin 1
ALBO "On" Voltage	1.2	1.5	2.1	V	Measured at Pin 16
ALBO Threshold	1.35	1.50	1.65	V	Measured differentially across Pins 4 and 5
Differential Threshold	-75		+75	mV	Threshold difference for polarity reversal at Pins 4 and 5
ALBO "On" Impedance		5	10	Ω	Measured at Pin 1
ALBO "Off" Impedance	20	50		kΩ	Measured at Pin 1
Comparator Thresholds					See Figure 9
Clock Threshold	82	86	90	%	% of ALBO threshold
Data Threshold	47	50	53	%	% of ALBO threshold
Clock Extractor					See Figure 10
Oscillator Current	10	14	20	μA	@ +25°C
Tank Drive Impedance		50		kΩ	
Recommended OSC. Q	100				
$I_{injection}/I_{OSC.}$	6.5	7	7.5		Ratio of current in Q_{1B} to current in Q_{1A}
Output Driver					See Figure 12
Low Output Voltage	0.65	0.75	0.95	V	Measured at Pins 8 and 9, I_L = 15 mA
Output "Off" Current		5	100	μA	V_{out} = 20V
Output Pulse					See Figure 13
Maximum Pulse Width Error			±30	n sec	
Rise Time			80	n sec	
Full Time			80	n sec	

Part VIII
Precision Linear Circuits: Trimming Techniques

The matching and tracking characteristics of monolithic components are often not sufficient for precision circuits such as the digital-to-analog converters with 10-bits or higher accuracies. In such cases, it is often desirable to employ a trimming technique, both to improve the circuit performance and to achieve higher yields. In most cases, such precision trimming can be employed at the wafer-level, i.e., prior to scribing and packaging of the IC chip, in order to reduce the reject rate at the finished and packaged product level.

The papers included in this part describe a number of differing approaches to trimming techniques for precision linear IC's. The first paper, by Erdi, describes a trimming technique using selective shorting of zener diodes. The next paper, by Comer, demonstrates a direct application of this technique to the design of a 12-bit D/A converter.

Another commonly used method of adjusting critical parameters of monolithic circuits is the laser-trimming technique for adjusting the values of critical resistors. This can be done by either actively trimming the resistor cross-section, or using a laser beam to cut out the shorting links across a resistor. This latter method is the subject of the third paper, by Price, which describes the application of laser-trimming to the design of a 10-bit digital-to-analog converter circuit.

ADDITIONAL REFERENCES

1. G.A. Bulger, "Stability Analysis of Laser-Trimmed Thin-Film Resistors," *IEEE Trans. Parts, Hybrids, Packaging.*, vol. PHP-11, September 1975.
2. R.C. Headley, "Laser Trimming is an Art That Must be Learned," *Electronics*, June 21, 1973.
3. R.B. Craven, "An Integrated Circuit 12-Bit D/A Converter," *ISSCC Digest of Tech. Papers*, pp. 40-41, February 1975.
4. A.P. Brokaw, "A Monolithic 10-Bit A/D Using I^2L and LWT Thin-Film Resistors," *ISSCC Digest of Tech. Papers*, pp. 140-141, February 1978.

A Precision Trim Technique for Monolithic Analog Circuits

GEORGE ERDI

Abstract—A technique for permanent adjustment of precision analog circuits at wafer test by selective shorting of Zener diodes is presented. Analytical details of the trimming procedure and a physical description of diode short-circuiting are given. The method is applied to a precision operational amplifier with input offset voltage reduced to 10 μV. The necessity of optimizing other related parameters is demonstrated. Practical considerations limiting wafer test accuracy are discussed. Circuit performance is summarized.

INTRODUCTION

THE character of the monolithic analog circuit is essentially established once it has emerged from wafer processing. Hybrid and modular devices, on the other hand, usually undergo some adjustment procedure before final assembly. These facts explain why monolithic IC's enjoy significant cost advantages, while hybrid and modular circuits have the potential for better performance. To bridge this performance gap, several trim techniques for monolithic devices have been developed in recent years, resulting in the adjustment and improvement of a critical variable.

Laser trimming has received the most attention of all of these schemes. Although most of the early technical problems are apparently being solved [1], laser adjustment sacrifices some of the cost advantages of the monolithic process. It does require a high initial investment for equipment that can only be applied to thin or thick film resistors (more expensive to manufacture than diffused resistors). In most cases the adjustment cannot be made at wafer test, but only after the device has been die-attached and bonded—further increasing costs.

Trimming with fusible links is a well-established procedure, primarily for programming memories. However, it also needs additional processing steps for the formation of the fuse. Several attempts have been made [2], [3] to use standard linear process metal (1000-Å thick aluminum) as fusible links. Although the production-worthiness of this scheme is difficult to judge, metal regrowth due to electromigration and/or thermal expansion [4] after the fuse is blown is a possible problem area. In addition, the current levels required are in the ampere range resulting in a fast deterioration of the wafer test probes and unsightly—and therefore often unacceptable—blown metal connections.

The selective shorting of emitter-base diodes (in the avalanche or Zener mode) to adjust monolithic circuits does not require extra processing, it can be implemented at the wafer test phase, with minimal increase in manufacturing cost. The shorting of diodes for programming circuits was first proposed in 1962 [5] and used extensively in digital circuits since then. Analog designs employing this trimming method have appeared only recently. Examples are an FET operational amplifier [6] developed concurrently with the precision op amp described below [7], and other unpublished, and therefore unsubstantiated, applications.

In the following sections a trim technique utilizing shorted diodes and its application to a precision operational amplifier are described. Details of the diode shorting procedure are delineated in the Appendix.

Manuscript received May 16, 1975; revised July 14, 1975. This paper was presented at the International Solid-State Circuits Conference, Philadelphia, Pa., February 1975.

The author is with Precision Monolithics Inc., Santa Clara, Calif. 95050.

OFFSET VOLTAGE ADJUSTMENT

Improvements in monolithic operational amplifier design have decreased the input referred error contribution of most parameters, such as bias and offset currents, common-mode and power-supply rejection, voltage gain, to the order of microvolts. One exception has been input offset voltage (V_{os}). Even a V_{os} of a few hundred microvolts will dominate the error budget, and thus represents an unacceptably large error term in many applications. External V_{os} adjustment by potentiometers is expensive and potentially unreliable. In addition, field-servicing is difficult, since the faulty device cannot be replaced by a substitute without extensive recalibration. Permanent V_{os} adjustment at wafer test eliminates all of the above difficulties.

The equivalent circuit of Fig. 1 represents an operational amplifier with a resistor-loaded input stage. Offset voltage is defined by

$$V_{os} = (V_{BE1} - V_{BE2})\big|_{V_{out}=0} = \frac{kT}{q} \ln \frac{I_1}{I_2} \frac{I_{s2}}{I_{s1}}. \tag{1}$$

I_{s1} and I_{s2} are saturation currents of $Q1$ and $Q2$, respectively.

If V_{os2}, the offset voltage of the second stage $\ll I_1 R_L$, then $I_1 R_L = I_2 R_R$ and (1) reduces to

$$V_{os} = \frac{kT}{q} \ln \frac{R_R}{R_L} \frac{I_{s2}}{I_{s1}}. \tag{2}$$

Hence, the causes of V_{os} are readily identified as mismatches between R_R and R_L, I_{s1}, and I_{s2}.

V_{os} can be altered by changing the ratio R_R/R_L, and, in particular, if $(R_R/R_L)(I_{S2}/I_{S1}) = 1$, V_{os} will be zero.

Offset trim at wafer test is accomplished by the additional circuitry of Fig. 2. Zener diodes Z_0 to Z_n are normally nonconducting. The differential stage is balanced, since nominally:

Reprinted from *IEEE J. Solid-State Circuits*, vol. SC-10, pp. 412-416, Dec. 1975.

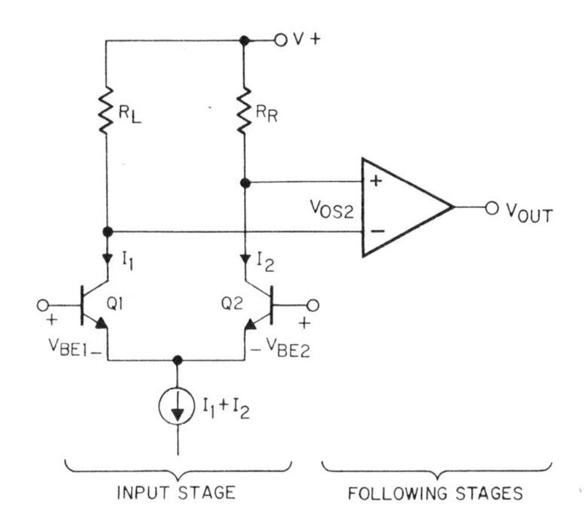

Fig. 1. Equivalent circuit of operational amplifier.

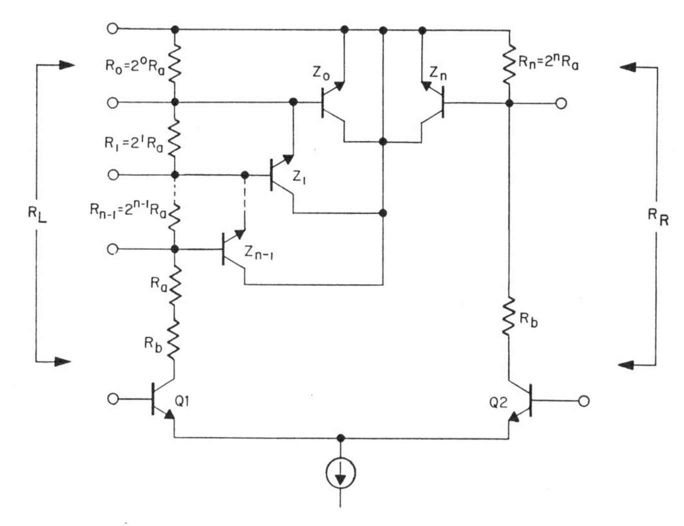

Fig. 2. Offset adjustment by selective shorting of Zener diodes.

$$\begin{aligned} R_L &= R_b + R_a(1 + 2^0 + 2^1 + \cdots + 2^{n-1}) \\ &= R_b + 2^n R_a \\ &= R_R. \end{aligned} \tag{3}$$

V_{os} is adjusted by the selective shorting of Z_0 to Z_n. Irrespective of the Z_i shorted, the resistance difference between the two sides, $\Delta R \equiv R_L - R_R \leqslant 2^n R_a$. If $2^n R_a << R_b$, and since $R_b \leqslant R_L$, the change in V_{os}, ΔV_{os}, from (2) is

$$\begin{aligned} \Delta V_{os} &= \frac{kT}{q} \ln \frac{R_L + \Delta R}{R_L} \\ &\simeq \frac{kT}{q} \frac{\Delta R}{R_L}. \end{aligned} \tag{4}$$

The following equations describe the maximum and minimum changes in V_{os} in the positive (ΔV_{os}^+ max, ΔV_{os}^+ min) and negative (ΔV_{os}^- max, ΔV_{os}^- min) directions. Because of the conclusions of (4), all equations can be reduced from a logarithmic to a linear form.

If $Z_0, Z_1, \cdots, Z_{n-1}$ are shorted,

$$\Delta V_{os}^+ \max = \frac{kT}{q} \ln \frac{R_b + 2^n R_a}{R_b + R_a} \simeq \frac{kT}{q} (2^n - 1) \frac{R_a}{R_b}. \tag{5}$$

If Z_n is shorted,

$$\Delta V_{os}^- \max = \frac{-kT}{q} \ln \left(1 + 2^n \frac{R_a}{R_b}\right) \simeq \frac{-kT}{q} 2^n \frac{R_a}{R_b}. \tag{6}$$

If Z_0 is shorted,

$$\Delta V_{os}^+ \min = \frac{kT}{q} \ln \frac{R_b + 2^n R_a}{R_b + (2^n - 1) R_a} \simeq \frac{kT}{q} \frac{R_a}{R_b}. \tag{7}$$

If all Zeners are shorted,

$$\Delta V_{os}^- \min = \frac{-kT}{q} \ln \left(1 + \frac{R_a}{R_b}\right) \simeq \frac{-kT}{q} \frac{R_a}{R_b}. \tag{8}$$

If the initial, unadjusted V_{os} varies between

$$\begin{aligned} -(\Delta V_{os}^+ \max + \tfrac{1}{2} \Delta V_{os}^+ \min) &\leqslant V_{os} \\ &\leqslant -(\Delta V_{os}^- \max + \tfrac{1}{2} \Delta V_{os}^- \min), \end{aligned} \tag{9}$$

then V_{os} can be changed by appropriately selected shorting of Z_i to

$$\tfrac{1}{2} \Delta V_{os}^- \min \leqslant V_{os} \leqslant \tfrac{1}{2} \Delta V_{os}^+ \min. \tag{10}$$

From (5)-(10) all negative V_{os} values can be improved $(2^{n+1} - 1)$ times, all positive offset voltages $(2^{n+1} + 1)$ times.

The discussion above is concerned specifically with operational amplifiers. However, the conclusions are quite general and can be directly applied to other types of analog circuits, such as comparators, instrumentation amplifiers, and analog multipliers.

Precision Operational Amplifier

As a specific example, a precision operational amplifier[1] was designed with $R_a = 230\,\Omega$, $R_b = 50\,\mathrm{k}\Omega$, $n = 3$. Assuming $kT/q = 26$ mV at room temperature, from (7) and (8)

$$\Delta V_{os}^+ \min \simeq -\Delta V_{os}^- \min = 120\,\mu\mathrm{V}.$$

From (5)

$$\Delta V_{os}^+ \max \simeq 840\,\mu\mathrm{V}.$$

From (6)

$$\Delta V_{os}^- \max \simeq -960\,\mu\mathrm{V}.$$

Therefore, from (9) and (10) offset voltages ranging from $-900\,\mu$V to $+1020\,\mu$V can be improved to $-60\,\mu\mathrm{V} \leqslant V_{os} \leqslant 60\,\mu$V. Offset voltage reduction is accomplished by dividing the untrimmed V_{os} range into sixteen equal, 120 μV wide, sorting sections, each one is covered by one of the sixteen shorting combinations available from Z_0, Z_1, Z_2, and Z_3.

Practical considerations govern the selection of both the maximum boundaries of adjustment, and the resultant V_{os}. The approximately ±1-mV range was chosen because most of the devices manufactured fall within these limits. In addition, those few units with V_{os} exceeding 1 mV are not considered

[1] Precision Monolithics mono OP-07 technical data, June 1974.

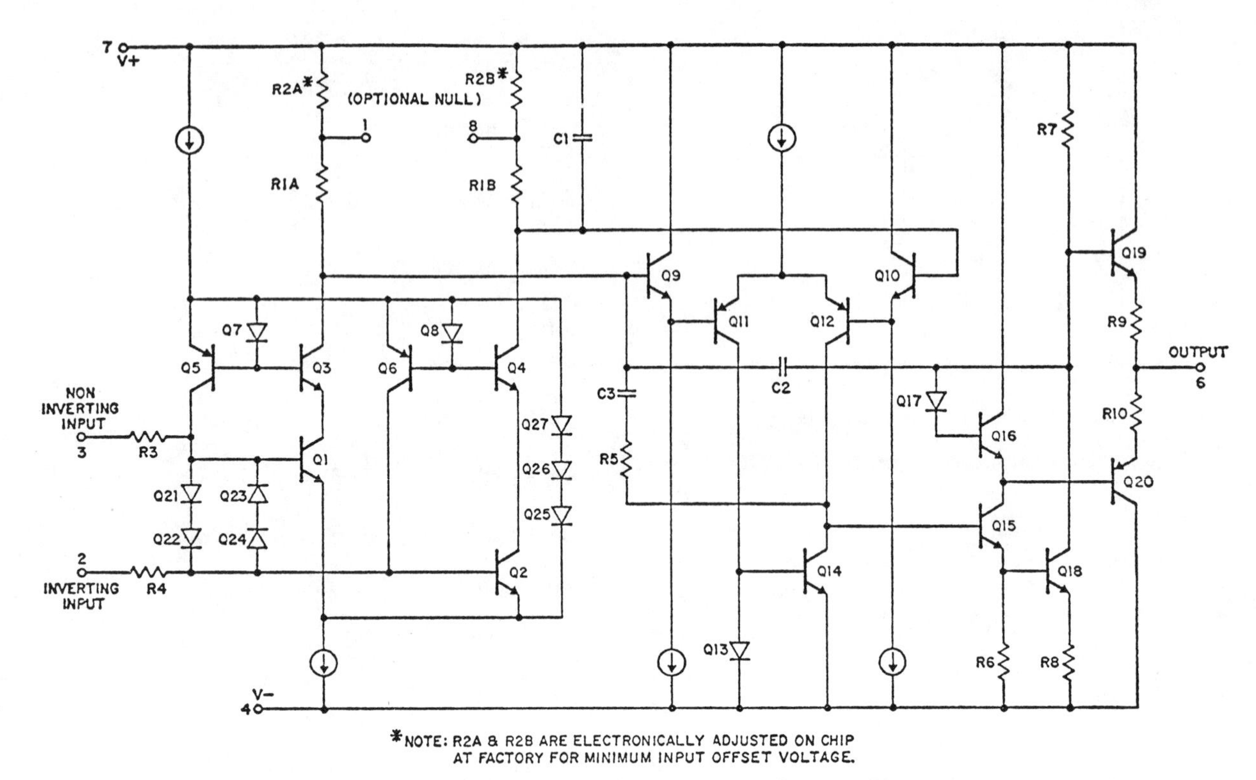

Fig. 3. Simplified schematic of complete operational amplifier.

to have adequate reliability for a final specification of less than 60 μV. In other words, the device must have some serious imperfection, leading to the relatively high-offset voltage. One more Zener diode could have been used to lower the trimmed limits to ±30 μV. However, several error terms, each amounting to 3 to 10 μV, would negate the possible benefits of a ± 30-μV reduced range. These are measurement inaccuracies at wafer test; changes through assembly, particularly due to die-attach; increased chip temperature in the packaged device; warm-up drift, since by necessity the amplifier is measured a few milliseconds after turnon at wafer test.

The procedure at wafer sort starts with the measurement of V_{os} by an automated tester; if V_{os} is within the adjustable range, the appropriate preprogrammed Zener diode combination is shorted. The heat generated in the fused Zener diodes (see Appendix) creates a pattern of thermal gradients across the chip. Although the isothermal layout with cross-connected input devices (Fig. 4) eliminates most of the temperature differentials, the measurement of V_{os} to microvolt accuracies would still be effected. The test sequence is organized to evaluate less sensitive parameters immediately after Zener shorting, and thus, several hundred milliseconds elapse before V_{os} is remeasured, allowing the gradients to decay.

Since the V_{os} of most devices is less than 60 μV, a prime grade with 25 μV of guaranteed offset voltage can be defined with reasonable yields. However, a 25-μV V_{os} specification is meaningful only if two other V_{os}-related parameters are optimized: V_{os} drift with temperature (TCV_{os}) and time (V_{os}/t). Even relatively good TCV_{os} or V_{os}/t performance, such as 2 μV/°C and 20 μV/month can render the initial offset voltage guarantee worthless with temperature and time variations. A monolithic operational amplifier [8], which has been in production for years, has excellent input specifications, including TCV_{os} and V_{os}. Therefore, its basic design was retained, with the trimming circuitry of Fig. 2 added to facilitate V_{os} adjustment.

A simplified schematic is shown in Fig. 3. The asterisks of $R2A$ and $R2B$ indicate that these resistors were altered at wafer test using the technique described above.

The three-stage amplifier is internally compensated with feedforward capacitor C_3. C_2 sets the dominant pole. Low bias currents are achieved with the bias current cancellation network of $Q1$, $Q3$, $Q5$, $Q7$ and $Q2$, $Q4$, $Q6$, $Q8$. Additional circuit design details are described in [8].

The photomicrograph of the device (Fig. 4) illustrates the symmetrical layout [9] which is essential for precision operational amplifier design. The cross-coupled input transistor connection on the left side of the chip cancels gradients due to wafer processing and thermal effects generated by variations in power dissipation in the driver and output stages located on the right side of the device. Fig. 5 shows an enlarged view of the trimming circuitry. The chip area assigned to the pads contacting the four Zener diodes is larger than the area of the diodes themselves, due to the fact that the Zeners are in a common isolation pocket. Since as many as eight pads could be employed to contact four shorting diodes, another advantage of the trim technique of Fig. 2 is the significant area reduction represented by the use of only four pads in addition to the ones needed for device operation.

Circuit Performance

The performance attained with this design is detailed in Table I. The 25-μV guaranteed offset voltage (60 μV maximum over the −55°C to 125°C temperature range) no longer

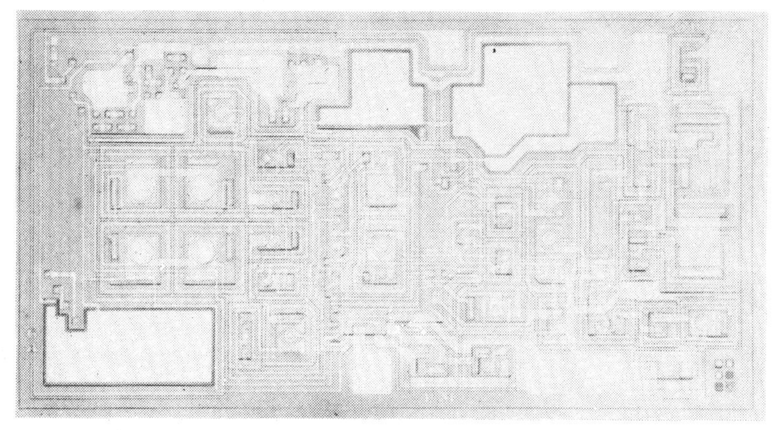

Fig. 4. Chip photomicrograph of precision operational amplifier (100 X 53 mils [2]).

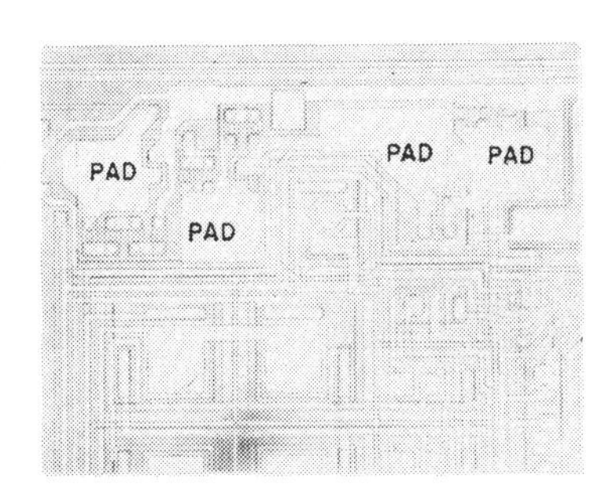

Fig. 5. Detailed view of trimming circuitry.

TABLE I

PRECISION OPERATIONAL AMPLIFIER PERFORMANCE @ $V_S = \pm 15V$, $T_A = 25°C$

PARAMETER	TYPICAL	MIN/MAX	UNITS
Offset voltage, V_{OS}	10	25	μV
drift with temperature	0.2	0.6	μV/°C
drift with time	0.2	1.0	μV/mo
Offset current, I_{OS}	0.3	2.0	nA
drift with temperature	5	25	pA/°C
Input bias current, I_B	0.7	2.0	nA
drift with temperature	8	25	pA/°C
Noise voltage 0.1 Hz to 10 Hz	0.35	0.6	μV p-p
Noise current 0.1 Hz to 10 Hz	14	30	pA p-p
Input resistance--differential	80	30	MΩ
Input resistance--common mode	200	--	GΩ
Common-mode rejection	126	110	dB
Power supply rejection	110	100	dB
Voltage gain	500	300	V/mV
Slew rate	0.25	--	V/μsec
Unity gain bandwidth	1.2	--	MHz

dominates the error budget. Indeed, a worse case 25°C dc analysis of the voltage follower circuit of Fig. 6 shows that the individual error contributions of V_{os}, bias current, voltage gain, common-mode and power supply rejection all are within 20 μV and 33 μV. Therefore, offset voltage has been successfully reduced to the level of other error sources.

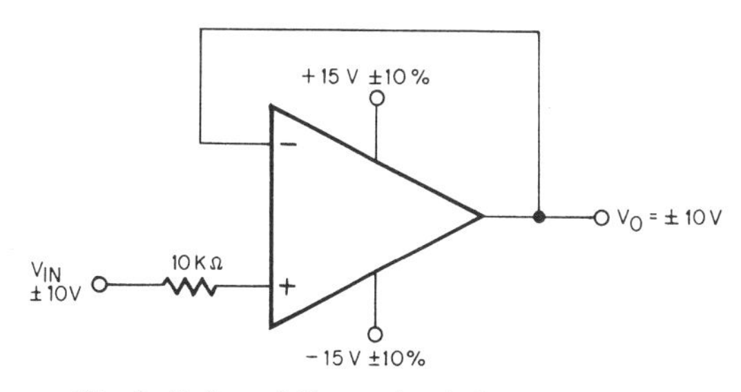

Fig. 6. Voltage follower circuit for error analysis.

Reliability

Reliability data for shorted Zeners are claimed [6] to be superior to other trimming methods. For the precision operational amplifier reliability is further enhanced by the fact that the shorted diodes only carry 8 μA, the input stage current of the amplifier.

In excess of 200 000 device-hours of operation at 125°C yielded only one failure. The cause was unrelated to the trimming circuitry. In general, reliability differences between the device and its predecessor without shorted diodes described in [8] are indistinguishable, providing additional proof of the basic soundness of the offset adjustment scheme.

Conclusion

A precision trim technique which has significantly improved the performance of a monolithic operational amplifier has been described. It is a general method; therefore it can be

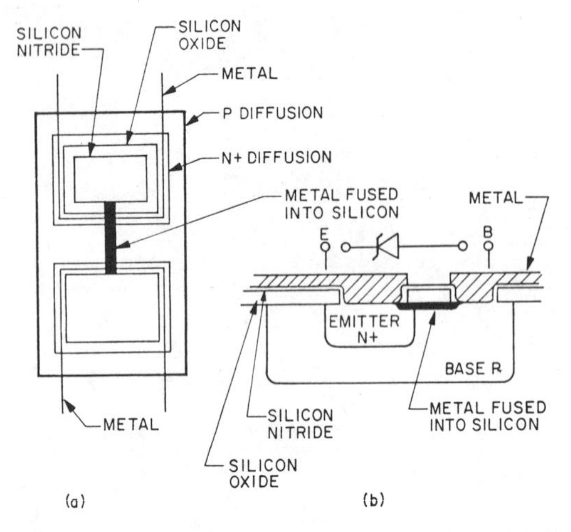

Fig. 7. Short-circuiting of Zener diodes. (a) Top view. (b) Side view.

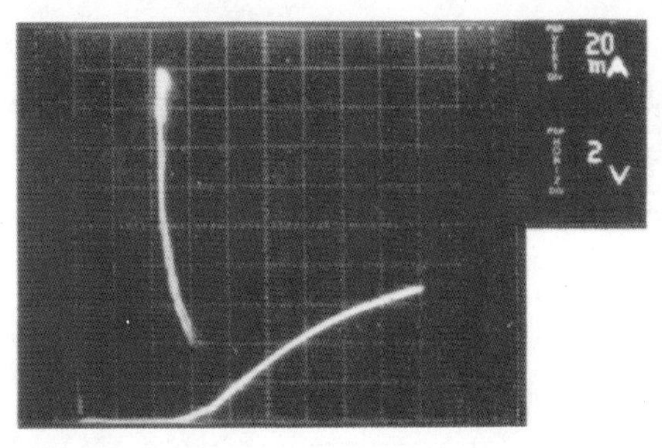

Fig. 8. Curve-tracer photograph of Zener diodes being shorted.

used to adjust other types of analog circuits. On wafer trimming, however, it is not a panacea: careful attention must be paid to other parameters, before the full benefits of the technique can be realized.

Appendix

When large current is passed through the emitter-base diode in the avalanche-mode, the localized power dissipation at the junction reaches such a magnitude that the junction will be destroyed. As shown in Fig. 7, a narrow link of metal fused into the silicon will be created. In addition, due to the large electric field present, metal will be swept across the silicon surface, beneath the oxide layer. In effect, a double short is produced: one as a result of the destroyed junction, the other due to the metal bridging. The typical resistance between the emitter and base contacts is 1 Ω. The curve-tracer photograph of Fig. 8 illustrates the chain of events required to short-circuit a Zener diode. Diode voltage is swept at a 120-Hz rate. Avalanche breakdown occurs at 6.3 V. Beyond this voltage, to 18 V, the diode behaves as a 170-Ω resistor, which is the resistance of the emitter and the 200-Ω per square base region between the contacts. At 18 V the instantaneous power dissipation exceeds 1.2 W. If the voltage is increased, the diode will switch into an oscillatory, thermal runaway condition. This state lasts less than a second, after which the Zener is destroyed and a 1-Ω resistor results. It is apparent from Fig. 8 that a necessary condition for shorting Zeners is that sufficient voltage and current has to be provided to exceed the "knee" of the curve at 18 V and 70 mA.

Using an automated tester, a current pulse train can be applied to the device [10], and as the avalanched junction is progressively degraded, the resistance can alternately be monitored. When a predetermined resistance is reached the pulse train is terminated.

The shorting method selected for the precision operational amplifier described above was to experimentally arrive at a set of conditions which will short the Zeners without fail considering all possible manufacturing variations. Application for 80 ms of a 25-V power supply, current-limited at 300 mA consistently shorted the Zener diodes.

Acknowledgment

The author wishes to thank J. Biran, J. Lundstrom, and S. Saylor for their assistance in testing the device and B. Dunaway for the chip layout.

References

[1] S. Harris and D. Wagner, "Laser trimming on the chip," *Electron. Packaging and Production*, pp. 50-56, Feb. 1975.
[2] W. F. Davis, "A five-terminal ±15-V monolithic voltage regulator," *IEEE J. Solid-State Circuits*, vol. SC-6, pp. 366-376, Dec. 1971.
[3] J. R. Butler and R. Q. Lane, "An improved performance MOS/bipolar op amp," in *1974 IEEE Int. Solid-State Circuits Conf. Dig. Tech. Papers*, 1974, pp. 138-139.
[4] I. A. Blech, "Recent experiments on electromigration and whisker growth in thin metal films," presented at the Western Sem. Material Science, Bell Laboratories, Palo Alto, Calif., 1975.
[5] J. E. Price, "Programmable circuit," U.S. Patent 3 191 151, June 22, 1965.
[6] J. DuBow, "Process yields low-cost FET op amp," *Electronics*, p. 154, Dec. 20, 1973.
[7] "Bipolar op amp matches specs of chopper stabilized units," *Electron. Des.*, p. 119, July 19, 1974.
[8] G. Erdi, "Instrumentation operational amplifier with low noise, drift, bias current," in *Northeast Res. Eng. Meeting Rec. Tech. Papers*, Oct. 1972.
[9] G. Erdi, "A low drift, low noise monolithic operational amplifier for low level signal processing," Fairchild Semiconductor, Appl. Brief 136, July 1969.
[10] J. D. Rizzi and L. D. Fagan, "Electrically alterable integrated circuit read-only memory unit and process of manufacturing," U.S. Patent 3 742 592, July 3, 1973.

A Monolithic 12-Bit D/A Converter

Donald T. Comer

Precision Monolithics, Inc.

Santa Clara, CA

MANY MONOLITHIC D/A converter circuits have been developed and reported in the literature[1-2]. However, the accuracy of the conventional implementations is ultimately limited by the degree to which the elements of the precision resistor ladder networks can be matched. To achieve the accuracy required for 10- and 12-bit converters it is common practice to use active laser trimming of thin-film resistors[3-4] or laser cutting of metal links to produce discrete correctional currents[5]. The approach to be discussed offers an alternative to these methods and is based upon selective shorting of zener diodes[6] to produce discrete correctional currents. The method is compatible with either trim at wafer sort or trim after final packaging. An extension of the basic idea allows after package programming of weighting currents to achieve either a binary or a binary coded decimal (BCD) coding. This allows the flexibility of converting devices which fail to meet binary linearity requirements to the less stringent BCD requirements.

The foregoing principles have been applied to produce a single chip 12-bit D/A converter on the 145 x 166 mil die shown in Figure 1. Other features of this device include thin film reference and span resistors which are similarly trimmed for full scale calibration. The trim circuitry utilizes approximately fifty percent of the chip area and is implemented in the upper half of the die photo shown in Figure 1. By trimming the first 6 bits of the 12 bit converter, very good yields to 0.5 LSB linearity are possible.

Figure 2 shows a functional block diagram of the circuit. The 12-bit DAC is implemented with an R-2R ladder for the first six bits, with the remainder current of the primary ladder being further divided by a slave ladder. The remainder current of the slave ladder is ultimately used to scale the trim correction currents in the trim network. The ladder resistors in both primary and slave ladders are implemented with thin film Si-Cr resistors with sufficient precision that the 6 least significant bits contribute a total linearity error which is acceptable. Thus, only the six most significant bits are trimmed.

The function of the trim network is to generate program selectable trim currents I_{T1} through I_{T6} which are directly proportional to the full scale current. This is accomplished by referencing trim current sources to the remainder current of the slave ladder.

Figure 3 shows the circuitry used to implement the trim currents for the most significant bit.

In this case binary multiples of a reference current, which is directly proportional to ladder remainder current, are selectable by shorting at the appropriate zener diode Z_1, Z_2, Z_3 or Z_4.

Binary multiples are created by using both multiple and split collector PNP devices. Since the desired correction may be either positive or negative, devices Q_1 and Q_2 form a current mirror which produces a current which may be either positive or negative depending upon whether zener Z_5 is shorted or left open. The output I_{T1} is then summed at the collector of the device setting up the most significant bit current. Likewise, trim correctional currents for 5 additional bits and full scale currents are provided. However, since the range of errors expected for the lower order bits are not as great as the most significant bit, progressively fewer selections are provided for the lower order bits.

The total number of trim currents provided for the 12 bit D/A converter is 31. The selection of the zener diodes to be shorted is provided by a matrix decoding scheme and current drivers capable of providing sufficiently high current to accomplish the aluminum migration necessary to produce a reliable short.

Since trimming is accomplished after the die has been packaged, the inputs to the matrix decoder which selects individual zener elements are provided through the 12 digital input pins to the converter using a threshold sensitive circuit. The 12 bit input lines are then used to control the inputs to a 6 x 6 decoding matrix which allows up to 36 discrete zener elements to be shorted.

In addition to providing selection to any one of the 31 trim zener elements, the 6 x 6 matrix allows selection of additional zener elements which are used to: (1) provide an expansion of trim ranges for all bits to allow a trade-off between trim range and trim resolution; (2) provide activation of a *lock-out* device to prevent further zener shorting after initial trimming is completed, and (3) provide activation of circuitry which changes bit current weighting from binary to BCD.

The combination of trim currents which are selectable after packaging with conventional bipolar D/A converter design techniques using Si-Cr ladder resistors, results in a single chip converter design with good yields to linearities heretofore unachieved with single chip designs.

Acknowledgments

The author expresses his thanks to J. Schoeff and D. Dooley who helped provide many of the innovative ideas that helped to create a 12-bit D/A converter on a single chip.

---

[1] Dooley, D.J., "A Complete Monolithic 6-bit D/A Converter", *ISSCC Digest of Technical Papers*, p. 50-51; Feb., 1971.

[2] Dooley, D.J., "A Complete Monolithic 10-bit D/A Converter", *ISSCC Digest of Technical Papers*, p. 12-13; Feb., 1973.

[3] Craven, R.B., "An Integrated Circuit 12-bit D/A Converter", *ISSCC Digest of Technical Papers*, p. 40-41; Feb., 1975.

[4] Holloway, P. and Norton, M., "A High Yield, Second Generation 10-bit Monolithic DAC", *ISSCC Digest of Technical Papers*, p. 106-107; Feb., 1976.

[5] Price, J.J., "A Passive Laser-Trimming Technique to Improve D/A Linearity", *ISSCC Digest of Technical Papers*, p. 104-105; Feb., 1976.

[6] Erdi, G., "A Precision Trim Technique for Monolithic Analog Circuits", *ISSCC Digest of Technical Papers*, p. 192-193; Feb., 1975.

Reprinted from 1977 *IEEE Internat. Solid-State Circuits Conf. Dig. Tech. Papers*, Feb. 16-18, 1977, pp. 104-105.

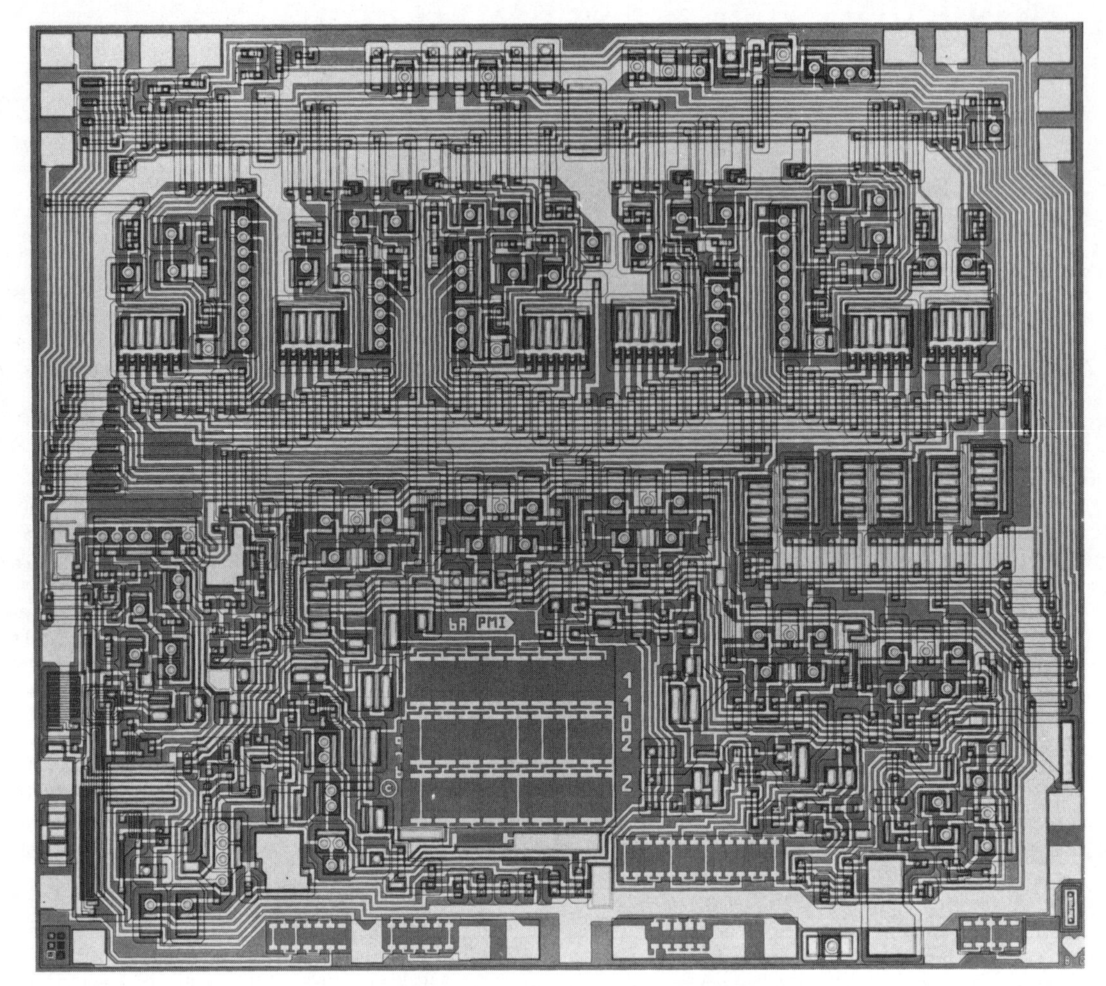

FIGURE 1 – Die photo of single chip 12-bit DAC.

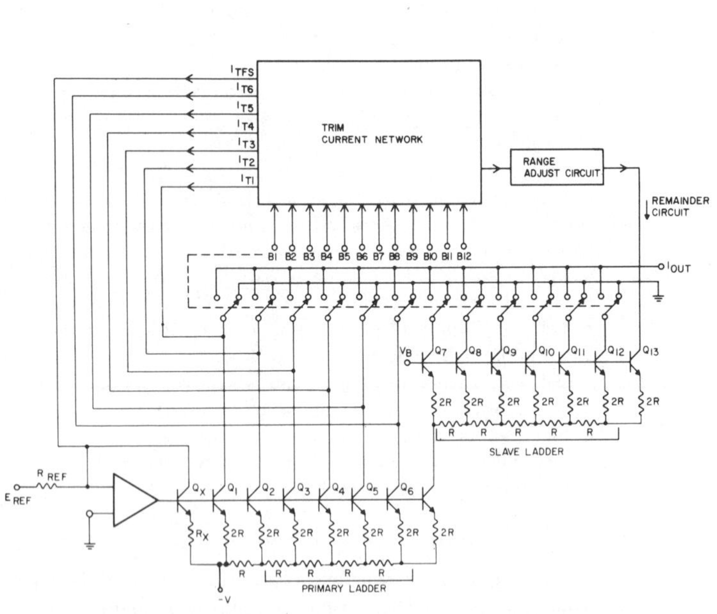

FIGURE 2 – Functional diagram of 12-bit DAC.

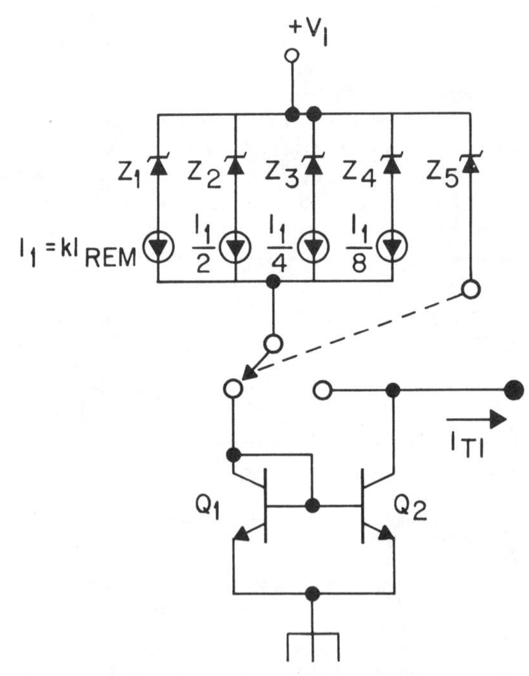

FIGURE 3 – Trim circuit for bit 1.

A Passive Laser-Trimming Technique to Improve the Linearity of a 10-Bit D/A Converter

JOHN J. PRICE

Abstract—The development of a new laser-trim technique is described. Metal links, which are arranged to provide incremental adjustments to critical circuit parameters, are cut by a laser. Trimming operations are monitored and controlled by means of a computer program. A comparison is made between currently used trimming methods. An application, to improve the linearity of a 10-bit D/A converter, is described. This trimming technique, which is implemented so that its effects track temperature and reference current variations, allows the converter to be adjusted for better than ±0.05 percent linearity by appropriately modifying the ladder termination voltages.

Introduction

WITH present day integrated circuit technologies, untrimmed wafer yields of high-accuracy circuits (such as a 10-bit D/A converter) are too low for the profitable production of a low-cost part [1], [2]. Quite often, some method of trimming the circuit to improve performance and achieve greater yield is employed.

The most common method of adjusting the critical parameters of monolithic circuits is to laser-trim a matrix of thin-film resistors. In addition to the expensive and difficult processing steps of depositing the thin-film material, there are problems associated with active[1] trimming of thin-film resistors. One problem, which can be caused by the localized heating of the resistor, is mechanical stress and its resultant resistor instability [3], [4]. The light of the laser beam and the heat it generates can also adversely affect sensitive circuitry near the trim and cause the overall circuit to operate in an abnormal manner [5].

An alternative to laser-trimming thin-film resistors is utilizing a laser to cut metal links that are connected to provide incremental changes in a circuit parameter [5]. Thus, standard diffused resistors, as opposed to thin-film resistors, may be used.

Thermally induced stresses in the trimmed material are of no concern because each link is either left intact or is completely opened. However, light and heat generated by the laser beam may still affect any nearby circuitry.

The trimming technique discussed here avoids the above problems because aluminum links are cut in a completely passive manner. That is, based upon measurements taken of the untrimmed circuit, the test computer determines which links should be cut and performs the entire trim operation without monitoring the effects as it proceeds.

Manuscript received May 24, 1976; revised August 2, 1976. This paper was presented at the International Solid-State Circuits Conference, Philadelphia, PA, February 1976.

The author is with the Motorola Integrated Circuits Center, Mesa, AZ 85202.

[1]Active trimming is defined as a feedback trimming method in which an output is continually monitored while trimming is performed. Trimming is stopped as soon as the output reaches a certain level.

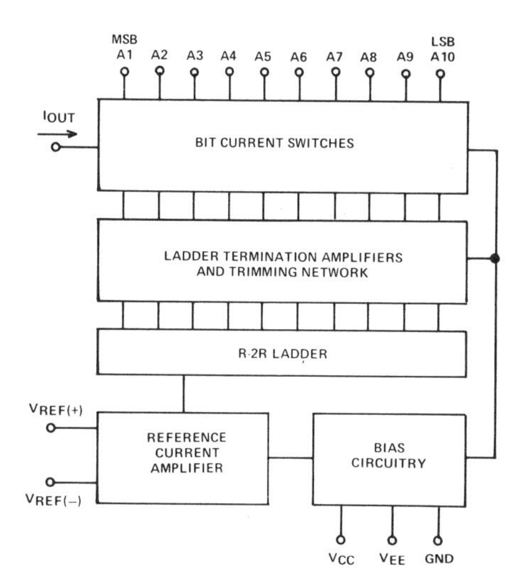

Fig. 1. Block diagram.

Circuit Operation

Fig. 1 is a block diagram of a 10-bit D/A converter that requires laser trimming. The reference current amplifier converts the user-supplied reference voltage to a reference current. The converter is classified as a multiplying type because the reference current is allowed to vary. The analog output current, therefore, is the product of two variables: the reference current and the input digital word.

The reference current is accurately mirrored and doubled to produce the full-scale current that is fed to the *R-2R* ladder. The ladder network divides the full-scale current into the required binary-related components. Ladder termination amplifiers of unity current gain provide low-impedance terminations for each leg of the ladder. Bit current switches steer the individual bit currents from either the supply or the analog output I_{out}, depending upon the value of the input digital word.

Linearity, or relative accuracy, is a measure of how closely the actual transfer curve follows an ideal straight line that passes through zero and full scale, as shown in Fig. 2. Nonlinearity results from nonideal current division in the *R-2R* ladder and is caused by mismatches in the ladder resistors and in the ladder termination voltages. By designing the circuit so that the voltage drop across the *R-2R* ladder is several volts, the errors from mismatches in the termination amplifiers can be made negligible. For this circuit, such errors are typically less significant by more than an order of magnitude than are the ladder resistor mismatches.

The five most significant bits of the *R-2R* ladder are shown

Reprinted from *IEEE J. Solid-State Circuits*, vol. SC-11, pp. 789-794, Dec. 1976.

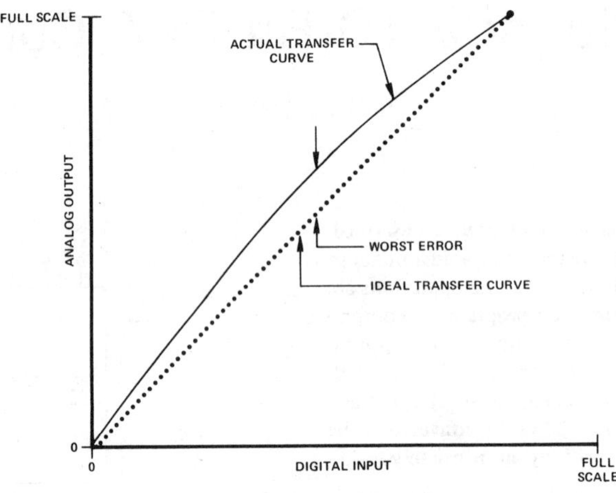

Fig. 2. Definition of linearity.

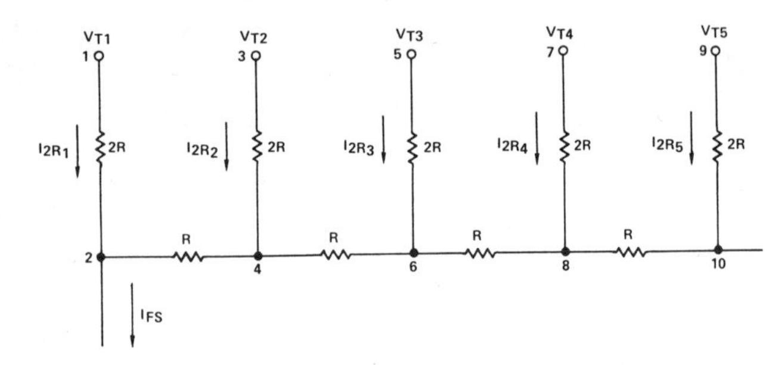

Fig. 3. The five most significant bits of the R-$2R$ ladder.

in Fig. 3. This ladder is capable of achieving the required binary ratios because the lumped resistance of the portion of the ladder to the right of any even-numbered node is the same as the resistance from that node to its respective termination amplifier ($2R$ in each case) [6]. Since nonlinearity is caused primarily by mismatches in the ladder resistors, the most common method of improving linearity is to adjust the value of each ladder resistor to the required matching tolerance. The feasibility of this approach, as applied to the trimming method of cutting metal links, is quickly dismissed when one becomes aware of the minute increments in resistor values that would be required to achieve the desired resolution in the linearity adjustments. Also, the network devised to provide those increments would have to be repeated on each resistor in the ladder in order to minimize initial nonlinearity; this would require a significant amount of die area.

Trim Technique

Rather than perform a trim to adjust the values of the ladder resistors themselves, mismatches in the ladder can be effectively cancelled by making slight adjustments to the termination voltages with respect to each other. As an example, if the $2R$ resistance of the most significant bit (MSB) (see Fig. 3) is low with respect to the other resistors (because of processing tolerances), its current will be higher than precisely half the full-scale current. By lowering the MSB termination voltage V_{T1} an appropriate amount, the current will also be lowered to its proper value.

At this point, let us consider the effect of typical resistor mismatches on linearity. The ratio of the current split at the MSB (node 2 in Fig. 3) is equal to the ratio of the two impedances connected to that node, which statistically would be approximately the same as the resistor-matching tolerance. A mismatch of 1 percent in the resistor values would result in an error of $\frac{1}{2}$ percent in the MSB current and an error of $\frac{1}{2}$ percent, but of opposite sign, in the current which continues on to the remaining bits. Since the MSB current is nominally one half of the full-scale current, its error contributes $\frac{1}{4}$ percent to nonlinearity, relative to full scale. A similar argument applies to the other bits, but since each successively less significant bit handles half as much current as the previous one, the typical error with respect to full scale is half that of the previous bit. Eventually a point is reached in the ladder where the errors contributed by all the less significant bits are normally well within the required linearity. Thus, only the more significant bits need to be trimmed. Using this rationale, as well as verification from a computer Monte Carlo analysis program, it was determined that only the five most significant bits require trimming to achieve the desired ±0.05 percent linearity.

A matrix can be calculated that describes the relationship between a change in any termination voltage and its effect on the voltages across all of the $2R$ resistors in the ladder. This matrix, designated (F), is

$$(F) = \begin{pmatrix} 0.5 & -0.25 & -0.125 & -0.0625 & -0.03125 \\ -0.25 & 0.625 & -0.1875 & -0.09375 & -0.04688 \\ -0.125 & -0.1875 & 0.6562 & -0.1719 & -0.08594 \\ -0.0625 & -0.09375 & -0.1719 & 0.6641 & -0.1680 \\ -0.03125 & -0.04688 & -0.08594 & -0.1680 & 0.6660 \end{pmatrix}.$$

The derivation of this matrix, which is straightforward and involves only resistive voltage division within the ladder, is not presented here. The change in current through any particular $2R$ resistor is simply the change in voltage divided by the resistance, or

$$\Delta I_{2R} = \frac{\Delta V_{2R}}{2R}. \tag{1}$$

By superposition, the total change in current through a $2R$ resistor is the sum of the effects from the changes in all the termination voltages. For an n-bit ladder,

$$\Delta V_{2R_1} = \Delta V_{T_1} F_{11} + \Delta V_{T_2} F_{12} + \cdots + \Delta V_{T_n} F_{1n}$$

$$\Delta V_{2R_2} = \Delta V_{T_1} F_{21} + \Delta V_{T_2} F_{22} + \cdots + \Delta V_{T_n} F_{2n}$$

$$\vdots$$

$$\Delta V_{2R_n} = \Delta V_{T_1} F_{n1} + \Delta V_{T_2} F_{n2} + \cdots + \Delta V_{T_n} F_{nn} \tag{2}$$

where

ΔV_{2R_i} is the change in voltage across the ith $2R$ resistor

ΔV_{T_i} is the change in the ith termination voltage

F_{ij} is the appropriate element in the (F) matrix.

If trimming is performed on only the five most significant bits, then ΔV_{T6} through ΔV_{Tn} remain zero. Equation (2) is then valid in this case if n is replaced by 5. In matrix form, we have

$$(\Delta V_{2R}) = (F)(\Delta V_T). \tag{3}$$

Modifying (1) to read

$$(\Delta V_{2R}) = 2R\,(\Delta I_{2R}) \tag{4}$$

and substituting into (3), we find that

$$2R\,(\Delta I_{2R}) = (F)(\Delta V_T). \tag{5}$$

As used in trimming, the (ΔI_{2R}) matrix will represent those changes in the bit currents that are required to eliminate nonlinearity in the five most significant bits. These values are determined by measuring the bit errors prior to trim. Since linearity is generally expressed as a percentage of full-scale current, the (ΔI_{2R}) matrix should be replaced with one that describes the needed bit current changes with respect to full scale. This matrix will be called (P). Since the full-scale current is twice the reference current, then

$$(P) = \frac{1}{2I_{\text{ref}}} (\Delta I_{2R}) \tag{6}$$

or

$$(\Delta I_{2R}) = 2I_{\text{ref}}\,(P). \tag{7}$$

Substituting (7) into (5), we have

$$4R \cdot I_{\text{ref}}\,(P) = (F)(\Delta V_T). \tag{8}$$

The unknown in the above equation is (ΔV_T) and is determined from

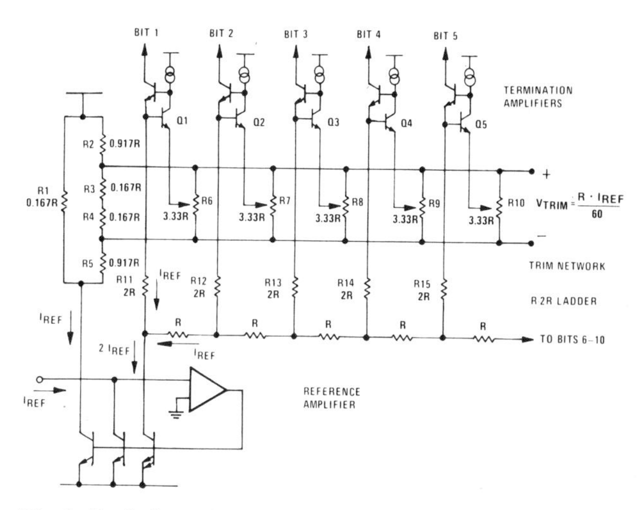

Fig. 4. Equivalent schematic of trim network and related circuitry.

$$(\Delta V_T) = 4R \cdot I_{\text{ref}}\,(F)^{-1}\,(P). \tag{9}$$

Equation (9) states that the required individual trim voltages, or changes in termination voltages, are a function of the reference current and the value of R in the ladder. Unless the circuitry which implements the trim can compensate for these two dependencies, linearity will not be maintained as the reference current varies, nor as temperature varies and the resistors change in value.

Trim Circuit Implementation

Fig. 4 shows the equivalent circuit of the reference amplifier, the R-$2R$ ladder, the termination amplifiers, and the network which derives the trim voltages. A current equal to the reference current I_{ref} is generated in the reference amplifier and is used to produce a voltage drop across resistors R_6 through R_{10}. This voltage is

$$V_{\text{trim}} = I_{\text{ref}} \frac{R_1 \cdot R_P}{R_1 + R_2 + R_5 + R_P} \tag{10}$$

where R_P is the lumped resistance of R_3, R_4, and R_6 through R_{10}.

The values of R_1 through R_{10} are shown relative to the value of R in the ladder. Using these values, we find that

$$V_{\text{trim}} = \frac{R \cdot I_{\text{ref}}}{60}. \tag{11}$$

Resistors R_6 through R_{10}, which are shown schematically as trimpots in Fig. 4, divide V_{trim} into increments that are appropriate for the required linearity adjusting resolution. Fig. 5 shows the layout of one of the trim resistors. The laser is programmed to cut all the links but one; therefore, the link that is left untrimmed functions as a center tap and can be adjusted to one of a finite number of settings.

At this point, we will assume that V_{trim} is to be divided into N incremental trim voltages as determined by the length of the resistor between the links. The value of N will be determined later. One increment, V_{inc}, has the value

$$V_{\text{inc}} = \frac{1}{N} V_{\text{trim}} \tag{12}$$

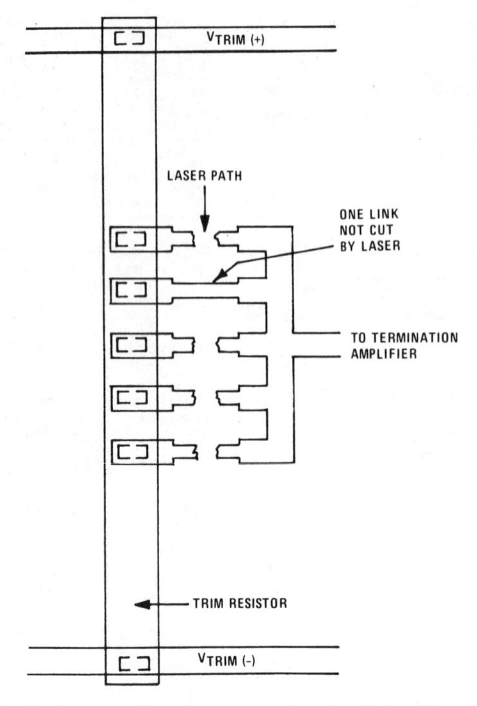

Fig. 5. Layout of a trim resistor showing links after trim.

or, combining (11) and (12),

$$V_{\text{inc}} = \frac{R \cdot I_{\text{ref}}}{60N}. \tag{13}$$

Using this relationship, it is now possible to calculate the number of incremental trim adjustments that each of the five most significant bits requires in order to minimize nonlinearity. The number of trim increments, designated inc, is simply equal to the required trim voltage divided by the value of one increment, rounded to the nearest integer:

$$(\text{inc}) = \frac{1}{V_{\text{inc}}} (\Delta V_T). \tag{14}$$

Combining (13) and (14), we have

$$(\text{inc}) = \frac{60N}{R \cdot I_{\text{ref}}} (\Delta V_T) \tag{15}$$

or, solving for (ΔV_T),

$$(\Delta V_T) = \frac{R \cdot I_{\text{ref}}}{60N} (\text{inc}). \tag{16}$$

Substituting back into (9) yields

$$\frac{R \cdot I_{\text{ref}}}{60N} (\text{inc}) = 4R \cdot I_{\text{ref}} (F)^{-1} (P) \tag{17}$$

or

$$(\text{inc}) = 240N (F)^{-1} (P). \tag{18}$$

Equation (18) demonstrates that the number of incremental trim voltages required on each bit of the circuit is a function of only one variable, the (P) matrix, or the necessary changes in bit currents. It is not a function of actual resistance values nor of the value of the reference current. Therefore, we have a completely passive trim which tracks variations in reference current and temperature.

Yet undetermined is the number of increments N into which V_{trim} is to be divided. Referring to the (F) matrix, note that the diagonal elements are fairly constant for all bits of the ladder. They represent the changes in voltage across the $2R$ resistors ΔV_{2R} for a unit change in the respective termination voltage ΔV_T. These values, which range from $\frac{1}{2}$ for the MSB to $\frac{2}{3}$ for the LSB, indicate that the incremental change in termination voltage V_{inc} has a nearly constant effect on linearity, regardless of which bit is being considered.

A trim increment size of V_{inc} allows the trim to be accurate to within half that value ($\frac{1}{2}$ V_{inc}). This is because the greatest error occurs when the required trim is exactly halfway between the two closest possible trim values. Linearity to $\pm\frac{1}{2}$ LSB for a 10-bit D/A converter corresponds to ± 0.05 percent of full scale. A trim resolution of $\frac{1}{20}$ LSB, corresponding to a trim increment of $\frac{1}{10}$ LSB, is therefore adequate to enable parts to be trimmed to within $\pm\frac{1}{2}$ LSB linearity.

Equation (18) can now be used to determine N, the number of increments into which V_{trim} is divided. Rewriting the equation to solve for N, we have

$$N = \frac{1}{240} (P)^{-1} (F) (\text{inc}). \tag{19}$$

Using the MSB for analysis, we set the first element of the (P) matrix to the desired effect of the trim increment on linearity: $\frac{1}{10}$ LSB or 0.01 percent of full scale. The other elements are set to zero. The first element of the $(P)^{-1}$ matrix will be the reciprocal of 0.01 percent, or 10 000. Multiplication of the $(P)^{-1}$ and the (F) matrices produces a vector with its first ele-

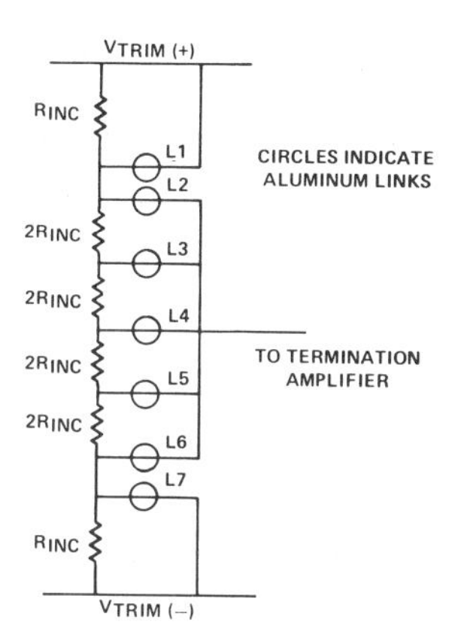

Fig. 6. Link reduction scheme as discussed in text.

TABLE I
TRIM VOLTAGES AS A FUNCTION OF LINKS NOT CUT FOR LINK REDUCTION SCHEME OF FIG. 6

ΔV_T RELATIVE TO THE UNIT V_{INC} DEFINED IN EQUATION 13	LINKS NOT CUT
+4.3	L2 L7
+4.0	L2 L1 L7
+3.4	L2 L1
+2.4	L3 L7
+2.0	L3 L1 L7
+1.5	L3 L1
+0.5	L4 L7
+0.0	L4 L1 L7
−0.5	L4 L1
−1.5	L5 L7
−2.0	L5 L1 L7
−2.4	L5 L1
−3.4	L6 L7
−4.0	L6 L1 L7
−4.3	L6 L1

ment equal to P_1^{-1} times F_{11}, or 5000. The (inc) matrix has the first element set to one and all the others set to zero because we are only considering the MSB. The product of the three matrices in (19) is a scalar and is 5000. Now (19) can be written

$$N = \frac{1}{240}(5000) \simeq 20. \tag{20}$$

Dividing V_{trim} into 20 increments provides the required resolution for trimming.

In order to determine the trim range for each bit, or the number of links each trim resistor will have, the Monte Carlo analysis program was again used. Based upon observed resistor and V_{BE} matching tolerances, the results predicted that the optimum capabilities for the five most significant bits are ±10, ±6, ±4, ±2, ±2 trim increments, respectively. The total number of links required for this trim capability is 48. A reduction in the number of links as well as the required area was found to be possible if the actual increments were made twice as large by dividing V_{trim} into 10 increments instead of 20. Special increments of the smaller value are added to each end of the resistor as shown in Fig. 6. By either leaving these smaller increments shorted or trimming one, the effective increment size is reduced by a factor of two. Table I illustrates the increments that are obtained this way. The two most significant bits are implemented with this technique, and the total number of links was reduced to 39.

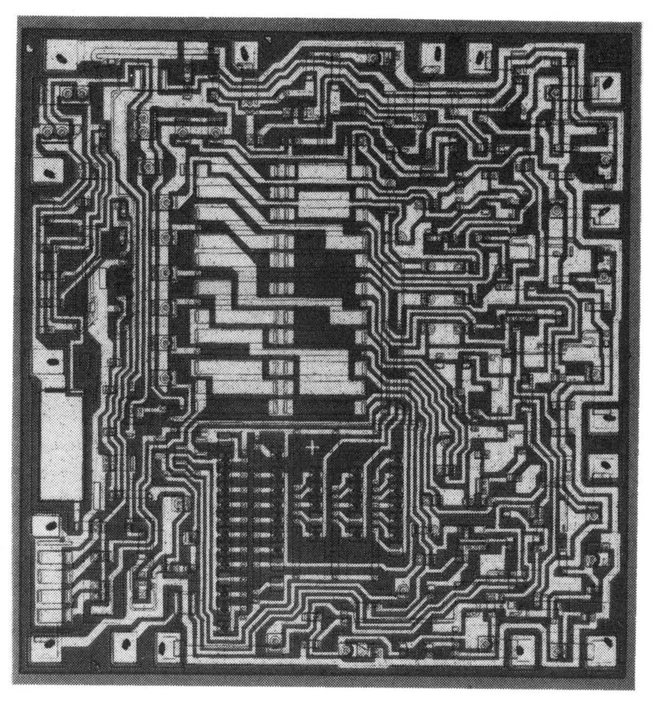

Fig. 7. Photomicrograph of 10-bit D/A converter.

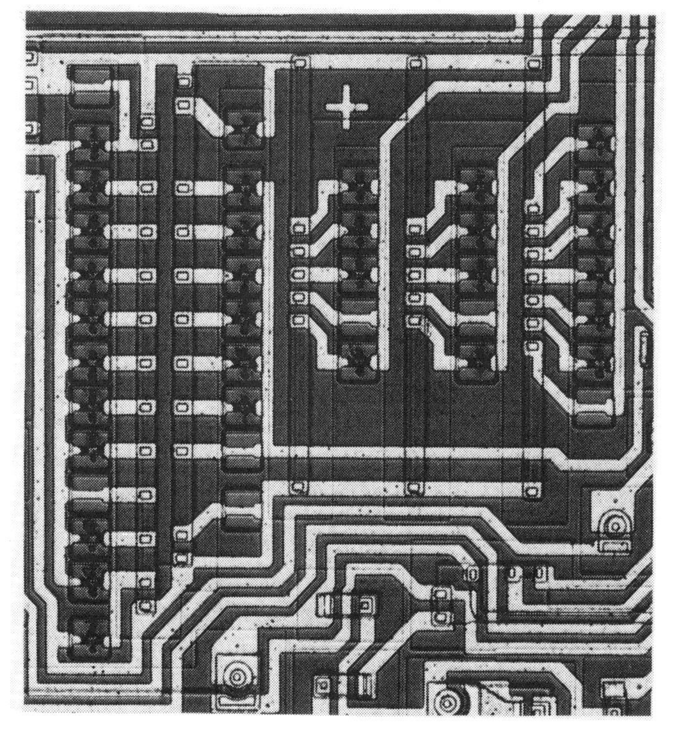

Fig. 8. Enlargement of trim area.

Fig. 7 is a photomicrograph of the 10-bit D/A converter. The R-$2R$ ladder is near the center of the die. The trim network is directly below the ladder and can be seen to take up less than 10 percent of the total area.

An enlargement of the trim area is shown in Fig. 8. The trim resistors are oriented vertically with the links extending horizontally. From left to right, the resistors are those associated with bits 1, 2, 5, 4, and 3. Note that the links which were cut are completely open to a width of approximately 1 mil (0.0025 cm), eliminating the possibility of regrowth of the metal.

Sources of Error in Calculating Trim

The preceding discussion has assumed perfectly matched devices to derive (18). It should be obvious that if such were the case, no trim would be required. It is important to determine how much error is introduced into (18) when such mismatches are considered. There are three sources of error which can affect the accuracy of the trim.

1) Resistor mismatches in the R-$2R$ ladder.

2) Resistor mismatches in R_1 through R_{10} of Fig. 4 and lack of tracking with the ladder resistors.

3) Ratio errors in the reference amplifier between the full-scale current ($2I_{ref}$) and the trim voltage setup current (I_{ref}).

Before a circuit is trimmed, it must be functional and it must meet an initial linearity requirement corresponding to a resistor matching of 0.5 percent typically. Resistors R_1 through R_{10} of Fig. 4 are not as optimized as the ladder resistors for matching, but can be expected to match between themselves and the ladder resistors by typically 1.0 percent. The error in the current ratioing in the reference amplifier is also maintained to 1.0 percent.

All of these errors combine to cause an error in the actual trim voltage increment and in the calculated number of increments required.

The term $R/60$ on the left side of (17) will be in error by the amount of error in the matching of R_1 through R_{10} in Fig. 4. The R term on the right side of the equation will be in error by the mismatches in the R-$2R$ ladder. The I_{ref} terms on both sides of the equation actually match only as well as the currents are mirrored in the reference amplifier. The combination of these errors can result in an error in the elements of the (inc) matrix by up to 2.5 percent. This error is most significant for a needed trim of a full ten increments. Thus the error can be 2.5 percent of ten, or 0.25 increments, which can affect linearity by only $\frac{1}{40}$ LSB.

Conclusion

A trim technique, involving aluminum links which are cut to produce incremental adjustments on circuit performance, has been discussed. Its advantages include the fact that circuits, utilizing standard processing with diffused resistors, can go through an inexpensive laser trim operation to compensate for processing variations that affect critical circuit parameters. The technique was applied to a 10-bit D/A converter in a way that enables the trimming to be passive since the required trim pattern is calculated from only the measured nonlinearity. The trim also tracks with input and temperature variations.

Achieved results, after trim, are better than ±0.05 percent linearity, with a typical drift of ±2.5 ppm/°C. These results agree well with the theory and the computer simulations used in design.

This trim technique is also being used to produce low offset operational amplifiers and precision voltage references. Many other circuits will no doubt be able to benefit from this development.

References

[1] D. J. Dooley, "A complete monolithic 10-bit D/A converter," in *ISSCC Dig. Tech. Papers*, Feb. 1973.

[2] H. H. Stellrecht and G. Kelson, "A self-contained 10-bit monolithic D/A converter," in *ISSCC Dig. Tech. Papers,* Feb. 1973.

[3] G. A. Bulger, "Stability analysis of laser trimmed thin film resistors," *IEEE Trans. Parts, Hybrids, Packaging*, vol. PHP-11, Sept. 1975.

[4] R. C. Headley, "Laser trimming is an art that must be learned," *Electronics*, June 21, 1973.

[5] M. L. Faber, Jr., "Algorithmic trimming on active circuitry," *IEEE Trans. Parts, Hybrids, Packaging*, vol. PHP-10, Sept. 1974.

[6] D. H. Sheingold, *Analog-Digital Conversion Handbook.* Norwood, MA: Analog Devices Inc., 1972, sect. II, ch. 1.

Author Index

Subject Index

Editor's Biography

Alan B. Grebene(M'65—SM'77) was born in Istanbul, Turkey, in 1939. He received his B.S.E.E. degree from Robert College, in Istanbul, Turkey in 1961; the M.S.E.E. degree from the University of California at Berkeley in 1963, and the Ph.D. from Rensselaer Polytechnic Institute, Troy, New York in 1968. During 1963 and 1964, he was employed as an integrated circuit research engineer at Fairchild Semiconductor Research and Development Laboratories, Palo Alto, California. From 1964 through 1965, he was on the technical staff at the microelectronics division of the Sprague Electric Company, North Adams, Massachusetts. From 1965 through 1967, he was a member of the Electrical Engineering faculty at the Rensselaer Polytechnic Institute of Troy, New York. He joined Signetics Corporation of Sunnyvale, California in 1968, where he was the Director of Circuit Research until 1971.

Since 1971, Dr. Grebene has been with Exar Integrated Systems, Inc., of Sunnyvale, California. He is one of the founders of Exar where he is the Vice Presisent of Engineering. Dr. Grebene has authored over thirty-five technical papers and holds seven U.S. patents in the area of integrated circuits and solid state devices. He is also the author of the book "Analog Integrated Circuit Design," published by Van Nostrand Reinhold Co., in 1972, and a contributor to the "Electronic Engineering Handbook," published by McGraw-Hill, in 1975.